D0161407

Boiler Operator's Handbook

2nd Edition

Boiler Operator's Handbook
2nd Edition

By
Kenneth E. Heselton, PE

THE FAIRMONT PRESS, INC.

CRC Press
Taylor & Francis Group

Library of Congress Cataloging-in-Publication Data

Heselton, Kenneth E., 1943-
Boiler operator's handbook / by Kenneth E. Heselton, PE -- 2nd edition.
     pages cm
 Includes bibliographical references and index.
ISBN 0-88173-730-5 (alk. paper) -- ISBN 0-88173-731-3 (electronic) -- ISBN 978-1-4822-5389-4 (Taylor & Francis distribution : alk. paper) 1. Steam-boilers--Handbooks, manuals, etc. I. Title.

TJ289.H53 2014
621.1'94--dc23

2014005475

Published by The Fairmont Press, Inc.
700 Indian Trail
Lilburn, GA 30047
tel: 770-925-9388; fax: 770-381-9865
http://www.fairmontpress.com

Distributed by Taylor & Francis Ltd.
6000 Broken Sound Parkway NW, Suite 300
Boca Raton, FL 33487, USA
E-mail: orders@crcpress.com

Distributed by Taylor & Francis Ltd.
23-25 Blades Court
Deodar Road
London SW15 2NU, UK
E-mail: uk.tandf@thomsonpublishingservices.co.uk

Printed in the United States of America
10 9 8 7 6 5 4 3 2 1

ISBN-0-88173-730-5 (The Fairmont Press, Inc.)
ISBN-978-1-4822-5389-4 (Taylor & Francis Ltd.)

While every effort is made to provide dependable information, the publisher, authors, and editors cannot be held responsible for any errors or omissions.

The views expressed herein do not necessarily reflect those of the publisher.

Table of Contents

Preface
To the Second Edition

In the seven years since I finished the first edition of this book I have had several opportunities to think about what else this book could include to give operators more intelligence and tools and therefore better operation of plants. Many ideas were generated by operators asking questions at classes that I teach regularly for operators to comply with regulations of the Maryland Department of the environment. Also, while working with the Testing Subcommittee for the ASME QFO standard and preparing test questions for the operator's exam I learned much about the perception of boiler operators. As a result there are a few changes in this edition that should be fruitful.

I've also included guidelines for operation of HRSGs and combined cycle systems because those plants are increasing dramatically with their higher efficiency and the increase in natural gas production.

I've also grown to understand an increasing need for knowledge about our environment and how our operations affect it. So, there's more stress in this edition on the environmental effects of boiler operation.

Because so many boiler plants include chilled water or can include refrigeration equipment for plant stores, and boiler operators are called upon to operate that equipment, I've also included a chapter on refrigeration systems. One section of the chapter addresses the environmental effects of inadvertent and intentional discharges of some of our refrigerants. It does not, however, include all the information that's essential for passing an examination to achieve EPA certification for air-conditioning and refrigeration technicians in accordance with section 608 of the federal clean air act. There are several short courses available for operators to take for that examination and they are normally accompanied by the exam.

Finally, because it's apparent that many Operators are expected to control and maintain heating and air conditioning systems the refrigeration chapter includes guidelines on air conditioning.

I do hope this new edition will help you better understand the importance of wise operation of boiler plants and associated facilities.

Ken

Acknowledgements

While many authors list people that have supported, guided, or assisted them in the preparation of their book I failed to do so with the first edition. That was primarily due to the request of Linda Hutchings at Fairmont Press to produce a boiler operator's handbook instead of what I had originally envisioned. The result was an concerted effort to produce a book that could be called a handbook within a limited time frame that I ran well over. I'm not certain that I have managed to improve on that target with this edition but I decided I had better recognize the many people that contributed to the book. At least Linda gave me some time to work on this second edition so I wasn't rushed and only submitted the material two weeks behind the agreed schedule.

Regrettably many of the individuals that contributed to my knowledge and experience with boiler plants are now dead. Regardless I'm compelled to name them because they were among what I call the original "steammakers." The knowledge and experience I gleaned from working and conversing with them is so great that I can't find words to describe it. They're listed in chronological order because any other means doesn't seem right. Chuck Sgamato, the second assistant engineer aboard the African Glade, my first ship sailing as a cadet, made sure that I understood the duties of every man in the engine room by doing them. He had me doing the dirty cleanup jobs, firing the boiler, changing oil burners, etc., all the way up to operating the throttle valves while maneuvering the ship in and out of port. There were many other ship's engineers, and crew members, that aided in my education but Chuck did so much in my first trip that he's the only one that I remember well.

When I quit sailing and took my first shore job at Hercules Incorporated Norm Lind, Walt Applegate, Tom Gamble, Lou Tori and others at Hercules provided the challenges I needed to develop design experience and a better knowledge of air conditioning.

After Hercules I spent twenty-two years with Power and Combustion and was allowed to screw up and fix my own screw-ups plus work on many projects that were unique and difficult and just plain fun to do. I owe a lot to the guidance of Russ Conklin and Ed Deacon and an excellent working relationship with Bob Jackson, Elmer Sells, Harry Deiter and many steamfitters and boilermakers that enjoyed doing an excellent job as much as I did. There's a bunch of us, including Len Stetz, Ray Kollman, Ray DiLiello, Dave Pugh, Mark Schneider and others that still get together a couple of times a year to reminisce.

There are many other people that I've worked with who contributed to my knowledge and the material in this book that I may forget but they include Charlie Gonnerman and others at Leach Wallace Associates, Ken Donithan, Lance Brown, and an innumerable number of boiler plant operators that shared their experiences with me.

Finally, I have to pay loving tribute to my beautiful wife, Sue, who allowed me in my few minutes of spare time to work on the book in it's first edition and to prepare this second edition. She's been supportive, loving, and helpful in immeasurable ways.

Introduction

This book is written for the boiler operator, an operating engineer or stationary engineer by title, who has knowledge and experience with operating boilers but would like to know more and be able to operate his plant wisely. It is also simple enough to help a beginning operator learn the tricks of the trade by reading the book instead of learning the old-fashioned way (through experience) some of which can be very disagreeable. The book can also be used by the manager or superintendent who wants a reference to understand what his operators are talking about. It's only fair, however, to warn a reader of this book that it assumes a certain amount of experience and knowledge already exists.

The day I mailed the contract for this book to the publisher I sat across a table from a boiler operator who said, "Why hasn't somebody written a book for boiler operators that isn't written for engineers?" I've tried to do it with this book, no high powered math and minimal technical jargon.

There are two basic types of operators, those that put in their eight hours on shift while doing as little as possible and those that are proud of their profession and do their best to keep their plant in top shape and running order. You must be one of the latter and you should take pride in that alone.

There is a standard argument that operators operate; they don't perform maintenance duties or repair anything because they have to keep their eye on the plant. That's hogwash. As an engineer with more than forty-five years experience in operating and maintaining boiler plants, I know an operator can't allow someone else to maintain and repair his equipment. It's imperative that the operator know his equipment, inside and out, and one of the best ways of knowing it is to get into it. The operator should be able to do the work or supervise it. Only by knowing what it's like inside can the operator make sound judgments when operating situations become critical.

As for keeping an eye on the plant, that phrase is nothing more than a saying. If you are a manager, reading this book because operators report to you, you should know this—the experienced operator keeps an **ear** on the plant. The most accurate, precise, sensitive instrument in a boiler plant is the operator's ear. The operator knows something is amiss long before any alarm goes off because he can hear any subtle change in the sound of the plant. He can be up in the fidley, and notice that a pump on the plant's lower level just shut down. Hearing isn't the only sense that's more acute in an operator, he "feels" the plant as well. Sounds, actually all sound is vibrations, that aren't in the normal range of hearing are sensed either by the ear, the cheek, or through the feet. Certainly an operator shouldn't be inside a boiler turbining tubes, while he's operating the plant but there are many maintenance activities he can perform while on duty. Managers with a sense of the skill of their operators will use them on overtime and off-shift to perform most of the regular maintenance.

Chapter 1, "Operating Wisely," is the guiding outline for an operator that wants to do just that. The rest of the book is reference and informational material that either explains a concept of operation or maintenance in greater detail, or offers definitions.

I hope this book gives you everything you need to operate wisely. If it doesn't, call me at 410-679-6419 or e-mail KHeselton@cs.com.

Chapter 1

Operating Wisely

If it were not for the power of the human mind with its ability to process information and produce concepts that have never existed before we would be limited to living out our lives like the other species that reside on this earth. We would act as we always have and never make any progress or improve our lives and our environment.

We could, of course, do only those things expected of us and be content with the rewards for doing so. Read on if you're not contented with simply being and doing.

WHY WISELY?

Actually I intended the title of this book to be "Operating Wisely" because there are many books with the title of "Boiler Operator's Handbook" available today. Some are small, some are large, and all have good information in them. If you don't already have one or two, I'm surprised. This isn't just another boiler operator's handbook. However, the publisher wanted to call it a boiler operator's handbook to be certain its content was properly described. Those other books describe the plant and equipment but don't really talk about operating, and in many cases they fail to explain why you should do certain things and why you shouldn't do others.

It's said that "any automatic control will revert to the level of competence of the operator."[1] It's clear that engineers can design all sorts of neat gadgets but they won't work any better than the operator allows. What they always seem to miss is the fact that they never told the operator what the gadget was supposed to do and how to make sure it does it. Lacking that information, the operator reverts to a strategy that keeps the plant running. Hopefully this book will provide you with a way to figure out what the engineer was trying to accomplish so you can make the gadget work if it does do a better job. In some cases you're right, the darn thing is a waste of time and effort, but hopefully you won't dismiss them out of hand anymore. New gadgets and methods are tools you can put to use.

Over the years I've observed operators doing a lot of things that I considered unwise; some were simply a waste of time, some did more harm than good, and others were downright dangerous. Most of those actions could be traced to instructions for situations that no longer exist or to a misunderstanding by the operator of what was going on. To learn to operate wisely you have to know why you do things and what happens when you do the wrong thing. This book tries to cover both. When you understand why you do things you're more likely to do them correctly.

When you have an opportunity to make a mistake, it's always nice to know how someone else screwed up. As Sam Levenson once said, "You must learn from the mistakes of others. You can't possibly live long enough to make them all yourself." Many mistakes are described in the following pages so you will, hopefully, not repeat them.

Two other reasons for this book are the environment and economics. If every boiler operator applied a few of the wise actions described in this book there would be a huge reduction in energy consumption and, as a result, a dramatic improvement in our environment. You can earn your salary by proper operation that keeps fuel, electricity, and water costs as low as possible while still providing the necessary heat to the building and processes. Wise people don't do damage to their environment or waste the boss' money. I hope to give you all the wisdom I gained over fifty years in this business so you can operate wisely.

PRIORITIZING

The first step in operating wisely is to get your priorities in order. Imagine taking a poll of all the boiler plant operators you know and asking them what is the most important thing they have to do. What would they list first? I'm always getting the reply that it's keeping the steam pressure up, or something along those lines. Why? The answer is rather simple; in most cases, the only time an operator hears from the boss is when the pressure is lost or everyone is complaining about the cold or lost production. Keep the pressure up and you will not have any complaints to deal with, so it gets first billing. Right? ... Wrong!

1

History is replete with stories of boiler operators doing stupid things because their first priority was continued operation. There are the operators that literally held down old lever acting safety valves to get steam pressure higher so their boat would beat another in a race. Many didn't live to tell about it. I recall a chief engineer aboard the steamship *African Glade* instructing me to hit a safety valve with a hammer when he signaled me; so the safety would pop at the right pressure. The object was to convince the Coast Guard inspector that the safety valve opened when it was supposed to. A close look at that safety valve told me that hitting it with a hammer was a dumb thing to do. Thankfully the valve opened at the right pressure of its own accord. That was an example of self endangerment to achieve a purpose that, quite simply, was not worth risking my life.

It's regrettable that keeping pressure up is the priority of many operators. Several of them now sit alongside Saint Peter because they were influenced by the typical plant manager or others and put the wrong things at the top of their list of priorities. Another operator followed his chief's instructions to hit a safety valve so it would pop several years ago. The valve cracked and ruptured, relieving the operator of his head. Without a doubt the superintendents and plant managers that demanded their now dead operators blindly meet selected objectives are still asking themselves why they contributed to their operator having the wrong impression. Despite how it may seem, your boss doesn't want you risking your life to keep the pressure up; he just loses sight of the priorities. The wise operator doesn't list pressure maintenance or other events as having priority over his safety.

So what is at the top of the list? **You** are, of course. An operator's top priority should always be his own safety. Despite the desire to be a hero, your safety should take priority over the health and well being of other people. It simply makes sense. A boiler plant is attended by a boiler operator to keep it in a safe and reliable operating condition. If the operator is injured, or worse, he or she can't control the plant to prevent it becoming a hazard to other people.

For several years a major industrial facility near Baltimore had an annual occurrence. An employee entered a storage tank without using proper entry procedures and subsequently succumbed to fumes or lack of oxygen. Now that's bad enough, but... invariably his buddy would go into the tank in a failed effort to remove him, and they both died. Rushing to rescue a fool is neither heroic nor the right thing to do; calling 911 then maintaining control of the situation is; so nobody else gets hurt. The operator that risks his life to save a friend that committed a stupid act is not a hero. He's another fool. Abandoning responsibility to maintain control of a situation and risking your life is getting your priorities out of order. While preventing or minimizing injury to someone else is important, it is not as important as protecting you.

Other people should follow you on your list of priorities. There are occasions when the life or well being of other people is dependent on a boiler operator's actions. There are many stories of cold winters in the north where operators kept their plants going through unusual means to keep a population from freezing. A favorite one is the school serving as a shelter when gas service was cut off to a community. When the operator ran out of oil, he started burning the furniture to keep heat up. That form of ingenuity comes from the skill, knowledge and experience that belongs to a boiler operator and allows him to help other people.

Next in the proper list of priorities is the **equipment and facilities**. Keeping the pressure up is not as important as preventing damage to the equipment or the building. A short term outage to correct a problem is less disrupting and easier to manage. It's better than a long term outage because a boiler or other piece of equipment was run to destruction. The wise operator doesn't permit continued operation of a piece of equipment that is failing. Plant operations might be halted for a day or week while parts are manufactured or the equipment is overhauled. That is preferable to running it until it fails—then waiting nine months to obtain a replacement. You can counter complaints from fellow employees that a week's layoff is better than nine months. There are several elements of operating wisely that consider the priority of the equipment.

Many operators choose to bypass an operating limit to keep the boiler on line and avoid complaints about pressure loss. Even worse, they bypass the limit because it was a nuisance. "That thing is always tripping the boiler off line so I fixed it." The result of that fix is frequently a major boiler failure. Operator error and improper maintenance account for more than 34% of boiler failures.

The **environment** has taken a new position on the operator's list of priorities within the last half century. Reasons are not only philanthropic but also economic. Regularly during the summer, the notices advise us that the air quality is marginal. Sources of quality water are dwindling dramatically. The wrong perception in the minds of the company's customers can reduce revenue (in addition to the costs of a cleanup) and the combination is capable of eliminating a source of income for you and fellow employees.

Several of the old rules have changed as a result. It is no longer appropriate to maintain an efficiency haze because it contributes to the degradation of the environment. The light brown haze we thought was a mark of efficient operation when firing heavy fuel oil has become an indication that you're a polluter. Once upon a time an oil spill was considered nothing more than a nuisance. I have several memories of spills, and the way we handled them, that I'm now ashamed of. You should be aware that insurance for environmental damage is so expensive that many firms cannot afford insurance to cover the risk. Today a single oil spill can destroy a company.

Most state governments have placed a price on emissions. At the turn of the century it was a relatively low one. The trend for those prices is up and they are growing exponentially.

You must understand that operation of the plant always has a detrimental effect on the environment. You can't prevent damage, but you can reduce the impact of the plant's operation on the environment. The wise operator has a concern for the environment and keeps it appropriately placed on the list of priorities.

Those four priorities should precede **continued operation of the plant** on your priority list. Despite what the boss may say when the plant goes down, he or she does not mean nor intend to displace them. Most operators manage to develop the perception that continued operation of the plant is on the top of the boss's list of priorities, that impression is formed when the boss is upset and feels threatened, not when she or he is conscious of all ramifications. Continued operation is important and dependent upon the skill and knowledge of the operator only after the more important things are covered.

Since continued operation is so important, the operator has an obligation many never think of, and some avoid. The wise operator is always **training a replacement**. If the plant is going to continue to operate there must be someone waiting to take over the operator's job when the operator retires or moves up to management. Producing a skilled replacement is simply one of the more important ways the wise operator ensures continued operation of the plant.

Right now you're probably screaming, "Train my replacement! Why should I do that, the boss can replace me with that trainee?" It's a common fear, being replaceable, many operators refuse to tell fellow employees how they solved a problem or manage a situation believing they are protecting their job. That first priority is not your job, it's your safety, health, and welfare. Note that protecting your position is not even on the list. When an employer becomes aware of an employee's acting to protect the job, and they will notice it, they have to ask the question, "If he (or she) is afraid of losing her (or his) job maybe we don't need that position, or that person."

Let's face it, if the boss wants to get rid of you, you're gone. On the other hand, if the boss wants to move you up to a management position or other better paying or more influential job and you can't be replaced readily, well… Many operators have been bypassed for promotion simply because there wasn't anyone to replace them. It's simply a part of your job, so do it.

Preserving historical data is a responsibility of the operator. The major way an operator preserves data is maintaining the operator's log. The simplest is getting the instructions back out of the wastebasket. If that information is retained only in the operator's mind, the operator's replacement will not have it and other personnel and contractors will not have it. Lack of information can have a significant impact on the cost of a plant operation and on recovery in the event of a failure. Equipment instructions, parts lists, logs, maintenance records, even photographs can be and are needed to operate wisely. It's so important I've dedicated a couple of chapters in this book to it.

Operating the plant economically is last and the priority that involves most of your time. The priorities discussed so far are covered quickly by the wise operator. You are paid a wage that respects the knowledge, skill, and experience necessary to maintain the plant in a safe and reliable operating condition. You earn that money by operating the plant economically. One can make a difference equal to a multiple of wages in most cases.

Note that the word efficiency doesn't fall on the list of priorities. It can be said that operating efficiently is operating economically but that isn't necessarily true. For example, fuel oil is utilized more efficiently than natural gas; however, gas historically costs less than oil. The wise operator knows what it costs to operate the plant and operates it accordingly. Efficiency is just a measure used by the wise operator to determine how to operate the plant economically.

Frequently the operator finds this task daunting because the boss will not provide the information necessary to make the economic decisions. The employer considers the cost data confidential material that should only be provided to management personnel. If that is the case in your plant you can tell your boss that Ken

Heselton, who promotes operating wisely, said bosses that keep cost data from their employees are fools. Show him (or her) this page. If an operator doesn't know the true cost of the fuel burned, the water and chemicals consumed, electrical power that runs the pumps and fans, etc., the operator will make judgments in operation based on perceived costs. And frequently those perceptions are flawed. I was able to prove that point many times in the past. Regrettably for the employer, it was after a lot of dollars went up the stack.

I have a few recollections of my own stupidity when I was managing operations for Power and Combustion, a mechanical contractor specializing in building boiler plants. When I failed to make sure the construction workers understood all the costs they made decisions that cost the company a lot of money. Needless to say, I could measure the cost of those mistakes in terms of the bonus I took home at Christmas.

You don't have to know what the boss's or fellow employee's wages are. They're not subject to your activities. You should know, however, what it costs to keep you on the job. Taxes and fringe benefits can represent more than 50 percent of the person's wages. Many of the extra costs, but not all, for a union employee appears on the check because the funds are transferred to the union. Non-union employers should also inform the operators what is contributed on their behalf. Even if the employer doesn't allow the operator to have that information, the wise operator should know that the paycheck is only a part of what it costs to put a person on the job. In addition to retirement funds, health insurance, vacation pay and sick pay there is the employer's share of Social Security and Medicaid; the employer has to contribute a match to what the employee has withheld from salary. There are numerous taxes and insurance elements as well. An employer pays State Unemployment Taxes, Federal Unemployment Taxes, and Workmen's Compensation Insurance Premiums at a minimum. If you have to guess what you really cost your employer, figure all those extras are about 50 percent of your salary.

Economic operation requires utilizing a balance of resources, including manpower, in an optimum manner so the total cost of operation is as low as possible. You might want to know even more to determine if changes you would like to see in the plant can reduce operating costs. That, however, is to be covered in another book.

To summarize, the wise operator keeps priorities in order and they are:

1.   The operator's personal safety, health and welfare
2.   The safety and health of other people
3.   The safety and condition of the equipment operated and maintained
4.   Minimizing damage to the environment
5.   Continued operation of the plant
6.   Training a replacement
7.   Preserving historical data
8.   Economic operation of the plant

Prioritizing in the Real World

Prioritizing activities and functions is simply a matter of keeping the above list in your mind. Every activity of an operator should contribute to the maintenance of those priorities. Only by documenting them can you prove they are done, and done according to priority. We'll cover documenting a lot so it won't be discussed further here. Following the list of priorities makes it possible to decide what to do and when.

Changes in the scope of a boiler plant operator's activities make maintaining that order important. Modern controls and computers that are used to form things like building automation systems have relieved boiler plant operators of some of the more mundane activities. We have taken huge strides from shoveling coal into the furnace to what is almost a white collar job today. As a result, operators find themselves assigned other duties. You may find you have a variety of duties which, when listed on your resume, would appear to outweigh the actual activity of operating a boiler. A boiler plant operator today may serve as a watchman, receptionist, mechanic and receiving clerk in addition to operating the boiler plant. As mentioned earlier, maintenance functions can be performed by an operator or the operator can supervise contractors in their performance. The trend to assuming or being assigned other duties will continue and a wise operator will be able to handle that trend.

Many operators simply complain when assigned other tasks. They also frequently endeavor to appear inept at them, hoping the boss will pass them off on someone else. Note that if you intentionally appear inept at that other duty it may give rise to a question of your ability to be an operator. An operator has an opportunity to handle the concept of additional assignments in a professional manner. One can view the new duty as something that can be fit into the schedule; in which case it increases the operator's value to the employer. A wise operator will have developed systems that grant him (or her) plenty of time to handle other tasks. If, however, you can't make the duty fit, you can demonstrate that the new duty will take you away from the work you must do to maintain the priorities and, pleasantly, inform the boss of the increased risk of damage or injury that could occur if you take on the new requirements. Should your boss insist you assume duties that will alter the priorities you should oppose it. Every place of employment should have a means for employees to appeal a boss's decision to a higher authority. Seek out that option and use it when necessary but always be pleasant about it.

It is during such contentious conditions that the value of documentation is demonstrated. A wise operator with a documented schedule, SOPs, and to-do-list will have no problem demonstrating that an additional task will have a negative effect on the safety and reliability of the boiler plant. On the other hand, documentation that is evidently self-serving will disprove a claim. The wise operator will always have supporting, qualifying documentation to support his or her position.

Another situation that produces contentious conditions in a boiler plant involves the work of outside contractors. Frequently the contractor was employed to work in the plant with little or no input from the operators. That's another way a boss can be a fool, but it happens. When a contractor is working in the plant, it changes the normal routine and regularly interferes with the schedule an operator has grown accustomed to. The wise boss will have the contractor reporting to the operator; regrettably there aren't many wise bosses in this world. Even if I'm just visiting a plant I still make certain that I report in to the operator on duty and check out as well. I always advised my construction workers to do it. Regardless of the reporting requirements the operator and contractor will have to work together to ensure the priorities are maintained.

The wise operator will be able to work reasonably with the contractor to facilitate the contractor getting his work done. Many operators have expressed an attitude that a contractor is only interested in his profit and treat all contractors accordingly. Guess what, the wise operator wants the contractor to make a profit. If the contractor is able to perform the work without hindrance or delay he will be able to finish the work on time and make a profit. If the contractor perceives no threat to the profit he contemplated when starting the job he will do everything he intended, including doing a good job. If the operator stalls and blocks the contractor's activity so the contractor's costs start to run over, he will attempt to protect his profit. If the contractor perceives the operator is intentionally making life difficult he may complain to the operator's boss as well as start cutting corners to protect his profit. A contractor can understand the list of priorities and work with the operator that understands the contractor's needs.

Dealing with fellow employees also requires demonstrative use of the list of priorities. The problem is not usually associated with swing shift operation because the duties are balanced over time. When operators remain on one shift it is common for one shift to complain another has less to do. Another common problem is the one operator that, in the minds of the rest, doesn't do anything or doesn't do it right. If you've got the priority order right in your mind you already know that number 6 applies; train that operator.

There's nothing on the list about pride, convenience, or free time. Self interest is not a priority when it comes to any job. You can be proud of how you do your job. You may find it convenient to do something a different way (but make sure your boss knows of and approves the way). You should always have a certain amount of free time during a shift to attend to the unexpected situations that arise, but no more than an hour per shift. Keep in mind that you are not employed to further your interests or simply occupy space. You can, and should, provide value to your employer in exchange for that salary.

Most employers understand an employee's need to handle a few personal items during the day. They'll tolerate some time spent on the phone, reading personal documents, and simply fretting over a problem at home. They will not, however, accept situations where the employee places personal interests ahead of the job. I've encountered situations where employers allowed their employees to use the plant tools to work on personal vehicles, repair home appliances, make birdhouses and the like during the shift. On the other hand I've encountered employers that wouldn't allow their people to make personal calls, locking up the phone. Limiting personal activity as much as possible and never allowing it to take priority over getting that list we just looked at should prevent those situations where, because the boss's good nature was abused, the employer suddenly comes down hard restricting personal activity on the job.

Your health and well being is at the top of the list primarily because you're the one responsible for the plant. Keep your priorities straight. Maintaining your priorities in the specified order should always make it possible to resolve any situation. The priorities will be referred to regularly as we continue operating wisely.

SAFETY

The worst accident in the United States was the result of a boiler explosion. In 1863 the boilers aboard the steamship *Sultana* exploded and killed almost eighteen hundred people. The most expensive accident was a boiler explosion at the River Rouge steel plant in February of 1999. Six men died and the losses were measured at more than $1 billion. Boiler accidents are rare compared to figures near the first of the 20th century when thousands were killed and millions injured by boiler explosions. Today, less than 20 people die each year as a result of a boiler explosion. I don't want you to be one of them. I'm sure you don't want to be one either. Safety rules and regulations were created after an accident with the intent of preventing another.

A simple rule like "always hold the handrail when ascending and descending the stair" was created to save you from injury. Don't laugh at that one, one of my customers identified falls on stairs in the office building as the most common accident in the plant. Follow those safety rules and you will go home to your family healthy at the end of your shift.

There are many simple rules that the macho boiler operator chooses to ignore and, in doing so, risks life and limb. You should make an effort to comply with all of them. You aren't a coward or chicken. You're operating wisely.

Hold onto the handrail. Wear the face shield, boots, gloves, and leather apron when handling chemicals. Don't smoke near fuel piping and fuel oil storage tanks. Read the material safety data sheets, concentrating on the part about treatment for exposure. Connect that grounding strap. Do a complete lock-out, tag-out before entering a confined space and follow all the other safety rules that have been handed down at your place of employment. Remember who's on the top of the priority list.

PPE stands for personal protective equipment and was an acronym that was readily ignored long before I was operating plants and for quite a while afterward. Because I wasn't required to wear ear protection I have a lot of trouble hearing lots of things today. Luckily I was smart enough to wear safety glasses and gloves when appropriate most of the time, but not always.

I'm sure you will want to be able to see and hear those grandchildren when they come along and you won't want them to have to observe that grandpa (or grandma as applicable) has deformed hands, burns or other injuries that can only be described as ugly. Wear that PPE, I wish the rules were enforced better when I was starting out because I would be able to see and hear better now and I wouldn't have to explain to my grand nieces and nephews (let alone the grandchildren I couldn't have) why my hands and face look the way they do.

Prevention of explosions in boilers has come a long way since the Sultana went down. The modern safety valve and the strict construction and maintenance requirements applied to it have reduced pressure vessel explosions to less than 1% of the incidents recorded in the U.S. each year, always less than two. On the other hand, furnace explosions seem to be on the increase and that, in my experience, is due to lack of training and knowledge on the part of the installer which results in inadequate training of the operator.

You must know what the rules are and make sure that everyone else abides by them. A new service technician, sent to your plant by a contractor you trust, could be poorly trained and unwittingly expose your plant to danger. Even old hands can make a mistake and create a hazard. Part of the lesson is to seriously question anything new and different, especially when it violates a rule.

What are the rules? There are lots of them and some will not apply to your boiler plant. Luckily there are some rules that are covered by qualified inspectors so you don't have to know them. There should be rules for your facility that were generated as a result of an accident or analysis by a qualified inspector. Perhaps there's a few that you wrote or should have written down. When the last time you did that there was a boiler rattling BOOM in the furnace a rule was created that basically said don't do that again! Your state and local jurisdiction (city or county) may also have rules regarding boiler operation so you need to look for them as well. Here's a list of the published rules you should be aware of and, when they apply to your facility, you should know them.

ASME Boiler and Pressure Vessel Codes (BPVC):
   Section I—Rules for construction of Power Boilers[a]
   Section IV—Rules for construction of Heating Boilers[a]
   Section VI—Recommended Rules for Care and Operation of Heating Boilers[b]
   Section VII—Recommended Rules for Care and Operation of Power Boilers[b]
   Section VIII—Pressure Vessels, Divisions 1 and 2[c] (rules for construction of pressure vessels including deaerators, blowoff separators, softeners, etc.)
   Section IX—Welding and Brazing Qualifications (the section of the Code that defines the requirements for certified welders and welding.)
   B-31.1—Power Piping Code
   CSD-1—Controls and Safety Devices for Automatically Fired Boilers (applies to boilers with fuel input in the range of 400 thousand and less than 12.5 million Btuh input)

National Fire Protection Association (NFPA) Codes
  NFPA—30—Flammable and Combustible Liquids Code
  NFPA—54—National Fuel Gas Code
  NFPA—58—Liquefied Petroleum Gas Code
  NFPA—70—National Electrical Code
  NFPA—85—Boiler and Combustion Systems Hazards
    Code (applies to boilers over 12.5 million Btuh in-
    put)

---

aRequires inspection by an authorized inspector so you don't
  have to know all these rules.
bThese haven't been revised in years and contain some recommendations that are simply wrong.
cRequires inspection by an authorized inspector so you don't
  have to know all these rules

That's volumes of codes and rules and it's impossible for you to learn them. They are typically revised every three years so you would be out of date before you finished reading them all. It's not important to know everything, only that they're there for you to refer to. Flipping through them at a library that has them or checking them out on the Internet will allow you to catch what applies to you. CSD-1 or NFPA-85, whichever applies to your boilers, are must reads. Some of those rules are referred to in this book.

Sections VI and VII of the ASME Code are good reads. Regrettably they haven't kept up to the pace of modernization. The rest of the ASME Codes apply to construction, not operation. You'll never know them well but you have to be aware that they exist.

As I said earlier, many rules were produced as the result of accidents. That is likely true in your plant. A problem today is many rules are lost to history because they aren't passed along with the reason for them fully explained. I'll push the many concepts of documentation in a chapter dedicated to it but it bears mentioning here. Keep a record of the rules. If there isn't one, develop it. The life you safe will more than likely be yours.

MEASUREMENTS

If you pulled into a gas station, shouted "fill-er-up" on your way to get a cup of coffee then returned to have the attendant ask you for twenty bucks and the pump was reset you would think you'd been had, wouldn't you? You might even quibble, "How do I know you put twenty dollars worth in it?" Why is it that we quibble over ten dollars and think nothing about the amount of fuel our plant burns every day? I'm not saying yours is one of them but I've been in so many plants where they don't even read the fuel meter, let alone record any other

measurements, and I always wonder how much they're being taken for. I also wonder how much they've wasted with no concern for the cost.

Any boiler large enough to warrant a boiler operator in attendance burns hundreds if not thousands of dollars each day in fuel. To operate a plant without measuring its performance is only slightly dumber than handing the attendant twenty dollars on your way to get coffee when you know there may not be room in the tank for that much. When I pursue the concept of measurements with boiler operators I frequently discover they don't understand measurements or they have a wrong impression of them. To ensure there is no confusion, let's discuss measurements and how to take them.

First there are two types of measures, measures of quantity and measures of a rate. There's about 100 miles between Baltimore and Philadelphia, that's a quantity. If you were to drive from one to the other in two hours, you would average fifty miles per hour, that's a rate. Rates and time determine quantities and vice versa. If you're burning 7-1/2 gpm of oil you'll drain that full 8,000-gallon oil tank in less than 19 hours. Quantities are fixed amounts and rates are quantity per unit of time.

The most important element in describing a quantity or rate is the units. Unit comes from the Latin "uno" meaning one. Units are defined by a standard. We talk about our height in feet and inches using those units without thinking of their origin. A foot two centuries ago was defined as the length of the king's foot. Since there were several kings in several different countries there was always a little variation in actual measurement. I have to assume the king's mathematician who came up with inches had to have six fingers on each hand; why else would they have divided the foot by twelve to get inches?

Today we accept a foot as determined by a ruler, yardstick, or tape measure all of which are based on a piece of metal maintained by the National Bureau of Standards. That piece of metal is defined as the standard for that measure having a length of precisely one foot. They also have a chunk of metal that is the standard for one pound. As you proceed through this book you'll encounter units that are based on the property of natural things. The meter, for example, is defined as one ten millionth of the distance along the surface of the earth from the equator to one of the poles. Regrettably that's a bogus value because a few years ago we discovered the earth is slightly pear shaped so the distance from the equator to the pole depends on which pole you're measuring to. Many units have a standard that is a property of water; we'll be discussing those as they come up.

Unless we use a unit reference for a measurement nobody will know what we're talking about. How would you handle it if you asked someone how far it was to the next town and they said "about a hundred?" Did they mean miles, yards, furlongs, football fields? Unless the units are tacked on we can't relate to the number.

With few exceptions there are multiple standards (units) of measure we can use. Which one we use is dependent on our trade or occupation. Frequently we have to be able to relate one to the other because we're dealing with different trades. We will need conversion factors. We can think of a load of gravel as weighing a few hundred pounds but the truck driver will think of it in tons. He'll claim he's delivering an eight-ton load and we have to convert that number to pounds because we have no concept of tons; we can understand what 16,000 pounds are like. Another example is a cement truck delivery of 5 yards of concrete. No, that's not fifteen feet of concrete. It's 135 cubic feet. (There are 27 cubic feet in a cubic yard, 3 × 3 × 3) We need to understand what type of measurement we're dealing with to be certain we understand the value of it. Also, as with the cement truck driver, we have to understand trade shorthand.

When measuring objects or quantities there are three basic types of measurement: distance, area, and volume. We're limited to three dimensions so that's the extent of the types. Distances are taken in a straight line or the equivalent of a straight line. We'll drive 100 miles between Baltimore and Philadelphia but we will not travel between those two cities in a straight line. If you were to lay a string down along the route and then lay it out straight when you're done it would be 100 miles long. The actual distance along a straight line between the two cities would be less, but we can't go that way.

Levels are distance measurements. We always use level measurements that are the distance between two levels because we never talk about a level of absolute zero. If there was such a thing it would probably refer to the absolute center of the earth. Almost every level is measured from an arbitrarily selected reference. The water in a boiler can be one to hundreds of feet deep but we don't use the bottom as a reference. When we talk about the level of the water in a boiler, we always use inches and negative numbers at times. That's because the reference everyone is used to is the center of the gage glass which is almost always the normal water line in the boiler. The level in a twelve-inch gage glass is described as being in the range of –6 inches to +6 inches. For level in a tank we normally use the bottom of the tank for a reference so the level is equal to the depth of the fluid

and the range is the height of the tank.

With so many arbitrary choices for level it could be difficult to relate one to another. That could be important when you want to know if condensate will drain from another building in a facility to the boiler room. There is one standard reference for level but we don't call it level, we call it "elevation" normally understood to be the height above mean sea level and labeled "feet MSL" to indicate that's the case. In facilities at lower elevations it is common to use that reference. A plant in Baltimore, Maryland, will have elevations normally in the range of 10 to 200 feet, unless it's a very tall building.

When the facility is a thousand feet or more above mean sea level it gets clumsy with too many numbers so the normal procedure is to indicate an elevation above a standard reference point in the facility. A plant in Denver, Colorado, would have elevations of 5,200 to 5,400 feet if we used sea level as a reference so plant references would be used there. It's common for elevations to be negative, they simply refer to levels that are lower than the reference. It happens when we're below sea level or the designers decide to use a point on the main floor of the plant as the reference elevation of zero; anything in the basement would be negative. The choice of zero at the main floor is a common one. Note that I said a point on the main floor, all floors should be sloped to drains so you can't arbitrarily pull a tape measure from the floor to an item to determine its precise elevation.

An area is the measurement of a surface as if it were flat. A good example is the floor in the boiler plant which we would describe in units of square feet. One square foot is an area one foot long on each side. We say "square" foot because the area is the product of two linear dimensions, one foot times one foot. The unit square foot is frequently written ft^2 meaning feet two times or feet times feet. That's relatively easy to calculate when the area is a square or rectangle. If it's a triangle the area is one half the overall width times the overall length. If it's a circle, the area is 78.54% of a square with length and width identical to the circle's diameter. A diameter is the longest dimension that can be measured across a circle, the distance from one side to a spot on the opposite side. In some cases we use the radius of a circle and say the area is equal to the radius squared times Pi (3.1416). When you're dealing with odd shaped areas, and you have a way of doing it, laying graph paper over it and counting squares plus estimating the parts of squares at the borders is another way to determine an area. A complex shaped area can also be broken up into squares, rectangles, triangles and circles, adding and subtracting them to determine the total area.

Volume is a measure of space. A building's volume is described as cubic feet, abbreviated ft^3, meaning we multiply the width times the length times the height. One cubic foot is space that is one foot wide by one foot long by one foot high.

I'll ignore references to the metric system because that's what American society appears to have decided to do. It's regrettable because the metric system is easier to use and there's little need to convert from one to the other after we've accepted it. After all, there's adequate confusion and variation generated by our English system to keep us confused. When it comes to linear measurements we have inches and yards, one twelfth of a foot and three feet respectively. Measures of area are usually expressed in multiples of one of the linear measures (don't expect an area defined as feet times inches however). For volumetric measurements we also have the gallon, it takes 7.48 of them to make a cubic foot.

Note that the volumetric measure of gallons doesn't relate to any linear or area measure, it's only used to measure volumes. That's some help because many trades use unit labels that are understood by them to mean area or volume when we couldn't tell the difference if we didn't know who's talking. A painter will say he has another thousand feet to do. He's not painting a straight line. He means one thousand square feet. We've already mentioned the cement hauler that uses the word "yards" when he means cubic yards. Always make sure you understand what the other guy is talking about.

When talking, or even describing measurements we will use descriptions of direction to aid in explaining them. While most people understand north, south, east and west plus up and down other terms require some clarification. Perpendicular is the same as perfectly square. When we look for a measurement perpendicular to something it's as if we set a square on it so the distance we're measuring is along the edge of the square. An axial measurement is one that is parallel to the central axis or the center of rotation of something. On a pump or fan it's measured in the same direction as the shaft. Radial is measured from the center out; on a pump or fan it's from the centerline of the shaft to whatever you are measuring. When we say tangentially or tangent to we're describing a measurement to the edge of something round at the point where a radial line is perpendicular to the line we're measuring along.

Another measure that confuses operators is mass. Mass is what you weigh at sea level. If we put you on a scale while standing on the beach, we would be able to record your mass. If we then sent you to Cape Kennedy, loaded you into the space shuttle, sent you up in space, then asked you to stand on the scale and tell us what it reads, what would your answer be? Zero! You don't weigh anything in space, but you're still the same amount of mass that we weighed at sea level. There is a difference in weight as we go higher. You will weigh less in Denver, Colorado, because it's a mile higher, but for all practical purposes the small difference isn't important to boiler operators. Once you accept the fact that mass and weight are the same thing with some adjustment required for precision at higher elevations you can accept a pound mass weighs a pound and let it go at that.

Volume and mass aren't consistently related. A pound mass is a pound mass despite its temperature or the pressure applied to it. One cubic foot of something can contain more or less mass depending on the temperature of the material and the pressure it is exposed to. Materials expand when heated and contract when cooled (except for ice which does just the opposite).

We can put a fluid like water on a scale to determine its mass but the weight will depend on how much we put on the scale. If we put a one gallon container of 32° water on the scale, it will weigh 8.33 pounds. If we put a cubic foot of that water on the scale, it will weigh 62.4 pounds.

Density is the mass per unit volume of a substance, in our case, pounds per cubic foot. So, water must have a density of 62.4 pounds per cubic foot. Ah, that the world should be so simple! Pure clean water weighs that. Sea water weighs in at about 64 pounds per cubic foot. Heat water up and it becomes less dense. When it's necessary to be precise, you can use the steam tables (page 353) to determine the density of water at a given temperature but keep in mind that its density will also vary with the amount of material dissolved in it.

In many cases water is the reference. You'll hear the term specific gravity or specific weight. In those cases it's the comparison of the weight of the liquid to water (unless it's a gas when the reference is air) Knowing the specific gravity of a substance allows you to calculate its density by simply multiplying the specific gravity by the typical weight of water (or air if it's a gas). One quick look at the number gives you a feel for it. If the gravity is less than one it's lighter than water (or air) and if it's greater than one it will sink.

Gases, such as air, can be compressed. We can pack more and more pounds of air into a compressed air storage tank. As the air is packed in, the pressure increases. When the compressor is off and air is consumed, the tank pressure drops as the air in the tank expands to replace what leaves. The compressor tends to heat the air as it compresses it and that hot air will cool off while it sits in the tank and the pressure will drop. We need to know the pressure and temperature of a gas to determine the density. The steam tables list the specific volume (cubic feet per pound) of steam at saturation and some superheat temperatures. Specific volume is equal to one divided by the density. To determine density, divide one by the specific volume.

Liquids are normally considered non-compressible so we only need to know their temperature to determine the density. The specific volume of water is also shown on the steam tables for each saturation temperature. Water at that temperature occupies the volume indicated regardless of the pressure.

We also use pounds to measure force. Just like a weight of, say ten pounds, can bear down on a table when we set the weight down we can tip the table up with its feet against a wall and push on it to produce a force of ten pounds with the same effect. Weights can only act down, toward the center of the earth, but a force can be applied in any direction. Just like we can measure a weight with a scale we can put the scale (if it's a spring loaded type) in any position and measure force; they're both measured in pounds.

Rates are invariably one of the measures of distance, area, volume, weight or mass traversed, painted, filled, or moved per unit of time. Common measurements for a rate are feet per minute, feet per second, inches per hour, feet per day, gallons per minute, cubic feet per hour, miles per hour and its equivalent of knots (which is nautical miles per hour, but let's not make this any worse than it already is). Take any quantity and any time frame to determine a rate. Which one you use is normally determined according to the trade discussing it or the size of the number. We normally drive at sixty miles per hour although it's also correct to say we're traveling at 88 feet per second. We wouldn't say we're going at 316,800 feet per hour. Be conscious of the units used in trade magazines and by various workmen to learn which units are appropriate to use. You can always convert the values to units that are more meaningful to you. The appendix contains a list of common conversions.

There are common units of measure used in operating boiler plants. Depending on what we're measuring we'll use units of pounds or cubic feet or gallons when discussing volumes of water. We measure steam generated in pounds (mass) per hour but feed the water to the boiler in gallons per minute. We burn oil in gallons per hour, gas in thousands of cubic feet per hour, and coal in tons per hour. We use a measure that's shared with

the plumbing trade which we call pressure, normally measured in pounds per square inch. Occasionally we confuse everyone by calling it "head."

We normally describe the rate that we make steam as pounds per hour and use that as a unit of rate abbreviated "pph." The typical boiler plant can generate thousands of pounds of steam per hour so the numbers get large and we'll identify the quantity in thousands or millions of pounds of steam. A problem arises in using the abbreviations for large quantities because we're not consistent and use a multitude of symbols.

We'll use "kpph" to mean thousands of pounds of steam per hour but use "MBtuh" to describe a thousand Btu's per hour. Most of the time we avoid using "mpph" both because it looks too much like a typo of miles per hour and because many people wonder if we mean one thousand or one million. A measure of a million Btu's per hour can be labeled "MMBtuh" sort of like saying a thousand thousand or use a large "M" with a line over it which is also meant to represent one million. I've also seen a thousand Btu's per hour abbreviated "MBH." The ASME is trying to be consistent in using only lower case letters for the units. It will be some time before that's accepted. This book uses the publisher's choice.

Pressure exists in fluids, gases and liquids, and has an equivalent called "stress" in solid materials. Most of the time we measure both in pounds per square inch but there are occasions when we'll use pounds per square foot. Pounds per square inch is abbreviated psi. The units mean we are measuring force per unit area. It isn't hard to imagine a square inch. It's an area measuring one inch wide by one inch long. Then, if we piled one pound of water on top of that area the pressure on that surface would be one pound per square inch. If we pile the water up until there was one hundred pounds of water over each square inch, the pressure on the surface would be 100 psi. It isn't necessary for the fluid to be on top of the area because the pressure is exerted in every direction, a square inch on the side of a tank or pipe centered so there's one hundred pounds of water on top of every square inch above it sees a pressure of 100 psi. The air in a compressed air storage tank is pushing down, up and out on the sides of the tank with a force, measured in pounds, against each square inch of the inside of the tank and we call that pressure.

When we're dealing with very low pressures, like the pressure of the wind on the side of a building, we might talk about pounds per square foot but it's more common to use inches of water. A manometer with one side connected to the outside of the building and another to the inside would show two different levels of water and the pressure difference between the inside and outside of the building is identified in inches of water, the difference in the water level. It's our favorite measure for air pressures in the air and flue gas passages of the boiler and the differential of flow measuring instruments.

There is another measure of pressure we use; "head" is the height of a column of liquid that can be supported by a pressure. I have a system for remembering it, well… actually I mean calculating it. I can remember that a cubic foot of water weighs 62.4 pounds. A cubic foot being 12 inches by 12 inches by 12 inches means a column of water one foot high will bear down on one square foot at a pressure of 62.4 pounds per square foot. Divide that by 144 square inches per square foot to get 0.433 pounds in a column of water one inch square and one foot high so one foot of water produces a pressure of 0.433 psi. Divide that number into one and you get a column of water 2.31 feet tall to produce a pressure of one psi. The reason we use head is because pumps produce a differential pressure, which is a function of the density of the liquid being pumped, see the chapter on pumps and fans.

Head in feet and inches of water (abbreviated "in. W.C." for inches of water column) are both head measurements even though a value for head is normally understood to mean feet.

Okay, now we've got pressure equal to psi, why do we see units of psig and psia? They stand for pounds per square inch gage and pounds per square inch absolute. The difference is related to what we call atmospheric pressure. The air around us has weight and there's a column of air on top of us that's over thirty miles high. That may sound like a lot but if you wanted to simulate the atmosphere on a globe (one of those balls with a map of the earth wrapped around it) the best way is to pour some water on it. After the excess has run off the wet layer that remains is about right for the thickness of the atmosphere, about three one-hundredths of an inch on an eight inch globe. Anyway, that air piled up over us has weight. The column of air over any square inch of the earth's surface, located at sea level, is about 15 pounds. Therefore, the atmosphere exerts a pressure of 15 pounds per square inch on the earth at sea level under normal conditions. (The actual standard value is 14.696 psi but 15 is close enough for what we do most of the time) If you were to take all the air away we wouldn't have any pressure, it would be zero.

A pressure gage actually compares the pressure in the connected pipe or vessel and atmospheric pressure. When the gage is connected to nothing it reads zero,

there's atmospheric pressure on the inside and outside of the gage's sensing element. When the gage is connected to a pipe or vessel containing a fluid at pressure the gage is indicating the difference between atmospheric pressure and the pressure in the pipe or vessel. Absolute pressure is a combination of the pressure in the pipe or vessel and atmospheric pressure. Add 15 to gage pressure to get absolute pressure, the pressure in the vessel above absolutely no pressure. If you would like to be more precise use 14.696 instead of 15. Atmospheric pressure varies a lot anyway so there's not a lot of reason to be really precise.

Later we'll also cover stress, the equivalent of pressure inside solid material, under strength of materials.

Viscosity is a measurement of the resistance of a fluid to flowing. All fluids, gases and liquids have a viscosity that varies with their temperature. Normally a fluid's viscosity decreases with increasing temperature. You're familiar with the term "slow as molasses in January?" Cold molasses has a high viscosity because it takes a long time for it to flow through a standard tube, what's called a viscometer. The normal measure of viscosity is the time it takes a certain volume of fluid to flow through the viscometer and that's why you'll hear the viscosity described in terms of seconds. A chart for conversion of viscosities is included in the appendix along with the viscosity of some typical fluids found in a boiler plant. More on viscosity when we discuss fuel oils.

It's only fair to mention, while we're discussing measurements, that there is something called dimensional analysis. Formulas that engineers use are checked for units matching on both sides of the equation to ensure the formula is correct in its dimensions (measurements). It ensures that we use inches on both sides of an equation, not feet on one side and inches on the other. Since I promised you at the beginning of the book that you wouldn't be exposed to anything more complicated than simple math (add, subtract, multiply and divide) I can't get any more specific than that. Just remember that you have to be consistent in your use of units when you're making calculations.

Not a real measurement but a value used in boiler plants is "turndown." Turndown is another way of describing the operating range of a piece of equipment or system. Instead of saying the boiler will operate between 25% and 100% of capacity we say it has a four to one turndown. The full capacity of the equipment or system is described as multiples of the minimum rate it will operate at. Unless you run into someone that uses some idealistic measurement (anybody that says a boiler has a

3 to 2 turndown must be a novice in the industry) minimum operating rate is determined by dividing the larger number into one. If you run into the nut that described a 3 to 2 turndown then the minimum capacity is 2/3 of full capacity. Divide the large number into one and multiply by 100 to get the minimum firing rate in percent.

We also use the term "load" when describing equipment operation. Load usually refers to the demand the facility served places on the boiler plant but, within the correct context, it also implies the capacity of a piece of equipment to serve that load. If we say a boiler is operating at a full load that means it is at its maximum; half load is 50%, etc.

A less confusing but more difficult measure to address are "implied" measures. Some are subtle and others are very apparent. A common implied measure in a boiler plant is half the range of the pressure gauge. Engineers normally select a pressure gauge or thermometer so the needle is pointing straight up when the system is at its design operating pressure or temperature. We always assume that the level in a boiler should be at the center of the gauge glass, that's another implied measurement. In other cases we expect the extreme of the device to imply the capacity of a piece of equipment; steam flow recorders are typically selected to match the boiler capacity even though they shouldn't be. The problem with implied measurements is that we can wrongly assume they are correct when they're not. Keep in mind that someone could have replaced that pressure gauge with something that was in stock but a different range. I failed to make that distinction one day and it took two hours of failed starts before I realized the gauge must be wrong and went looking for the instruction book. Yes, I've done it too.

Probably one of the most common mistakes I've made, and that I've seen made by operators and construction workers, is not getting something square. All too often we'll simply eyeball it or use an instrument that isn't adequate. The typical carpenter's square, a piece of steel consisting of a two foot length and sixteen inch length of steel connected at one end and accepted as being connected at a right angle works well for small measurements but using it to lay out something larger than four feet can create problems. I say "accepted as being square" because I've used more than one of them to later discover they weren't. Drop a carpenter's square on concrete any way but flat and you'll be surprised how it can be bent. On any job that's critical, always check your square by scribing a line with it and flipping it over to see if it shows the same line. Of course the one side you're dealing with has to be straight. Eyeball-

ing (looking along the length of an edge with your eye close to it) is the best way to check to confirm an edge is straight.

For measures larger than something you can check with that square you should use a 3 by 4 by 5 triangle; the same thing the Egyptians used to build the pyramids. You lay it out by making three arcs as indicated in Figure 1-1. You frequently also need a straight edge as the reference that you're going to be square to, in which case you mark off 3 units along that edge to form the one side, that's drawing the arc to find the point B by measuring from point A. An arc is made 4 units on the side at point C by measuring from point A then another arc of 5 units is made measuring from point B and laying down an arc at D. Where the A to C and B to D arcs cross (point E) is the other corner of the 3 by 4 by 5 triangle and side A to B is square to A to E. The angle in between them is precisely 90 degrees.

The beauty of the 3 by 4 by 5 triangle is the units can be anything you want as long as the ratio is 3 to 4 to 5. Use inches, or even millimeters, on small layouts, and feet on larger ones. If you were laying out a new storage shed you might want to make the triangle using 30 feet, 40 feet, and 50 feet. It's difficult to get more precise, even if you're using a transit.

Another challenge is finding a 45 degree angle. The best solution for that is to lay out a square side to get that 90 degree angle then divide the angle in half. Figure 1-2 shows the arrangement for finding half an angle. Simply measure from the corner of the angle out to two points (C and D) the same distance (A to B) then draw two more arcs, measuring from points C and D a distance E, and F identical to E to locate a point where the arcs cross at G. A line from A to G will be centered

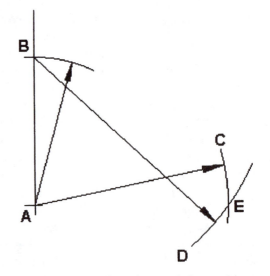

Figure 1-1. Creating a right angle

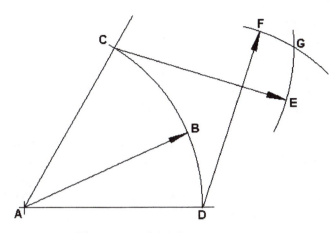

Figure 1-2. Dividing an angle

between the two sides, splitting the angle. If you started with a 90 and wanted to split it into three 30's, measure off F at twice the length of E then shift around to get two points that are at 30 and 60 degrees. The same scheme will allow you to create any angle.

WATCH OUT FOR WADITWs

Huh? What's a WADITW? You should know because you've heard it many times and it's used regularly by a lot of people. It's another mnemonic (an abbreviation which assists the reader in determining the meaning because each letter represents a word). You really must watch out for these because they're not wise; think twice when you hear it and back off when you say "We Always Did It That Way."

A few years ago I was called upon to assist in the design of an installation of new burners plus burner management and controls on three existing boilers in a central heating plant. Observation of the plant's operation and a review of the fuel records revealed that the maximum load was less than two-thirds of the capacity of one boiler so I suggested that the design include only two boilers and reducing the capacity of them to reduce turn-down losses. The operator's instant response was that I couldn't do that because "We Always Run Two Boilers In the Winter." Not exactly a WADITW but it's the same concept. After several exchanges and their being convinced that I was wrong, they decided to prove their point and did so by collecting data. They recorded

outdoor air temperature and oil consumption for several hours then produced that data at our next meeting indicating it proved their point.

I charted that data and the results are shown in Figure 1-3. Note that an increase in fuel usage with lower temperatures is obvious; then note the difference between the markers. The round symbols are for readings with two boilers in operation but the square ones are for one boiler in operation. Not only were they able to operate one boiler at colder temperatures, they burned less fuel doing so. Note that on the coldest hour, 16°F, they only burned 300 gallons of fuel while operating one boiler but on a warmer hour, 22°F, they burned 420 gallons. Imagine that, a difference in fuel of 120 gallons per hour! At current prices today, near $4 per gallon, that's $480 per hour more for fuel. This is only one example of how operators can make a difference—when they operate wisely.

TRENDS AND CHARTING DATA

Figure 1-3 and the accompanying story is only one example of several situations where graphing data managed to prove a point which Owners and Operators simply didn't understand until they looked at a graph. It's interesting that most of time I'm not sure about the advantages of looking at data until I have generated a graph using that data. It's also not uncommon that I surprise Owners and operators when the graphics revealed

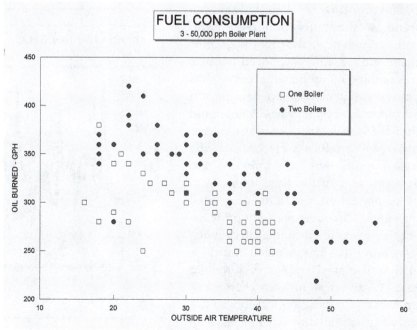

Figure 1-3. A Comparison of boiler operation

that they had a misconception about their operations just like the one mentioned above.

New control systems and data acquisition systems typically have graphic capability. They are typically referred to as trends. When you have the opportunity to use the systems to look at your data take advantage of it. Trending data which produces a graph on the monitor or display in your plant on a regular basis simply provides one more visual aid that you can use to catch problems early or detect changes that you would not have noticed otherwise.

Lacking on the fancy instrumentation doesn't restrict you in creating graphics that serve as a visual aid. You can do what I have done for years. Leaf through the logbook and enter the data into one of the spreadsheet programs then use that program to produce a graph. That's how the graphic of Figure 1-3 was produced. I typically use Lotus 123 most of the time but will also convert it to produce one in Excel for customers that don't have Lotus.

Hopefully the discussion about the situation that produced the graphic in Figure 1-3 is a clue. If you don't graph your data, someone else, perhaps like me, will produce one that contradicts your perception of your facility's operation. In other words, someone could prove that what you are doing is wrong! I don't know who said it but you should remember the saying "if you always do what you've always done, you'll always get what you always got." Make your next step in operating wisely be spending time looking at trends and graphs of your data.

According to the dictionary a trend is "the general course or prevailing tendency" and it's the graphic that's most readily produced because it is nothing more than a record of data values, frequently called data points, all taken at specific times. You should find that the instrumentation systems today will produce a graphic showing data for different time intervals and you can specify the time intervals to be displayed on a monitor or printed on paper. Of course we only print the graphic when it's meaningful, showing something that has changed. You look at that displayed graphic on a regular basis to detect significant changes or undesirable trends. Typical indications include a sudden drop in condensate returns (sometimes indicated by a sudden

increase in makeup) which frequently allows connecting the change to another event that occurred at the same time. Undesirable trends can include a constant decrease in evaporation rate, an increasing stack temperature, and a gradual decrease in condensate temperature, all of which are indicative of something going wrong and a need to discover what's causing it.

There are, however, load related reasons for undesirable trends and situations where changes in load cloud the issue, producing oscillations in a trend graphic that prevent detecting trends. Many times those deviations can be resolved by filtering the data so you're only looking at a trend graphic for a specific load on a boiler, plant, or other specific equipment or system. At other times you may want to actually plot the data relative to the load. Figure 1-4 is one example of a trend graph for outside air temperature and fuel use for a large office building during a cool week. The fuel use obviously increases as the temperature drops. Later we'll look at another way to compare these values using degree days. The effect of night set-back when the building is closed from 7 pm to 6 am and weekends reveals the reasons the fuel use is not consistent for changes in outdoor air temperature alone. The peaks of fuel use around 6 and 7 in the morning of the first five days are due to recovering from the night set back. They're missing the last two days because that was the weekend. The set back also reduces the fuel use over the weekend. You'll also notice the increases in load associated with people entering and leaving the building when air infiltration

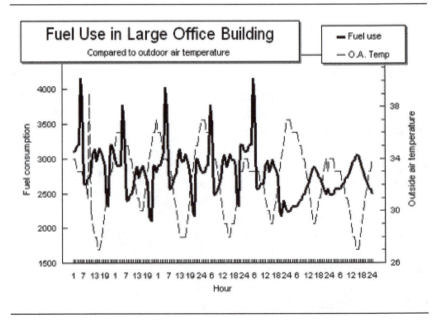

1-4. Fuel use trend for an office building

is greatest. There are many activities that can have an effect on load that you'll have to pick out to understand your operation. In addition to the ones mentioned here there's the effect of wind, solar gain (sunny or cloudy), unusual activities such as a conference that increases use of doors, hot water heating and such things as kitchens, laundries, and other heat users.

FLOW

Here's a concept that always raises eyebrows: You can't control pressure; you can't control temperature; you can't control level; the only thing you can control is flow. Before you say I'm crazy, think about it. You maintain the pressure or temperature in a boiler by controlling the flow of fuel and air. You maintain the level by controlling the flow of feedwater. Pressure, temperature, level, and other measures will increase or decrease only with a change in flow. An increase in flow will increase or decrease the value we're measuring depending on the direction of the flow.

That's usually my first statement in response to operators' questions about their particular problem in maintaining a pressure, temperature or level. It always brings a frown to the operator's face and I continue relating it to their specific problem until that frown turns into a bright smile. They don't get an answer to their problem from me; they get an introduction to the concept of flow and how it affects the particular measure they are concerned with so they can see for themselves what is causing their problem. It's a fundamental that, once grasped, will always serve an operator in determining the cause of, and solution to, a problem with control.

If you don't buy it you simply have to think about it for a while. Read that first paragraph again and think about your boiler operation and you'll eventually understand it. There's absolutely no way for you to grab a pressure, temperature, or level and change it. Any description you can come up with for changing those measures always involves a change in flow.

Now that you have the concept in hand, let's talk about how you control flow to maintain all those desirable conditions in the boiler plant. You have two means for controlling flow. You can turn it on and off or you can vary the flow rate. When you're changing the flow rate we call it "modulating" and the method is called "modulation." To restore the level in a chemical feed tank you open a valve, shut it when the level is near the top, and you add chemicals to restore the concentration; that's on-off control. A float valve on a make-up water

tank opens as the level drops to increase water flow and closes to decrease flow as the level rises; that's modulation. There is, of course, more to know and understand about these two methods of control but they'll be addressed in the chapter on controls; we need to learn a lot more about flow itself right now.

Accepting the premise that all we can control is flow makes it a lot simpler to understand the operation of a boiler plant. Every pound of steam that leaves the boiler plant must be matched by a pound of water entering it or the levels in the plant will have to change. Water wasted in blowdown and other uses like softener regeneration must also be replaced by water entering the plant.

The energy in the steam leaving the boiler plant requires energy enter the plant in the form of fuel flow. If the steam leaving contains more energy than is supplied by the fuel entering then the steam pressure will fall. Some of the energy in the fuel ends up in the flue gases going up the stack so the energy in the fuel has to match the sum of the energy lost up the stack and leaving in the steam. The sum of everything flowing into the boiler plant has to match what is flowing out or plant conditions will change. An operator is something of a juggler. You are always performing a balancing act controlling flows into the plant to match what's going out.

A boiler operator basically controls the flow of fluids. The energy added to heat water or make steam comes from the fuel and you control the amount of energy released in the boiler by controlling the flow of the fuel. Gas and oil are both fluids because they flow naturally. Operators in coal fired plants could argue they are controlling the flow of a solid but when they look at it they'll realize that they're treating that coal the same way they would a fluid. The only other flow an operator controls is the flow of electrons in electrical circuits, another subject for another chapter—electricity. Controlling those flows requires you understand what makes them flow and how the flow affects the pressures and temperatures you thought you were controlling.

All fluids have mass. Fuel oil normally weighs less than water. Natural gas weighs less than air but it still has mass. We can treat them all the same in general terms because what happens when they flow is about the same. Gas and air are a little more complicated because they are compressible, their volume changes with pressure. In practice the relationship of flow and pressure drop are consistent regardless of the fluid so we'll cover the basics first.

Flow metering using differential pressure is based on the Bernoulli principle. Bernoulli discovered the relationship between pressure drop and flow back in the seventeenth century and, since it's a natural law of physics, we'll continue to use it. In order for air to flow from one spot to another, the pressure at spot one has to be higher than the pressure at spot two. It's the same as water flowing downhill. The higher the pressure differential the faster a fluid will flow. If you think about the small changes in atmospheric pressure causing the wind, you know it doesn't take a lot of difference in pressure to really get that air moving. Bernoulli discovered the total pressure in the air doesn't change except for friction and that total pressure can be described as the sum of static pressure and velocity pressure.

The measurement of static pressure, velocity pressure, and total pressure is described using Figure 1-5. The static pressure is the pressure in the fluid measured in a way that isn't affected by the flow. Note that the connection to the gage is perpendicular to the flow. The gage measuring total pressure is pointed into the flow stream so the static pressure and the velocity pressure are measured on the gage. What really happens at that nozzle pointed into the stream is the moving liquid slams into the connection converting the velocity to additional static pressure sensed by the gage. There is no flow of fluid up the connecting tubing to the gauge. The measurement of velocity pressure requires a special gage that measures the difference between static pressure and total pressure. With that measurement we can determine the velocity of the fluid independent of the static pressure. A velocity reading in a pipe upstream of a pump, where the pressure is lower, would be the same as in a pipe downstream of the pump (provided the pipe size is the same).

If you've never played in the creek before, go give it a try to see how this works. Notice the level of water leaving a still pool and flowing over and between some rocks. Put a large rock in one of the gaps and you'll reduce the water flow through that gap but that water has to go somewhere. The level in the pool will go up, probably so little that you won't notice it because the water flow you blocked is shared by all the other gaps and the only way more water can flow is to have more cross-section to flow through. I think I learned more about hydraulics (the study of fluid flow) from playing in the creek in my back yard than I ever learned in school. You could gain some real insight into fluid flow by spending some time observing a creek. That's a creek, now, not a large deep river. All the education is acquired by seeing how the water flows over and through the rocks and relating what you see to the concepts of static, velocity, and total pressure.

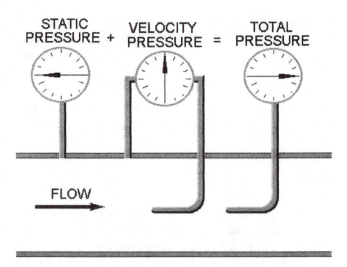

Figure 1-5. Static, velocity, total measurements

PRESSURE DROP AND FLOW

There's another thing about flow that's important to understand; a change in pressure drop is proportional to the square of a change in flow. You'll recall from high school math that you multiply a number by itself to square it. In my boiler operator training classes I use the diagram (Figure 1-6) of a sprinkler on the lawn hooked up to a hose after a tee with a pressure gage that reads 10 psig and ask the operators what the gage should read when the flow is twice as much as shown. Almost invariably I get the answer of "20 psig" but I also get a bunch of wild guesses and it was two years before I got the right answer after hundreds of classes. If I want two times the flow the pressure drop has to increase by four times (two squared is four) and the gage has to read 40 psig.

I also demonstrate to those classes that fractions can be used to compare different situations. Imagine a

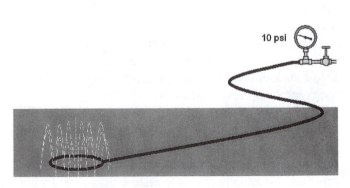

1-6. A lawn sprinkler example

gas-fired boiler where the gas head pressure is 16 inches of water column at full load. What's the pressure going to be at three-fourths load? Squaring three-fourths gives you nine-sixteenths which permits you to answer the question by saying 9 inches of water column. Note that nine-sixteenths is much closer to one-half than it is to three-fourths. If the firing rate is decreased to one-half then the pressure drop is one-fourth and the gas head pressure would be only 4 inches of water column. At one tenth load the pressure drop is one one-hundredth of full load. That should help explain why many of your flow instruments fail to record accurately at flows less than 10%; the pressure drop is so small that its value is lower than the pressure fluctuations associated with noise in the piping. I trust you can now see that, if you don't understand this fact of nature, you can make some very erroneous decisions about your operation.

The square root chart in the Appendix is provided so you can use percentages to obtain more precise values. Divide the full load pressure drop you note on gages between two points in a system (actually using the same gage connected at two points would be better, so you can eliminate gage calibration error) to get a 100% (full load) pressure drop value. For any other pressure drop divide it by that full load value to get a percent pressure drop. The result of the division will always be less than one and more than zero. Then draw a horizontal line from that value to the intersection of the curve and a vertical line down to the bottom scale to read the percent of full load flow. To compare one state to another use the one with the higher pressure drop as if it was full load and perform the math again to determine the flow at the lower pressure drop in percent of the value of flow at the higher pressure drop.

WHAT COMES NATURALLY

Observing everything in nature helps you understand what's going on in the boiler plant. Most of our engineering is based on learning about what happens naturally then using it to accomplish purposes like making steam. The formation of clouds, fog, and dew all conform to rules set up by nature. By observing them we learn cause and effect and can make it work for us. We can be just like Newton, sitting under the apple tree and being convinced, by an apple dropping, that there's such a thing as gravity and we can use it to do some work for us. You can see how it works, then relate it to what's happening in the boiler plant.

Many natural functions occur in the boiler plant and by observing nature we can get a better understanding of what's going on. Steam is generated and condensed by nature, we experience it by rain falling and noticing the puddles disappear when it's dry. Fire occurs naturally and we can see what happens when the fuel and air are mixed efficiently (as in a raging forest fire) and not so efficiently (our smoldering campfire). We can observe the hawks spinning in close circles in a rising column of air heated by a hot spot on the ground or air deflected by wind hitting a mountain. Even though we can't see the air, can understand buoyancy or how an air stream is diverted.

Buoyancy is also evident in a block of wood floating on water. The wood is not as dense as the water so it is lifted up. The hot air the hawks ride is not as dense as cold air so it floats up in the sea of colder air around it. The movement of air and gases of different densities is important in a boiler plant, we refer to it as "natural draft," movement of air that naturally occurs because air or gas of higher temperatures is lighter than colder surroundings and rises.

We can see the leaves and twigs in a stream spin off to the side indicating the water is deflected by a rock in the stream. We can see the level of the water increase beside the rock revealing the increase in static pressure as the velocity pressure is converted when it hits the rock. That conversion of velocity pressure to static pressure is how our centrifugal fans and pumps work.

When something happens that doesn't make sense try to relate it to what you observe happening in nature. That's how I arrive at many solutions to problems.

WATER, STEAM AND ENERGY

At almost every hearing for the installation or expansion of a new boiler plant there is the proverbial little old lady in tennis shoes claiming we don't need the plant because it's much easier and cleaner to use electricity. We have to explain to her that almost all the electricity is generated using boilers, even nuclear power. Each time I'm questioned about why the facility needs a boiler plant I think of how history was shaped by the use of boilers. If it were not for the development of boilers, we could still be heating our homes with a fireplace in each room; imagine the environmental consequences of that!

Most people know so little about the use of water and steam for energy that it's important to establish an understanding of the very simple basics, which is what I'll attempt to do in this section. Although you may feel you understand the basics you want to read this section because there are some simple shortcuts described here that can help you.

Water is the basis for heat energy measurement. Our measure of heat energy, the British thermal unit (Btu for short) is defined as the amount of heat required to raise the temperature of water one degree Fahrenheit. We engineers know that's not precisely true at every condition of water temperature but it's good enough for the boiler operator. As for the energy in steam, well it depends on the pressure and temperature of the steam but, for all practical purposes it takes 1,000 Btu to make a pound of steam and we get it back when the steam condenses.

If you want to be more precise, you can use the steam tables (Page 353) A few words on using those steam tables is appropriate. Engineers use the word "enthalpy" to describe the amount of heat in a pound of water or steam. We needed a reference where the energy is zero and chose the temperature of ice water, 32°F. That water has no enthalpy even though it has energy and energy could be removed from it by converting it to ice. So, the enthalpy of water or steam is the amount of energy required to get a pound of water at freezing temperature up to the temperature of the water or steam at the outlet of the boiler. Since we use freezing water as a reference point, the difference in enthalpy is always equal to the amount of heat required to get one pound of water from one condition to the other.

Did I forget to mention that steam is really water? Some of you are going to wonder about my sanity in making such a simple statement but I've run into boiler operators that couldn't accept the concept that the water going in leaves as steam. Steam is water in the form of gas. It's the same H_2O molecules which have absorbed so much energy, heated up, that they're bouncing around so frantically that they now look like a gas. The form of the water changes as heat is added, it gets hotter until it reaches saturation temperature. Then it converts to steam with no change in temperature and finally superheats. There is, for each pressure, a temperature where both water and steam can exist and that's what we call the saturation point or saturation condition.

Most of us are raised to know that water boils at 212°F. That's only true at sea level. In Denver, Colorado, it boils at about 203°F. Under a nearly pure vacuum, 29.75 inches of mercury, it boils at 40°F. The steam tables list the relationships of temperature and pressure for saturated conditions. Since a boiler operator doesn't need to be concerned with the small differences in atmospheric pressure the table shows temperatures for inches of mercury vacuum and gage pressure. If you

happen to be a mile high, like Denver, you'll have to subtract about 3 psi from the table data. Any steam table used by an engineer will relate the temperatures to absolute pressure.

What is absolute pressure? If you must ask you missed it in the part on measurements, flip back a few pages.

Provided the temperature of water is always less than the saturation temperature that matches the pressure the water is exposed to, the water will remain a liquid and you can estimate the enthalpy of the water by subtracting 32 from the temperature in degrees Fahrenheit. For example, boiler feedwater at 182°F would have an enthalpy of 150 Btu. It takes 970 Btu to convert one pound of water at 212°F to steam at the same temperature so you're reasonably accurate if you assume steam at one atmosphere has an enthalpy of 1,150 Btu (212–32+970). If we sent the 182°F feedwater to a boiler to convert it to steam, we would add 1,000 Btu to each pound. Just remembering 32°F water has zero Btu and it takes 970 Btu to convert water to steam from and at 212°F is about all it takes to handle the math of saturated steam problems.

We do have other measures of energy that's unique to our industry. One is the Boiler Horsepower (BHP). With 1,000 Btu to make a pound of steam and the ability to generate several hundred pounds of it the numbers get large and cumbersome, so the term Boiler Horsepower was standardized to equal 34.5 pounds of steam per hour from and at 212°F. Since we know that one pound requires 970 Btu at those conditions a boiler horsepower is also about 33,465 Btu per hour (34.5 × 970), more precisely it's 33,472. It's important here to note the distinction that a Boiler Horsepower is a rate value (quantity per hour) and Btu's are quantities. We abbreviate Btu's per hour "Btuh" to identify the number as representing a rate of flow of energy.

Another measure of energy unique to our industry, but not used much anymore, is Sq. Ft. E.D.R. meaning square feet of equivalent direct radiation. It's also a rate value. It was used to determine boiler load by calculating the heating surface of all the radiators and baseboards in a building. There are two relative values of Sq. Ft. E.D.R. depending on whether the radiators are operating on steam or hot water. It's 240 Btuh for steam and 150 Btuh for water. There are rare occasions when you will encounter the measure but its better use is to relate what happens with heating surface. If a steam installation were converted to hot water, it would need an additional 60% (240/150 = 1.6) of heating surface to heat the same as the steam. Flooded radiators can't produce the same amount of heat as one with steam in it even though the water is at the same temperature.

The rate of heat transfer from a hot metal to steam and vice versa is always greater than heat transfer from a hot metal to water. It's because of the change in volume more than anything else. Take a simple steam heating system operating at 10 psi (240°F). Check the steam tables and you'll find a pound of water occupies 0.01692 cubic feet and a pound of steam occupies 16.6 cubic feet. As the steam is created it takes up almost 1,000 times as much space as the liquid did. That rapid change in volume creates turbulence so the heating surface always has water and steam rushing along it. It's about the same effect as you experience when skiing or riding in a convertible, you're cooler because the air is sweeping over your skin. When the steam is condensing it collapses into a space one one-thousandth of it's original volume and more steam rushes in to fill the void. That's the mechanism that improves heat transfer with steam, not the fact that steam has more heat on a per pound basis.

Steam may have more heat per pound but those pounds take up a lot more space. One cubic foot of water at 240°F contains 12,234 Btu but one cubic foot of steam only contains 69.88 Btu. Say, that provokes a question. Why don't we only use hot water systems because water can hold more heat? The best answer is because we would have to move all those pounds of water around to deliver the heat. To deliver the heat provided by one pound of steam would require about 200 pounds of water. Steam, as a gas, naturally flows from locations of higher pressure to those of lower pressure, we don't have to pump it. The rate of water flow is restricted to about 10 feet per second to keep down noise and erosion. Steam can flow at ten times that speed. Nominal design for a steam system is a flowing velocity of about 6,000 feet per minute. If you found that confusing, check the units, there are 60 seconds in a minute.

Hot water is a little easier to control when we have many low temperature users. A hot water system has a minimal change in the volume of the water at all operating temperatures. For that reason we will pay the cost of pumping water around a hot water system in exchange for avoiding the dramatic volume changes in steam systems. Never forget that there is a change in volume in a hot water system; to forget is to invite a disaster. Water changes volume with changes in temperature at a greater rate than anything else, almost ten times as much as the steel most of our boiler systems are made of; see the tables in the appendix. Unlike steam it doesn't compress as the pressure rises so the system must allow

it to go somewhere. The normal means for the expansion of the water in a hot water system is an expansion tank, a closed vessel containing air or nitrogen gas in part of it. Modern versions of expansion tanks have a rubber bladder in them to separate the air and water. The bladder prevents absorption of the air into the water. The air or nitrogen compresses as the water expands, making room for the water with a little increase in overall system pressure. Tanks without bladders normally have a gage glass that shows the level of the water in the tank so you can tell what their condition is.

A hot water system will also have a means to add water, usually directly from a city water supply. Most have a water pressure regulator that adds water as needed to keep the pressure above the setting of the regulator. A relief valve (not the boiler's safety valve) is also provided to drain off excess water. Older systems can be modified and added to the extent that the expansion tank is no longer large enough to handle the full range of expansion of a system. In some newer installations I've found tanks that were not designed to handle the full expansion of the system. Those systems require automatic pressure regulators to keep pressure in the system as the water shrinks when it cools and the relief valve to dump water as it expands while the system heats up. The tank should be large enough, however, to prevent the constant addition and draining of water during normal operation. A good tight system with a properly sized expansion tank should retain its initial charge of water and water treatment chemicals to simplify system maintenance.

All hot water systems larger than a residential unit should have a meter in the makeup water line so you can determine if water was added to the system and how much. Lacking that meter a hot water system can operate with a small leak for a long period of time during which scale and sludge formation will occur until you finally notice the stack temperature getting higher or some other indication of permanent damage to the boiler or system.

Steam compresses so there is seldom a problem of expansion with steam boilers unless you flood the system. However, since steam temperature and pressure is related when using steam at low temperatures we frequently get a vacuum and air from the atmosphere leaks in. We will say a vacuum "pulls" air in but it really doesn't have hands and arms that can reach out to grab the air. The atmospheric air is at a higher pressure so it will flow into the vacuum. In those cases where we have a tight system the vacuum formed as steam condenses will approach absolute zero so the weight of the air outside the system will produce a differential pressure of 15 psi which can be enough to crush pressure vessels in the system. To prevent that happening low temperature steam systems usually have vacuum breakers to allow air into the system. Check valves make good vacuum breakers because they can let air in but not let the steam out. Thermostatic steam traps and air vents are required to let the air out when steam is admitted to the system. If installed and operated properly low pressure steam systems can work well because the metal in the system will be hot and dry when the air contacts it so corrosion is minimal.

To know how much heat is delivered per hour you determine the difference in enthalpy of the water or steam going to the facility and what's returning then multiply that difference by the rate of water or steam flowing to the process. The basic formula is (enthalpy in less enthalpy out times pounds per hour of steam or water). In the case of water there's a little problem with that formula because you normally determine flow in water systems in gallons per minute. Well, just like the others, there's a simple rule of thumb; gpm times 500 equals pounds per hour. One gallon of water weighs about 8.33 pounds and one gpm would be 60 gallons per hour so 8.33 × 60 equals 499.8 and that's close enough. Since the difference in enthalpy is about the same as the difference in temperature for water, heat transferred in a hot water system can be calculated as temperature in minus temperature out multiplied by gpm times 500.

For steam systems it's simply 1,000 times the steam flow in pounds per hour if the condensate is returned. There are times when the condensate isn't returned because a condensate line or pump broke or the condensate is contaminated. That's common in a lot of industrial plants because it's too easy for the condensate to be contaminated so it's wasted intentionally. In those circumstances you have to toss in the heat lost in the condensate that would have been returned. What you're really delivering to the plant under those conditions is the heat to convert the water to steam plus the energy required to heat it from makeup temperatures to steam temperature.

There are also applications where the steam is mixed with the process, becoming part of the production output. An example is heating water by injecting steam into it. The amount of heat you have to add to make the steam is the same as the previous example but the heat delivered to the process is all the energy in the steam.

The one problem many boiler operators have is grasping the concept of saturation. Steam can't be generated until the water is heated to the temperature corresponding to the saturation pressure. Once the water is at that temperature, the temperature can't go any higher as long as water is present. At the saturated condition any addition of heat will convert water to steam and any removal of heat will convert steam to condensate. The temperature cannot change as long as steam and water are both present. When the heat is only added to the steam then the steam temperature will rise because there's no water to convert to steam. Whenever the steam temperature is above the saturation temperature it is called superheated.

Superheated steam doesn't just require addition of heat. If you have an insulated vessel containing nothing but saturated steam and lower the pressure then the saturation temperature drops. The energy in the steam doesn't change so the temperature cannot drop and the steam is superheated. In applications where high pressure steam is delivered through a control valve to a much lower pressure in a process heater the superheat has to be removed before the steam can start to condense. The heat transfer is from gas to the metal, without all the turbulence associated with steam condensing to a liquid. It isn't as efficient as the heat transfer for condensing steam. Process heaters can be choked by superheated steam where the poor gas to metal heat transfer leaves much of the surface of the heat exchanger unavailable for the higher rates of condensing heat transfer. That's right, your concept that superheated steam would be better just went out the window.

So why superheat the steam? We superheat steam so it will stay dry as it flows through a steam turbine or engine. Without superheat some water would form as soon as energy is extracted. The water droplets would impinge on the moving parts of the turbine (a familiar concept would be spraying water into the spinning wheel of a windmill) damaging the turbine blades. In an engine it would collect in the bottom of the cylinder. In electric power generating plants it's common to pipe the steam out of the turbine, raise its temperature again (reheating it) then returning it to the turbine just to maintain the superheat.

When we're generating superheated steam some of it is needed for uses other than the turbine so we don't want it superheated. In that case we desuperheat it. Heat is removed or water is added to the superheated steam for desuperheating. When water is added, it absorbs the heat required to cool the steam by boiling into steam. In most applications superheat cannot be eliminated entirely because we need some small amount of superheat to detect the difference between that condition and saturation. As long as we have a little superheat, we know it's all steam. When it is at saturation conditions, we can't tell how much water is in the steam.

Understanding saturation is the key to understanding steam explosions. When water is heated to saturation conditions higher than 212°F, as in a boiler, it cannot exist as water at that temperature if the vessel containing it fails. Under those circumstances the saturated condition becomes one atmosphere and 212°F as the water leaks out. A portion of the water is converted to steam to absorb the heat required to reduce the temperature of the remaining water to 212°F. How much steam is generated is determined by the original boiler water temperature but every pound of water converted to steam expands to 26.8 cubic feet. The rapid expansion of the steam is the steam explosion.

Let's do the math for a heating boiler operating at 10 psig. The 240°F water has to cool to 212°F releasing 28 Btu per pound. It can only do so by generating steam at 212°F which contains 1,150 Btu per pound. One pound of steam can cool 41 pounds of water (1,150 ÷ 28). The volume of 42 pounds of 240°F water at 0.01692 cubic feet per pound (0.71 cubic feet) becomes 41 pounds of water at 212°F (0.01672 × 41 = 0.685 cubic feet) and one pound of steam (26.8 cubic feet) so the original volume of water expanded 38.71 times (0.685 + 26.8 = 27.48 ÷ 0.71) and it happens almost instantly.

Other situations involving steam at saturation are described in the discussion of equipment where it must be understood.

THE STEAM AND WATER CYCLE

For most of you who are operating commercial, industrial, or institutional boiler plants the concept of the steam and water cycle is a foreign concept. Understanding a cycle becomes important when you're generating power with the steam. By generating power I mean generating electricity or powering a mechanical device normally with a steam turbine. In order for this edition of the book to be more meaningful for plant operators I'll attempt to present some descriptions of steam and water cycles.

For almost any plant there is a steam and water cycle; the exception being hot water heating boilers where steam is not generated. In hot water heating plants the cycle is simply heat in and heat out. It is added to the water in the plant boilers, the water is transferred by circulating pumps (or natural circulation) to radiators, convectors, kitchen equipment, etc. where the water is cooled by the users of the heat, then the water is returned to the boiler to be heated again.

Flow arrows are seldom shown on a cycle diagram

because it's typically assumed that the fluids flow from higher pressures (higher in the diagram) to lower pressures. I included arrows to guide the novice. Valves, both isolating and control are not shown, nor are steam traps always shown. They are understood to exist in components that use the steam and where required to isolate systems for maintenance.

The simplest steam and water cycle exists in a low-pressure heating system. Water is pumped into the boiler, is heated to saturation temperature, then converted to steam. Almost all of the energy added to the water and steam in the boiler is latent heat. We refer to that latent heat as the latent heat of evaporation; the energy used to convert the water to steam. The steam then leaves the boiler and flows through piping and control valves to radiators, convectors, kitchen equipment, etc. where the steam is condensed. Those users of the heat primarily use the latent heat of condensation. There may be some heat transferred to the user by cooling the condensate. The condensate is then returned to the boiler completing the cycle. Figure 1-7 is an example of a steam cycle diagram for a conventional heating plant, perhaps a school, a restaurant, or an apartment building.

I can hear it now! "Hey, Ken, what's so simple about this diagram?" I couldn't resist throwing in some things to make you think about your facility. When steam is used to heat things at temperatures close to, at, or above 212°F the temperature of the condensate from that heating equipment is much higher and any drop in pressure will result in some of the condensate flashing into steam.

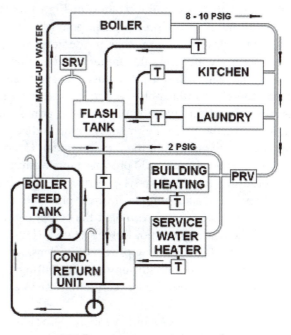

1-7. Low pressure steam cycle

That's very true for condensate that's formed in the main steam piping. The cycle diagram in Figure 1-7 is provided to explain how some little changes in the cycle can make a difference. First, I'll provide a few explanations for the diagram. Steam is represented by two lines, sort of like along a highway, where the steam flows down the middle of the lines. Water, being much heavier, is shown in black between the two lines. The boxes containing a capital "T" represent the steam traps for the systems. "SRV" represents a safety relief valve (not a safety valve) it's one that relieves a fluid proportional to the pressure above its set point. "PRV" represents a pressure reducing valve or a complete pressure reducing station.

What's going on here is a system that doesn't waste heat like so many I've seen. It contains a flash tank that does what it's supposed to do—unlike so many that are installed and do nothing or simply separate the flash steam from the condensate and throw it away. Some of the condensate that's drained from the steam piping will flash into steam. Condensate from the main header drains, kitchen and laundry is bound to be hotter than 212°F and once it leaves the steam traps some of it will flash into steam. By piping that condensate to the flash tank that flash steam is recovered at a lower pressure and can be used for heating the building in the winter and service water (domestic hot water at 120°F) in the winter and summer. Note that the condensate from different pressure sources is always piped independently to the flash tank so flashing condensate from the higher pressure sources doesn't block the flow of condensate from lower pressure sources. I do not like the use of check valves that, supposedly, prevent high pressure condensate entering low pressure equipment.

If the flash steam is more than the demand for heating and service water the safety relief valve (in this system set at 2.5 psig) will open and waste the excess steam because there's no place else to use it. Under normal operating conditions there is no waste to atmosphere because all the flash steam from the flashing higher pressure condensate is used in the 2 psig system. Condensate from the flash tank would produce more flash steam when entering the condensate return unit and that's why it is piped to a subsurface diffuser (pipe with holes drilled in the bottom) so it can heat the colder condensate from the heating and hot water systems. The steam condenses as it heats the low pressure condensate. The system practically eliminates any steam venting from the condensate unit and boiler feed tank (unless a trap has failed).

Why all that to save a little vented steam? Never… never… never… think of vented steam as "a little." Every pound of vented steam represents about 1,000 btu wasted. Some systems waste condensate that has to be replaced by cold makeup water and heated to saturated steam in the boiler. In our low pressure plant example that's 1,180 btu per pound. Part of the wise operator's job is to constantly monitor vents for wasted steam and try to do something about it.

In plants where the steam is primarily used to generate power the steam and water cycle becomes more complex. Some of these are referred to as supercritical boiler plants because the steam is generated at pressures exceeding 3,190 psig where the density of water and steam are the same. Density is simply a value representing the weight of a substance in pounds per cubic foot. Because there's no difference in density those boilers do not have a level indicator or gage glass because there's no water level. After the water is converted to steam in the boiler it passes through the superheater, more tubes exposed to the boiler flue gases or radiant heat from the fire where it's heated further increasing its temperature. Steam heated to temperatures higher than the saturation temperature is superheated steam. The superheated steam is piped to a steam turbine or steam engine where the energy in the steam is converted to power. The conversion of energy to create power is associated with a drop in steam pressure and temperature. Each horsepower of mechanical energy is produced by removing 2,545.1 btu per hour from the steam plus an allowance for losses due to inefficiency.

Many industrial plants produce superheated steam in higher pressure boilers to generate power with steam turbines that exhaust at a nominal high-pressure for use in the facility. The turbines are referred to as "back pressure turbines." I have worked on plants that exhaust steam at 15 psig after a turbine where the steam was used mostly for evaporators in the process plant and other facilities with turbine exhaust pressures as high as 150 psig for distribution to all users in a chemical plant.

Now compare that low pressure plant with an electrical utility like the one depicted in Figure 1-8. The plant consists of two boilers and two turbine sets. Each set of turbines (H.P. for high pressure, I.P. for intermediate pressure, and L.P. for low pressure) in this plant are all driving one shaft connected to their respective electricity generator.

Steam generated in the boiler is piped to headers either external to or within the boiler that connect to superheater tubes which raise the temperature of the steam from saturation to a maximum that's normally determined by the strength of the metal the superheater tubes are made of. Utility plant superheaters are typi-

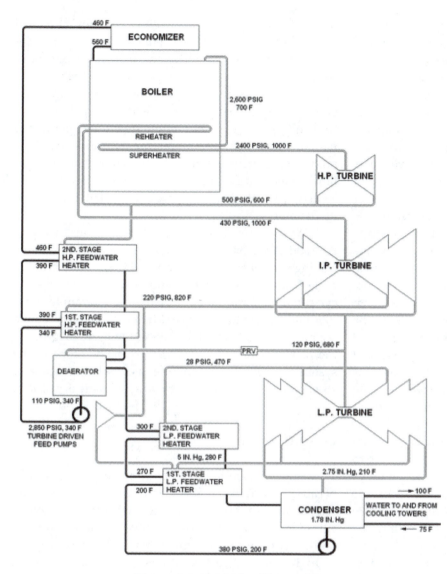

Figure 1-8. Typical utility cycle

extracted in high pressure turbines reduces the temperature of the steam so much that generating any additional power with that steam would result in the steam reaching saturated conditions where droplets of water would form, strike the rapidly rotating turbine blades, and damage the turbine. So utility plants typically have reheaters, additional tubes in the boiler, that are exposed to the flue gases before they reach the boiler section. There the steam from the high pressure turbine is heated again to increase its superheat before continuing through the intermediate pressure turbine. The reheater is "convective" because the superheater shields it from the radiant heat of the furnace and it is heated only by the flowing flue gas. Each "stage" of a turbine has different inlet and outlet pressures. So the reheater operates at a lower pressure than the boiler and superheater but can have an outlet temperature as high as the superheater.

As pressure decreases the volume of steam increases. To accommodate the increasing volume of the steam the turbine casing would have to grow in size substantially, so much so that it would become extremely large and extremely expensive to build. Economics in place when the plant was designed determine an optimal size for the casing and turbine blades for each stage of the turbine. At selected stages steam is piped from the turbine for auxiliary uses removing enough steam to keep the turbine from becoming excessively large. That steam is normally referred to as extraction steam or bleed steam. In some industrial plants steam supplied to the facility is extracted with the remaining steam continuing through the turbine to the condenser. An additional feature of some turbines in that application incorporate what is referred to as a goggle plate that throttles the flow of steam to the rest of the turbine to maintain the pressure of the steam supplied to the facility. In electric utility plants extracted steam is used to power auxiliary turbines, comfort heating of the facility, and heating the feedwater at different stages between the condenser and the boiler inlet.

cally radiant meaning they are exposed to the radiant energy from the flame in the boiler. Several methods are used to control the superheat temperature. An "interpass" desuperheater could consist of redirecting a portion of the superheated steam between two sections of the superheater through coils in the steam drum where the boiler water is heated by the superheated steam so the steam is cooled then mixed with the remainder before passing through the next section of superheater tubes. Another interpass method is injecting boiler feedwater into the stream of superheated steam between sections where the steam is cooled by evaporating the added boiler water. Controlling the flow of the steam through the drum desuperheater or the water to the injection desuperheater controls the temperature of the steam at the outlet of the superheater.

In a typical electrical utility plant the energy

Steam leaving the last stage of the turbine enters the condenser where the latent heat is removed and the steam converted to water. The condenser can be air cooled which normally consists of multiple heat exchangers with fan forced atmospheric air used to cool them. Air cooled condensers are only used where water comes at a premium or is in very limited supply; most condensers are water cooled. Water cooled condensers can use water from cooling towers but can be from a river, lake, or the ocean which normally provides the highest plant efficiency because that water is the coldest. The colder cooling water reduces the pressure on the steam side of the condenser because the colder surfaces lower the saturation temperature of the steam. The pressure in the condenser is always less than atmospheric so that pressure is referred to as a vacuum. The vacuum, typically expressed in inches of mercury, can be estimated from the steam tables by adding ten to twenty degrees to the cooling water temperature. In the example of Figure 1-8 you'll note the temperature of the condensate is higher at the discharge of the condensate pumps. That's because the condensers are designed to sweep the steam leaving the turbine over the droplets of condensate before turning up to contact the tubes that condense the steam.

The steam condensate is removed from the condenser by condensate pumps that force the water to flow from the condenser to the deaerator. On its way the condensate will first pass through low-pressure heat exchangers that use extraction steam and operate at a partial vacuum, draining their condensate into the condenser. Condensate from auxiliary systems that heat equipment at low temperatures, such as space heating, is typically drained into one of the low-pressure heat exchangers.

When the condensate reaches the deaerator it is mixed with some extracted steam to heat it to saturation conditions. That steam source is commonly augmented with steam from a pressure reducing station supplied by a higher pressure source to maintain the plant's heat balance (explained later). When water is heated to saturation any gases that are absorbed in the water cannot remain absorbed because the water is about to boil. The gases form bubbles in the water, which is agitated to force the bubbles out of the water and out a vent at the top the deaerator. The heated, deaereated water is collected in a storage tank which is either part of the deaerator or as separate vessel normally designed to hold enough feed water to keep the plants boilers operating for 20 minutes.

Boiler feed pumps draw water from deaerator storage and pump it to the boilers through high-pressure heat exchangers, typically, an economizer, and feed water control valves. Newer plants may operate without feed water control valves by using variable speed control of the boiler feed pumps to change the feed water flow. Some plants, especially supercritical, may use booster pumps to help increase feed water pressure. Controlling flows of extraction steam are one of the ways a utility plant boiler operator earns a salary. Maximizing electricity generation by ensuring maximum flow of steam through each turbine stage produces more Kwh output and that's the product the plant is selling to its customers.

At least most of the water, sometimes a liquid and sometimes steam, is circulated within the system leaving the boiler then returning to it; that constitutes a cycle. The wise operator should be able to describe the cycle in his or her plant. If there isn't a diagram for your plant laying around or buried in a file somewhere then try making your own. Displaying the cycle diagram in the plant not only informs visitors of the scope of the system but serves as a reminder as to what an operator should monitor to make the cycle as efficient as possible.

COMBUSTION

Most of our fuel that we use is called "fossil fuel" because its origin is animal and vegetable matter that was trapped in layers of the earth where it became fossilized, breaking down, for the most part, into hydrocarbons. Hydrocarbons are materials made up principally of hydrogen and carbon atoms. It's the hydrocarbon portion of fossil fuels that generates more than 90% of the energy we use today, from the propane that fires up your barbecue to the coal burned in a large utility boiler to make electricity. The normal everyday boiler plant that you're operating also burns hydrocarbons but we concentrate mainly on four forms, natural gas, light oil, heavy oil, and coal.

The principal difference in these fuels is the hydrogen/carbon ratio and the amount of other elements that are in the fuel. Despite the fact that our typical hydrocarbons vary from a gas lighter than air to a solid they all burn the same, combining with oxygen from the air to release energy in the form of heat. It's not necessary to know how it does it, only to understand that certain relationships exist and generally what happens depending on changes you make or changes that are imposed on you by the system. If you look at a number of what we call "ultimate analysis" of fuels you'll discover that the fuel gets heavier with an increase in the amount of carbon in the fuel and lighter as hydrogen increases. There are other factors but let's just discuss simple combustion first.

If you were ever in the Boy Scouts, you were taught the fire triangle. To create a fire you need three things, a fuel, air, and enough heat to get the fire going. You also probably discovered that you can stack up a campfire (you'll discover I love campfires) using pieces of wood about four inches in diameter and over a foot long and even though you have a lot of fuel there with air all around it you can't start the darn thing with a match. Obviously there's fuel and air so the problem is not enough heat. To get that fire going you have to have some kindling, smaller and lighter pieces of fuel that will continue to burn once you heat them with a match and they produce more heat to light those big sticks you put on the campfire.

Once the fire gets going, the heat generated by those big sticks burning is enough to keep them going and light more big sticks as you stack them on the fire. If you pull the fire apart, isolating the big sticks from each other, the fire will go out. Now we have a very good lesson on the relationship of fuel and heat in a fire. As the fuel burns it generates heat and some of that heat is used to keep the fuel burning and some is used to start added fuel burning. When the fire is compact, where a good portion of the heat it generates is only exposed to the fire and more fuel the fire will be self supporting. If the fire is spread out where all its heat radiates out to cold objects the fire will go out.

The fuel in the furnace of a boiler burns at temperatures in the range of 1200 to 3200°F which is usually more than enough to keep it burning and heat up any new fuel that's added to the fire. Modern furnaces, however, are almost entirely composed of water-cooled walls which absorb most of the radiant heat of the fire. Despite that high temperature a fire in a modern boiler is barely holding on and it doesn't take much to put it out. That's why we need flame detectors, which are covered in a later chapter.

All of our fuels are principally hydrocarbons, material containing atoms of hydrogen and carbon in various combinations with varying amounts of other elements. The reason hydrocarbons are important is they release energy in the form of heat when they burn. We call the burning of the fuel the "process of combustion." That's because we engineers have to use big words, we say combustion instead of burning to give the action a name, burn is a verb, combustion is a noun. It really isn't that complicated a word and most operators have no problem using it.

We use different adjectives for combustion including partial, perfect, complete, and incomplete to describe different results when burning fuels. Partial combustion means we burned part, but not all, of the fuel. Incomplete combustion is basically the same but the difference is we intentionally have partial combustion and incomplete combustion is undesirable. Perfect combustion is an ideal condition that is almost never achieved. It's when we burn all the fuel with the precise amount of air necessary to do so. Of course we engineers have to use a fancy word to describe that condition, and it's "stoichiometric" combustion. Complete combustion burns all the fuel but we always have some air left over.

Every fuel has its air-fuel ratio. That's the number of pounds of air required to perfectly burn one pound of fuel. The air-fuel ratio of a fuel is principally dependent on the ratio of carbon to hydrogen in the fuel, the amount of hydrocarbon in the fuel, and, to a lesser degree, the air required to combine with other elements in the fuel. Note that this is a mass ratio, not related to volumes, but it can be converted to a volumetric ratio (cubic feet of air per cubic foot of fuel) provided we specify the conditions of pressure and temperature to define the density of the fuel and air. The air-fuel ratio for a fuel can be determined from an ultimate analysis of the fuel (Appendix L, page 380).

The air required for the fuel is not consumed completely, only part of the air is used, the oxygen. I'm sure you know that atmospheric air, the stuff we breathe, contains about 21% oxygen by volume. We engineers get more precise and say it's 20.9% but for all practical purposes 21% is close enough. What's in the other 79%? It's all nitrogen, what we call an "inert" gas because it doesn't do much of anything except hang around in the atmosphere. When we get to talking about the air pollution we create when operating a boiler you'll discover it isn't entirely inert. That little tenth of a percent we engineers consider contains a lot of gases, mostly carbon dioxide, that don't really do anything in the process of combustion either so we can say they're inert.

It's a good thing that air has that 79% nitrogen because it absorbs a lot of the heat generated in the fire and limits how fast that oxygen can get to the fuel. It's considered a moderator in the process of combustion because it keeps the fuel and oxygen from going wild; without it everything would burn to a crisp in an awful big hurry.

You should recall an incident in the early days of the manned space flight program where three astronauts were burned to death in a capsule during a test while sitting on the ground. At that time they were using pure oxygen in the capsules, a small electrical fire provided enough heat to get things started and, without the nitrogen to moderate the rate of combustion, the inside of the capsule was consumed by fire in seconds. We do have flames that burn fuel with pure oxygen, the space shuttle's engines do it and the typical metalworker's cutting torch uses it, but those applications have a limit on their burning imposed by consumption of all the fuel and the moderating effect of the nitrogen in air surrounding those operations. Keeping those cutting torch oxygen tanks properly strapped down in the boiler plant is important because they're a source of pure oxygen that could produce a rapid, essentially explosive, fire in the plant where we aren't prepared for it.

The appropriate title for this part should be combustion chemistry but I know what would happen. Mention the word "chemistry" and a boiler operator's eyes glaze over and they look for a route of escape. Hey, if we wanted to be an engineer or chemist maybe we would study chemistry, we're not engineers or chemists so don't bother us with that stuff. Okay, I understand the feelings and I remember them but you have to understand what's happening in that fire to know how to operate a boiler properly. I'm not going to present anything that's far out, no confusing calculations or any of that stuff, it's really quite simple and you'll find you can understand it and use that understanding to become a wiser operator.

Any fossil fuel has only three elements in it that will combine with the oxygen in the air and release heat. All of a sudden combustion chemistry is not so complex is it? Actually there are only four reactions that you need to know. (Combining of materials to produce different materials is a reaction). Let's start with the easy one first; hydrogen in the fuel combines with oxygen in the air to produce di-hydrogen oxide (H_2O). Yes, you're right, that's really what we call H-two-O and it's water.

Of course the heat generated by the process produces water so hot that it's steam so we don't see liquid water dripping from a fire. I like to say hydrogen is like the best looking girl at the dance. She always gets a partner. Hydrogen will mug one of the other products of combustion if necessary to get its oxygen. To date nobody has been able to find any hydrogen left over from a combustion process because it always gets its oxygen to make water. You're assured that all the hydrogen in the fuel will burn to water if combustion is complete. If it isn't complete, the hydrogen will still be combined with some carbon atoms to produce a hydrocarbon, sometimes it isn't any of the hydrocarbons that the fuel started out as, it can be an entirely different one.

Carbon, in complete combustion, combines with the oxygen in the air to make carbon dioxide, CO_2 for short. We say "C-oh-two" basically reading off the letters and number. That's one atom of carbon and two atoms of oxygen. You'll probably recall that it's the fizz in soda pop and what we breathe out. Actually our bodies convert hydrocarbons to water and carbon dioxide. We just do it slower and at much lower temperatures than in a boiler furnace. Since carbon is the major element in fuel, we make lots of carbon dioxide in a boiler. Next in quantity is water. Now, that brings up an interesting point, if we're making carbon dioxide and water, both common substances that we consume, then what's the problem with boilers and the environment? We'll get to that but, for the most part, firing a boiler is natural and it produces mostly CO_2 and H_2O which aren't harmful.

Notice that I had to say "in complete combustion" in the lead sentence of that last paragraph. If we have incomplete combustion, the carbon will not burn completely. Instead of forming CO_2 it forms CO, carbon monoxide. That's the colorless, odorless gas that kills. The person deciding to commit suicide by sitting in his running car in a closed garage dies because the car engine generates CO and he breathes it. That CO is trying to find another oxygen atom to become CO_2 and it will strip it from our bodies if it can. That's what happens, it robs us of our oxygen and we die of asphyxiation.

The last flammable (stuff that burns) constituent in fuel is sulfur. Sulfur combines with the oxygen in the air to form SO_2, sulfur dioxide. There isn't a lot of sulfur in fuel but what's there burns. And, that's it! Three elements, Carbon, Hydrogen, and Sulfur combine with oxygen to produce CO_2, water, and SO_2 and heat is generated in the process. Now, hopefully, I can show you the chemical combustion formulas and they'll all make sense. When we use numbers in subscript (small and slightly below normal) that indicates the numbers of atoms (represented by the letter just in front of the number) in a molecule. Numbers in normal case indicate the number of molecules. Atoms, represented by the letters, combine to form molecules. Many gases, oxygen is one of them, are what we call diatomic; that means it takes two atoms to make a molecule of that gas. All fuels are made up of atoms of hydrogen and carbon, it's the mix of atoms to form the molecules of the fuel that produces the different fuels we're used to. In other words, it's the combination and number of hydrogen and carbon molecules that determines if the fuel is a gas, an oil, or a solid material like coal. Here's the list of basic combustion chemistry equations.

$$C + O_2 => CO_2$$
+ 14,096 Btu for each pound of carbon burned.

$$2H_2 + O_2 => 2H_2O$$
+ 61,031 Btu for each pound of hydrogen burned

$$S + O_2 => SO_2$$
+ 3,894 Btu for each pound of sulfur burned

$$2C + O_2 => 2CO$$
+ 3,960 Btu for each pound of carbon burned

C is Carbon, one atom

CO is a molecule of carbon monoxide, containing two atoms

CO_2 is carbon dioxide, one molecule containing three atoms

H_2 is a molecule of hydrogen, consisting of two atoms

H_2O is a molecule of water, consisting of three atoms

O_2 is a molecule of oxygen, consisting of two atoms

S is an atom of sulfur

SO_2 is a molecule of sulfur dioxide, three atoms

The rules of the equations are rather simple. You have to have the same number of atoms on both sides of the equation. Try counting and you'll see that's the case. You see, we don't destroy anything when we burn it. It's one of the natural laws of thermodynamics that's called the law of conservation of mass. It may appear that the wood in the campfire disappeared but the truth is that it combined with the oxygen in the air to form gases that disappeared into the atmosphere along with the smoke. Every pound of carbon is still there. It's just combined with oxygen in the CO and CO_2. I know it doesn't make sense that we get energy without converting any of that matter to the energy but that's the case. At least nobody has been able to find a difference in weight to prove it.

You'll also notice that we don't get much heat from the carbon when we make CO. That's one sure way to know you're making any significant amount of it. When I was sailing, we sort of used that fact to tune the boilers. Once we were at sea we pushed the boilers to generate as much steam as possible to turn that propeller with the turbines. Every rotation of that big screw got us 21 feet closer to Europe or 21 feet closer to home, depending on which way we were going, and the more rotations we got the faster we got there. We would push the fans wide open then increase fuel until we noticed our speed wasn't increasing. Usually what happened is the speed would drop off. That was a sure indication we were making CO so we would back off on the fuel a little and that was the optimum point for firing.

Why did the speed suddenly drop off? Notice in the formulas that one oxygen molecule produces only one molecule of CO_2 and two of CO. There's another natural rule that says all molecules at any particular pressure and temperature take up the same amount of space. Since we double the number of flue gas molecules when we make CO the gas volume increases. The increased gas volume produces more pressure drop through the boiler which restricts flue gas flow out. Since the gas can't get out as fast, less air can get in and there's less oxygen so we make more CO. The result is a generous generation of CO until the heat input has dropped to where there's a balance between the pressure drop from more CO and the reduced generation of CO as the air input is decreased. Try it some time… carefully. Just decrease your air or increase your fuel at a constant firing rate and watch the steam flow meter. When the CO starts forming you'll see the steam flow drop off.

Maybe it's a little late, but I think this is a great time to discuss how fuels are produced. It's because the methods used in creating those fuels are partially occurring in our fire in our boiler and by talking about both at the same time it may make more sense why I would insist you know how some fuels are made. Coal is not necessarily made but is simply dug up and transported to the boiler plant right? Not really, some of it is put through a water washer, some of it is treated by exposure to superheated steam, and a small amount of it is ground up fine and mixed with fuel oil to create another fuel. Natural gas and fuel oil also go through preparation processes. Natural gas is normally put through a scrubber after it's extracted from the ground to remove excess carbon dioxide and sulfur compounds.

For all practical purposes the gas flowing up the large pipelines from Louisiana and Texas to all us consumers on the east coast doesn't have any sulfur in it to speak of. If it did the sulfur might react with the oils in the big compressors the pipeline companies use to pump the gas north and make those oils acid. Once the gas arrives at a gas supplier in the northeast sulfur is added back into the gas in the form of mercaptans, chemical compounds that give gas its odor so we can detect leaks. Those mercaptans contain sulfur.

Fuel oil whether it's number 1 (kerosene), 2 (diesel), or any of the heavier grades (4, 5 & 6) all come from crude oil, the oil that's pumped from the earth or gushes when it's under pressure. The crude oil is "refined" in a refinery to separate the different fuels, and a lot more, from the material that comes out of the ground. One big fraction of crude oil is gasoline. In fact there is such a big demand for gasoline that some of the other products are re-refined by different processes to make more gasoline to satisfy our love for driving around in automobiles. The basic process of separating the different components from crude oil is distillation where the oil is heated until the lighter portions including naphtha, gasoline, and others evaporate.

A good portion of our kerosene and light fuel oil (Number 2) is produced by distillation. Some of that and heavier parts of the crude oil are heated further and exposed to catalysts (materials that help a reaction occur) to "crack" them, breaking more complex hydrocarbons down into lighter, less complex ones. That's what happens when the fuel is exposed to the heat of the fire, it's distilled and cracked until it becomes very simple hydrocarbons that readily react with air to burn. It's argued, with some degree of accuracy, that only gases burn and the heat has to convert the fuel to a gas before it will burn. All that distillation and cracking takes some time and that's why a fuel doesn't burn instantly once it's exposed to air.

Now let's try something just a little more complicated. Let's burn the major portion of our natural gas. It's mostly methane, which is represented by the formula CH_4. The same rules for formulas apply. To burn the methane we need a couple of oxygen molecules, O_2 from the air. One molecule of the O_2 combines with the carbon to form CO_2 and the other combines with the four hydrogen atoms to make two molecules of H_2O. The equation is:

$$CH_4 + 2\,O_2 => CO_2 + 2H_2O$$
$$16 \qquad 32 \qquad\quad 28 \qquad 20$$

The numbers under the groups of molecules in the equation represent the atomic weights of the different molecules. I'm sure you know that metals have different weights, aluminum being a lot lighter than steel so you can easily agree that carbon, hydrogen, and oxygen have different weights. You'll also be pleased to know that even I don't remember the atomic weights, it's not necessary to, so you can relax, you don't have to remember the numbers, only the concepts. Atomic weights have no units, they're all relative with oxygen assigned an atomic weight of 8 as the reference because it's the standard we use to measure molecular weights. Hydrogen has an atomic weight just slightly more than one and we use one because it's close enough for what we're doing. Carbon has an atomic weight of twelve and that's all we need to see the total balance of the combustion equation for methane. One carbon plus four hydrogens gives methane a molecular weight of 16 (12 + 4). The two

molecules of oxygen consist of four atoms so its weight is 32 (4 × 8). The CO_2 is 12 + 2 × 8 and the two water molecules are twice (2 × 1 + 8). The law of conservation of mass means that we should have as much as we started with and, sure enough, 16 + 32 is 48, the same as 28 +20.

We engineering types use this business about weights to get an idea of the amount of energy in the fuel. Remember earlier we said we could make 14,096 Btu for every pound of carbon we burned? Well, in the case of methane 12/16ths of it is carbon, and that will provide 10,572 Btu per pound of CH_4 (12 ÷ 16 × 14,096). Similarly, the 4/16ths of hydrogen will produce 15,257 Btu (4 ÷ 16 × 61,031). Add the two values to get a higher heating value of methane of 25,829 Btu per pound. Now I know that doesn't meet with your understanding of how we normally measure the heating value of natural gas. We say natural gas produces about 1,000 Btu per cubic foot, right? That's because it's always measured by volume, in cubic feet. However, the measurement is also always corrected for the actual weight of the gas because it's the mass that determines the heating value, not the volume.

Whenever an engineer wants to know exactly how much flue gas will be produced by a fuel, precisely what the air to fuel ratio is for that fuel, and how much energy we'll get from the fuel we ask for an "ultimate analysis" of the fuel. That analysis tells us precisely how much carbon, hydrogen, sulfur, etc. is in the fuel. An ultimate analysis also includes a measure of the actual heating value. The worksheet in the appendix on page 382 is used to determine the amount of air required to burn a pound of fuel and some other information we use as engineers.

I still haven't really explained why the big sticks on that campfire didn't start burning right away. In addition to the fact the big heavy stick sucks up all the heat from the match without its temperature going high enough for it to burn it has to do with something we call flammability limits. If you add enough heat to any mixture of air and fuel some of it will burn. What we really have to do is come up with a mixture of air and fuel that will not only burn, but will produce enough heat in that process that it will continue to burn. I really wonder if I'll ever stop finding situations where I can't get a fuel and air mixture to burn. After fifty years in the business you would think I could always get a fire going, not just campfires, fires in a boiler furnace. Throw in enough heat and some fuel and air and it should burn, right? Well, I can honestly say "no" because I've been through several bad times trying to get a fire going with no success. This is one of those situations when you can, hopefully, learn from my mistakes and not get as frustrated as I have trying to get a fuel to burn. There are two rules. First, the fuel and air mixture has to be in the flammable range and secondly, you need a fuel rich condition to start. The hard part for those of us designing and building boiler plants is to make certain we have those conditions.

What's the flammable range? It just happens to be the same thing as the explosive range. It's a range of mixtures of fuel and air within which a fire will be self supporting, not requiring added heat to keep the process of combustion going. To be perfectly honest with you, every time we fire a burner we're producing an explosive mixture of fuel and air. It doesn't explode because it burns as fast as we're creating it. If it doesn't burn and we keep creating that mixture the story is a lot different. Eventually something will produce a spark or add enough heat to start it burning. Then the mixture burns almost instantly and it's that rapid burning and heating to produce rapidly expanding flue gases that we call an explosion.

A graphic of a typical fuel's flammability range is shown in Figure 1-9. At the far left of the graph is where we have a mixture that's all air, no fuel. On the far right is where we have all fuel and no air. The quantities of fuel and air in the mixture vary proportionally along the graph as indicated by the two triangles. The thin line in the middle of that band is the stoichiometric point, the mixture that would produce perfect combustion. Mixtures to the left of the stoichiometric point are called lean mixtures because they have less fuel than required for perfect combustion. They can also be called air rich. Mixtures to the right are called fuel rich because there is more fuel in the mixture than that required for perfect combustion. Keep in mind that we're looking at pounds of air and pounds of fuel, not volume. The flammability range is the shaded area and it's only within that narrow range of mixtures that a flame will be self sustaining.

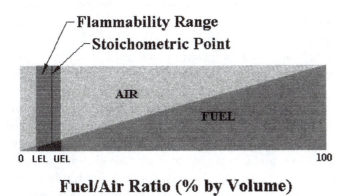

Fuel/Air Ratio (% by Volume)

Figure 1-9. Flammability range

At either end of the flammable range, which we also call the explosive range, are the two limits of flammability. The one where flammability will be lost if we add any more air is called the lower explosive limit, LEL for short. The one where too much fuel prevents sustained combustion is called the upper explosive limit, UEL. If you think about it, it's essential that we have this flammability range. Otherwise the sticks would burn as you carried them back to put on the campfire; actually everything would burn up. On the other hand, that narrow range of mixtures keeps me humble and could do the same to you. It isn't as easy to get a fire going in a furnace when you consider that you have to get the fuel and air mixture within that narrow range. You get to bypass most of the experiences we engineers have because we make sure it works before you get your hands on it.

Getting the mixture in the flammable range isn't the only criteria when it comes to combustion in a boiler furnace. The only way that flame will burn steady and stable is if it begins at the UEL. In other words, the point where ignition begins is where the fuel and air mixture pass from a really fuel rich condition into the explosive range. I can still recall looking through the rear observation port into a furnace full of pulverized coal and air, so much that it looked like a fog in there. I could see the bright flame of the oil ignitor burning through the fog but the darn coal wouldn't light! Needless to say I was very uncomfortable looking at that mixture of fuel and air and wondering whether it might suddenly light.

Many a boiler failed to light because there wasn't that fuel rich edge right where the ignitor added the heat to light it off. Usually it's due to the mixture being too fuel rich and the ignitor not reaching the point where the UEL is to get things started. In other situations the fire is lit and the heat from the fire manages to force ignition into the fuel and air entering the furnace until the fire reaches a point that's way too fuel rich and the fire goes out. Then, because the furnace has some heat, the fuel and air mix again to reach the flammable range and the mixture lights again and burns back toward the burner again. We call it instability, you typically call it "run like hell."

Here's where I always tell boiler operators that you shouldn't always do what you see the service engineer doing. It's standard practice for service engineers to manually control the fuel going into the furnace when lighting a burner they just adjusted. They do it because they aren't certain about the mixture and have their hand on the valve to control it, usually shutting the burner down faster than the flame safety system would. Once they get it right, they usually let it light off the automatic valves. Of course I should say that applies to service engineers that worked for me at Power and Combustion. In some instances a service engineer will leave a job that doesn't light off properly; as far as I know we never did.

I always tell this story because it introduces another term in a manner that operators understand. One of the reasons Power and Combustion provided quality boiler and burner installations was the interaction between the design engineer (me) and the technicians in the service department who performed the work in the field. They never hesitated to show me how I had screwed up or call when they had a problem they couldn't resolve. In the 1980's my service manager at Power and Combustion was a gentleman named Elmer Sells. Elmer and I got along well because we're both hillbillies, natives of the Appalachians, I grew up in western New York State and he grew up in West Virginia. We were into the start-up phase of a project to convert three oil fired boilers at Fort Detrick to gas firing.

I got a call from Elmer asking that I come out to the plant to look at a problem they had. When he called he used that West Virginia drawl that normally meant he figured he had me, so I knew I was in trouble before I even left. I arrived right after lunch time and found Elmer standing next to the largest boiler, a four burner unit rated at 140,000 pph. Working that WV drawl he informed me they had just purged the boiler and he would like me to try to light off the bottom left burner.

As I climbed up the ladder to the burner access platform I noticed the observation port on the burner was open so I stood off to the side of the burner while I started it. The gas-electric ignitor started fine but there was a little delay after I opened the last main gas shut-off valve. The burner ignited, the boiler shuddered, and a tube of flame shot out of that observation port about six to nine feet long. I had my finger on the burner stop button immediately but realized the burner was operating normally. Then I turned to look down at Elmer who was standing there with his hands clasped behind his back while rocking back and forth on his toes and heels. He dropped his broad smile and said, again with that WV twang "little rough, ain't it?" I agreed and realized what I had done wrong so we set out to correct the problem. Today those burners light off quietly and smoothly. The lesson to be learned here is any roughness on light off is just another form of explosion and shouldn't be tolerated.

In recent years I've encountered facilities where the contractor that placed the equipment in operation couldn't establish a smooth light off and left the job in-

forming the owner that it was "just a puff" that occurred as the burner started. Don't ever let anyone convince you that a puff is anything other than an explosion. A puff is simply an explosion that did no or limited damage. Every puff you experience should be considered a warning and is not be tolerated because sooner or later whatever is causing the problem will get worse and you will experience an explosion that does some serious damage.

What causes explosions, including puffs? It is the direct result of an accumulation of a flammable mixture. Make no mistake about it, when you're burning a fuel you are creating an explosive mixture because there is no difference between a mixture of fuel and air that will burn and an explosive mixture. The reason we can safely fire a boiler is we burn the explosive mixture at the same rate that we create it. It's only when the mixture doesn't burn and accumulates that we have an explosion. We control the combustion by controlling the rate of burning. When an accumulation ignites it burns at a rate dictated by nature and that's a lot faster than our normal fire, so fast that the products of combustion expanding can create a pressure wave which will create a force of 18 to 70 psig. The explosions we experience and call a puff were simply small accumulations of an explosive mixture which did not produce pressure high enough to rupture the furnace.

It's not always possible to avoid a puff or rough light off. They occur when burner systems fail to repeat the conditions established when they were set up. Material can plug orifices, linkage can slip, regulator springs can soften and many times a combination of minimal factors can combine to prevent a smooth light off or burner operation. If you experience a puff you should consider it a warning sign that something is going wrong and do something about it. If your sense of what has been happening with your burner is sound, you may be able to correct the problem yourself but you should keep in mind that more than 34% of boiler explosions are attributed to operator error or poor maintenance; make adjustments only when you are confident that you understand what is causing the delayed ignition. If you aren't certain, it's much wiser to call for a service technician that has experience with burner adjustments.

I think it's important that a flame begin within the throat of the burner where heat radiating from the refractory throat provides ignition energy. I normally don't see a stable flame on a burner without a good refractory throat. A boiler just south of Baltimore had a furnace explosion in 1993 that was due to the improper adjustment of the burner such that the UEL was established so far out in front of the burner that it would not light the first two or three tries; an accumulation of unburned fuel brought the mixture into the explosive range on the next attempt and the boiler room walls flew out into the parking lot. That incident and several others I've investigated justifies my instructions to all boiler operators. The best thing I can tell you at the end of a chapter on combustion. You can push the reset push-button on the flame detector chassis two times and only two times, never take a chance on strike three.

I can't leave the subject of combustion without touching on the latest buzzwords that has EPA's attention and, therefore, every State's department of air quality. Combustion optimization is simply the process of adjusting the air to fuel ratio on a boiler to get the most heat out of the fuel. The environmental engineers also want it to be while generating the smallest amount of emissions. For many a small plant a service technician comes in once or twice a year (the typical state regulation requires a combustion analysis at least once a year) and he "tunes up" the boiler. From all I can tell that's the EPA's perception of it. Those of you with more sophisticated controls and oxygen trim have automatic combustion optimization, the controls are constantly adjusting the fuel to air ratio.

THE CENTRAL BOILER PLANT

Steam and hot water are used for building and process heating because the conversion of our fossil fuels (coal, oil, natural gas) and biomass (like wood and bagasse) to heat is not a simple process. Water and steam are clean and inexpensive and are excellent for transferring energy from one location to another. It is also relatively easy and inexpensive to extract the heat from the steam or hot water once it has been delivered to where the heat is required. Boilers made it possible to centralize the process for converting fuel to heat so the heat could be distributed throughout a facility for use. One boiler plant in a large commercial or industrial facility can serve hundreds or even thousands of heat users. The central plant concept is the most efficient way to deliver heat to a facility.

Many will question that statement, I know. If central plants are so efficient then why are so many facilities installing local boilers and doing away with the central plant? The answer is false economy. Many of our central plants are at the age where all the equipment and piping are well past its original design life and should be replaced. Replacing the central plant with several small

local boilers is seen as a way to reduce the capital (first) cost. We can install one gas pipe distributing fuel to all those local boilers at a much lower cost than installing insulated steam and condensate or hot water supply and return piping.

However, the cost of several small boilers with a combined capacity exceeding that of the central plant puts a considerable dent in the distribution piping saving. Those are not the principal reasons for the switch; the main reason central plants are abandoned is the contention that all those little local plants, operating a low steam pressure or with hot water below 250°F don't need boiler operators present. The justification is eliminating the high wages of boiler operators. There's the main source of the false economy. Installing many more boilers to maintain will reduce the cost of qualified operators. Ha!

The most recent study I'm aware of is one by Servidyne Systems Inc., & the California Energy Commission which claims "a well trained staff and good PM program has potential of 6% to 19% savings in energy." If the staff is eliminated then an increase in cost of 6.3% to 23.4% is possible because they are not there to maintain that savings. A little plant with a 500 horsepower boiler load could see energy cost increases in terms of 2013 dollars of $46 to $171 per day in fuel alone. That's considerably less than the numbers quoted in the first edition of this book because fracking has increased the production of natural gas dramatically thereby forcing the price of gas down. I have to admit that a plant has to have an average load closer to 1500 horsepower to justify the expense of Licensed Operators by fuel savings alone but there's much more to be saved by knowledgeable operators and, in my humble opinion, disposing of a central plant isn't justified if it's load is greater than 500 boiler horsepower.

Fuel prices in January of 2001 were triple the 1999 cost and they're increasing again as I write this. So, you see, decentralizing almost any existing plant will save on labor but burn those savings up in fuel. That doesn't consider the additional cost of maintaining several boilers instead of two or three. By the time all those local boilers start needing regular maintenance the people that decided to eliminate the central plant have claimed success and left. The facility maintenance bill starts to climb to join the high fuel bills associated with all those local boilers.

Now someone's going to claim that the local boilers are more efficient because they're operating at low pressure. That's not true. Nothing prevents a high pressure steam plant with economizers generating steam more efficiently than a low pressure boiler when the feedwater temperature is less than the saturation pressure of the heating boiler. A typical central plant in an institution will have 227°F feedwater to cool the flue gases but local heating boilers will be about 238°F. Since the flue gases can be cooled more by the high pressure plant the central plant boiler efficiency will be higher.

Add to the higher efficiency of a central plant the ability to burn oil as well as gas and the purchasing price advantage for the fuel, the most expensive cost when operating a plant, is also lower. Today's time of use pricing has almost eliminated the deals we got for interruptible gas. In the 1990's when firm gas was about $5 a decatherm interruptible gas was about $3.50. You could save 30% on the price of gas by allowing the supplier to call for you to stop burning that fuel at any time. The ability to burn fuel oil allowed you to take advantage of an interruptible gas contract. Today it's not interruptible, but you pay a much higher price than oil when gas is in short supply.

Running fuel oil supply and return piping to a lot of local boilers is usually abandoned as a first cost savings. Besides, who will be around to switch them? There are automatic controls for switching fuels but the geniuses that decide to abandon a central plant must be afraid of them. With time of use prices someone needs to compare them for oil and gas to decide when to fire oil. In the winter of 2001 I had a customer capable of firing oil that fired gas at prices of $10 to $11 a therm when oil cost only about $7.50; they burned up a difference in less than two months that would have paid a boiler operator's salary for a year. The only way a central plant can cost more to operate than a lot of local boilers is if the heat loss from the distribution piping is excessively high. However, it takes a lot of quality installed distribution piping to produce enough heat loss to justify a lot of local boilers. If your management is considering shutting down your central plant lend them this book so they can ask the right questions of whoever is pushing for it.

I was always encouraging customers to install boilers in their central plants with higher pressure ratings. The cost differential for a boiler capable of operating at 600 psig instead of 150 psig is not that great compared to the value of the potential for adding a superheater and converting the boiler for generating electricity later. Very few chose to heed those suggestions and today they're regretting it because distributed generation is the big thing. A plant that generates power with the same steam that's used in the facility produces that electricity at a fraction of the cost of an electric generating station. Usually 80% of the energy in the fuel a simple boiler plant uses is converted to useful energy in the facility; less than 40% of the energy in a conventional utility steam plant gets converted to electricity. All facilities that dumped their central plants for a multitude of little boilers also dumped their ability to make power economically.

Distributed generation is a new buzzword that basically means electricity is generated in many locations (instead of large centrally located power plants that are usually long distances from the users of the power). By having several small plants distributed throughout an area transmission lines lose less power and don't have to be so big.

ELECTRICITY

If there's anything that boiler operators pretend to know nothing about it's electricity. I have met several boiler operators that would send for an electrician to change a light bulb. To choose to know nothing about it is to doom yourself to becoming a janitor, with pay to match. Not only are we in an age where electricity powers our controls but we're coming into the age of distributed generation where every decent sized boiler plant will be generating electricity. It's essential that the boiler operators of tomorrow know enough about electricity to use it, generate it, and occasionally troubleshoot a circuit.

The current trend is toward engine and gas turbine cogeneration. That's where the fuel that's normally burned in the boiler is fired in the engine or gas turbine instead. The engine or turbine generates electric power and the steam or hot water is generated by the heat from the exhaust of the engine or turbine.

Some visionaries like to think we'll all be running with fuel cells in the future. Fuel cells generate electricity by reversing the electrolysis process. I trust you'll remember that day in chemistry lab in high school when you put two wires into water with an inverted test tube over each and watched gases form at the ends of the wires with the bubbles rising to collect in the test tubes? That was electrolysis, breaking water down into its two elements, hydrogen and oxygen. A fuel cell combines hydrogen and oxygen to form water and generate electricity. Heat is also generated in the process and that's what would be used to generate our steam and hot water. Fuel cells have advantages like no moving parts, other than fuel and cooling fluid pumps, so they are very reliable. We might all be using them today if it weren't for one simple problem. They can't generate electricity using the carbon in the fuel. Any fuel cell using a typical hydrocarbon fuel like natural gas basically burns the carbon.

Whether it's an engine, a gas turbine, a fuel cell, or a very conventional steam turbine driving an electric generator you will eventually be operating one because all plants will have them. So, ... now's the time to get an adequate understanding of electricity.

I'm not going to use all the hydraulic analogies we engineers typically try to use because I think they are just confusing. Electricity is different but it isn't a dark and mysterious thing that is beyond the understanding of a competent boiler operator. There are only two basic things you have to know about electricity and the rest falls into place.

For electricity to work there has to be a closed circuit. A circuit is a path that the electricity flows through. Break the circuit anywhere so it is not a closed path and electric current can't flow through it. The second thing is that there has to be something in that circuit that produces electrical current. If electric current isn't flowing through the circuit the circuit isn't doing anything. That's it, create a circuit to make electricity work and break the circuit to stop it. When the path is complete so current can flow we call it a closed circuit. Whenever there's a break in it we call it an open circuit. To be fair I should also explain that a "break" is typically undesirable whereas the "open" is a normal interruption in the circuit.

You pull the plug on the toaster that's stuck and belching black smoke while incinerating the last slice of bread that you planned on having for breakfast and you opened the circuit. Actually, you opened it in two places, the plug does have two prongs. When you turn the light switch off you opened the circuit. In most cases opening a circuit consists of moving a piece of metal so there is a gap between it and the rest of the metal that forms the circuit. In almost every case where we use electricity we use metal wire and metal parts to form the circuit. Sometimes, as with the toaster plug, you can see the open. In other situations, as with the light switch, you can't see the open because it's enclosed in plastic to protect you and it.

When mother nature is dealing with electricity metal is not a requirement. At some time in your life you had to walk across a carpet on a cold dry winter day, reach for the doorknob and get surprised by a spark jumping from your finger to the knob. We call that static electricity but there wasn't anything static (as in standing still) about it. As you walked along the carpet

your shoes scraped electrons off the carpet which then collected in your body. When you reached for the doorknob the electrons passed through your finger, through the air, into the doorknob. Another way mother nature shows us how she handles electricity is lightning. In those cases electric arcs form where the electricity just flows through the air, just like the static spark off your finger traveling to the doorknob.

Those two natural examples imply that a circuit doesn't have to be like a circle (so the electrons can continue to flow around it) but the truth is that they are. The electrons you dumped to the doorknob eventually bleed through the door, hinges, door frame and into the floor to get back to the carpet. The discharge of lightning is dumping electrons dragged to the earth by the rain drops back up to the clouds in the sky. Those rather fast and furious discharges of electricity are not the kind of thing we want to do in the boiler plant. Note that it's called a "discharge" which means the electric charge is eliminated, at least until it builds up again. Once you've recovered from that spark between your finger and doorknob you will not get shocked again, provided you didn't move around the carpet some more.

A battery is like having stored electrons. The difference is a battery contains chemicals that react to replace the electrons when you start discharging it. You can discharge a battery by running the electrons through a light bulb, as in a flashlight, or, as I sometimes do when carrying some spares around, by shorting the battery. I do that when the keys in my pocket manage to touch both ends of the battery. I have some rechargeable batteries in which the chemical process is reversed to restore the charge. A battery will keep restoring the charge until the chemicals all change then we call it "dead." There's not much difference between a dead battery and a dead electrical circuit except that the battery just can't produce enough electrons to raise the voltage and a dead circuit can have full voltage someplace.

It's important to realize that an electrical circuit that isn't doing anything can still have a charge of electrons stored someplace ready to surprise us just like when we reached for the doorknob. The problem with electric circuits is they have the capacity to store a lot more electrons than our shoes can rubbing the carpet and it's current that kills. The voltage you build up walking across the dry carpet is a lot higher than most electrical circuits, it takes a lot of voltage to make electrons jump that gap between your finger and the doorknob.

You'll recall there was this earlier chapter on flow? Electricity is no different. You control the flow of the electricity, those little electrons have to flow for something to happen. Voltage is nothing more than a reference value like steam pressure. The electric company, or you if you're generating it, produce enough electron flow to keep the voltage up just like you produce enough steam flow to keep the pressure up. Most electric flow control is on-off; you close the switch and open it to control the flow. You may have a dimmer on one or more lights in your home, they control the flow of electrons to dim the lights. At other times the equipment is designed to automatically control the flow.

I've managed over forty years to deal with electricity but I have to admit that I still don't really understand what happens with alternating current. I base all my operating judgment on principles for direct current and a little understanding of alternating current. I trust you can do the same, you don't have to be able to design electrical systems, only understand how they work and how to operate them. Of course you can troubleshoot them to a degree if you understand how they work.

I even use the basic Ohm's law on AC circuits to get an idea of what's going on. I know it isn't a correct analysis but it's good enough for me. You know Ohm's law, it's really mother nature's law, Ohm is just the guy that realized it. The voltage between any two points in a circuit is equal to the value of the current flowing through the circuit times the resistance of the circuit between the two points. $V = IR$ where V stands for voltage, I stands for current in amperes, and the R represents resistance in ohms. If you know any two of the values you can determine the third because current equals voltage divided by resistance and resistance equals voltage divided by current.

Ohm's law is a lot of help when troubleshooting electronic control circuitry. Most of our control circuits today use a standard range of four to twenty milliamps to represent the measured values. For example, a steam pressure transmitter set at a range of 0 to 150 psig will produce a current of 12 milliamps when the measured pressure is 75 psig. If we aren't getting a 75 psig indication on the control panel and want to know why we can take a voltmeter and measure voltage at several points in the circuit to see why. Start with the power supply, it should be about 24 volts if it's a typical one. That gives you a starting point and you can use one side of the power supply, whenever possible, to check for voltage at other points in the circuit.

The voltage drop across the transmitter should be more than half that of the power supply because all the transmitter does is increase or decrease its resistance; to control the current so it relates to the measured steam pressure. If there isn't much voltage drop across the transmitter then there's a problem elsewhere in the circuit. I'll frequently check for a voltage drop between each wire before it is connected to the transmitter terminal and a spot past the screw that holds the wire because poor connections are frequently a problem. 24 volts DC can't push current through a loose or corroded connection and corrosion is always a problem in the humid atmosphere of a boiler plant. I've fixed many a faulty circuit by just tightening screws without even checking the voltage.

A voltmeter or even a light bulb in a socket with two wires extended can be used to check the typical 120 volt control circuit. Just make sure you don't touch those test leads on the light to anything that could be higher or lower voltage. If the resistance between two points is zero, or nearly zero, then there's no voltage and your meter or test light will show nothing. If the circuit is open between the two points you put your test leads on you will get a reading or the light will shine. The circuit will not operate because the meter or light doesn't pass enough electrical current.

In the days of electro-mechanical burner management systems I added a light to a control panel, down in the bottom door, and labeled it "test." The light was connected to the grounded conductor and a piece of wire long enough to reach anywhere in the panel was connected to the light and left coiled up in the bottom. All an operator had to do was pick up the coiled wire and touch it's end to any terminal or other wire in the panel to find out if the wire or terminal was "hot." The idea was to allow the operator to pick up that lead and troubleshoot the system when he had problem.

Most of the time that provision was eliminated from the design after the submittal to the owner. Why? It was a combination of Owner management being convinced that an electrician was the only one that could troubleshoot electrical circuits or they had trade restrictions which required that work be done by an electrician. Frequently it's assigned to a trade identified as an instrument technician. I've discovered, however, that most electricians are totally lost in a burner management control system and few instrument techs understand them. Set up your own test light so you have it when you need it.

The need for troubleshooting burner management systems has decreased considerably with the introduction of microprocessor based systems. Many of them include a display that will tell the operator what isn't working (failure to make a low fire start switch on start-up being a very common one) and they're simply more reliable than all those relays and that extensive wiring.

Just the same, you should be able to do it. Read the drawings and sequence of operation until you understand how your system works then review it every year so you will have most of it in your head when the need to solve a problem comes up.

What good was that test lead? Well, all you had to do was touch the end of it to one of the terminals or wires in the system (while holding the insulation on the wire so you don't light up) and see if the test light comes on. If the light comes on then there's a closed circuit up to that point. If it's not on then you know there's an open somewhere between the power supply and that terminal. When one terminal is hot and the next one isn't you can look on the drawing to see what's connected between the two. If it's supposed to have a closed contact at the stage you're looking at then you go out into the plant to find the device to see what's wrong with it. The device could be broken or it could be valved off (although there aren't supposed to be valves between a boiler or burner and the limit switches). It could be something as dumb as a screw vibrated out and the switch flopped over, something that really screws up mercury switches.

Figure 1-10 is an example of a circuit test. For some reason the fan motor starter (FMS) coil isn't energizing to start the fan. The test light is connected to the grounded conductor. That upside down Christmas tree symbol is normally used to represent a ground wire connection. The clip on my test light allows me to slip it over any grounded metal part of the boiler plant for testing but there's times when something accessible and metal isn't grounded even if it should be so don't count on it. The other side of the light is connected to a test lead with an insulated handle and extended metal probe. Mine was recovered from a broken meter that someone threw away. Always check your test light by tapping a hot lead to ensure it's working before testing. You'll note the test lead is indicated to be touching ter-

minal 4 in the panel. To quickly isolate a problem start testing at the middle of the circuit, and terminal 4 is in the middle in this example. If the light works then the break or open is closer to the grounded conductor and you can forget checking everything before terminal four. If the light doesn't work then the break or open has to be between the hot lead and terminal four. Once you've figured out which side contains the open you can test at a point near the middle of the part of the circuit that you just isolated. You should pick terminal 3 or 5 depending on which end of the circuit must have the break because they're conveniently in the panel. If the light didn't work at 4 or 3 then check the wire connected at the on/off switch because it's also conveniently in the panel. You've isolated a problem when you find the spot where the light doesn't work when checking after a device and does before it. Of course that doesn't mean that there isn't a problem with something else in the circuit that's between the problem you found and the left connection of the FMS. It's slightly more complicated because other parts of a circuit must be activated first but once you determine where the circuit is open it's a lead to finding out what's preventing operation.

If a fuel safety shut-off valve should open, but doesn't, you can check its terminal (when the burner management system indicates it should be energized) to see if it's getting power (light on). If it isn't then you can check back through the panel circuitry to find what's open. Keep in mind that you only have ten or fifteen seconds to do that most of the time and you'll have to go through several burner cycles until you spot the problem. If the output terminal is energized then you'll have to check the power at the valve to be certain it's not a loose or broken wire between panel and valve motor.

I used to take it for granted but got stung so many times that now I always check to be certain a burner management system is properly grounded. Lack of a ground can produce some very unusual and weird conditions. Anytime you see lights that are about half bright or equipment running that's noisy and just not normal look for lack of a ground or an additional one.

Exactly what is a ground? It's anything that is connected to a closed circuit to mother earth. In most plants there is a ground grid, an arrangement of wires laid out in a grid underground and all interconnected to each other and the steel of the building to produce a grounded circuit. At your house it's your water line and possibly also separate copper rods driven straight into the ground. A ground wire is any

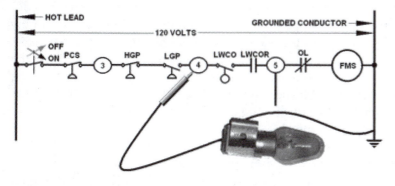

Figure 1-10. Testing a control circuit

wiring connected to the ground.

Don't confuse a ground wire with a grounded conductor. Ground wires are there to bleed stray voltage to ground, not to carry current. A grounded conductor is a wire that carries electrical current but is connected to a ground wire. All the white wires in your house should be grounded conductors. If you took the cover off your circuit breaker panel you should see that they're all connected together in there and also connected to a wire that is attached to your water line (the ground wire).

All the steel in a building, the boilers, pumps, piping, etc., should all be connected to a ground. In cases like the building steel or pumps and piping the electricians will call them "bonded." Bonding and grounding is the process of attaching everything that could carry electrical current (but shouldn't) to the ground below the building. At sometime in your career you should have an opportunity to do what I've done, three times. You're working around a pump or something and step back or drop a tool and knock the grounding conductor loose. There's more in the section on maintenance that addresses that.

With everything connected to a ground the difference in voltage between any wire and ground should indicate the voltage of the system the wire is in. System voltages do vary though and you shouldn't get excited if the voltage seems a little off. The common 120 volt system will vary from a low of 98 to a high of 132 although they typically fall in the 115 to 120 range. 480 volt systems usually range from 440 to 460 volts between leads at the motor.

We never give it much thought but you should always know another location where you can disconnect the power to a circuit. Remember the toaster? The reason you pulled the plug out of the wall was the toaster control didn't work. There's usually a button or lever we can push or flip to release the toast and turn the toaster off but sometimes it gets jammed. That's a regular for me because I like the whole grain large loaf bread and those slices are always getting stuck in the toaster. Well, just like the toaster, you should be able to identify another means of shutting down every piece of electrical equipment in the plant.

Usually you just push a button labeled stop and that's all you have to do. The stop button moves a metal bar away from two contacts to open the control circuit which stops current flowing through a coil that holds the motor starter contacts closed. The coil releases the motor starter contacts and the motor stops. The question is, what do you do when a) the push-button contacts don't open? b) the insulation on the two wires leading to the push-button in a conduit placed too low over a boiler melts and the wires touch each other (what we call a short)? c) a screwdriver somebody left in the motor control center dropped onto the terminal board for the starter shorting out that same push-button circuit? d) Humidity in the electrical room promoted corrosion on the metal core of the coil until the portion holding the motor contacts rusted to it so the motor contacts stay closed even when there's no power to the coil? e) two or more of the motor starter contacts fused together and will not release even though the coil isn't holding them shut? (I could go on with a lot of other scenarios) What do you do? Make sure you know where to flip a circuit breaker or throw a disconnect in case something like that happens.

Keep in mind that disconnects are not normally used to break circuits. They're the devices that have copper bars that are hinged at one end and slip between two other pieces of copper that press against the bar to produce a closed circuit. If you pull one of those to shut a motor down expect some sparks. You wouldn't normally do it because those copper bars aren't designed for arcing and they'll melt a little wherever the arc forms. When you do have to do it, do it as fast as possible.

Speaking of arcs... you know, that spark between your finger and doorknob and the lightning are arcs: they can be hazardous to you and the equipment. Every motor starter and circuit breaker is fitted with an "arc chute." It's constructed of insulating material and designed to help break the arc that forms when you're opening a circuit. You won't see them used on common 120 volt or lower circuitry because that's not enough voltage and seldom has enough current to produce a sizable arc. Normally the arc chute has to be removed to see, let alone get at, the main circuit contacts to inspect and maintain them. You'll recognize them after peeking into several starters and breaker cabinets. Whatever you do, make certain it's put back!

When somebody leaves the arc chutes off, and it happens frequently, the arc that forms when the contacts open lasts longer and does serious damage to the contacts because all the current in the arc tends to leave through one point and that point gets so hot that the metal melts and tries to follow the current producing a high spot on the contacts. The next time the contacts close that high spot is the only place contact is made and the metal is overheated because all the current for the motor has to go through that one little point. It melts and the coil pressure pushes the contacts together squeezing that melted part out until enough metal is touching on the contact to reduce the heat. Then the contact is fused closed and it won't necessarily open when the coil is de-energized. That's when you're running around trying to find another way to shut the damn motor down!

If only two of the contacts fuse together or something happens to one of the three circuit wires for a three phase motor it runs on only one phase. We call that single phasing because current can only flow one way at a time between two wires. Three phase motors can operate on one phase if the load is low enough but it will destroy the motor in a short period of time.

Due to the considerable number of injuries from arc flashes there are now new regulations for maintaining electrical equipment to prevent injuries from arc flashes. If you have an opportunity to take a course in arc flash protection it's advisable to do so. You'll discover that there are requirements for (PPE) personal protective equipment to be worn when working with energized electrical components. While it's improbable that you'll have a problem in normal operation it is very advisable to leave the handling of high-voltage (600 V and over) to electricians trained in arc flash prevention.

Three phase motors use three electrical currents that flow between the wires. If they aren't balanced the motors can run hot and fail early. Your motor starter terminals should be checked regularly (every two or three years) and after any maintenance to be certain that the voltage is balanced. Use a meter to measure the voltage on each pair of leads, L1 to L2, L2 to L3, and L3 to L1. That big L, by the way, stands for "line" meaning line voltage, the supply voltage. The difference between the average difference and the lowest or highest measurement shouldn't exceed five percent. If there is a big difference in voltage you should get an electrician to check everything in the plant.

Most boiler plants use MCCs (motor control centers, Figure 1-11) to control and distribute electrical power around the plant. You may even have more than one MCC in your plant, something I always recommend, for improved reliability. These are more economical to install than local motor starters at each motor but they can also be a source of unique problems in operation. Individual motor starters use their own power source for control power. Control power is electricity used to feed the motor starter coil that, when energized, pulls in the contacts that close the circuit to feed power to the motor. A typical MCC will have one control power transformer that supplies control power for all the motor starters in the MCC. It's necessary to rotate or pull on the handle on the door of a particular section of the MCC because the handle operates disconnects for the three phase power before the door can be opened. In some MCCs

that will also disconnect the control power but in others it will not. To access the starter it will normally pull out because they're on a drawer assembly and that will disconnect the control power. Because the control power can still be on and the three phase power disconnected you can encounter a situation where everything indicates the motor is running when it isn't. Always be aware that indicating lights on any electric motor starter or MCC can provide false indications. Also, a boiler can try to start without the fan running so disconnect and tag out control power for any boiler before disengaging the starter at the MCC.

The typical MCC indicating light arrangement provides a red light to indicate a motor is running and a green light to indicate it is shut down. I've always had problems with that concept because I think of green to mean it's going. Accept the fact that the electrical engineers use that color arrangement to indicate that you should stop when you see that red light before doing anything that would shut down that motor unless, of course, you want to shut it down.

That's about all I know about three phase motors that is worth telling an operator. The current has to flow in all three wires for it to work and the current isn't flowing through each wire at the same rate and the voltage isn't the same in any wire at any particular instant in time. Don't do anything that could result in one wire having an open circuit when the others don't.

Figure 1-11. Motor control center

Speaking of motors, that's one of the few things I haven't destroyed… yet. I can proudly say that I haven't burned up a motor. We won't talk about all the other things I've managed to destroy. You can, however, burn up a motor if you don't treat it properly. The common method is starting and stopping one. Motors are rated for "continuous duty," "intermittent duty," and "severe duty." You might think that had something to do with where they were located or how many hours the run a day but it doesn't. Continuous duty motors are designed to operate continuously but only be started once or twice an hour. Intermittent duty motors are designed to start and stop a little more frequently and severe duty motors are designed to be started and stopped all the time. So, if you have a small boiler with a level controlled feed pump that starts and stops all the time it should have an intermittent or severe duty motor.

When a motor is started the electricity has to bring it from a dead stop up to speed and that takes a lot of energy. It's sort of like pushing somebody's car when they're broke down (does anybody do that anymore?) It takes a lot of push to get it moving. A motor has what we call high inrush current, in other words a lot of electricity flows through it when it starts. All that energy heats up the motor because it isn't as efficient as it is when it's up to speed. If you stop it, then start it up again right away the heat is still there and added to. So don't start and stop continuous duty motors a lot. Sometimes we have some problems getting a boiler started and repeatedly start and stop the burner blower. If there's a selector switch on the panel that lets you run the fan constantly that's a better thing to do than let it continually start and stop.

One operating technique I was taught was starting a centrifugal pump with the discharge valve shut. It won't hurt the pump, at least not right away, and preventing any fluid flow reduces the load of the pump while the motor is coming up to speed. Once the motor is up to speed you open the discharge valve so fluid can flow. That only works on centrifugal pumps.

You can also overload a motor. One of the things I always used to do when designing boiler plants was specify a pump or fan be supplied with a motor that was non-overloading. In other words, it was oversized so no matter what we did operating it, we couldn't overload it. Now I know that oversized motors are very inefficient so I try not to do that (oversize them). Since we're all working toward more energy efficient installations you will have more opportunities to burn up a motor than I ever did!

DOCUMENTATION

The importance of a boiler plant log, SOPs and disaster plans has already been stressed. Since I measure the quality of care a plant receives by its documentation I thought it important to let you know what I believe should be documented in a boiler plant.

Okay, that's a fair question, what is documentation? It's all the paperwork. Frequently I get a comment from an operator that goes something like "If I wanted to do paperwork I would have got a desk job!" It's not so much doing it, if you think about it the only paperwork you do regularly is filling out the logs. Since the logs are your proof of what you did they're always part of an operator's job. SOPs, disaster plans, and the rest that I'm about to cover are primarily one time deals with maintenance as required. You prepare them once and revise them when necessary.

Maintaining documentation can make a big difference in plant operation. Occasionally I get a call to visit a customer to attempt to determine who made a piece of equipment, what size is it, and where they can get another one. Of course those situations are always crisis ones because whatever it is just broke down and they need it desperately. Frequently I'll be in a plant collecting data for a new project or to troubleshoot a problem and discover the nameplate on a piece of equipment is either (1) covered with eight layers of paint, (2) scratched and hammered until it's beyond recognition, or (3) simply missing… and the plant will not have one piece of paper that describes it. Look around your plant at every piece of equipment and imagine what's going to happen if it falls apart when you need it!

Just a couple of weeks ago I was in a plant with pumps that were so corroded you couldn't even read the manufacturer's name and markings formed in the casting, let alone the nameplate. They had no paperwork on those pumps and no spares. If one broke down they would have no idea where to find a replacement for it. They couldn't even go to their local pump shop and get something that would work because they had no idea what the capacity or discharge head of the pump was. There's an old saying in the construction industry that applies to everyone, it's short and sure, "Document or Disaster."

Not only do you need plant documentation, it has to be organized. I insist the design for every project have an equipment list and a bill of materials and that they be correct. When the job is done those documents become the index for the operating and maintenance instruction manuals. I've had customers who didn't seem to care if they had them and others who requested as many as eighteen copies. Of course the ones that asked for all those copies never managed to have one in the plant when I visited it later!

My method is to assign every piece of equipment in the plant a 3 digit equipment number beginning with 101. Drawing number 02 for every job is the equipment list where every piece of equipment is described along with a common name, manufacturer's information (including shop order, invoice, and serial number) and performance requirements. Drawing number 01, by the way, is a list of the drawings. When equipment or systems are added to the plant the 02 drawing for that job becomes an extension of the first, etc. When they're properly prepared on 8-1/2 × 11 paper equipment lists are an invaluable, single and readily accessible information source.

I also produce an alphabetical index for equipment which references the number so the information can be found in the equipment list.

Material is identified by a bill of material number that consists of a drawing number and the bill of material item number from that drawing. My drawing numbers were all two digit (I never made more than 99 drawings for a job) so you can tell a number is a bill of material number because it has two digits followed by a dash and the item number. It tells you where you can find it on a drawing (the drawing number) and where it's described (in the bill of material on the drawing). If there isn't a drawing describing some material (for example, there's no creating a drawing of water chemicals) I make up a drawing that is nothing but a list of those materials.

What's the difference between equipment and material? If I can define it in the space for a material item on a drawing it's material. When it takes more than one or two lines to describe everything I need to know about it, it's equipment. It's also equipment when you need an instruction manual to use it.

I want the equipment number marked on the equipment, and some materials, to facilitate reference and I stamp every page of the O & M Instructions with the number before I put them in the binders. Everything is then arranged and stored by the numbers. I've encouraged every plant I work in to take that format and extend it to identify everything in the plant.

Most plants will find my numbering method works for them. Large facilities may find it is easier to use four digit equipment numbers where the first digit segregates items (0_ _ _ _ for general equipment, 1_ _ _ _ for Boiler 1, etc., and drawing numbers get much larger

as well. If possible, form a scheme for yourself and use it to identify equipment and material so you can find something when you want it and you have a rationale for where the paper is stored in a filing cabinet.

Someone's bound to ask, why use numbers? Why not just arrange alphabetically by the equipment name? The answer is, if you are a very small plant then you can use alpha. However, any reasonable size of boiler plant is going to have a lot of equipment and it may take several file drawers to store all the information. Every time you add something to the plant with a numbering system that material goes to the last space in the last drawer in the file, the next consecutive number. If you add something with an alpha arrangement you will have to insert it somewhere in the middle and move all the rest of the material about to make space for it. Numbering devices and using an index to locate the number is easier to manage.

Each equipment file also needs to have references to repairs and maintenance history, spare parts, and other pertinent information. Since repairs and maintenance are ongoing the easiest way in a paper system is to have a sheet in each equipment file which has a line for each activity. The sheet might look something like this:

101—Boiler 1—Maintenance and Repair History

Original installation and start-up complete—October 11, 1993

Annual Inspection—July 18, 1994
Annual Inspection—July 22, 1995
Replaced fan motor—August 12, 1995
Annual Inspection—June 30, 1996
Annual Inspection—July 11, 1997
Annual Inspection—July 17, 1998
Annual Inspection—June 23, 1999
Annual Inspection—July 21, 2000
Replaced burner—October 11, 2000
Plugged three tubes—January 22, 2001
Annual Inspection—June 30, 2001
Replaced probed on low water cutoff—August 21, 2001
Replaced steam pressure switches—August 30, 2001

As you can see, this brief history of repairs and maintenance can easily fit on one sheet of paper to cover several years. To know more about, say... why the three tubes were plugged, you would simply look at the maintenance and repair logs for January 22, 2001. It's also obvious that this requires some discipline on your part, the item has to be added to the equipment record. It's so much easier with a computerized system and equipment numbers.

Today it's easiest to use a computer to maintain your records, just be sure you back it up. You can identify the location of the instruction manual by file number and drawer number or other reference. The digital processing allows you to insert information for a piece of equipment in a record without having to move everything about. Actually it's moved, it's just that you don't do it, the computer does. You can also find maintenance and repair information and other data related to a piece of equipment by simply searching those files for an equipment number.

Even though the matter of filing is facilitated by the computer you should still use equipment numbers. A number is unique to the computer but it can't always pick out differences in alpha references that we all use. For example, your data files could have references to Boiler No. 1, boiler #1, Blr. 1, boiler 1, and Number one boiler all entered by different people and sometimes even by the same person. The computer doesn't realize all those references mean boiler 1, and some information could be lost in the depths of the data files.

With little plants I like to see everything stored together, the original specification, the manufacturer's paperwork, maintenance and repair records, parts lists, record of parts on hand and where they're stored. When all the documentation for a piece of equipment is stored in one spot you can find information quickly and, quite importantly, when you dispose of the equipment you can pull the paper from only one spot to discard it or move it. If the equipment was replaced you can replace the documentation readily as well. You shouldn't have to sift through tons of paper that describes pieces that were thrown out years ago; it seems I'm always doing that.

In the first edition of this book I made a mistake, no question about it. You should never have all of your eggs in one basket and you should definitely not have documentation all stored in one place without a copy somewhere else in the event of a disaster. I imagine I didn't think much of it when I wrote the book because I had always retained a copy of our instruction manual and the drawings for every job we ever did at Power and Combustion and the drawings were original vellums kept in a fireproof safe. There were several occasions when customers lost their original data due to fire or water leakage and we were able to replace their material with copies. So, everything I produced was stored in two completely isolated locations. You should make it a point to have a second copy of all of your documentation stored in a separate location. Information is kept only on a computer should have backup files stored somewhere else in a separate remote building, on an isolated server, or in a cloud.

Okay, we have a need for documentation, a means of keeping it in order, now what do we have to keep? Here's a list of equipment items that is as complete as I can make it. You won't always need everything but none are unnecessary. The best thing to do is keep everything because you never know when a piece of information is valuable until you can't find it!

• An equipment list, arranged in numerical order with a description of each piece of equipment. A name for the equipment; manufacturer, manufacturer's model number, a copy, rubbing or photo of the nameplate, model number, serial number, National Board and State Numbers for boilers and pressure vessels, capacity, maximum allowable pressure, maximum operating temperature, minimum operating temperature, maximum and minimum ambient temperatures for operation and storage, voltage requirements, power or amp draw, weight dry, weight operating, overall dimensions.

• Original specification and/or purchase order for the equipment.

• Manufacturer's Data Report Forms and all Repair Forms (boilers and pressure vessels).

• The Manufacturer's Operating and Maintenance Manual.

• PIDs (Process and Instrumentation Diagrams) These drawings show the intended flow of all the process fluids (water, steam, gas, oil, etc.) in the plant and the instruments that are used to measure, indicate, and record the values of those fluid flows. Frequently they will have the range of flow for each fluid. A steam line may show values like "0 to 25,000 pph" so you'll know what the range of flows are. It can also show pipe sizes.

• Lubrication records, what lubricants are required and when the equipment was lubricated or lubricant was changed. Include tribology reports.

• Maintenance and repair records. Either a reference to the date of repair (see above) so details can be found in the maintenance and repair log or a description of the work and when it was done.

- Spare Parts List furnished by Manufacturer (including updated lists when they change part numbers and prices)

- List of spare parts on hand and the location where they are stored.

Of course you'll also need a material list. In small facilities that can just be the bills of materials on the drawings. When you have more than ten to twenty drawings for the plant that begins to get cumbersome. A prepared material list, again you could use a computer, can consist of a number of pages in a three ring binder (my preference) with pages for each drawing bill of material (could be a copy of the original drawing) and an index that helps me find the more important ones. The advantage here is that you can change the information in the notebook to reflect replacements and not have to alter the original drawings. When you replace a valve you can edit the material list to include the manufacturer and figure number of the valve you put in. The figure number on the drawing may identify a valve that's no longer available or the original manufacturer could be out of business.

All those documents should be prepared initially by the engineer and contractor that built your boiler plant. They're something you should have if you don't and, if you don't, you should take the time to create. Once you have them, all you should do is keep them current and add maintenance history. Now, it's time to talk about documentation that has to be produced by the operators.

STANDARD OPERATING PROCEDURES

It's so regrettable that many boiler plants have lost valuable knowledge and experience that was developed over the years of the plant's operation. I'm always amazed that people have an attitude that is expressed in statements like "if Charlie ever retires this plant is in a lot of trouble." The problem isn't just Charlie's retirement, if he dies tomorrow the plant is in a lot of trouble! The message that's really being passed with those comments is that Charlie knows a lot about the boiler plant and he's the only one that knows it. You may think a lot of Charlie, you may rely on him for help on a regular basis, but the truth is that Charlie is a selfish SOB that intends to take his knowledge with him when he leaves the plant and doesn't give a damn about what happens to it or anyone else working there after he leaves.

Maybe everyone thinks he's great right now but different words will be used when he's gone and someone has to do what Charlie has always done. Charlie may do something a certain way because he remembers how someone (maybe himself) got hurt doing it another way. If he leaves the plant and takes that knowledge with him it's highly likely that equipment will be damaged, the plant will be shut down, someone will get hurt, or, god forbid, someone dies—because nobody knows what Charlie knew. I don't want any Charlie's in my boiler plants and I'm constantly warning chiefs about his type. Don't be a Charlie, help document your SOPs and keep them up to date.

Standard Operating Procedures (SOPs) are known, followed and disregarded, changed and updated but seldom written down. It's the lack of SOPs in written form that make Charlie and his kind bad boys in my book, and in the books of people that later suffer from the lack of knowledge that Charlie had. Charlie has SOPs, the problem is they are all in his head. That doesn't do anyone a damn bit of good when Charlie is gone.

As far as I'm concerned very operator owes it to his fellow employees and successors to keep a written set of SOPs, keep them up to date, and be certain that they are complete enough to be followed properly. When a bad experience demonstrates you did it the wrong way that should result in a change in SOPs so nobody else has to have that bad experience. I always suggest a footnote be added in the SOP that reads something like "To avoid failure experienced on (date)" so new and future operators will be able to look up the history of that incident in the log should they question the SOP. Documenting the operation that works well is one way to ensure that the experience is normally a pleasant one and you (and everyone else) avoids the unpleasant ones.

If it were a simple matter to write down steps to follow for each operation in a boiler plant and they always worked then this book wouldn't be necessary. Hell, operators wouldn't be necessary. No two plants, no two boilers, function exactly the same and the only way you determine how to handle those variances is with experience. The manufacturer's instructions for operating the equipment are almost always inadequate because they can't (nor do they even try to) foresee the unique situations that surround their equipment when it's installed in your plant. Don't expect the chapters that follow to be complete either. I list the general activity and identify some things you should know to perform the activity wisely but I don't know what your plant is like and I can't write your procedures either. You and your fellow operators (if any) are the only ones that can produce a quality document of SOPs for your plant.

Of the many reasons I get from operators that claim they can't prepare their own SOPs a lack of skill in handling the English language is one of the weakest. "Aw, Ken, I can't write procedures, I don't write well at all." That's not a good excuse, you write it down in the same words you would use to explain it to another operator, there's no difference between saying it and writing it, you're trying to document an operation, not write a Pulitzer prize winner. I can't write worth a damn but I felt obligated to put what I do know down in this book.

If your SOP doesn't read well, that's tough, what's important is the message and not how it is expressed. Some operators have been concerned with the appearance of their writing and used the services of someone with more language skills to help. Be cautious and read their editing out loud because they don't know squat about operating a boiler plant and can change meaning. I remember reviewing a lovely looking document for one plant. One of the operators was married to a teacher and she typed it all for them. It contained the words "make sure you fill the blower with water before turning the burner control switch on." The correct word was "boiler," it must have been misspelled in the original form; and, hopefully, nobody is stupid enough to try to fill the forced draft blower casing with water before turning the burner on but... If you have access to help with writing your procedures feel free to use it but don't expect someone else to do your job. The final text should be understandable to you and other operators. I won't tell you some of the things that were in an SOP rewritten by an operator's sibling that happened to be in marketing for a toy company. It was humorous reading, actually entertaining, but it didn't serve the purpose at all.

Plain old lined paper in a three ring notebook will do the job. It's not necessary to have the SOPs typed but print if you're doing them by hand, too many people have trouble reading someone else's writing. Someone in the plant may be able to type them for you after they're written down and checked but, like using creative prose assistance, check it afterward. I'm a strong proponent of putting a computer in every boiler plant so the operators can use it to record log data, analyze plant performance, plan maintenance and document maintenance activities, etc., so using a word processor on it to produce your SOPs is a good thing to do. It just isn't important that it be so fancy. Some advantages include the ability to change a sentence or paragraph without having to type a whole new page, indexing, and all the other niceties of word processors. If you can get them

on a computer that's the best deal, just make sure you have backups and at least one up-to-date printed copy. To make sure you're dealing with the current document the date of the last revision of each page should always be written on the bottom in what we call a footer.

I also recommend some form of review of SOPs. If you're the only one writing them you should add your initials to the bottom of each page. If you are one of several operators all of you should initial a page when it's created or revised. The implication of the initials is that you read the page and agree that it is the way you operate; so read before initialing.

Your SOPs should include all the operating activities in the boiler plant and in other areas of the facility that you are responsible for. They can include related items such as, how shifts rotate, which shift is responsible for operating certain equipment or in certain areas of the plant, what equipment a particular shift is responsible for maintaining, etc. Your SOP should contain your description of each of the operating modes that will be discussed later in this book along with all the detail associated with operating each piece of equipment. Some may have to contain special provisions for specific pieces of equipment such as modifying flow loops for different hot water boilers because the piping arrangement produces different situations at each of the boilers, even though the boilers are identical.

It wouldn't hurt to have your SOPs reviewed by a competent engineer. Two years ago I was asked to represent the family of a young man who was killed in a boiler explosion and, although I didn't perform the investigation myself (I suggested they use a friend of mine who lived much closer to the scene of the accident), the result of the investigation was that the operators had bypassed all the operating limits before starting the boiler. What won the case for the family was the fact that the written SOPs instructed them to do exactly that. Recall what I said about WADITWs?

Your SOPs can modify the order of operations where it is more convenient for you. An example would be where the order of opening valves is reversed (without consequence) because the operator would have to go from one level to another then back again to open them in the normal order. They should recognize additional valves, drains, vents, switches, disconnects, and circuit breakers that are particular to your plant or added over time. I must have made ten trips up and down three flights of ladders on one ship trying to determine why I wasn't getting steam to an evaporator and finally found a valve had been added in the piping to accomplish a major repair; it was closed. Opening and closing it wasn't in the SOP for starting that evaporator, the SOP wasn't updated to recognize the change but someone had started closing that valve. I scribbled "make sure valve is open on third deck beside Boiler 2" in the margin under start-up and "leave that damn valve on third deck open" under the shutdown description.

When you get into writing your SOPs you'll discover why some of us engineers like to put pretty brass tags on valves to label or number them. Then there's little or no confusion as to which valve is which and writing the SOP is easier. So, don't hesitate to tag valves. If the boss is too cheap to go for the brass tags there are alternatives, including using a magic marker and writing the number on the wall next to the valve.

SOPs can also include standard maintenance procedures which, even though they're maintenance, not operating activities, are performed by the operating staff and, when included in one document, show the extent of activities performed by the operators. If you are in a large plant with separate maintenance staff there should be another document for maintenance activities and someone should check for coordination of the two to ensure that all procedures are documented, there are no duplications and no conflicting procedures.

Once you have a set of SOPs the difficult work begins, You have to keep them up to date. After initial preparation of your SOPs and for a week on each anniversary of their completion you should think about each function as you perform it and ask yourself "Is this procedure in the SOPs? Am I doing the job the way it's described?" If the answer to either is "no" then you need to get your SOPs up to date. Be very attentive to any construction going on in the plant because that work may change your SOPs or require you to create some new ones.

Don't make them and forget them. I would estimate that every fifth plant I visit for the first time has written SOPs that are completely out of date. Only four months ago an operator exclaimed "of course we have SOPs, they're right here" and proudly showed me a notebook that described coal unloading, coal firing, ash handling, etc. The problem was the plant had been converted to oil ten years ago and gas three years later.

When projects involve such things as adding a new boiler, replacing the burners, replacing a pump, adding new controls or technology such as VSD's (variable speed drives) changes in your SOPs are a foregone conclusion. If you prepare an initial draft of the SOP for the operation prior to project completion it gives you time to think about how you're going to operate that new or modified equipment. Look in the manufacturer's

instructions for keys to successful operation and mentally rehearse the operation before it's time to do it. After you've gone through start-up and a few normal operations of the project you can edit your SOP to account for things you learned during the start-up and operation.

If you don't have SOPs or they're not up to date don't put off creating or correcting them. When you have a highly skilled and experienced Charlie in your plant bounce them off him and make certain you have captured as much of his knowledge as possible in those documents so you're not wishing he was there after he's gone.

Finally, know and follow your SOPs. When I evaluate a plant and its operators I frequently pick out a procedure and ask the personnel to run through it, describing what they would do while I stand there with the copy of the written procedure. It's tough on 'em! First of all, they can't grab the procedure and read it (I have it in my hand) and secondly, if they don't follow it I will know every step they missed. Pretend I am coming to check out your plant every quarter and review your knowledge of your written procedures.

DISASTER PLANS

Preparing disaster plans has become a big deal since the tragedy of 9/11 but I've been promoting the development of disaster plans for a boiler room ever since I spent 92 hours resolving a ground fault in the main propulsion system of a ship in the middle of the Atlantic Ocean. We would have completed the recovery in a lot less time and been far more confident of what we were doing if someone had prepared a plan for such a failure. Sometimes it's unpleasant to consider what we would do if something happened but if we don't prepare we may find ourselves running around in circles like Chicken Little (an old children's story)

Let's face it, if steam pressure is lost you are going to hear about it even if it isn't your fault and there's nothing you could do about it. That's a given and it's easy to explain away a disaster but there's no explaining when you aren't prepared to handle it. Just as you develop SOPs for new installations, by imagining what you would do to operate, you develop disaster plans for situations that you can imagine happening. Preparing may make you the hero someday in the future, not because you did something brave, but that you did something wise, planning what to do in the event of a disaster.

First the plans have to consider what to do if a disaster is happening and what you can do to limit the damage. Plans for fire are essential, especially if your facility does not have sprinklers. Even if you have sprinklers you have to consider what you would do if they were not available, as in loss of all water. Pick spots at ten foot spacing all over the plant, imagine a fire starting at that point then decide how you would fight it with and without water supply. Of course you're going to have some duplicate situations; you'll have areas where a fire is impossible (don't bet on it though, even concrete can burn) so you can simply refer to plans for those other locations. In some cases you have to consider protecting a bigger potential loss (like fuel oil storage tanks) before fighting the actual fire.

Look at the equipment in the vicinity and pay special attention to electrical conduits because it's possible for a small fire in one location to completely shut down the plant. For some dumb reason, probably because it was cheaper for the contractor to do it, many of the plants I know of have all the control wiring for the entire plant run through one spot. They're extremely vulnerable. Pay special attention to what you would do with a fire in the control room, if you have one, or at the control panels. Once you've developed plans for fires that start you can work on plans for fires that get out of control and, finally, how to restore operations after a fire. This exercise typically leads to some decisions to reduce vulnerability to a fire by adding sprinklers, relocating systems (especially wiring) and duplicating some services to make a fire more survivable.

A good appendix to put together for your disaster plan manual is a list of every piece of equipment in your plant with a source for that equipment. In the case of critical parts that are known to break down regularly you probably have that equipment in your parts inventory and can simply indicate "parts" in the manual. Other devices that are too expensive to keep as spares or are not likely to break down are the ones that you need sources for. Sources can be a rental company, stocking parts distributor, or the manufacturer. Include contact names, phone numbers, fax numbers, e-mail addresses and travel directions (in case you have to go get it) for each potential supplier. This list has to be maintained and kept up to date. Don't neglect anything when preparing your list, it should include such items as transformers, transfer switches, distribution panels, fuel oil storage tanks, large valves and pipe fittings that are not the standard stock item for your local suppliers.

Some disasters we don't expect to happen do. Total loss of the plant is one possibility. I've seen boiler rooms practically flattened by an explosion. In another plant the building was untouched but all three boilers had their casings blown off by a simultaneous combustibles explosion. The disaster plan for such an incident would include a list of suppliers of rental boilers that have capacity and pressure ratings to match your plant, contact names and phone numbers, two sets of prepared directions for the contractors on truck routes to deliver the boilers and set them up (two in case the primary site is unusable), in addition, a design for piping to connect the boilers to existing service connections, with alternates for each source and each service pipe.

It's best to have plans broken down by area, here's what we will do to set up a temporary plant in area A and here's the one for area B. Each plan should include an option for temporary water treatment facilities, deaerator, etc. if needed. It's best to include options for the ability to use some existing equipment in a plan that considers what to do if the entire plant is lost.

I'm going to give you a list of disasters which you can address by preparing a disaster plan that you would follow in each event. You will discover that throwing up your hands and walking away is your first impression but after you have had time to think about it that isn't the only solution. Even with total disasters you should have a plan for what to do when they happen. Try developing a plan for each of these disasters where the conditions described relate to your plant:

- You are experiencing heavy rain; flooding is occurring all over the place; the nearby stream is over it's banks and threatening to enter the boiler room; your relief can't get in; oil delivery is out of the question; you have a natural gas supply line over that stream that's starting to catch debris and back up the water; the roof drains are plugged with leaves so the roof is flooded and water is running down all the walls.

- All the weather just described happened up river from you and all of a sudden the water is pouring into the boiler room because the river overflowed.

- A tornado just swept through the plant; all the windows are blown out, the roof is gone and rain is coming in; the insulation was swept off several hundred feet of distribution piping supplying an area where steam supply is critical; the stack for your largest boiler was buckled over by the storm.

- It's an unusually hot summer; temperatures in the upper levels of the boiler room are so high that motor starters located there are tripping as if the motor was overloaded. You lost some ventilation fans; you can't stand to be in the boiler room for more than ten minutes at a time; insulation on steam lines that were soaked by an oil leak are smoking; the control room air conditioning isn't making it so you're perspiring all over the log book as you try to record all the systems that are shutting down from overheating.

- You are experiencing heavy snow, well beyond normal such that you're trapped in the plant, your relief can't get in; oil delivery trucks can't get there for a day or two; the roof of the boiler room is buckling under the weight of the snow; the atmospheric vents for gas systems and the oil tanks are buried in a snow drift; combustion air openings are plugged or plugging with snow.

- Today is the third day of sub zero weather and systems that were supposed to keep operating in the cold are beginning to freeze up. For you in the south, it only has to be the first day of sub freezing weather.

- The electrical power is out and you were just told by the electric company that it's down for at least a day. Two subsidiary considerations are when it's below freezing and when it's extremely hot.

- Consider loss of city water supply due to a city line rupture. You just got told it will be at least twenty-four hours before you can expect water pressure but you have to keep the plant going and you need makeup water.

- Boiler No. 1 (or the lowest number that's still around) just blew up shredding all piping and wiring within six feet of the boiler; steam, water, chemicals, fuel gas and/or fuel oil are spilling into the area; you can't hear a thing because the blast just destroyed your ear drums temporarily. Repeat this consideration for each boiler in the plant.

- Your plant is next to a chemical complex that makes a hazardous gas; they have an alarm system to indicate a gas release and it's been blowing for five minutes which is a fair indication that it's not a drill.

Almost every operator that looks at this list complains "C'mon, Ken, that's not fair! These things don't happen every day, how can I plan for them?" Sometime later they're realizing what they can do, and you should be doing the same thing. Prepare disaster plans and don't be afraid to imagine the almost incomprehensible. At least now, after 9/11 I don't have to explain that part to people.

LOGS

Recording data in a log has been addressed in prior sections but the maintenance of logs is so critical to operating wisely that it deserves a section of its own. I have a multitude of stories that reflect on the performance of plants and operators and almost every one involves a failure to maintain an adequate log. A few describe how maintenance of a log favored the operators and the plant. I won't bore you with all the stories but I will provide some direction in how to avoid cost, embarrassment, and injury through the dedicated maintenance of logs.

Logs are tools. They contain information that allows the operator to make better decisions. In many cases they are the only records of a plant's operation and the activity therein. By looking at the log an operator can determine if a current condition of pressure or temperature is consistent with what existed at another time under similar conditions; a valuable check on the memory which can, and frequently does, fail. Mine does.

The wise operator knows the value of his log. By maintaining an adequate log the operator is demonstrating his skill, protecting the interest of his employer, and developing a database as a resource for evaluating the performance of his plant which allows him to improve on the plant's performance. There are many sources of information available to an operator today but the one resource that continues to be a reliable source of information is the log.

Modern plants are equipped with computers, recorders, electronic devices called data loggers and other means of recording data but those devices do not record everything. The electronic devices may not retain information, some only retain data for twenty-four hours. Frequently the traditional boiler plant log is abandoned in the mistaken belief that all that modern instrumentation eliminates the need for a log. All too frequently those plants realize, after a serious incident, that belief was ill founded. A major, or even a minor, incident can destroy electronic data to leave the plant and operator with no historic data for reference or evidence.

The typical boiler plant at the turn of the century should have a log "book," not a three-ring binder or loose pages. A bound book with consecutively numbered or dated pages is the best type of log book. Contrary to what one might believe, handwritten paper logs have survived many of the worst boiler plant incidents, being lost only when the entire plant was destroyed. Others have survived a plant burnt down although the edges of every page was burned back.

Most importantly, if ever required as evidence in court, it should survive scrutiny. A judge or jury will be confident that the document wasn't tampered with or altered, believing the document is factual and representative of what the operator recorded. Loose pages and electronic data can be altered readily without evidence of that alteration so they are not considered a legal record. When you are facing a law suit it's too late to create a log. And, in today's litigious society it's foolish to think that you'll never be sued.

On the other hand, your maintenance of a log could support your employer's claim against a contractor or manufacturer or even be the basis for a claim by your spouse in the event you're injured or killed. A log is more than just a piece of paper you have to fill out, it's every operator's responsibility to maintain one.

The best log today is a combination of electronic data, printed records and handwritten logs. The handwritten log can contain data that isn't stored electronically or it can include that data as an original source that is subsequently entered into an electronic database by the operator. There is no need to put all data on a single piece of media.

As technology continues to develop, an electronic database will eventually eliminate the handwritten log. An electronic log that could eliminate the handwritten log should consist of a non-erasable media (such as a CDR) with provisions for the operator to record all pertinent data in concert with electronic data storage. The log should be duplicated in another location to preserve it and should also be on non-erasable media. One or more could store the electronic data normally captured by recorders and data loggers while another could store data entered by the operator. Password control can provide the equivalent of the operator's signature. Unless the data are secure and duplicates exist at a location outside the plant where they're not exposed to the same opportunities for damage don't abandon a paper log.

Types of Logs

A boiler plant log can consist of many documents and devices that, as a group, constitute the log. Typical documents that form a log as of the writing of this book include:

Operator's log—A paper document that contains consecutive dated entries made by the plant operators to describe activity on their shift or watch. The log can contain a record of data readings recorded by the operator along with a narrative on activities undertaken by the operator, a record of visitors, contractors, and others that visited the plant, work performed by contractors, problems encountered, etc. Of all documents this one must be arranged to survive as a legal document of what occurred in the boiler plant. It should not be alterable nor altered absent of signature. If an operator decides to change what he has written in the log he should do so according to prescribed procedures discussed later.

Water treatment log—A paper document that contains a record of water analysis and water chemical additions. This document could be part of the operator's log if desired but normally consists of forms prepared by the water treatment service organization.

Maintenance and repair log—Documents that constitute a record of maintenance and repair of everything in the plant. This log should be arranged to facilitate locating the information. There's more on this log in the documentation and maintenance sections.

Visitor's log—A paper document recording the signatures of visitors to the plant. Normally unnecessary unless the plant has a great number of visitors on regular occasions which would clutter the operator's log.

Contractor's log—A paper document recording the signatures of contractors working in the plant. Normally unnecessary unless the plant has a great number of contractors regularly working in the plant so the information would clutter the operator's log.

Recorder charts—All charts from recorders are a part of the plant log. They provide a continuous record of pressures, temperatures, and levels that would normally be recorded at intervals in the operator's log. These are normally paper documents that show values for pressures, temperatures, and levels over a twenty-four-hour period or a week. Some recorder charts span a month and strip charts can easily hold data for three months.

Modern recorders are provided that store the data on floppy disks or CD's but these have their limits and their survivability is questionable. See previous comment on digital data.

Creating Your Log

Many plants simply visit the nearest stationary store to purchase journal binders. These are fabric-covered cardboard bound books with lined and numbered pages. All data are entered by the operator according to standard operating procedures. That is the least expensive approach to producing a log but not necessarily the best method. Anything larger than a small heating plant should consider using a custom log book.

Why a custom log book? There are basically five reasons. First, it saves an operator's time. Second, it provides a consistency not available with a journal, even with well-developed SOPs for log entries. Third, it ensures data are recorded consistently over time. Fourth, it invites contributions of a professional to assist in the development of the log to ensure all important information is recorded. Fifth, a custom log provides a sense of professionalism that isn't associated with the journal type.

A preprinted log can provide assigned spaces for entering much of the data and recording normal activities. Every log must have space for an operator's narrative. The operator's narrative is that written portion of the log normally referred to as notes. It contains a description of what happened in the plant in the operator's own words. Custom preprinted logs also incorporate the feature of a carbon copy. Every other page is perforated at the binder so it can be removed and carbon paper is used over that page to produce a duplicate that can be removed every day to another location. That copy is also used by the manager to perform more detailed analysis and note comments by the operators that require the manager take action to correct deficiencies or have work performed that isn't within the purview of the operators.

I promote an unusual log format—Bound paper operator's logs that are maintained by the individual operators and a computerized log which provides the electronic database for the plant. Contents of the operator's log is entered in the database. Thus, the best of both worlds are possible, there's an original document prepared in the operator's handwriting and an electronic database the operator transfers his information to. It also allows some independence on the part of the operator and will reveal the lack of understanding of an unqualified operator.

What to Record, Why and When

Despite the installation of recorders there is lots of important data in a boiler plant that is not recorded other than in the operator's log. The content also depends on the provision of other logs; data can be recorded in different binders that, combined, form the plant's log. The amount of data recorded is dependent on factors such as personnel responsibilities, the type of plant and the importance of plant reliability and efficiency. For that reason a full evaluation of the log by a professional or an in-depth review by a facility's operators and management personnel should be conducted to ensure the log contains all the data necessary for the plant. Frequently operators and management are not aware of the value of certain data. For that reason the following recommended list is included with a rationale for why that data should be recorded. If you can't justify a professional review, this list should help you produce an adequate log.

When to record data depends on the type and size of plant. A small heating plant may have limited visits by operating personnel and choose to record data once a week. There is a dramatic exposure to additional expense for fuel and water and serious damage to equipment that is seldom considered with that timing. A household heater receives more attention than those plants because the residents note deviations in temperature or noise. A boiler installation in any building should be checked at least daily by someone that is competent in checking the plant and recording and interpreting data.

Probably one of the most serious exposures for limited operator attendance is in our country's schools. It is not in the least unusual for parents to discover, only after asking the children, that the temperatures have been irregular in their school for several weeks, or even an entire season. In our schools the attendant is typically the janitor who, without training, can define his attendance to the boiler as storing his mop bucket in the boiler room. A qualified person should check the boiler plant and record readings twice a day while school is in session. That same rule applies to apartment and office buildings. Plants with boilers larger than 300 horsepower and supplying critical loads such as hospitals and nursing homes should have a qualified person check the boiler plant three times daily as a minimum.

High pressure boiler plants are commonly required to have a licensed boiler operator in attendance but that is not the case in every state; many times the presence of a boiler operator is a function of a union contract rather than state law. When an operator is in attendance recording data hourly is a common practice. The actual written log, however, may only include a record of data by shift or on a four or two-hour interval. There is little value to hourly data other than requiring the operator to be within the vicinity of each piece of equipment every hour. It's a matter of professionalism, operators with a sense of being a professional enter data in the log every hour to demonstrate that they're watching the plant.

Suggested Matter and Data to Record

Here is an abbreviated list of things that should be documented in the boiler plant log along with some good reasons for maintaining a record of the values or information. It's arranged in alphabetical order and many of these items won't apply to your plant so you wouldn't include them in your log.

Air heater outlet air temperature: Monitoring the heated air temperature along with flue gas inlet and outlet temperatures provide an indication of fouling of the heat transfer surfaces, leakage past seals or through corroded tubes, and other performance problems with the air heater.

Annual inspection: The operator's narrative should record the annual (bi-annual or fifth year in certain jurisdictions and with certain types of pressure vessels) inspection of the boilers and pressure vessels in the plant. Inspections are required by law in every state so documenting that it happened is imperative. Don't rely on the inspector, some of whom have been known to lose paperwork. The record should include the name of the National Board Certified Inspector and any findings that inspector relates to the operator.

Blowdown heat exchanger drain temperature: This data provides a means of calculating the cost of heat lost to blowdown. The temperature is an indicator of the performance of the heat recovery system and blowdown/makeup relationship. The drain also dumps to a sanitary sewer which, by Code and law, can't be higher than about 140°F so it's also a record of compliance.

Boiler inlet water temperature: For steam boilers it is an indication of heat lost in feedwater piping or heat added by feedwater heaters and economizers. For hot water boilers it is an indicator of load, required for output calculations. The inlet temperature for fluid heaters and vaporizers serves the same purposes.

Boiler outlet water temperature: Hot water heating boilers are typically controlled to maintain this temperature. It is required for output calculations.

Boiler water flow: Hot water boilers, especially certain types of HTHW boilers, require a controlled flow of water. The value is required for output calculations and should also be monitored for reliability because minimal flow should trip a limit switch.

Booster pump pressure: See condensate pump pressure.

Burner gas pressure: The gas pressure at the burner is indicative of input and should be monitored for consistency relative to load. Increases in gas pressure relative to load are indicative of plugging of or damage to the gas burner. Decreases are indicative of failure or damage to the gas burner.

Burner oil pressure: The oil pressure at the burner is indicative of input and should be monitored for consistency relative to load. Increases in oil pressure relative to load are indicative of plugging of or damage to the burner gun or the atomizing medium controls. Decreases are indicative of failure or damage to the burner gun or atomizing medium controls.

City water temperature/pressure: See makeup water

Combustibles: Monitoring the combustibles content of the flue gas can lead to early detection of burner problems and fuel air ratio control failure. Larger plants may actually control air to fuel ratio using combustibles and monitoring that value is very important to them.

Combustion air temperature: Frequently this is also the boiler room temperature. The combustion air temperature is the base for a boiler heat loss efficiency determination.

Contractor's activities: The operator's narrative should describe which contractors were in the plant, when they were there, how many men, and what they were working on. It wouldn't hurt to list the names of each of the contractor's employees. Less needs to be recorded if there's a contractor's log. Even if there is, the operator's log should note the presence of the contractors as well, use a simpler record such as "XYZ Contractors on site at 8:20 a.m.—seven men."

Condensate pump pressure: Also called booster pumps, these lift condensate to the deaerator and the discharge pressure relative to plant steam load and deaerator pressure is indicative of the condition of the spray valves in the deaerator. The discharge pressure of condensate return pumps, not necessarily in the boiler plant, can reveal steam blowing through traps connected to the same header.

Condensate tank temperature: The tank temperature is a first indicator of excessive trap failures. Once the temperature exceeds 200° a trap inspection is warranted. When makeup and condensate are blended in the tank, the temperature can indicate the percentage of returns. An upward shift in temperature of those tanks indicate trap problems.

Deaerator pressure: Small variations in the deaerator pressure relative to feedwater temperature or plant steam load can indicate problems with the deaerator.

Draft readings: The draft readings are seldom recorded by electronic equipment but they are indicative of the internal conditions of a boiler. Variations in draft readings are frequently subtle, occur over extended periods of operation, and are load related so the operator can miss a significant change. Variations relative to load can indicate fireside blockage, loose baffles, loss of refractory baffles and seals.

Drum pressure: For high pressure steam boilers the drum pressure is indicative of load because of the drop through the non-return valve, and the superheater when equipped. The drum pressure also permits a more accurate calculation of blowdown losses.

Feedwater pressure: Changes in heating plants with cycling feed pumps indicate problems with the pumps or piping. Changes in plants with feedwater flow control valves are relative to boiler load.

Feedwater temperature: The amount of steam a boiler can generate is dependent on feedwater temperature. Lower temperature feedwater will reduce the capacity of the boiler to generate steam. It has an effect on evaporation rate and overall plant performance. The temperature is also indicative of deaerator performance.

FGR: See recirculated flue gas.

Flame signal strength: Upsets in burner conditions and soot or moisture accumulations on the flame detector are indicated by changes in the flame signal strength. Monitoring it can preclude a sudden unexpected boiler outage. Gradual degradation of the flame detector can be monitored for guidance in replacement beyond the normal one year.

Fuel oil meter reading: The totalizer should be read at the beginning or end of the shift to track how much fuel was burned each shift. These data are essential for calculating evaporation rate and fuel inventory maintenance. A fuel oil meter reading should be taken for each boiler whenever possible to determine the boiler performance. If there is no meter then fuel tank level readings have to be used to determine consumption.

Fuel oil supply temperature: Measured at the inlet of the pumps it provides an indication of the temperature in the tank(s) for inventory management and detecting leaks in UST's (underground storage tanks). When burning heavy oil the temperature after the heaters is monitored to confirm heater operation. Temperature to the burners is critical for proper atomization and it can vary with oil deliveries because the viscosity of the delivered oil can change.

Fuel oil tank levels: Required for fuel oil inventory management and detecting UST leaks.

Gas fuel meter reading: The totalizer should be read at the beginning or end of the shift to track how much fuel was burned on that shift. These data are essential for calculating evaporation rate and comparing with the gas supplier's meter readings. A gas fuel meter reading should be taken for each boiler whenever possible to determine the boiler performance. If the only meter available is the gas supplier's meter, it should be read to monitor consumption relative to steam generated, heat output, degree days, or other measure of performance.

Gas supply pressure: The pressure of the gas supplied to the plant is monitored to confirm the gas supplier's delivery promise. Gas supply pressure should also be monitored for possible loss of supply. Gas pressure supplied to each boiler, after the boiler pressure regulator, must be maintained constant and at a prescribed value for accuracy of boiler gas flow meters and/or air fuel ratio. Changes in the gas pressure supply pressure to boilers with parallel positioning controls can alter the air fuel ratio and must be monitored to prevent unsafe operating conditions.

Happenings: Anything that happens which is not normal should be documented. An operator's comment that he heard what sounded like a gunshot proved beneficial in a later court case. Happenings must be recorded consistently to support the credibility of a single incident report in court.

Header pressure: In high pressure steam plants this is the pressure that is controlled. Changes indicate problems with controls, excessively large load changes and inadequate boiler capacity.

Low water cutoff tests: (See testing)

Makeup water meter reading: The principal source of contaminants in boiler water is the makeup water. If makeup is consistent and there is no leakage of untreated water into the system (such as a domestic hot water heating coil break) the water chemistry should be consistent. A sudden decrease in makeup is an indication of an external coil break that can be returning untreated water to the boilers. Monitoring makeup water permits extending time between water chemistry analysis. The quantity of makeup has a significant impact on energy consumption. Every gallon of cold, say 50°, makeup water that replaces 180° condensate requires more than 1,000 Btu of additional heat input. I consider this one essential, even in the smallest of plants and regardless if they are steam or hot water.

Makeup water pressure: This is a value that is seldom monitored because operators take the continuous supply of city water for granted. Someday the city will disappoint you. Monitoring the pressure from wells is more critical.

Makeup water temperature: Determines heat required for makeup, see makeup water meter reading.

Oil supply pressure: Main oil supply and boiler oil supply pressures must both be monitored. Variation in oil supply pressure is indicative of problems with fuel oil pumps, tank levels, variations in oil viscosity or quality. Changes in burner oil supply pressure can upset fuel air ratio.

Operating hours: Recording the amount of time a piece of equipment is operating can permit output and input calculations as well as a record of the amount of time the equipment has been running. Logging equipment start and stop times or operating hour meter readings are invaluable for plant performance analysis and maintenance scheduling.

Outdoor air temperature: Preferably the high and low outdoor air temperature should be recorded. The outdoor air temperature is a prime indicator of heating and ventilation loads. Taking the high and low temperatures for a day permits calculating Degree Days for the facility location. Sophisticated recording devices can record the time the outdoor air temperature is within a given range to provide bin data when desired. Bin data are records of the number of hours the outdoor temperature was within a certain range, and they allow very accurate evaluation of heating plants.

Oxygen: Monitoring and maintaining a minimum oxygen content of the furnace gases is one good practice for maintaining efficiency. Usually, however, the analysis is made of the stack gases. Recording oxygen readings can reveal problems with air to fuel ratio controls, damage to boiler casings or burner problems. When available it should be recorded regularly.

Primary air temperature (coal firing): Too high a temperature will result in pulverizer fires, too low a temperature will result in pulverizer plugging because the coal is not dried adequately. The temperature of the primary air (leaving the pulverizer) when compared to the entering air (air heater outlet temperature) is indicative of coal condition, moisture content, and/or pulverizer condition.

Recirculated flue gas temperature: This temperature should be monitored for changes that indicate fan seal leakage and stratification in boiler outlet ducts.

Reheater steam flow and inlet and outlet pressures and temperatures: On boilers equipped with reheaters these data are required to determine the heat absorbed by the steam. Reheater outlet temperature also has to be monitored like superheater outlet temperature.

Softened water pressure: Comparing the pressures at the inlet and outlet of the softener is a simple measure for determining the cleanliness and quality of the resin bed. Higher pressure drop through a softener can limit the capacity of the makeup water supply.

Stack gas oxygen: see oxygen

Stack gas combustibles: see combustibles

Stack temperature: This list is in alphabetical order but stack temperature is undoubtedly one of the most important data points to record. Monitoring stack temperature is like monitoring a human's temperature. Stack temperature is the most important indicator of boiler health so it should be recorded as frequently as possible. Stack temperature varies slightly with load so load related temperatures should be monitored to indicate scale accumulation, fireside accumulation, baffle failures, improper air fuel ratio and other problems.

Steam flow indication: If the plant load varies considerably during a shift, say more than ten percent of operating boiler capacity, recording the indication of steam flow consistent with the other data readings is desirable to maintain a correct relationship for evaluation.

Superheater outlet pressure: This pressure should be recorded because, combined with the outlet temperature, it is used to determine the amount of heat added to the steam. Variations (relative to load) in superheater pressure drop can indicate superheater leaks or blockage that is otherwise undetectable.

Superheater outlet temperature: The damage associated with an excessive superheater outlet temperature requires constant monitoring of the superheater outlet temperature. The superheater outlet temperature combined with the outlet pressure is required to determine the amount of heat added to the steam.

TDS: The total dissolved solids content of the makeup, condensate, boiler feedwater, and boiler water

should be monitored at a frequency adequate to detect problems and any time a problem with water chemistry is indicated.

Testing: Regular testing such as testing operation of the low water cutoffs on steam boilers should have a check box where, by checking the box, the operator indicates he performed that operational test. An initial box, where the operator's initials indicate who did it is appropriate when more than one person is on the shift. Most other tests, conducted infrequently, such as quarterly lift testing of a steam boiler's safety valves can be included in the operator's narrative. Tests that should be recorded, and their frequency, include:

- Combustion analysis—Frequency is subject to State Environmental Regulations but should be performed at least quarterly for boilers that operate continuously and any time the efficiency of combustion is questioned.

- Flame sensor tests—each month for gas and oil fired boilers.

- Hydrostatic tests—for boilers, annually. For unfired pressure vessels, bi-annually except for compressed air storage tanks which may only be tested every five years. Note that these are common time frames, your jurisdiction may require a higher or lower frequency. For any pressure vessel or piping system a test should be conducted after the vessel or piping is opened for inspection or repair.

- Low water cutoff tests—each day for steam boilers, each shift for high pressure steam boilers, semi-annually for hot water boilers. Testing of the low water cutoff is imperative since fully one third of boiler failures are due to low water.

- Safety valve lift tests—each quarter for steam boilers operating at less than 400 psig, annually for hot water boilers.

- Safety valve pop tests—each year for steam boilers and hot oil vapor boilers. Alternatively record replacement with rebuilt safety valves. The boiler inspector normally governs the performance of these tests because many boilers have more than one safety valve and the seals have to be broken (and replaced by the inspector) to test the second valve.

Water analysis—depends on the plant. High pressure steam boilers with highly variable loads and makeup water requirements should have water analyzed every shift. Other high pressure plants may test water daily. For steam plants where makeup water is limited and consistent, condensate returns cannot be contaminated, and makeup water is metered, weekly analysis should do. For hot water boiler plants with limited leakage and when makeup water is metered monthly analysis should be adequate. Monitoring the makeup is the key, analysis should be checked immediately when makeup usage changes abruptly, either up or down.

Water pressure/temperature: See Makeup, Boiler, Feedwater.

Visitors: Unless frequent visitors suggest having a visitor's log the operator's log should record all visitors to the plant.

Log Calculations

The logged record of a boiler plant's operation should include calculations of fuel consumed (absolute minimum), steam generated or MBtu output, and percent makeup as a minimum. These are fundamental values that, if not monitored, can allow plant performance to decay until it becomes a serious problem.

Other calculations that can be incorporated into a log include evaporation rate or heat rate, a degree day calculation and steam generated or heat output per degree day or according to a degree day formula. Reconciliation of fuel oil inventory (including shrink or swell of oil in outdoor above ground storage tanks) to account for variations in inventory is recommended for oil burners. Reconciliation of boiler fuel flow meters with gas service meters is invaluable for monitoring the quality of the gas service instrumentation as well as in plant instruments. Calculation of the plant heat balance will permit determining how much steam was delivered to the facility.

Chapter 2

Boiler Plant Operations

We cheer the football quarterback that throws the winning touchdown, the baseball player that hits the last inning home run and the jockey that rides the leader over the finish line. Inside boiler plants around the country are other heroes. He demonstrates skill and experience as he flawlessly lights off a boiler, brings it up to pressure and puts it on line. That's controlling thousands of horsepower with explosive energy that exceeds the imagination of most of us. She moves swiftly to respond to a cacophony of alarms, swinging valve handles and pressing buttons in a long practiced dance to restore operations to normal and the noise to the low roar we're used to.

If it were not for the experience, training, and skill of today's boiler operators we could be learning of the thousands of accidents and significant number of injuries and loss of life that was normal a century ago. They are operating equipment with a lower designed margin of safety and more complex limits on operation than their predecessors ever dreamed of.

Operations are covered in this chapter without discussion about the equipment. That's not true for operation of refrigeration systems, gas turbines, and HRSGs where their operation is described along with the equipment in general and without the detail covered here. This chapter does, however, contain guiding information that can apply to those other systems so it's wise to read it again to see how it may relate to those other systems after you're familiar with them.

OPERATING MODES

There are many different modes of boiler plant operation. The one normally dealt with is "normal operation" when the plant is generating steam (vapor) or heating water (fluid) and all the operator need do is monitor it in the event something goes wrong. The other modes of operation require an operator act to change the condition of the plant.

No book can provide a specific set of instructions to perform those activities because every boiler plant is different. The following are guidelines to use for writing your own procedures if they don't exist and to check them in the event they do.

VALVE MANIPULATION

If it weren't for the fact that piping systems are normally built with generous safety factors I would consider the operation of valves one of the most critical skills for a boiler operator. It's still a critical skill, there's just other ways that you can kill yourself and other people more readily because the piping can take the punishment some of us manage to hand out. At some time in your life as a boiler operator you're bound to discover this because you'll be torn between standing there and doing your job and running like hell because all the piping in the plant is shaking around and making banging sounds that make you think it's going to blow apart at any minute.

After fifty years I've grown accustomed to it and start checking out the plant to find where the problem originated while everyone else is running out the doors. That doesn't mean that someday I won't run, only that I've experienced the normal hammering enough to know when the piping will survive… if I stop it in a reasonable period of time! Most of the time those banging and shaking incidents are due to improper operation of a valve.

Sometimes the problem isn't involved in operating the valve, it's because it didn't work or was left in the wrong position. One such incident happened after starting up a new boiler plant and while I was operating it during construction. Steam wasn't needed at night and I was the only one there so I just shut off the boilers, opened a header drain and left the plant. The following morning as the boilers came up the whole main steam system started banging and thrashing about. After everything quieted down again, which took a while, and I ended up draining what seemed like an awful lot of water out of the header I finally realized the drain valve I opened the night before was plugged. A vacuum had built up in the system and drawn condensate into everything. After I dismantled the drain piping, cleaned it, and the valve, I vowed I would make sure more than one drain or vent was open to ensure a vacuum didn't build up in any steam piping I shut down.

A similar instance created havoc when a steam line on a bridge flooded due to a vacuum. You see the bridge was temporarily supported during the original hydrostatic test of the piping because it was never designed to support the flooded steam line. Guess what happened!

When manipulating valves on steam piping it's important to remember that a cold line is either full of air or water, it's rare for it to contain a vacuum. When shutting down a steam system the space occupied by the steam has to be filled with something when the steam condenses, either air or water; unless you're in a plant that injects nitrogen into cooling steam piping. Water setting in any piping system will descend to the lowest level if allowed. Air can compress in piping to preclude admission of steam or water. Steam at pressures less than 15 psig is lighter than air and steam at 15 psig (actually a tad lower than that) and above is heavier than air. It's one reason we keep a high pressure boiler vent open until the pressure is above 25 psig and vent low pressure boilers until we're carrying a load, counting on the flow of steam to sweep the heavier air out of the boiler.

Air can be trapped high or low in a steam system depending on the pressure and it can create pockets where piping is suddenly heated as the air is displaced. Some air is desirable in water systems to serve as a cushion to absorb the shock of sudden changes in flow. There's always a standing length of piping at the top of any water system. It's there to trap air for that purpose. In your house it'll be in the wall behind your medicine cabinet.

Modern plumbing systems use a special fitting with a seal so the air can't be absorbed in the water to lose the cushion. Plumbers used to know that the solution to a hammering sound in the customer's pipes every time a valve closed was to drain, then refill, the system to restore that air cushion. Of course some of them made a pretty elaborate thing of it so they could charge more to perform that simple act. Draining and refilling the water piping in your house is usually all you have to do to eliminate pipes banging every time you close a faucet.

Every time you fill or drain a system you should follow a prescribed procedure that's proved successful for your plant. If it's a new plant you'll have to develop the procedures so you should think about how you've done others and apply your experience in producing a prescribed procedure for each piping system in the new plant. There's no sense in busting it before you even get it started.

The first step in filling a system is opening vents and drains. Keep in mind that they're never empty, usually they're filled with air and it's necessary to get it out. When shutting down a system you have to open the vents and drains so the liquid can drain out and the air can fill the space left by condensing steam. Speaking to the latter, it's always important to open some vents first a little steam escaping proves to you that the valve is open.

Once you've closed a main steam valve to a piping system the pressure will drop quickly and a vacuum could be generated before you get a vent or drain valve open. Open the vents first and let a little steam escape because it's safer. On large systems it may take several vents and drains to admit air fast enough to prevent pulling a vacuum. Any system containing large pieces of equipment (deaerators, tanks, heat exchangers, etc.) should be monitored closely as you shut them down to ensure a vacuum doesn't happen because the equipment isn't necessarily designed for a vacuum and atmospheric pressure can crush them. Fail to do it and you'll appreciate that the day you suck in a heat exchanger that costs several thousand dollars to replace.

Simply draining water without venting a system can also create damaging vacuums. Anytime the column of water in the piping gets over 35 feet it can create as pure a vacuum as steam. Draining a water system without venting tanks on upper floors can result in all those tanks being crushed by atmospheric pressure because the water draining out left a vacuum.

It boils down to knowing the fluid you're dealing with, what's in the piping, and what will happen when you open or close that valve. Filling any large system, whether with water or steam, should be done with a valve installed for that purpose. Normally it's a small valve mounted on the side of the shut-off valve (Figure 2-1) but it can also be piped as a bypass or even consist of a simple drain and hose bib where you should connect a hose from the supply to fill the water piping. The problem is that sometimes (Okay, I'll be honest… frequently) we engineers don't think about it and put in a bypass or fill valve that is so small it will take hours to fill the system. On the other hand, I've seen systems where the fill valve was the largest. Just forgive us dumb engineers and take the time to fill the system or, if you have to fill it regularly, put in a larger bypass or fill valve (like the additional one in Figure 2-1). Please note that I don't encourage you to leave the insulation off the valve and piping.

Some operators choose to crack the main isolating valve to speed up the filling process. Before I continue I

Figure 2-1. Warm-up bypass valves

want to make sure that term is clear. I remember an apprentice that we called Tiny who happily trotted off to follow my instructions to crack a ten-inch steam valve in the upper level of a boiler room. I was very grateful that he decided to get some clarification and leaned over the rail on an upper platform (with the twelve-pound maul he had in his hand showing) and yelled down "Mr. Ken, exactly where do you want me to hit it?"

To crack a valve means to open it until the disc lifts off the seat (creating a small opening or crack for the fluid to flow through). A ten inch steam header shut-off valve should have something like a three inch globe bypassing it to allow warm-up of the steam main. Try as hard as you can and you still won't be able to crack a valve that large without producing a significant surge in steam flow. I encountered a valve that took two turns of the wheel to close it back off after I cracked it open and the resulting jump in steam flow lifted the boiler water level in the boiler to the point it tripped on high level.

Another common and dumb trick is filling a hot water boiler by opening the main shut-off valves so you drop the pressure in the whole system and steam starts flashing off at all the high points then collapses as pressure is restored.

Regardless, you should always crack any valve as the first stage of opening it. When the valve is larger than two inch wait a moment or two to see what happens while preparing to spin it shut again; if you have to. Then you can wander off whistling and looking around, playing the innocent party, if systems start hammering and banging because you changed the pressure in a system too fast. The important thing to remember here is that it will do the same thing the next time so change your operating mode to eliminate that action thereafter.

Always open and close valves slooowly until such time as you know you can get away with spinning them. Even then, don't spin valves. Someone else may see you doing it and follow suit anytime they're directed to operate one. Once upon a time I was a cadet on my first ship and spun a boiler non-return valve open just like I observed the second assistant doing. The difference was he had done it while there was less pressure in the boiler. I did it when the boiler pressure was considerably higher than the rest of the system, oops! (I know what it sounds like when boiler water is lifted out of the boiler, bangs around in the superheater, and then hits the first stages of the turbine; …it isn't a pleasant sound.)

Despite my yelling at them about the same thing year in and year out, I still find steamfitters putting valves way up in the air where you need a ladder, and sometimes to act like a monkey, to get at the darn valves to operate them. A valve is installed in a piping system so someone can shut it off when necessary and anything higher than four feet off the floor is a pain to operate. I design systems with piping dropped to pressure reducing stations, distribution headers, etc. to put valves at operating level only to find later that the contractor convinced the owner (who doesn't do the operating) that money could be saved by rearranging the piping a little. How frequently do operators expose themselves to potential harm by climbing up to get at a valve just so someone could save a few bucks on an installation?

In other instances the contractor simply put the valve where a pipe joint was needed. If you're associated with new construction do your best to get valves located where they're convenient. If they aren't convenient and you have to operate them more than once a year then ask for an extension. A chainwheel or extension rod is going to cost the owner something but all you have to do is mention the cost of the workmen's compensation claim if you fall while trying to operate that valve. Don't let them get cheap either, ask for the chainwheels with the built in hammers that help drive the valve open whenever it's larger than three inch. Use oversized chainwheels otherwise. Push the issue, think of yourself and remind your employer, if you're all alone in the boiler plant and fall while climbing to reach a valve it's going to cost a lot more than installing an extension rod or chainwheel.

I don't understand why but I haven't run into any

operator that knew the proper procedure for operating a lubricated plug valve before I explained it. That funny looking knob that sticks out of the square where you put the handle isn't a giant grease fitting that takes an equally large grease gun. It's just a screw and when you turn it the movement presses a small amount of grease into the valve. The grease isn't soft flowing material either, it's very thick and stiff; when you replace it you turn that fitting all the way out so you can put in a stick of grease.

You should give that fitting a quarter turn every time you operate a lubricated plug valve unless you're operating it several times in a shift, in which case you give it a turn a shift. I've had several steamfitters tell me that a lubricated plug valve is no good because "they always leak." I don't understand where they get that, it's the only valve that you can stop leaking in service. When you turn that plug screw you're driving that stiff grease in between the metal parts of the plug valve to seal it. Unless nobody has operated the valve for years, so the grease has hardened and doesn't flow uniformly into the valve, it will always seal. That's one reason Factory Mutual first chose the lubricated plug valve for fuel safety shut-off service, what we commonly refer to as an "FM Cock" because they should never leak if operated properly.

With the exception of those lubricated plug valves all valves do leak. Some soft seated valves can last what seems like indefinitely but an operator should always be conscious of the fact that a valve can leak and should never, even with lubricated plug valves, rely on a valve holding right after it was closed. Sometimes indications like pressure dropping can give false assurance that a valve isn't leaking so you should always wait until conditions have stabilized, cooled down or heated up as the case may be, before taking the position that a valve is closed tight. Also keep in mind that zero pressure measured by a gage at the high point of a system (or a gage with a water leg that's compensated for it) doesn't reveal the pressure at the low point of a system which could have several feet of static fluid pressure on it.

A system isn't down and without pressure until all the vents and drains have been opened and, to be absolutely certain, the lowest drain valve passed some fluid when it was opened (to prove it really was open and the connecting piping wasn't clogged) and, finally, no fluid is leaving it. If there's a possibility of gas lighter than air entering the system (like natural gas) test for it at the high point vent and a high point closest to the potential source of that gas before declaring a system isolated. Also, don't count on a valve holding if it held last time. I've had many experiences with random leaks through valves; they leak one time and not several oth-

ers or never leak, except occasionally. Hmmm... wasn't that a statement typical of an engineer?

When isolating systems (see more under lock-out, tag-out) it's always advisable to ensure that you've double protection in the event one of the valves fails or leaks; if there's another one in the line close it. A vent or drain between the two valves will release any leakage to atmosphere instead of into the system that's isolated. Resilient seated valves (butterfly, ball, globe, and check) can seal initially then leak later if upstream pressures increase.

An important consideration in valve operation is the use of a valve wrench. If you don't have any valve wrenches in your plant then make some and hang them where they're convenient. You don't slap a pipe wrench on a valve handle to open or close the valve. I've been in many a plant where the chief engineer would fire anyone caught doing it. The pipe wrench is designed to grip a pipe by cutting into it; using one on a valve handle will create sharp slivers and grooves in the handle's metal which can tear through leather gloves and cut up the hand of the next person that tries to operate the valve.

Make some valve wrenches, all you need is different sizes of round stock, a vise to bend it, and for larger sizes a torch to heat the metal so you can bend it. Never put the portion you grip in the vise so it remains smooth. The standard construction (Figure 2-2) includes drilling a hole for a hook for hanging the wrench near the valve for use when you need it. Valve wrenches, by the way, are not for closing valves, only for opening them. Those chief engineers I mentioned would also ream you out if they caught you using a valve wrench to force a valve closed.

One last comment on operating valves. It's a matter of courtesy that has almost been abandoned since I was an operator. When you open a valve you always close the handle back down one half, then back one quarter, turn. That way anyone coming along behind

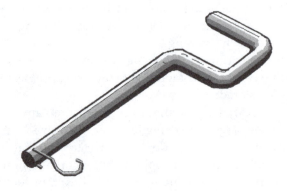

Figure 2-2. Valve wrench

you will be able to tell immediately if the valve is open because they'll try to close it and it will make at least a quarter turn toward closed. If you leave the valve jammed open someone can think it's closed because it doesn't spin that quarter turn. I was so used to that practice, and still believe in it, that I'm regularly foiled by someone leaving a valve jammed open. Thank goodness the important ones have to be rising stem so I can tell their position by looking at them.

NEW START-UP

There is a significant difference between starting a boiler plant that is new and one that has been in operation. Hundreds of wiring connections, pipe joints, and other work went into preparing the boiler and there's bound to be a few unforeseen problems as the start-up proceeds. These guidelines should help you achieve a smooth start-up. They should also be used after any maintenance that resulted in opening a system.

First, have a written procedure prepared, not an outline. Each step should be described along with who is responsible for the action. In many cases it will be the installing contractor's responsibility to produce this document but you should check it for completeness and accuracy. Imagine the start-up proceeding and try to imagine all the things that could go wrong as well when preparing or checking a written procedure. The following should be addressed by the written procedure.

Preparing for Operation

Be certain the safety shutdown push-buttons, switches, valves, and other devices are in place and operational. Test each one if possible and refuse to continue the procedure if one is not present or not operational.

Check all electrical circuits for shorts and grounds before energizing them. Make sure all equipment and piping is electrically grounded before admitting fluids into the plant. Energize all electrical circuits before admitting fluids into the plant to ensure they can be powered up. Test all electrical emergency trips and shutdown devices. De-energize circuits before admitting fluids.

Prior to closing a boiler or pressure vessel inspect it to ensure there are no personnel, tools, or other things inside that shouldn't be there. In Amsterdam in 1967 I almost closed a boiler with ten shipyard workers napping in its furnace.

Small boilers can come set up from the factory to reduce the chances of a problem on initial start-up. It's rare that a boiler to be attended is factory tested and even then you can't be certain that the conditions in your plant are identical to the conditions in the factory. So, the initial start-up of a boiler requires a careful approach to lighting the initial fire. You should ensure air flow, make certain it's linear on modulating boilers and establish safe light-off conditions before thinking about starting to fire.

The codes require a minimum amount of building opening to admit fresh air for combustion but I've found that it's frequently overlooked. If you're starting up the only boiler in the plant it's possible there's no way for combustion air to enter that boiler room. If the boiler is an addition to an existing plant the likelihood that someone paid attention to the requirements for combustion air is even more remote.

A basic rule is two openings consisting of one square inch in each opening for every 1,000 Btuh of boiler input and a minimum of 100 square inches for small boilers. Larger installations allow 4,000 Btuh per square inch. One opening should be high up in the building and the other near the floor. Prior to starting a new boiler the availability of fresh air should be confirmed and the openings should be labeled "combustion air, do not cover."

I've ventured into many a building where the air openings were blocked because the operators could feel a draft. Then they couldn't understand why their boiler was smoking. Once you've confirmed the fresh air source make sure you have linear air flow on any modulating boiler; refer to the chapter on tune-ups for establishing linearity.

Make sure each fluid system is closed and ready to accept fluids before opening shut-off valves. When preparing to admit liquids identify vent valves and make certain they are open, you can't put much water in a boiler plugged full of air. If the fluid is admitted through a pressure reducing station position a person to monitor the pressure in the system.

Position observers to detect leaks in the piping and equipment. Be certain that observers are capable of seeing all drains leaving the plant to ensure hazardous or toxic materials don't escape. Ensure the person controlling the valve(s) admitting the fluid is in contact with all observers and can shut the valves immediately if a problem arises. Ensure personnel are positioned to close vent valves as the system is filled.

Have I said it before? Look at the instruction manuals. Know how much fluid is required to fill the system and estimate the filling time. It's another way to ensure you know the fluid is going where it's supposed to. Wondering where all that fuel oil went several minutes after the tank should have been full is not a comfortable feeling. Whenever possible have means of detecting the level as the system fills so you will know what's happening.

Fill Systems

Fill the system slowly. Whenever possible use bypass valves even though the filling may be slower than desired. The person attending to the valves controlling fluid entering the systems should not leave that post and close the valves immediately upon instructions of, or any sounds from, any observer. I must add that the valve operator should announce at regular intervals after closing a fill valve. We once stood around waiting for a boiler to fill for more than three hours when we finally checked with the apprentice that was stationed on the valve. He closed it when someone shouted "hold it" and that's how it had been for three hours.

Observe vent valves and close them as fluid reaches them. After the system has filled operate the vent valves again to bleed off any air that may have been trapped and then migrated to the vents. When filling systems with compressible gas use testers and bleed the system at the high or low points accordingly (high points for systems where the fluid is heavier than air, low points for fluids lighter than air).

Allow the systems to reach supply pressure or controlled pressure slowly while diligently looking for leaks. Compressed gases (including air) will expand explosively if the container ruptures so your plan should provide for small increases in pressure with hold points at regular intervals to check for leaks and any signs of distortion of the vessel or piping that could be caused by the pressure. A hold point, by the way, is when you have reached a certain time or condition in an operation where you planned to hold everything while checking that the procedure is happening as planned and all safety measures have been taken. In many cases they're described in the SOP as a hold point.

Hydrostatically test each system after it is filled following the procedures described for pressure testing. As with filling there should be a person assigned to control the pump or valve that is pressurizing the system.

Check electrical circuits that are connected to the systems during hydrostatic tests to ensure the liquid did not introduce an undesirable ground. Check them again after all test apparatus is removed and normal connections reinstated.

Finally, make certain that all the tests performed are documented. A note in the log saying "tested Boiler 2" isn't adequate. The documentation should contain values that demonstrate you really did it. The log should read "Tested Boiler 2 to 226 psig by the boiler gage."

Every time I'm told something was tested and I ask for the pressure, voltage, resistance, and I don't get numbers I doubt it was done. Yesterday I checked on one of those general statements and found it was a lie, the tests hadn't been done because there's no way the results would meet the requirements.

Start Makeup Systems

Once all pressure testing is completed, begin operation of the systems in an orderly manner. Water softeners, dealkalizers, etc. should be placed in service to condition water to be fed to a boiler system. Provide means to drain water until water suitably conditioned for the boiler is produced. If the installing contractor was sloppy you'll find yourself flushing mud, short pieces of welding rod, and lunch bags containing leftovers out of the line and flushing will become a major project. I can remember one job where we had to cut the pipe caps off the bottom of drip legs to get the large rocks out.

Establish Light-off Conditions

The combustible range (see fuels) is so narrow that it really is difficult to establish conditions to create a fire in a furnace. Today's modern boilers which surround the fire with (relatively) cold surfaces don't provide heat or reflect it back to help maintain a fire making firing difficult if conditions are not correct. On fixed fire boilers (no modulation) check the instructions for any measurements that will help you establish the proper air flow or conditions for the combustion air. On modulating boilers set the air flow at a low fire (minimum fire) condition.

If there is no other means of determining where to set the air flow I start at maximum on fixed fire units and 25% on modulating units. Yes, maximum is easy to set and no, 25% isn't that hard to determine. If you don't have a manometer make one by taping some clear tubing to a yardstick (actually that's a better manometer, I always have problems with the tubing coming off my fancy purchased one); leave a loop of tubing hanging off the low end to hold the water. You just need a way to measure the air flow and pressure drop across anything in the flow path is adequate.

Set up your manometer on a ten to one slope (Figure 2-3) so every inch on the ruler is a tenth of an inch in actual pressure. Position the end of the tubing at the inlet of the forced draft fan or air inlet then fill the manometer with water until the level is at zero. Run the fan to high fire (maximum) and record the reading on the manometer. Recall that pressure drop is proportional to

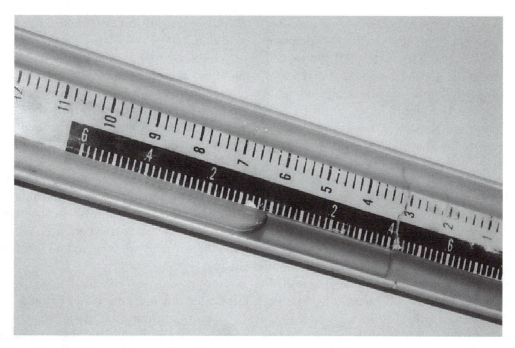

Figure 2-3. Manometer on slope

the square of flow so the measurement when you are at 1/4 flow (25%) will be 1/16 of the reading at high fire. Run your modulating controls down to the bottom to see if the manometer reading is about 1/16 of what you got at high fire. If it isn't check that manual again; some boilers are only rated for a 2 to 1 turndown so low fire is 50% and the differential pressure reading would be 1/4 of the high fire value.

If you must adjust linkage, and that's very possible, remember to check for any changes to the high fire reading after you've ensured the controls will stroke (go from high to low and back) without binding any of the linkage. Once you've established light-off combustion air flow you can set up the fuel or fuels.

Setting up fuel oil at low fire should be a snap. The only problems with it could be an improper piping design which, among other things, doesn't include any fuel oil return. As far as I'm concerned if you install a boiler without a fuel oil return line you're setting it up for a furnace explosion! With a fuel oil return line you can set up your oil conditions without creating a fire.

Before opening the oil valves, make sure the oil atomizer is not in the burner or open a joint at the hose or tubing to the burner so you know you're not dumping oil during this process. Who says the safety shut-off valves don't leak?

Once again check the instruction manual, this time you're looking for a burner oil pressure at light-off. That's either operating pressure for a fixed fire burner or a specific pressure for a modulating burner. Pressure atomizing burners will follow the rules for flow and pressure drop but air and steam atomized burners don't. If you can't get the information from the manual use the pressure that's half the range of the pressure gauge for fixed fire burners, 1/5 of that for modulating pressure atomizing burners, and 1/16 of it for steam or air atomized burners. Half the gauge is explained in the chapter on measurements.

Once you've established the required oil pressure for light-off you can set it. Close the fuel oil recirculation control valve; a globe type valve in the fuel oil return line at the boiler. Position the controls at low fire on modulating boilers. That can be as simple as holding the "decrease" push-button on a jackshaft controlled boiler to several adjustments on a pneumatic control valve. Slowly open the fuel oil supply valve while observing the burner supply pressure gauge. Slowly is because you could interrupt the flow of oil to another (operating) boiler and shut down the plant. (That's said by a dummy that did it more than once!) The burner pressure should suddenly jump to oil supply pressure because the recirculating valve is closed and there is no flow through the piping.

Now you know why we want to be sure no oil is going to the atomizer, if the safety shut-off valve is leaking there will be oil dripping or spraying out of the burner yoke or the opening we created. Needless to say, if the safety shut-off leaks we stop start-up and call the manufacturer. Assuming there are no leaks we have now pressure tested the burner piping at operating pressure (or did you actually hydro test it?) and we can continue with the setup. Crack the recirculating control valve then slowly open it until you've established light-off pressure at the burner piping (after the firing rate control valve) then continue slowly opening the supply valve while adjusting the recirculating control as necessary to maintain the pressure.

Nope, you're not done. Establishing a pressure for light-off isn't that simple. Remember the chapter on flow? You aren't so concerned with pressure as you are with flow and establishing the pressure doesn't prove the flow. Use the oil flow indication on full metering systems or take two oil meter readings at a set interval to determine gpm to determine the flow; it should be the design flow for fixed fire boilers and 20% to 50% (depending on turn-down capability) for modulating boilers.

If there's no meter I will count quarter-turns of the recirculating control valve on another identical boiler, match that position and establish light-off pressure by adjusting the control valve. Barring any other means of setting it I'll listen to the recirculating control valve and set the low fire pressure while the squeal through the recirculating control valve sounds familiar. After you have established a final position for the control valve you can set the recirculating control valve to produce a pressure that matches operating pressure at low fire for a good smooth light-off.

Since we don't recirculate gas you can't guarantee a light-off position by measuring the flow. We can establish the pressure. For fixed fire units it's a matter of setting the pressure regulator. The pressure regulator on a modulating burner should be set for the design supply pressure. We'll get light-off pressure refined when we perform the initial light-off.

Light-off pressure is not necessarily low fire but it usually is. Some burners will operate at lower flows than that required for light-off and your plant may have operating conditions where it is imperative to establish a low fire position independent of light-off. If that's the case, your control provisions should includes means of proving the light-off conditions.

Low fire is typically the light-off condition on most boilers. It's imperative that the low fire conditions are fixed and reliable because many upsetting situations could produce unstable fires and explosive conditions otherwise. The fuel flow control valves should never shut, and I do mean never! Their minimum position should be set mechanically so something has to break before they shut. That way any upset in the controls,

including broken linkage, should establish a low fire condition.

It pays to look at your equipment to see how it will fail, if linkage comes loose and can fall to open the fuel control valve add weights so it will close to minimum fire instead. The air flow controls should also rest on a mechanical stop at low fire so the dampers never shut, unless they leak so much at closed that low fire air flow is still achieved. I've run into a few new full metering systems where the designer or contractor felt a mechanical stop was unnecessary, establishing minimum fire using control signals; of course most of those discoveries were on plants that had experienced a boiler explosion! I was there to find out why and no low fire stops is usually one reason.

Fill Boiler and Test Low Water Cutoffs

Before starting a fire in the boiler, fill it with water to a low level in the gauge glass, about an inch. Make sure the vent valve on the top of the boiler is open so air can get out to let the water in. When the water is heated from cold to boiling it will have swelled so much that the level will rise to over the middle of the glass. In unusual boilers it's sometimes necessary to drain some water before the boiler reaches operating temperature because the boiler has a large volume of water compared to the room for expansion in the steam drum. You'll have to drain water to keep it in sight in the glass.

From this point on you have to keep an eye on that water level. When the water level is visible in the gauge glass it's time to test the low water cutoff. Proving a low water cutoff works on a new boiler is doubly important because there are so many ways to defeat those devices. The cutoffs should be tested without operating any bypass buttons or similar provisions to ensure they operate properly. Their failure is a primary reason for boiler failures.

Be sure to test the low water cutoff properly; simulate a loss of water due to evaporation by draining the water column or cutoff chamber slowly so the water level drops gradually to the cutoff setting. If it doesn't shut the burner controls down, don't continue the start-up until it's fixed.

Prove Combustion Air Flow

After the boiler is filled with water it's time to start a burner cycle which always begins with establishing and proving air flow through the burner and furnace. In very small boilers, like your home hot water heater, airflow is a function of combustion and is not proven. In most boilers, however, it amounts to starting a fan which will produce a measurable air flow that can be proven. Proof typically consists of a fan motor starter interlock contact and an air flow switch.

Note that I said air "flow" switch, on many systems a simple pressure switch is used and pressure doesn't prove there's flow. Too often I see boilers with a simple windbox pressure switch used to prove combustion air flow. It's contacts will close when the fan runs and open when the fan is shut down because a pressure switch simply compares pressure at the point of connection and atmospheric pressure. If one of those switches is giving you difficulty (they seldom do) you can usually get it to function by closing the burner register. I'm not saying you should do that, you shouldn't; there's no air flow through the burner when the register is shut... but the switch is made!

I've seen many installations where the operators have pulled similar tricks to get the boiler operating or keep it operating. Air flow should be proven by a means that's independent of such conditions and my favorite method is using the differential pressure across a fixed (not adjustable) resistance somewhere in the air flow stream. I'll mention some methods later in the book.

Purge the Boiler

Once air flow is proven we "purge" the boiler. A purge is a constant flow of air through the boiler that must occur long enough to ensure any combustible material is swept out the stack so it can't be ignited by the starting burner. On an initial start-up some math has to be done to determine the purge timing and the flow rate may have to be established. Your state law and frequently insurance company requirements dictate the flow rate and timing of a purge. These are the more common requirements: Single burner boilers can be purged at the maximum combustion air flow rate unless they are coal fired. Multiple burner and coal fired boilers purge air flow requirements vary but the basic rule is 25% of full load air flow.

Single burner fire tube boilers must purge for sufficient time to displace the volume of the setting four times. Single burner water tube boilers must purge for sufficient time to displace the volume of the setting eight times. Multiple burner and coal fired boilers must purge for sufficient time to displace the volume of the setting five times and for at least five minutes.

So how to calculate the purge air timing? First calculate the volume of the setting. The setting is everything from the point where combustion air enters enclosed spaces leading to the furnace to the exit of the stack. For all the fans, ductwork, air heaters, burner

windbox and similar parts the inside is mostly air so you can determine its volume by simply measuring the outside and multiplying length, width and height to get the volume. Do the same thing for the boiler. The manufacturer's instruction manual will list the weight of the boiler empty and flooded so you can calculate the volume of water, steam, steel, and refractory then subtracting that to get the volume of the gas space in the boiler. Divide the dry weight by 500, the approximate weight of a cubic foot of steel to determine the steel volume and divide the difference between flooded and dry weight by 62.4 to determine the volume of water; subtract the results from the outside volume of the boiler. The total gives you the volume of the setting.

For single burner oil and gas fired boilers you can use the required combustion air flow rate for full load air flow, see the section on fuels. If the boiler fans cannot be operated at full load air flow on a purge determine the actual purge air flow rate (as a percent of full load) using the processes described for estimating the minimum air flow. For multiple burner and coal fired boilers use 25% of the full load air flow as a purge rate. Now you have a volume in cubic feet and a rate of flow in cubic feet per minute. Divide the volume by the flow rate and you know how many minutes it takes to displace the volume of the setting which is one air change. Multiply that result by the required number of air changes (4, 5 or 8) to determine the purge timing.

Maybe you aren't starting a new boiler but you would like to know what the required timing for your existing unit is. The means is described and it's always a good thing to know. For many of you the result is going to be a surprise. The required purge timing is usually a lot longer than what the boiler is originally set up for. I'm the only engineer I know that actually performs those calculations to determine the required purge timing.

It wasn't a big deal in the days of pneumatic timers where an operator could reach in the panel and readily turn the timer back... most of them did it. When we started installing microprocessor based (programmable controller) systems near the end of the twentieth century some of our customers got very excited; the operators couldn't reach into the control system to change the program memory and shorten the purge. As a result, they had some very long purge times to go through after a power interruption or any boiler trip. There's more on this subject in the section on why boilers blow up. Try to live with the legally binding number if you can.

Once you've done the math, calculated the correct purge time and set the controls for it, use the purge to clear the boiler every time before attempting ignition.

It provides a lot of time to think about why the boiler tripped or how it did as you continue with the start-up; valuable time if you use it.

I had a few service technicians working for me that thought it a nuisance and always shortened the purge time (remember, all they had to do was spin a knob on a timing relay to change it). Their skill and experience allowed them some leeway in breaking the rules and, luckily for them, they normally got away with nothing more than a few singed eyebrows when they stretched it too far. You don't have that skill and experience so don't play Russian Roulette with a boiler explosion. Do a complete purge.

A purge must be proven before you start timing it and the purge conditions must be proven during the entire purge period. Purge proving is one thing very few systems do well and you should assure yourself that the system on your new burner really proves a purge air flow exists. I insist on installing a purge air flow proving device that actually measures the air flow but I discovered that can be defeated (see why they fail) so I'm confident no automatic control system can really prove the purge air flow.

The boiler operator should be the final authority on purge air flow and ensure the automatic system's acceptance of the condition is correct. On small boilers the typical proof is a fan running and the controls at high fire position. As the boiler size increases, a device to monitor flow should be provided and one that measures air flow just like I suggested for combustion air flow is best. A proven purge is imperative for safe boiler operation. Many of the explosions and regular puffing I've experienced were the result of an inadequate or non-existent purge (see that chapter on why they fail) so don't be satisfied with anything less and don't trust the sensors entirely.

Open Fuel Supply, Prove Lightoff Conditions:

When you're satisfied the purge system is working properly you can open manual fuel shut-off valves to bring fuel up to the safety shut-off valves. Don't open the burner shut-off valves yet. The piping should be checked to ensure the fuel is up to the safety shut-off valves and there are no leaks before proceeding. You should also perform a leak test of the fuel safety shut-off valves (see maintenance) to ensure they're working properly before proceeding. I know, they're new valves; but I also know that valves leak, even new ones!

After that final check you can install the oil guns or gas guns that you intentionally left out so no fuel could get into the boiler. It's a good habit to get into, always remove the guns when the boiler is not to be fired and you can readily remove them. That way you have some degree of confidence that fuel can't possibly get into the boiler. It's better for a leak to appear at the front where you can see or smell it than to quietly create an explosive condition inside the boiler. The most expensive boiler accident to date, a billion dollars worth, was the result of leaking fuel. If you can't break a fitting to show leakage at the front then you should check an idle boiler regularly when there is (or could be) fuel in the burner piping.

Once a purge is complete modulating boilers should be positioned for light-off. Because most boilers light off at low fire we commonly refer to this as the low fire position, not the light-off position and I will follow suit. You should be aware that light-off position and low fire do not have to be the same. Once a burner is operating it can usually remain stable at firing rates lower than rates required to achieve a smooth light-off.

Where loads can require a boiler to operate at very low firing rates on occasion, and it's more desirable to keep a boiler going, separate minimum (low) fire and light-off positions may be established. In those instances the position switches have to prove the settings are high enough for ignition as well as low to minimize input during light-off.

Low fire position switches have always proven to be difficult to maintain and set because the low fire is at the minimum stops described earlier. I've never quite understood why they're such a problem because the position proving switch(es) do not have to be set right at the minimum position. I do know that many technicians try to do just that but it's simply easier than doing the more logical thing which is to determine an acceptable upper limit for light-off and adjust the low fire switch accordingly. The acceptable upper limit is determined by increasing the firing rate until lightoff gets rough. That's above the upper limit and you back off it a little.

There's plenty of room for switch adjustment on a multiple burner boiler because low fire has to be established by an independent means of control. The minimum stop on the main fuel flow control valve should be set so the flow produces a pressure slightly less than desired at low fire with one burner in operation. The additional flow can then be provided by the minimum fire controls. I always use minimum fire pressure regulators which bypass the main fuel control valve to maintain a certain minimum pressure in the burner header regardless of the number of burners in operation. Setting the main control valve with its minimum stop to produce almost enough flow for one burner helps make it possible to keep the boiler from losing all burners in the event the minimum fire pressure regulator fails. You'll probably encounter multiple burner boilers without minimum pressure regulators, and experience the difficulties of operating without them. All multiple burner boilers should have them, one for gas, one for oil, for more reliable operation.

A low fire position is not a certain solution to problems when lighting off a boiler. Fixed fire boilers light off at full fire so there is no switch or adjustment to be made and they can experience rough light off. A rough lighting is due to creation of a fuel and air mixture that is outside the flammable range (see the chapter on fuels) which finally lights when a proper mixture is established.

When firing gas it's usually because the mixture is fuel rich due to gas leaking past a regulator. A quantity of gas is trapped between the regulator and safety shut-off valves at a higher pressure than normal. When the shut-off valves open the result is a flow of gas larger than normal for a few seconds until that buildup of gas bleeds off. On oil fired boilers the gun can start empty with fuel mixing with the air in the gun to produce too lean a mixture. On the next try, if the gun isn't purged, the mixture can be fuel rich. A rich or lean condition can be created depending on operation of atomizing medium controls. If the boiler doesn't have smooth ignition start looking for short term surges or sags in fuel pressure and flow when compared to conditions after a stable fire is established.

There's usually a lot of room for low fire variations because most boilers have fan dampers that simply can't close enough to produce a minimal excess air condition at low fire. Those dampers leak so badly that low fire is usually established with the dampers in what could be considered a closed position and excess air is still 200 to 300%. A good variable speed drive will provide lower excess air at low fire but the flow usually has to be controlled to overcome problems with changes in stack draft producing significant changes in the air flow.

Remember, it's always important that low fire be a stable condition. With multiple burner boilers where the code limits low fire air flow to 25% of full load air flow that can be difficult because air fuel ratios can change with number of burner registers open. Set procedures must be established to get the air flowing at the correct amount through the burners to be started; and those procedures must then be followed religiously.

Establish an Ignitor

With low fire (actually light-off) position determined it's time to actually get a flame going in the burner. Except for very small boilers that involves the operation of an ignitor. Most boilers will be equipped with a gas-electric ignitor. Small boilers frequently use nothing more than an electric spark to light the fire, that's because their burner is the size of an ignitor or smaller. You can also run into some with oil-electric ignitors and a few with high energy electric ignitors and other unique methods. The bulk of boilers use a gas electric ignitor and we'll stick to that mode. Many of you may choose to call an ignitor a "pilot fire" or "pilot light" but I'll simply talk about ignitors and you're free to use any of those labels.

Since ignitors use an electric arc to start the gas or oil fire it's appropriate to make certain the flame sensor doesn't think the electric arc is a fire. Begin by closing all the manual fuel shut-off valves including the ones that supply fuel to the ignitor. Next, go through several partial ignition cycles to see if the spark is detected. What's an "ignition cycle?" That's everything you and the Burner Management and combustion controls do from starting of the fans to igniting the main fuel on the first burner fired, including a purge.

On multiple burner boilers we also have a "burner ignition cycle" which includes waiting then trying to light that burner. If the flame scanner "picks up" (the system indicates the scanner recognizes a flame which is probably from another burner) your burner supplier has a problem and you shouldn't continue to operate the boiler until the problem is fixed. Sometimes you can correct the difficulty by re-sighting the scanner (adjust it so it points in another direction) but if you do you should perform this check regularly to ensure the adjustment hasn't failed to prevent sensing a spark as a flame.

We also talk about discrimination and it's very important in multiple burner boilers. The flame scanner for a burner should not detect the fire of any other burner. If it does, it can improperly indicate the ignitor, or main flame, of its burner is on and allow the fuel valves to remain open when, in fact, there's no fire there. To prevent this, and any false indication of a fire, a Burner Management system will normally lock out when a fire is detected that shouldn't be there. That's anytime a flame is detected but the fuel safety shut-off valves aren't energized. Even with a single burner boiler you should make sure that works. Slip the scanner out of the burner assembly during the purge period and expose it to a flame, the burner should lock out.

For almost all ignitors the trial time is ten seconds. We call it the pilot trial for ignition (PTFI). That means that ten seconds after the ignitor gas shut-off valves open the scanner must detect a flame to permit continued operation of the boiler. If the ignitor flame isn't detected the valves should shut and the Burner Management should "lock out." When you're satisfied that the system passed the spark test check the timing and make sure that the system locks out.

Open the ignitor manual valves when the spark test and check of PTFI is complete. Then see if you can establish a proven ignitor flame. Once the ignitor is proven the Burner Management allows at least ten seconds for main flame trial for ignition (MFTI) before shutting down and we can use that period to check the ignitor fire and do some other things. You don't have to do this every time you start the burner, only after maintenance or adjustments have been made that could allow the scanner to see a spark as a flame. That includes a change in scanner alignment or simply removing and replacing the burner; you can't be certain it's back in exactly the same place.

You should be satisfied that the ignitor lights quickly (not just before the end of its ten second trial) and burns with a clean and stable fire. If the ignitor isn't stable you can't expect it to do a good job of lighting the main fire. It should be bright and ragged looking because there's lots of excess air there. You don't want it snapping and breaking up like fire from a machine gun where there are bursts of fire.

An ignitor gives us an opportunity to check operation of the boiler safeties and, during initial start-up, maintain a minimum input into the furnace to slowly dry-out the refractory. Drain the water from the low water cutoffs during the main flame trial period without pressing any bypass push-buttons to be certain the cutoffs shut the burner down; I have encountered systems that do an excellent job of alarming a low water cutoff but didn't trip the burner.

Use the cycling of the fuel safety shut-off valves to check fuel safeties as well (see the chapter on setting safety switches). Every safety and limit switch should be operated to ensure they will actually shut the burner down.

Start Refractory Dry-out

The life of refractory in a boiler is almost entirely dependent on how it was treated on the initial start-up. By performing a controlled, slow warm-up of the boiler you can ensure a long life for the refractory. Slam the fire to it and you can count on repairing refractory every time you open it until you break down and do a complete refractory replacement. I like to use the ignitor to begin a dry-out. It requires some temporary wiring and a relay in most cases (you simply energize the ignitor fuel valves (not the spark) in place of the main fuel to

keep the ignitor going.

Operating on ignitor will provide a very slow warming of the boiler, so slow that it may seem like it's doing nothing; give it a day if you do it. Only when it's apparent that the ignitor can't bring the temperature up anymore remove any temporary wiring to restore normal ignitor operation and allow main burner operation. A critical temperature during refractory dry-out is 212°F because at that point you start making steam out of any water that's in the refractory. The steam, expanding rapidly, can erode the refractory as it seeps out into the furnace. If you raise the temperature rapidly through that temperature the steam generation can be so great that it creates pressure pockets in the refractory to force it apart, creating voids and cracks that will be repair items for years to come. That's why long-term operation on ignitor can be beneficial to a new boiler, drying out that refractory so slowly that erosion, cracks and voids are dramatically minimized.

Of course, all of this is a waste for boilers with refractory that's already been fired, right? Wrong! How do you know what weather conditions that boiler was in traveling to your site? Treat your new boiler as if the refractory was soaking wet and you'll never regret it. Treat it as if you should be able to run it to high fire right away and plan on a lifetime of refractory repairs. Once you've reached the limits of ignitor operation you're ready to establish a main flame and prepare for a combination of refractory dry-out and boiler boil-out.

Repeat the operation to dry-out any major refractory repairs as well. Refractory is one of those things that can't be guaranteed because the manufacturer and installer have no way of knowing how the dry-out was handled. You want it to remain intact so give it the tender loving care it deserves.

Establish Main Flame

Having spent a day or two on initially drying the refractory and testing ignitor operation we're ready to light that main burner. This is not a time to be faint of heart or careless and quick. Although most small boilers come factory tested so you have some reason to believe it's set right for main flame ignition that's no guarantee. On many boilers you'll find that particular burner arrangement is being fired for the first time ever, so nobody knows what the right settings are. I frequently see operators slowly opening the burner manual shut-off valve after the automatic valves open as a burner starts; that's because they saw the technician do the same thing on the initial start-up. I'll explain later why you shouldn't do that but now, on initial start-up, that's what you have to do.

I say you can't be timid or quick in this operation because you don't want to create a flammable mixture that doesn't light right away. If you open the manual valve too slowly you will allow so little fuel in that the mixture at the burner will always be too lean to burn. However, the fuel can settle or rise and accumulate to create a mixture that's just right, waiting for you to finally get ignition. If you open the valve rapidly you can shoot right past the point where the mixture is right and into a fuel rich condition that won't burn; that only happens if the controls admit too much fuel, and they have half a chance on an initial start-up. That fuel in its rich condition can mix with some air in the furnace to produce a flammable mixture and accumulate in preparation for an explosion, suddenly lighting when you don't expect it.

If you're going to be the one operating that valve do a few practice runs before doing it for real and time yourself. You should operate a plug valve or butterfly valve from closed to open in approximately five seconds. That gives you enough time to stroke through all the potential mixture conditions within the trial for ignition period without going so fast that you miss the proper point of ignition. When lighting off on oil you're usually using a multiple turn valve that's really open enough for low fire in two turns so practice getting it two turns open in five seconds. Also train yourself to close the valve at the same speed.

Keep in mind when you're operating that valve that there is a delay involved; the fuel has to displace the air in the burner piping and burner parts before it can enter the furnace and start mixing with the air. An assistant watching the fuel gauge can read off pressures to you so you can get an idea of where you are. You want to stop opening the valve the instant you see that main fire light and be prepared to close it a little or open it a little more depending on your perception of the fire. If the fire is bright and snappy, an indication that it's air rich, you should open the valve more. If the fire is lazy, rolling, and smoking to indicate it's fuel rich you should close down on the valve.

If you didn't get a fire then allow a full purge of the boiler. It's not uncommon to have several attempts at starting that first fire. My service technicians were artists and knew what they were doing but they always upset me by shortening the purge time so they could get back to trying the main burner faster. Please don't do that! Every missed fire leaves an accumulation of fuel in the boiler that can produce a healthy explosion when it's lit by the next operation of the ignitor. Always, please, allow for a full purge; and ...if you saw a smoky

fire purge it twice to get all that fuel out of there before trying again.

Use the purge period to think about why you didn't get a fire. It might be because the gas piping was full of air and you forgot to purge it. It might be that you simply forgot to start the oil pump (in which case, why did the low oil pressure switch not prevent an attempt at ignition?) or you forgot to open a fuel or atomizing medium valve. Maybe you saw a little light burning indicating you didn't have enough fuel or a lot of smoke indicating you had too much so you can adjust the controls accordingly. One problem with steam and air atomized burners is not enough fuel but it's not apparent because the steam or air is breaking it up. Make some corrections then, after a purge, try again until you get it going.

Now for an operation that many service technicians fail to do, mainly because it can take some time and several light-offs, do a pilot turndown test. It's a process where you prove to yourself that, if the ignitor fire has decayed to the point where it can't light the main burner, the scanner will prevent an attempt at ignition on a faulty pilot. Throttle the gas supply to the ignitor until you note a drop in the flame. Make sure the ignitor can light the main flame. Continue dropping the pressure and checking to be sure the main flame ignites. If, during the process, the scanner fails to detect the ignitor flame and the Burner Management locks out the test is complete.

That seldom happens, what usually happens is the ignitor fails to light the main fire. Now you have to repoint or orifice the scanner so it will not detect the ignitor flame when it isn't adequate to light the main burner. Matching that scanner position or orificing so it also allows reliable detection of the main flame can frequently be a problem. You have to do it, however, or the system can be forced to repeatedly attempt light-off of a main flame with an inadequate ignitor and the results have been very devastating in some installations.

Now that you have managed to light a main burner you want to establish proper firing conditions so you can repeat them for every light-off. If you managed to open the manual valve completely without changing the condition of the fire you're past the need to balance the manual valve and controls. If not, then note the burner pressure and close down on the main fuel control valve a little (or adjust the minimum pressure regulator) then open the manual valve a little to restore the pressure and repeat the process until the manual valve is wide open. Once you know what the proper conditions for start-up are the only reason for operating the manual valve is when you question the ability of the system to repeat those conditions. You should get a smooth light-off every time, once you have it set.

Now that you can get a main flame it's a good time to review the process. Open valves admitting fuel to the furnace only after purge and low fire position interlocks are proven. Open the valves in the main fuel only after a pilot (ignitor) flame is proven. Prove the purge limits prevent completion of a purge cycle when combustion air is blocked by blocking it. Ensure the purge requirements are not satisfied when the burner register(s) is(are) closed, when the fan inlet is blocked to the degree the required air flow cannot achieve the specified flow rate and when the boiler outlet is similarly blocked. Ensure the burner start-up cannot continue after purging until the low fire position is proven. Admit main fuel only after observing a stable and adequate pilot flame exists and extinguishes at the end of the main flame trial for ignition period (unless there are separate pilot and main flame sensors where you assure the main flame sensor does not detect the pilot flame). Purge the boiler completely according to the code after each test or failure to produce a main flame. Don't alter flame trial timing of the control.

Boil-out and Complete Dry-out

This normally only applies to a new boiler. You may have to boil-out a boiler after tube replacements or complete dry-out of some refractory repair so follow the sequence when necessary. The entire process is skipped for normal operation of a boiler.

Some boilers will have pipe caps or plugs in casing drains where moisture can escape during dry-out. They should be removed for this period of operation.

Normally the boiler is simply filled with treated makeup water or feedwater before this stage. Once the process begins that will have to change. Boil-out chemicals should be as prescribed by your boiler water treatment supplier or the boiler manufacturer. Be certain you don't have conflicting requirements. Handle those chemicals with extreme care and using all the required protective clothing and equipment; they're a lot tougher than normal chemicals. They should be added right before you start the boil-out and dry-out and removed as soon as the boil-out is done.

Burner operating time should be limited until the boiler is operational and you've completed refractory dry-out and boiler boil-out. When it's possible to operate the boiler on main flame, make the first step a combined refractory dry-out boiler boil-out procedure. Neither function can be performed without having an effect of the outcome of the other. Procedures supplied by the boiler manufacturer should be followed or the se-lected procedure should be submitted to, and approved by, the manufacturer.

The contractor may say "we always did it that way;" but that doesn't make it right; insist on a written document. Be certain to remove brass, copper or bronze parts exposed to the boiler water because the caustic water can damage them. In many instances that includes the safety valves. Replace them with overflow lines run to a safe point of discharge where any liquid that passes through can be collected and treated. Be prepared to dispose of the boil-out chemicals after the process is completed. Sometimes it's necessary to interrupt the dry-out procedure to dump the boil-out chemicals, flush, and refill the boiler. Have a procedure in place for re-establishing the dry-out. Be prepared to commence normal water treatment immediately after the boil-out.

Don't rush these steps, pushing activity along at this point can damage the boiler in a manner that will last its lifetime. Have adequate personnel on hand for the maximum period required because it is not unusual to start and stop the boiler frequently during the initial phase of a dry-out. It's also possible for the procedure to take much more than an eight hour shift. On any large boiler it's common for it to take more than a day.

Normally the dry-out and boil-out are performed with controls in manual for minimal adjustments as necessary to obtain a clean burning fire. You started the dry-out before beginning boil-out and will probably end up finishing boil-out before the dry-out is complete. That's because you don't produce any steam pressure to speak of while boiling out so the temperature is only a little over 212°F when the boil-out is complete.

You'll have to let the boiler cool some before draining the boil-out chemicals and refilling it but there's no harm in dropping them while the boiler is hot. There are two arguments about dropping boil-out water; one is solids will stick to the metal and bake on so allowing the water to cool is best, the other is they will retain the solids while hot but drop them out if they're allowed to cool so dumping the water hot is best. I happen to believe the second argument but always look for recommendations of an appropriate temperature to drain the water from the boiler and chemical manufacturers.

The boil-out water is considerably more caustic than normal boiler blowdown so you should provide for proper disposal of that water, neutralizing it before dumping it in the sanitary sewer or employing a licensed hauler to dispose of it.

Once the boil-out chemicals are drained the boiler water must be treated. The boil-out removed all the varnish and grease that was covering the inside of the boiler and protecting the metal from corrosion. It also removed that material so it couldn't burn on to produce a permanent scale on the boiler heating surfaces. From completion of boil-out on those surfaces have to be protected by proper water treatment.

After boil-out is complete, the safety valves and other materials removed for the boil-out should be replaced. This can also produce an interruption in the dry-out of the refractory and require a gentle reheating before continuing.

Refractory dry-out is complete when the temperature of the refractory at any point has gradually raised to something higher than atmospheric boiling temperature. That's usually 212°F but can be lower (203°F in Denver, Colorado). Some people will accept termination of water flowing out of casing drains, others are more elaborate. The minor expense of some thermocouples located at certain points in the refractory and monitoring them is the best way to determine a dry-out is complete.

The controlling temperature is the temperature of the refractory closest to the outer wall of the boiler not the surface of the refractory in the furnace.

Boiler Control Adjustments

Now that I have spent some time training boiler operators in tuning boilers I've come to the realization that I was somewhat cavalier in explaining tuning of a new boiler in the original edition of this book. Tuning a boiler is a complicated process and potential for an untrained individual to blow up the boiler is much higher than I thought back then. While it's far more dangerous with an older boiler the potential still exists with a new one. Therefore, I must insist that, as a boiler operator, you should not attempt burner adjustments without a significant amount of training and hands-on experience.

The contractor, or the boiler manufacturer's service technician should tune up a new boiler on startup. That's because the boiler is the contractor's property until such time as the new owner accepts it. If you make any adjustments and the boiler blows up you bought it. So, leave them at it. I have had the opportunity to investigate boiler explosions that occurred during startup under the supervision of a contractor or a manufacturer's representative with sufficient regularity to justify that instruction.

Transition to Automatic or Manual Control

Another requirement for a new boiler is establishing a smooth transition from light-off to automatic operation. This is normally accomplished without any trouble on boilers with jackshaft type controls and isn't a factor on fixed fire units. Making the transition with full metering controls is another matter. Normally there is an interface between the combustion controls and the Burner Management systems which allows the Burner Management system to control damper and valve positions to satisfy requirements for purge and light-off (low fire) positions. At some point after a successful ignition of the main fuel the interface lets the automatic controls take over. A stable, safe, and smooth transition between light-off and automatic operation requires more than a simple switching from one to the other.

To begin with, a cold boiler with modulation shouldn't be released to automatic control immediately. There's enough thermal shock for a boiler to experience going from relatively cold (even in what we would call a hot boiler room) to firing at low fire where the steel is less than a millimeter from hot flue gases over 1,000°F. If the controls simply shift to automatic that temperature difference will readily double. Limiting thermal shock as much as possible is important to extending boiler life so provisions to prevent the controls running to high fire right after ignition is important. The simplest approach is you set the controls in manual before the boiler starts and make sure that the manual signal is adjusted to low fire. Other approaches include low fire hold systems and ramping controls.

Low Fire Hold

A low fire hold consists of provisions to keep the burner at low fire until the boiler is near operating temperature. The normal arrangement is a pressure switch or temperature switch similar to the operating and high limit controls but with an electrical contact that's normally open. The pressure or temperature has to reach the switch setting before the contact closes to allow automatic operation. The switch has to be set lower than the normal pressure or temperature modulating controls so the burner isn't affected by the low fire hold system after the boiler is up to operating conditions. Sometimes during emergencies you'll have to bypass the low fire hold controls or the boiler will not get hot until spring. Be certain you can operate in manual to over-ride low fire hold controls.

With the typical jackshaft control the switch prevents an increase in firing rate above light-off position until the pressure or temperature is reached. An automatic low fire hold is very important for modulating boilers that are controlled by a thermostat. A few warm days could prevent the boiler operating until it was dead cold; the low fire hold will prevent the rapid heating of that boiler on high fire with severe thermal shock.

When the outdoor temperature is swinging from warm to cold the amount of time the boiler is held at low fire is almost proportional to the average heat load, it will be less as the average temperature drops and the delay before release to modulation will decrease. Unless you are always on hand to control the warm-up of a boiler you should have low fire hold controls. One final note, on some steam boilers where operating pressures are low you might want to use a temperature switch for low fire hold because pressures can swing more significantly generating control problems.

You really don't want to suddenly switch from light-off position to modulating because the controls will simply run the burner right up to high fire when it isn't necessary. If you're controlling the boiler manually you should allow it to come on line while at low fire. Then, when it seems to have reached its limit, gradually increase the firing rate until the load is up to normal operating conditions, then switch to automatic.

When the boiler is unattended ramping controls function the same way and are recommended for high pressure steam boilers that start and stop automatically. They control the rate of change of the firing rate so it gradually increases at a constant rate (like going up a ramp) until it's at high fire or, more normally, the set pressure is reached and the automatic controls take over. A ramping control should only function on the initial transition from light-off to automatic, or from low fire hold to automatic. The transition rate should be adjustable and you should set it so the rate is as slow as possible to minimize thermal shock. Pneumatic and microprocessor based systems are described in the section on controls.

Test Safeties

Never forget that the safety valves are the last line of a defense against a boiler explosion and test them as soon as possible. First, do a lift test on steam and high temperature hot water boilers when the pressure has exceeded 75% of the set pressure of the valves. Hot water boiler safeties can usually be tested before firing by applying city water pressure.

As soon as possible in the start-up of a new boiler run a pop test of steam and high temperature hot water boilers. A pop test is described later.

Boiler Warm-up

Your boiler manufacturer should have indicated a warm-up rate in the instruction manual. A problem with it is normally there's no way for you to determine if you're actually doing it. If it were critical for temperatures below 212°F then the boiler should be equipped with thermometers. Normally it is a psi per hour rate that you can track. On large boilers it's not at all uncommon to have to stop and start the burners to limit that warm-up rate. Most boilers smaller than a quarter of a million pounds of steam per hour can be allowed to warm up at the low fire rate.

Fixed fire boilers are absorbing the maximum heat input every time the boiler is fired so they have to be started and stopped to reduce the warm-up rate. If that's required, other than on initial start-up, the manufacturer should provide automatic provisions for it.

Multiple burner boilers can be warmed up slowly by only operating one, or a portion of the burners. The burners should be switched regularly, according to the manufacturer's instructions or every fifteen minutes to one half hour so the heating is more uniform. Always start another burner before extinguishing the one it replaces so you don't have to purge the unit. A purge is blowing cold air over the metal you just heated to produce a sudden swing in its exposure to temperature. That could produce stress cracks in the metal that you don't want to have. A boiler should be limited to the number of starts and full stops it is exposed to. When the manufacturer recommends limiting stops and starts it's for high pressure boilers with very thick metal that is more susceptible to damage from stress due to temperature variations across its thickness.

Full Metering Switch to Automatic

Simply switching from light-off position to firing rate control, whether it's manual or automatic, can be rough with a full metering control system. The fuel and air controls are pre-positioned by the interface with the Burner Management system and may be lower or higher than the position that produces flow rates acceptable to the control system. The result is what we call a "bump" as the controls are suddenly allowed to react to the difference and make some rather abrupt, and usually excessive, changes in valve or damper positions in an effort to establish the required flows.

On almost any pneumatic or electronic (not microprocessor based) controls you can also experience problems with reset windup, where the controls detect an error and try to correct it, but can't, so the controller output continues to increase or decrease until it reaches zero or maximum possible output. The outputs are outside the control signal range (such as 3 to 15 psig where the signal can drop to zero or climb to 18 psig—the standard supply pressure. Similarly a 1 to 5 volts range can be a negative voltage and go as high as 12). In either case there is no response to controller action until the control signal winds back into the normal control range.

Modern microprocessor based controls have anti-windup features and procedureless, bumpless transfer (manual to auto and vice versa) features that eliminated the problems with earlier pneumatic and electronic controls. It's possible the system designer didn't properly configure those features and you can still experience bumps on transfers.

A fuel control valve should be positioned at a minimum (mechanical) stop where fuel flow after ignition is more than the controller's set point. If it isn't, the controller would wind up to maximum output (and it has lots of time to do it before a main flame starts) so the fuel valve would suddenly swing open when the controls are released to automatic. If the flow is a little higher than the controller's set point, reset windup (in this case it would wind down), there's simply a delay in response. However, there may not be sufficient time between main flame ignition and transfer to automatic for the controls to wind down and excessive fuel feed could still occur.

If the controls do wind down before transfer they will have to recover and once the fuel valve starts to open it swings open more than it should. To overcome those strange actions the interface between Burner Management and combustion controls should actually adjust the set points to achieve purge and light-off conditions so the controls are controlling all the time. The ramping controls should help overcome that problem with reset wind-up on light-off. Bumps off low fire and maximum fire can occur during normal firing and are discussed in the section on controls.

Collect Performance Data

Among the many things that don't need a "break in period" a boiler should top the list. From the moment the first fire lights in the furnace a boiler begins breaking down. So, collecting performance data on a new boiler as soon as possible is essential. It gives you a record of what the boiler was capable of when it was new, completely clean, and not broken. While I know some examples of where problems with a new boiler had to be corrected before it was accepted a boiler is normally at its best when it's new. Should it happen that changes have to be made in a new boiler this data should be collected once the boiler is accepted. This data provides a basis for comparison to actual operating conditions during the life of the boiler and is very helpful in detecting the source of problems such as inefficient operation.

Given some time to think about this since the first edition of this book was published I have come up with a list of data that should be collected once a new boiler

is ready to go into service. The data should be collected as soon as possible and, when necessary, before performance testing. The following list is not necessarily complete. You may note certain things about your new boiler that are not indicated or documented and choose to record those as well. The more data you collect the more likely you will be able to detect problems with the boiler during its life.

Data should be collected at each firing rate (each screw on a jack shaft cam as shown in Figure 2-4 or each 10% interval of control signal) and include much more than readings of all boiler mounted instruments, pressure gauges, thermometers, etc. On steam boilers the drum level using either a signal level, a temporarily mounted ruler, or simply the number of nuts at the side of the gage glass should be recorded. For hot water boilers the water level in the expansion tank should be recorded even if it is adjusted automatically. I have seen many new boilers supplied without any pressure gauges on the fuel supply system. Despite that you should find that required test ports permit connection of a gage to collect that data. This is especially important when you have multiple boilers or other equipment using a fuel gas supply which would prohibit taking fuel meter readings. If the boiler does not have a fuel meter I would recommend purchasing one and installing it right away so that data could be recorded and collected continually.

Not all boilers are fitted with draft gages but that should not stop you hooking up a manometer, measuring, and recording the draft (sometimes a pressure reading) at the fan discharge, the burner wind box or burner head, in the furnace, at intermediate points within the convection bank (where accessible), the boiler outlet, economizer outlet (when equipped with an economizer), the air heater outlet (when equipped with an air

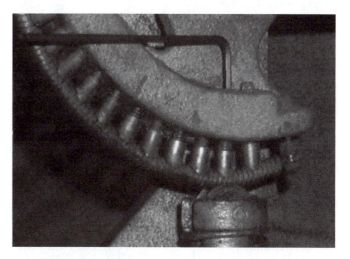

Figure 2-4. Adjustment screw on control valve

heater) and the inlet to the stack.

Sometime in the future you should be applauded for taking readings of shell or casing temperature. The best way to do this today is with an infrared camera that is calibrated to accurately indicate temperatures. There are contractors that provide an infrared survey service that could be used. There are less expensive methods including use of an infrared thermometer or as simple as holding a thermometer against the casing or shell with a piece of insulation. That last method has to be described along with the recorded data because the insulation does restrict heat flow and produces a higher temperature than actual so the same method must be used for comparison at a later date. I have detected considerable differences in the temperatures of casing surfaces on water tube package boilers and recommend taking several measurements of those surfaces. I also insist that new water tube boilers be fitted with insulated drum head covers and their surface temperatures should be recorded as well. If your boiler (new or not) isn't fitted with insulated drum covers measurements of the drum head surfaces should suggest an early modification to your new boiler to include them.

Photographs of the flame in the furnace for each fuel as well as each load can be used for comparison in the future. I carry a piece of blued glass to reduce the light intensity when photographing a fuel oil or coal fire because my camera is typically blurred when photographing flames of those fuels.

If the boiler is fitted with automatic controls your recorded data should include each controller's set point (whether manually set, fixed, or remote), process variable value and/or percent, and the controller's output signal, normally in percent.

Acceptance Testing

The final step in start-up of a new boiler should be the performance of an acceptance test. Data should be collected and recorded at the firing rate where efficiency is guaranteed by the manufacturer and, if it is a modulating boiler, at no less than three other firing rates (maximum, 75%, 50%, and 25% being common). All data collected should be carefully recorded and stored in a binder for future reference. If it is a new plant the performance of all equipment should be documented at the various firing rates. Occasionally a plant is started when there is no place to use the steam and no way to perform the test until other installations are in place. The installing contractor then requests a delay in testing until a load is available. When that occurs collect data at firing rates which can be handled. Nearly identical readings

at a later date will prove the boiler wasn't abused while waiting for a load.

Acceptance tests vary. ASME PTC-4.1 the "Steam Generating Units" power test code provides three means of testing a boiler for acceptance. However, a test in conformance with that test code is an expensive proposition requiring continuous documented operation of the boiler for a period of 8 to 12 hours. It's justified for a boiler designed to generate more than 60,000 pounds of steam per hour but not for a small 50-horsepower boiler that only generates 1,700 pounds per hour. There are other simpler and acceptable means for testing boilers; the important element is having established one acceptable to owner and manufacturer before the boiler is purchased. In the unlikely event the boiler fails to perform the manufacturer is then committed to make it right.

For small boilers, I always recommend testing for one hour at each load point and, with that exception, testing using the "heat loss method" of PTC-4.1. That way you have a formal acceptance test but not the expense of long runs. It shouldn't require any overtime because you should allow it to settle for half an hour before each test to establish test conditions, and you can do six a day. On a boiler with two fuels that would mean no less than four days just running acceptance tests. I always wonder what some engineers were thinking when they said the start-up should take a couple of days, it takes more time to check operation and tune the boiler than it does to test it.

A final acceptance test when a boiler is field erected is very important. A contractor can build a boiler wrong and many have. What about the boiler that is factory tested? I would still run an acceptance test of the final installation. The cost of a boiler is a small fraction of the cost of fuel it will burn in its lifetime; on average—ten times the price of the boiler each year. A small difference in performance can represent a considerable sum. I actually estimate the cost of a 1% difference in efficiency for a particular installation and use that value, with the vendor's knowledge, in evaluating boiler offerings.

Other start-up activities that may be associated with a new plant are covered in the following descriptions.

DEAD PLANT START-UP

All personnel working on the site within the boiler plant and equipment rooms should be notified of the plant startup so they are not surprised and more importantly not injured as, the system comes up to temperature. Confirmation from all those people indicating they are prepared should be formally documented before the boiler is fired. One time one of my service techs started an ignitor in a furnace where the boilermakers were working. He lived to tell about it, but it took a lot of apologizing to those boilermakers to keep them from lighting him off.

Normally when we say a plant is dead we mean dead cold. There's no heat in a boiler or any auxiliary equipment associated with normal operation. We'll also loosely use the term to describe a plant that has a hot, or warm, boiler but isn't maintaining normal operating pressure. A dead plant start-up is returning a dead plant that had been operating to operating condition. It's not uncommon to return a plant to service that was shut down for the summer or a protracted business slump. It's also occasionally necessary to return a plant to service after a loss of electric power or water supply that forced it to shut down. The operations mentioned here are assumed to occur after a plant was laid-up according to the procedures described later. Some activities also apply to simply returning a plant to normal operation.

Remove sorbent from the boiler, deaerator and other closed vessels, install new gaskets and close manholes. Check all personnel, tools, etc. are out before closing the vessels. Fill fluid systems as described in New Plant Start-up. Everything from leaves to birds can find their way into air and gas openings to block them while a plant is shut down. Check to confirm stack clean-outs, vent openings and air inlets are clean. Confirm the vent valve on the boiler and the free-blow drain are open. If the burner on the boiler was dismantled or repaired the steps in New Plant Start-up should be followed to ensure proper burner operation. As soon as possible compare initial operating data with current operating conditions to ensure there have been no significant changes in the boiler's performance.

Record oil tank levels, fuel gas, steam, and water meter readings to establish values at start-up. Leakage, testing, and other activities may have changed the meter readings from the shutdown or last recorded state.

A cold boiler should be returned to operating conditions slowly. When starting a boiler in a dead plant it's advisable to bring the served facility up with the boiler. That increases the time it takes to raise pressure on the boiler and the facility to allow for gradual heating. Open all valves that lead to the facility only after confirming all drains and vents in the facility have been closed or are manned by trained observers.

In steam plants this process normally creates a flood of returned condensate as pressure builds so provisions for handling it should be provided. Lower the operating level of the boiler feed tank or deaerator and condensate tank beforehand if possible. If that's not possible, close isolating valves for make-up to those tanks and manually maintain the lowest reasonable level until pressure in the facility is near normal.

For hot water installations the system should be flooded, the expansion tank level confirmed, and circulating pumps started to generate at least minimum flow in the system. This may require a walk-through of all equipment rooms to ensure the systems are ready to circulate water. Any equipment still receiving maintenance should be adequately isolated using proper lock-out and tag-out procedures.

Lock controls in manual at low fire. Starting a dead plant or boiler should provide a very slow increase in temperature until the boiler's contents are above 220°F. That minimizes damage to the refractory from pockets of absorbed moisture; a sudden increase in volume as liquid changes to steam will build up pressure inside the refractory and rupture it. It's sometimes necessary to repeat an initial dry-out because the refractory got wet or refractory repairs were performed while the plant was down.

Performing operational tests of the boiler's operating limits during the initial firing of the boiler will provide frequent interruptions to the heat. That will reduce problems with the refractory and provide early reassurance that the safety and operating limits are functioning properly. A wise operator will not only confirm limit operations but record it in the log book.

As soon as steam is evident at the boiler vent, operate vents in the facility to remove air from the steam distribution system. If the system has automatic air vents it's a good idea to operate a few manual vents anyway to ensure the automatic vents are working.

In high pressure steam plants close the free blow drain valve only after steady steam flow is certain. The purpose is to prevent any condensate accumulation over the non-return valve that would slug over into the steam piping when an interrupted flow is re-established.

Close the boiler vent valve when the pressure is up to 10 psig on heating boilers or 25 psig on power boilers. Allowing a loss of steam until those pressures are reached helps ensure all the air is removed from the boiler.

If the boiler feed tank is fitted with a steam heating sparge line it should be placed in operation after the boiler vent valve is closed. If it is a coil heater it may be allowed to come up with the plant.

Open the vent valves on a deaerator wide before

admitting steam and gradually open the steam supply to the deaerator only after there is a constant flow of water to the boiler. Any sudden surges in water flow could rapidly produce a vacuum in the deaerator. Also avoid any rapid changes in facility steam consumption that could cause a drop in steam pressure. If a vacuum is formed the deaerator and its storage tank could be damaged. Once the deaerator pressure is up to normal, open the isolating valves wide so the steam pressure regulator can function and close the vent valve to its normal throttling position.

Test the low water cutoff before reaching normal operating pressure and after the pressure is high enough for the boiler to return to firing. That's normally when pressure exceeds 6 psig for heating boilers, 30 psig for power boilers. Lift test the safety valves when the pressure is above 75% of the safety valve set pressure. They could have corroded shut during the shutdown period.

At some point low fire will not be adequate for pressure to continue to rise. Increase the firing rate manually in small increments (less than 10%) and allow the pressure to stabilize before increasing it again. Initially all the condensate will stay in a steam system because the pressure will be below atmospheric wherever automatic vents aren't operating properly or don't exist. Condensate will not return until there is enough pressure differential to push it back to the boiler plant. At several points during the start-up the pressure differential will accelerate condensate returning; the slow steps will limit the rate at which that happens. Wait until the pressure is at, or slightly above, normal operating pressure to switch control to automatic.

After an hour or so of automatic boiler operation the normal operating levels of the condensate tank and deaerator may be restored if they were lowered for the start-up. Increase the level gradually to avoid any damage associated with a rush of cold inlet water. If your timing is right you shouldn't have an inrush because the vessels will be filled by the condensate stored in the system. Make a point of noting the amount of condensate returned to provide better guidance in an SOP for the next dead plant start-up.

With steam generation stabilized, draw water analysis and determine setup of chemical feed and blowdown controls. Open cooling water valves to any quench system. Open valves to put the continuous blowdown heat recovery system into operation. Vent the flash tank until steam has been flowing out the vent for ten to fifteen minutes so you don't push air into the deaerator. Alternatively, leave the deaerator vent wide open until the blowdown system is in normal operation.

Record the start-up activity in the log and begin monitoring the plant as required for normal operation. It's very important to note all problems that came up, changes in operating procedures that were required to accomplish the start-up or correct problems, and the conditions at various times during the process with the times noted. That data can be used to compare with the original SOP for dead plant start-up and modify it to improve the process.

Notice that I didn't say shorten the process. Usually when starting up a dead plant you have time because many other operations won't even be contemplated until you have steam or hot water flowing normally. A slow start-up ensures minimal stress from thermal shock and avoids the pitfalls of rushing to get the job done.

On the other hand, when the plant is being restored after an unscheduled interruption, you can take the shortest reasonable time based on experience with prior start-ups. If called upon to rush you should already know which boiler to select for it—the one that needs the most refractory repairs anyway. Selectively damaging the plant under emergency conditions, such as restoring heat to a hospital or nursing home where it's critical, is part of a well prepared disaster plan.

NORMAL BOILER START-UP

After that initial plant start-up we begin to relax and, regretfully, can get too casual about a boiler start-up. We tend to forget that the equipment deteriorates with age and use to the degree that something could go wrong. A certain amount of that should be addressed each year right after the annual inspection when, because we had the boiler apart, we should start it up as if it were new. We should also pick up a few other good habits that take that wear and age into consideration.

Close circuit breakers as needed to apply power to the burner management control at least 24 hours prior to starting a fire in the boiler. Flame sensors can deteriorate and provide false flame signals but may operate normally when they are first energized. The long warm-up ensures the sensors are properly checked by the burner management system during start-up.

A normal boiler start-up assumes other boilers in the plant are operating and the boiler to be started has not had maintenance or other work performed on it. If there was work performed, review the recommendations for new plant and dead plant start-up to determine if there's anything you should check or test before proceeding. Make sure the vent valve is open. If the stop valve at the steam header is closed the free blow drain valve should be open.

Set firing rate controls to manual and low fire. Make one quick trip around the boiler to be certain it isn't open and all valves are in the proper positions before starting it. Open the fuel block valves slowly to ensure you don't upset fuel supply to operating boilers. When firing oil, check an oil burner assembly then and insert in the burner. If oil is steam atomized, open isolating valves to admit steam to the inlet of the burner steam shut-off valve. If oil is air atomized, start compressor and admit air to the inlet of the burner air shut-off valve. Check to be certain normal operating fuel supply pressures have been established. Blow down the gauge glass and water column while observing the water level in the glass to assure yourself the boiler contains water.

Turn the burner management control on to allow a burner to start. On multiple burner boilers and in older single burner plants it may be necessary to initiate a purge and burner ignition. When the pilot flame is proven, gradually open the atomizing steam or atomizing air shut-off valve at the burner. This ensures that any fuel oil that may be transported to the burner by the atomizing medium will be exposed immediately to the ignition energy of the ignitor and burned at nearly a normal rate. Opening the valves earlier can inject a slug of oil into the furnace that would subsequently vaporize to produce an explosive mixture in the furnace and ignite when the pilot comes on. Open the fuel shut-off valve, if the atomizing medium didn't produce a fire, to start the main burner. If the atomizing medium dumped in some fuel that produced a fire it's best to repeat the purge. Sometimes you're so slow at opening the steam or air valve that you don't have time to get fuel on. That's okay, wait until it has purged again. This is a normal start-up and you aren't in a hurry.

Shortly after the burner has started and is operating normally, close the burner manual valve. The burner management system should detect a flame failure and initiate a boiler shutdown. Only if the boiler shut down, reset the burner management system for another start. Open the burner manual valve after the burner management system indicates an ignitor flame is proven. Having restored operation, check the low water cutoffs by blowing each one down and confirming the burner management system shuts the boiler down. Do not use any bypass push-button while testing the cutoffs at this time, you want a full operational test. Once the boiler is up and operating it may not shut down for months; this is the one and only, best and truest time to confirm that the flame failure and low water safety systems all work. Repeat the

low water cutoff tests if it's necessary to shut the burner down to control the rate of heating of the boiler.

Close the boiler vent valves when the pressure is up on heating boilers or 25 psig on power boilers. If the non-return valve on a high pressure boiler is closed, open it so steam will flow to the free blow drain. If the second steam stop valve was left open, open the free blow drain to drain the boiler header and leave the non-return closed.

Allow the pressure to increase while observing it closely. The burner should shut down when the operating pressure or temperature control setting is reached. Once that operation is proven, test the high pressure or high temperature limit switch by temporarily installing a jumper on the terminals of the control switch. The high limit should shut the boiler down before the safety valves open or the temperature of a heating boiler exceeds 250°F. It should also lock out to prevent continued operation. Allow the pressure to fall until it is below the operating pressure then reset the controls so the burner can be started again. Remove the jumper from the control switch terminals.

Once you've proven operation of the low water cutoff and the boiler pressure or temperature control and limit switches you can run through successive tests of each combustion air and fuel limit switch. Proving the operation of the low combustion air flow switch can produce a condition of flammable mixtures in the boiler so you must be careful with that one. In some cases you will have to simply adjust the switch setting to simulate a condition, not the best of tests, but at least you will have done something to ensure it operates. I've attended testing programs where many of the fuel and air limit switches didn't function when the operator thought they would.

- With the firing rate set at minimum fire, reduce combustion air flow by slowly sliding a blank over the inlet of the forced draft fan while someone watches the fire. The minimum air flow switch should trip before the fire gets smokey or unstable. Take care that the blank doesn't affect the switch sensing the air flow, use another method of reducing air if it does.

- Increase gas pressure to the burner while watching the fire, again at minimum fire. The high gas pressure switch should trip before the fire gets smokey.

- Decrease gas pressure to the burner while watching the fire, again at minimum fire. The low gas pressure switch should trip before the fire becomes unstable.

- Decrease oil pressure to the burner while watching the fire, again at minimum fire. The low oil pressure switch should trip before the fire becomes unstable.

- If the oil is heated at the boiler you can check operation of high and low oil temperature switches (if present) by adjusting the oil temperature while observing the fire. This takes time due to the thermal inertia of the system so be prepared for that. If the fuel is heated at a common supply point the testing should only be done when you will not interrupt the operation of other boilers.

What happens if the burner doesn't trip on low water cutoff, flame failure, or high pressure limit? You secure it, note the failure in the log, and notify your superiors that it isn't working properly. A boiler with malfunctioning safety controls should not be placed in operation.

Open the boiler isolating valve on a heating boiler when the boiler pressure is reasonably close to the header pressure. It's best to open the second stop valve on high pressure boilers when the pressure in the boiler is within twenty pounds of header pressure. The minimal difference in pressure limits steam wire drawing the second stop valve seats and makes it easier to open the valve. When there's a bypass built into, or around, the second stop valve you can use it to pressurize the boiler header. The normal way is to open the non-return valve when ready to put the boiler on line to build up pressure in the boiler header. In either case, always be certain the free blow drain valve is open and blowing steam to ensure yourself there's no condensate in the header that would suddenly enter the plant steam header.

After steam is flowing to the header, as indicated by a steam flow recorder or a drop in boiler pressure as the non-return valve lifts, close the free blow drain of a high pressure boiler.

Once the boiler is "on-line" which means it is delivering heat to the facility, record the fuel and steam or other output meter readings. It's one of the little things I ask operators for that I never get an answer to—"how much fuel does it take to bring that boiler up and onto the line?" If yours is one of those plants that change boilers frequently it may be a very important question because there's considerable amount of fuel used to do that and an associated amount of energy lost when a boiler is taken off line and left to cool.

The final step in a normal boiler start-up is to establish its manual firing rate or place it in automatic control. Since you should still be at low fire, this can require increasing the firing rate manually until the desired firing rate is reached. If you intend to place it in automatic you should increase the firing rate until you notice that it's about the same as the same sized boiler that's already on automatic before switching to auto. Simply throwing the switch to auto isn't the appropriate way to do things because the boiler controls could swing for some time before they are stable again.

EMERGENCY BOILER START-UP

Emergencies come in two forms, instantaneous and impending. If you've done your job as far as observing the equipment is concerned, regardless of who maintains it, you shouldn't face too many of either. There are some emergency situations that are beyond our control, such as power failures, but we should have plans for them; right? Instantaneous emergencies involve an immediate shutdown of the plant or an operating boiler such that you can't supply steam to the facility served by the boiler plant. Impending emergencies are the ones where you know it's only a matter of time until you can't supply that steam.

Impending emergencies involve things like the severe squeal of a fan belt or motor bearing on operating equipment that tells you it's bound to fail very soon. It can also be clouds on the horizon and the sound of thunder when you know the power is going to fail because you never seem to make it through a thunderstorm without a power failure. Natural occurrences from flood to excessive heat to deep snow and forest fires seldom come without a warning so they should be impending emergencies, something you know is coming.

With your disaster plan in mind you can take appropriate action. And that's why you make the plans, so you don't have to stop and think about it. The less time you take to act the more time is allowed for warming up a boiler and other things that you don't want to rush unless you have to. If you've actually rehearsed those disaster plans you will find yourself surprisingly comfortable with what's going on.

Steam may be such a precious commodity in your plant that you maintain a boiler on hot standby. In that case, start the standby boiler so it can be brought on line. Once the problem with the boiler that tripped is resolved you can put it on standby or restore it to service.

Frequently the reason for a boiler shutdown can be determined, and corrected, quickly so it can be returned to service. Many times it's resolved quickly and the boiler is returned to service even before anyone else notices you have a problem. When that can't happen, then it's time for an emergency boiler start-up.

Any emergency that results in the shutdown of a boiler should be responded to with an instant evaluation of the condition of that boiler. If you're confident that it cannot be returned to operation or are not sure why it went down the first step would be to start another boiler, if you have one, so it's warming up. Starting the other boiler takes time from finding the problem with the unit in operation but it also allows for a more gradual warm-up of that boiler in the event the one that was running can't be restarted.

Of course, if you know it went down because of a short power interruption, or other cause you know will not prevent a restart, there's no reason to start that other unit. If there's water and steam pouring out of the boiler that shut down or large gaping holes in what used to be square casing you know there's no hope for the boiler that went down and all you can do is secure it.

I define an emergency boiler start-up as one that requires operation of a boiler from a dead cold condition in as little time as possible. There are things you can do to limit the damage to the boiler in that process and actually accelerate the start-up time. Which ones are available to you will determine what you do. You can use these suggestions as guidelines to prepare your own disaster plan that describes an emergency boiler start-up.

Frequently heating boilers are allowed to sit idle with their steam valves open. This frequently gives the operator an impression that the boiler is ready to go because there's pressure on it. Nothing could be further from the truth and I discourage that practice because it injects a considerable temperature swing in the shell of the boiler right at the water line. Steam at the surface is at saturation and hot, the original boiler water and condensate below the surface can be much cooler. Even systems that drain the condensate from the bottom of the boiler do not correct for the fact that the majority of the water in the boiler is relatively cold.

Power boilers will always be considerably colder than normal steam condition. The principle concern in an emergency start of a boiler is the development of stresses in the boiler metal associated with rapid heating of the boiler. Whether it's a low pressure firetube or a large watertube doesn't matter much, both have thick steel parts in contact with the boiler water that have to be heated to normal saturation temperature and the time spent in doing that will determine the extent of damage by thermal overstress.

Rather than heating all the water in the boiler you can bring warm or hot water in to help accelerate the warm-up. That's especially true when the boiler is the only one you're firing. Temporarily shutting off the makeup water and operating the boiler blowoff valves to drop level so the heated water from the boiler feed tank or deaerator displaces much of the cold water in the boiler will both add to the heating of the boiler and provide some movement of water to help transfer that heat to the thick parts of the boiler metal. Once you've about drained a boiler feed tank you can restore the makeup. Let a deaerator sit until you're producing steam then bring the makeup on real slowly.

In an emergency you want to push the envelope as much as possible without damaging the boiler. Your disaster plan should have been developed after some testing that determines what firing rate provides the fastest warm-up of the boiler within the limits recommended by the boiler manufacturer so you can immediately set that firing rate to get the fastest possible warm-up.

If your normal procedure for warm-up includes shutting the burner down, don't do it. I've never been a proponent of that activity other than for refractory dry out. If you think about it, the operation of the burner followed by a purge produces dramatic swings in the metal's exposure to temperatures on the fire sides. I think you will do less damage to the boiler by firing continuously, although at low fire, than cycling the burner on and off. In multiple burner boilers, where you're only operating one or a portion of the burners during warm-up, operation should consist of firing another burner before shutting one down as explained in the first discussion on start-up.

Then, of course, there's the matter of how serious the need for steam is. Loss of steam for blanketing chemical reactions may be more critical than damage to the boiler. In a hospital during a disaster where every operating room is handling emergency surgery maintaining steam for sterilization is a must. In such situations your disaster plan can call for ignoring the manufacturer's recommendations so you bring a boiler up to operation as fast as possible.

At some point you've established how much time it will take to recover and documented it in your disaster planning. That information should be supplied to the facility served by the boiler plant so they're aware of it when preparing their disaster plans. Some facilities may reply with the question "is that the absolutely quickest you can do it?" In most cases you can answer "No, but it will expose (multiply your steam generating capacity in pounds per hour by $20 or the boiler horsepower by $700) of boiler to probable failure to do it quicker. Now they have a time and a dollar value for doing it faster, usually you won't get any more questions.

Note that I didn't mention refractory. If the boiler has been laid up properly there's no reason to believe serious damage to the refractory could occur during an emergency start-up. Very old boilers and coal fired units may have sufficient thicknesses of refractory that it's a concern and your plan should address those conditions when they exist. Finally, log it all!

NORMAL OPERATION

I hear it so frequently: "all that boiler operator does is sit on his butt and read the paper" or words to that effect. Of course, you may have the same perception of others; does a night watchman do anything? a librarian? how about us engineers? Remember that old Indian proverb: "never criticize a man until you've walked a mile in his moccasins." I always address that first quote with the following inquiry and offer the list that follows it so the person knows what a boiler operator does on a normal shift.

What is the most sensitive, precise, and accurate sensor in a boiler plant? I always wait for the person I ask to bring up a few answers. Occasionally they answer the operator's brain but I can say that's wrong. It's the operator's ear. Think about it… Even before the pressure gages drop or the alarm goes off you know when something goes wrong; you hear it! When I'm asked what is necessary to eliminate personnel during the evening or night shifts I always manage to get the inquirer's mental gears turning by explaining that and asking how much they're willing to spend to get a system that approaches the ability of an operator listening to his plant.

Since the original edition of this book something has happened that gives me another response to people that question the need for a boiler operator. IBM created the computer system they named Watson to serve as a contestant on the television program jeopardy and it actually won. I believe the estimated cost of that system was well over $5 million. Of course Watson could hear and make decisions but it cost a lot more than the services of a boiler operator. I doubt if anyone is ready to spend more than $5 million to replace a boiler plant operator.

Now, in addition to always being on top of everything going on in that plant, what does an operator do?

During any typical working day in a steam plant a boiler operator will spend no less than 4 hours plus 1 hour per operating boiler and 1/2 hour per idle boiler to:

- Note in that newspaper the weather forecast for his next shift and predict the steam load to see if another boiler must be started or one stopped to accommodate that load. Transfer the number of local degree days from the paper to the log. Review communications from the prior shift, the chief engineer and plant engineer to see if facility operations will change the load and plan accordingly. In production facilities, review the production schedule for the same purpose. In some cases today's operator checks the standing orders and production schedules on the plant's Intranet to determine the boiler load.

- Check each boiler in operation to note water level, steam pressure, feedwater pressure, fuel pressure, fuel temperature, stack temperature, draft, casing color and temperature, firing rate, position of control linkage, security of control linkage connections, condition of air inlets, temperature of blower bearings, temperature of blower motor and its bearings, signs of vibration at blower or its motor, flame signal strength, flame appearance, flue gas appearance, and detect signs of leakage.

- Check each idle boiler to note water level, internal pressure, position of vent valve, stack temperature, draft conditions, casing temperature, position of control linkage, security of control linkage connections, condition of air inlet, furnace and boiler pass conditions, and look for signs of leakage.

- Check auxiliary equipment and systems to note salt storage level, brine level, softener in service, other pretreatment equipment as applicable, condensate tank level, deaerator level, deaerator pressure, condensate temperature, feedwater temperature, condition of deaerator vent gases, temperature of condensate pump bearings and the pump's motor and motor bearings, temperature and condition of the condensate pump seal and seal flushing flow, temperature of the boiler feed pump bearings and the pump's motor and motor bearings, temperature and condition of the feed pump seal, continuous blowdown discharge temperature, flash tank pressure, blowdown drain temperature, chemical feed tank levels, fuel oil supply pressure, fuel oil service pumps, motors, and bearings when firing oil, fuel gas supply pressure, fuel tank levels, and look for signs of leakage.

- Draw representative samples of boiler water and test the water for partial alkalinity, total alkalinity, phosphate residual, sulfite residual, chlorides, iron, total dissolved solids, and other concentrations as dictated by the water treatment supplier. Draw representative samples of condensate and test for hardness, pH, iron, total dissolved solids, and other concentrations as dictated by the water treatment supplier. Draw multiple samples of condensate and test when necessary to isolate hardness leakage. Draw representative samples of the boiler feedwater to test for pH, chlorides, total dissolved solids, and other concentrations as dictated by the water treatment supplier. Draw samples of raw water and test for hardness and total dissolved solids. Draw samples of softened makeup water and test for hardness, repeating frequently near ends of softener runs to detect breakthroughs.

- Record, in the boiler plant log, many of the levels, pressures and temperatures described above, maintenance activities described below, unusual activities and events, and observations of conditions that are precursors to failures. Record water, fuel and steam flow meter readings. Calculate and record evaporation rate and fuel consumption per degree day then evaluate the results to identify changes or upsets in system operation and quality of control adjustments. Calculate percentage of returns and compare with history to detect system leaks and upsets.

- Perform normal operating activities including: Test the low water cutoffs on each operating boiler each shift for three shift operation and at least twice each day. Calculate effect of changes in raw water hardness on softener capacity and adjust softener regeneration rates accordingly. Adjust the continuous blowdown rates at operating boilers to maintain dissolved solids concentrations, iron, alkalinity, or whatever is the controlling factor. Adjust the chemical feed pump rates to restore normal water chemistry for each concentration. Clean fuel oil filters when firing oil. Operate boiler soot blowers as required. Adjust firing rate controls to maintain normal operating pressures and/or cycling controls to maximize cycle time according to the

load. When indicated, sample and test boiler flue gases to evaluate firing conditions then adjust fuel to air ratio accordingly.

- Provide escort for visitors, inspectors and contractors. Note work being performed by contractors and service providers, inspect their work where required. Receive shipments of fuel oil, water treatment chemicals, maintenance parts and other materials. Document all visitors, contractors, deliveries, etc., in the log.

In addition to the daily activities described above, perform weekly activities including: Inspect air inlet louvers and screens for blockage, clean as necessary. Restore full levels to all lubricating oil reservoirs in pumps, blowers, fans, air compressors, etc., using the required lubricant. Check salt elutriation conditions and adjust brine feed accordingly. Draw representative samples of the boiler feedwater and test for dissolved oxygen. Take direct level readings and check for water incursion in fuel oil storage tanks. Perform bottom blowoff of operating boilers (this activity normally requires the presence of two operators).

In addition to the foregoing, perform monthly activities including: Lift test safety valves on all operating steam boilers. Conduct slow drain test of low water cutoffs. Test flame detectors. Check along all fuel gas piping elements with leak tester. Check fuel gas regulator vents to detect diaphragm leaks, vent valve vents to detect leaking vent valves. Inspect all piping in plant for loss or dislodging of insulation. Inspect stack cleanout for accumulation of debris, clean as required. Changing and cleaning of filters is usually performed on a monthly basis but each one is staggered to provide a level load of work as much as possible.

Annually the operators should prepare each boiler for the internal annual inspection by the National Board Commissioned Inspector. During that process the operators should inspect the boiler internals on the water side to assess their performance in maintaining water quality and on the fire side to detect any soot accumulation, refractory damage or dislodging, seal damage or loss, and other problems that might change the heat transfer rates in the boiler. At least two people are needed for inspections to satisfy confined space requirements.

Biannual, five-year, and ten-year inspection and maintenance cycles need to be considered as well. Programs for greasing motors and driven equipment can be scheduled in a manner that spreads this work out rather than doing it all at once.

Annual tests that should be performed by the boiler operators include: Leak testing of fuel oil safety shut-off valves, regulators, and vent valves. Calibration checks of gauges and thermometers. Removal and replacing of safety valves where the insurance inspector requires rebuilding, normally on a five year per valve basis.

All the above assumes a bare bones boiler plant. There is always additional equipment and systems that need to be monitored and maintained on a regular basis and service the facility and/or the boiler plant including (but not limited to) domestic hot water heaters, air compressors, cooling towers, chillers, air handling units, etc. Adding the monitoring, maintenance, and water conditioning for those systems can easily consume another operator's time for a normal day.

SAFETY TESTING

Both the National Board of Boiler and Pressure Vessel Inspectors and ASME have recommended that safety valves be lift tested monthly on boilers operating at pressures less than 400 psig and pop tested annually. For higher pressures they recommend testing based on operating experience. Since most of you are operating boilers at pressures lower than 400 you should be testing the safety valves. Whenever I bring this point up I see the operator's eyes begin focusing on the ceiling because they don't do it. Since those safety valves are the last line of defense to prevent a boiler explosion checking them at a reasonable interval is very important. We're only talking steam boilers here. Safety relief valves on hot water boilers are another story and I'll address them in maintenance.

As to the frequency, I've come to the conclusion that monthly lift testing isn't such a good idea. That's because plants will set a schedule for testing on, say the first of each month or the first Monday of each month and here's what happens: The little old lady in tennis shoes that lives nearby notes the schedule and is pushing for local officials to be on hand to hear all that dreadful noise that's upsetting and deafening her each month. I suggest testing quarterly, every three months, and at random hours on a random day because the longer period and randomness reduces complaints from the neighbors. And, since most of you aren't testing regularly a quarterly schedule for lift testing ensures more frequent checks of the safeties than you were doing.

What exactly is lift testing? It's raising the lever on the safety valve to open it, thereby proving that the mechanism will allow the disc in the safety valve to come off its seat to release the steam. If it doesn't lift that means the valve isn't operational and preventing a steam explosion due to high pressure. How it's done is rather important because, and my ears should, perhaps, be burning because some of you are saying to yourselves "Hey, Ken, If we test the valve it's going to leak." I get that complaint from contractors all the time. Well, if it's done right it will not leak. Also, it won't leak provided there isn't anything blocking its operation or that can clutter it up. Your round performed before lift testing should include a close inspection of the safety valve outlets, discharge piping, drip pans, and drains to ensure they're all clear. A safety valve pops open and pops closed with no feathering or weeping because it's either open or it's shut and the transition is so fast that there's no way wire drawing can occur. Wire drawing is a term used to describe the cutting of a valve seat, disc, or both by a small steam leak where the high velocity of the leak erodes the metal.

According to ASME a safety valve should be lift tested only when the pressure in the boiler is higher than 75% of the valve setting. I've told many an operator it's because the lifting lever will break if you try lifting the valve without the steam pressure to assist it. That's not necessarily a true statement but I use it to support the contention that they should only lift test the valve when the boiler pressure is as high as normal operation will allow. Another consideration, one that may have resulted in the contention by many that the valve will leak, is the load on the boiler. If there are only two safety valves then lift testing when the load is higher than 50% is producing a temporary load greater than 100%. That means the steam and water separation in the drum will exceed what the manufacturer designed it for and it could promote foaming and priming which could carry boiler water droplets into the safety valve at high velocity to produce damage from impingement and possibly produce deposits of boiler chemicals on the valve disc and seat that dry immediately as the steam becomes superheated; that's how they can be made to leak. So, if you have two safety valves you shouldn't lift test them unless the load on the boiler is less than 50%, with three valves it's 66%, four or more valves 75%.

Now, I want to say something about how to do a lift test. The problem is you have to lift that lever and it's on top of the boiler. It's not convenient to get to and, to be perfectly honest, I don't like to be near a valve that automatically pops wide open. I'm always concerned for the safety of the Operator and climbing up on the boiler, a ladder, or even an access platform, and lifting that lever doesn't seem safe to me. Therefore, I strongly recommend you go to the hardware store and get some lightweight welded chain, machine screws and nuts that fit through the chain and two pulleys per valve that will carry the chain. Figure out where you're going to put it before you go to the store then add twenty feet per boiler to have enough chain to do the job. You may also need some clamps to attach to building steel, screw eyes for wood structures, or lead anchors and eye bolts to attach to a concrete ceiling to suspend the pulleys. Connect one end of the chain to the eye that's at the end of the safety valve lifting lever (that's what the eye is there for) and run the chain through the pulleys so it drops to a reasonable height above the operating floor near a walkway. Finally, to prevent the chain running out on you, you need something at the other end to offset the weight of the chain between the two pulleys. I always suggest making a combination operating handle and weight that looks like the one in Figure 2-5. Yes, it looks sort of like a stop sign and that's intentional. It should have a red background with white border and lettering to add to that effect. That's because you don't want some visitor to come wandering by and wondering what happens if he pulls it. Making it about 2½ inches wide makes it relatively comfortable in the hand. It you make it from at least 10 gauge steel with the edges ground smooth is should be very comfortable to operate the lifting lever with it. Finally, mount it so it's above the top of the head of the tallest Operator in the plant and low enough for the shortest one to reach it. You can make a wire hook from an old coat hanger wrapped to a pipe off to the side of the walkway to store it to keep it from knocking heads or catching on something you're carrying down that aisle.

Oh! I didn't say how to do a lift test? Well all you do is check the pressure, the boiler load, and make sure you're grabbing the handle that's attached to the safety

Figure 2-5. Safety lift grip

on the correct boiler, pull on it and let go of it. Repeat for each safety on the boiler allowing a few seconds between lifts to let things stabilize before doing the next one. What should happen is you'll hear the steam belch out of the valve and then stop—pop and stop. In the very unlikely situation (which normally occurs only when you haven't tested the valves regularly) that you hear steam weeping through the valve after a lift test you should operate it again as soon as possible to blow whatever got under the seat out.

Pop testing of a safety valve is normally done as part of the semi-annual inspections on high-pressure boilers and during annual inspection of another boiler in low pressure plants. That operation is described in the section on annual inspection.

What about safety testing on hot water boilers? Well, you can test the safety valves on HTHW boilers using the same rules as steam boilers. Always make it a point to ensure everyone is clear of the vents on the boiler room roof before testing because the discharge will include water as well as steam. If problems occur when testing (like hot water splashing all over) the valve discharge piping is not arranged properly.

Regular hydronic and hot water heating boiler safety relief valves shouldn't be tested by lifting them. Normally the system pressures are nowhere near 75% of the valve setting and that's one reason why. When asked what to do about them I only recommend that they be removed from the boiler during each annual or biannual inspection and examined by looking at the inlet and outlet. If you see scale buildup on either side replace the valve. Hot water boilers are typically equipped with PTV valves (pressure and temperature relief) The temperature relief is accomplished by a small diameter cylinder that hangs out of the bottom of the valve and contains a fluid or wax that expands with temperature and will force the valve open at the set temperature. Unlike steam safety valves they do not have to be installed vertically and must be installed so that cylinder is in the heated water.

A hot water boiler explosion occurred several years ago at a school in Oklahoma with several children and some teachers injured. It was because a contractor didn't understand the application of a PTV valve and changed its mounting. Another hot water heating boiler generated steam in a school in Baltimore in 1999 so the water was pushed back through the cold water line by the expanding steam to flash out a toilet as a little girl flushed it with disastrous consequences for her. Those valves have to installed according to the manufacturer's instructions and inspected regularly.

IDLE SYSTEMS

For some strange reason people think a boiler plant that's shut down during the summer or an air conditioning system that's shut down during the winter doesn't need any attention. The contrary is true, they need more attention because it's during those periods when the equipment isn't operating that they normally incur the most damage. Before you say I don't know what I'm talking about consider this: most of the rusting and corrosion in heating systems occurs during the summer when the boilers are shut down and a typical reason for catastrophic failure of a chilled water system is freezing when it's shut down. Idle equipment deserves just as much attention as operating equipment.

Idle boilers should be warm (see the section on standby boilers) or laid up wet or dry. Concerns with warm boilers include checking to ensure they're really warm; the temperature of the water at the bottom of the boiler should be the same as the water at the top of the boiler. Boilers that are not up to operating pressures and temperatures can weep enough to promote high rates of localized corrosion so casing drains should be checked daily to ensure there's no evidence of the boiler weeping excessively.

Idle boilers require more attention because an operating boiler is generating inert gas; it's less likely to explode than an idle boiler. The fuel oil and gas supply shut-off valves should be checked to ensure they're closed and supply pressures after them down to zero. Gas fired boilers should be checked by sniffing at an observation port or other sampling means to ensure there isn't any gas leaking into the boiler.

The most expensive industrial accident incurred to date was the result of gas igniting after leaking into an idle boiler at the River Rouge Steel Mill in February of 1999. The result of that boiler explosion was six dead, several injured and over a billion dollars in damage. If the boiler is oil fired the oil burner should be removed or the oil supply piping disconnected from the burner and plugged so no oil can leak into the furnace. Separate ignitor gas supplies should also be isolated and checked.

The ash pits, bunkers and furnaces of coal and solid fuel fired boilers should be checked for accumulation of anything that could create problems including water, trash, rodents and sleeping contractor employees. Speaking of contractors, an idle boiler should be covered to prevent damage from contractor operations above and around it and panels and fan inlets should be sealed to keep construction dust from entering them.

I like to leave power on a burner management panel and control panels so the indicating lights, transformers and the like keep the enclosures dry. Alternatively you should check for operation of panel heaters or temporary lights installed for that purpose. You can't be certain that there's sufficient power to keep the panels dry so simply open the panels once a week to check for condensation; any rusting or discoloration says you need heaters in them.

You don't want to discover your boiler is full of holes when you try to start it up in the fall so, if the boilers are in wet layup the water should be tested for sulfite content and pH weekly and corrected if the analysis shows the levels to be inadequate for proper storage. Boilers without stack caps should have the stacks covered if they are above the boiler and stack base access doors opened if they aren't so you can be certain rain isn't entering the boiler and corroding it. Sometimes that isn't easy to do so it's more important to see to it that any rain that falls dries out quickly by providing, and regularly confirming, good ventilation over the metal surfaces and up the stack.

During the winter an idle boiler can freeze up if the plant is sealed so much that combustion air from operating equipment is drawn down the stack of the idle boiler. That's why I say stack temperatures should always be recorded, even on idle boilers. Stagnant water piping and the like can also freeze if the cold outside air that's always drawn into a boiler plant for combustion happens to flow over that piping or equipment.

Chillers, cooling towers, and other air conditioning equipment plus any equipment or piping system that contains water should be drained completely when it's idle. If it's not possible to drain a system completely then it should be filled with an antifreeze solution that's guaranteed to prevent freezing at the lowest known temperature at your plant. If neither of those options are available to you then you have to be concerned with freeze protection, checking every piece of idle equipment regularly during the winter months to be certain it's not freezing.

Some freezing is due to us engineers, I'll admit. I recall one installation where the engineer designed louvers for combustion air in the wall of a boiler room where the air drawn in traveled right over the chiller; since it was inside the boiler room it was supposed to be warm and plant personnel failed to drain it. At the beginning of the cooling season they got an expensive surprise. Remember that story because you can't forget that standing water in the boiler plant can freeze if cold air is drawn over it, including water in idle boilers.

Your water supply piping is susceptible to freezing

because the water is already cold and it won't take much more to start freezing it. There's been more than one boiler plant shut down in the winter because cold drafts froze their city water line solid. Don't take an indicating light's operation as proof that electric tracing is on, put your hand on the covering. If it isn't warm slip a thermometer under the lagging and if necessary push it through the insulation to the pipe (be careful with pointed thermometers that you don't penetrate the tracing).

Salt storage tanks are usually idle but they can overflow at any time. Brine can also freeze. An idle softener can freeze if exposed to a cold draft and can contribute to salt leaking into the effluent (another one of those engineering terms, it's the treated water leaving the softeners) if it isn't checked while it's idle.

Idle condensate and boiler feed pumps can freeze up. That's why it's important to rotate them regularly. That's rotate, not bump. When you bump a pump you simply push the electric motor's start and stop buttons one after the other so the motor turns over. The problem with bumping any rotating equipment is it tends to stop turning right where it stopped last time. Any rotor suspended between bearings will tend to sag over time and if left in, or returned to, the same position every time the sagging increases.

To rotate a pump you should turn it by hand. Sometimes that means temporarily removing a coupling guard or reaching under it. The final key is to turn it $1°$ turns so it's 90 degrees off its last position. Rotate it once a month and it will only be in the same position one fourth of the year. All rotating equipment, anything run by an electric motor, gas or diesel engine or steam engine or turbine including the drives should be rotated monthly. By maintaining a schedule of the rotating equipment and rotating one a day or one a week (depending on how many you have) all the equipment in a facility can be rotated on that monthly schedule.

Idle piping systems also deserve some attention. The first lesson of idle liquid piping systems should be to ensure there is always one way for the liquid to expand out of the piping system. If you valve off a piping system to the extent that the liquid is trapped inside, the piping will be exposed to considerable swings in pressure as the liquid is heated and cooled. The liquids that enter a boiler plant are typically colder than the plant so it's very easy to isolate a cold liquid which will expand when heated. If that liquid is completely trapped the only way it can expand is to stretch the pipe and you better believe that it can do it.

Expanding liquid normally raises the pressure to the point of failure of a gasket or packing at a valve stem

and operators will consider it a simple leak. If, however, you fix all the leaks the pressure will eventually split the pipe because expanding heated water can produce as much force as freezing ice.

The best example for this is a typical run of hydronic piping where the temperature of the water ranges from an installation temperature of 70°F to a maximum operating temperature of 250°F, 100 feet of the piping heated from 70 to 250 will increase in length by 1.3 inches but the water in the piping will extend its length 67.44 inches. Since the water isn't very compressible it will try to stretch the pipe that far, with drastic results if that water doesn't have anywhere to go.

The simple solution for idle systems is never isolate them completely. If you have to, then install provisions for expansion or a relief valve on them that discharges the liquid to a safe location. A favorite spot for this problem is the short length of piping between two fuel oil safety shut-off valves. The engineer's solution is a relief valve connected to that piping and discharging to the oil return line. If you don't have one of those you should have a branch line with a small valve for leak testing closed with a nipple and pipe cap. Remove the cap and open the valve each time the boiler is shut down for an extended period then close it back up after a little air has gotten in. That little bit of air should not create a problem at the burner because it should pass through while the ignitor is still operating.

Fuel oil in idle piping exposed to the heat of a boiler room can gradually break down to form heavier hydrocarbons and gases that produce the equivalent of air pockets in piping. That doesn't necessarily create a problem for the piping but pumping that fuel with its pocket of gas to a burner can create a flame out (there's not enough energy in the gas to keep the flame going) and subsequent re-ignition of the fuel oil to produce a furnace explosion. Always recirculate oil to eliminate any gases long before starting a burner on fuel oil. Fuel oil piping can also be a hazard if it is fully isolated

I have seen four-inch water piping reduced to less than 3-inch internal diameter in a matter of months because it was idle. Despite chlorination and other forms of water treatment microbes manage to survive. Given stagnant water and a minimal source of nutrients (food to eat) they can thrive. Not only do those microbes construct rather solid homes on the inside of the pipes they also generate waste that can be very acidic or caustic to corrode the piping. Just recently I have seen a large number of articles in engineering magazines on the problems of MIC (microbe induced corrosion) which, in many cases, is comparable to oxygen pitting because the microbes concentrate under a little growth on the inside of the pipe and emit the acids and alkalis that attack locally.

Normally the solution for idle cold water piping is simply opening a vent or drain valve to refresh the water in the idle piping once a week. Microbes can't survive in water above 140°F and don't do well in water much warmer than 120°F. Water lines that are in the upper levels or a boiler room shouldn't have a problem with microbial growth because of the heat but would suffer from oxygen pitting if you regularly added oxygen rich water to them.

Oxygen is another problem in water piping, not as persistent as in boilers but the cold city water usually warms up in idle pipes in the boiler plant and raising the temperature of the water reduces its ability to absorb oxygen so some of it is released to produce the damage we know as oxygen pitting. (See deaerator operation for more on oxygen problems). If the piping is to be idle for long periods of time it should be drained and kept dry. That way, both microbes and water borne oxygen can't do damage to it. A dry line will develop a very thin coat of rust that will protect it.

If you can't keep the pipe dry then adding chemicals to the water or filling the piping with nitrogen to inert it are options. A nitrogen inerting system consisting of a regulator and safety valve on a portable cylinder should maintain the inert status for several months. You only need to maintain a few inches of water column as pressure in that idle piping. Nitrogen can find some pretty small places to leak through and maintaining high pressures will result in wasting a lot of nitrogen.

Vent and bleed lines for gas pressure regulators, gas pressure limit switches, and the bleed of double block and bleed shut-off valve systems are basically idle piping. The vent lines from a regulator or pressure switch is there to provide a direct connection for atmospheric pressure on the diaphragm of the control valve plus convey fuel gas to a safe location in the event the diaphragm leaks. The bleed line is used intermittently to dump the gas trapped between the two safety shut-off valves. Those lines should always be treated as gas lines even though they may contain air most of the time. The condition of the terminations of gas system vents and bleeds, normally a screened fitting, should also be checked on a regular basis to ensure they aren't blocked.

An ear to the line can detect a good sized gas leak. They should also be checked by stretching a rag over their outlet (or a union just inside the building when the outlet is inaccessible) and soaking it with soapy water. Bubbles indicate a leak. They should be checked whenever there's reason to believe they could be leaking or on annual inspection. I'm reminded of when my service technicians made repeated visits to a plant in an attempt to locate an intermittent gas leak. They eventually discovered the rubber disc of a bleed valve had been cut by the sharp seat of the valve and occasionally buckled to block the valve partially open while the boiler was operating.

Fuel oil tanks that aren't in use should be full except for one that may be filling. That way you minimize the exposure of the metal in the tanks to air and its corrosive properties. You also limit the contact of air with the oil. Full, of course, doesn't mean up to the brim; you always need some freeboard (space between the liquid level and the top) to allow for expansion. I thought I had the matter of expansion down and filled fuel oil tanks up to the very top once. The oil was normally delivered hot (good old bunker C) so it would shrink into the tank. I discovered later that the particular shipment I received was colder than normal so it expanded instead of shrinking and I got to spend a day cleaning up the fuel spill I created. See fuel oil in the section on consumables for more on the wise use of fuel oil storage.

Propane and fuel oil storage facilities have a bad habit of becoming garbage dumps. In the fall leaves accumulate in the diked areas around the oil tanks and on the ground around the supports of propane tanks. That's fuel for a fire from an inadvertent spark or cigarette that could produce a disastrous fire and possibly an explosion. Water can accumulate in diked areas or simply form ponds that stand on metal pipes, supports and tanks to promote their corrosion. Every day shift should visit the fuel storage locations for the express purpose of identifying hazards and eliminating them. Raking leaves and mopping water may not be in the job description but you are responsible for those facilities and should take any action necessary to protect them.

A very important piece of equipment that's idle most of the time is an emergency generator. Many plants test them on a regular schedule but they deserve attention between tests to detect any problems that might arise. There are probably many items and systems in your plant that weren't included in this discussion but deserve your attention when they're idle because they're critically necessary when you need them. You have to identify them and make certain your SOPs include procedures to check on them.

SUPERHEATING

The recent deregulation of electricity has resulted in more superheated steam boilers to permit plants to generate electric power so you better know the important requirements of superheater operation. The first and foremost rule is the superheater has to have steam flowing through it to absorb the heat getting to the tubes or the tubes will overheat and fail. Water in superheater tubes doesn't help, it can block flow in some tubes to permit them overheating or suddenly blow over at high velocities to create water hammer damage.

The following guidelines should ensure proper start-up of a superheated boiler. Note that some boilers, HRSGs in particular, can have special requirements so be sure to read that instruction manual. When the boiler is equipped with a reheater you should have to adjust valving to direct steam from the boiler through the reheater and open the reheater vents and drains. When starting up a boiler with a superheater make sure all vents and drains on the superheater are open. Similarly, check that all reheater vents and drains are open.

As soon as a reasonable flow of steam is evident at the boiler vent, close it to develop maximum flow through the superheater. When superheater drains appear to be blowing clear with no moisture present (a slight gap between the pipe and the cloud of water droplets) close down on the drain valves to increase flow through the whole superheater. Similarly choke down on any intermediate vents. Constantly observe the superheater outlet temperature, paying close attention after any change in firing rate, number of burners or ignitors in service, and other activities that can change flue gas flow past the superheater. Close drains and vents except for the final superheater vent valve once you have the turbine rolling over. Close the superheater vent valves after the turbine is carrying a load. Close the bypass valve to the reheater as well, confirming reheater flow from and to the turbine before closing the reheater vent valve.

During operation note the superheater, and reheater if equipped, outlet temperature on a regular basis. There are many things that can go wrong to produce a problem with overheating the superheater or reheater that aren't necessarily going to be associated with changes in sound. If the turbine trips, open the superheater vent valve before trying to reset the turbine trip valve. If the boiler has a reheater establish flow through it as well.

Fooling around with a trip valve without superheater flow is dangerous. There's no steam flow so the superheater outlet temperature indication will fall even though the metal a few feet inside the boiler is overheating. It's very embarrassing and quite scary to see the superheater outlet indication peak well above design temperature after you get the trip valve opened back up.

If your plant makes it a practice to lift check the safety valves then do so with caution, waiting until the boiler has settled down after lifting each drum safety.

Open the superheater vent first before starting to take a boiler off line, that's first before anything else. If other boilers are serving the load any reheater will have to be set up to maintain steam flow as well. Whenever possible keep serving the load after shutting off the fires to keep the flow up, allow the turbine to drop off with the boiler so you maintain maximum possible superheater steam flow. Don't open the other vents and drains until the boiler is down to 25 psig when you're ready to open the drum vent. There are so many variables in superheater and reheater design today that I can't begin to ensure you these procedures are the best for your plant. Be certain you follow manufacturer's instructions.

Some superheaters are equipped with gas bypass dampers inside the boiler so you can control the superheat temperature to a degree. Others will have an intermediate desuperheater that injects feedwater into piping connecting two sections of a superheater to drop the temperature coming out of the first stage and you may find desuperheaters on reheaters. Some of these devices can produce a false sense of security by producing safe superheat readings at the boiler outlets but the temperatures upstream of the desuperheaters or in parts of a superheater that aren't affected by the dampers go too high. In any kind of upset operating condition check as many temperatures as you can and don't bet on the lowest reading being the right one, always figure the highest reading is the right one.

Desuperheaters are used to increase the supply of desuperheated steam (the added water evaporates and becomes part of the steam). When the steam is used in heat exchangers and similar apparatus desuperheating reduces the amount of heating surface required in the heat exchanger. They should always leave a little superheat in the steam so you know there's no water racing down the piping looking for an elbow to run into. When you're operating a superheated steam plant you have to know what the saturation conditions are for every service and what are the maximum temperature ratings of the equipment and piping.

SWITCHING FUELS

Any boiler plant of a reasonable size should be capable of burning more that one fuel. It provides the owner or user with an alternative fuel in the event the supply of one is interrupted. It also provides a basis for negotiating price with the suppliers. Most boiler operators, don't make the fuel supply or price decisions but they should be prepared to choose, and choose wisely, which fuel to burn.

In most northern states the operator is informed by a phone call when to switch from natural gas to oil firing. Their natural gas is purchased in accordance with a special contract so the supply is "interruptible." It's a method that benefits the gas supplier and the consumer. The large pipelines that transport gas from the southern states, principally Texas and Louisiana, have a maximum capacity. The pipeline owners want to optimize the use of those pipelines. They are limited by the pipeline capacity to the customers that are supplied "firm" gas. Those firm gas customers don't use much, if any, during the summer and when outdoor temperatures are mild so there's always room in the pipelines for more gas to flow except on very cold days. By selling interruptible gas the pipelines make use of that extra room in the pipeline. The purchaser gets a discount, paying less for interruptible gas, and that's why both benefit. The only compromise for the purchaser is a switch to an alternate fuel when notified by the supplier of an interruption.

Once you're familiar with your plant you will know an interruption is coming most of the time. On rare occasions the supplier may have to take a pipeline out of service for maintenance or repair and will require an interruption but most of them are due to load (see Know your Load, page 93). Most of the time a weather forecast will forewarn you that you will have to stop firing gas and change to an alternate fuel. You'll also know about when you will receive a call that allows you to switch back to gas.

Here's an appropriate word of caution when considering a fuel transfer. There's no such thing as a "flick of the switch" fuel transfer. I've had to observe the cleanup from a couple of boilers where someone thought it was that simple. Most boilers have to shut down and go through a regular boiler start-up to change from one fuel to the other. The idiots that believe in "flick of the switch" end up blowing up boilers.

You might even have a plant that automatically switches from gas to oil and vice versa. You'll have what is called an "automatic interruptible gas service" controlled by an automatic interruptible system (AIS). Those consist of a set of controls in a panel, normally sealed by the gas supplier, that sense outdoor air temperature and control the boilers to automatically switch fuel. These are typically small heating boiler plants where only one boiler is required to carry the peak load

and a short interruption in steam supply or a dip in steam pressure or water temperature is not considered a problem. At a prescribed cold temperature the controls stop boiler operation then automatically restart it on the alternate fuel. When the temperature rises to a higher value the boiler is stopped then restarted on natural gas.

Today there's another reason for switching fuels and it's more important for the boiler operator to become involved because it relates to the fast paced financial situations of today. Many gas contracts today do not set a fixed price for gas. The price varies according to any one or more sets of rules or price indices. A typical index is "well-head price" meaning the price of the gas where it is extracted from the ground. Currently that price is set for each month but it could easily be set hourly in the future.

The boiler operator may have to watch the Internet on a computer in the control room to be prepared to switch fuels when the gas price goes high enough. 2000-2001 produced some significant swings in natural gas pricing with prices ranging from $2.97 to $10.81 per Decatherm when fuel oil cost was about $7.12 per Decatherm. There were a few plant chiefs called upon to answer why they continued firing natural gas when it was cheaper to fire oil.

Pricing is the principle reason for fuel switching but loss of service is another. During an earthquake buried gas piping is typically interrupted. I've also experienced interruptions due to contractors digging into the gas mains and gas piping breaks from flooding that washed the line out. Sudden ruptures can also interrupt your gas service so having an alternate oil supply is a way of recovering from those situations.

As with AIS the simple way to switch fuels is to shut the boiler down then restart it on the alternate fuel. One of the reasons AIS is seldom utilized today is many people didn't manage to get that right. There were several failures in the 1970's associated with systems created that simply switched fuel valves (the flick business). The installer or designer didn't understand that could result in a loss of flame with continued admission of fuel and a subsequent explosion. So, unless your system is specifically designed as one of the two "on-the-fly" switching systems I'm about to describe, shutting down then starting on the second fuel is your only option.

One favored method of fuel switching is the "low fire changeover" method. The alternate fuel system (for the one presently not firing) is placed in service to bring the fuel supply up to the safety shut-off valves. The operator also makes certain the manual burner shut-off valve for the alternate fuel is closed. The controls are switched to manual and firing rate is reduced to minimum fire. The operator then begins the changeover by turning a selector switch on the control panel to "Dual" or "Changeover" so the burner management system will energize both sets of fuel safety shut-off valves. The operator then throttles the manual burner shut-off valve for the fuel being fired and slowly opens the manual burner shut-off valve for the alternate fuel. When observation indicates the alternate fuel is firing the operator spins the alternate fuel's manual burner shut-off valve open while simultaneously closing the valve for the fuel that was firing. The selector switch is then turned to the alternate fuel position so the burner management system will close the original fuel safety shut-off valves. The controls are adjusted to bring firing rate back to slightly above the rate before the changeover until pressure or temperature in the boiler is near normal before switching back to automatic firing rate control.

The designers of burner management systems incorporate additional logic in their systems to ensure a low fire changeover is performed properly. That logic requires the low fire interlock be maintained while the selector switch is in the position to admit both fuels. They frequently add a timing sequence that limits the time when both fuels can be admitted. If the selector switch remains in the two fuel position for more than a few minutes the boiler is shut down. I don't like those standard provisions because logic is complex and the time limit produces a sense of urgency in the operator that may cause her or him to make a mistake.

In low fire changeover systems I have designed (keep in mind that I really don't like this approach to switching fuels) I allow the operator to initiate it by turning the selector switch. The control logic then knows controls have to be in manual and at low fire so the logic switches controls to manual and low fire. The operator doesn't have to do it. Once the low fire position is established the control energizes the ignitor and waits ten seconds for it to be established.

Gas is normally admitted at the perimeter of the burner while oil enters at the center; rather than accept one will light the other I use the ignitor which is designed to light both. After ten seconds, the normal PTFI (Pitot trial for ignition) timing, the alternate fuel safety shut-off valves are opened. Then, after the normal MFTI (main flame trial for ignition) timing the ignitor and original fuel are shut down. Manual control of the fuel flows is not required but the operator may do it. The controls should be set such that excess air at low fire is at least 150 to 200%. During the period both fuels are firing the excess air would be 25% to 50%; that doesn't guarantee complete combustion but it will assure a stable flame exists. Once the operator observes the stable firing of the alternate fuel and turns the selector switch to the alternate fuel the controls are released back to automatic. Switching to Manual and manual adjustment of the firing rate controls is optional. I also inject ramping controls mentioned earlier. If the selector remains in the two fuel position for more than a minute after both fuel valves are energized the system shuts down the alternate fuel and returns to automatic. There's no reason to shut the boiler down.

Low pressure heating systems and similar applications that do not have a critical steam pressure or water temperature requirement can accept shutting down and restarting a boiler so the simple stop and restart method is satisfactory for them. The low fire changeover method manages to eliminate the loss of heat input during the purge period to reduce pressure or temperature loss but some drop is associated with holding operation at low fire. In my experience any facility that can't afford a drop in pressure or temperature has two other means of switching fuels that will, unlike the previous methods, ensure a reasonably constant maintenance of pressure and temperature. Smaller plants will have a spare boiler that can be brought up on the alternate fuel and placed on line. Larger facilities normally don't have spare boilers so a means of switching fuels on operating units while maintaining pressure or temperature is required.

Larger facilities will have full metering combustion control which allows dual fuel firing to maintain pressure or temperature. Dual fuel firing is simply operating with both fuels at once. When equipped with a full metering system the two fuel flows are measured, their values added and the total fuel flow measurement is used by the controls to maintain a proper air-fuel ratio. The alternate fuel is started at low fire with its control in manual. The ignitor is brought on, then the alternate fuel, and the boiler simply fires both fuels. Once the operator observes a stable alternate fuel the controls are adjusted to bring the alternate fuel up manually until the automatic control has reduced the original fuel firing rate to low fire. Once the original fuel is at low fire the operator switches its control to manual and transfers control of the alternate fuel to automatic. Finally, the original fuel valves are de-energized to complete the transfer.

This method has been successfully applied on multiple burner boilers with capacities of 250,000 pph. When applied to multiple burners the second fuel is started one burner at a time to limit control upsets. An interlock requires all burners be firing on both fuels before the alternate fuel firing rate can be increased above low fire. Safety shut-off valves for the original fuel are tripped in unison when at low fire; a sudden increase in excess air will not produce an abnormal furnace environment with a good control system.

Many times I hear the argument that switching at load is dangerous. As I said earlier, I don't like low fire switching and I consider shutting a boiler down, then starting on the alternate fuel, a little more dangerous. There's a reason most boiler explosions occur on light-off. You're creating an explosive mixture then trying to get it to burn instantly. When a boiler is operating you have a fire so low fire changeovers or dual fuel firing don't involve that opportunity for an explosion. You're also producing an inert gas while you're firing so any injection of fuel that isn't burned is surrounded by inert gas instead of air and it can't burn. (There's reasons to be cautious about this when you have boilers with a common breeching)

The low fire changeover method requires significant quantities of excess air so there is air there for any introduced fuel to burn if it isn't ignited immediately by the existing fire. That's a bit of a problem because the existing fire isn't very stable and all that excess air makes it even more unstable. Bringing on a second fuel when dual fuel firing with full metering controls results in the combustion air increasing as the fuel starts flowing to the furnace. The fire of the existing fuel is above minimum to produce more heat and is more stable than it would be at low fire. (Low fire position is normally determined to be when the fire is stable; anything lower being unstable)

The method available to you for switching fuels should be documented by a detailed SOP for that operation because it is always possible for something to go wrong to produce an explosive condition.

Finally, practice it. Before an operator is compelled to switch—it happens when the gas company called and he or she can't reach the chief or anyone else for help— that operator should have done it under supervision at least twice each way. It's also advisable to practice it in the early fall, before cold weather sets in, so everyone has the memory of it refreshed.

STANDBY OPERATION

Whenever I bring up my opinion of standby operation it provokes conversation. Before you sit down to write me a note or call to tell me I'm full of it, please read this whole section. You may just agree with me that firing a boiler to keep one on standby is inefficient, bad for the boiler, and nothing more than an indicator of an operator's (or an operator's boss's) lack of confidence in the equipment and/or the operator's skill. If you still disagree after reading this section you should also review your logs to see what has happened. You should find that boiler operation is highly reliable, more reliable than the electrical service, and should be treated that way.

Boilers do shut down unexpectedly and loss of pressure or temperature will happen. You should find your logs document that the shutdowns were primarily due to loss of electrical service and an unexpected boiler failure is rare to nonexistent. So, I ask you, "why do you continue firing another boiler to keep it hot just in case the operating unit fails?"

Ever notice that you can't break a wire by bending it once but you always can by bending it repeatedly? As far as I'm concerned you are probably doing more damage to your standby boiler by running the pressure up regularly than you would if you poured the fire to it to get it up to pressure from a dead cold start the one or two times in its life that was necessary.

A well maintained plant where equipment is tested regularly and maintained properly will not have boiler failures and has no need of keeping a boiler on standby. The damage to the boiler and the fuel and electricity costs for keeping it hot normally outweigh any advantage of keeping it hot by regularly warming it up. On the other hand, the maintenance of pressure or temperature may be so critical that loss of a boiler is unacceptable. In the 1980's I had one customer with a simple formula: if the pressure dropped from 240 psig (normal operation) to 230 psig it cost the plant a quarter of a million dollars. A standby boiler isn't the solution in those cases, it's having a sufficient number of boilers on line so loss of any one will not prevent maintenance of pressure or temperature.

There is simply no way I can justify the concept of keeping a boiler on hot standby by firing it regularly. The only means of maintaining a hot standby that I will agree with are (1) installation of convection heaters and (2) blowdown transfer. By installing a heating coil in the bottom drum of a boiler or installing a heat exchanger, circulator and piping connecting the blowoff and feedwater to heat the boiler water using steam from operating units you can keep a boiler hot enough that it can be brought on line as fast as one that's fired to keep it warm.

Blowdown transfer uses the continuous blowdown from operating boilers to keep an idle boiler hot. Depending on the amount of blowdown it's possible to keep more than one boiler in hot standby without firing them. Either of these methods doesn't apply heat to the refractory so some minor refractory damage may incur if a standby has to be brought on line immediately but the pressure parts will be uniformly heated and the boiler will come on line quickly without danger of stress cracking.

Now, quit heating up a boiler to maintain a standby. It wastes fuel, it increases environmental pollution, it's bad for the equipment, and it's a waste of your time.

I've discovered that plants which seem enamored with the concept of standby boilers also like to rotate them frequently. They're kept on standby so it's easier to rotate them. There's also a bit of confusion regarding the status of a boiler on standby that should be cleared up; it seems to happen frequently in plants with multiple heating boilers. Just because the pressure gauge shows the same pressure as operating boilers doesn't mean the boiler is hot. Steam from the operating boilers will flow to an idle boiler. A power boiler with a leaking non-return valve can hold a head of steam.

The problem is that pressure and temperature is only above the water line; everything below can be dead cold, and in one case was actually freezing. For the same reasons that water circulates in a boiler when it's firing it will stagnate when it isn't. I commonly come across boilers that show pressure where I can reach down and touch the bottom drum or a portion of the shell and find it cold. A boiler in that situation is not a hot standby, it's a bunch of thermally distorted steel. Any rapid changes in the water level can result in stress cracking of the drum or shell and tube sheets.

Systems that simply drain the condensate off at the surface of these boilers maintains an artificial state that is dangerous. Those boilers should either be allowed to flood, so they're all cold, with the condensate removed in a section of piping above the boiler, or isolated and put in lay up properly. It's not too expensive to replace a piece of piping compared to replacing a boiler.

ROTATING BOILERS

The act of rotating boilers, sometimes called alternating although I prefer that label be used to refer to automatic rotation, is the operation of boilers in a manner that assures that all the boilers have the same amount of operating time. It has been common practice and many facilities have alternating controls that ensure every boiler take its turn at operating. Why is it so important to make certain that all the boilers have an equal amount of use to improve the certainty that they all start having break downs at the same time?

Like the old rule of thirds (page 111) I recommend you operate your plant so one boiler has half the total operating hours and another has one third of the total operating hours. The boiler with the most operating hours will experience problems giving you a good indication when to maintain, rebuild, or replace parts to ensure the problems aren't repeated on the other two. You'll also have two boilers with less wear than one and one with less wear than the other two. If you only have two boilers one should have twice as many hours as the other.

Another perpetuated bit of foolishness is alternating systems that are constantly switching boilers. Either each time a boiler cycles or every day. Heating up a boiler takes energy and switching to another results in all that energy being lost. Why waste it every day? If one boiler is too big for the load (it is cycling) why would you operate two to double radiation losses? Rotate the boilers on at least a quarterly schedule so they get at least three month's rest before you start them up again. Start-ups always put a strain on a boiler, why strain them any more frequently than necessary?

Oh, that's right, you would have to lay up the boiler properly if you didn't use it regularly. Collect some data, do a little math and you'll discover that it's costing the owner a considerable amount of money to keep two boilers running when one is adequate. Lay one up for a summer, or a year. The little bit of work it takes to do the job right will pay off in lower fuel bills that you can take credit for.

BOTTOM BLOWOFF

Some of you will argue this point because you've used it for everything, everything but the only purpose for bottom blowoff. Its only purpose is to remove sludge, scale, and sediment that collects in the bottom drum of the boiler. There is a prescribed procedure for it with some variations depending on the type of bottom blowoff valves that are on the boiler. Some of my customers don't perform a bottom blowoff... ever. That's because their water pretreatment and chemistry methods don't create any accumulation in that bottom drum and the little bit that does collect is removed with each cleaning for annual inspection.

Yes, you may open the bottom blowoff valves to drain the boiler for its annual internal inspection (biannual for some of you) but draining the boiler is not the same as performing a bottom blow. Other reasons for opening those valves are simply not acceptable. The bottom blowoff valves are not there to regulate the water level; if the water continuously runs high then get the level controls fixed. The bottom blowoff valves are not there to lower the concentration of solids in the boiler water, that's what the continuous blowdown system is for. Continuous blowdown removes water with the highest concentration of solids and, when diverted to a blowdown heat recovery system, waste very little energy.

They are definitely not for maintaining boiler operation; I had a hard time believing an operator was blowing his boiler down every fifteen minutes so enough cold water was added to prevent the boiler cycling off; he was wasting fuel, water, and his own energy to keep the boiler from doing something normal. Operating the bottom blowoff valves without concern for operating conditions can interrupt boiler water circulation to result in an eventual failure. Use them only for their intended purpose.

The first and principal consideration for a bottom blow is to make certain you are in control of it. I prefer they be done at the change of shift so two operators are there to do it. You can do it yourself if, and only if, you can see the gauge glass while you're operating the valves. There are very few boilers set up so you can do that and it's still a good thing to have another person on hand in case something goes wrong; I once had a blowoff valve stick open.

Whatever you do, don't consider the blowoff an option to test the low water cutoff. I see that done regularly and ask the operator the question "What are you going to do the day the low water cutoff doesn't work?" Oh, I get a lot of assuring answers but the only right one is that operator will finally decide that something's wrong, close the blowoff valves and walk to the front of the boiler to see the gauge glass empty and, as in one case related to me, look into the furnace to see all the tubes glowing red! If there's no one there to keep an eye on the gauge glass don't blow the boiler down until someone is. Watch the glass every second until the entire process is complete.

A bottom blow removes a considerable amount of water in a very short time and can change the natural circulation in the boiler. Unless the manufacturer's instructions specifically state that a bottom blowoff can be performed below a certain load never perform a bottom blow without shutting down the burners. Never blow a boiler with loads above the limit prescribed by the boiler manufacturer either.

A bottom blow can temporarily stall flow in risers resulting in high concentration of solids and scale formation in those tubes to promote subsequent failure. Water tube boilers are very susceptible to that form of damage. There should be written procedures in any plant for performing a bottom blow and they should be complied with.

Since the purpose of a bottom blow is to remove solids from that mud drum you want to have enough water flowing out to flush it well so the first step in preparing for a bottom blowoff is to either temporarily raise the drum level controller setpoint, use manual control, or bypass the feedwater valve to raise the boiler water level up to within a couple of inches of the top of the glass. That provides the maximum reservoir of water for a good flush of the mud drum.

Open the first valve (more later on which valve gets opened first) then crack (see valve manipulation) the second valve to allow some water to slowly drain out of the boiler and heat up the blowoff piping and flash tank or blowoff tank. When the level in the gauge glass has dropped an inch, open the valve completely to provide full flow to flush the mud out of the boiler. Then, when the level is about two inches from the bottom of the glass close the valves. Restore the setpoint or automatic control to establish normal water level.

Continue to monitor the level until it returns to normal and check it frequently for about an hour afterward. I like to blow down the water column a few times at two minute intervals after the bottom blow; if any mud was left in the boiler it was loosened and will show as color in the fresh water in the gauge glass. If I see some color then I know I have to blow down more frequently; usually when that happens I knew it was coming because the water supply or some other factor that would increase solids accumulation in the boiler had changed.

As to which valves to operate first; forget the arguments about the valve closest to the boiler, that's seldom the criteria. It depends on the valves. If the two valves are identical Y pattern globe valves then the closest is opened first and closed last so all the erosion is concentrated on the valve furthest from the boiler; however, such arrangements are unusual.

The most common mistake I see is associated with the systems that have one slow opening valve and one quick opening valve. The operators tend to believe the slow opening valve should be opened first and closed last so you can give the boiler a real quick blow with that quick opening valve; after all, that's why they put it on there, right? Nope, the quick opening valve is there because you can open it quickly without anything flowing.

With the slow opening valve you don't produce

sudden changes in flow and you crack the valve to warm up the piping slowly. I can still remember watching one operator whip a valve open to immediately fill cold blowoff piping with hot boiler water. Then I watched the little puffs of steam where the cracks in the piping had formed from repeated shocks of that nature rise up between his legs (he was straddling the blowoff piping). He obviously had no concern for the life of the family jewels. It's bad enough to hit cold piping with 212°F water, let alone water well over 350°F.

Seatless blowoff valves (Figure 2-6) must be operated in a manner based on their arrangement. The piston assembly in the valves creates a void as they're opened and closes one as they're closed. The valve closest to the boiler in this picture is opened last and closed first. The piston in the second valve in line creates a void in the blowoff piping when it's opened, drawing back some air or water, and pushes it out as it's closed. If the second valve were closed first the piston in the valve closest to the boiler would act to compress the water between the two valves as it's closed.

I've watched operators do that, many times using a valve wrench because they had to apply enough force to squeeze the water out the packing or gasketed joint, and they always complained because the valve was so difficult to close. If they're unlucky they'll compress the water and produce pressures so high that the gasket will blow out of the flange and hit them in the head or, more likely, those family jewels.

Figure 2-6. Seatless blowoff valves

So you should only use bottom blowoff valves to remove sediment or drain the boiler and operate them properly so you don't thermal shock them and the piping too much. If you do use them to drain the boiler be sure to close them off once it's drained and you're ready to open the boiler. It's very embarrassing when you blow a lot of dirty water into a boiler you just drained because you forgot to close the valves. It's downright dangerous too. The valve closest to the boiler in Figure 2-6 should be removed and the remaining valve locked and tagged closed before anyone enters the boiler.

CONTINUOUS (SURFACE) BLOWDOWN

Only we old-timers use the term surface blowdown. Both continuous blowdown and surface blowdown are basically the same thing. In the early years of steam power we used a lot of reciprocating steam engines and they required lubrication. As a result lubricant ended up in the condensate and was returned to the boiler. We did have systems to remove the lubricant but there's no reason for discussing them today. Today a reciprocating engine would use Teflon seals that can operate without oil lubricants. We used the surface blowdown to remove the oil and similar contaminants from the surface of the boiler water.

Today, we should always use the term continuous blowdown. The purpose of continuous blowdown is removing concentrations of minerals dissolved in the boiler water. It should be noted that with the advent of chelants and polymers that solids and suspended as well as dissolved minerals are also removed by continuous blowdown. When we were using phosphates to remove dissolved minerals, by converting them to a nonadherent sludge, that sludge had to be removed by bottom blowoff. With current modern chemical water treatment very little bottom blowoff is required.

Why is it connected to the boiler near the surface if it isn't surface blowdown? It's because that's where the steam and water rising in the boiler separated so the water at that point has the highest concentration of solids, more than anywhere else in the boiler.

Unlike bottom blowoff continuous blowdown should be continuous as its name implies. One reason is there are limits to the concentration of TDS (total dissolved solids) in the boiler water, normally based on operating pressure. If we blowdown intermittently then we have to maintain an average concentration of TDS that is lower than the limit. With continuous blowdown we can hold the TDS concentration near the limit. That reduces the amount of blowdown and the amount of energy removed from the boiler in the blowdown water.

Maintaining a continuous blowdown also makes it possible to use the heat in the water removed from the boiler to preheat the makeup water that has to be added to replace water removed by blowdown. That's accomplished with blowdown heat recovery systems. There are two types of blowdown heat recovery systems depending primarily on boiler operating pressure. One simply includes a heat exchanger arranged for counterflow with the continuous blowdown on one side and makeup water on the other. These are only employed with low-pressure steam systems and HTHW. For high-pressure steam systems the blowdown heat recovery equipment consists of a flash tank and a heat exchanger. The high-pressure system blowdown heat recovery occurs in two stages. The high temperature boiler water enters the flash tank which operates at a much lower pressure than the boiler so some of the blowdown water flashes into steam. That steam is used in the deaerator to help preheat the boiler feed water. The remaining water then passes through a heat exchanger to preheat the makeup water.

Automatic continuous blowdown control systems come in two types. The least expensive and more commonly used has a sensor to detect the TDS in the boiler water and opens the blowdown control valve at set time intervals for short period of time to allow the water in the boiler to reach the sensor. Then the valve is held open until the TDS in the boiler drops below the controller setting. This is basically an intermittent controller. The other type modulates the blowdown control valve to maintain the TDS at set point. When the plant is fitted with a continuous blowdown heat recovery system intermittent operation of the blowdown is undesirable. With the intermittent type control the solution is to install a small control valve bypassed by a manually set valve so the Operator can adjust the manual valve to handle the bulk of the blowdown and allow the automatic control to cycle the automatically controlled valve to maintain the TDS level. That produces a constant flow of blowdown with small intermittent increases in the flow to better match a constant flow of makeup. You increase or decrease the constant flow to keep the automatic valve open about 50% of the time.

Normally the operator doesn't have to be concerned with the economics of blowdown heat recovery but this is one system that engineers fail to understand. Several engineers have told me that a blowdown heat recovery system can't be justified economically. The problem is they're only looking at the heat recovery. If you don't have blowdown heat recovery system then you are wasting a valuable resource. With the exception of low temperature hot water heating boilers boiler blowdown is discharged to a flash tank where some of the energy in the high temperature water is used to convert some of the water to steam that is vented to atmosphere. The remaining water, now at 212°F is conveyed to a drain. The Plumbing Code clearly restricts the temperature entering a drain to 140°F so we use high quality drinking water to cool it by dumping more water down the drain. We refer to it as quench water and it's normally high quality drinking water that goes down the drain to sewer along with the blowdown. The typical blowdown heat recovery system cools the blowdown to about 90°F so quenching with fresh water isn't required. I can't tell you how much water costs in your particular area because the cost differs substantially. Where I live in the Baltimore Metropolitan area we're blessed with a rather low cost of around four dollars per thousand gallons of water. That charge is a combination of water and sewer charges and a large proportion of it recovers the expense of processing the sewage. I am aware of jurisdictions where the combined water and sewer cost is over $25 per thousand gallons. Because the engineers don't think about the fact that blowdown heat recovery eliminates that water waste they fail to provide a suitable blowdown heat recovery system in many boiler plants. When wasted drinking water costs are considered I have yet to fail to justify a blowdown heat recovery system. And, because potable water is such a valuable and rapidly depleting resource I insist on having one anyway.

So, don't disparage that system; make sure it's operating properly and, if it isn't, point out to the powers that be how important it is. The amount of blowdown is determined by dividing the average TDS of the boiler feedwater by the average TDS level maintained in the boiler. The result multiplied by 100 is percent blowdown. Determine how much is flashed into steam by dividing the enthalpy of the boiler water less the enthalpy of saturated water at the pressure in the flash tank by the enthalpy of steam at the pressure in the flash tank. Multiply the result of that calculation by 100 to determine percent of flash steam then subtract that value from 100 to determine the percent going down the drain or through a heat exchanger. The heat lost to blowdown is determined by dividing the difference between the boiler water enthalpy and the feedwater enthalpy by the difference between steam enthalpy and feedwater enthalpy. That result times 100 is percent heat in the blowdown. Finally, to determine the quantity of quench water divide the difference between the temperature of the water after flashing and 140°F by the difference between 140°F and the average quench water temperature. The result times 100 is the percent of quench water and it's not unusual for it to be more than 100%.

Sorry, the math isn't done; we need some values to justify blowdown heat recovery or the cost to repair it. The cost is determined by calculating the heat losses and their value then adding the cost of quench water. Since we calculated everything in percent it's easy to figure out the cost of the heat in the blowdown, multiply the annual fuel bill (or your estimate of it if management doesn't have the sense to keep you informed of it) by the heat in blowdown multiplied by boiler efficiency. Divide by 100 each time you use a percentage. If you have a flash tank that recovers the steam multiply that result by the percentage of water after flashing divided by 100. To determine the cost of quench water requires knowledge of how much steam you made. Use meter data or one of the estimating methods described in the section on instrumentation in the chapter on controls. Multiply the pounds of steam made per year by the percent blowdown divided by 100, correct for flash steam used as just described, then multiply by the percent of quench water to determine the pounds per year of quench water. Multiply that value by the cost per pound of water, your combined water and sewage charge.

ANNUAL INSPECTION

The annual inspection is a standard requirement except for some jurisdictions. Either the State or your insurance company will require you arrange for a National Board Commissioned Inspector to inspect your boilers. The very limited number of incidents with boilers can be attributed to that one requirement more than any other. Normally inspection is a maintenance activity but every year you should also have the inspector stop by for an operating inspection. The inspector should visit to observe the boiler in operation and require you demonstrate the operation of certain safety devices.

Used to be the inspector wanted to see those safety valves operate, some still may. To make it possible to test the safety valves you will be asked to temporarily jumper the high pressure safety switch or adjust it to a value above the safety valve settings. If other boilers are on line to carry the load you may also close the boiler's isolating valve(s) so the other boilers and piping systems are not affected. The inspector will also require you connect his, or her, test gage to the connection adjacent to the boiler's pressure gage; the inspector's gauge connection is required by code.

The boiler is then operated in manual control to raise the steam pressure until the safety valve lifts or the inspector refuses to let the pressure go higher, or you do. If the boiler is larger than 100 horsepower it will have two safety valves and the inspector can ask you to break the valve seals of the valve with the lower setting and gag it shut so the higher set valve can be tested. After the higher set valve operates you remove the gag and the inspector replaces the valve seals. The code requires the valves open within a certain percentage of the pressure their nameplate indicates. If one of the valves fail the test the inspector will require it be sent out for repair or be replaced.

Notice I said "used to be." To reduce their costs many insurance companies have changed their requirements to reduce the amount of time an inspector is on site. It takes some time to set up the boiler, raise the pressure, and let it fall. In some cases they'll accept a lift test (see maintenance) of the safety valves. Many insurance companies are now simply requiring the valves be sent out to an authorized shop for rebuilding at five year intervals.

An authorized shop would be one that has received authorization from the National Board to use the "VR" (for valve repair) symbol stamp issued by the National Board. However, manufacturers who hold an ASME Certificate of Authorization "V," "HV," or "UV" (depending on the valve) Code Symbol Stamp can also rebuild safety valves.

I'm not suggesting you accept those changes. If your insurance company will not let the inspector observe actual lift tests and reseal the valves then suggest to your employer he get another insurance company. Rebuilding safety valves isn't an inexpensive proposition and an owner typically ends up buying a spare set to switch out because the rebuild takes several days. I have one customer that simply buys new valves because they cost less than rebuilding. It's simply false economy again, save some time for an inspector and spend much more than the inspector's time on new safety valves and rebuilding.

I believe the trend is apparent and indicates that the slack in testing of safeties is allowing more incidents. 2002 data[2] show slightly more than 2% of boiler and pressure vessel "incidents" could be attributed to failure of a safety valve. That's more than twice what it used to be. Pop tests of safety valves should be performed every year. There's no guarantee that they will pop when they should just because you can lift them.

After the safety valves are tested you should remove the jumper or reset the high pressure switch then demonstrate that it opens at or near its setting and below the set pressure of the safety valves.

The inspector should also expect you to demonstrate a functional test of the low water cutoff, either by an evaporation test or a "slow drain" test. The evaporation test consists of boiler operation with the boiler feed pump off or feedwater control valve closed so no water is fed to the boiler. As the water evaporates the level drops until the low water cutoff shuts the burner off. A slow drain test is used when there is little or no steam demand. The blowdown valves are opened to drain the boiler slowly until the low water cutoff functions.

When performing these tests you should not take your eye off the gage glass or have someone else watch it. Fully one third of all boiler failures are due to low water condition according to National Board data. That means those low water cutoffs fail; that's why you're performing a functional test of each one.

The inspector can also require you demonstrate the function of other safety interlocks. Specific tests are required by code depending on the size of your boiler and State laws can include other requirements. ASME CSD-1 has a checklist requirement. NFPA-85 contains a list of mandatory tests. The National Board promoted adoption of those Standards in the mid 1990's and most jurisdictions have adopted them. You will find, however, that not all inspectors are up to speed on those Standards.

In many cases the inspector will simply require you show you have documented evidence that you conducted the tests. As far as I know the National Board has not added a requirement in the inspection code that says the inspectors have to observe any of those tests.

I have every respect for anyone who carries a commission as a National Board Inspector. However, I'll use an old saying that those of you that also grew up on a farm will understand; "There's a rotten apple in every bunch." There are inspectors that will sit at their home office and fill out inspection reports. There are those that will come to the plant but, other than walking past them, never really look at the boilers. They'll spend all their time in the chief's office drinking coffee and talking. If you have one of them, quietly report what you observed to the chief boiler inspector of the state or commonwealth.

I know the feeling such a suggestion provokes—it's none of my business; we keep our boilers up so it doesn't matter; I like the guy and don't want to get him in trouble; I might be found out and lose my job... Think of it another way. Think of the people that are going to

be injured by the failure of another boiler that inspector doesn't properly inspect. Imagine that boiler is in the building where your children go to school! It's a subject near and dear to my heart because there isn't enough monitoring of inspectors. I have hard and unpleasant experience with such situations. I know of one little girl that was severely burned and... That's all I'll say on a subject I could rant on for another ten pages but I won't. I'll just trust you to do the right thing.

Testing of safety valves and inspection of the boilers by inspectors is essential in reducing our exposure to a boiler failure. We certainly don't want to return to conditions that existed in the first decade of the last century when millions were injured and thousands died from boiler failures.

It's the benefit of a third-party inspection with no responsibility to the owner of the boiler that makes the system as good as it is. Every boiler inspector is well trained and tested before receiving a commission as a National Board Inspector. You should take advantage of their training and skills during every inspection, calling their attention to changes or conditions that you question. Never treat them as someone you have to hide things from. That's exposing yourself. After all, who's going to be closest to that boiler if it does explode?

OPERATING DURING
MAINTENANCE AND REPAIR

You have some additional duties when a contractor or other employees are working in the plant on maintenance or repair activities. Concerns are protecting the health and welfare of those workers, making certain they don't do damage to the plant, and making certain they don't disrupt normal operations inadvertently.

You may be required to start and secure boilers to provide access for the workmen to the equipment or parts of the plant. It can be as simple as operating to reduce temperatures where they are working above a boiler. It could also be as complicated as generating steam required for the contractor's operations. It isn't uncommon to isolate sections of piping for work. Whatever the activity and regardless of who does the work the operator should be the final authority for accessing any system and that should be made perfectly clear to anyone that enters the plant.

Frequently the chief or manager of the boiler plant takes the attitude that an operator should have no authority over contractors working in the plant. If that happens with you it's an indication of a lack of trust in your skill but can also be an indication that the chief can't relinquish authority appropriately. You should go to that superior and explain that you are not comfortable operating a plant when others can do things without your knowledge and consent. Make certain he or she understands that your interest is in the safe operation of the plant and they should make certain the contractor works with your approval.

That's not an excuse to be dictatorial and unwavering. I've known operators that seemed to enjoy the power they had over contractors and saw to it that they didn't interrupt the operator's schedule, regardless. If the owner is paying the contractor to work on a time and material basis the contractor won't complain a bit. Every minute the contractor's employees stand around waiting for you to give them approval or shut down a system simply means more time and more profit for the contractor. Treat every one of them as if they were working on time and material.

Probably the most difficult thing for the operator to remember during these periods is the requirement that everything done is recorded in the log. In the unlikely, but probable, revelation of problems later—either as a result of the workmen's activities or because they failed to do something—the log provides a documented history of the work for reference. Believe it or not, I served as an expert witness for a customer whose boiler operators failed to record a contractor blew up each of their new boilers on two different days. I do hope you're not that lax in maintaining your log.

There's frequently an air of distrust between boiler plant operators and contractors working in the plant. Without going into the reasons for it, because I don't understand it anyway, I just want to mention that a log entry that reveals that distrust through nonspecific statements or general comments will not satisfy the requirements of a court. An owner whose operator made entries like "contractor XYZ is breaking everything" and "the stupid contractor broke it" couldn't get the jury to accept it. The jury couldn't get past the implication that the operator logged an opinion rather than fact.

All log entries regarding a contractor's activities should be factual and devoid of comment. Log entries should indicate what was done, who did it, and when it was done, nothing more. It's very important you do it because there may be nobody else there to see it—forcing a later conclusion that what you're testifying happened, without a log entry, may be nothing but your imagination. I know one time a simple seven word entry "Cliff working on Boiler 3 control panel" later proved to recover a rather expensive burner management chassis

that Cliff had simply removed and taken with him.

Whenever possible there should be checklists prepared for any repair or maintenance work in the plant. That's so it can be consistent with normal operating procedures. Otherwise what is a normal activity could be made unsafe. Many a contractor has decided a line has no pressure or contents and started working on it without realizing it could suddenly be filled with boiler water (bottom blowoff).

That also provokes the thought that operating procedures may have to be changed to accommodate work in the plant. Despite the fact that the blowoff lines should be locked out and tagged out when working on them people make mistakes or bad assumptions. A notice for the day regarding operation of bottom blowoff should also be prepared by the chief or maintenance manager so operators know the piping will be worked on.

When contractors are working in the plant you should be in relatively constant observation of their activities. You can't fail to enforce the owner's safety rules and regulations, informing the contractor when the rules are violated and reporting any refusal to comply. If a contractor's employee is injured as the result of a hazard addressed by the safety rules and that employee was not informed of the rules the owner could be found liable for the person's injuries.

Make sure safety rules are complied with but don't help the contractor comply. The contractor should do confined space testing before contractor's employees enter a confined space. The contractor should perform the lock-out tag-out so all you should have to do is add your lock when everything is proven out.

The best projects for repair, retrofit, or maintenance in a plant by a contractor exist when the operator and contractor work together. By preparing a schedule and working to it you will help the contractor get done and get out of your plant as soon as possible. When several people are in a plant and their goals differ that situation produces many opportunities for things to go wrong. If the contractor and operator share a goal of limiting interference to plant operation and getting the work done readily and quickly then there is less likelihood of problems cropping up. Remember what I said back in that first chapter on priorities.

CODE REPAIRS

I'm sure you're skilled at repairing several things. Based on my own experience I believe you have to be knowledgeable and capable of many repairs yourself.

That doesn't mean that you have the credentials that are required to perform certain repairs. Even what are called "repairs of a routine nature" which appear to require little skill are still required by Law (because your State, Commonwealth, or Province, has adopted the National Board Inspection Code) that not only addresses the requirements for inspection of your boilers but includes requirements for the repair of your boilers. Any repair within the environment of a boiler or pressure vessel requires the work be performed by an organization (which can be a one man shop) that has Authorization to use the National Board "R" Symbol Stamp or ASME Authorization to use the applicable Code Symbol Stamp. Without those credentials you're not qualified to perform the repair. You should also be wary of Contractors that may have convinced some purchasing agent that they can do the job cheaper but don't have those credentials. It costs a considerable amount of money to satisfy the National Board Inspectors that review firms to determine if they are qualified to do the work and that expense has to be included in the cost of the work. The firms have to repeat qualification every three years so it doesn't matter how long they've been in the business. If the Contractor can't produce the Code Symbol Stamp or a copy of their authorization to use it they shouldn't be working on your equipment either.

You should be impressed by the skill of the welders and boilermakers that work on your boilers and pressure vessels. Once you've seen the result of a window weld to replace a blistered or bulged boiler tube in a watertube boiler you'll be convinced that they're the right people for the job. If, however, it looks sloppy, inconsistent with anything you've seen before, or just plain wrong in your mind ask to see their Authorization or the Code Symbol Stamp and report them to the State, Province, or Commonwealth's Chief Boiler Inspector if they can't produce it. That should get rid of any unqualified contractor. I think the days of Contractors buying their authorizations is long gone, at least for now, so you shouldn't have to accept shoddy or inadequate repairs.

Maybe you don't know what to look for in terms of a good weld. What you can do to inspect welding is described in the chapter on maintenance.

PRESSURE TESTING

The most catastrophic incidents within a boiler plant are due to sudden releases of steam and water under pressure. To help ensure the equipment, piping, etc. is capable of operating without rupture, regular pressure testing is performed. Pressure testing is normally limited to hydrostatic testing but that's not always possible. The procedures you use should be consistent to ensure the systems are safe for operation under pressure and not damaged while pressure testing. I'll cover hydrostatic testing first, because it's common and preferred.

As with filling there should be a person assigned to control the pump or valve that is pressurizing the system. Be as certain as possible that you have removed all air from the system. A system is usually air free if the water pressure increases rapidly once everything is closed. If the pressure doesn't jump to city or system pump pressure there may be air. Once you've started the hydro pump look at the gage. If pressure isn't jumping up with each cycle of the pump then there's still air in it; get it out. If the system ruptures with compressed air in it the air and water will pass out through the point of failure with dramatic force.

Hydrostatic tests should be conducted with water between 70°F and 120°F for reasons of safety, that temperature range is also required by code. Normally hydrostatic test pressure is 150% of the maximum allowable pressure or the setting of the safety valves.

Newer equipment is manufactured with steel at higher rated pressures than when I was designing their installation and, because higher stresses are allowed, the requirements for a Hydrostatic test pressure may be 125% of the design pressure instead of 150%. There's still a lot of equipment out there that was tested at 150% and I still insist that it be tested at that pressure after any repair or for a preventative maintenance check. When questioned by the repair or maintenance crew I always use the same answer, "The boiler / pressure vessel had to pass a 150% hydro when it was made. Don't you want to prove that it is as good as new?" And, when discussing it with a repair contractor I always add "You did repair it so it's as good as new, right?"

Of course you can't just apply 150% test pressure to a system without concern for what's attached to it. Many pressure switches, transmitters, etc., can't withstand the hydrostatic test pressure so they have to be disconnected. That includes some thermal wells and temperature switches and sensors so be certain they're okay or remove them for the test. It's all that cumbersome removing stuff and putting it back that many contractors wish to avoid so they'll try to get away with a lower test pressure.

Many times even boiler inspectors permit testing at normal operating pressures but I consider that foolish because the system can fail and allow pressure to reach the settings of the safety valves plus the valves can stick a little resulting in higher pressures. We tested a large number of compressed air storage tanks for an installation in the 1980's at the request of their inspectors. It's a good thing we did it hydrostatically. Eleven of them failed, four at pressures below the safety setting and one just slightly above normal operating pressure.

A hydrostatic test, done properly, will not result in injury if the containment fails; a little water will run out and the pressure will drop instantly. A boiler in operation doesn't fail that pleasantly. Which would you rather have, a rupture (consisting of a leak of cool water) due to a hydrostatic test and when you're looking for it or an explosion of steam and boiling hot water (or worse) when you least expected it? Testing at anything less than the standard test pressure is providing false hope that the containment won't fail in operation.

That's why I said "help ensure" back in that first paragraph. Pressure testing a vessel at 150% of its maximum allowable working pressure still doesn't mean it can't fail at lower pressures. During operation temperatures of boilers and many pressure vessels are substantially higher than the maximum hydrostatic test temperature. Those higher temperatures introduce additional stress into the vessel and can contribute to failure of one that just passed a 150% hydro. It's even more likely to fail in service if it passed a hydro at normal operating pressures.

One reason you always have somebody at the pump or valve controlling the application of pressure is to release it immediately if a problem is detected. Another is to make certain that the pressure doesn't exceed the chosen test pressure. If a manufacturer (who has to test at 150%) exceeds the test pressure by more than 6% the engineering of the vessel must be repeated to ensure it was not subjected to excessive stress during the hydro. There's no excuse for letting the test pressure run above the 150% so don't do it. Ensure the pressure in the system never exceeds the specified test pressure by more than 6%. If it does, note it in the log and notify the manufacturer to determine if any damage was done by exceeding the test pressure.

Check electrical circuits that are connected to the systems during hydrostatic tests to ensure the liquid did not introduce an undesirable ground. Check them again after all test apparatus is removed and normal connections reinstated.

When testing is performed pneumatically (with air) the test pressure should not exceed 125% of maximum allowable working pressure. Also, the pressure must be increased in steps with inspections for leaks at each step. The rapid expansion of the air in the event the

vessel ruptures could do serious damage. That's why flooding a vessel with water for a hydrostatic test is so important, the water pressure will drop instantly with a rupture but any air in the system will expand to push the water out with considerable force.

A sound test requires the source of pressure be disconnected and the pressure observed for a period of time to ensure there are no leaks. Occasionally the pressure will increase or decrease as the testing fluid heats or cools. If leaks are found, drain the system for repair and repeat the test when the repairs are complete. Note that any air test requires precautions and should only be used when there's no option.

A special test not normally performed is a boiler casing test. It ensures there are no significant leaks of the products of combustion from the boiler into the boiler room. The test requires blocking the stack, preferably at a point outside the boiler room, and the burner opening into the boiler. The actual test pressure should not exceed the manufacturer's rating for the casing or any ductwork connected to the boiler that is also included in the test. The best way to apply pressure is using the test setup shown in the Figure 2-7 which, by it's construction, serves as a gauge for the test and a way to prevent exceeding the test pressure. When some bubbles rise through the loop the test pressure is achieved. Once the pressure is reached the air supply is disconnected and the level drop observed. It shouldn't drop more than one inch per minute after bubbles stop rising through the column. If leaks are indicated drop the pressure, insert a lit smoke bomb through the capped connection, and reinstate test pressure to locate the leak. Reduce air line to 1/2 inch and other piping to match the size of the observation port if it is less than two inches.

Note that the 25-inch water leg is selected for boilers designed for a maximum casing pressure of 25 inches water column. Many are only capable of 10 inches so the leg should be shorter. I normally specify a 25-inch pressure rating and that's why the system in Figure 2-7 shows it.

The application of a smoke bomb is necessary to spot leaks in the casing. It's normally done for a replacement casing job and I have yet to see one done where a couple of smoke spurts didn't point out a spot where the boilermaker missed a little stretch of casing weld.

This test only applies to boilers with casings designed to operate under pressure. A person should remain, hand on air valve, at the test apparatus whenever the compressed air connection is open. Be certain to remove blanks and any combustible sealing material (caulking) when the test is completed.

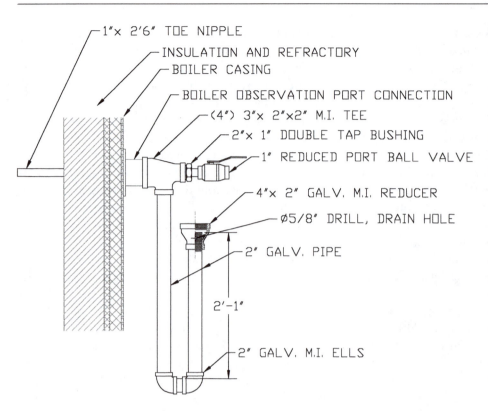

Figure 2-7. Casing test assembly

LAY-UP

When a boiler will not be used for an extended period of time (more than a week or so) it is important for operators to be certain that boiler is maintained in such a manner to prevent corrosion or other damage while the boiler is inactive. The operating activities that prepare the boiler for an extended period of inactivity is called laying it up. There are two means of boiler lay-up, dry and wet; as the names imply, it depends on whether the boiler contains water or is drained.

Wet lay-up is the common method because it is used for short term lay-up and does not require as much preparation to put the boiler into lay-up and restore it to operating condition. The first step in laying up a boiler is to shut it down and allow it to cool completely. During the cool down period some circulation of boiler water occurs and it's the best time to measure boiler chemistry and establish water conditions for lay-up. The sulfite content of the water should be doubled compared to normal (60 ppm vs 30 ppm) and alkalinity raised to the maximum value (pH of 11) so the boiler internals will be protected from corrosion. During a short term lay-up the only other provision that is made is raising the water level to the top of the drum to minimize the internal surfaces that are exposed to air.

For a long-term wet lay-up the entire boiler drum

should be protected from contact with air so it should be flooded. I recommend the installation of an expansion tank on the boiler to maintain a flooded condition. The expansion of the water can be determined from values in the steam tables, the difference between dry and flooded weight of the boiler, and the normal range of boiler plant temperatures (40°F to 135°F) to size the expansion tank. A tank with a capacity of 3% of the boiler (in gallons) should work in most situations. Best is a bladder tank connected to a branch connection off the boiler vent with another vent valve to bleed water when chemicals have to be added or the pressure adjusted. Starting with a tank drained of air until there's no pressure over the bladder will allow the pressure in the boiler to raise to 15 psig when the tank is half full. An alternative method is to install a bucket on a pipe nipple, set it up on the vent valve and add or remove water to maintain the level in the bucket. Water should be added by introducing additional sulfite using the chemical pump and maintaining 60 to 120 ppm in the chemical pump's storage tank.

Long-term wet lay-up requires addressing the condition of the fireside of the boiler. When it will be down for more than a month it's advisable to seal the stack or block the boiler breeching at a point inside the boiler room. The daily swing in temperature and humidity can produce conditions that promote condensation of water in atmospheric air on the surfaces of the boiler because the water and steel are colder at some times. By restricting air flow you reduce the potential for condensation but you don't eliminate it.

Once the air in the boiler is confined you can use silica gel as explained for dry lay-up or simply add a little heat with lights or a short length of tubing using condensate, blowdown water, or steam to raise the temperature of the air in the boiler to a couple of degrees above the water temperature so it's never condensing on the surfaces.

Dry lay-up, as the name implies, is achieved by draining the boiler. It is not that simple however. Left exposed to air and the varying temperature and humidity around a boiler plant there will be significant deterioration of the boiler's interior unless protected. To prevent corrosion the boiler should be free of moisture. After the boiler is drained all drain valves should be closed, the drum covers or inspection openings opened and dry air blown through the boiler to remove any remaining moisture. Checking the exhaust air with a hygrometer until the humidity in the boiler is less than 10% or 5% above the humidity of the drying air is recommended. Then, insert a package of silica gel with a corrosion proof drain pan under it and close the boiler completely. The air will simply compress and expand in the boiler as it heats and cools so there is no reason to install an expansion tank. The silica gel must be checked twice a year to ensure it is active. Any moisture found in the drain pan should be removed.

The fire sides of the boiler have to be considered for long-term lay-up. The connection to the stack and the combustion air inlets should be blocked off. The enclosed spaces should be dried and maintained with a silica gel dryer as described above.

Normally boiler control panels, motor starters, etc. can be maintained by simply leaving the power on the panels. The indicating lights in the panels should supply sufficient heat to lower the internal humidity and prevent corrosion from moisture. If the panels are exposed to the weather addition of some light bulbs inside to lower the humidity is recommended. Wiring two 100 watt lights in series will produce about 25 watts of heat but the likelihood of one of the bulbs failing is very low. Add lights to panels that do not have any. Motors for combustion air fans, boiler water feed or circulating pumps can be heated by applying reduced voltage to the windings or using heaters that are supplied for such a purpose.

Always refer to the manufacturer's instruction manuals for suggestions or requirements for lay-up. Regardless of how the boiler is laid up its condition must be monitored on a regular basis, preferably weekly, to ensure it is not deteriorating. All you're normally doing is making sure the seals are intact (nobody opened it and left it) and there's no external signs of corrosion or other problems. Initially test water during wet lay-up on a weekly basis to ensure it has sufficient sulfite to remove any oxygen. When testing reveals consistent retention of the sulfite extend the testing interval until a period with a drop of 10 ppm is experienced and incorporate that testing interval in your SOPs. Check silica gel inside the furnace on a monthly basis and inside the boiler on a semi-annual basis.

All too often I've seen a boiler abandoned to the ravages of weather, etc., simply because the plant had no need of it. Later, when they attempted to sell it, the condition was so bad they couldn't and their only option for removing it was to pay for its removal. Even if you don't need the equipment, preserve it. Someone may need it and, if it's in good shape, the owner will get enough for it to pay for its removal. Otherwise you may be looking at that rusting hulk until the day you retire.

When the whole plant is put in lay-up these guidelines can be extended to other equipment. Special consideration should be given to a long-term lay-up. Valves and Pumps with packing should have the packing removed and replaced with fresh material heavy in graphite. Packing that was in use and allowed to dry will harden and be almost impossible to remove later. Pumps that have mechanical seals can't be reliably preserved but you could try disassembling the seals, coating the sealing surfaces with a mineral oil and reassembling them. Pumps containing oil and such materials that lubricate without freezing can simply be isolated after filling with liquid that is confirmed water free and not prone to form acids while stagnant. Pumps containing water should be drained completely, close their supply and discharge valves, then use the vent and drain connections to blow dry air through them and dry them completely before isolating.

TUNE-UPS

While I prefer the title "tune-up" the PhD's at the EPA who like to use longer words have chosen to use the title "Combustion Optimization" for the same thing. This edition is revised to include a section for Combustion Optimization in the Chapter on maintenance because it's a normal maintenance function. It's included in the startup of a new boiler and operation of the plant during a tune-up is covered here.

Along the east coast of the US I've found that it's uncommon for a boiler operator to be expected to perform the tune-up of a boiler. A few plants do their own tune-ups but use other personnel with labels like Instrument Technician to do the work. Rarely is it done by a licensed boiler operator. Tune-ups should be performed on an annual basis and whenever there's reason to believe the controls are out of tune and it is always the boiler operator's role to identify a problem with the controls that require a tune-up.

Another factor in tune-ups are the requirements of the local environmental office, whoever is responsible for enforcing the Clean Air Act. Many states now require a tune-up be performed each year. That is, however, not as frequent as I believe they should be done. I've documented many cases where performance of a tune-up as soon as evidence of mis-operation exists will pay for itself in as little as a couple of weeks. The larger your plant is, basically the more fuel you burn, the sooner a tune-up will pay for itself. The important thing is that the operator monitor operation to determine when it's needed independent of regular intervals.

Sometimes the evidence is rather apparent, smoke pouring out the stack or frequent flame failures, but that's the extreme and an operator should detect problems long before it gets that bad.

I personally believe a plant should use a contractor for tune-ups because the contractor's employees are doing the job at a higher frequency so their equipment is maintained in calibration, their skill level is higher, and they aren't distracted by other things going on in the boiler plant. A contractor can afford to invest in high tech equipment for tune-ups when doing several a month.

That same equipment is too expensive for a plant that only needs to use it once or twice a year. That doesn't mean that a contractor is always the best option. I've also encountered many situations where the contractor considers the tune-ups as fill-in jobs and pulls the employee regularly to handle emergencies so the tune-up loses the continuity that's required to ensure it's done properly. The single biggest problem with operators doing tune-ups is they get pulled away to handle other situations and if the contractor's operation is the same that's a disadvantage to using that contractor.

Is a tune-up necessary right now? That's a question a boiler operator has to ask whenever plant operating conditions indicate it. Monitoring of evaporation rate or heat rate and other conditions can indicate a tune-up is necessary. Of course the operator has to be aware of situations that can create a problem that could be wrongly attributed to controls (like blocking of plant air entrances) and correct them first.

Something coming loose and shifting position from vibration or for other reasons should also be sought out before committing to a tune-up. An employer will get very upset if the cost of a tune-up is revealed to be something other than a problem with the controls. I remember one chief that got peeved when he discovered the operator called for a regular tune-up just because he got lonely and wanted the company of the contractor's technician.

A number of things must be considered in association with a boiler tune-up and some of them are best accomplished by the operator. To tune a boiler it's necessary to create stable firing conditions for at least a short period of time so the technician can collect data that are all relative to that firing rate. This can mean anything from operating the subject boiler in manual, while using another to handle load, to controlling steam dumped to atmosphere to produce a constant load.

An operator can be so involved in simply maintaining the firing condition that there's not time to collect the data and that's another reason for using a contractor. In many cases there are problems creating the load conditions for tune-ups because there isn't enough load. Wasting steam may seem like a logical solution but if the plant normally operates with high condensate returns wasting steam may be impossible because the water pretreatment system can't produce enough water to waste as steam. That's why, in some cases, boiler tune-ups are restricted to the winter.

When a boiler is tuned up in the summer the data and adjustments at high fire may be made by temporarily running the firing rate up to grab readings which isn't the same as establishing a stable condition so performance at those rates may be a lot different than the report indicates. A boiler plant log should always include a note to the effect that a tune-up was achieved by grabbing readings so the assumption that it was a normal load tune-up is not made.

I will argue that it isn't necessarily important to fire a boiler at or near full load to tune it up with a full metering combustion control system. When properly configured a full metering system can be set up with a few readings, preferably at loads to at least 50% of maximum firing rate because the variables associated with load are corrected for by the system with one single exception assuring that the maximum firing rate according to air flow is.

A final note on tune-ups. They are not a final fix. As the boiler continues to operate the linkage, fan wheel, and everything else is subjected to friction and wear. With jackshaft type parallel positioning controls everything in the plant can alter the burner's air to fuel ratio.

I've been told that all you have to do is to repeat a tune-up every year, whether it needs it or not, and you find your readings are still the same. If you do that, give me a call, I want to see that boiler! It's always possible that something can slip, wear, or change in some manner during normal operation and you'll have to repeat the tuning process to restore efficient and clean firing before the year is up. When that happens it's best to treat the time between tune-ups as the required interval unless a couple of repeat runs prove that one time was a fluke and you can go back to annual tune-ups or whatever interval your equipment sets for you.

AUXILIARY TURBINE OPERATION

Contrary to popular belief auxiliary turbines are not there just in case you lose electric power. I frequently hear an operator complain that the turbine driven auxiliaries are a waste of time because they would lose everything on a power outage anyway. While it's true that an auxiliary turbine will operate without electricity their more important function is reducing operating cost while contributing to the heat balance of the plant.

The auxiliary turbines are an optional source of power and the wise operator will make best use of them because, operated properly under the right conditions they can reduce the cost of powering the auxiliary equipment by about 75%. I should also note that, if you run an auxiliary turbine under the wrong conditions you can increase the cost of powering the equipment by 1000%.

There's no easier way I know of to get rid of a new boss that doesn't know anything about boiler plants and proves to be intolerable. I'm not suggesting you operate auxiliary turbines improperly to bump up operating costs and get rid of a boss, but it is one trick I've seen used.

There's that term again, exactly what is a heat balance? In it's truest sense a heat balance is the result of calculations that determine exactly where heat goes in a boiler plant with the balance meaning heat out equals heat in. The more common reference is the balance of heat into and out of a deaerator which could leave a lot of you out when you don't have a deaerator.

If you have a sparge line in a boiler feed tank and heat the boiler feedwater by injecting steam into that line you're operating with something similar but seldom use enough steam in that feed tank to effectively run a turbine.

Maintaining a heat balance is operating a deaerator and auxiliary turbines to get the most efficient use out of the steam going to the deaerator. When steam flows through an auxiliary turbine some energy is extracted from it to drive the pump, fan, or other auxiliary device. The exhaust steam then flows to the deaerator where it is used to preheat and deaerate the boiler feedwater. That steam condenses as it mixes with the feedwater delivering virtually all the heat left in it to the feedwater which is then fed to the boiler.

For all practical purposes (by ignoring the little bit of heat lost from the piping and equipment through the insulation) all the energy in deaerator steam is recovered and returned to the boiler. If it happens to flow through a turbine on its way to the deaerator and produce a little power, the cost of generating the power is only the little bit of heat lost by the steam as it passes through the turbine.

When compared to the typical electrical utility plant where 60% of the heat from fuel ends up lost, your auxiliary turbines are super efficient. Despite their economies of scale, burning cheap coal, etc., the utility can't make power as inexpensively as you can with auxiliary turbines. That's why you can typically power a piece of auxiliary equipment for one fourth of the cost of doing it with an electric motor.

If, on the other hand, you run too many auxiliary turbines so you're dumping steam out the multiport (relief valve) to atmosphere you're wasting all the energy that should have gone to the deaerator and it costs more than ten times as much as electricity. The trick is to operate the turbines so you're putting as much as possible through the turbine without pushing any out the multiport.

The best auxiliary turbines to use are boiler feed pump turbines. They require power proportional to feedwater requirements and deaerator steam is proportional to feedwater requirements. Forced and induced draft fans are second best. Regrettably turbines don't use steam proportional to their power output, they need a certain amount of steam to overcome friction and windage (like fighting the wind, it's losses associated with the rotor of a turbine whirling in the steam) so the steam consumption of an auxiliary turbine isn't perfectly proportional to its power output.

There is a reasonable degree of proportionality that is evident when you look at the Willians line for a particular turbine. The Willians line is a line on a piece of graph paper that shows the relationship of steam consumption to turbine power output and it looks something like that shown in Figure 2-8. Since there is a fixed amount of energy needed just to keep it spinning there's some point where the turbine's steam requirement per gallon of boiler feedwater pumped exceeds the requirement for heating steam at the deaerator. When operating a feed pump turbine below that point some of the steam is wasted, when operating above that point the deaerator needs more steam than the pump does.

Your basic task is to determine the boiler load closest to that point then operate one or more auxiliary turbines accordingly; run the turbine whenever you can without wasting steam. If you have more than one turbine driven feed pump you have to determine the boiler load above which you can run two turbines. If the turbine drives are of different sizes and there are some for other services (like condensate pumping or driving fans) you have to learn how to juggle them for making the most use of the auxiliary steam going to the deaerator.

When you do have many auxiliary turbines of different sizes using the Willians lines in their instruction manuals will help you determine ways to mix them for maximum utilization. When you have an option of changing turbine nozzles (note the two lines in Figure 2-8) you determine when the extra nozzles are needed by when the turbine seems to be inadequate to power the pump. Note the feedwater flow or steam flow when that occurs so you can determine when to adjust turbine nozzles.

Boiler feed pump turbines can help maintain the heat balance because they're equipped with controls. These vary from constant speed controllers which will vary steam usage as the water flows change to special control loops for maintaining a constant feedwater pressure or constant differential between feedwater and steam headers. As the boiler load increases the pump horsepower has to increase to pump more water. The increased load will tend to slow down a speed regulated turbine so the controls open the steam valve more to restore the speed. Similarly the steam supply to the turbine is increased to maintain feedwater header pressure or water to steam differential as load increases.

Very large auxilliary turbines may actually have control linkage that opens and closes turbine nozzles. Those systems will open one nozzle control valve entirely before starting to open the next so only a small quantity of steam is throttled. That increases the efficiency of the turbine and improves the ratio of feedwater to turbine steam demand.

The steps in starting up and shutting down auxiliary turbines are all pretty much the same. The first task is deciding which one to start. You then set up it's driven equipment the same way you would in preparation for starting one powered by a motor. The turbine casing vents and drains should be open but check that they are. Check oil levels in the turbine bearings or sump, any reduction gear, and on the driven equipment. If the turbine is fitted with an electric motor driven lubricating

oil pump start it to start oil circulating through the bearings. If it's possible to get at the shaft, rotate the shaft a quarter turn every five minutes while it's warming up to help ensure uniform heating.

Damage to auxiliary turbines is normally due to alignment problems associated with thermal imbalance so take your time to ensure the casing and rotor are uniformly heated. Large auxiliary turbines can have some very thick metal parts, especially around the nozzle blocks and shaft seals so the larger the turbine, the more time you give it to warm up.

When a bypass is provided on the exhaust valve crack it to start warming up the casing. Admit only enough to get steam at the vent then throttle down the vent so the air is pushed out the drains. If you don't have a bypass then crack the exhaust valve. Leave the vent open enough to dispel air that's heated by the steam. Don't leave it wide open. With a wisp of steam coming out there should be enough pressure to push air out the drains. The steam from a typical 100 to 150 psig supply (or higher) is about half the density of air when dropped to atmospheric pressure. It's so light that you need some push to force the air out the turbine casing drains.

Since most auxiliary turbines operate with exhaust pressures of 15 psig or less the steam will always be less dense than the air. You want to be certain the entire casing is flooded with steam so the rotor and casing are heated uniformly. As the casing warms less steam will be used to heat it up so the drains will begin blowing more and more steam. Throttle the drain valves to limit steam waste but be sure to keep them open enough to drain all the condensate.

When there's little to no condensate evident at the drain valves open the exhaust valve; at this point the steam has nowhere to go and isn't condensing so the casing pressure should be close to exhaust line pressure.

Open the drain valve above the steam supply shutoff valve to drain any accumulated condensate above the isolating valve then throttle it until you're primarily draining condensate. If there's a bypass on the steam supply valve crack it to bring steam up to the trip valve. Once the supply line is dry, open the supply valve. If there is no drain at the trip valve body don't open the supply until you're ready to start rolling the turbine. While that supply piping is warming up open outlet then inlet valves of any turbine bearing coolers, throttling the inlets if the coolers are lacking temperature controls.

I've received many complaints that my timing is off here because heating up the casing will heat up the oil in the bearings. That's true, and I want it to. If you open the cooling water to the bearings first the oil may

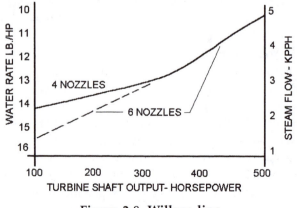

Figure 2-8. Willans line

still be colder than design operating temperature when you start rolling the turbine over and you may have insufficient lubrication because the oil is too cold. By using the casing heating to warm up the oil you ensure it's at the right temperature for operation before you start rolling the turbine. It's the kind of consideration you need to include in your SOPs but I've never run into a turbine that overheated oil while warming up the casing.

Once you're certain the casing and the steam supply piping is warm and dry and the oil is up to operating temperature it's time to start rolling the turbine. Sometimes you have to run the trip valve down (turn it as if closing it) because someone tripped it earlier and didn't reset it. If the valve doesn't seem to be opening try that first; there's a spring loaded trip mechanism that shuts the valve by releasing the yoke screw and you have to turn the valve as if to close it until the trip mechanism is reset.

Open the supply shut-off valve. Crack the trip valve and continue slowly opening it until the turbine starts to turn over. The minute you see the shaft start moving stop opening the valve and close it back down to maintain a slow rotation of the turbine.

If your ears suddenly hurt because of a loud screeching noise shut the trip immediately and back up in the start-up process because you forgot to open the exhaust valve or there's another valve in the exhaust piping that's closed or throttled. Auxiliary turbines are equipped with what we call a sentinel valve. It's expensive to put a full capacity relief valve on every turbine casing in case someone forgets to open the exhaust valves so sentinel valves are used. They're like a relief valve but they don't have much capacity; they just let enough steam out to make one loud squeal that's designed to shake up any operator that forgot to open all the exhaust valves.

This is prior to the most critical stage of auxiliary turbine operation and where things can go very wrong so it's important to take your time and allow the turbine to gently roll over for a while. You've just started steam flowing in the exhaust piping and any pockets of condensate should be slowly flushed out during this time. If there are known areas where the piping may have pockets of condensate and they're equipped with drain valves those valves should still be open.

If there's a reduction gear between the turbine and driven equipment you want to give it time to warm up and get the oil properly distributed over the gears and bearings. Some will have heaters to keep the oil hot enough while the turbine is down, some will have coolers, and some have nothing but a sump full of oil. Let the turbine roll slowly until all the temperatures are in the normal operating range and you're absolutely certain you don't hear any screeching, bumping, or grinding in the whole assembly. It doesn't hurt to use the screwdriver at the casing with handle in your ear trick to listen for any unusual sounds while a turbine is slowly rolling over. Open valves for cooling water to any reduction gear or oil coolers on the driven equipment.

The final step before bringing the turbine up to speed is checking the trip. Normally there is some linkage between the turbine and the trip valve and all you have to do is push gently on the lever closest to the turbine shaft to trip the valve. Some turbines will have a means to manually operate the trip. Make sure it works then reset and open it again to restore normal rolling.

When you're satisfied that the turbine is rolling over without problems and the overspeed trip should work you can start bringing it up to speed. First make certain that you and anyone around you are not in line with the rotor. If it flies apart and pieces penetrate the casing you don't want to be in the way.

You want to open that trip valve real slow. A fair amount of energy is required to overcome the inertia of the rotor, driven equipment, and any gears to get them moving but once they're moving it doesn't take much to keep the speed up. If you bring the turbine up to speed too fast it will overspeed and sometimes that trip just doesn't act fast enough. If the turbine has a tachometer you should watch it and slow the opening of the trip valve as normal speed is approached.

The turbine speed controls should eventually take over control of the steam flow. Once that happens you can run the trip valve the rest of the way open. If the turbine is equipped with a process control (like feedwater header pressure or feedwater to steam pressure differential) that valve or controller should take over. Resist the temptation to bring a turbine up on one of those controllers, especially if they're in automatic. Neither the controller nor the manual signal output can control the steam flow as well as you can with your hands on that trip valve.

If you were starting a centrifugal pump it's time to open the pump discharge valve. Open it slowly so the turbine controller has an opportunity to respond to the increased load.

Once the turbine is up to speed and carrying load you can close the vent and drain valves, provided you don't see any condensate coming out. If the exhaust line from the turbine is routed up from the casing connection then the casing should have a steam trap to continuously remove condensate. Make certain that such traps are really working by temporarily opening a manual casing drain about five to ten minutes after you closed it; you should get nothing but steam.

Stop any electric driven oil pump and observe oil pressures to ensure the turbine's pump is satisfactorily providing proper lubrication pressures. Some electric pumps will automatically stop as the turbine's oil pump generates a higher pressure.

What if you have to bring one up in a hurry? I hope you never do have to because the potential for damage to an auxiliary turbine by rapid starting is very high. If you are in an operation that must be able to bring a turbine up quickly then you should have condensate traps on the casing and steam supply drains, an automatic air vent at the top of the casing, and means to rotate the turbine regularly, either automatic or prescribed manual means, so it's always ready. When starting one of these units always check by opening a free blow drain to ensure the casing is dry before starting the turbine. They make a lot of racket and exhaust steam piping and the deaerator can get pretty rattled if you start that turbine with any accumulation of water in the casing.

When shutting down and the turbine has an electric oil pump make certain it is running. Begin to shut down the turbine by slowly closing down on the trip valve. The steam supply shut-off should be open or shut, not throttled, so there's no erosion or wire drawing to cause it to leak. Make certain that the load served by the driven equipment is handled by another turbine or motor driven device as the turbine you're shutting down starts slowing noticeably.

When the turbine has slowed a little more close the discharge valve of any pump powered by the turbine to ensure a hung up check valve doesn't allow reverse flow to start driving everything backwards. Throttle down on the trip valve until the turbine is gently rolling over and allow it to continue rolling for twenty minutes to one half hour. This slow rolling allows the turbine parts to cool from operating conditions to exhaust temperatures and slow cooling is desirable for the heavy metal parts.

After that cool down period close the trip valve and high pressure steam supply valve then immediately open all the drains a couple of turns. If you're going to start it back up again in a few hours leave it under exhaust pressure. Otherwise, after the turbine stops rolling, close the exhaust valve, open the vent and drain valves completely and once the rotor stops turning, stop any electric oil pump.

It's a little complicated, it takes time, you have to handle small handwheels in tight spaces around the turbine because the guy that piped it never thought about operating. Proper operation of auxiliary turbines can make a real difference in the overall operating cost of a boiler plant. Wise operators know that and operate them wisely.

POWER TURBINE OPERATION

Power turbines can operate similar to auxiliary turbines in that their exhaust consists of steam under pressure that is used for other purposes than simply powering the turbine. These are referred to most commonly as back-pressure turbines with the only limitation on their exhaust pressure being the capability of the turbine casing to withstand the pressure. A common exhaust pressure for back-pressure turbines is 150 psig. These turbines are commonly used to generate electric power from superheated steam produced by the boilers and their exhaust used to service the facility. The amount of steam generated and passing through the turbines is typically controlled to maintain the facility service pressure by throttling the steam to the turbines.

Unlike auxiliary turbines that may have simple ring lubricated bearings or a startup lubricating oil pump that is supplanted by a shaft driven oil pump power turbines normally have a complete lubricating oil system that includes pumps to pump the oil to the bearings and gears of the turbine and the equipment it powers. We never shut the main lubricating oil system down on the ship other than for maintenance of that system. It has gravity tanks, a large reserve of oil in tanks installed much higher than the turbine that use the static head to produce the pressure for feeding the oil to the turbine, and oil pumps that lift the oil from the sump of the turbine and gear casing back up to the gravity tanks. The pumps always deliver more oil than the turbine and gear require so there is an overflow line from the tanks that drops oil back to the sump. The piping has an illuminated observation window at the operating level and it is always a comforting thing to see that oil flowing through that port back to the sump. It means we have a reserve of oil that would keep the turbine and main gears lubricated for a period of time even if electric power is lost. Shore-side turbines may rely on emergency generators to operate the lubricating oil pumps but some may have a similar arrangement to provide enough lubrication to allow the main turbine to come to a stop without losing lubrication in the event of loss of power.

On the ships, to prevent distortion of the rotor shafts we always engage a "jacking gear" driven by a small electric motor that powers one of the turbine rotors and the gearing to drive the other rotor and the main shaft. That jacking gear constantly rotates everything taking about 10 minutes for rotating the main shaft and propeller over one revolution. Many power turbines are equipped with the jacking gear and these can be manually engaged or automatically engaged. Even if there is no vacuum or steam on the turbine the jacking gear should be operated continually when the turbine is out of service. On board ship we always had a sign hanging on the two throttle control valves (ahead and astern) that indicated that jacking gear was engaged.

Power turbines can be used to drive large pieces of equipment, everything from sewage treatment plant pumps to chillers. Every electric utility plant whether burning hydrocarbon fuels or generating steam with nuclear energy use power turbines to generate electricity exclusively. In order to maximize power output and efficiency those plants use condensers. Some plants with back-pressure turbines will also use condensers. Condensers can be air-cooled or water-cooled but most are water-cooled. Water cooling is preferred because the temperature of the water is normally much lower than air temperature. The water used for cooling the condensers can be drawn from a well, a city water supply, a river, a lake, or the sea. When using well water or city water a closed loop system utilizing cooling towers dramatically reduces that water consumption.

Condensers for power turbines operate at a vacuum. The vacuum in the condenser is related directly to the temperature of the cooling air or cooling water. The lower the temperature the higher the vacuum and therefore, the higher the pressure drop through the turbine and correspondingly higher power output. Operating a turbine with a condenser introduces a number of pieces of other equipment to maintain the vacuum.

Unlike auxiliary turbines power turbines with condensers have low casing exhaust pressures and, therefore, lower casing temperatures. The ships I operated normally maintained vacuum in the main condenser at all times exclusive of requirements for maintenance and long-term shutdowns of the turbine. This had very little effect on steam consumption and stabilized casing temperatures. Temperatures throughout the rotors (we had a high pressure turbine rotor and another rotor that handled low pressures for propulsion and reversing, cannot be expected to be stable because the pressures along those rotors and the corresponding temperatures vary from boiler outlet conditions to the saturated conditions in the condenser.

Shaft seals on a power turbine will consist of those exposed to pressure and they normally incorporate piping connecting to shaft seals exposed to lower pressures or vacuum. This permits recovery of the steam that would leak through seals exposed to high pressures while preventing air entering seals exposed to a vacuum. Normally the turbine will have a regulator installed to maintain a constant pressure in the seal piping.

The first steps in putting a back-pressure power turbine in operation is to ensure the lube oil pumps and jacking gear are running, admit cooling water to the seal steam condenser, and open the bypass of the turbine exhaust valve to admit steam to the turbine casing to start warming it up.

The first steps in putting a condensing power turbine in operation is to ensure the jacking gear is running, start the fans on an air-cooled condenser or the cooling water pumps for a water-cooled condenser, and admit makeup steam into the turbine seal system.

Once turbine seal steam pressure is established the next step to take is to start the vacuum pumps or air ejectors. The typical vacuum pump uses water to seal it and as the pump rotor turns if forces changes in the water level so the water acts like pistons to pump the air and other non-condensible gases out of the condenser. Steam from the condenser is condensed into the water in the vacuum pumps which typically makes up for any water that's lost in them. There is normally some automatic means of providing makeup water to the pumps to accommodate leaks and evaporation and the Operator should make sure the makeup isolating valves are open. The Operator should also open isolating valves for any lube oil coolers on the vacuum pumps.

On shipboard we utilized steam powered air ejectors to produce the vacuum in the condenser. Shoreside plants can also use steam ejectors. These typically consist of two steam jets that pump the air by simply accelerating it with a flow of steam. Two jets are used in series to increase the vacuum in steps. The jet steam and any steam drawn from the condenser is recovered by condensing it. Condensate from the condenser is normally used to condense the steam in the air ejector condenser. The air ejector condenser contains two separate condensing chambers to serve the two steps in vacuum. The recovered condensate from the higher pressure condenser drains to the lower pressure condenser through a loop trap and the condensate from the lower pressure condenser flows to the main condenser through another loop trap. A loop trap is simply piping that loops down and back up so the difference in pressures between the two chambers the loop connects is offset by static head in the loop.

Before starting a steam jet air ejector its cooling water has to be established which requires starting at least one main condensate pump and to ensure continuous flow a recirculating valve is opened so some condensate runs back to the main condenser. Then the steam jets can be placed in service by opening their steam valves. When I was sailing that recirculating valve was manually controlled and for reasons I'll never understand, so oversized that it was an absolute pain, requiring frequent adjustments. Most of them were too far open at one quarter turn which represents about five percent of the disc travel

While the casing is pressurizing on a back-pressure turbine or vacuum is building on a condensing turbine steam piping upstream of the throttle valves can be pressurized. Drains in that piping must be opened to remove condensate. Once the turbine is up and running the superheated steam will not produce any condensate in the piping so there are no steam traps attached to that supply piping. The drain valves are always operated manually. Ensuring there is no water collected over any one of the control valves or in the piping is necessary to prevent a slug of water banging on the turbine blades. I only screwed this up once and will never do it again. I was bringing up a boiler at the same time that I was preparing the main turbine for operation and didn't open the isolating valve from the boiler until we suddenly got a bell indicating a need for full astern operation. What we referred to as a "bell" was the ringing of a bell on a large circular dial marked in different pie shaped sections with "Finished with Engines, Stop, Slow Ahead, Slow Astern, Full Ahead and Full Astern" providing instruction from the bridge as to how they wanted the engine (the main turbine) operated. The First Assistant Engineer (the one that normally operates the engine during maneuvers) spun open the astern throttle valve and things went awry. Sensing a drop in pressure in the main header the controls increased the firing rate on the two boilers but, since I forgot to open the stop valve from one, only one provided steam. The pressure in the boiler with the closed valve climbed as the one connected to the header dropped and its water level dropped as the one serving the header went up. I rushed up and opened that stop valve but there was some water laying on top of it and I'll never forget the clatter as it raced toward the turbine. Luckily there was no damage to the turbine or I might have lost my license. So, make damn sure all those drains are clear of condensate and all the right valves are open well in advance of turbine operation.

The main exhaust valve of a back-pressure turbine can be opened as soon as the casing pressure matches the main exhaust pressure but it must be opened before admitting steam to the nozzles. Then, after the jacking gear is disengaged steam can be admitted to either type of turbine to start rotation.

It isn't necessary to establish a complete vacuum before starting operation of the turbine. As long as there's cooling water flow and evidence that the vacuum systems are working the turbine can be started. Before starting that turbine always disengage the jacking gear. Admitting steam to the nozzles may automatically force the jacking gear to disengage but I would rather do it manually. Admitting steam with the gearing engaged won't turn the turbine but increasing steam in an effort to get turning can result in damage to the jacking gear.

As with auxiliary turbines you should admit enough steam to achieve rotation and allow the turbine to slowly rotate and come up to operating temperature. The higher the temperature of supply steam the longer it will take to bring the turbine up. As with many other elements of operation always read the instruction manual for the turbine before you operate it and pay special attention to this part of the operation. We actually rotated the turbine (and the propeller) forward and backward initially to evenly warm up both ahead and astern turbine components.

An over-speed trip is provided on power turbines and it should be tested during startup similar to auxiliary turbines. I've know these to fail with disastrous results so be very careful in testing the trip to be certain it will work when it's needed. Regardless, always bring the turbine up to speed slowly so you don't over-speed because there is no load on the turbine to prevent it. At some point the controls for the turbine should automatically take over to control the speed and you can spin the trip valve the rest of the way open.

Once an electric generator turbine is up to speed it must be electrically connected to pick up a load. On the older ships I operated we had all direct current power and it was rather easy to raise the speed until the voltage on the generator was above the main line and throw the breaker to engage the generator I just started. It's another story with alternating current generators because you have to match the alternating current frequency. There are panel mounted devices that inform you when the generator is synchronous with the line but you really want the generator you just started to be running just a little faster so it picks up some load as soon as it's engaged. Then, you have to throw that breaker at the precise instant when the generator and line are in synch. This is one more place where reading and understanding all the instructions are very important. Your plant's SOPs should always address a turbine startup.

Chapter 3

What the Wise Operator Knows

To know is to perceive or understand clearly and with certainty. Knowledge is based on training, experience, and the ability to use that training and experience to develop perceptions of outcomes that haven't occurred. When you are in control of a facility that has the potential to level a city block under the worst of circumstances that certainty becomes very important.

KNOW YOUR LOAD

The product generated by a boiler plant is steam, hot water, or similar products that deliver the heat to the facility served by the boiler plant. The load is the rate at which heat must be delivered to the facility served by the boiler plant. Your normal concern (remember the priorities) is to maintain steam pressure or supply (return) water temperature. Do you know your load?

When I ask that question I seldom get an answer. When I'm more specific by asking for a peak load, low load, weekend load, winter load, or summer load the result is usually the same. Most of the time the operator moves to a recorder or log book to try to derive an answer from there. I've never understood why operators didn't know how much heat the facility required at a particular time because they have to know it to operate the plant properly. You have to know your load.

Let's face it, when it's late Friday evening near the middle of October and the weather forecast calls for a stiff cold front coming through before the end of your shift you better know whether or not you will have to start another boiler. You can't always count on the chief leaving instructions either. You have to know your load.

Your heating load is one of the first things you need to know because the weather is fickle and changes without notice. Maybe your plant is simply a heating plant so it's the most important load for you to know about. On the other hand you could be in a production facility where the weather has a minimal effect on your total load. Regardless, it's a load you should be aware of and be able to quantify.

The amount of heat needed to maintain temperatures in a facility is a function of the difference between the temperature in the facility and the outdoor air temperature. For more than half a century we have used Degree Days as a measure of the heating load, normally on a month to month basis. Degree Days are, as the units imply, degrees multiplied by days. They are calculated for a particular day by subtracting the average outdoor temperature during the day from 65°F. A typical example would be a day with a high of 50°F and a low of 40°F where the average is 45°F and the Degree Days are 20 (65-45). Why use 65°F? If you think about it you never really need to turn the heat on until the temperature drops below 65°F so it's reasonable to say that the heating requirement for a 65° day is zero. The numbers for each day are combined to provide the number of Degree Days for a period of time.

The numbers for all the days in a heating season (normally October 15th to March 15th) are added up to provide the number of Degree Days in a season. We engineers talk of a geographical region in terms of their seasonal degree days. We'll also compare degree days for one heating season to an average that's based on a collection of data over more than a century.

You may still find reports of the number of degree days in the newspaper and on your fuel and electric bills. Some utilities now list the average temperature for the month which may also be converted to degree days. The number of degree days is about equal to the number of days in the month multiplied by the difference between the average temperature and 65.

Today we will typically preface Degree Days with the word "heating" because there is an effort to establish a comparable value for Cooling Degree Days. In September and May you have to read the paper carefully to ensure you're reading heating degree days. It could be a hot month that produced more cooling degree days so that's what they report.

Problem is, Degree Days are reported after the fact so they're not available for predicting a boiler load. However, the same logic can be used to predict load. Whether your plant is strictly for heating, or provides heat for other purposes as well, you can determine a heating load based on outdoor air temperature. We have the 65°F value for zero load and there are published extreme temperatures, data are provided in the appendix for locations throughout the United States and Canada, that will allow you to determine what temperature matches full load or 100% heating load.

Your local air conditioning equipment salesman can tell you what the design low is in your area. You can also select your own number because your site could be as much as 5 degrees warmer or cooler than the nearest reporting station. If you have several years of logs to check back through you should be able to find the typical coldest temperature. Don't use one or even four extremes, they're so uncommon that nobody expects you to satisfy heating requirements for such temperatures. It's also unlikely that those temperatures will produce the predicted load because they're normally of short duration, only that cold for an hour or two, and the mass of the building will limit the effect on your load.

Using my home town of Joppa, Maryland, I can calculate my instantaneous heating load readily using the outdoor temperature. The extreme low for Joppa is 5°F, one degree cooler than the Baltimore airport, so the range of temperatures for heating at my home is between 5 and 65°F where the load is zero at 65°F and 100% at 5°F. To determine the percentage of load for a given outdoor temperature all I have to do is divide the difference between 65° and the current outside air temperature by 60. My heating load is 50% at an outside air temperature of 35°F.

All you need do for your location is determine the range by subtracting the extreme low from 65. You get the current Degree Day value by subtracting the outside air temperature from 65. Your percent load is the Degree Day value divided by the range times 100. Remember that you convert a number to a percentage by multiplying the result by 100. For an outdoor temperature of 42°F in Joppa my load is calculated as 65 less 42 divided by 60 to get 0.3833 which times 100 gives me 38.33%. That's how you determine a common heating load. Simply checking the weather forecast in the paper or from the radio or television will let you know what the load will be. I do hope you understand that I'm not implying you should only listen to a radio or watch television during your shift, you need those ears on the plant.

Of course the truth is that very few plants have a simple heating load. Boiler plant output is usually used for other purposes, a common one being hot water heating. Hospitals have sterilizers that run year round. Kitchens or cafeterias in the building can introduce substantial loads independent of outdoor temperatures too. However, they also require considerable ventilation so much of that load is outdoor temperature related. The heat in your steam or hot water can be used for many things that aren't related to outside air temperature.

In most systems used just for heating you'll find the loads are rather consistent in the summer and you can call that value a base load or summer load to which you can add the heating load. I've been able to generate formulas for steam loads that are very consistent for apartment buildings, nursing homes, and similar loads. The formula becomes the base load plus a factor times the number of degree days. Each base load and degree days should be for a specific period of time and degree days should be for a specific time frame (hour, day, month).

When generating a formula for heating load it's important to realize that the actual steam load at any one time will seldom match the formula due to everything from people opening and closing doors to the kitchen starting up in the morning while everyone's getting up and taking a hot shower. My experience is that the actual load will swing 25% of the maximum heating load in a typical heating plant. If you generate a formula to use, the actual load should be equal to the formula value plus or minus 25%.

Why produce a formula? Because boiler operators have to deal with us engineers and you can't convince an engineer of much without some supported documentation. So, by having a formula that represents your plant load conditions you can convince an engineer that you do know what you're talking about.

Here's how you do it. Keep track of your load using steam flow or Btu meter readings, fuel meter readings or tank soundings, preferably recorded each day. Also record the average temperature or number of degree days each day. You can use a properly installed (in the shade and away from sources of heat) high/low thermometer and average those two readings to have an accurate value for your site if the nearest airport isn't consistent with you. Eventually you'll have to convert average temperatures to degree days by subtracting them from 65. Any negative values should be converted to zero. Once you have some data you can start determining the value of the formula. If you haven't been collecting data it will take you a year to collect enough data to produce a reasonably accurate formula.

Once you have data you begin by determining your base load. During the months of July and August, when it's never cold, you can correctly assume that there's no heating requirement and the average steam generation, Btus, or fuel consumption is representative of the base load. For the few of you that live in the far north, you'll have to take the average of those readings on days when the outdoor temperature never got below 65°F.

If you're computer literate and can use a spreadsheet program then determining the formula is rather easy. If you aren't capable of doing that, try to get help from a friend that is. Should those options fail, get a cheap calculator and go at it. Create a table of values using your recorded data. In the first column put all the degree day readings. You can precede that one with such values as average outdoor temperature or the low and the high if those are the values you recorded then use them to calculate the degree days. In the second column record the steam generation, Btus or fuel use for that day. For the third column, calculate the heating load by subtracting the base load value from the value in the second column. If any of the results are negative, substitute a zero for that result. For the fourth column, calculate the heating ratio by dividing the heating load value of the third column by the number of degree days in the first column.

The values in that fourth column should all be close to each other. If you run into one, or some, that seems to be significantly different and you can't resolve it, cross out that row of data. After eliminating several rows from one set of plant data I finally realized that they were every seventh one and I was looking at data taken on weekdays where the fuel use covered the weekend. Simply dividing the odd result by three made the data useful. Count the number of rows of good data (each daily set of readings) and write the number down at the bottom of the page.

Add up all the values in the fourth column and divide by the count of good data rows to get an average of the values in the fourth column. Your load formula can now be determined as equal to the base value plus a factor times degree days and the factor is that average value. To get an idea of how accurate it is you can use it to calculate another value (put it in the fifth column) then compare that to the steam generation, Btus or fuel use in the second column. When using monthly data I find I'm normally within 5%, daily data are within 10% and hourly data are within 25% of the actual values. Continuing to record data and adjust the base and factor values improves the accuracy.

I use those formulas to compare the performance of a building at different times. Adjusting for the number of degree days corrects for variations in outdoor air temperature. It helps me detect when something went wrong in a boiler plant or the degree of improvement in efficiency a particular installation provided. You can use the formula to predict loads and to detect problems with the plant.

There's also another factor that changes your heating load and influences other uses of the heat you generate and that's the people load. The use of the facility will determine most of the people load. A nursing home or prison will have a relatively constant people load because the people are always there and doing the same thing. Apartment buildings will have a more variable people load, one of the more difficult to determine. College dormitories are another story because all the students are on the same schedule; if you know the schedule the loads are predictable despite the fact that they will vary considerably. Simply picture all the students rising at the same time to get ready for class, taking showers and washing then vacating the building; they will create a short-term peak load during that time. If the building was equipped with night set-back thermostats the load swing will begin with the warm-up and end with the students leaving for class.

When people are present your loads will be higher and when they're absent they'll be lower. In an office building, for example, everyone but the cleaning staff goes home in the evening so you don't have to heat the building to a comfortable 75°F at night. In that case you may have all the thermostats set back to 55°F. Under those circumstances your peak heating load isn't based on 65°F, it's based on 55°F. The difference between the thermostat set point of 75 during the day and the 65°F base we use for calculating degree days is covered by the people themselves (an office worker puts out about 550 Btuh of heat); then there's the equipment they're using (computers, etc.), and the lights.

People have other effects on heat load depending on what they're doing. When everyone is arriving for work in the morning they manage to pump a lot of the building heat out and the cold in when passing through doors. I know one building where they set the lobby thermostat for 85°F about an hour before starting time so they store some heat in the area to offset all the cold air that comes in with the arriving workers.

Store heat? Yes, everything can store heat to one degree or another. You have to raise the thermostat setting to 75°F in that office building well before the workers start arriving or it will still be 55°F when they arrive and you won't hear the end of it. It takes time for the temperature to return to 75°F because the air in the room has to warm up the walls, floors, ceilings, furniture, etc., from 55° to 75°. How fast it warms up depends on the weight of the materials and their specific heat, the amount of heat required to raise the temperature of the substance one degree Fahrenheit. The appendix has a table of specific heats for various materials.

When the outdoor temperature is mild the materials in the building may never get to 55°F before the thermostats are reset in the morning. When it's very cold out the temperature of walls and other surfaces exposed to the outdoors will drop quickly and may be cooler than the 55°F. Because partitions, floors and ceilings, furniture, etc. cooled slower, they might still be warmer and help offset the effect of the colder walls. Warm-up loads can be higher than heating loads if ventilation is not controlled. Unless the thermostat settings are timed to compensate for the variation in storage temperatures you may get some complaints in cold weather or waste heat in milder weather.

Ventilation loads are primarily people loads. For all practical purposes a facility has to introduce 20 cfm (cubic feet per minute) of fresh outside air for every person in the facility. There are more specific requirements that vary with the Jurisdiction but that is a good rule of thumb. Many older facilities may still be set for ventilation rates as low as 5 cfm per person so it pays to check the actual values before trying to determine the heating load they create. The amount of heat required for ventilation air is easy to determine, it's the total of ventilation air in cfm multiplied by a constant of 1.08 and the difference between the outdoor air temperature and room temperature. As an example, for 100 people you need 2,000 cfm of 0°F outside air which requires 162,000 Btuh (2,000 × 1.08 × (75 – 0). If you recall our earlier discussion that's equivalent to about 162 pounds per hour of steam. Note that we used 75° not 65° because we can't count on the heat from people, etc. to cover that portion of the load.

In areas containing a high concentration of people (movie theaters, stadiums, office buildings) the ventilation load can be the largest single load of the facility. The core of a building, in the middle where there are no outside walls, and floors and ceilings separate them from other occupied spaces, the ventilation air can produce a heating load that would not exist without it. If your facility has large changes in the number of people from day to night or over weekends you should see swings in load due to changes in the ventilation air.

Of course many older buildings don't adjust ventilation air depending on building occupancy. Yours may be one that continues full flow ventilation at night when There are only a few people, if any, in the building. If you have a way of closing that off at night (you'll never be able to get zero ventilation) you'll save a lot on heating all that air unnecessarily.

Modern facilities are using a combination of security and air conditioning controls to determine how many people are in the building and adjusting ventilation loads accordingly. Another method is measuring the carbon dioxide content of return air which indicates how many people are in the building or a certain area of the building. The new technical name for that is demand controlled ventilation. If you don't have the advantage of one of those specialized controls you'll probably have systems like time clocks that set the ventilation at a minimum when people aren't supposed to be in the building and adjust them to a value for full occupancy the rest of the time.

Any of those controls should be set for minimum ventilation air during the period when the building is warming up in preparation for occupancy. That way you avoid the ventilation load while handling the warm up load to limit the load on your boilers. It also makes no sense to heat up cold outside air to warm up walls. The ventilation should increase for a short period before people start entering the building to flush out the stale air.

Except for some process requirements the hot water heating load is largely a function of people activities. People have a direct relationship with hot water needs for cooking, showers, and washing. Each of those hot water uses is sporadic, occurring at specific (sometimes inconsistent) times so they're more on and off than a constant load. There are several means of producing hot water and satisfying the irregular loads so there's a section in this book devoted specifically to hot water heating. When the hot water is heated by many heat exchangers throughout the facility you have little control of those loads and you'll have to monitor plant loads to determine their effect.

An unusual load that I encountered at one chemical production facility a few years ago is a rain load. I was collecting nameplate data at one of the boilers and found myself almost run over by the operator who was suddenly rushing around trying to get that boiler operating. Once he had it on line I asked what the rush was all about. "It's about to rain" was his simple reply. That plant experienced a 30,000 pph increase in boiler load every time it rained! Many district heating plants experience a delayed rain load which is due to rain leaking into the manholes and tunnels containing the steam lines. It's a load that indicates inadequate or ineffective maintenance and shouldn't be as significant as that one plant. You may have one and it shouldn't be difficult to identify it.

Finally, there are production loads. These are requirements for heat to warm raw materials for production, to convert the product to another form (like melting it) or steam actually injected into the product to alter it. They can include tank heating and heat tracing where heat is used to keep the product in tanks and piping hot enough that it will flow or remain a liquid. Those heating requirements are independent of actual production. I like to treat those requirements like heating loads with a higher base temperature.

An asphalt plant, for example, may operate at 500°F to keep the asphalt a liquid and that temperature is so high that swings in outdoor temperature between 0°F and 100°F, an extreme winter to extreme summer outside air temperature would produce a variation between 100% and 80% [(500–100) ÷ (500–0) = 80%] If they're significant you can treat them the same as heating loads by using the product temperature instead of 65°F.

That's a way to determine production heating requirements which will exist as a load independent of the amount of product made. Actual production loads can be related to production output. It's one reason that boiler operators should know how many widgets or pounds of product the plant makes and be informed of how many are planned for production during the next shift.

Some production facilities produce a negative load. These include plants with waste heat boilers that can generate steam or hot water from exothermic reactions (chemical reactions of the product that generate heat). A boiler operator can be called upon to control those boilers. For the most part they conform to all the rules described for regular boilers in this book but each one can have unique characteristics or operating features and the operator should make sure he fully understands all the manufacturer's and process designer's instructions for their operation.

Except for simple heating plants the operator has to learn the contribution of each type of load and monitor loads to determine how much each one contributes to the total load. The simple mathematical relationships described here should help to explain some of the variations in loads you experience to provide a way to determine what the load will be when plant operations change.

You should be able to tell how much change in load will be associated with a change in outdoor air temperature, a change in production rates, shutdown of any particular part of the plant, and short-term swings associated with personnel activities. At the bare minimum you should know what your maximum, minimum, weekday, weekend, holiday, and total plant shutdown loads are. Once you know your load and know your plant you can begin operating wisely.

KNOW YOUR PLANT

I'm always amazed at the boiler operators that don't know their plants. I've been in plants with an operator that had been there 15 years and had him reply "I don't know" to what I thought was a simple question. I would be very embarrassed if someone asked me what steam pressure I normally operated at and I had to respond that I didn't know. More than half of the operators asked that question immediately wander over to the nearest pressure gage to look at it before responding. More than eighty percent of the operators of hot water plants can't tell me what the normal boiler water temperature is. I always say "it wasn't meant to be a trick question, I just wanted to know."

You shouldn't be asking yourself the same question now. You should know certain things about your plant and be able to respond to one of us dumb engineers without hesitation. We really don't ask trick questions. When I look at a pressure gage and it reads somewhere between 120 and 125 psig I have to ask the question because it could be either one of those values. Here's a quick list of common questions, see how many you can answer without looking them up:

1. What's your normal operating pressure/temperature?
2. What pressure/temperature are the safety/relief valves set at?
3. What's the capacity of each boiler?
4. What's your normal feedwater/return temperature?
5. What fuels do you fire?
6. What's the capacity of your fuel storage?
7. Where does your fuel come from? Are there alternate suppliers?
8. What is the turndown for each boiler?
9. What's your electrical power (208/230/460, 3 phase)?
10. How reliable is your electric power? (How many interruptions and their length in an average year)
11. What's your normal compressed air supply pressure?
12. What's your peak load? Peak day? Peak Hour?
13. What's your normal winter load?
14. What's your normal summer load?
15. What's your minimum load?
16. What's your water supply pressure?
17. What's the normal hardness of your water supply? Of alternate water supplies?
18. Where does your water come from? Do you have an alternate supply for water?
19. How many boilers do you run in the summer?
20. How many boilers do you run in the winter?
21. How frequently do you switch boilers?
22. What's your condensate return system leakage/

percentage?
23. What's your normal condensate temperature?
24. Is your condensate return pumped?
25. What does your blowdown drain to?

In addition to those questions I frequently aim my laser pointer to produce a red dot on a vertical pipe, one that comes up through the floor then continues to penetrate the ceiling, and ask the operator what the line is for and where it goes. While that is usually a question I want the answer to it's occasionally used when an operator gives the impression he knows it all. After forty years of learning boiler plants I know which of those piping systems are obscure. You should test yourself in this regard. Can you look at each pipe in your plant and name its contents, source and destination? No, you don't have to be able to do that to answer the questions of some dumb engineer like me, you need to know so you can react quickly and responsibly if that piping fails.

Since most of my operating was aboard ship we had another criteria for knowing the plant. The engine room aboard a ship is always at the bottom and there aren't any windows. If there's a skylight it's so small and far away it doesn't provide any light at operating levels. In the event the electric generator tripped we had to know how to get around in pitch dark. Most of us carried a working flashlight at all times but I don't see that in the typical land based boiler plant.

How about it? Especially you guys that work the night shift. Can you get around the plant safely in the dark? Trying it with your eyes shut is one way to test that skill. Be careful, however, that you don't put yourself in the position of falling down a stair or tripping over something. It's better to do something as goofy as walking around the plant with your eyes shut when someone is there that can call the ambulance if you land on your face. It may be goofy, but it might also save your life one day.

There are a lot more questions about your plant that you don't have to have immediate answers for because they're not asked frequently and, to be honest, you don't have to know the answer to operate. You do need to know a lot that you can't memorize and there's no need to commit it to memory; all you need to know is where to find the information. You should know the location of historical documents, logs, maintenance records… basically where all the paper and spare parts are stored and how to find something in that maze of paper or shelves of boxes. The next best thing to knowing an answer is knowing where to find it.

MATCHING EQUIPMENT TO THE LOAD

When we discussed priorities in the first chapter of this book the last was listed as the one you would spend most of your time on, operating the plant economically. Without a doubt, matching the equipment to the load is the easiest way to do that. I find so many boiler plants operating with two boilers on line and not enough load to keep one running constantly. I've also been in plants with four boilers on line looking at a load less than the capacity of one of them. When I make those statements I get a "so what" look from the boiler operators or the standard WADITW[3] response. Based on what I have seen, we should be able to conserve about 20% of the energy used in institutional heating plants in this country by simply matching loads.

Let's look at the example of two low pressure heating boilers operating when one could carry the load easily. My observations indicate the load is typically less than half the capacity of the one boiler. Radiation losses, normally 2% of input (or less) at full load, account for 11.5% of the input at the lower load; off cycle losses of the boiler that isn't firing account for another 1/2 to 2% depending on effective stack height; purging losses are doubled; demand charges for electricity when the two boilers just happen to be running at the same time; and the additional time an operator spends attending to an operating boiler all add up to a considerable additional cost for operating two boilers where one would do and that's ignoring the fact that cycling losses are doubled when the load is less than low fire capacity of one boiler.

Demand charges are calculated by the electric power company for medium and large installations. Maximum demand is determined by a separate meter that constantly measures the electrical load and keeps track of the maximum average electric load during a 15 minute period in each month. The utility bill includes a charge for that demand and it's not small change, $12 per kW which is equal to about $9 per horsepower. Any activity that produces a higher demand simply boosts that charge and any temporary operating condition that produces that demand creates the charge for the entire month. In some areas the utilities charge for the highest demand in the prior six months.

Running two feedpumps when one will do is not only boosting the demand charge, it's using electricity as well. Although it's not advisable to stop one feed pump before starting another to avoid a bump in the demand charge you can wait until the air compressor stops (so you know it won't run for a few minutes) to switch over pumps. A drop in demand of ten horsepower while the air compressor is down will reduce the demand while you're starting a thirty horsepower feed pump to switch over. That little bit of attention to the electrical demand could save your employer as much as $90 on the monthly electric bill.

In those days when all we had were coal fired plants conventional wisdom called for boilers to be of three sizes, one that could handle full load, one two thirds that size, and one a third of full load size. The two smaller units served as backup for the larger one and the variation in size ensured a closer match to steam load. Coal fired units didn't provide the turndown we have on modern boilers and cycling a coal fired boiler on and off left an operator awful tired at the end of a day.

There are many of you that will have a plant with only one boiler, one feed pump, etc. so your choices are limited or non-existent in operation. That doesn't prevent you noting your operation and estimating what could be saved if you had another, smaller boiler to carry the normal loads.

You should be able to justify the installation of a smaller boiler in any plant where the boiler cycles at the average winter outdoor air temperature. Cycling boilers are very inefficient and many times a much smaller replacement produces fuel savings that pay for it in a heating season.

For the rest of you, I'm betting that you can make a significant difference in the fuel and electricity consumption of your plant by doing your best to match the equipment to the load. For many of you it will simply be a matter of realizing there is a difference and acting to reduce the costs. Many others will find it's a matter of changing old habits and rationale.

The graph in Figure 3-1 is provided to show how you can make wiser decisions about boiler operation especially when you know what your equipment can do. This plant started out with only one boiler then grew until two were required. After a few years Boiler three was added and shortly thereafter Boiler four. The graphs show the evaporation rate (pounds of steam generated per therm of natural gas) for the boilers and a combination firing of Boilers one and two. For all practical purposes the evaporation rate

is representative of boiler efficiency.

The curve for Boiler one is typical for boilers manufactured in the middle of the 20th century in that the operating point of highest efficiency is not at maximum firing rate. A number of elements in the construction of those boilers combined to produce a curve similar to that shown. Boiler two, three and four all exhibit properties consistent with later designs of package boilers. The curve for the combination of firing Boilers one and two together is generated by adding the individual values for the two boilers with a specific trimming of the firing rate for Boiler one so operation is at maximum efficiency. Using the bias adjustment of the boiler master for Boiler one to limit its steam generation to 38,000 pounds per hour a higher evaporation rate for loads over 76,000 pounds per hour is achieved

It's obvious that Boiler one is the most efficient at loads less than 30,000 pounds per hour, Boiler two is the most efficient at loads between 30,000 and 48,000 pounds per hour, and three or four are most efficient at higher loads. However the plant was operated without this knowledge for several years. When the plant only had two boilers and total load was less than the capacity of Boiler one the Operators chose to fire them alternating every two months. Therefore there were periods of operation when considerably more fuel was burned than necessary. If it were not for the fact that the plant load was almost always over 15,000 pounds per hour the Operators should have elected to run Boiler one at low

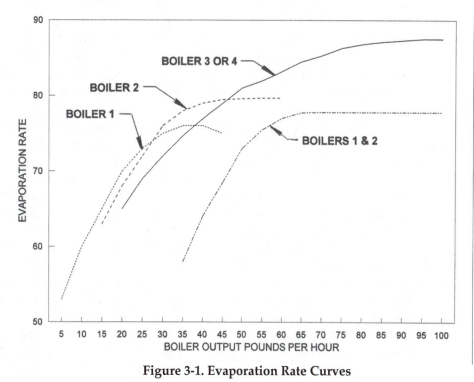

Figure 3-1. Evaporation Rate Curves

loads where it was the most efficient.

Once Boiler three was installed, a necessity to match increased plant load, the Operators chose to fire it alternating with a combination of Boilers one and two. The significant discrepancy between those two operating setups was obvious to the plant engineers and an investigation was conducted to determine what to do about it. It was determined that the principal reason for the difference was that Boiler three was equipped with an economizer. Due to the age of the first two boilers a decision was made to purchase another new boiler with an economizer rather than installing economizers on the older boilers. Part of the investigation included generation of the graphs. Now the Operators know that for certain periods of operation including the Christmas shutdown it's advisable to operate the oldest boiler because it's the most efficient for low steam loads.

Evaluation of the plant's steam loads later led to the decision to abandon, dismantle, and scrap Boiler two because plant operations almost never produced steam loads that were best served by that boiler and maintenance of it wasn't economical.

This example should serve to prove to you that knowing your load and knowing your equipment can make a significant difference in the cost of operation of your boiler plant and a graphing your data helps make it evident. Decisions to simply alternate firing boilers without regard to load performance cost that manufacturer hundreds of thousands of dollars.

Now that you know your plant and know your load you will make decisions that reduce the impact on operating costs. Frequently operators will decide to put another boiler on line whenever the load on one approaches 70%. That immediately converts operating conditions from one boiler operating at its maximum efficiency to two boilers operating near minimum efficiency at 35% load. Radiation losses are doubled with no change in load and all losses associated with lower firing rates are encountered. Knowing the load, being able to forecast its changes, and knowing what your boiler can do will frequently prevent putting that other boiler on the line until the load will exceed 100% of what is on line. Establish values based on experience and don't hesitate to experiment to see what the best matches are.

Matching equipment to load isn't restricted to the boilers. I don't know how many plants I visit where the scheme is to operate one boiler feed pump for each boiler on the line. Since feed pumps have to be capable of delivering water at the boiler safety valve pressure it's not uncommon for them to have significant capacity relative to normal operating pressures. As a result you should never associate the number of pumps in operation with the number of boilers. They deserve their own set of rules, established by experience and observation.

Many operators don't realize that there's a lower limit to efficient operation of water softeners. Once the flow in a softener, or any ion exchange bed for that matter, drops below a set value (usually 3 gpm per square foot of flow area) they begin channeling. The water tends to bypass much of the resin and its capacity isn't used. Operators can allow a lot of scale forming hardness to sneak through their softeners if they run too many of them in parallel.

If everyone in your plant is doing their best to conserve that valuable condensate you will have reduced the demand for makeup water and may have reduced it to the point that your softeners start channeling. You'll have to watch the softeners closer if you're down to one because it might start regenerating automatically when you're not looking. That will shut off your supply of makeup.

Some plants are constantly having trouble with condensate loss. It's either due to contamination indications or leaks. In those cases it's better to have the technician that services those softeners modify the programming to limit the softeners on line unless the pressure drop through them gets too high. It's a matter of adding a differential pressure switch so another softener will come on line when needed. He should also add a bypass switch that permits you to manually put a softener in service.

Whenever I visit a plant and find more than one piece of equipment operating I do a quick check of the loads to see if the loads and equipment match. I should note that this also applies to chillers and devices other than boiler systems. It is always a cheap way to give a customer a return on his investment in my time because I can usually show a considerable savings for doing nothing but shutting off some of the operating equipment.

In mid summer of 2001 I visited a plant where the gas booster was running constantly when the gas supply pressure was more than adequate to serve the boiler load. The owner had his operators shut the booster down and bypass it. Of course they had to check it when the temperatures got colder to determine when they might need it. I encouraged them to establish an SOP to check it out by running it temporarily every fall so they would be capable of putting it in service should they need it in the winter but, to the best of my knowledge, it hasn't run since. That wasn't just a case of matching load, it was a case of recognizing there was no load.

You shouldn't confuse matching loads and reacting to changing loads however. I was in one plant that started up a boiler every morning to handle the warm-up as the night set-back thermostats switched back up. An hour or two later the boiler was shut down until the next morning. First of all, that's rough on the boiler and it's really shortening its life. It's also wasting a certain amount of energy because what it took to heat the boiler up is lost up the stack before the next morning. If an operator is doing his job of checking all the operating limits when a boiler is started then that daily start-up would be rough on the operator; most don't seem to bother.

Short-term operation for an intermittent peak load shouldn't be considered unless there are problems with the steam pressure or supply water temperature drop associated with that load. In other words, it's okay for the steam pressure or water temperature to drop a little when everything starts heating up in the morning. The drop will limit the heat flow to the load because there's a smaller temperature difference and everything will eventually recover. Don't hesitate to try it. Let the pressure or temperature drop. A slight dip in conditions on an operating boiler is much less damaging than running a boiler up from cold.

If the pressure or temperature dips can't be tolerated you'll learn quickly what average night-time temperature signals that limit so you can have more boiler capacity in operation when it's necessary.

I also want to mention those plants where nobody seems to notice or care what the boiler to load relationship is. It's not at all uncommon for me to find a two boiler hot water plant where both boilers are always operating. In most of those plants the boilers were each sized to carry the full load and the operators discovered they could shut one down and never worry about having enough boiler capacity. The cost of fuel to simply keep a boiler hot can be considerable so they also found that they saved the owner a lot of money. Of course you have to shut at least one valve when you shut down that hot water boiler. Otherwise the hot water flowing through it will heat up a lot of air that's lost up the stack.

Don't think you have to run a proportion of boilers to match the load. I've been in many plants with four boilers, any one of which could carry the full load of the facility served. They'll run one or two boilers in the summer and three in the winter whether they need them or not. They're also usually the plants where the boilers are regularly switched so they will all wear out at the same time.

EFFICIENCY

There are so many definitions of efficiency and many operators (and most engineers) are confused as to which is which or simply assume they are all the same. I shall attempt to define the many different labels of efficiency and to clarify what they actually represent. I've even created a couple of definitions because I'm certain there's a need for them.

The first point of confusion involves the definition of boiler efficiency. It can be officially defined as one hundred times the heat absorbed in the steam and water divided by the heat energy added by the fuel and other sources of energy. That's the definition established by the ABMA and the one most of us accept as the true definition. Those other sources of energy include electric power supplied to the fans, and pumps that are integral to the boiler. If all of those values we engineers call "inputs" are accounted for then we get a correct value of efficiency. However, it's the one that is seldom used.

The energy added to the water and steam is the "output" of the boiler. There can be multiple outputs that have to be considered. If the boiler has a reheater the energy added to the steam that flows through the reheater is an output in addition to the water that is evaporated to produce steam and the energy added in the superheater.

Note that the official definition of boiler efficiency considers output to include all the heat absorbed by the water and steam. That includes the heat added to the water that's lost in blowoff and blowdown but not the heat lost in steam for sootblowing. Since the boiler's output that we get to use doesn't come from blowoff or blowdown water or sootblowing steam how can it be counted as output? It's counted because the boiler manufacturer has no control over the quality of water used to make steam and no control over the fuel fired and how cleanly it's fired. The boiler manufacturer is concerned with the heat that's transferred through the tubes.

Soot blower operation to maintain boiler conditions is one of the reasons that a boiler efficiency test in accordance with ASME PTC-4.1 (Steam Generating Units Power Test Code) is supposed to be run for a minimum of twelve hours. The Test Code does account for the sootblower steam because it's required to keep the heat transfer surfaces clean.

Several years ago the ABMA (American Boiler Manufacturer's Association) agreed to guanantee efficiency at only one firing rate and, unless otherwise specified by the customer, set it at full load. If you have some efficiencies listed at other firing rates in your boiler documentation you'll notice that those others are labeled "predicted performance" and only the full load is guaranteed. The problem with that wisdom is the boiler seldom, if ever, operates at full load. Whenever you have input, suggest that any new boiler you purchase be guaranteed for performance at a load you will have, say 50% or 75%. That doesn't violate the ABMA's rule. Today some chiller manufacturers, and possibly by the time this book is printed some boiler manufacturers may, guarantee the part load operating efficiency of their equipment.

Occasionally you will see a boiler efficiency guaranteed at something around 50% to 75% load. That is probably a sales tactic because the maximum operating efficiency of a boiler is typically in that range. As the load and firing rate decreases the volume of flue gas decreases. The heating surface, on the other hand, stays the same. Therefore the flue gas spends more time in contact with a proportionally larger heating surface so more heat is transferred.

You should notice that when you create your own performance documentation because the stack temperature will drop as you reduce firing rate from full load. Somewhere lower the efficiency will start to drop off because the flue gas is channeling so only a small portion of it is contacting the heating surface. As the firing rate decreases it becomes more difficult for the fuel and air to mix completely so excess air must be increased to prevent CO and efficiency suffers further. The radiation losses also become more significant as the load decreases. All these factors influence the operating efficiency of the boiler to different extents at different loads.

Heat loss efficiency is determined by backing into the value. An efficiency is considered to be the output (what you get out of it) divided by the input (what you put into it) with the result of the division multiplied by 100.

$$\text{Efficiency} = \text{Output} \div \text{Input} \times 100$$

The loss, and in the case of a boiler it's a loss of heat, is the difference between the input and output. Therefore, the output is equal to the input less the heat losses. By substituting input less losses for the output in the formula we get a formula that doesn't include output at all.

$$\text{Efficiency} = (\text{Input} - \text{losses}) \div \text{Input} \times 100$$

If we can calculate the losses as a percent of the input then all we have to do is subtract the percent

losses from 100 to get percent efficiency. Surprisingly it is easier and far more accurate to determine some of the heat losses as a percent of the input so determining efficiency using the heat loss method is the most widely accepted method.

The Power Test Code (PTC-4.1) provides a structured basis for calculating boiler efficiency by two methods, input-output and heat loss. All the larger boilers we installed while I worked for Power and Combustion were tested using both methods in a modified form of the Power Test Code. The cost of performing those tests in strict accordance with the Code could not be justified for even the larger boilers (up to 200,000 pounds per hour of steam) that we installed. The primary modifications we made to the Test Code included shorter test runs (three hours instead of the required eight to twelve) and less frequent measurements (every twenty minutes instead of every ten) so we could get two test runs in within one day and with only one man collecting data. Of course in those days we used an actual Orsat analyzer which took some time to operate, not one of those nice electronic analyzers we have today.

An examination of the results of the hundreds of test runs we made revealed a typical deviation in the input-output efficiency of as much as five percent while the heat loss results were normally within one percent. That's why I can say, with a reasonable degree of confidence, that the heat loss method is very acceptable.

I always get a kick out of some organizations indicating that they conducted hundreds of boiler efficiency tests. During my twenty years at PCI we only ran about two hundred boiler efficiency tests using that modified approach to the Test Code. Each test did consist of several test runs so I can say we made hundreds of test runs. Those were formal tests that included a printed report with all the calculations, records of collected data, and fuel analysis. They were not boiler tests conducted in strict accordance with the Test Code but they were a lot closer than what some people call a boiler efficiency test.

I don't consider a strip of narrow paper with a list of analysis values, temperatures, and a calculated boiler efficiency representative of a boiler test. Some firms that claim they've done hundreds of tests haven't included one fuel analysis. Unless you have the fuel analysis the test is simply flawed because the hydrogen to carbon ratio of fuels varies considerably. The modern flue gas analyzer contains programmed calculations based on an assumed fuel analysis and the odds that your fuel and the values used by that program are identical are slim to none. The results are only representative and based on an assumed fuel. They're sufficiently accurate to determine relative efficiency over the load range and to compare the boiler performance to another boiler burning the same fuel but if you use those results to challenge the boiler manufacturer's higher prediction you'll lose the argument. Calculations in Appendix L permit determination of boiler efficiency using the heat loss method and a fuel analysis for those purposes.

The most common value used today is what we call "combustion efficiency." When the technician visits your plant to do your annual combustion optimization (typically required by EPA [Environmental Protection Agency] or its equivalent in your State) or you draw stack samples that allow a calculation of boiler efficiency that's combustion efficiency. It's basically a heat loss efficiency that assumes a fuel analysis and determines the energy lost up the boiler stack. It's the one that is printed on that little strip of paper by the analyzer. Assuming the analyzer was properly calibrated the value is a reasonable indication of your boiler efficiency when it is adjusted for radiation loss.

That's because the stack loss is the largest single loss associated with boiler efficiency and the analyzer does a pretty good job of determining it.

It isn't much but radiation loss has to be considered in addition to that combustion efficiency. The manufacturer will provide you with a value of radiation loss, equal to a percent of input at a prescribed boiler load. All you have to do is determine its impact at the actual load. Divide the manufacturer's predicted loss by the percent of boiler load and, if the predicted loss is at a load other than 100%, multiply the result by the percent load for the prediction. In most cases the manufacturer's prediction is at 100% load so you only have to divide the predicted loss by the percent load. A few examples should suffice:

- A boiler with a predicted radiation loss of 3% at full load is tested and found to have a combustion efficiency of 79% at a 50% load. The radiation loss at that load is 6% (0.03 ÷ 0.5) so the operating boiler efficiency is 73% (0.79 less 0.06)

- A boiler with a predicted radiation loss of 2% at 80% firing rate is tested and found to have a combustion efficiency of 80% at full load. In this case the operating boiler efficiency is 81.6% (0.8 +0.02 ÷ 1 × 0.8)

- A boiler with a predicted radiation loss of 1.5% at 75% firing rate is tested and found to have a com-

bustion efficiency of 78% at a 40% load. In this case the operating boiler efficiency is 73% (0.82 +0.015 ÷ 0.4 × 0.75)

Why bother with the radiation loss? To ignore it is to invite some crucial errors in operating decisions. Radiation losses are, for all practical purposes, constant regardless of firing rate so their proportional effect varies with load. My favorite example is a plant with an old HRT boiler and a newer cast iron boiler. Since the HRT furnace was substantially hotter it was easier to get low excess air with a newly installed burner than possible with the cast iron boiler at the same loads. The predicted full load radiation loss for the HRT boiler was slightly more than 8% while the cast iron boiler had a predicted radiation loss of 4%. At the normal load of 50% the combustion efficiency of the HRT has to be 8% higher than the cast iron boiler to overcome the higher (16% versus 8% of actual input) radiation losses. The operators were firing the older boiler because combustion analysis indicated it was 5% more efficient. Evaporation rate data later proved they couldn't rely on their combustion efficiency.

For years we have settled on the concept of boiler efficiency being relative to the higher heating value (HHV) of the fuel fired. The advent of combined cycle and cogeneration plants has resulted in the return of lower heating value (LHV) to our definitions. There is a significant difference in the values expressed by these two references, with an efficiency at the LHV always being significantly higher than an efficiency at the HHV. In those rare applications where a CHX is applicable, an LHV efficiency could be greater than 100% because the system recovers heat the heating value doesn't acknowledge as existing. LHV doesn't include the heat that could be extracted if the water in the flue gas was condensed. When I talk efficiency I'm talking HHV, you'll have to be aware that someone can use the LHV.

Can a boiler efficiency be greater than 100%? Logic says the answer is no but by the definition of some efficiency labels some of them can. My favorite example is the Nevamar project we did back in 1974. That system used heated air off a process as combustion air. It contains a small amount of hydrocarbons with negligible heating value but can, when one particular process is operating, produce 360°F combustion air. When supplied to the one boiler with an economizer and a stack temperature of 303° it can produce results in the accepted definitions that exceed 100%. That, by the way, is efficiency at the HHV.

If we considered the true and full definition of boiler efficiency we would have to include the heat in that combustion air as an input. However, simple input output efficiency calculations only include the heating value of the fuel. They're used to avoid measuring the energy added by fan motors and pump motors along with that hotter combustion air. Combustion efficiency calculations will show a negative loss because the temperature of the hotter air is subtracted from the temperature of the flue gas.

For reasons I don't understand everyone concentrates on boiler efficiency when it doesn't change very much and has little to do with the overall "plant efficiency" which the boiler operator should be attending to. This is a bigger problem when there is so much confusion over what boiler efficiency really is. Two identical boilers in different plants can have the same boiler efficiency and combustion efficiency but one will produce less usable energy than the other because it has a higher blowdown rate. The energy absorbed by the water and steam in the boiler (ASME definition) includes the heat added to the blowdown water. Two plants with identical boilers and loads can have different plant efficiencies simply because one plant doesn't have water softeners so it must blow down more. Maybe they both have softeners but one has very little condensate return; it must heat the makeup water to replace that condensate and blow down more. Those and other variations can produce plants with boilers having an 80% efficiency operating with a plant efficiency as low as 40%. Take a plant with a mismatch between equipment and load and that plant efficiency can be as low as 20%.

So what's "Plant Efficiency?" It's the amount of heat you deliver to the facility, the usable heat you generate, divided by the energy used in the plant. What you deliver to the facility is your output. I like to use energy in steam or hot water going down the pipe to the plant less the energy in the condensate or return water. That way my output is what the facility is using. The energy used in the plant includes electric power in addition to fuel.

A kilowatt-hour is 3,413 Btu. Multiply the kWh on your electric bill by that number to know how many Btus were added by electricity. If you're firing gas and want to deal in therms then multiply the kWh in your electric bill by 34.13 to convert the electricity use to therms. If you're larger and use decatherms or millions of Btu multiply it by 3.413. With identical units you can add your electrical and fuel energy inputs to the plant to get the total energy used.

If you deliver steam to the facility and get nothing back you're a 100% makeup plant and the energy you're delivering is all in the steam. Look for the enthalpy of

the steam in the steam tables in the appendix, subtract the enthalpy of the water supplied to the plant and multiply by the number of pounds of steam produced to get an output in Btu. Divide by 100,000 to convert to therms and one million for decatherms or million Btu.

If you're getting condensate back, you'll have to meter it or subtract makeup and blowdown from steam output to determine the quantity of it. Use the enthalpy in the steam tables for water at the condensate temperature. Multiply by pounds of condensate returned to get Btu. Adjust that result to match your output units and subtract from the steam output to get plant output.

Maybe you're generating electricity too, use the conversion and add that to your output.

For hot water plants determine the water flow rate. Hopefully it is constant. Convert gallons per minute to pounds per hour then multiply by the number of hours in the day, week, or month you're evaluating. One gpm is approximately 500 pounds per hour so multiplying gpm by 500 is close enough. The time period is determined by how you measure your fuel usage. If you're relying on the gas billing it's usually the month and you'll use 720 or 744 hours depending on the month (except February which will be 672 or 696). Once you have the number of pounds you were pumping around you multiply it by the temperature difference of the water. After all, the definition of a Btu is the amount of heat required to raise the temperature of water one degree.

You'll have to use an average temperature for return water (or supply water if you control on the return temperature) to calculate the output. Since the loads swing, a Btu meter, which constantly performs that calculation, should be an integral part of your plant so you can measure your output.

That's it, plant efficiency is your output divided by input. You can calculate it regularly or use some of the rate measurements we're about to cover. So, what do you do with it? You compare it! By measuring your plant efficiency you're developing a measure that will allow you to determine, first and foremost, if the plant performance is consistent, increasing, or decreasing. You want to produce the highest efficiency or highest rate of output per unit of input that you can. It's called burning less fuel and using less electricity while still satisfying the load.

So, you measure it to determine where you are. You'll discover that running one boiler instead of two makes a big difference. You'll find out when you shut down the continuous blowdown heat recovery system that it costs a lot more to operate without it. However, continuous blowdown saves more money in water than it does in fuel.

Now I hope you're beginning to see where you can make some difference. All that attention to the tuning of the boiler to get optimum boiler efficiency is not as productive as making certain that the energy converted to steam and hot water is used efficiently.

Plant efficiency deserves all our attention because it is the sole purpose of the boiler plant to deliver heat to the facility. I'm careful to point out that when I say "facility" I mean the buildings, production equipment, etc., served by the boiler plant. The facility itself is involved in the energy equation under these conditions because it can contribute to the performance of the boiler plant. It does so primarily by returning condensate and, in some cases, generating some of the steam or producing some of the heat.

A facility can also waste much of the heat energy produced in the boilers to increase fuel and electricity consumption. It may not be your responsibility to reduce that waste but you should be monitoring and documenting it for the benefit of the owner so it can be reduced. To identify your own overall performance, calculate the plant efficiency as defined. To get a measure of the facilities performance, compare fuel used to production quantities (production ratios) heating degree days, or a formula you develop that accounts for the load variations.

You can also keep track of the difference in energy returned by the facility. It can make a difference. If the third shift is assigned cleanup and discovered that the hot condensate did a better job of cleaning than the heated domestic water you would catch them doing it. After all, condensate is distilled water and it will dissolve a lot more than city water.

Which efficiency should you use? Well, I've already said plant efficiency is the one you should monitor for overall plant performance. For comparing boilers use what I call the boiler operating efficiency which is basically combustion efficiency with an accounting for radiation loss.

Blowoff and blowdown losses as explained earlier are functions of water treatment and operation, not boiler efficiency. They have to be accounted for in Plant Efficiency because the heat lost to blowoff and blowdown isn't delivered to the facility. Steam generated that's used in the deaerator isn't delivered to the facility nor is steam used to heat the plant.

For all practical purposes every piece of equipment has an operating efficiency that is separate and distinct from predicted efficiency. We seldom manage to operate equipment at its designed capacity so we should be aware of what it's efficiency is at the actual operating conditions. When we lower steam pressure, or raise it, we've changed operating conditions for the boilers, economizers, boiler feed pumps and system steam traps. An increase or decrease in pressure will alter the pressure drop in steam mains to amplify the change at the steam utilization equipment. In some cases we'll have charts or graphs that will predict the efficiency at the new condition. Some, like pump curves, do so with an accuracy that we can use. We may have to measure performance of other equipment to determine if the change is beneficial or detrimental.

In some cases operating efficiencies are described using terms other than percent. Chillers, for example, will list the kilowatts per ton values at different loads. In those instances the important thing to know is whether the ratio should be increased or decreased to increase efficiency. As operators we don't have to know the value precisely, we only need to know whether we want to increase it or decrease it. In the case of kW per ton we want to decrease it. In the next section we'll discuss some of these parameters which are much easier for boiler operators to use.

At the risk of being accused of trying to generate too many new terms I'll stick my neck out and talk about "cycling efficiency." It isn't addressed in any of the literature and is not given the attention it deserves. I've discovered it's very important and have developed an analysis method to determine it. It's surprising how many boilers are out there serving a load only by cycling. Very few of them are in boiler plants manned by operators but you may have to attend to one.

Whenever the load on a boiler is less than that boiler's output at low fire the boiler has to cycle to serve the load. All the time it sits there it's radiating heat, that radiation loss that's only a few percent at the most at high fire but may be 10% or more of the input when it's cycling. When the pressure or temperature control switch contacts close the boiler starts, warms up, and serves the load until the pressure or temperature control switch contacts open. Every time it's off the boiler loses heat to the load and air drafting through it. When it starts the boiler loses heat as the purge air cools it down. Those heat losses, purge air cooling and off cycle cooling become very significant as a percentage of the input. Cycling efficiency accounts for all those losses.

Now most engineers will tell you that it really doesn't matter much because the boiler input is very low when it's cycling. That's true, but a boiler that is serving a load at 5% of capacity may be operating at a cycling efficiency of 30% or less which means it burns more than three times as much energy in fuel as it delivers to the facility. Now consider the fact that so many boilers are outsized so they're running at those low loads most of the time and that cycling efficiency becomes meaningful.

Uh oh! Used another word that isn't in the dictionary. Outsized means the boiler is no longer the right size for the facility. With added insulation, sealing up air leaks, adding double glazing, and other activities to conserve energy we have decreased the load so much that the boiler is now too big for it. It got outsized! I can't guarantee that it wasn't too big to begin with because that's usually a fact, but calling it outsized doesn't raise the hackles of engineers like telling them they oversized it does.

When a modulating heating boiler is cycling at temperatures that are halfway between the winter design low and 65°F cycling efficiency has to be determined because it's so low that replacing that boiler with one that's the right size (perhaps you've heard of right-sizing, it's been the rage) fuel savings will pay for it in one or two heating seasons. Use that half the load and cycling determination to identify boilers that are cycling excessively and get an engineer to do an evaluation to determine if the boiler should be replaced.

Perhaps the boss won't go to the expense of hiring an engineering firm to do it but you might suggest he contact an ESCO (Energy Service Company) and invite them to look into it. ESCO's install modifications to plants to reduce energy consumption and get their money back from the savings with no money layout by the owner.

PERFORMANCE MONITORING

Calculating boiler efficiency may not be considered part of the duties of a boiler operator. Monitoring and optimizing plant performance is. To make it simple we use values that are less complicated to determine, and easier to understand and work with. Of course you have to understand how they're calculated, and whether you want the results to be higher or lower to indicate an improvement in performance or they're a waste of time.

If you want to work in terms of efficiency the previous section provides guidelines to do just that. Don't be surprised if you get numbers that seem out of place but don't accept them as true either. It's simply unrealistic to believe something can operate at more than 100% efficiency, even if the calculations are accurate.

The best method for evaluating steam boilers is evaporation rate. Divide the quantity of steam generated by the gallons of oil or therms of gas burned to get it. Don't, as one plant in Missouri did, simply put 122 in the column on the log for evaporation rate because that's what it is. In that instance, and in many others, I found the operators put a value in the log that the chief wanted so everyone was happy. It wasn't anywhere near the actual value which could be calculated from the other entries in the log.

In the case of the Missouri plant I upset everyone because I did the math and showed the actual value was around 108 pounds of steam per gallon of oil and that two of the three operators managed to run the plant so their value was 105 while one managed to maintain 114. Once the other two were clued in as to what they were doing wrong, and settled down, the average went to 114. There were sound reasons why the plant couldn't manage an evaporation rate of 122 but, since the manager wouldn't accept anything less, the operators put what he wanted in the log book.

Evaporation rate can be used to compare boilers to each other and to performance at other loads and at other times. It's comparable to a boiler efficiency as far as variations is concerned. A change in evaporation rate should be relative to a change in combustion efficiency.

Of course that doesn't come close to monitoring plant efficiency. For that you have to compare the delivery rate, how many pounds of steam you deliver to the facility divided by the amount of fuel burned in the same time frame.

The actual value of the number itself isn't important. The object of calculating these rates is to see if they've changed and, if so, did they change for the better. Whatever you use it should be treated as a flexible number with a goal of increasing or decreasing it depending on how you calculate it. The concept is exactly the same as monitoring your gas mileage on your car where the miles per gallon dropping off indicates there's something wrong or you just did a lot of city driving you normally don't do. Changes in the rate can be an indication of improved performance or changes in the load.

Evaporation rate provides a value very consistent with boiler operating efficiency and delivery rate is consistent with plant efficiency so they are good parameters to measure, log, and compare to monitor your performance and the performance of the plant.

Evaporation rate can indicate problems that can't be determined by combustion analysis or other methods of monitoring boiler efficiency because the latter are instantaneous readings. Frequently combustion analysis are performed while the boiler controls are in manual and the service technician has adjusted them to optimum. That can be a significantly different condition when compared to operating at varying loads in automatic.

Okay, so you have a steam plant but no steam flow meter. Well, you're not unusual. There are still ways of determining the amount of steam generated. A simple one in many plants is achieved by installing a twenty dollar operating hour meter on the boiler feed pump motor starter. This will work in all cases where the pumps are operated to control the boiler water level. The pump has a listed capacity in gallons per minute which, when multiplied by 60 gives you gallons per hour then multiply by 8.33 (or the actual density) to get pounds per hour. Multiply differences in hour meter readings times the pump capacity, 60 minutes per hour and density to determine how many pounds of steam you made then divide that by the amount of fuel burned to get evaporation rate. If you have a lot of blowdown then calculate it's percentage, subtract that from 100, divide the result by 100 then multiply that result by the meter reading to get steam generated.

Oh, it's a hot water plant; well, that's a little more difficult. If the water flow through the boiler is constant a recorder for the water temperatures will provide you with an average temperature difference and you can multiply that by the water flow to determine how many Btu's went into the water. If the boiler water flow varies you'll need a Btu meter that calculates the heat added based on flow and temperature. Any decent sized plant will have a Btu meter that makes that calculation.

Check out your situation, since a Btu is the amount of heat added to one pound of water to raise the water's temperature one degree you just have to get the degree rise and number of pounds figured out. Number of pounds times temperature rise gives you heat out and dividing that by fuel used provides Heat Rate. Since most hot water plants are heating plants you may find you can get along with a degree day ratio.

Plant efficiency can also have a relative parameter that's easy to calculate. In many cases it's not so easy but we'll get to that later. If the plant is used solely for heating then you can use a degree day ratio. Divide the quantity of fuel burned by the number of Degree Days in the same period. You will probably find that the ratio changes with load so you should always compare gallons per degree day or therms per degree day to periods with the same or a similar number of degree days. That value is the opposite of evaporation rate, you want to keep it as small as possible.

If the boilers are also used to heat hot water, the hot water use is reasonably consistent with variances that are insignificant compared to the heating load so you can treat it as a constant value. Refer back to that earlier discussion on knowing your load.

If your boiler is serving an industrial plant you have the potential for a variety of plant efficiency comparisons. There are pounds of product per pound of steam, a very common measure, and complex calculations that vary depending on the industry, method of production, and product manufactured. Usually these plants are large enough that process steam metering is justified so you can work with a Plant Rate, pounds of steam delivered to the plant divided by the quantity of fuel consumed.

No fuel meters? If you're firing oil then all you need do is sound the tanks regularly and after every delivery. If you're firing gas the gas company always has a meter you can use. If firing coal there has to be some way to get an idea of the weight burned.

In plants that are so small that the price of a fuel meter isn't justified the boilers usually fire at a fixed rate so another twenty dollar operating hour meter connected to the fuel safety shut-off valves will give you a reading. You can go to the trouble of determining how many gallons or therms were burned but a formula as simple as hours of operation divided by degree days will give you a performance value you can monitor. Put another operating hour meter on the feed pump and you're comparing fuel input to steam output. Don't bother with all the other math, just divide the difference in readings of one meter by the difference in readings of the other.

Always make sure the ratios you use are quantities divided by quantities or flow rates divided by flow rates. I sometimes think we should use a different word for some of these ratios because A "rate" implies flow when, in fact, it has nothing to do with flow rate in this context.

Keep in mind that, unlike your car, the boiler plant is in operation 8,760 hours a year so a little change in fuel consumption represents a significant change in cost of operation. Monitoring the performance using one of the several ratios available to you will allow you to make those little differences in plant performance that can amount to significant reductions in operating cost.

MODERNIZING AND UPGRADING

There are two ways of looking at modernizing and upgrading. An operator either arrives for work one day to find contractor's personnel swarming around the plant or the operator simply sits and dreams of what would be nice to have. Occasionally there is some blend of the two but, for the most part, operators only get to experience one or the other. There are ways to become more involved in any modernization or upgrading of your plant. Even if you can't get involved you should respond to an upgrade professionally.

When we were looking at a project for Power and Combustion I tried to make time to get to the plant to discuss the modifications with the operators. Usually that visit benefited us because the operators were always willing to reveal the skeletons in the closets that might come out to bite us during the performance of the project. In many cases I managed to learn what wasn't working and what had been a problem so I could modify the design to correct or eliminate those things.

It's recent encounters of that nature with operators that convinced me this book was something that was needed. I encountered operators totally opposed to the concept of the project and for many of the wrong reasons. In some cases the operators simply misunderstood and in others they had a perception that was erroneous. I've learned to treat perceptions much differently than I used to because a perception is reality to the person that has it and in many cases I can't confuse them with facts—because they've made up their mind. I guess that's the first suggestion I can come up with when you're faced with some plant modernization or upgrades, don't close your mind to it and insist it'll never work.

If you are one of those people that chooses to decide it will never work, I'll watch out for you. I have first hand experience with operators proving their point by what I would call sabotage. If you do decide to insist it'll never work then I'm going to try to be on your side. I've learned through some very bad experiences that when an operator says it'll never work, it won't. I know that because the operator makes damn sure it won't work. That operator is in the position to prove his or her prediction.

I've also learned that a lot of engineers dismiss an operator's contention and put the project in anyway, figuring the operator will learn to live with it once it's demonstrated that it does work. Most of the time it does work, but only until the engineers and contractors leave. I'm not accusing any boiler operators of anything, it's what happens because nobody bothers to spend enough time with the operators to show them it does work and how they should operate it.

If only an operator would be honest enough to say "Hey, I don't understand it and if I don't understand it I won't be able to keep it running" instead of saying it won't work. Try it if this situation comes up, you may find that you're respected more for your honesty than

your knowledge and, hopefully, you'll get the training you need.

Why do so many of us buy another Chevy or another Ford or another whatever it is we're driving? It's a matter of comfort, we're used to that make of car and the one we have has treated us well so we go buy another one. Occasionally someone will see another make and decide that next time I'll buy that one because it looks, seems to perform or whatever better than what we have. Of course if you're like me you would love to have a Corvette; it's just that we can't afford one. When it comes to boilers there aren't any ads on television or in the paper that tell us what else is out there and that's a problem.

There are ads for boilers in trade magazines and ways of learning of other makes of boiler and burner and you should take advantage of that. I once had the misfortune of winning a contract to replace an old HRT boiler with a rotary cup burner run by an operator that had never seen anything else and was insistent that he get the same equipment, just new. It didn't matter to him that the old boiler was very inefficient and the burner was illegal, he knew them and that's what he wanted. The toughest part of that job was getting that old timer to even look at the brochures and instruction manuals for modern equipment. When I finally decided to incur his wrath by telling him point blank that he wasn't going to get his old boiler and burner back and he had better try learning about the new one his response was unexpected. He shrugged his shoulders, said "okay" and reached for the instruction manual. That was a success story only because there was no way to satisfy his desires.

I've seen many a boiler plant rebuilt to look just like it did simply because the boiler operators wanted the same thing they had. I've seen antique equipment with promises of very expensive parts and service bills installed as new. I've seen boilers so old and inefficient that they should have been replaced years ago fitted out with new burners and controls. I've seen more bad engineering performed because it was the will of the boiler operators than for any other reason and, I'm ashamed to admit, did some of it myself because there was no other alternative but walking away from the job.

Many engineers and contractors are more than willing to give the operators what they want. It's easy for them to copy what's there. It doesn't take any imagination and it doesn't really require any engineering. I know that millions of dollars of fuel go up the stacks of plants that were expanded, supposedly modernized, or upgraded with no improvement in performance all because the operators had no vision. But, because the higher-ups in the organization didn't know and wouldn't oppose their operators, their requirements were met. I hope you don't repeat that error.

I'll cover one more point on this side of this subject and then quit making some of you feel guilty. The reason many operators object to any changes in a plant is they feel their job is threatened. I've seen many situations where plant changes were made intentionally to reduce personnel. There's no guarantee that it will not happen to you, regardless of the fact that eliminating operators can't possibly save money because plants left to their own will not operate as efficiently. I've only seen a couple of instances where money was truly saved and it was because the operators originally didn't do anything but show up for work.

In today's market there's no reason to fear being put out of a job. Qualified, experienced boiler operators are becoming a rare commodity. You may have to change jobs but you won't be out of work long. I really doubt if you will be laid off with any plant upgrade or modernization because you're interested enough in doing a quality job to purchase this book. No wise employer will get rid of a wise operator. Just last Tuesday an employer told me frankly that he had to eliminate the steam plant but he was going to keep all his employees by transferring them.

If you know the equipment you're operating is inefficient, always breaking down, costing too much to maintain, etc., then you might just be able to demonstrate to your employer that it would pay to replace it. The typical employer is concerned first for the reliability of the plant and secondly for its cost of operation. Actually I'm not certain that many of them really realize how much it's costing them to run their plant; many of them never think about the sum total of all the monthly fuel bills.

Anyway, you should be aware of how your operation compares with others and what's available to improve the operation of your plant. That requires obtaining information on how other plants perform and what's available to improve the operation of yours. I've never attended a NAPE (National Association of Power

Engineers) meeting because I spend enough time with ASME, ASHRAE, AEE and others but I still believe every boiler operator should belong to that association, its an association for boiler plant operators.

Attending the regular meetings of your local chapter of NAPE will give you an opportunity to talk to other operators and learn what they're doing. There are also a considerable number of publications, mostly magazines, that target decision makers in boiler plants and similar facilities and a lot of them provide the subscription at no cost; the advertisers pay for the cost of publishing them. If you join NAPE you'll probably get a lot of invitations to free subscriptions to the magazines. That association and others like it are your best resource for information. Use them to increase your knowledge about the industry and you'll be prepared for whatever comes down the road. You'll also be knowledgeable enough that your opinions will be welcomed in any planning for modernization or upgrades in your plant.

Even if you don't have a say in the modernization or upgrading of your plant you do have a part to play. The first and most important thing to do is listen. I wish I could learn to follow that piece of advice myself, I seldom listen long enough; I allow my mind to start winding up before I hear the whole story and then stick my foot in my mouth. It's hard, I know it's hard, but whenever you try to just listen and say nothing until asked you're a lot better off. You'll learn what's going on and you'll gain insight into what will happen.

Right after listening comes reading. I've said it before and I'll continue saying it, the wise operator is the one that reads the instruction manual. I've had experience with manufacturer's service engineers that didn't read their own instruction manual and enjoyed laughing at them when the supposedly dumb boiler operator pointed out the solution to their problem in their own book. Every piece of equipment is unique and has its own unusual features, sometimes just to make them different from everyone else's, and those features should always be in the instruction manual. There will come a time when you will be expected to operate that new stuff and you better be prepared.

Chapter 4

Special Systems

A working knowledge of steam systems makes it possible to understand the use of special and unique systems and heat exchange materials because all the rules of heat and flow don't change with a system or the fluids used as the heat exchange medium. This section provides a little insight into some of the special systems that a boiler operator can encounter and may be called upon to operate.

SPECIAL SYSTEMS

You can always read the instruction manual but just in case you happen to encounter one of the special systems found in some boiler plants I thought I would touch on them here. You may never encounter one but if you do, at least you'll have an idea what you're dealing with before you open the instruction manual.

VACUUM SYSTEMS

In the chapter on energy we touched on what happens in steam systems with temperatures below 212°F but there are systems that are designed to operate with a vacuum. Vacuum pumps (Figure 4-1) intentionally produce a vacuum by removing air from the piping system, both the original air on start-up and air that manages to leak in. Condensate flows to the vacuum system which is operating as the lowest pressure in the system and is pumped out to the boiler feed tank or deaerator. The system shown in Figure 4-1 is a common one that produces a vacuum by pumping water through a water jet that acts as an ejector to pump the air out of the system. The vacuum system allows users of the heat to operate at lower temperatures, maybe a necessity in some situations where there's a concern for someone touching a radiator and the problem is solved by operating at 25 inches of mercury where the steam temperature would be 134°F.

You won't run into many vacuum systems because they've been declared unworkable by many engineers and boiler operators. A singular big problem with them

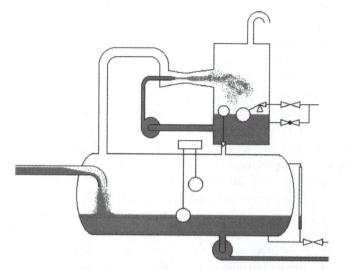

Figure 4-1. Vacuum pumps for condensate system

is air leakage which is impossible to locate during normal operation and even when you can pressurize the system they don't show up because a drop of water or piece of scale can prevent water leaking out but will allow air to leak in. Once air leaks start they tend to get worse because the air dries out the joint sealing compounds. Technology could probably provide us with a joint compound that could maintain a seal in a vacuum system but the horse has already escaped the barn.

Another problem I encounter regularly with vacuum systems is someone works on the system with no knowledge that there's a vacuum pump back at the boiler plant and they put in a vent. Now you're assured of a leak because someone created it and it looks perfectly normal. I find open vented condensate return units on vacuum systems regularly. If someone does this to you the simple solution is to connect the vent to the steam line instead of atmosphere when the tank can take the steam pressure. You'll also have to install a valve so you can service the unit and put a liquid trap in the overflow line to block it. The water in the trap tends to dry out so you have to have a way to refresh it as well.

Since there's so few of these systems around I'll just suggest you use the manufacturer's instruction manual as a guide and other information in this book that should help you understand what's happening with them and how your SOPs, etc., should address them.

HYDRONIC HEATING

Much of this book addresses the steam generating boiler plant and, while much of what we cover applies to water heating as well, there are many considerations in a water plant that are not a concern in a steam plant. Hydronic is just a word we use to differentiate low pressure hot water heating systems from other types of boiler plants. I tend to use whichever label is selected by the people I'm dealing with, hot water one minute and hydronic the next but that's simply to make the other people comfortable by using their label.

Unlike a steam plant a hydronic system can be shut down without admitting air to prevent a vacuum. For that one reason hydronic systems should last at least twice as long as a steam system under otherwise equal operating conditions. How long is that? About 60 years.

It's the system of choice today for residential boiler applications and most commercial buildings because it doesn't require as much attention as a steam system. Properly maintained it will require a minimum of make-up, almost nothing at all when new, and therefore need little attention to chemical treatment. With all that said, there's some reason to wonder why anyone even considers having an operator in a hydronic heating plant but I think I answered that question already.

You don't have to admit air to a hydronic system like you do steam because the change in volume from operating to idle is not significant. That doesn't mean that changes in volume are no concern for the operator. The problem with most hydronic systems is due to changes in volume that aren't accounted for in various stages of operation. Close off a section of steam system and the steam will condense leaving a vacuum that might permit atmospheric air to crush some thinner walled vessels attached to the system, that's all that will happen. Of course one of those vessels could be a $60,000 stainless steel heat exchanger! That happened.

Hydronic systems will also produce a vacuum as the water cools so you should expect air in that piping if you isolate it. Hot water and steam piping is usually strong enough that it can withstand the vacuum and nothing happens. Close off a section of chilled water piping in a building so that water is trapped and you have another story. As the chilled water heats it expands to build up pressure rapidly. It will rupture the piping if it can't leak out somewhere. Unlike steam and air water isn't compressible. The best thing to do is close only enough valves to stop flow, not so many that the system is completely isolated. When isolating for maintenance, open some vents as soon as the system is isolated.

Hydronic heating systems must have provisions for thermal expansion. When you heat water from a nominal building temperature of 65°F to an operating temperature of 180°F each cubic foot of water in the system will swell by almost 3%. That's not a lot percentage wise but when you consider the total volume of a heating system that can be several hundred gallons. A plant that's waterlogged (all elements full of water) can experience extreme swings in pressure associated with the expansion and contraction of the water. An expansion tank is provided in a hydronic heating system to reduce pressure swings to a tolerable range.

The tank can be an open type, located above the highest point in the system at a height adequate to maintain the desired system operating pressure. The top of the tank is open to atmosphere and the gage pressure at any point in the system is a function of the height of the water. The tank has to be large enough to accept the expansion of the water in the system without a considerable change in level because the system pressure will change about 1 psi for every 2.31 foot change in tank level.

Sometimes the tank is too small to handle full expansion and the water overflows from the tank as it expands. A float valve can be added to replenish the water when the system cools. Open tanks are used infrequently and normally only in systems using ethylene or propylene glycol and rust inhibitors for freeze and corrosion protection. The only time I've encountered these tanks they're on cheap systems in locations that contained glycol and received very little maintenance. The principle problem with an open tank is it allows oxygen to get into the water with corrosion as the outcome.

Closed expansion tanks can be a simple pressure vessel or be fitted with a neoprene or Buna-N bladder that separates the water in the system from the air that provides the expansion cushion. Pressure maintenance in systems with closed expansion tanks is established by controlling the air pressure over the liquid and/or the amount of water in the system. Some systems use nitrogen instead of air to eliminate the oxygen as a source of corrosion of the tank and system. Tanks without bladders are usually epoxy coated internally, that's why they have those "do not weld" stencils that someone painted over several years ago. (That was a another snicker generator, a comment that indicates what some people manage do to destroy a plant, hopefully you're much wiser)

Most plants are served by an expansion tank that can take the full swing of expansion from an idle condition to design operating temperature. A few plants, however, either due to space or price limitations, or as a result of expansion of the building and adding boilers without changing the expansion tank, will not have enough room in the expansion tank. All systems are normally fitted with a make-up pressure regulator that admits city water to maintain a certain minimum pressure in the system and a relief valve that will drain off water when the pressure builds.

Open and simple closed expansion tanks are fitted with a gauge glass so you can see the water level and know what's going on. Bladder type tanks do not provide any indication of level unless special instruments are provided or you have a good ear and can get to the tank to tap your knuckles on it. I prefer a simple closed tank because, in addition to knowing what's happening in the system by looking at the water level, you can add a low water cutoff to any tank mounted above the boiler for primary protection in the event of a loss of water.

The tank low level cutoff can't work alone because steam can be generated in the boiler to displace water in the tank so you don't get a low water indication at the tank. That's why you need a low water cutoff on the boiler and why a low system pressure alarm switch, shutdown if the plant isn't attended, is a necessity as well.

Unlike steam plants the fluid in a hydronic heating system doesn't move around on its own. You'll find I swap the words water and fluid around when talking of hydronic systems. That's because many of them use a glycol mixture, not just plain water. The glycol changes the boiling point of the fluid so you need another set of tables besides the steam tables but they otherwise work the same.

Steam will readily flow from one point to another with a very little difference in pressure. A hydronic heating system is full of water with the only pressure variation being the elevation at a particular point. There may be a little thermosyphoning going on where lighter hot water is lifted up as heavier cold water drops down to displace it but it's never enough for heating any reasonably sized system. You might find what we call a gravity system in a house where the pipes are large enough to allow the liquid to move around but I doubt if you'll see it elsewhere. So, for most installations there's no pressure differential to force the heated water out of the boiler and to the load.

That's why every hydronic heating system has circulators. Circulators are pumps that push the water around the hydronic heating system. They're not sized to fill the system, nor capable of pushing the water up to the highest level in a system. They are selected to overcome the resistance to flow through the system at the designed flow rate and that's all they do. If there

is any large volume of air in the system it will create differential pressures that can prevent or limit system flow (Figure 4-2) because the pump wasn't designed to overcome that differential. The pump in Figure 4-2 was designed to pump the water around the system. Once air accumulates in the radiator to produce a condition where the water drains to the boiler the pump has to push the water up to the radiator and frequently doesn't have the ability to do it. Opening the vent on the radiator allows the pressure in the expansion tank to push the water up to displace the air. Air in water systems can create all sorts of problems.

If there's not enough in the system, and a makeup regulator isn't provided, you will have to add water to the expansion tank manually in order to to restore the operating level.

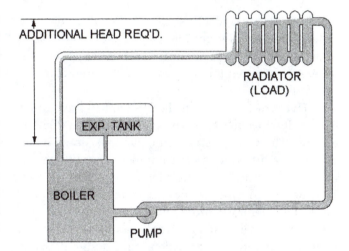

Figure 4-2. Differential produced by air in hydronic system

One neat thing about hydronic systems is they're easy to measure. Given the definition of a Btu all you need to know is the temperature in, temperature out, and the flow rate to know how many Btu's a boiler is putting out or how much a particular piece of equipment is using. That's true at any instant anyway. It's another story when you want to get average or total readings.

The flow rate has to be close to the rating of the circulator There are pressure drop curves (Figure 4-3) in the instruction manuals for most equipment so you can read the pressure drop through a coil and read the flow off the curve. I prefer a differential gauge but using the same gauge on both connections will give you a fairly accurate differential; just reading both installed gauges assumes they're identically calibrated and they almost never are. Two weeks ago I saw two gages on the same

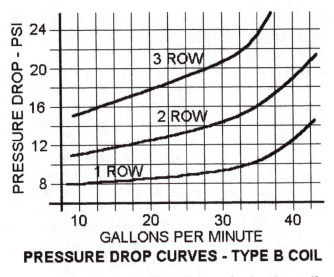

Figure 4-3. Pressure drop curves for heating coil

line read 30 psig and 21 psig, I wonder which one was right? You'll usually get the reading off a coil table in gpm so multiply by 500 (to convert gpm to pph) and the difference between the inlet and outlet temperatures to get the Btuh.

Hydronic systems in the US tend to have much higher flowing pressure requirements than systems in Europe. The Germans in particular look down on us because we introduce so much unnecessary differential in our systems and it wastes a lot of motor horsepower.[1] That's a matter of initial design. In many systems I've found the operators throttle down on a valve here and there to resolve heating complaints until the whole system is operating at a fraction of it's design flow and in other situations they adjust valves open enough that flow through some systems prevents flow in others.

Building owners don't like to hear that their distribution system is totally upset and they have to bring in a balancing company to put everything back in order, a task that is very expensive relative to building size. I'm not telling you to leave the darn thing alone, If you believe a small adjustment will solve a problem then try it; just count every turn or partial turn of that valve and log it so you can always return and put it back where it was. It's preparing to dig yourself out of a hole.

Sometimes the flow control valves in hydronic systems or piping loops themselves accumulate mud and sludge because the flow is slow enough to allow the sediment to drop out. What should happen is the accumulation reduces the size of the flow stream so velocity increases until a balance is reached where no more material accumulates. In the initial years of a building system that sediment accumulation can reduce the flow through the loop so it's necessary to open a throttling valve a

little to return to the design flow.

If you're going to do it, do it right and use the measuring device (you may have to rent it) and the flow sensing taps on the valve and restore the design flow which should be shown on the piping drawings. While you're at it, check some of the other valves in the same area to be certain you didn't alter their flow rates, taking readings on them before and after you make the adjustment on the one. See the chapter on flow.

Sometimes it's just a matter of blowing sediment out. Before we had balancing records for systems I would recommend opening a valve on each loop after noting its position then counting the quarter turns and restore its position afterwards. The temporary jump in flow would flush out that particular loop and may return its operation to normal.

Hydronic systems need blowdown just like steam systems. You shouldn't have a lot of sludge and sediment in a system. The problem is—there's always a little bit of it; water contains solids and we add chemicals to treat it so there's some in the water. It will be swept along in the areas of the piping that have higher velocities and settle out in the areas that have the lowest velocities.

Systems with sections designed for future expansion include piping larger than necessary for current operation so the velocity in those sections will be considerably lower than individual unit loops and other parts of the system. A unit loop is piping from supply headers to return headers that serves one piece of equipment that uses the heat.

When you have future service connections they are the ones you should use to blow down occasionally to flush the mud and sediment out because that's where it will settle (in addition to the bottom of the boiler). If you don't clear them occasionally the sludge will build until it can be swept up in chunks by the flowing water and jammed into a smaller distribution or unit loop, then you'll have a real problem to fix.

As for how frequently you blow down a hydronic system, it depends on how much of what quality water you add to the system. I always recommend installation of a meter on the makeup water supply for a plant because that will be your guide to how much water you've added. Then it's simply a matter of knowing the quality of the water to see how much mud, sludge, etc. you added along with that water.

The mud and sludge which is dirt that entered with the makeup water and sludge created by the water treatment to remove scale forming salts doesn't leave with a water leak unless the leak is a big one. Usually the leak is in the form of steam. If you heat water to 220°F a lot will flash off as it drops in pressure at a leak and flow out as pure steam. All the mud and sediment that was in that water stays in the system. It's one reason leaks aren't as much of a problem, the remaining mud and sludge plug the leak.

It's safe to say you can blow down a new system once a month as long as makeup is minimal. Remember that blowing down removes water so you will have to add makeup water and more treatment chemicals with it to replace what you blew down. Watching the first gush out the drain valve will be the clue to frequency. Normally a hydronic system should be tested for TDS (see chemical water treatment) just like a steam system and the blowdown should be managed to keep TDS below a prescribed value (usually 2500 ppm). However, if you see a slug of mud (the water will be discolored) for more than ten seconds you're not blowing down frequently enough, increase the frequency. No sludge, decrease it.

TDS is dissolved solids, not settled solids so there's a distinct difference and unlike a steam system (where everything solid stays in the boiler because it can't become a gas and leave with the steam) the settled solids tend to pick many points in the hydronic system to accumulate.

Don't believe that old lie that you don't have to do any water chemistry testing and maintenance in a hydronic system. Even systems with zero leaks have problems with the water chemistry changing as it reacts with the metals in the systems and any air it comes in contact with. It's essential to maintain the proper pH of the system and a supply of Nitrite or Sulfite to prevent corrosion due to oxygen getting in. (See water treatment.)

If you have system leaks that must be replaced by makeup water then that water has to be treated. As systems grow older the number of leaks tend to increase, despite good maintenance practices; and the water treatment program has to improve to handle the large volumes of makeup water. Many hydronic systems are equipped with nothing to pretreat the water (see water treatment) so more chemicals are required and in many cases adding pretreating equipment is justified.

In my experience the major concern with hydronic boilers is preventing thermal shock. Be sure to read the chapter on thermal shock in the section on why boilers fail. It's particularly important when the plant has more than one boiler because you have to avoid sending a slug of cold water from an idle boiler into a system operating on another boiler and avoid dumping hot water into a boiler that's cold.

Most hydronic heating plants permit firing the boiler without any water flow through it so the boiler can be warmed up without pumping it's cold contents into the system piping. There might be situations and conditions where you have slugs of cold water in the piping even though the boiler is up to temperature and careful manipulation of the boiler's isolating valves is required to warm up that piping.

It's best to crack open one of the two valves (return or supply) connecting the boiler to the system before starting the boiler to maintain consistent pressures throughout the system. Leaving one valve open when a boiler is out of service but not isolated for repair or other purposes is not a bad idea. The selected valve should be in a position where thermosyphoning will not generate any thermal shock, sometimes warming the boiler up with a valve open allows thermosyphoning to warm up piping to avoid thermal shock. Since every plant is different you should develop an SOP that allows starting and engaging a hydronic boiler with minimal thermal shock.

Arrangements of hydronic boilers in multi-boiler plants come in two forms. Parallel installations (Figure 4-4) are most common and can be used with any number of boilers. Serial installations (Figure 4-5) are less common and the number of boilers is limited to two or three. In parallel installations each boiler handles a portion of the system water and care is recommended to ensure the water flows to each boiler uniformly.

In some parallel installations the system water is left flowing through each boiler so a boiler that is shut down acts as a radiator, wasting heat to the air that is drawn through it by stack effect to actually cool the system water. If you can't do anything else about this type

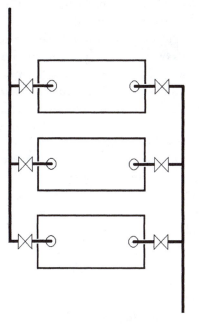

Figure 4-4. Hydronic boilers in parallel

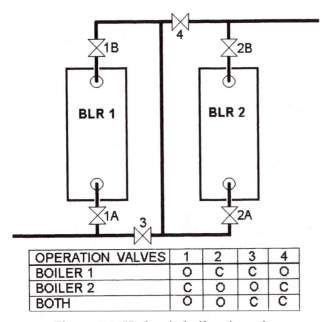

OPERATION VALVES	1	2	3	4
BOILER 1	O	C	C	O
BOILER 2	C	O	O	C
BOTH	O	O	C	C

Figure 4-5. Hydronic boilers in series

of arrangement put a cardboard blank over the combustion air inlet to minimize the airflow due to draft. The hot boiler will still waste heat to the boiler room as radiant losses and some thermosyphoning of the air will occur in the stack so it's not the best solution.

Closing one of the valves (supply or return) on an idle boiler will eliminate the heat losses but it will change system and boiler flows and those effects have to be considered. Some boiler plants have a bypass line between the supply and return headers that simulates the pressure drop of one boiler so you can open it after closing off a boiler to restore the flow rates in the operating boiler and system to normal.

When operating with less than the full complement of boilers on line and bypassing around or through others be aware that the system supply temperature will be less than the boiler outlet temperature because it is mixed with the return water flowing through the idle boilers or bypass. Some plants use a header temperature control so the idle boilers or bypasses doesn't change the hot water supply temperature. It will require higher temperatures in the operating boiler.

If you have a common header temperature control it should be on the return. These systems usually have a proportional control so the firing rate of the boiler will be proportional to the difference between return temperature and the set point (desired return temperature). The return temperature will be held near the set point but the supply water temperature will vary depending on the blend of firing and idle boilers or bypasses. It won't hold a constant return temperature either because there's a

delay in response to changes in the boiler firing rates.

Checking the temperatures and a little math will allow you to determine what percentage of the water is flowing through the operating boiler. When waters of two different temperatures are mixed the resulting temperature is dependent on the quantities of water at each temperature. The percentage of water flowing through a boiler will equal the difference between the mixed water temperature (Tm) and the return temperature (Tr) divided by the difference between the boiler outlet temperature (Tb) and the return temperature times 100; boiler water flow as % of total = (Tm-Tr) ÷ (Tb-Tr). This formula comes in handy when you want to know how much water is in each part of a mixture.

You can also use the basic formula for energy to determine how much heat is lost in an idle boiler, the temperature at the outlet will be lower than the temperature at the inlet. As in all cases where you're comparing differences in gauge or thermometer readings it's a good idea, where possible, to switch the devices so you have a different reading from the same instrument.

Series operation of hydronic plants requires the piping arrangement allow for total flow through each boiler and means for isolating the boiler which requires three valves, two valves to isolate the boilers and one for bypass as shown in Figure 4-5. The water is heated first in one boiler then its temperature is raised further in the second boiler. These systems commonly use a header temperature controller to regulate the firing rate so the two boilers fire at the same rate. When the boilers are controlled independently the modulating controller for the first boiler has to be set lower than the second one so it doesn't take all the load.

Without the common controller you will find yourself constantly adjusting the controller set points (or firing one boiler on hand) to fire the two boilers evenly. An alternative to the common controls is using the position of the second boiler as a controller for the firing rate of the first boiler, simply adding another rheostat to the modulating motor of the first boiler and installing a selector switch will allow both single and two boiler operation.

BOILER WATER CIRCULATING PUMPS

Boiler water circulating pumps are common on hot water heating boilers and consist of a small pump that delivers a fraction of the total flow through the boiler. The purpose of the pump is to help eliminate thermal shock. They pump enough water from the boiler outlet back to the inlet to produce a blend of return and circulated water to raise the inlet temperature well above the return water temperature. These should be started before the boiler and normally have proof of operation switches that prevent boiler operation if the pump doesn't operate.

If you live long enough and move around you might just run into one of these on a water tube boiler, they are rare. Some boilers were manufactured with the requirement that the circulation of water in the boiler had to be assisted by a boiler water circulating pump. The pump typically takes suction from a drum or header at the bottom of the boiler and pumps into a header a little higher. It's difficult to tell if the pump was provided as an element of the design of the boiler or added once it was revealed that natural circulation couldn't provide sufficient flow through the risers. I'm only aware of one situation where the downcomers were a restriction. On that boiler, a new and unique design, the manufacturer didn't provide any downcomers.

Because boiler water circulating pumps are designed to pump hot boiler water you may learn by reading the instruction manual that the pump should not be started until the boiler has been warmed up to a certain temperature or pressure. That's because the pump motor was sized for handling the hot boiler water with assistance in producing a differential from natural circulation so operating it with colder water would result in overloading the pump motor. You should also find the pump will have cooling for the lubricating oil or direct cooling of the bearings using an external source of cooling water because of high operating temperatures. There should also be cooling provided for the pump shaft seals and, of course, pump operation should require proof of cooling water flow before the pump motor can be started. You should find that there is a differential pressure switch used to prove the pump in operation incorporated into the burner management system to safely shut down the boiler in the event of the failure of the boiler water circulating pump.

HTHW BOILER PLANTS

High temperature hot water (HTHW) plants have all the characteristics, features and problems associated with hydronic systems. The defined difference is an HTHW plant operates with water temperatures higher than 250°F. The typical HTHW boiler plant has design conditions of 400 psig and 400°F. HTHW plants also have some other unique characteristics that are not found in the typical hydronic system. In most HTHW plants the boilers are called HTHW generators. They differ considerably in construction and operation. The typical HTHW generator (Figure 4-6) is a once through boiler.

Okay, I still call them boilers; because they are boilers. They're just unique boilers and that's why we call them generators. I've often wondered if they were called "generators" in an effort to exclude them from the requirements of the boiler construction codes but I have never researched it. They don't have drums and the headers are usually small enough that someone could argue that the code doesn't apply. Those generators require water flow through them to operate because they don't store any hot water, their water volumes are very low. The controls will include low water flow switches that prevent burner operation and will shut the burner down if the water flow in the boiler is too low.

The controls require Btu calculation with measurement of the return water in order to ensure the outlet temperature is close to steady. Flow through the boiler consists of several parallel circuits and the tubes are frequently orificed at the headers to ensure proper distribution of water. It stands to reason that a tube designed for water flow on a once through basis will have a real problem if steam is generated in it because the larger volume of steam will fill the tube. Once steaming starts in one of those boilers failure due to overheating rapidly follows.

Each HTHW generator is commonly fitted with its own circulating pump (standby circulating pumps are normally shared) to ensure adequate water flow. Normally, there are separate pumps used to circulate the high temperature hot water through the system.

The circulators (in an HTHW plant I've always heard them called circulating pumps) have to pump water much hotter than the standard pump. Even though they are installed to pump the water into the boiler like hydronic circulators they are exposed to temperatures that are so high the oil or grease in the pump bearings

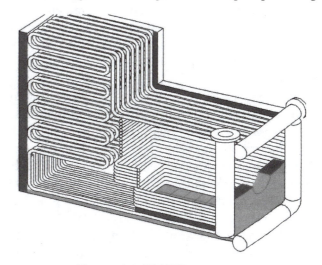

Figure 4-6. HTHW generator

could be overheated. The pump seal or packing would also be exposed to those high temperatures and few can handle it. Any leakage of the hot water along the shaft would start flashing into steam and that could do serious damage to shaft and seal or packing.

To prevent problems with the seals or packing the circulating pumps are normally fitted with sealing fluid systems. Where the seal or packing is exposed to the suction side of the pump sealing fluid is commonly drawn off the pump discharge. Some may extract water using a Pitot tube inside the discharge of the pump so the velocity pressure is used to generate the differential to move the sealing fluid. In others it may be necessary to have a seal pump draw water off the system and produce the differential necessary to force the water through the sealing fluid system. Newer pumps may be fitted with a special impeller on the shaft inside the seal housing that pumps liquid through the cooler and back to the seal.

Sealing fluid systems typically consist of two elements, a strainer to remove any particulate that might damage the pump seal, packing, or shaft, and a cooler to reduce the water temperature to values that the seal or packing can accommodate. After the sealing fluid passes through the strainer and the cooler it is returned to the pump to flow over the seal and back into the pump and, in the case of packing provide the little leakage that separates the packing and the shaft. In the case of packing it's supplied to a lantern ring (see pumps). Proper control of the cooling of the sealing fluid is required to ensure the fluid isn't overcooled to cause thermal shock.

The expansion tanks for HTHW plants are occasionally called accumulators. They can serve the typical expansion tank role but can also become a storage space for the hot water. To limit corrosion problems at the high temperatures they are always pressurized with pure nitrogen instead of air, although a true accumulator might be pressurized with steam and can contain electric heating coils to build up the steam pressure on a system start-up and to maintain pressure when the system is shut down.

It's common for the low water cutoffs to be mounted on the accumulator because the generators don't have any point where a low water level can be detected. To avoid thermal shocks in the system the makeup water is added to the accumulator where there's a considerable volume of water for it to mix with before it hits any metal.

Preventing thermal shock is even more of a problem in HTHW boiler plants. Most HTHW plants have more than one boiler (unlike the hydronic plant that typically has one) and the higher temperature operation requires careful management of the system when starting a boiler and putting it in service. The temperature differences between atmospheric and operating conditions are significant.

You should be careful so you don't suddenly expose metal at 80°F to high temperature water at 390°F. In some circumstances that's difficult to do but operations that mix the two fluids (hot and cold) to gradually warm up a boiler, pump, or piping system can be managed. Steps in bringing a boiler on line and taking one off line can get very involved because the pumping and piping arrangements have to be reconfigured to ensure even distribution of the load on the boilers.

I have encountered plants with piping arrangements that restricted single boiler operation during periods of low load to a particular boiler because the system arrangement didn't permit isolating the other boilers. In another plant where the facility load had increased significantly the design did not permit operating two boilers to carry the load because there was no way to arrange the piping to parallel the boilers. It's possible for HTHW boilers to operate in series but its uncommon and the piping arrangement has to provide for it.

Unlike low pressure hydronic plants HTHW boiler systems seldom have accumulators large enough to hold all the expansion of the system from atmospheric to operating conditions. A large pressure vessel designed to hold several hundred gallons of water is very expensive so they are occasionally reduced to a size that provides a cushion on the operation instead of allowing for complete expansion and contraction.

Those larger plants are equipped with provisions to fill the system as it cools from normal operating temperatures and tanks that allow steam to flash off and recover the remaining hot water as the system expands. In some cases the requirement for expansion tanks to accommodate normal operating temperature swings is so great that even smaller tanks with operating and standby provisions for fill and drain are installed instead, a lower pressure or open storage tank being used to prevent wasting the treated water as the system heats and cools.

Any HTHW system requires makeup water pumps to force the makeup water into the system. The pressure in a city water supply just isn't adequate. Lack of electric power in these plants can't be tolerated because the liquid in the system will cool and shrink to require makeup. A drop in pressure will result in steam flashing in some systems and driving water to others with much noise and pipe rattling. The emergency electric generator is very important and some plants even have engine driven makeup pumps as a backup.

There is one more point I would like to make about

HTHW plants. I consider them to be far more dangerous than any other kind of boiler plant. The heated water contains a lot of energy and any rupture of a piping system or a piece of equipment will result in a steam explosion. The rupture of an HTHW pipe will discharge almost 100 times as much steam as a steam pipe with steam at the same temperature. The number and location of exit doors from a HTHW boiler plant should greatly exceed those for a steam plant and any control room should have at least one exit that leads directly outdoors.

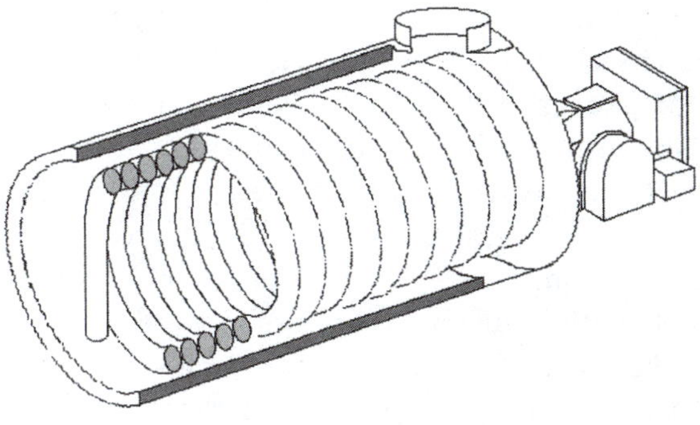

Figure 4-7. Fluid heater

ORGANIC FLUID HEATERS AND VAPORIZERS

Organic fluid is basically oil, hydrocarbons that are used as heat transfer fluids because they have much lower vapor pressures than water. What that means is they can be heated to higher temperatures before they evaporate. Organic fluids are available that will remain a liquid and not evaporate at temperatures as high as 800°F at atmospheric pressure. By and large these materials function the same as water and steam, they simply evaporate and pressurize at much higher temperatures.

Organic fluids are used to produce high temperatures without the expense of handling high pressure. A system can be designed to operate at 500°F (a common maximum operating temperature) and pressures not exceeding 30 psig where a steam or HTHW plant would have to operate at almost 900 psig. Both liquid and vapor systems are considered high pressure plants because the temperature is always higher than 250°F. The boiler is a power boiler even if the operating temperature is below 15 psig. A fluid heater is basically the same as a hot water boiler and a vaporizer is very much like a steam boiler, the principal difference is the operating temperature.

The typical fluid heater (Figure 4-7) looks a lot like a common firetube boiler from the outside and many operators confuse them with a firetube boiler. They're actually water tube boilers. What looks like an outer shell is a casing. The tubes form one continuous coil surrounding the furnace and in many cases are two coils to produce a secondary pass surrounding the furnace pass. Unlike a firetube boiler, flow has to be proven in these units before the burner is started and flow must be maintained or the burner should be tripped.

Other significant differences between steam and organic fluids include flammability, especially when they are heated to such high temperatures. If a water or steam boiler has a leak the tendency is to put the fire out. If an organic heater or vaporizer has a leak the tendency is to add to the fire. Almost any plant with organic heat-

ers will also have a steam boiler that must be in operation in order for the organic device burners to function because the steam is used to quench any fire that might occur in the organic device.

Normally a thermocouple in the outlet or stack is monitored and any rapid increase in temperature automatically results in burner shut down and opening of the steam quench valves. A few small units are fitted with compressed CO_2 extinguishing systems to avoid the provision of a steam plant but it takes a lot to put out an organic heater fire. Once it takes off, any leak adds enough fuel to melt more of the boiler metal to allow a bigger leak and bigger fire.

The higher temperature fluids tend to have high pour points. That means they don't flow well, if at all, at normal atmospheric temperatures and the system will freeze up on shut down. Fluid systems for those high temperature fluids use steam tracing to warm up the organic fluid enough that it can be circulated in the system in order to get it started.

One operator I know is very happy that he's operating the fluid heaters at his plant. He told me he's happy because "I don't have to fool around with water treatment." While it's true that organic fluids don't need the attention of a water plant, because the systems are designed to retain the fluids and vapors so there's no to little makeup, the fluids do break down and regular sampling and chemical analysis is still required.

Over a period of time the fluid can break down and has to be replaced or reconditioned. Scale as we know it in water based systems isn't a problem but carbon can build on the inside of tubes just like scale if the boiler is fired too hard, fluid flow is lost, or the fluid begins to break down, and that can eventually result in a tube failure. A tube failure can result in the entire heater melting down so there is a concern for proper operation to prevent carbon formation just like there are concerns for scale formation in a water boiler.

Monitoring the pressure drop across the liquid

side of a fluid heater is critical to detecting a buildup of carbon in the tubes. Monitoring is not as simple as reading the gauges at the inlet and outlet then subtracting the difference. Since viscosity changes with temperature you need to have a record of pressure drop at different average temperatures so you have relative pressure drops for comparison. You want to be as precise as possible with your measurements because you want to catch the carbon formation the instant it starts.

Even a very thin coating of carbon is so rough it can produce a significantly rough surface on the inside of the tubes so the pressure drop increases significantly. That's usually not a problem because the circulating pumps are normally positive displacement types that will continue to force the designed flow of fluid through the heater. When carbon builds up failure tends to be instantaneous because the increased pressure drop is handled until the pump motor is overloaded and trips out. Systems with centrifugal circulating pumps are uncommon because the viscosity variation with temperature has a significant effect on the flow in the system and the performance of the pump.

I did have one customer that solved his problem temporarily by installing a larger motor on the pump. It was nearly impossible to tell what was going on in the system because none of the pressure gauges worked. When they finally got some gages in place a high pressure drop was detected across the heater and they had to shut the whole plant down to retube it.

Any organic fluid system should be checked throughout its entire length at least once a shift with special attention paid to any signs of leakage. The insulation is typically calcium silicate in order to handle the high temperatures, and it's also very thick, so a slow leak can penetrate a lot of insulation (store a lot of fuel) before it's detected. System leaks are dangerously close to becoming fires and they must be caught before they become a fire; there is no steam quenching on the piping like there is in the furnace.

Since most organic fluid systems are used in petrochemical and similar industrial production plants immediate shutdown to repair a leak could result in thousands of dollars of production loss so you may be compelled to simply monitor a minor leak and be prepared to extinguish any fire that results until the entire facility can be economically shut down. It's one of those situations where the operator has to consider multiple risks and the cost of each; any leak that can't be made up, or becomes extensive to the degree it's a dramatic hazard, requires a shut down.

Shutting down a fluid system takes time so the growth of a leak also becomes a factor to consider. The fluid has to be circulated long enough to allow the heater to cool until it will not carburize the fluid left standing in it. It also has to be cooled enough so it will not spontaneously ignite when exposed to air, then the fluid must be drained from the system back to storage until the level is below the point of the leak. Some facilities don't have sufficient storage to completely drain their systems and require a supplier's empty truck, on rental, to hold the fluid as it's drained.

Organic fluid heaters and the occasional vaporizer make some chemical processes possible only because they can produce high temperatures at low pressure. A common application is in the asphalt industry where the product must be heated to high temperatures so it can flow readily. All the rules for high pressure boilers apply and every plant will have unique and special provisions that the operator should know. Among all plants these are the ones where the SOPs must be memorized because lack of rapid and proper response to an upsetting condition can lead to hazardous conditions or long-term shutdown of the facility.

SERVICE WATER HEATING

Service water is the term currently used by ASHRAE to describe what I always called domestic hot water heating. Heating of water for cooking, showers, baths, washing, etc., is not the same as heating water for closed hydronic building heating systems so we'll use the term "service water" to describe it.

Service water heating systems are frequently ignored. I didn't think about it much at my home because I have an electric hot water heater and it managed to operate trouble free for thirty years. Finally, the plastic dip tube failed, disintegrating into thousands of little pieces that fouled every faucet and toilet tank float valve until I was so frustrated I called the water company to complain about the junk they put in the water. It was a little embarrassing to have them tell me it was probably the dip tube then discover that was the case. Anyway, I figure my new electric hot water heater and it's dip tube will outlive me.

I wish you were all that lucky. It won't happen very often. Service water heaters do not enjoy the presence of chemically treated water to prevent scale and corrosion and most of them have such problems. I remember one area where the well water contained so much calcium sulphate that it would form heavy scale if the water temperature was increased by 6°F. There's nothing you can do to the water to prevent scale formation or corrosion so the equipment has to be made for the service and you will have to operate and maintain it properly to provide continued operation.

Service water heaters usually have much lower rates of heat transfer than steam and heating boilers to reduce scale formation. They are also fabricated for the application, some of them are glass lined with glass coated heating surfaces. We can't treat the water to make it non-corrosive so we have to protect the heater from corrosion.

The equipment sold in your area is usually suitable for service water heating of water used in the area. Don't do like a friend of mine that thought the hot water heater prices were too high in his new neighborhood and transported one from his old neighborhood in another state. He saved a lot on the heater, but it didn't last a year.

Small electric and gas or oil fired service water heaters require more attention in a commercial application than the ones in your home because they get more use. You should have a schedule for blowing them down on a regular basis to remove any mud, scale, or other debris that may accumulate. Regular checking and recording of the stack temperature is also a must for the fired heaters because that can indicate problems with scaling; as scale forms it insulates the heating surfaces requiring higher flue gas temperatures to do the heating. There's also some checking and adjustment required for storage water heaters to keep everything working right.

Since entering semi-retirement I've encountered a fair number of projects involving problems with storage water heating. I've also encountered many installations where someone felt they solved the problem by installing instantaneous water heaters. If you think an instantaneous water heater is an appropriate solution to any problems you may be having with your hot water system I urge you to reconsider.

Instantaneous hot water heaters do just what they say they'll do, heat water quickly, primarily as it is used. Except for facilities where the instantaneous hot water heating load is less than about 25% of the lowest plant loads those heaters can be a real problem for smooth and reliable boiler operation. It's also hard to believe an instantaneous heater is anywhere near efficient because they're capable of heating more water than is normally heated so they only operate a fraction of the time allowing considerable off cycle losses.

The amount of hot water used is a function of the activities of the occupants of the buildings. The curve in Figure 4-8 is based on ASHRAE data[4] indicating the typical hot water consumption for a family over a 24-hour period. It's obvious that an instantaneous hot

water heater has to be able to produce the quantity of hot water drawn between 7 and 8 in the morning but is required to produce a fraction of that load for the rest of the day.

Note that the chart is based on gallons per hour and does not show instantaneous flows that could easily exceed the values shown. In my home I can draw water at the rate of 920 gallons per hour, about eight times the maximum rate shown on the chart. However, since my bathtub has a capacity limit of approximately 200 gallons I would draw hot water at that rate for no more than a few minutes. An instantaneous heater with a capacity of at least 920 gallons per hour would be required to ensure a continuous supply of heated water. However, with a 200-gallon storage tank I am able to fill the tub and satisfy other household requirements with a heater that can heat water at the rate of 10 gallons per hour. Unless, of course, I intend to fill the tub more frequently than once a day.

Since most of us do not take 200-gallon baths that example is improbable. It does, however, do well to explain the difference between instantaneous and storage water heating. The best system will always consist of a proper mix of water heater and storage that handles the load without excessive cycling of the water heater. See the discussion on cycling boilers for reasons why excessive cycling is a problem.

When your hot water loads are large and variable a modulating burner on an instantaneous hot water heater will reduce cycling or eliminate it. Instantaneous heaters with modulating burners can only eliminate cycling if the burner's turndown capability exceeds the variation in hot water usage. As you can see from the figure, that would require a burner with a turndown better than 20 to 1. Such burners are very expensive so

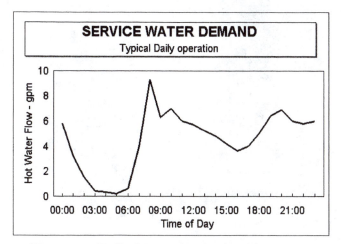

Figure 4-8. Daily hot water consumption curve

cycling is a normal condition.

In case you haven't already figured it out, I dislike steam powered instantaneous hot water heaters because they produce load swings in the summer that prevent smooth and constant operation of the boilers. Now that I've made my position clear (that storage is a necessity) it's time to talk about operation and control of hot water heaters.

Figure 4-9 is a graphic of a boiler and storage tank system typical of that used in a large apartment building. Cold city water enters the system at the bottom center of the graphic where it can either enter the circulating pump or the storage tank. Service water is drawn off the top of the tank. The arrow at the bottom right side of the tank represents flow of water circulated through the system to maintain hot water in the piping distribution system.

This combination of heater and storage will cycle but it has the advantage of extended cycle operation and a fixed firing rate for the burner that makes it efficient, but still simple to operate and maintain, if you know what you're doing.

A service water boiler deserves the same attention as a heating boiler on initial start-up. Before the system is started the owner, design engineer or installing contractor (depending upon the requirements associated with installation) should contact the owner's insurance company or the authority having jurisdiction (normally the state, county or municipality) to obtain a boiler certificate (or a document of similar title) which authorizes the owner to operate the boiler. There may be provisions in the jurisdiction to exempt certain equipment but any requirements should be determined before placing the system in service. Normally the boiler is subjected to a

visual inspection by a National Board Certified Inspector before the certificate to operate is issued.

Initial operation of the burner should be achieved under the supervision of a technician trained in the proper set up of a fired piece of equipment. That technician should produce a "start-up sheet," a document that includes, as minimum: The name, address, and phone number of the technician's employer, the technician's name and signature, and the date the initial start-up was performed; a record of the actual settings of the operating limit (OL) and the high limit (HL) temperature switches and an indication that their operation was confirmed; a record of the setting of the pressure-temperature relief valve and a record that its operation was confirmed; a record of the burner performance while firing including, but not necessarily limited to: stack temperature, flame signal measurement, percent oxygen in flue gas, carbon monoxide level of flue gas, if measured, smoke spot test recording (oil only) if measured, gas consumption rate (gas firing), temperature of water at the boiler inlet during normal operation, temperature of water at the boiler outlet during normal operation, pressure at the inlet of the system, pressure at the discharge of the pump or other location between pump and boiler, and position of the throttling valve. The start-up sheet should be retained as a part of the original documentation for the system and referenced on each subsequent start-up (after shutdowns for maintenance or other purposes) to ensure the conditions do not differ substantially from the original start-up conditions.

All openings into the boiler and tank should be checked to ensure the system is closed and will not lose water unintentionally when placed in service. Before closing openings the internals should be inspected to ensure there are no loose parts, tools, personnel, or anything else inside the system that does not belong there.

Valves and some spigots are opened to vent air and admit water until the system is flooded and at city water pressure. It is important to note that, if the city water supply to the inlet shown in the graphic is separated from the city water supply by a check valve or backflow preventer, an expansion tank or similar provision is required to prevent an increase in the system pressure when the water expands as it is heated.

Disconnects, circuit breakers, and control switches are closed (in that order) to permit system operation. The circulating pump should start first, followed by the burner. The start-up sheet should be checked as soon as operation stabilizes to ensure the conditions do not differ substantially from the original start-up conditions.

Since the first edition of this book was written the

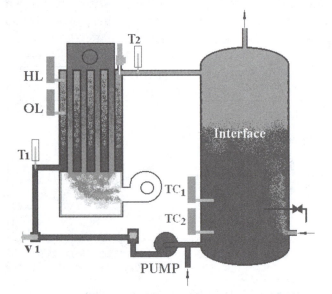

Figure 4-9. Service water heating system

Codes have changed and in every state and province all hot water boilers must have a low-water cutoff. If your boiler is equipped with a low-water cutoff its operation should be confirmed during fill-up and before the burner is fired. Since service water heaters are seldom fitted with a gage glass this test assumes a water level. Turn control power on before filling the boiler; the burner should not start. Then constantly monitor it to note that the burner starts to fire once the level reaches the low-water cutoff and closes its contacts. Then, immediately shut off the water supply and open the drain. If the low-water cutoff is operating properly it should immediately stop burner operation. If it doesn't secure the burner yourself and see to it that it's fixed. Secure the drain turn the power off and fill the system completely venting at all points before restoring power.

When stable operation is achieved the throttling valve (TV) should be adjusted to achieve the desired outlet temperature as indicated by the thermometer (T2) at the boiler outlet. Throttling of that valve is normally required to restrict the rate of water flow through the boiler to get the desired hot water temperature. If the valve is open too far the flow exceeds the design flow rate and boiler outlet water temperature is too low. If the valve is throttled too much the boiler will heat the water excessively and the burner will start short cycling on the operating limit (OL).

Think about it, what's a Btu? If the heater is fired at a constant rate (most are) then there is a consistent output in Btu. Since the water flow is constant (the tank is a detour for any water that isn't used in the system) the water temperature rise should be constant.

Provided the demand for hot water does not exceed the capacity of the boiler, hot water will enter the tank faster than it flows to the building. Therefore some of the water heated by the boiler remains in the tank, mixing with and displacing the cold water. Once the volume of the tank above the inlet pipe from the boiler is filled with hot water an interface forms between the hot and cold water because the cold water is denser than the hot water.

Boiler operation continues and hot water displaces the cold water in the tank until the level of the interface drops to the level of the lower tank temperature control switch (TC2) to terminate heating operation. The opening of contacts on the lower tank temperature control switch interrupts operation of the pump and burner to complete a heating cycle.

During the period when the circulating pump and burner are shut down the building is supplied by hot water from the tank. The weight of the check valve on the pump discharge provides sufficient differential pressure to prevent flow of water through the boiler during this period. Sometimes the valve is fitted with a spring rather than using weight. Don't put another type of valve in its place or it may not work.

As the hot water flows out the top of the tank it is replaced by cold water entering the bottom of the tank. The interface level raises until it is above the level of the upper temperature control switch (TC1). Contacts on TC1 close to start the pump. Auxiliary contacts on the pump motor starter close to bypass the TC1 Contacts so the pump will not stop when the TC1 contacts open. The auxiliary contacts also permit burner operation.

Whenever hot water demand does not exceed the capacity of the boiler the system continuously repeats the operation described above. The pump and boiler start, heat a volume of water equal to the volume of the storage tank between TC1 And TC2, then stop and wait until that volume of hot water is consumed.

When service water demand exceeds the capacity of the boiler the difference between hot water demand and boiler capacity is made up by hot water flowing out of the storage tank and cold water entering the tank. The tank supplies all the hot water until the level is above TC1 then the hot water from the boiler and the hot water stored in the tank combine to serve the hot water demand.

Whenever the service water demand exceeds the capacity of the boiler the elevation of the interface increases. Provided the high demand does not continue until the hot water stored in the tank is consumed the boiler will continue to fire until the storage tank is once again filled with hot water down to the level of TC2, completing a boiler operating cycle.

Under unusual circumstances of sustained high demand for hot water the reserve in the storage tank is consumed. Thereafter the water leaving the system will be a mix of cold water passing up through the tank and hot water produced by the boiler. Hopefully this will never be the case in your plant. If it frequently is, suggest a larger tank, larger boiler, or a combination because there's a hazard associated with it that is not desirable.

You may wonder why there are two temperature switches on the tank. Tests I performed indicate the interface in a storage tank has a temperature gradient of 5 to 10 degrees per inch depending on turbulence. A system with a single temperature control would cycle on and off frequently as the interface rises and falls during each cycle. Each time the burner starts and stops a purge is performed that, despite its purpose of safety, cools the boiler and water with purge air. Provision of two temperature controls properly spaced (more on that later) can significantly reduce losses and wear and tear associated with burner and circulating pump cycling.

City water temperature can vary significantly with the season depending on the water source. If all water is supplied from wells then the temperature varies less. When the water is stored in reservoirs or lakes and towers the temperature can vary between 35° and 65°. If the boiler operates at a fixed firing rate, as most do, the outlet temperature of the boiler will vary with the season. During burner operation the operator should note the temperature on the outlet thermometer (T2) regularly and adjust the position of the throttling valve (TV) to restore the desired tank temperature (±5°F) at least monthly. To increase the temperature the valve is closed some, to lower the temperature the valve is opened further. Make the adjustment when the boiler operation has stabilized then wait a few minutes to see the results before adjusting the valve further.

Normally the boiler operating limit (OL) and high limit (HL) do not function. However, when the boiler operates for extended times during periods of high demand the operating limit could open its contacts because the temperature gradient in the boiler changes. The operating limit should not be adjusted to the point that it controls the boiler (starting and stopping it) during normal operation.

There is no provision to adjust the pressure in the system. It should follow the supply water pressure. The safety relief valve should not be tested to determine if it operates. Operating personnel wearing proper protective equipment should raise the lifting lever of the safety relief valve every three months to confirm that the valve mechanism is free and the water flow passages are not blocked. Testing of the safety relief valve should be recorded in the log.

The purpose of the high limit is to prevent overheating of the boiler in the event the circulating pump fails or operating personnel inadvertently close a valve in the piping that prevents flow through the boiler. Its adjustment should be noted, lowered into the operating range to ensure it functions to interrupt burner operation, then restored to the original setting on an annual basis. The test of the high limit should be recorded in the log.

The bacteria blamed for the deaths of several members of the American Legion in Philadelphia is frequently found in water supplies. When exposed to warm water in a confined environment it can flourish. It's not the only one that can cause problems. The interface in the hot water storage tank always contains a level of water at the optimum temperature for that bacteria to grow and multiply. I suggest you sample water from the interface for presence of Legionella at quarterly intervals after initial start-up and, if none is discovered, annually thereafter. Annual testing should coincide with heavy rains in the summer where the bacteria is most likely to enter your system.

The process of checking for Legionella consists of drawing a sample and sending it to a laboratory for analysis. It requires a water sampling connection installed in the storage tank at the location indicated, just below the level of TC1. If the sample connection is above the return line inlet it should penetrate the tank as shown to ensure a sample of the interface is drawn. To ensure the operating personnel are not exposed to the bacteria (in the event it is there) they should wear protective equipment recommended for this operation. A sample bottle should be placed such that the sample piping extends into the bottle to the bottom to minimize splashing and generating aerosols while sampling. The sample should be drawn in the late afternoon or early evening when demand is normally low and immediately after the pump and boiler start operating (when the interface is near the level of the sample line.

If the laboratory test indicates Legionella is in the interface it should be flushed from the storage tank. Connect a hose to the sample valve outlet and extend it into a drum containing sufficient sodium hyperchlorite (Clorox™) to super treat a drum full of water. Turn off the pump circuit breaker immediately after it starts to prevent pump and boiler operation temporarily then, after a few minutes of drawing hot water from the building system, open the sample valve and close the pump circuit breaker. When hot water is flowing to the drum the sample valve can be closed because the complete interface was flushed to the drum. Repeat the procedure until a laboratory test of the interface does not show Legionella.

Even if Legionella does form in the storage tank interface it should not contaminate the hot water delivered to the building unless the storage tank temperature is too low or hot water demands result in all the storage in the tank being consumed. In the latter case the interface flows into the building's hot water distribution system. Operating the system to maintain hot water in storage at 180°F for more than one half hour should kill all bacteria except what's in the interface. Blending valves should be installed to provide the maximum 120°F water for hand washing, bathing, etc.

I've looked at a few service water heaters where thermal shock was determined to be the cause of their failure. Thermal shock is observed by anyone pouring liquid into a glass of fresh ice. The ice cracks instantly, even when the liquid is very close to freezing. Iron, steel, and brass boiler parts are more malleable and slightly stronger than ice so the effect is not as dramatic, but it does happen.

Boiler damage due to thermal shock is normally the result of repeated heat/cool cycles. Damage occurs when the metal is over-stressed because the surface is cooled or heated at a rate that exceeds the heat flow through it. As a result one surface is at a different temperature than the one opposite it. The differences in thermal expansion result in compressive stress at the hottest surface and tensile stress at the coldest surface. When the difference in stress reaches the breaking point of the metal then tiny micro cracks form in the colder surface. Repeated exposure to the heating and cooling expands the cracks until leaks are evident. Thermal shock can also be associated with rapid changes in firing rate but most service water heaters are designed to accommodate the changes associated with their on/off operation.

You would think that a hot water heater with normal temperature differentials of 140°F would be damaged regularly by thermal shock if even smaller temperature differentials are a problem. They don't because the overall temperature differential is distributed along the length or height of the boiler. The boiler in Figure 4-10 would normally have 40°F water entering the bottom (at T1) and 180°F water leaving the outlet (at T2) with the temperature between those two levels varying almost linearly from top to bottom. The high temperature differentials between the products of combustion and the water in the boiler do not produce a significant temperature difference across the thickness of the metal because the heat flows through the metal much faster than through the thin film of flue gas between the metal and the products of combustion. The temperature differential across the metal is normally less than 30°F.

Thermal shock occurs when a liquid in contact with the metal is quickly displaced by other liquid at a temperature significantly lower or higher than the original liquid. The direct contact with the metal parts and turbulence associated with the rapid replacement of the liquid heats or cools the metal surface rapidly, faster than the heat transfer through the metal itself.

So what caused the damage to the boilers I mentioned earlier? What can cause thermal shock? Well, in the case I first examined, the temperature control was different. Instead of installing a temperature switch that penetrates the storage tank at a level above the water inlet (as shown in Figure 4-9) the contractor provided a "strap-on" aquastat. That is a temperature switch with a bare thermal sensing bulb that is simply clamped to the outside of a tank or pipe to sense the temperature.

In that case, the bulb was clamped to the pipe where the cold water enters the tank.

Each time the system filled the storage tank until hot water flowed out of the storage tank into the piping and into the bottom of the boiler for a short period until the temperature controller finally responded to the change from cold to hot water. When the circulating pump started again the hot water was immediately displaced by cold water. The thick metal at the bottom of the boiler was repeatedly subjected to swings between hot and cold water entering the boiler which resulted in cracks around the bottom of the boiler shell.

As you can tell, simply heating hot water isn't as simple as it sounds. There's even an unusually different attitude about scale formation among people that maintain these devices. Why? They manage to get away with a considerable amount of scale because water temperatures are so low. It's a common practice to allow scale to build in one of these heaters (keep in mind, you can't treat it because it has to be potable where someone could drink it) until you can hear the loose scale (they call it lime deposits) rattling in the bottom of the heater where steam is forming under the material and then collapsing as it contacts the colder water.

Since water is not concentrated in a service water heater you would not expect it to form scale except under unusual conditions, but it happens regularly. It's not uncommon for scale to form on the heat transfer surfaces to the point that the heater capacity is less than demand and you can't make enough hot water. I can recall one location where the solids content of the water was so high that a mere 6°F increase in water temperature was all that was required for scale formation. The best solution for these applications is water softeners but that's not always accepted by the powers that be so you should be prepared to clean a service water heater regularly as part of its maintenance when the calcium and/or magnesium content of the water is high.

WASTE HEAT SERVICE

As far as I'm concerned these are the best boilers; the cost of fuel, the single largest cost for any other kind of boiler plant, is zero! That one great benefit also encourages us to put up with some unique and sometimes hazardous flows that contain the heat we extract with the boiler. I think the one most hazardous I've seen is a sulfur dioxide stream from firing pure sulfur to make sulfuric acid. Knowing what you know about problems with sulfur in conventional fuels should make you appreciate the special requirements for one of those boilers.

A waste heat boiler will always have a lot more heat exchange surface than a fired boiler because there is no radiant heat transfer. It's safe to assume a waste heat boiler will have twice the heating surface of a conventional boiler for the same capacity. It isn't uncommon to encounter a waste heat boiler with finned tubes to provide additional heating surface so you will not necessarily encounter boilers with twice the number of tubes. Depending on the source of the heat the boiler can incorporate an economizer section to preheat the feedwater and can be of once through design. The materials of construction may include materials that don't conform to the requirements of the Rules for Construction of Heating Boilers (Section IV) or Rules for Construction of Power Boilers (Section I) because the liquids or gases that are the source of the heat would destroy those materials. In those cases the boilers are constructed in accordance with the Rules for Construction of Pressure Vessels (Section VIII) as an "unfired boiler" which allows use of exotic materials including stainless steels, Inconel, and others.

The largest, physically, waste heat boiler I ever encountered was one I helped design and, as far as I know, is still in service in Wilmington, North Carolina. It is twenty-four feet in diameter, over ninety feet tall and generates about 25,000 pph of low pressure steam. The largest in capacity is a unit that looks more like an economizer and only preheats boiler plant makeup water, 120 million Btuh. They come in a variety of sizes and configurations that are so variable that there's no describing them all and their operation varies significantly depending on the conditions of the fluid flow stream the heat is coming from.

A low water cutoff is a required element for any boiler and they should always be provided on waste heat boilers unless the temperature of the fluid stream is less than about 750°F where the metal will not overheat. I can recall one system where a contractor installed a waste heat boiler connected directly to the exhaust of a steel annealing furnace which exhausted a heating stream at about 1800°F. The new boiler was melted down two days after installation because the water source failed. If the temperature is high enough there should always be a way of diverting the waste heat stream to prevent overheating the boiler. In some cases there is no diversion of the waste heat stream but it's possible to add air to dilute it until the boiler metal can withstand the temperature. With those exceptions any waste heat boiler should be treated like a normal boiler.

ONCE THROUGH BOILERS

I am aware of one manufacturer of once through boilers in the commercial to industrial range and many electric utility boilers are designed for once through operation. The boiler feed pump(s) or a boiler water circulating pump is used to produce the differential pressure required to force the water and steam through the boiler. Similar to HTHW generators these boilers can contain multiple circuits of tubing with inlet and outlet headers but many have one single circuit to maintain a point of conversion from saturated to superheated steam. Feed water is forced into the tubes and steam which may be saturated or superheated leaves the tubes. All electric utility boilers of this design generate superheated steam. Needless to say, those boilers must use absolutely pure water because any dissolved solids would build up in the boiler tubes as scale.

Once through boilers that generate saturated steam also incorporate a steam and water separator at the outlet so only steam is delivered to the facility. The water that exits the boiler with the steam is separated and returned back to the inlet of the circulating pump. The one boiler design I know of has only one tube formed into a long coil that surrounds the boiler furnace and may have additional tubing extending the coil as a convection section. A safety feature on these boilers, that doesn't exist on other types, consists of a switch activated by the thermal expansion of one of the boiler tubes. If there is insufficient water flow through the tube its coil expands as it is heated to trip the switch.

Flow has to be proven before lighting a fire in one of these boilers so treat the low water flow switch with the same concern and testing as a low water cutoff.

ENGINES AND EMERGENCY GENERATORS

At the risk of being called overly emphatic I will repeat that you should never operate equipment until you've read the instruction manual. These guidelines only apply if you can't find the manual, something that happens with surprising regularity.

The most common engine in a boiler plant is normally the emergency generator. For reasons that still escape me ships, from other nations, which I visited during my tenure in the merchant marine were almost always powered only by engines. I enjoyed taking tours of those ship's engine rooms to see diesel engines so large that you could stand inside one of the cylinders and fuel injectors that were so large it took two men to carry them. Ships today can have engines rated at 60,000 to 80,000 shaft horsepower. Despite their large size their requirements for operation are no different than small engines.

Environmental regulations developed under EPA's Tier 4 requirements have changed the procedures for operating engines. Those requirements apply to all land based equipment whether stationary or mobile. Some of you are bound to be familiar with requirements for adding DEF (diesel exhaust fluid) into a separate reservoir of your diesel engine powered pickup truck. New emergency generators and stationary diesel engines will also have that requirement. They will be fitted with CDPFs (catalyzed diesel particulate filters) that capture the engine's particulate emissions, which are principally unburned hydrocarbons, then burn off the carbon while simultaneously reacting to break down nitrogen oxides using a catalyst alone or the DEF. To achieve the emission limits now established engines will also have control features added to augment PCV and EGR valves. Because the implementation of Tier 4 regulations are progressing as I work on this edition I can't tell you about all the modifications to normal engines that will become normal in the coming years. Once again, read your instruction manual.

It's not necessary to describe putting an engine driven generator online or disconnecting it from the line because the operation is basically the same as it is for steam turbine powered generators. The only concern with an engine is that operators and automatic controls can connect an engine driven generator before it's warmed up and some instability will result in fluctuating frequency and voltage. Except for actual emergency operation it's advisable to allow the engine to idle for a period of time so it reaches normal operating temperatures before putting it online.

Engines can be used to power pumps, fans, blowers, and any number of pieces of mechanical equipment associated with a boiler plant in addition to the more common emergency generator. Engines have hazards associated with them that make them almost as dangerous as a boiler. Blow by, which is the term used to describe the leakage of a fuel air mixture in an engine cylinder past the piston rings and into the crankcase, combined with the aerosol produced by the mechanical agitation of the lubricating oil can be counted on to produce a combustible, hence explosive, mixture in the crankcase. Those spring loaded covers over the access openings for the crankcase have the dual purpose of first relieving "puffs" then immediately closing to prevent the admission of air which could produce a larger puff or a true explosion as air is added

to the fuel. Recall what I said about boilers in that a puff is simply an explosion that didn't do any damage and is very strong warning that a more extensive explosion could follow. Normally the blow by consists principally of the products of combustion so the crankcase should contain a reasonably inert gas.

Of the two types of engines diesels are the most common in boiler plants. Occasionally you might be called upon to operate a gasoline engine that introduces additional concerns. I'm sure you've been told that you should not store gasoline in your house or attached garage. Some of you may have a hot water heater in your garage but noted that it is installed on a platform. All this is because gasoline evaporates at normal atmospheric temperatures and the vapor is heavier than air so if gasoline vapors leak from a storage container, or your automobile, and mix with the air in the space it will produce a combustible mixture and the mixture will gravitate to the lowest space in your garage. The solutions to these problems in your home consist of elevating the hot water heater and not storing gasoline inside even if it's in an approved container. When it's necessary for the use of portable gasoline engines in a boiler plant it's not always possible to keep the engine outdoors. It is, however, normally possible to take the engine powered equipment outside for filling and starting. During normal operation when gasoline is drawn from the tank a small orifice in the top of the tank admits air and no gasoline vapor is emitted from the tank. Operation of an engine inside a building without providing additional openings for ventilation may result in that engine running fuel rich so it's exhaust is discharging gasoline vapor into the building. Make sure you have adequate ventilation for those engines and whenever possible means to discharge the exhaust outdoors.

Procedures for starting an engine should always include checking the lubricating oil level first. Before you argue with me about checking it every time before starting keep in mind that someone else could have decided to change the oil, drained it, and failed to get around to refilling the engine. Yes, that's happened thousands of times. When lead acid batteries are used to power a starter for the engine it's also a good procedure to check the battery water level as well. The typical emergency generator has radiator cooling and the level of the coolant in the radiator needs to be checked, otherwise the cooling systems for the engine must be in operation before starting. Then, whenever possible after starting the engine, wait for it to warm up before connecting to the load. Closely monitor cooling water and lubricating oil temperatures after connecting to the load to ensure they stabilize before leaving the engine. Frequently engines are started with compressed air or hydraulic fluid pressurized by compressed air and restoration of the pressure should be checked before leaving the engine.

During normal operation of an engine you should check for proper lubricating oil level, pressure drop through air filters and fuel filters, then log cooling and lubricating oil pressures and temperatures every shift. I'm reminded that there are engines that don't permit checking oil level while they're operating, like your car, they have to be shut down to check the oil level. Regardless, this puts you next to the engine where your ears and your brain are put to work to detect any abnormal sounds and their meaning.

Remember that the air used for combustion must come from outdoors. I've always questioned the application of air filter sensors (these usually provide a green or red indication) that are simply connected to the air piping between the filter and engine. On large engines I would specify the sensor detect the difference between outside air and the piping so failure of automatic louvers and the like to operate properly would produce a red indication.

It is always a good idea when shutting down an engine to remove the load, if that's possible, and allow the engine to cool down from load conditions. Alternatively reduce load to cool the engine. Stopping it immediately could result in carburization of oil in areas where the oil is trapped instead of flowing through. When the engine is equipped with an auxiliary oil pump it should be started before shutting down the engine and stopped only after the oil temperatures entering and leaving the engine differ by less than 5°F. Then, treat the cooling system similarly but don't shut it down until after the auxiliary oil pump is shut down.

Like boilers, idle engines deserve to be checked regularly. Potential problems include loss of starting power, freezing cooling water, and moisture condensing and accumulating in the crankcase to name a few.

It isn't unusual for emergency generators to be treated as peaking generators. In other words the owner of the emergency generator has a contract with the local electric utility that pays the owner to operate the generators during periods of peak electric load. This arrangement is beneficial to both parties. The agreement does normally include provisions for penalty should the owner fail to operate the equipment when it's requested by the utility. While someone could imagine that peak electric loads occur in the summertime it's also possible in the winter. PCI installed a couple of large emergency generators for a customer under those circumstances a number of years ago. Luckily the owner had purchased the generators and

associated equipment so PCI was not responsible for the design. On an extremely cold winter night the year after the installation, when temperatures drop below 11° F, the utility called but the engines would not start. We were called to resolve the problem and discovered there were no heaters for the fuel oil or the lubricating oil. It took a combination of halogen lights and welding machines to heat the oil and get the engines running. So, even if you don't have that agreement to deal with you should consider the potential for an extremely cold start of your emergency generator and be assured that it will start and run and carry a load.

GAS TURBINES

One type of gas turbine, that I'm used to and would be installed in a boiler plant, is referred to as aeroderivative. The turbine assembly consists of a typical jet airplane engine with its exhaust discharging into a few stages of gas turbine blading connected to the power output shaft. All other GTs (gas turbines) consist of a combination of compressor, burners, then turbine mounted on the same shaft as the compressor and connected to a generator or other piece of driven equipment. I have only seen a few dozen and, for all practical purposes, they were all different so, once again, the instruction manual becomes very important.

Unlike an engine most gas turbines are cooled by air flow and the lubricating oil. The lubricating oil is, in turn, cooled by air or water. Many of the aeroderivative plants have staged oil cooling where the airplane engine's oil is cooled by another lubricating oil system. For air cooling some of the air compressed by the compressor flows around the combustion chamber(s) for cooling then mixes with the burner exhaust to reduce the temperatures entering the turbine section. More modern engines have cooling air from the compressor diverted through the turbine blades to cool them. The air leaves the blades through multiple orifices at the tip of the blade to flow over the external surface of the blade to help protect the blade from the hot flue gases.

It isn't uncommon to require fuel gas compressors for natural gas-fired turbines. Despite the large supply of natural gas at the time of preparing this edition many gas turbines exist that are designed to fire fuel oil as well as natural gas. The oil nozzle may look a little Strange (Figure 4-10) but you should be used to enough variations now that you won't appear to be confused when confronted with something like this.

Gas turbines, like engines, need to be started. At

Figure 4-10. Gas turbine oil nozzle

minimum they must be brought up to a speed that produces enough compression of the air to produce a differential pressure across the turbine that, when combined with burning of the fuel can generate enough power at the turbine shaft to power the compressor and the rotating resistance of anything attached to the shaft of the gas turbine. This is normally done with an electric motor; or, the electric generator attached to a gas turbine can function like a motor to get it started. Before everything starts

turning, however, lubrication has to be established and proven.

Maintenance of lubricating oil temperatures during a GTs operation is very important because the oil can be too cool as well as too hot. Other temperatures that should be monitored closely are compressor exhaust temperature, burner exhaust temperature, and turbine exhaust temperature. The first two must be within limits for continued operation and the last as low as possible for maximum turbine efficiency.

During operation the pressures, temperatures, and gas velocities through the GT vary as indicated in Figure 4-11. Atmospheric air accelerates toward the inlet of the compressor (1-2) accompanied by a drop in static pressure created as some of the static pressure of the air is converted to velocity pressure. In the compressor (2-3) the pressure is increased and velocity decreases as the air is compressed because compression reduces the volume of the air. Temperature of the air rises from the work of compression just as it does in diesel engines. Between the compressor and the burner (3-4) some pressure is lost as the air is directed through different passages with provisions to increase turbulence for mixing with the fuel. Velocity there can decrease when some of the compressed air is extracted to feed through the turbine blades for cooling them. Inside the burner

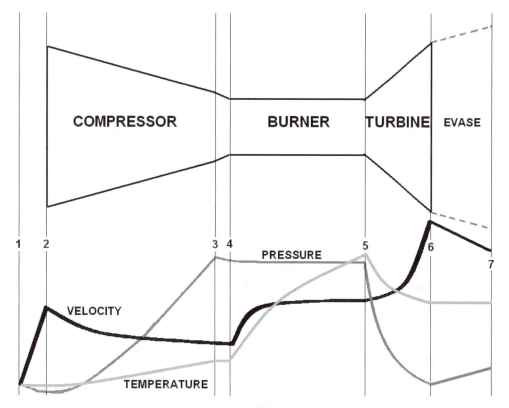

Figure 4-11. Gas turbine PTV curves

(4-5) pressure remains about the same and velocity and temperature increase. Mass flow also increases a small amount due to the addition of the fuel. Air to fuel ratios are considerably higher than those in a boiler because all the heat from combustion is retained in the flue gases and more air is required to absorb the heat and keep the temperature from exceeding the tolerances of the metal of construction. I'm familiar with turbines that run 400% excess air but more modern designs continue to reduce that value to reduce the energy used in the compressor for higher GT efficiency. In the turbine (5-6) pressure and temperature drop as the gases convert energy to the turbine blades to rotate the shaft. Velocity increases as the gases expand through the turbine stages. A GT turbine only has reaction blading (see steam turbine descriptions) because there are no nozzles. Finally, at the outlet of the GT (6-7) an evase (gradual expansion of ductwork) serves to convert some of the velocity pressure to static pressure. That allows for a slightly higher differential pressure across the turbine and the turbine outlet pressure reading to be lower than atmospheric for turbines without heat recovery and lower than the inlet pressure to heat recovery equipment.

A process used to boost power output of GT generators when summer electrical loads are high is misting. Immediately after the turbine inlet filters spray nozzles inject water into the air stream entering the compressor. This cools the inlet air by evaporation of the spray water to produce increased mass flow of combustion air along with the additional moisture so the GT can burn more fuel and produce more power.

FUEL CELLS

While I've read several articles I have to honestly say that I have never laid hands on a fuel cell. Unless I have an opportunity to work with one before I have submitted this edition of this book I can only say that you should really study the instruction manual.

I do know that the cell itself generates power by reversing electrolysis, something you should have seen, or even performed in a lab, in high school. That's where you

place two electrodes under water each within an inverted test tube of water and force a direct current of electricity through the water. Hydrogen is generated at one electrode and oxygen at the other and those gases collect in the bottom of the inverted test tubes because the electricity breaks down the H_2O into its two components. The fuel cell combines oxygen in the air and hydrogen in the fuel to generate electricity, a reverse of electrolysis.

HRSGs AND COMBINED CYCLE PLANTS

HRSGs are connected to the discharge of a gas turbine. It's not uncommon for people to use that abbreviation when discussing simple waste heat boilers. The typical HRSG captures the heat remaining in the exhaust of a gas turbine generator and supplies superheated steam to feed a steam turbine also connected to a generator and that is the typical combined cycle plant. A typical combined cycle plant consists of gas turbine driven electric generators with a HRSG at the exhaust of each and steam from the HRSG powering a steam turbine. The steam cycle isn't quite the same as a typical utility plant like the one shown in Figure 1-8 because the combined cycle plant is typically a number of HRSGs supplying one or two steam turbine generators.

A single stage combined cycle plant is schematically represented in Figure 4-12. The steam turbine and condenser plus associated auxiliaries are the same as for

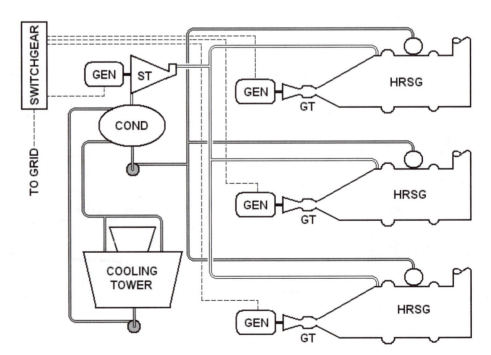

Figure 4-12. Combined cycle plant schematic

any steam plant. Each gas turbine (GT) drives an electricity generator (GEN) and the hot exhaust is converted to superheated steam in the HRSGs and the steam powers the steam turbine (ST) to drive its generator. The steam turbine exhaust is condensed in the condenser (COND) and is pumped back to the HRSGs to be converted to steam again. A cooling tower is normally provided to cool the condenser water but condenser water can also come from an adjacent body of water. Multiple GTs and HRSGs permit turndown without sacrificing efficiency and multiple nozzle blocks on the ST allow it to operate at varying loads with minimal loss in efficiency. I have seen plants with as few as two and as many as twelve HRSGs.

The HRSG is, for all practical purposes, a boiler plant combined in one package and is shown schematically in Figure 4-13. It consists of (1) a GT exhaust duct that conveys the turbine exhaust to, and distributes the exhaust into the first stages of the boiler portion. A duct burner (2) provides additional heat to the boiler section. The superheater (3) raises the temperature of the steam as described for utility plants before continuing to the steam turbine. A HRSG could contain a reheater and although I haven't seen one I can imagine a HRSG having a duct burner between a superheater and reheater to increase reheat. The boiler (4) generates the steam that supplies the superheater from the feedwater heated by the high

pressure economizer (5). After leaving the economizer the flue gases enter a lower pressure boiler (6) with its own economizer (7) that boiler generates steam for the deaerator (8) which is typically mounted integral to the HRSG and is confused by some as another steam drum. Before discharging up the stack (9) the flue gas could be exposed to a condensate heater that preheats a mixture of returned condensate and makeup water. Feedwater is fed to the two boiler stages by independent feedwater pumps. The main feed to the main boiler can also be heated by an external high pressure feedwater heater using bleed steam from the steam turbine.

Note that some of the boiler tubes and all economizer tubes are finned. The provisions of fins are dependent upon the operating flue gas temperatures (omitted where they would be burned off) and possibly eliminated on the last row of boiler tubes, which serve as downcomers.

HRSGs are optimized in design to recover the heat of gas turbines within the smallest possible footprint and lowest combined cycle cost (initial and operating costs combined over the life of the unit). The designer tries to extract as much heat as possible from each stage which requires close attention to the pinch points. Pinch points are where the flue gas temperature approaches the temperature of the fluid it's heating. If the two temperatures are the same heat will not flow from the flue gas to the water

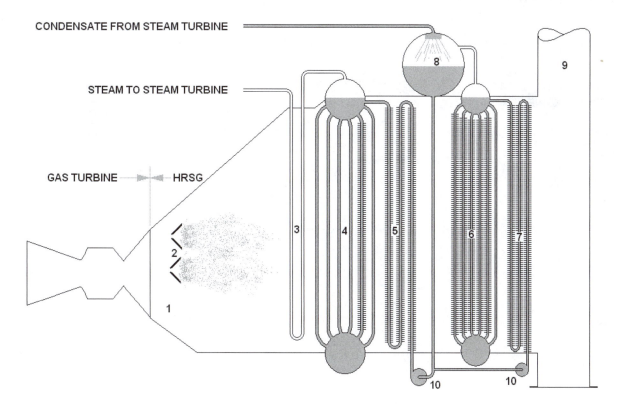

Figure 4-13. HRSG schematic

or steam. The temperature profile of a typical HRSG is shown in Figure 4-14 and pinch points are indicated.

Since the HRSG is designed for these operating conditions there are situations during startup and shutdown that require special control, especially to prevent steaming in the economizers which would result in water hammer in them along with potential damage to the economizers, feedwater control valves, and the drum internals. At times it's necessary to maintain flow through the economizers by returning some of the economizer outlet water to the deaerator. High pressure in the deaerator is normally prevented by a relief valve and vacuum by a vacuum breaker. Return of some feedwater to the deaerator from the low pressure economizer can also be used to help stabilize the deaerator pressure.

The duct burner is used primarily to control superheat temperature and the HRSG is designed so superheat will normally be lower than design with the duct burner out of service. Since the duct burner is in a stream of hot gas with 400% excess air the only air added to it may be some for the ignitor. Being a radiant source in an otherwise simply convection flow the duct burner principally alters the superheat. The mass flow through the HRSG isn't altered as much by simple addition of fuel so the effect on steam generation isn't as significant as increasing firing of a conventional boiler.

The literature (*Power Magazine* among others) is constantly revealing new techniques and features of combined cycle plant operation and their use is increasing rapidly. It's not just the availability of natural gas, it's also the fact that these plants are more efficient, converting more of the energy in the fuel to electricity than conventional boiler plants. That's more than 40% and almost 50% conversion efficiency.

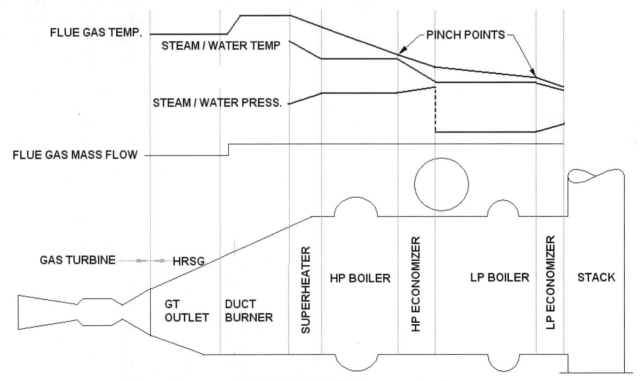

Figure 4-14. HRSG PTV profile N ### needs values

Chapter 5

Refrigeration and Air Conditioning

With this edition I have added refrigeration, and air-conditioning, because many boiler plants have refrigeration equipment within the plant, attached, or in an adjacent room or building and the boiler plant operator is expected to operate and help maintain that equipment as well. Several states and municipalities include questions on refrigeration systems and air-conditioning in their boiler operator exams. If this surprises you, it shouldn't, that's because refrigeration is very similar to boiler operation, and refrigeration is simply transferring heat from one location to another. Advances in SCADA systems and centralized monitoring of facility equipment have also served to drop responsibility for the operation of the systems in the lap of the boiler operator because the access terminals are in the boiler control room.

Refrigeration has many applications. Principal uses include food preservation and comfort cooling but refrigeration can also be used in production processes. Of the latter the most interesting one I've encountered is making soap. While I was at PCI we had the opportunity to install a system designed to operate at -25°F at Lever Brothers. At that time Dove bars were made in the plant in Baltimore and we were helping install support systems for a new product line. The bars are formed in a press at a speed so fast that the heat of working the bars would melt the soap without the cooling. When I was working at Hercules I worked on the air-conditioning for a new Hystron fibers plant that made Trevira™, a synthetic silk. I had to come up with a design for air cooling the synthetic fibers that were squeezed out of the plastics extruder quickly enough to prevent their sticking to each other as they were twisted into a single strand. I also designed some 130 Ton air handling units for that plant.

A ton of refrigeration is a rate, not a quantity. The derivation of value is the amount of heat that can be absorbed by one ton of ice melting in 24 hours. I like to tell a story, that's undoubtedly fiction, where it started with a theater owner deciding he could get more customers in the summer by having a cool theater and negotiated with a contractor to cut blocks of ice from a nearby lake and store them under straw (as an insulator) in a huge pile for use during the summer. The theater owner drew outside air over blocks of the ice to cool it before blowing it into the theater and the theater needed tons of ice. Others employed the same provisions and the standard unit of measure became a ton. It takes 80 Btus to melt a pound of ice so a ton is 80 x 2000 ÷ 24 = 12,000 Btuh. Just as we quantify electric power in kilowatt hours a quantity of refrigeration is Ton hours.

REFRIGERANTS

What's a refrigerant, why don't I just say Freon™? Why is because there are many refrigerants and materials we wouldn't normally think of as refrigerants, including water. There are a considerable number of refrigerants and the one used in any particular system is dependent upon a number of factors.

Of course water can be a refrigerant. I love to point out what I understand to be a true point in history. It should make the boiler plant operator more amenable to dealing with refrigeration. The first air conditioning system installed by Carrier himself was in a printing plant in Brooklyn, New York, in 1902 and it was a steam powered system using water as a refrigerant. Before you say "yeah, sure it was" look at your steam table in the Appendix. All we have to do to get water cold enough for air conditioning is to expose it to a vacuum at 29.75 inches of mercury and it will boil at 40°F. Carrier created a vacuum over a tank of water using steam jets to remove the evaporated water plus any air and non-condensable gases. The water in the tank boiled until the heat absorbed by the evaporating water cooled the remaining water to 40°F. The cold water was pumped to coils that absorbed the heat in the air in the printing plant and simultaneously condensed excessive moisture in that air to lower the humidity in the printing plant. The slightly warmer water was then returned to the tank where the heat absorbed from the printing plant air was removed by evaporating some of the water. Why was a printing plant air conditioned first? Simply because paper properties change with temperature and moisture content and keeping it consistently spaced between printing rolls containing different colors of ink, a process called registration, required maintaining the temperature and humidity in the plant.

Many other fluids can be used as refrigerants but they have undesirable properties. Some known to cause cancer were used in early systems and others, such as propane, are not used because they're flammable or heavier than air. Along with water another natural substance used as a refrigerant is ammonia. It's an excellent refrigerant but it's also flammable, corrosive, and poisonous. If it leaks from the refrigeration system it can catch on fire, ruin materials stored in the refrigerated spaces and kill people. That's why it's only used in facilities that aren't normally open to the public and monitored constantly for leaks.

Then, in the later half of the 20th century we discovered that Freon (a trade name of the DuPont Company that became more or less a generic label for refrigerants) was not as safe as we thought. It wasn't flammable nor poisonous and it was lighter than air so it wouldn't pool inside a basement like propane does but, being lighter than air it rose up in the atmosphere until it got to the stratosphere then disassociated releasing chlorine atoms that break down ozone. Located about 30 miles above the surface of the earth the stratosphere contains an ozone layer that works like a filter to screen out harmful ultraviolet radiation from the sun. By the time the chemistry of the action of most refrigerants (CFCs, short for chlorofluorocarbons) on the ozone layer was recognized the ozone depletion was so extensive that there was a "hole" in the ozone layer over the North Pole. International efforts to limit the release of refrigerants and development of new refrigerants that lack chlorine have reduced the damage to the ozone layer and it is now recovering.

Part of the solution is heavier than air refrigerants that will not reach the upper atmosphere. However, they can accumulate in spaces containing the refrigeration equipment with the potential of suffocating people. Should you happen to go to work in a plant that uses one of those refrigerants you should find a SCBA (self contained breathing ap-

paratus) in a cabinet immediately inside the door. Do not fail to read the instructions and, preferably, try out the apparatus so you know how to operate and use it. Encourage other operators to do the same because they may have to use it to reach you and carry you out of the space in the event of an incident.

Figure 5-1 is a tabulation of our more common refrigerants including replacements for refrigerants that are no longer allowed to be used in new equipment. You will note that there are color codes for most refrigerants. Containers of the refrigerant are painted that color in an effort to ensure a system is not charged with the wrong refrigerant or refrigerants are not mixed. Most refrigerant systems will fail to operate properly when exposed to a mixed refrigerant. There are, however some refrigerants which are intentional mixtures; these are known as binary refrigerants, a mix of two.

The question mark for replacement of R-22 is there because; at the time of writing this there was no consensus on a suitable replacement for it. Unlike the CFC refrigerants which can not be made anymore the HCFC R-22 is scheduled to be replaced and no longer manufactured sometime between now and the year 2030. R-134a has become the standard for all applications that used to use R-12. The orange containers of R-11 and containers of R-123 are actually drums because they are a liquid at normal atmospheric conditions; in use the evaporator for those refrigerants operates in a vacuum.

R-404a is one of the binary refrigerants which have a range of saturation temperatures between two conditions known as dew point and bubble point. Because it's a mixture of two refrigerants with different saturation condi-

Application	Temperature Range	Refrigerant	Color	Uses	Replacement
High temperature	50°F to 90°F	R-22	Green	Comfort Cooling	?
Medium temperature	32°F to 50°F	R-12	White	Coolers	R-134a
Low temperature	0°F to 50°F	R-502	Purple	Freezers	R-507 & R-404a
Ultra-low temperature	-50°F to -250°F	R-13 & R-503		Cascade	R-403B
Cryogenics	-250° to -400°	R-728		Expendable	None
Low-pressure chillers	42°F to 55°F	R-11	Orange	Chilled water	R-123

Figure 5-1. Information on some common refrigerants

tions one portion boils off at a slightly lower temperature than the other, creating bubbles of the lower temperature refrigerant then that vapor is superheated a few degrees until the other portion reaches its saturation temperature. The span between the dew point and the bubble point is referred to as "temperature glide." Handling of binary refrigerants requires special procedures, so once again, it's important to read that instruction manual. There are also ternary blends, three refrigerants blended together. Some blends are azeotropic, they have no temperature glide and act just like a single refrigerant.

R-728 is actually liquid nitrogen it's expendable and one of the only refrigerants that can be vented legally. R-13 is carbon dioxide, can be vented, but is normally contained because simply venting a refrigerant gets expensive and now it's considered a pollutant as a "greenhouse gas."

Venting of refrigerants other than nitrogen and carbon dioxide is illegal and perpetrators can be subjected to a fine of up to $27,500.00. That's one of the many things you should learn about handling refrigerants when you take a course in that activity to obtain your certification as a technician for handling refrigerants to comply with Section 608 of the Federal Clean Air Act of 1990. That class will teach you about stratospheric ozone depletion, the Clean Air Act, refrigerants and oils, and recovering refrigerants. Then you will have to take an examination and pass it before receiving your certification. I suggest you take the course and get certified if you're working in a refrigeration plant so you are qualified to handle refrigerants and understand fully what to do in that process. You may feel you're contented with simply operating the equipment but your employer may choose to employ someone else that has that certification.

Oil for refrigeration equipment is special because it has to work in an environment where it's surrounded with refrigerant. The lubricating oils for refrigerants can be specific to a particular refrigerant or to a group of them and extreme care is required in handling them. Almost all refrigerant oils are hygroscopic, they suck up water, and since water isn't desirable in a refrigerant system it's important that those oils are handled and stored in sealed containers so they can't soak up water from the air.

THE REFRIGERATION CYCLE

I covered steam cycles near the beginning of this book and hope to show that the refrigeration cycle isn't significantly different. As shown in Figure 5-2, the ba-

sic refrigeration cycle has four stages. Once a couple of differences are cleared up you should have no trouble understanding what's going on. The four basic stages are evaporation, compression, condensation, and throttling.

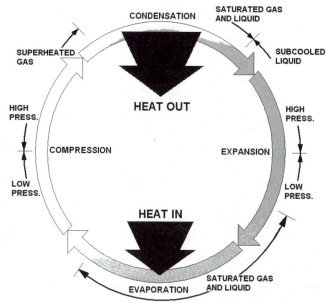

Figure 5-2. Basic Refrigeration Cycle

Evaporation

Evaporation, as the title implies, is converting a liquid to a gas, just like making steam in a boiler. The refrigerant absorbs heat by evaporating and that's how the evaporator removes heat from another substance to cool it. The principal difference is the refrigerant is exposed to pressures in the evaporator where the saturation temperature of the refrigerant is lower than the substance to be cooled. Heat flows from the substance to be cooled through the tubing or casing of the evaporator into the refrigerant causing the refrigerant to boil. Despite some descriptions you may have heard a refrigeration system doesn't push heat, it always flows in one direction, from a higher temperature to a lower temperature. By reducing the evaporator pressure the boiling temperature of the refrigerant is lowered until it's colder than the substance to be cooled.

Why do I use the word substance? It's a catch-all word to replace what I would have to say, which is—a gas, a liquid, a liquid that upon removal of heat becomes a solid, a solid, or another refrigerant. That's because refrigeration is not only associated with changing the temperature of a substance, it can also change the state, from gas to liquid and liquid to solid. We cool air and dehumidify it, cool water and freeze it, and that's only the simple ones.

Compression

Now you might argue that Carrier system didn't compress the refrigerant, it pulled a vacuum. Another look at the steam tables should reveal the volume of the pound of steam at 40°F is 2423.7 ft.³ per pound of steam but at zero psig it's 26.8. The steam Jets compressed the steam removed from the tank at a compression ratio of 90 to 1. Now the next question may be why don't we always use water for a refrigerant? The answer to that question is that other liquids that boil at lower temperatures under pressure can be used so we don't have to create a vacuum. Absorption chillers, discussed later, do use water for a refrigerant.

With any fluid increases in pressure increase the saturation temperature of the fluid. In order to eliminate the heat that was absorbed in the evaporator we compress the fluid to raise its saturation temperature to a value higher than the substance the heat is rejected to. The many forms of refrigerant compressors are covered later in this chapter.

It takes power to compress a gas and the energy of compression is added to the gas. Therefore heat is also added to the refrigerant in the compression stage.

Condensation

The compressed gas loses heat to a substance at a temperature lower than the saturated temperature of the gas when it condenses. The heat is transferred from the refrigerant through tubing or shell into the substance that absorbs the rejected heat. At this point it should be evident that all we are doing is transferring heat, the same thing we do within our steam and water cycles. The construction of condensers is dependent upon the refrigerant and the substance accepting the rejected heat.

Throttling

Since condensers operate at a higher pressure than evaporators we require a means of controlling the flow of refrigerant from the high pressure condenser to the low-pressure evaporator to ensure the compressed hot gases do not enter the evaporator. We also have to ensure that liquid does not leave the evaporator to enter the compressor because, unlike gas, liquid is not very compressible. Therefore, a means of controlling refrigerant flow between the condenser and the evaporator is required. Many throttling systems utilize means of sensing temperature to control the refrigerant flow while others use liquid level and two are simple orifices.

REFRIGERANT SUPERHEAT AND SUB-COOLING

Just as we superheat steam by adding heat after all the liquid has been boiled away refrigerants are superheated. The heat added in the compression stage by the work done on the refrigerant to compress it increases the refrigerant superheat. In refrigeration the temperatures are simply much lower. Refrigerant in an evaporator will continue to absorb heat by evaporating the liquid until it runs out of liquid. In most evaporators the throttling controls limit the admission of liquid such that none exits in the evaporator and, because the gas temperature is still lower than the substance being cooled the gas absorbs heat and that additional heat raises the temperature of the gas, superheating it.

Terms need to be clarified so there's no confusion so I will repeat two definitions. A superheated gas is gas at a temperature higher than its saturation temperature at the pressure of the gas. Superheat is the difference in temperature between the temperature of the gas and its saturation temperature. A refrigerant at 83 psig where the saturation temperature is 40°F may be a superheated gas at 50°F and the gas has a superheat of 10°F. The same refrigerant compressed to 275 psig where the saturation temperature is 120°F can be a superheated gas at 130°F and once again have 10° of superheat. Normally the superheat after the compressor is greater than 10°.

While superheating occurs in the evaporator, sub-cooling occurs in the condenser. In the condenser all of the gaseous refrigerant is condensed, the liquid is still exposed to temperatures lower than the saturation temperature and heat leaves the liquid so its temperature is lower than saturation. Liquid leaving a condenser is normally a sub-cooled liquid. The refrigerant described in the previous paragraph could be condensed at 275 psig then cooled to a temperature of 110°F. In that case sub-cooling is 10°F.

It's important to understand these conditions because lack of superheat or too much superheat can result in damage to compressors. Inadequate sub-cooling can result in poor operation, or damage to, the throttling device. Inadequate sub-cooling also permits liquid flashing to vapor before the refrigerant reaches the evaporator thus restricting the flow of the liquid refrigerant.

Subcooling is accomplished by removing heat from a liquid that has just condensed from a vapor to a temperature lower than the saturation temperature. Subcooling is normally accomplished in the condenser but other provisions and equipment can be utilized to subcool the liquid. A system can, for example, include a heat exchanger that uses the cool vapor coming out of the evaporator to help cool the liquid refrigerant. The cool gas is simply superheated a little more as it absorbs the heat from the liquid. It's also possible to have an independent subcooler to cool the liquid refrigerant.

Subcooling is necessary to ensure the liquid remains a liquid until it has exited the throttling device. If the liquid is not subcooled sufficiently it can breach saturation conditions as the pressure in the liquid tubing drops due to friction or changes in elevation. When that happens small bubbles of refrigerant vapor form in the liquid line and, because the vapor uses up more space than liquid, result in increased velocity in the liquid line for more pressure drop and generation of more vapor.

For several years, while my wife Sue served on our County Council, we attended a Maryland Association of Counties conference in Ocean City Maryland. We stayed in a condominium on the Bayside of the island adjacent to the convention center. The condo's heating and air-conditioning consisted of heat pumps for the individual units scattered around on the ground just outside the condominium deck. Each year I wondered how well those units worked for the condos on the third floor. The change in elevation (about 30 feet) would reduce the pressure in the liquid line by almost 15 psi (density of liquid R-22 being approximately 71.8 pounds per cubic foot at 96°F) which would require the refrigerant liquid to be at least 6°

cooler to overcome that differential. If the pressure drop in the liquid refrigerant tubing introduced another 15 psi the combination would approach the normal design of 10° subcooling. Then I would remember that since the liquid line ran up the building on the inside it would be cooled by the conditioned room temperature. That might not help in a taller building.

Measuring superheat and subcooling

Unlike our boiler plants refrigeration systems do not always have pressure gauges and thermometers mounted in the piping. When there are gauges or thermometers it's an indication that the reading should be monitored and logged regularly. Usually a set of portable gauges, called a refrigeration gauge set as shown in Figure 5-3 and a clip on thermometer are used to determine superheat and sub-cooling. The gauge set in your facility will normally have the saturation temperatures printed on the gauge face along with the pressure indication. If the temperature is not indicated on the gauge set then you will need a refrigerant card like the one shown Figure 5-4. Most of those cards list several refrigerants so you don't have to carry a lot of them, only remember to use the right column. The gauge set has three hoses connected and terminating in special fittings that match fittings provided for their connection on the refrigeration equipment. Sometimes, but not always, the fittings are keyed so that they cannot be connected improperly. The color of the hoses is always a key to how they should be connected. The blue hose fitting gets connected to the vapor line at the outlet of the evaporator and the red hose is connected to the hot line somewhere between the compressor and the condenser. Those hoses lead to the gauges to provide an indication of pressure and the corresponding saturation temperature of the refrigerant at the point where the hose fitting is connected. The third hose, typically yellow, the one in the middle, is separated from the other two by the valves and is typically used for a connection to a bottle of refrigerant. That hose is used to add refrigerant to, or remove refrigerant from, the system. The color of that hose can be keyed to the refrigerant. Gauge sets exist that can only be used for specific refrigerants; R134a is one. Their fittings will also be specific matches to prevent contamination of a refrigerant system by a different refrigerant. The clip on thermometer should be clipped onto the refrigerant piping as close as possible to the pressure gauge connection so a true superheat, the difference between saturation and actual temperature, can be determined.

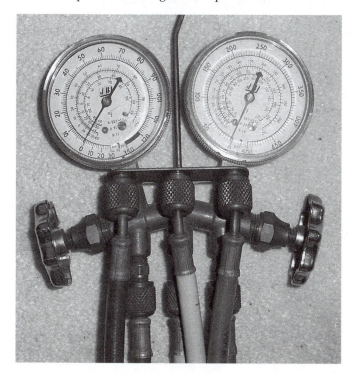

Figure 5-3. Gauge set

SPORLAN — PRESSURE-TEMPERATURE CHART at SEA LEVEL

TEMPERATURE °F — REFRIGERANT (SPORLAN CODE)

PSIG	YELLOW 12 (F)	GREEN 22 (V)	GREEN 124 (M)	BLUE 134a (J)	PURPLE 502 (R)	TEAL AZ50 or 507 (P)	WHITE 717 (A)
5*	-29	-48	3	-22	-57	-60	-34
4*	-28	-47	4	-21	-55	-58	-33
3*	-26	-45	6	-19	-54	-57	-32
2*	-25	-44	7	-18	-52	-55	-30
1*	-23	-43	9	-16	-51	-54	-29
0	-22	-41	10	-15	-50	-53	-28
1	-19	-39	13	-12	-47	-50	-26
2	-16	-37	16	-10	-45	-48	-23
3	-14	-34	18	-8	-42	-46	-21
4	-11	-32	21	-5	-40	-44	-19
5	-9	-30	23	-3	-38	-41	-17
6	-7	-28	26	-1	-36	-39	-15
7	-4	-26	28	1	-34	-38	-13
8	-2	-24	30	3	-32	-36	-12
9	0	-22	32	5	-30	-34	-10
10	2	-20	34	7	-29	-32	-8
11	4	-19	36	8	-27	-31	-7
12	5	-17	38	10	-25	-29	-5
13	7	-15	40	12	-24	-27	-4
14	9	-14	41	13	-22	-26	-2
15	11	-12	43	15	-20	-24	-1
16	12	-11	45	16	-19	-23	1
17	14	-9	46	18	-18	-21	2
18	15	-8	48	19	-16	-20	3
19	17	-7	49	21	-15	-19	4
20	18	-5	51	22	-13	-17	6
21	20	-4	52	24	-12	-16	7
22	21	-3	54	25	-11	-15	8
23	23	-1	55	26	-9	-14	9
24	24	0	57	27	-8	-12	11
25	25	1	58	29	-7	-11	12
26	27	2	59	30	-6	-10	13
27	28	4	61	31	-5	-9	14
28	29	5	62	32	-3	-8	15
29	31	6	63	33	-2	-7	16
30	32	7	65	35	-1	-6	17
31	33	8	66	36	0	-4	18
32	34	9	67	37	1	-3	19
33	35	10	68	38	2	-2	19
34	37	11	69	39	3	-1	20
35	38	12	71	40	4	0	21
36	39	13	72	41	5	1	22
37	40	14	73	42	6	2	23
38	41	15	74	43	7	3	24
39	42	16	75	44	8	4	25
40	43	17	76	45	9	4	26
42	45	19	78	47	11	6	28
44	47	21	80	49	13	8	29
46	49	23	82	51	15	10	31
48	51	24	84	52	16	11	32
50	53	26	86	54	18	13	34
52	55	28	88	56	20	15	35
54	57	29	90	57	21	16	37
56	58	31	91	59	23	18	38
58	60	32	93	60	24	19	40
60	62	34	95	62	26	21	41
62	64	35	97	64	27	22	42
64	65	37	98	65	29	24	44
66	67	38	100	66	30	25	45
68	68	40	101	68	32	26	46
70	70	41	103	69	33	28	47
72	71	42	104	71	34	29	49
74	73	44	106	72	36	30	50
76	74	45	107	73	37	32	51
78	76	46	109	75	38	33	52
80	77	48	110	76	40	34	53
85	81	51	114	79	43	37	56
90	84	54	117	82	46	40	58
95	87	56	120	85	49	43	61
100	90	59	123	88	51	45	63
105	93	62	126	90	54	48	66
110	96	64	129	93	57	51	68
115	99	67	132	96	59	53	70
120	102	69	135	98	62	55	73
125	104	72	138	100	64	58	75
130	107	74	140	103	67	60	77
135	109	76	143	105	69	62	79
140	112	78	145	107	71	64	81
145	114	81	148	109	73	66	82
150	117	83	150	112	75	68	84
155	119	85	152	114	77	70	86
160	121	87	154	116	80	72	88
165	123	89	157	118	82	74	90
170	126	91	159	120	83	76	91
175	128	92	161	122	85	78	93
180	130	94	163	123	87	80	95
185	132	96	165	125	89	82	96
190	134	98	167	127	91	83	98
195	136	100	169	129	93	85	99
200	138	101	171	131	95	87	101
205	140	103	173	132	96	88	102
210	142	105	175	134	98	90	104
220	145	108	178	137	101	93	107
230	149	111	182	140	105	96	109
240	152	114	185	143	108	99	112
250	156	117	188	146	111	102	115
260	159	120	192	149	114	105	117
275	163	124	196	153	118	109	121
290	168	128	201	157	122	113	124
305	172	132	205	161	126	117	128
320	177	136	209	165	130	120	131
335	181	139	213	169	133	124	134
350	185	143	217	172	137	127	137
365	188	146	221	176	141	129	140
380	192	150	224	179	144	132	143
400	197	154	229	183	148	136	147

* Inches mercury below one atmosphere

SPORLAN — PRESSURE-TEMPERATURE CHART at SEA LEVEL

TEMPERATURE °F — REFRIGERANT (SPORLAN CODE)

Note: The refrigerant blends in this chart are read as both BUBBLE POINT and DEW POINT (vertical labels on the chart); in the mid-pressure overlap region both values are printed in a cell (shown below as two numbers).

PSIG	PINK MP39 or 401A (X)	SAND HP80 or 402A (L)	ORANGE HP62 or 404A (S)	LIGHT BROWN 407C (N)	REDDISH PURPLE FX-10 or 408A (R)	BROWN FX-56 or 409A (F)
5*	-23	-59	-57	-48	-54	-22
4*	-22	-58	-56	-45	-52	-20
3*	-20	-56	-54	-42	-51	-19
2*	-19	-55	-53	-39	-49	-17
1*	-17	-54	-52	-36	-48	-16
0	-16	-53	-50	-34	-47	-15
1	-13	-50	-48	-31	-44	-12
2	-11	-48	-46	-29	-42	-9
3	-9	-45	-45	-27	-39	-7
4	-6	-43	-41	-24	-37	-5
5	-4	-41	-39	-22	-35	-2
6	-2	-39	-37	-20	-33	0
7	0	-37	-35	-18	-31	2
8	2	-36	-33	-17	-29	4
9	4	-34	-32	-15	-27	6
10	6	-32	-30	-13	-26	8
11	8	-30	-28	-12	-24	9
12	9	-29	-27	-10	-22	11
13	11	-27	-25	-8	-21	13
14	13	-26	-23	-7	-19	14
15	14	-24	-22	-5	-18	16
16	16	-23	-20	-4	-16	17
17	17	-21	-19	-3	-15	19
18	19	-20	-18	-1	-13	20
19	20	-19	-16	0	-12	22
20	21	-17	-15	1	-11	23
21	23	-16	-14	3	-9	25
22	24	-15	-12	4	-8	26
23	25	-14	-11	5	-7	27
24	27	-12	-10	6	-5	29
25	28	-11	-9	8	-4	30
26	29	-10	-8	9	-3	31
27	30	-9	-7	10	-2	32
28	32	-8	-5	11	-1	34
29	33	-7	-4	12	0	35
30	34	-6	-3	13	1	36
31	35	-5	-2	14	3	37
32	36	-4	-1	15	4	38
33	37	-2	0	16	5	39
34	38	-1	1	17	6	40
35	39	0	2	18	7	41
36	40 / 30	1	3	19	8	43
37	42 / 31	2	4	20	9	44
38	43 / 32	3	5	21	10	45 / 30
39	44 / 33	3	6	22	11	46 / 31
40	45 / 34	4	7	23	12	47 / 32
42	46 / 36	6	9	25	13	48 / 34
44	48 / 38	8	11	26	15	50 / 36
46	50 / 40	10	12	28	17	38
48	42	11	14	30	19	39
50	44	13	15	31	20	41
52	45	14	17	33	22	43
54	47	16	19	34	24	45
56	49	18	20	36	25	46
58	50	19	22	37	27	48
60	52	21	23	39	28	50
62	53	22	25	40	30	51
64	55	23	26	42 / 30	31 / 30	53
66	56	25	27	43 / 32	32 / 32	54
68	58	26	29	44 / 33	33 / 33	56
70	59	27	30	46 / 34	35 / 34	57
72	61	29	31 / 30	47 / 36	36 / 36	58
74	62	30	33 / 32	48 / 37	38 / 37	60
76	64	31	34 / 33	49 / 38	39 / 38	61
78	65	32 / 30	35 / 34	51 / 39	40 / 40	63
80	66	34 / 31	36 / 35	41	41 / 41	64
85	69	37 / 34	39 / 38	44	45 / 44	67
90	73	40 / 37	42 / 41	46	48 / 47	70
95	76	42 / 40	45 / 44	49	50	73
100	78	45 / 43	48 / 47	52	52	76
105	81	48 / 45	50 / 50	54	55	79
110	84	50 / 48	52	57	57	82
115	87	50	55	59	60	84
120	89	53	57	62	62	87
125	92	56	59	64	65	89
130	94	57	61	66	67	92
135	96	60	64	69	69	94
140	99	62	66	71	71	96
145	101	64	68	73	73	99
150	103	66	70	75	76	101
155	105	68	72	77	78	103
160	108	70	74	79	80	105
165	110	72	76	81	81	107
170	112	74	78	82	83	109
175	114	75	80	84	85	111
180	116	77	81	86	87	113
185	117	79	83	88	89	115
190	119	81	85	90	91	117
195	121	82	87	91	92	119
200	123	84	88	93	94	121
205	125	86	90	95	96	123
210	127	87	92	96	97	125
220	130	91	95	99	100	128
230	133	94	98	102	104	131
240	136	97	101	105	107	137
250	140	99	104	108	109	137
260	143	102	107	111	112	141
275	147	106	111	115	116	145
290	151	110	114	119	120	149
305	155	114	118	123	124	153
320	159	118	122	126	128	157
335	163	121	125	130	131	161
350	167	125	129	133	135	165
365	170	128	132	137	138	169
380	174	131	135	140	141	172
400	178	135	139	144	145	177

* Inches mercury below one atmosphere

Figure 5-4. Saturation card

EVAPORATORS

Evaporators absorb heat by evaporating the refrigerant. Ignoring, for the moment, the pressure drop associated with the friction of flow through the evaporator the heat is absorbed at a constant temperature, the saturation temperature of the refrigerant at the pressure in the evaporator. With the exception of chillers all evaporators are intentionally operated to evaporate all the liquid before it leaves the evaporator. Once all the liquid is evaporated the refrigerant, as a gas, absorbs additional heat to become superheated. The absorption to superheat is intentional in order to prevent liquid entering the compressor and damaging it. The superheat is maintained automatically in most systems by the throttling device.

The evaporator most people can relate to is the one that's visible in the window air-conditioning unit when you remove the cover and filter so you can clean the filter. You should recall that what you see are rows of tubing wrapped with aluminum that looks shredded or the tubing runs through sheets of aluminum. We commonly refer to those as fins. We also refer to it as extended surface. Heat transfer from a solid to a boiling liquid and through highly conductive metal is much faster than it is between a metal surface and air. Therefore an evaporator to cool air is designed with a lot of surface area where the metal contacts the air. An evaporator in a window air conditioner cools the air and removes moisture by condensing it. The velocity of the airflow over the cooling surface has to be restricted to prevent blowing the droplets of moisture off the coil and into the circulating fan or the room. The additional surface compensates, to a degree, for the loss of turbulence that would be provided by higher airflow rates. The fins are also designed to accelerate the drainage of the condensed water off the heat transfer surfaces.

That water cooler you stop to drink at has an evaporator that's a coil of two tubes, one inside the other, with the refrigerant inside the inner tube and the water that you're drinking in the annular space between the two tubes. Heat transfer rates from metal to flowing water is higher than it is to a gas like air so fins aren't economically justified on something so small that only requires a little extra length of tubing to make up the difference between heat transfer to boiling liquid and heat transfer to flowing liquid.

If you entered a walk-in freezer you would probably note few if any fins on the cooling coils. That's because fins would hinder defrosting and maintaining cleanliness. A freezer's cooling coils not only condense moisture from the air but freeze it on the coil surface. As that ice builds up it serves as an insulator restricting the flow of heat. Timers, or devices monitoring either the superheat of the refrigerant or temperature of the fins on the coil, are used to initiate a defrosting cycle. Defrosting of freezer coils can be accomplished in two ways. One way is using an external heating source, normally found in frost-free refrigerators, where an electric heating element is used to melt the ice on the coils while the refrigeration compressor is temporarily shut down. You may have opened the freezer side of the refrigerator and heard the hissing sound of melted water dripping on the heating element. In larger commercial freezers defrosting is achieved by first shutting off the supply of liquid refrigerant to the evaporator then admitting hot refrigerant gas coming from the discharge of the compressor. Drain pans under those coils are designed to separate the ice that falls off the coils from the melted water so the water is drained. Sometimes maintenance is required to remove the ice. Fins on the cooling coils of a walk-in freezer would not permit effective defrosting because they would hold the ice in place. However, fins on the cooling coils of your refrigerator are effectively defrosted by the heater. In both the walk-in freezer and your refrigerator flow of air over the coils while defrosting is reduced by shutting off air circulating fans.

It pays to check the condition of the drains that remove the water melted off the coils after a defrost cycle to ensure they are flowing clear. The drains in a typical household freezer drop the water into an external pan positioned at the condenser cooling air outlet where the heated air can evaporate the water. The tubing between the pan installed in the freezer to collect the melted ice water and the external pan or drain can get plugged with chunks of ice to block flow. That results in the pan inside the freezer overflowing to make the bottom of the freezer into an ice rink, a dangerous condition for a walk-in freezer.

Heat transfer between metal and flowing water occurs at a rate much faster than between metal and flowing air. Therefore you will seldom see fins on tubing used for cooling water or freezing water. Chillers frequently have a form of fin that will be covered later.

When making ice the standing water and the ice itself restrict heat flow. That requires additional surface which has to be smooth to permit removal of the ice. Ice makers can use one of the two methods described for defrosting freezers to melt a thin layer of the ice at the surface of the freezer to allow removal of the ice.

You might see what looks like a P trap used in plumbing in the piping of an evaporator or the gas piping between the evaporator and the compressor. It has a specific purpose when the compressor is higher than the evaporator. It accumulates coalesced oil until the trap is full and the gas pushes it up the piping to the compressor. In most systems some lubricating oil circulates and those little traps make sure it doesn't get trapped in the system and always returns to the compressor.

FREEZING AND ICE STORAGE

I doubt that there are many of you who can remember an icebox. The name sort of says it, an icebox was a wooden box with doors and an opening in the top where the Iceman dumped a chunk of ice every few days. Before refrigeration the only way we had to obtain ice, other than in the winter, was to draw on reserves of ice cut from the lake surface and stored under a huge pile of straw during the summer. No, I'm not that old, but my grandmother still had her icebox so I know what they look like. The ability to make ice and freeze things is one of the major contributions to better health and longer living that we enjoy today.

On the other hand, freezing of water has contributed to considerable damage to refrigeration equipment, piping, and other equipment that contains water. I'm sure you know that ice expands as it cools; a contradiction of most substances because they shrink as they cool and expand when heated. It's a good thing in actuality because, by expanding, the density of ice is less. Frozen water floats. If it were not for this phenomenon earth could be an ice planet because the sunlight couldn't get at the ice to melt it. An operator has to be conscious of the fact that ice expands when cooled so he prevents damage to piping and equipment by preventing its temperature dropping below 32°F or draining the equipment and piping.

When making ice the operator also has to consider the fact that ice is a solid and heat transfer through ice is by conduction only. The rate of flow of heat through a solid, like ice, is proportional to the thickness of the solid; the thicker the solid the slower the heat transfer.

Freezing water requires removal of 80 Btu from each pound of ice and that ice will absorb the same amount of heat when it's melted. Ice, therefore, increases the capacity of the system to absorb heat within 1¼% of the mass of water. Storing ice, principally for air-conditioning systems, permits the installation of smaller equipment for cooling chilled water that is used in most facilities. Ice storage can also significantly reduce electricity cost for cooling when time-of-use electricity rates

are in effect by operating ice-making equipment at night when electricity rates are lower. Proper use of ice storage to limit peak electrical costs, and to absorb heat loads that exceed the capacity of the chilled water system, can significantly reduce the owners operating costs while ensuring the occupants of the building are always comfortable.

Ice storage systems don't work like the icemaker in your refrigerator. In order to absorb the heat the ice and chilled water have to be in contact and the rate of heat transfer between the two is proportional to the surface area where the water contacts the ice. In other words, you cannot simply create one huge block of ice then melt it. The only system I'm aware of that works well is made by Baltimore Aircoil and consists of a heat exchanger enclosed in a tank; the outer surface of the tubing of the exchanger is exposed to water that flows through the tank and ice is formed on the outside of the tubes by cooling using a refrigerant, brine, or glycol water mixture inside the tubes. The ice builds up on the surfaces of the heat exchanger when making ice and is melted by water pumped through the tank that chills that water. To ensure an even build up of ice on the surfaces of the heat exchanger air is blown into the tank to bubble up through the water thereby increasing turbulence to help improve heat transfer.

COMPRESSORS

The compressor takes in the refrigerant after leaving the evaporator and pumps it up to a higher pressure to raise the saturation temperature to a point where the heat absorbed by the evaporator can be dumped to another substance in the condenser. A low-pressure with a correspondingly low saturation temperature is maintained in the evaporator because the compressor removes the refrigerant vapor from the evaporator. Work is performed on the refrigerant by the compressor to squeeze it up to a higher pressure and the energy of the compression increases the heat in the refrigerant increasing its superheat. Because many compressor motors are cooled by the refrigerant, friction and heat from the motor windings typically adds to the superheat as well.

High evaporator pressures or high temperatures at the outlet of the compressor are indicative of inefficient compression. A noisy compressor, that is one noisier than normal, may indicate liquid flooding through the evaporator to the inlet of the compressor. It could also indicate loss of lubrication in the compressor. Because a compressor is a piece of mechanical equipment it requires lubrication. And, since the moving parts are in direct contact with the refrigerant, lubricating oils for compressors are specific to the refrigerant used. With the exception of centrifugal compressors some of the lubricating oil is broken up into small droplets forming an aerosol that travels with the refrigerant through the system. The oil is cooled and occasionally heated by the refrigerant itself so oil coolers are not always required.

There is no dipstick in a refrigeration compressor so other means are required to determine the level of the oil. Some compressors, like the one in a window air-conditioner and your car, don't require a lot of oil and because the systems are sealed, an adequate amount of oil is assumed. The wise operator always looks around and under compressors, evaporators, condensers and associated piping to detect any accumulation of leaking oil. First of all it's an indication of a possible refrigerant leak and secondly it can indicate a potential failure of the compressor. Larger compressors are usually fitted with a sight glass in the compressor crankcase that permits observation of the oil level. An idle compressor can have a high oil level due to accumulation of oil drained from connecting piping as well as portions of the compressor itself. On my one trip as a ship's refrigeration engineer I noticed entries in the log of my predecessor that mentioned frequent topping off and draining of lubricating oil from the compressors. I did note a higher oil level on compressors that were shut down but the level always returned to normal shortly after the compressor was placed in operation. That's because the oil filled the cavities in the compressor that had drained back to the bottom and some of the oil was circulated into the system. Simply allowing what happened naturally to occur saved me all the work of adding and removing oil.

While it seems strange every refrigerant compressor has a heater in it. The purpose of the heater is to boil off refrigerant that condenses in the crankcase or oil sump. When the compressor is operating the heater is normally turned off because the operation of the compressor heats the oil. When the compressor is idle, or out of service, refrigerant tends to migrate to the compressor crankcase, condense, and mix with the oil. This can happen if the heater doesn't work and the result will be higher crankcase oil level. This situation is problematic because as soon as a compressor starts operating the lower crankcase pressure drops below the saturation pressure of the liquid refrigerant in the oil and it boils. The boiling refrigerant mixes with the lubricating oil to create foam which results in slugs of lubricating oil entering the valves and cylinders of the compressor. Under

the right conditions liquid refrigerant can be carried into the cylinders as well. So, always check to see if the oil heater is working before starting a compressor and check that sight glass immediately after starting the compressor for foaming. A small amount of foaming that doesn't raise the oil level out of the glass would be acceptable. Compressor controls are typically fitted with a timer to prevent the motor starting after power is interrupted for sufficient time to allow the heater to boil off any liquid refrigerant that's migrated to the crankcase. That's why your home air conditioner compressor won't start right away after you over-adjusted the thermostat to stop it.

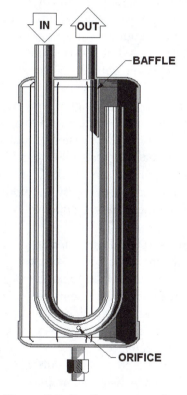

Figure 5-5. Suction accumulator

The description for evaporators mentioned the traps for accumulating and re-injecting oil into the flow of refrigerant gas. That has the potential of sending a slug of oil into the compressor that could be carried directly into one of the cylinders resulting in damage to the compressor. The device shown in Figure 5-5 is called a suction accumulator and is installed in the vapor piping near the compressor inlet to protect the compressor from slugs of refrigerant or oil. A slug of oil, refrigerant, or a combination thereof is trapped in the accumulator. The small hole in the riser tubing allows a gradual re-injection of oil into the flow of refrigerant to return it to the compressor and the superheat in the gas should vaporize any accumulation of liquid refrigerant. The operation of the accumulator means oil level can fluctuate and it's wise to observe the oil level for a while before reacting to a slightly higher or lower level by adding or removing oil.

Compressors are classified as open, hermetic, or semi-hermetic. Virtually all small compressors are hermetic with the motor enclosed within the sealed compressor casing. Since the motor is enclosed within the compressor casing the motor is cooled by the refrigerant flowing over it. A continuous flow of cool refrigerant while the compressor is operating is essential to prevent the motor overheating. Small compressors like the one in your refrigerator, window unit, and residential air-conditioners will not have means of monitoring conditions to detect a loss of refrigerant flow. However, another sense attributable only to a person, can provide an indication of a problem: your sense of smell. If the motor is not cooled adequately the higher temperature will result in vaporizing whatever managed to accumulate on the outside of the compressor casing producing a hydrocarbon emission that you recognize. Regular rounds by an experienced operator can usually result in detection of a problem and action to rectify it before the equipment fails even if there are no gauges or thermometers to read.

Medium sized and large compressors can be of any type but, because most of them are constructed to permit maintenance they are called semi-hermetic. Flanged and threaded connections permit opening the compressor enclosure for maintenance.

Large compressors are either semi-hermetic or have the motor mounted outside of the compressor casing in which case they are referred to as open compressors. Because your automobile air conditioner is driven by a belt and pulley from the engine it's an open compressor. An open compressor requires a shaft seal which can become a repeating maintenance problem. There are restrictions on the amount of refrigerant that can leak from a system and shaft seals are normally the only source of leaks. If you have an open compressor and repeated seal failures it's an indication that they are not installed properly, the compressor and motor are not in alignment, or the flow of refrigerant and/or

lubricating oil that lubricates and cools the seal has been interrupted.

Reciprocating Compressors

Reciprocating compressors are the most common type of compressor. One in your car, a window unit, your refrigerator or freezer typically contains two pistons and cylinders. Figure 5-6 is typical of reciprocating compressors used for home air conditioning units. I have operated and maintained compressors with as many as 16 cylinders, units similar to the one in Figure 5-7a & b which has 8 cylinders. Note the head spring in the section which reveals that the head of each cylinder in these compressors is free to move within the cylinder, compressing that head spring whenever liquid might be drawn into the compressor cylinder. A louder "thump" (actually it's more of a bang or rattle) is a sign that's happening to one of them. Reciprocating compressors function by the movement of the piston and cylinder to first, increase the volume inside the cylinder which allows the gas in the cylinder to expand until the pressure in that cylinder is lower than the pressure at the compressor inlet. Then gas from the evaporator flows into the cylinder as the moving piston makes room for it. Gas enters the

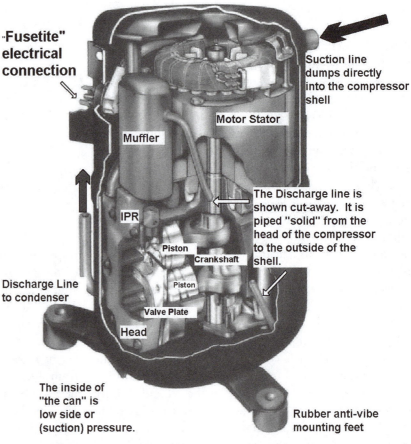

Figure 5-6. Small Reciprocating Compressor

compressor cylinder through an intake valve or valves. My best description for the valves of a refrigeration compressor is they look like popsicle sticks. They are guided within the casting of a cylinder head and are either a spring themselves or guided by springs that gently push them toward the seat. The flow of gas pushes the valve away from the opening and, when flow stops, the spring force pushes the valve back onto its seat. As the piston approaches the bottom of its stroke it slows down then reverses due to the rotary motion of the piston connecting rod at the crankshaft. Then, as the piston rises in the cylinder, reducing the volume, the gas is compressed until it is at a pressure slightly higher than

the pressure of the gas at the outlet to overcome the force of the spring on the valve and the discharge valve opens. For the remainder of the upward stroke gas is pushed out of the cylinder into the discharge piping. The volume of gas delivered with each rotation of the shaft isn't a function of the cylinder displacement because the gas has to expand and compress. The amount of gas pushed through the compressor is considerably less than the volume displaced by the piston in the cylinder.

Small compressors are started under load. To allow larger compressors to get up to speed before starting to pump refrigerant unloading systems are used. Unloading a cylinder is accomplished with a pin that pushes up against each suction valve to hold it open. The pin is connected to a small cylinder containing a spring as shown in Figure 5-8. When the compressor is shut down the springs push the pins up to hold the valves open. After the compressor gets up to speed the oil pump builds up pressure in the oil passages of the compressor and the small unloader cylinders to force the pins down and allow the suction valves to operate. You can tell when the unloaders operate because there will be a significant difference in the sound of the compressor as it comes up to speed. The unloaders allow a compressor to be powered by a standard duty motor instead of a high torque motor. Many modern compressors use a solenoid and spring to control the pins electronically.

In addition to reducing startup torque unloaders allow multiple cylinder compressors to handle varying loads without starting and stopping. A controller senses suction pressure, loading and unloading cylinders as required to maintain a set suction pressure. With older hydraulic actuated unloaders each one (sometimes pairs) can be set to operate at a different suction pressure allowing a wider range of evaporator temperatures or in a tighter range when temperature maintenance is more critical. If you're near the compressors regularly you should be able to hear the unloaders cutting in and out. If you no-

Figure 5-7a. 8-cylinder compressor

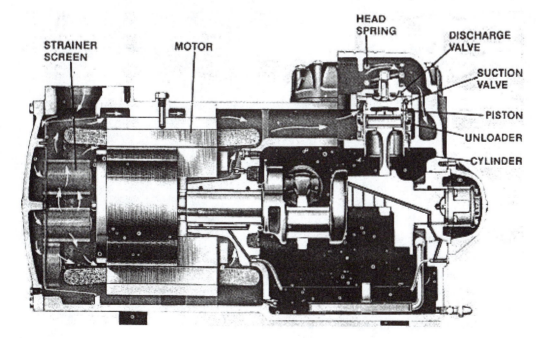

Figure 5-7b. 8-cylinder compressor section

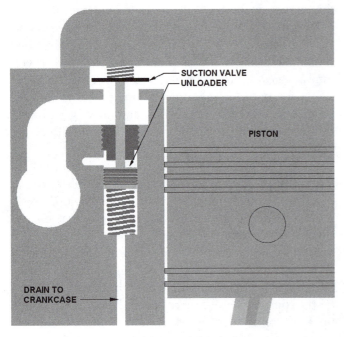

Figure 5-8. Unloader

tice that one of two or more compressors is constantly running with fewer pulsing sounds, indicating the compressor is totally unloaded, you should look at what happens if you shut it down to save energy and wear and tear on the compressor.

Vane Compressors

The only service I have seen for these compressors is automobile air-conditioning and small systems with very low pressure differentials. The compressor shaft has vanes set in slots around its circumference that are spring-loaded to press them against the compressor casing as shown Figure 5-9. The eccentric setting of the shaft inside the casing changes the position of the vanes which slip in and out of the slots so the spaces between the vanes change volume as the shaft rotates. The space between each vane works almost like a cylinder in a reciprocating compressor.

Scroll Compressors

Scrolls are new technology dependent upon modern material technology. Compression is achieved in a scroll compressor by the two pieces shown in Figure 5-10 one of which is stationary while the other wobbles eccentrically creating crescent shaped gaps between the walls that are moved along the scroll as the walls separate then move together. The result is pockets of refrigerant entering the scroll at the outside and being squeezed along the scroll to its discharge point in the center. The scroll pieces are mounted in the top of the compressor

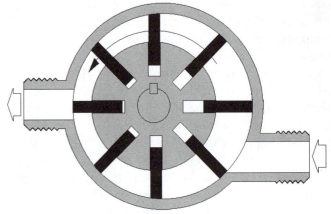

Figure 5-9. Vane compressor

(Figure 5-11). Figure 5-12 shows a series of the changes in position of the two scroll pieces to demonstrate how the compressor grabs a volume of gas and gradually compresses it. Usually a scroll compressor can be distinguished from a reciprocating compressor by shape, the scroll being taller and smaller in diameter for the same capacity. Scroll compressors at the time of writing this edition cannot be made larger than 25 tons. There

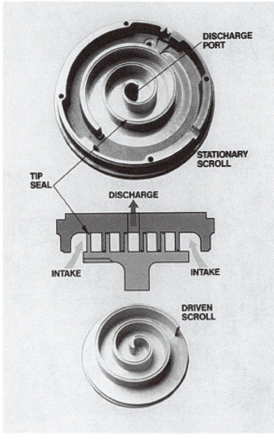

Figure 5-10. Scrolls

**Figure 5-11.
Scroll compressor**

**RIGHT: Figure 5-12.
Scroll progression**

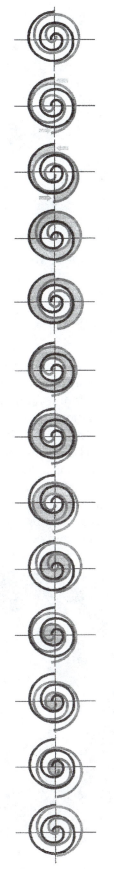

is no way to unload a scroll compressor. Scroll compressors have an advantage of lower sound power levels because they are smoother operating. To a degree scroll compressors can be turned down using variable speed drives or two speed motors for a better match to cooling loads. My home's heat pump contains a two speed compressor and in real cold periods is switching between high and low speed rather than starting and stopping.

Hot Gas Bypass and Defrosting

Many reciprocating and scroll compressors, especially the smaller ones, are on/off operation only. The typical refrigerator, freezer, icemaker, wine cooler, and window & auto air-conditioners can be operated in this manner because swings in temperature can be tolerated. On occasion the load can swing considerably and/or temperature deviations are

unacceptable and a means is required to keep the compressor in operation constantly. That's because the delay time between starting and stopping is typically around 15 minutes for those designs of compressors to ensure the oil in the crankcase is free of liquid refrigerant. The solution to the problem is hot gas bypass.

I've only encountered one situation where hot gas bypass to keep the compressor running was essential. An entryway for the plant laboratory at a plastic fibers plant was to be used to temporarily store pallet loads of production samples shipped to the lab for testing. The spools of fiber were not light and the laboratory technicians were not happy with a 15 minute wait time before cooling restarted. A pressure reducing valve and connecting tubing that delivered compressor discharge vapor to the inlet of the evaporator maintains the evaporator pressure to keep the compressor running.

Hot gas bypass can be confused with hot gas defrosting because they're piped the same. The difference in control operation will let you tell which system it is.

Screw Compressors

Screw compressors consist of two screw shaped rotors that are interlocked inside a casing. The two screws form cavities between the screws and the casing (Figure 5-13) that are sealed where the two screws meet and compression is achieved by capturing a gulp of refrigerant vapor in a cavity and that cavity progressing from the inlet where the pressure is low to the outlet. Theoretically the vapor would progress along the screws at suction pressure until the cavity is opened at the discharge where high pressure vapor would flow into the cavity and, like a piston, compressing what was trapped. However, leakage between and along the screws serves to raise the pressure in the cavity as it travels along so the sudden compression when the pocket of gas reaches the discharge is not as violent as it could be. A certain degree of unloading is provided with screw compressors through the movement of a plug laying along the screws. That same plug also provides a means of partially unloading the compressor for start up.

Lubrication is very important for screw compressors because the oil helps seal the clearances between the screws and the casing to reduce refrigerant leakage at the same time as it creates a wedge of oil between the moving parts. A typical screw compressor will have provisions for removing oil from the compressor discharge vapor and injecting it back in at the inlet to the screws. Large screw compressors will also have gears mounted on the two screw shafts that maintain screw alignment without exposure to gas leakage and they must be lubri-

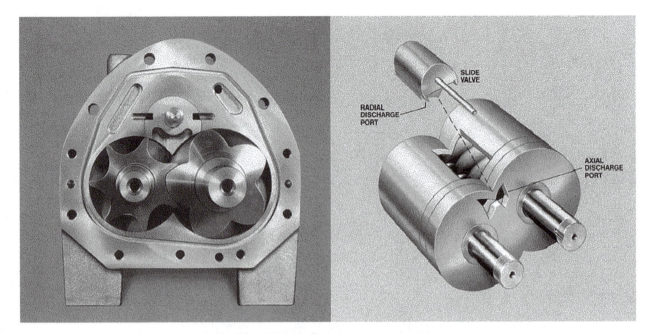

Figure 5-13. Screw compressor

cated as well. Some screw compressors will have a shaft mounted oil pump while others will rely on a separate electric pump at all times.

Although I've mentioned that a plant noise level should be restricted so you don't need to wear earplugs or earmuffs I have to admit I have never been near a screw compressor without one or the other. Screw compressors are inherently noisy; so noisy that you should never go near them without ear protection.

Centrifugal Compressors

If you've ever wondered why a facility would utilize chilled water for cooling and air-conditioning purposes instead of individual local refrigeration equipment it's because chillers, which are discussed later, are so efficient. Refrigeration using a centrifugal compressor has energy requirements that are fractional compared to other types of compressors. With the advances in design that include variable speed drives and magnetic bearings a centrifugal compressor today has half the operating cost of one from 20 years ago and those still produce cooling at half the power cost of local refrigeration equipment. A centrifugal compressor is a high tech piece of equipment and must be built in large sizes to be economical. A centrifugal compressor is very much like a blower (their forms are discussed in Chapter 10 under fans and blowers).

Considerable turndown is achieved with a centrifugal compressor using variable speed drives. However you may find yourself operating an older one that has a

fixed speed motor and inlet vanes (Figure 5-14). The inlet vanes are typically operated by a small motor mounted outside the compressor housing. They reduce flow through the compressor by closing off on the inlet while simultaneously creating a whirl in the flow of the gas in the direction of the impeller rotation to further reduce the motor horsepower required at the reduced flows.

Centrifugal compressors can be open or hermetic but are normally hermetic when powered by an electric motor. You may encounter one that's powered by a steam turbine or diesel engine. The principal use of

Figure 5-14. Centrifugal inlet vanes

centrifugal compressors is water chillers and you should read about those later to learn more about the operation of centrifugal compressors.

CONDENSERS

Once the compressor has increased the pressure of the refrigerant high enough to produce a saturation temperature above the temperature of the substance that receives the heat (heat removed in the evaporator and added by the compressor) the condenser serves to transfer the heat from the refrigerant vapor to that substance. Operating pressures and temperatures in the condenser are dependent upon the substance used to cool the condenser and its temperature. The condenser has to remove the superheat, simply cooling the refrigerant gas until it can start to condense, then condense it.

Because heat transfer from a gas to metal is a fraction of the rate of heat transfer from a condensing liquid to metal a fair portion of the condenser does nothing but desuperheat the gas. Once the gas temperature drops to saturation temperature the gas condenses to a liquid which drizzles over the heat exchange surfaces and drops to the bottom of the condenser. Many condensers are designed so the hot superheated gas flows around and through the condensed droplets because a small portion of them will be evaporated in cooling the superheated gas and the rest will be heated a bit because the liquid in contact with the heat exchange surfaces is cooled below the saturation temperature of the refrigerant. The evaporative cooling of the gas by the droplets is more efficient than the cooling of the metal surfaces.

Finally, the condensed liquid must be subcooled in most systems so it will remain a liquid until it reaches the throttling device. The amount of liquid in a system serves to balance the gas to liquid ratio in order to compensate for variations in load and operating conditions. Your refrigerator, window unit, automobile, and house air conditioning systems hold varying amounts of liquid refrigerant in the bottom of the condenser depending on those factors. The amount of subcooling is an indication of how much liquid is retained in the condenser. A small amount of subcooling can indicate insufficient refrigerant and a large amount would indicate excessive refrigerant. A measure of subcooling is a good measure of the refrigerant in a system when the liquid is stored in the condenser but the subcooling has to be measured at a consistent condenser pressure. Systems with receivers, discussed later, have other measures.

Free Cooling

When it's possible to obtain a source of fluid that's colder than what you want to cool, the refrigerant cycle can function without the compression and expansion (throttling) stages. The refrigerant gas will migrate (a term used to describe the flow of the gas because it's a natural thing) to the coldest spot in a system because that's where it will condense if the pressure in the system is below the saturation pressure that matches that temperature. By assembling the equipment so the condenser is above the evaporator the gas will condense in the condenser, giving up heat to a substance and the liquid will drain to the evaporator where it can cool another substance by boiling. When equipment is arranged to do this it's called free cooling because no energy is input to compress the gas.

Free cooling can also be accomplished with parallel systems that simply use the colder substance to remove heat from whatever it is you want to cool. One example would be using cold river water to create chilled water. To be truly effective these systems have to allow warmer chilled water temperatures. That's perfectly acceptable because when the river water is cold it's cold outside and the warmer chilled water will still do the job. Free cooling can also be utilized in more complex systems to pre-cool chilled water or cooling tower water before mechanical equipment is used to achieve the necessary water temperature.

Heat Pipes

Devices that exchange heat between incoming ventilation air and indoor air exhausted to make room for the ventilation air can perform free cooling. Devices call heat pipes consist of a number of finned tubes sealed at the ends, installed in a split casing, and sloped appropriately, can automatically transfer heat from warm indoor exhaust air to hotter outside air and, with airflow reversed, heat cold outside air with warm indoor exhaust air. Each heat pipe is a refrigeration system unto itself, the half at the lower end of the slope absorbs the heat by evaporating the liquid refrigerant that drains down into it. The evaporated gas then travels up the pipe to the high end where it is condensed, thereby giving up its heat.

Air Cooled Condensers

You are familiar with many air cooled condensers. You have one in your car, your refrigerator, your freezer if you have one, and your home air conditioning to name the principal ones. These have to be designed for operating temperatures of the cooling air at maximum temperature one would expect to encounter. Most of these are designed to condense the refrigerant using air at 120°F. The auto manufacturers cannot take a chance making vehicles that would not, on some day, pay a visit to the deep South or, as Sue and I did last winter, visit Death Valley. Heat transfer between metal and air is poor compared to metal to condensing vapor so you will find most air cooled condensers have fins in tight contact with the condenser tubing to increase the surface for heat transfer between the metal and air. Provisions are also made to force the flow of air over the transfer surface to provide turbulence at the interface between the two substances.

A window air conditioner has a unique extra feature to help condense a refrigerant. The condensate from moisture removed from the air-conditioned space is drained to a well in the casing where it is picked up by a slinger mounted on the condenser fan and thrown onto the condenser tubes and fins. It's cold water and it absorbs over 1000 Btu per pound as it evaporates to help condense the refrigerant. It's why you don't always have that experience of cold condensate dripping on your shoulders as you enter an establishment with a window unit above the door.

Some air cooled condensers may be called upon to operate in cold winter temperatures. Some examples are condensers for refrigeration equipment cooling the core of an office building, refrigerated food cases, or cooling a production process. In very cold weather the colder air would result in low pressures in the condenser. That would reduce the power consumption of the compressor

but it would also reduce the pressure differential across the throttling device and restrict flow of the refrigerant for inadequate cooling. Simply shutting down the fan(s) can reduce heat transfer to raise condenser pressure to an acceptable value but air cooled condensers in colder climates can also include dampers that can control airflow over the fins and coils to maintain a condenser pressure.

Condensing Units

This term is a label for a certain arrangement of refrigeration equipment, specifically a combination of a compressor and a condenser. It's normal to find condensing units outside residences, small office buildings, and grocery stores. Typically these are reciprocating or scroll compressors in the same housing as an air cooled condenser. A typical air-cooled condensing unit is shown in Figure 5-15. Liquid and suction lines connect the condensing unit to the evaporator and throttling device located inside the building. The condensing unit normally includes a receiver and suction accumulator and can contain multiple compressors with control switches set for different pressures to bring them on and off in stages. When used for refrigeration equipment they're also fitted with condenser pressure control switches that are staged to control the fans so the air flow is regulated to maintain a reasonably constant pressure on the liquid line returning to the evaporator. If there were no pressure control then the condenser could get so cold that there would not be enough pressure to push the liquid to the evaporator. I had, on occasion in the winter, found

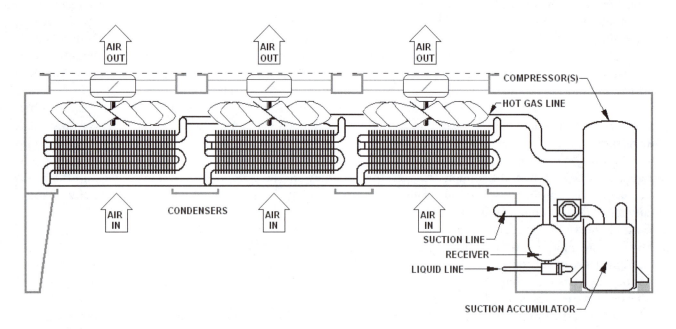

Figure 5-15. Air-cooled condensing unit.

it necessary to cover two of the fan discharge screens to restrict airflow enough to maintain an adequate condenser pressure. Wax coated cardboard hinged on one side with some duct tape did a good job of it but the cardboard had to be flipped back over in cold weather if the fan of that section was operated.

Regular inspections of condensing units are required because so many things can happen to them. During each inspection of the condensing unit you should always look for oily spots around the unit base and connecting piping; those spots would be refrigerant oil indicating a system leak. Sometimes the units are mounted on the roofs of buildings which requires a roof hatch and/or ladder for access and it's not uncommon for the latter to be one that's carried on a truck. These are less susceptible to damage than units mounted at grade level but at the same time are the ones that are frequently overlooked during normal inspection rounds. Problems with rooftop condensing units normally fall in the range of blockage of the coils and fins of the condensing unit by airborne contaminants. Cottonwood tree and oak tree pollen is a common problem when the trees are blooming and leaves can block the coils in the fall.

Newer condensing units mounted at ground level are exposed to other hazards. All new units are similar to a residential unit (Figure 5-31) because the three sided coil and vertical air discharge make them more efficient. A common problem is the accumulation of leaves, seeds, and small branches that fall onto the condensing unit and managed to drift past the fan to settle behind the condenser. The bottom tubes of the condenser then act as a receiver because tubes higher up have to serve as that portion of the heat exchanger dedicated to subcooling. The result is an apparent loss of refrigerant because more liquid is held in the condenser. Always look through the fan to check for accumulations that can cause this problem. It's not unusual for the discharge screen to be readily removed for access to remove any accumulations; however, always turn the unit off and disconnect power to the condensing unit before you reach your hands and arms through the fan to remove any accumulation. Don't forget to put the discharge screen back on before restoring power and control. The inlet side of the condenser can also be plugged up with mown grass and debris drawn into the fins by the airflow with the assistance of a lawn mower, weedeater, or wind. More frequent inspections in the spring when the pollen and seeds are falling and in the fall when leaves are falling should be made to ensure the condenser inlet is clean. Yard workers prove to be a hazard for most condensing units because the fins on the air cooled condens-

ers are readily damaged by the plastic strings of weed eaters; I even investigated one problem site where the Yard workers had elected to write their initials into the fins using a weed eater; needless to say the culprits were easy to identify. The grass and weed eater damage has the same effect as debris accumulation on the condenser outlet. Unless the damage is so extensive that the fins are torn, bent fins on an air cooled condenser can be repaired readily with a device called a fin comb. Even if none of these obvious conditions of condenser blockage are evident a regular annual cleaning of the condenser coils is recommended.

Water Cooled Condensers

Every source of water for a water cooled condensers can be counted upon to be colder than air except in the winter. Even water cooled by a cooling tower is colder than the temperature of the air. Heat transfer between metal and flowing water is considerably higher than the heat transfer between metal and flowing air. So, water cooled condensers are preferred because they permit lower saturation temperatures inside the condenser which reduces the horsepower requirements of the compressor.

You may ask, "Okay, why isn't my home air conditioner water cooled?" Well, if you live in the Southwest, it actually may be water cooled. We call them evaporative condensers (Figure 5-16), and they can be found almost anywhere. The refrigerant is piped or connected with tubing to coils in a casing where water drips on them to absorb heat as the water evaporates and is carried off by air drawn through the evaporative condenser. It requires a pump to circulate the water from the catch well at the bottom of the condenser to the top where it's distributed over the tubes. A makeup water flow control valve is required to add water from a water supply to replace the water that is evaporated. And, a small amount of water must be removed by blowdown to limit the concentration of solids in the circulated water. The operation and maintenance of an evaporative condenser is very similar to cooling towers which are discussed later.

Construction of a water cooled condenser is highly dependent on the quality of water used to absorb the rejected heat. The condensers for my shipboard refrigeration plants used seawater that is highly corrosive and contains all forms plant and animal life. It had to be filtered before the section of the circulating pump that forced salt water from the ocean through the filter, the pump, condenser, and condenser water flow control valve then back to the ocean. The shell, heads, and tubes of the condensers were made of high-quality brass and bronze. Sacrificial magnesium anodes electrically connected to the condensers provided additional protection against corrosion. Plants using seawater or brackish water would be constructed similarly but, if large, use electrically powered cathodic protection for corrosion. Growth of algae, bacteria, bivalves, etc., on the heat transfer surfaces are also concerns for those condensers. The pressure drop through and temperature differential of these condensers has to be monitored closely to identify conditions of organic growth hindering heat transfer.

Well water will have the highest dissolved solids content of any supply of condenser cooling water. Normally, when the well water is not circulated, it will not present a problem. However, I have seen sources that will form scale on heat transfer surfaces after a temperature rise of only a few degrees. Regular sampling and testing of the TDS of the well water should allow you to detect conditions that could contribute to scaling. The important thing to do with high solids water is to maintain a minimum rate of water flow through the condenser and avoid concentrating solids content by blowdown or other means.

Water drawn from rivers, lakes and reservoirs will normally have lower dissolved solids content but can have a considerably higher organic materials content. Water circulating through a condenser and the cooling tower will also pick up organic materials. When those waters are used in condensers they will require regular cleanings to remove the organics, algae, and bacterial growth. Monitoring of the pressure and temperature differentials under these conditions is imperative.

Condenser Pressure Control Valves

To maintain the pressure in a water cooled condenser above a minimum value self-contained pressure control valves modulate the flow of the water leaving the condenser. A capillary containing a refrigerant charge is connected to a bellows in the valve assembly (Figure 5-17). The pressure in the bellows is opposed by a spring that pushes on the valve stem to force the valve to close down and decrease water flow through the condenser. An increase in pressure on the refrigerant side of the condenser results in an increase in the pressure in the bellows on the valve to force the valve

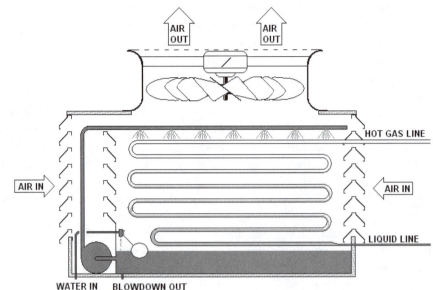

Figure 5-16. Evaporative condenser

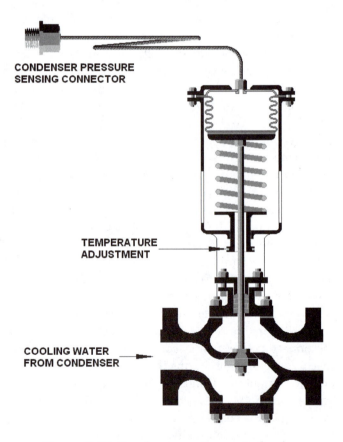

Figure 5-17. Condenser water control valve

to open increasing the cooling water flow. Control of the condenser pressure is maintained within a narrow range of pressure dependent upon the design of the spring and that range is adjustable up and down by adjusting the pressure on the spring.

Because this control valve is a proportional device and water temperatures can change due to season, storms, etc. the pressure in the condenser will vary. Under extreme conditions you might choose to adjust the setting of the valve to keep the range at the lowest possible values. That's because maintaining a low-pressure in the condenser reduces the amount of effort on the part of the compressor so it costs less to operate. The only time I ever considered it important to make that adjustment was when we were entering or leaving the Gulf Stream where seawater temperatures could swing by as much as 50°F. You might find it advisable to tweak the setting to increase the range between winter and summer. The control valve may completely block flow when the condenser is not in use and pressures inside it are low so it's always a good idea to have a small valved bypass around it that can be operated to maintain a minimum flow which helps discourage organic growth on the water sides. Organic growth can be a problem for any water cooled condenser so it's always a good idea to maintain some flow through them to discourage organic growth.

Why not just let the temperature of the condenser water drop in a refrigeration system? Primarily it's because the condenser can get so cold that there's not enough pressure differential to force the liquid through the throttling device. Some modern chillers and refrigeration systems may take advantage of cold water (or cold air for that matter) to condense the gas coming from the evaporator without compressing it by operating like a heat pipe. Those applications require a compressor bypass valve, a throttling device bypass, and because the flow of refrigerant is due to convection currents, the condenser has to be mounted higher than the evaporator.

THROTTLING DEVICES

A throttling device is required to separate the high pressure of the condenser from the low-pressure of the evaporator to ensure maintenance of the saturation conditions in each of those devices. The types of throttling devices vary considerably, from the very simple to quite complex, and can incorporate new high-technology methodology. The throttling device on the equipment you operate was selected based on economic factors. Most of them will operate with a minimum of attention for years. That doesn't mean that you should ignore them; you should be constantly checking

for symptoms of failure of the throttling device.

Frequently the throttling device is referred to as a metering device because it regulates the flow of refrigerant. Since I interpret the word "metering" to refer to items that actually measure flow I've avoided use of the word. Another label is "expansion valve" but I choose to limit the use of that title to the devices that are normally labeled with those words.

Failure of the throttling device to throttle sufficiently can result in liquid flooding the compressor resulting in damage to valves, the bearings, broken piston rods as in Figure 5-18, and occasionally result in crankcase rupture. On the other hand, excessive throttling will result in high superheat temperatures so the compressor overheats, burns up valves, burns up the motor, or melts parts as shown in Figure 5-19. If the throttling device isn't working properly insufficient flow will result in lower capacity cooling because heat transferred to the boiling liquid is much higher than heat transfer to a gas.

Figure 5-18. Broken compressor piston rods

Figure 5-19. Overheated compressor parts

If there isn't enough liquid entering the evaporator to ensure most of its internal surface is exposed to boiling liquid, overall heat transfer will be reduced.

In some of the graphics that follow you will note the formation of bubbles within the stream of liquid leaving the throttling device. This always occurs because the temperature of the liquid at the outlet of the condenser is always higher than the saturation temperature maintained in the evaporator. Also, the ambient temperature is usually higher than the evaporator temperature so the liquid refrigerant cannot be cooled by air around the liquid piping. A very small portion of the liquid vaporizes absorbing heat from the rest of the liquid to lower its temperature to the saturation temperature in the evaporator.

Capillary

The throttling device for many small units is a simple capillary. An extended length of small diameter tubing between the outlet of the condenser and the inlet of the evaporator restricts the flow of refrigerant (Figure 5-20). As the pressure drops within the refrigerant liquid it reaches a point of saturation and vapor starts to form. The bubbles of vapor take up more space within the narrow capillary so the velocity has to increase which results in an increase in pressure drop. It only takes a small amount of vapor, just a few bubbles, to produce a significant difference in pressure drop. The result is a stable flow rate through the capillary. Capillary tubing diameters are tabulated for each refrigerant and capacity.

Adjustable orifice

An adjustable orifice produces a pressure drop by restricting flow (Figure 5-21). Unlike a capillary the pressure drop is generated right at the valve. As the name implies an adjustable orifice is manually adjusted. It requires an operator adjust it while observing evaporator pressure and outlet temperature. They should only be used when the load on the evaporator is very stable and consistent.

Automatic Expansion Valve

An automatic expansion valve simply maintains a constant pressure in the evaporator, (Figure 5-22). In addition to the one shown these can be manufactured with an external sensing line so the pressure that's controlled is at a different location than the valve outlet. An automatic expansion valve can be used to control the refrigerant flow into an evaporator that has a means of separating the liquid and vapor so only vapor leaves the evaporator. That's normally accomplished with a retention space for separating the liquid and vapor plus enough room that gas velocity doesn't carry liquid droplets with it.

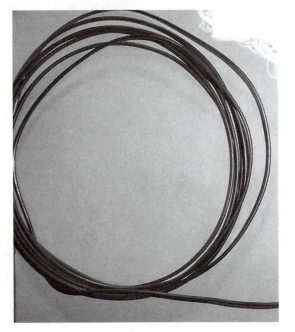

Figure 5-20. Capillary

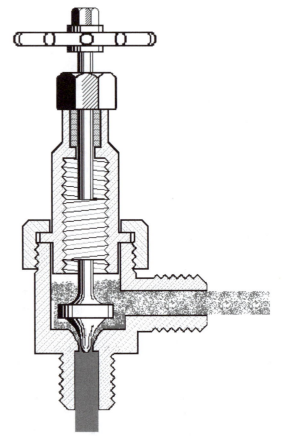

Figure 5-21. Adjustable orifice

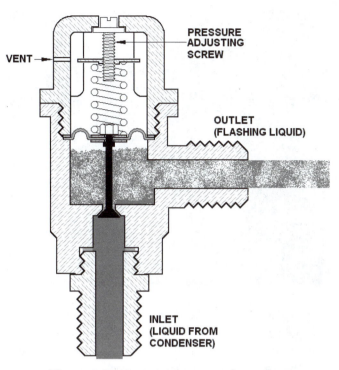

Figure 5-22. Automatic expansion valve

Figure 5-24 shows a cutaway of a TXV. The thermal bulb and diaphragm at the top of the valve, connected by a capillary, produces a force within the valve assembly equal to the saturation pressure in the thermal bulb times the area of the diaphragm. That force is opposed by the actual pressure at the evaporator inlet or outlet. The two pressures correspond to two different saturation temperatures and the superheat is the difference in

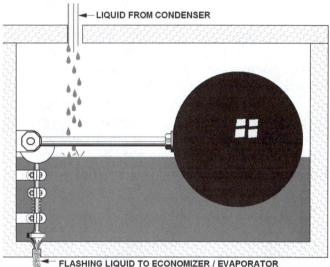

Figure 5-23. Float valve

Float Valve

A float valve restricts the flow of refrigerant through the valve to liquid (Figure 5-23). Liquid from the condenser enters the float chamber and accumulates until the increased level increases the buoyancy forces on the float to overcome the force on the valve disc that's imposed by the pressure difference between the float chamber and the evaporator over the area of the valve disc. The size of the float and the diameter of the orifice under the valve disc are selected to achieve a constant flow through the valve during normal operation. Similar to the automatic expansion valve it can be used to control the refrigerant flow into an evaporator with liquid vapor separation but it simply drains the condenser so there will be fluctuations in the level of the refrigerant in the evaporator. See chillers for more on float valves.

Thermostatic Expansion Valve

Thermostatic expansion valves are the most common means of controlling refrigerant flow. "Thermostatic" implies a combination of temperature and pressure control. The common abbreviation for the thermostatic expansion valve is TXV. The valve orifice is changed in size to vary the flow of refrigerant to maintain a constant value of superheat at the outlet of the evaporator. By ensuring a fixed value for superheat liquid surging into the compressor or inadequate flow of liquid into the evaporator is prevented.

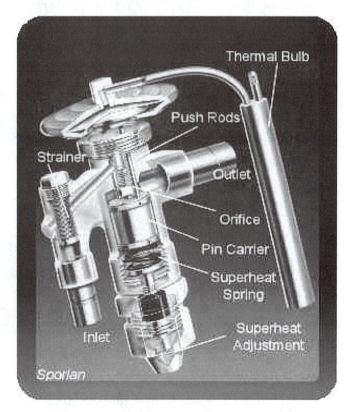

Figure 5-24. TXV

those two. The difference between the two forces on the diaphragm is opposed by a spring. When the difference increases, due to an increase in superheat, the spring is compressed to open the valve further and increase the flow of liquid into the evaporator. Conversely, a decrease in superheat will result in the valve closing down to reduce liquid flow.

In some applications where the pressure drop of the liquid and gas flowing through the evaporator is low the evaporator pressure is sensed only at the valve outlet. For most installations however, the lower diaphragm chamber is connected to the evaporator outlet with an external sensing line, commonly referred to as an equalizing tube, which is shown in Figure 5-25. The TXV outlet pressure sensing connection is a small orifice in the valve body and the flow through it is dumped off so fast through the sensing line to the evaporator outlet that the pressure on the bottom of the diaphragm is very close to the evaporator outlet pressure. This installation ensures the TXV is controlling by the actual superheat at the outlet of the evaporator.

The TXV has a strainer at the inlet that is accessible for cleaning. It cannot be removed for cleaning without removing the refrigerant from the system. Periodic cleaning is not required; the principal purpose of the strainer is to catch droplets of solder and contaminants in the liquid line upon startup of the system or products carried down the liquid line from breakdown of the filter dryer.

The TXV also has a cap that can be removed to access an adjustment of the spring pressure that will

Figure 5-25. TXV installation

change the setting of the superheat. This should be a "set and forget" adjustment. If you have reason to believe that the valve is not working properly there are many things to check before adjusting the superheat setting. Understanding the operation of the valve should aid you in determining what the problems might be. A common problem is vibration loosening the mounting of the thermal bulb.

Proper mounting of the thermal bulb is shown in Figure 5-26. The temperature of the bulb is maintained by heat transfer through contact of the bulb and tube because heat transfer through the metal is faster than

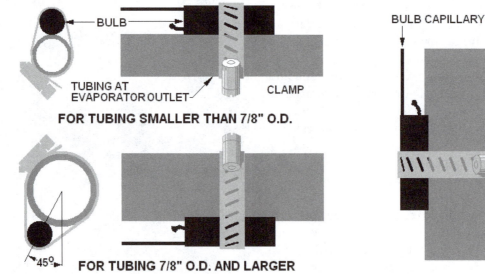

Figure 5-26. Mounting of thermal bulb

heat transfer from the bulb to air. While it isn't necessary it's a good idea to insulate the bulb and suction line. On large tubing the bulb is mounted at the bottom of a horizontal tube because boiling liquid will run along the bottom and superheated vapor could be at the top at low loads. The bulb is offset from the very bottom of the tube to eliminate the insulating effect of the thin film of oil that flows along with the refrigerant. The bulb is mounted with the capillary connection on top of vertical tubes to retain liquid in the bulb. If the connection were on the bottom vapor would form in the top of the bulb to reduce heat transfer. Normally the diaphragm and capillary contain some vapor because they are at a higher temperature than the bulb. The vapor in the capillary and diaphragm is usually superheated and doesn't materially affect the operation of the TXV.

Another possible problem with TXV operation is crimping of the capillary or the equalizing line. Crimping of the capillary can block the transfer of bulb pressure to the diaphragm or restrict the flow to increase the response time of the valve. Crimping of the equalizing line will result in a higher pressure on the underside of the diaphragm to reduce the superheat. If the evaporator has a high pressure drop that could result in liquid surging to the compressor.

Technological advances and production methods have resulted in the introduction of electronic expansion valves. These are controlled by the electronic controls which must sense evaporator outlet pressure and temperature to determine the positioning of the valve. I haven't had an opportunity to work with one of these and I would expect the cheaper ones to be a solenoid that simply opens and closes (repeated clicking should be noted) to control the refrigerant flow.

What might appear to be improper operation of a TXV may be attributable to problems with other pieces of the system. Evaporators and condensers can be fouled with dust and debris changing system performance so much that the TXV cannot correct it. A worn out compressor may not produce enough differential. There are other elements of a typical system that could contribute to a problem including the filter-dryer and other controls.

Regardless of the type of throttling device it is important to remember what it's supposed to do, what harm can come to the equipment if it doesn't do what it is supposed to do, and most importantly, how to monitor conditions to detect potential problems with the throttling device. For most systems that means making regular checks of superheat and subcooling.

MISCELLANEOUS COMPONENTS OF A REFRIGERATION SYSTEM

A refrigeration system consisting of an evaporator, compressor, condenser, and throttling device contains everything necessary to transfer heat from a substance we want cool to a substance that can accept the rejected heat. However, there are many additional components included in systems for protection of the system, ease of maintenance, or to serve diverse loads.

There are also specific labels for certain parts of a refrigeration system. Among those labels are specific names for different sections of piping or tubing. We use the word line generically to identify piping or tubing. The line between the evaporator outlet and the inlet of the compressor is called the suction line. The line connecting the compressor discharge to the inlet of the condenser is called the hot gas line. The line connecting the outlet of the condenser to the inlet of the throttling device is called the liquid line. Since there is seldom any significant length of line between the throttling device and the inlet of the evaporator there's no specific label for that portion. In many pieces of equipment it's actually a sort of Christmas tree that feeds the evaporator with a lot of smaller lines.

Liquid Observation Port

In many older refrigeration systems you will normally find a fitting on the liquid line before the throttling device that contains a glass window so you can look into the line. Most of the time you should see nothing because light passes straight through the liquid. If you have one you should always check it, because it only takes a glance, to be sure you do not see bubbles forming in it. The presence of bubbles indicates the liquid is not sufficiently subcooled and some of the liquid is vaporizing before it reaches the throttling device. Normally bubbles indicate a loss of charge but they can also be an indication of inadequate subcooling in the condenser.

Filter dryer

Practically every system made includes a filter-dryer. A filter-dryer is always installed in the liquid line before the throttling device. The filter-dryer (Figure5-27) performs those two functions. It contains an element that consists of fabric or paper filtering media combined with silica gel. Its purpose is to trap any rust, flaking, loose solder from installation or repair, and material from wear in the compressor to prevent plugging of the small orifices in the throttling device. Also, the desiccant, normally silica gel, absorbs water that gets into the

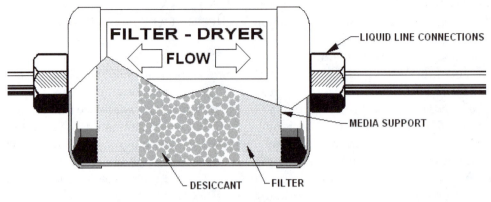

Figure 5-27. Filter Dryer

refrigerant through leaks in compressor seals and introduction of water from connected test equipment. Many filter-dryers are designed for bidirectional flow (note the two arrows on the label) but others must be installed in the right flow direction.

Occasionally a filter-dryer will be provided in the suction line before the inlet to the compressor. These have to be larger to handle the volume of vapor.

Since most refrigeration systems operate under pressure, in-leakage of water is uncommon. If your gauge set is kept dry and filled with the refrigerant you shouldn't introduce any water while testing. Therefore, replacement of the a filter-dryer is seldom required and most of them are hermetically sealed so replacement consists of replacing the entire unit. Most large systems will have a flanged cover to permit replacement of the element only.

Receiver

Refrigeration systems that serve a variety of loads with different temperature requirements in multiple evaporators require something like a surge tank to accommodate variations in system volume that occur. A receiver is a pressure vessel that contains liquid and vapor of sufficient volume to handle the fluctuation in the vapor/liquid ratio. Large receivers are normally ASME stamped pressure vessels and are subject to regular inspections by a National Board Commissioned Inspector at intervals of 2 to 5 years depending upon the State, Commonwealth, or Province in which they are located. Aboard ship they're inspected by the Coast Guard. A common feature of a receiver is a glass port to permit observation of the liquid level. Some receivers have two glass ports so you can shine a flashlight in one and look into the other to get a better view of the liquid level.

The purpose of a receiver is to absorb variations in the vapor/liquid ratio where changes in the liquid level are common and normally don't have to be corrected.

Adding and removing refrigerant in an effort to maintain a constant liquid level is ignoring the purpose of the receiver and creating a lot of work for the operator. You should actually know changes in receiver level are associated with a change in load. Sometimes turning on a load will result in a drop in refrigerant level and sometimes the opposite will occur. It all depends on things like elevation of the evaporator with respect to the receiver, the temperature to be maintained at the evaporator, and the length and run of the connecting piping. It's simply a matter of whether the liquid will end up in the receiver or be trapped in the evaporator and connecting piping.

Isolating Valves

Manual shutoff valves are provided on large refrigeration systems to permit isolation of compressors, condensers, and loads (evaporators) to place them in service or to remove them from service. Normally the valves are only used to secure loads that do not need refrigeration and to isolate compressors and condensers for maintenance and inspection. Refrigeration isolating valves are designed to prevent leakage of refrigerant. That can be accomplished by valves with a bellows seal instead of packing as shown in Figure 5-28. I'm more familiar with packed valves as shown in Figure 5-29. The handle of those valves also serves as a cap. In order to open or close the valve you remove the cap, flip it over and position the square hole in the top of the cap over the square end of the valve stem then turn it to open or close the valve. After operating the valve always return the cap, making sure the gasket is still inside it, and tighten it to prevent loss of refrigerant that could leak through the packing.

The King Valve is a label identifying the valve located in the liquid line at the outlet of the receiver. Closing the King Valve prevents liquid flow from the receiver to all loads served by that system. Continued a vacuum which might result in ambient air containing moisture entering the system should it be leaking. I believe I said elsewhere in this book that a system can be tight under pressure but leak under a vacuum because a piece of rust or a defect in construction can produce the equivalent of a check valve at the leak site.

If you are pumping down the system for maintenance, that requires opening the system, a separate independent vacuum pump should be used to recover the remaining refrigerant and evacuate the system. A recovery system should also be used to remove refrigerant from a compressor, condenser, or receiver before opening them for inspection or maintenance of the refrigerant side of the system. Recall that evacuating a system for maintenance and opening a system require you be licensed by EPA to handle refrigerants.

Load Control Valves

When a refrigeration system serves multiple loads control valves are installed in the liquid lines to start and stop the flow of liquid refrigerant to each load. These are normally solenoid controlled valves actuated by a thermostat. Aboard ship we had evaporators in many spaces. Some were used all the time such as ships stores, food freezers, refrigerators, and other equipment in the galley. Then each deck in each hold could have coils for refrigeration or freezing of the cargo in that space. The Santa Louisa, my first ship as a licensed ship's engineer carried bananas back from South America and we refrigerated all the holds that contained those bananas. Facilities ashore such as food distribution centers, large supermarkets, and the like can have central refrigeration condensing units serving freezers and refrigerated boxes.

Defrost Valves and Lines

In order to defrost evaporators used in freezing applications some systems include a defrost valve and piping or tubing that transfers refrigerant from the hot gas lines to the inlet of the evaporator to melt ice accumulating on the evaporator. A defrost valve is identical to a load control valve with the exception of size (larger because it handles vapor) and location; a defrost valve is always connected downstream of the throttling device. The defrost valve admits the hot gas into the evaporator so it has a higher temperature rating. When defrost controls call for defrosting the load control valve is closed, any circulating fans within the freezer space are shut down, and the defrost valve is opened to admit hot gas to the evaporator. Once the ice is melted off normal operation is restored.

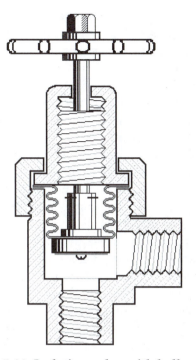

Figure 5-28. Isolating valve with bellows seal

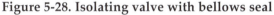

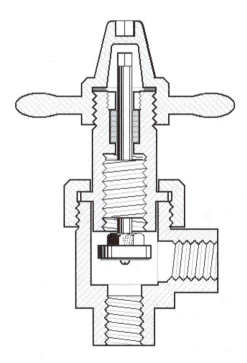

Figure 5-29. Isolating valve with packing

operation of a compressor then permits a "pump down" of the system transferring almost all of the refrigerant to storage in the receiver. Temporary overriding of the system controls can be used to lower the suction pressure and all elements between the King Valve and the compressor inlet until it is almost at atmospheric, zero psig. When pumping down you should avoid pulling

Back Pressure Control Valves

If the system has various loads back pressure control valves can be used to maintain a higher saturation pressure in evaporators that do not maintain freezing temperatures. That prevents the need for defrosting those evaporators. Back pressure control valves are also used in temperature sensitive applications where the product being cooled is damaged by colder temperatures. Back pressure control valves simply restrict the flow of gas (vapor) from the evaporator to keep the pressure, and therefore the related saturation temperature, above the preset minimum value. The throttling device controls the flow of refrigerant independent of the back pressure control valve. I've had to reset the valves several times with changes in cargo.

Limit and Operating Controls

Loss of refrigerant, accidental crimping of tubing, a dirty condenser, an iced up evaporator, and loss of condenser air or cooling water flow can result in damage to the equipment, especially the compressor. Limit controls are provided on systems to prevent damage to equipment in the system by shutting down the compressor whenever operating pressures or temperatures exceed preset limits. These devices are either mounted at strategic locations in the system or are incorporated in remote sensors wired to the system controls. You can think of each of them as a device operating a simple switch that turns a compressor on or off. Process refrigeration may include other devices not listed here and you should, I'll say it again, read the instruction manual so you understand their purpose and operation.

The most common limit device is a low-pressure switch which is connected near the outlet of the evaporator and opens an electrical circuit when the pressure in the evaporator falls below the switch setting. When evaporator loads are low it works more like a control switch. A typical situation would be a windshield defrosting condition in your automobile in the wintertime because the air conditioning system is operated even if the temperature inside the vehicle is below the setting of a thermostat. Of course that assumes you have a thermostat to control your automobile air-conditioning.

In larger systems a second low pressure switch may be included, called a "low-low" or "limit" switch and that switch would require manually resetting it to restore system operation.

Operation of the low-pressure switch when a system is not maintaining the set temperature is indicative of insufficient refrigerant. Checking you should find that evaporator pressure remains low in the system while cycling on and off on the low-pressure switch because there is inadequate refrigerant in the system. It's readily resolved by adding refrigerant but doing so requires that EPA license and you may need to bring in a licensed technician to do it. Checking for spots of oil that are indicative of a refrigerant leak before calling that technician can help you reduce the cost of his services because he'll know the extent of work necessary if you spot a leak. A leak means a considerable amount of work has to be done to return the system to normal.

A high pressure switch is normally connected to the hot gas line near the compressor discharge. You may hear this referred to as a high head pressure switch because they normally sense the pressure at the compressor head. In systems with multiple compressors you will normally find a combination high and low gas pressure switch mounted at each compressor with connections inside each compressor's isolating valves for exclusive control of that compressor. One purpose of the high-pressure switch is to shut down the compressor before the motor is overloaded. The high-pressure switch should always be set at a value equal to or lower than the maximum design pressure of the compressor discharge, hot gas piping, condenser, and liquid line. A high-pressure switch also prevents accidental operation continuing when a discharge isolating valve is closed—just in case you forgot to open it. Regardless, you should make a habit of checking to ensure the valves are open before trying to start the compressor. It's a bit difficult with packed valves because you have to remove the caps but my experience tells me it's always advisable to check, especially if you're not the only one operating that equipment. The combination high and low pressure switch also shuts down the compressor when you forget to open the suction isolating valve.

A high pressure switch may be provided on the condenser in systems with multiple compressors and condensers that can be isolated. In the event of a failure of a reversing valve they're also used on heat pumps. You might also encounter a high temperature switch that senses the temperature of the refrigerant in the compressor head and shuts down the compressor if that temperature gets too high. Low system pressure, loss of lubrication, or loss of refrigerant flow through the compressor could produce high head temperatures without a high head pressure.

Low oil pressure switches are provided on almost all large compressors. They are occasionally incorporated with the combination high and low refrigerant pressure switches into one assembly. Equipment such as reciprocating compressors may have an integral oil pump that builds up pressure after the compressor starts just like your automobile. In that case provision includes a contact that bypasses the low oil pressure switch for a few seconds during startup. If you have a compressor that starts then stops almost immediately check the oil level again. Do not repeatedly try to start a compressor that will not continue to run because bearing damage can occur. Since refrigerant flow cools the motor and in-rush current heats up the windings during starting you should give it a chance to cool off. During annual inspection simply lifting a wire from the bypass circuit should prove to prevent the compressor operating. If the compressor is fitted with an oil pressure gauge make it a point to observe the switch operation and record the pressure at which the contacts close. Note also that oil pressure switches are actually differential pressure switches that compare the oil pressure to the suction pressure (typically in the compressor crankcase) because the flow of oil is determined by that differential.

Large compressors can also have high and low oil temperature switches. The oil temperature switches will prevent compressor operation if the oil temperature is too high or low. Don't confuse an oil temperature switch that controls the heater with a high oil temperature switch, especially if you decide to set the switch by markings on the switch.

You may be wondering if the compressor could continue to operate completely unloaded. What you'll discover is that unloading is never 100% complete. Either the compressor is set up with one or two of the cylinders always in operation or throttling devices like the adjustable plug of a screw compressor doesn't unload everything and the inlet vanes of a centrifugal compressor simply leak. Remember that most motors are cooled by refrigerant flow; therefore every compressor should be protected by high and low pressure switches and a flow switch wouldn't hurt.

Testing of high and low pressure switches should be performed on an annual basis. On individual compressors throttling the suction and discharge valves while observing suction and discharge pressures on your gauge set should serve to confirm operation. As with other operating limits you should record actual values whenever possible in the log where you describe the testing. Be certain you've restored normal valve positions and any other alterations suggested for testing by the instruction manual before leaving the equipment even if the switches are shown not to work.

Water flow switches are used to prevent operation of the compressor whenever water flow through the condenser or evaporator is not proven. These are typically differential pressure switches that measure the pressure at the inlet and outlet of the heat exchanger and open contacts to shut down the compressor whenever the water flow is so low that there is a minimal pressure drop. Their operation should be tested annually by closing down on the outlet valves of the condenser and / or evaporator water circuits and attempting to start the compressor. An established number of turns open for each valve (degrees, percent, or notches on a butterfly valve) should be determined and included in your SOPs as starting points. You should be able to use a test light at the control panel terminals to detect switch operation when the equipment has both switches. You can then gradually open the outlet valve and note when the first switch in the circuit closes, record the inlet and outlet water pressures then do the same thing for the other. Using a test light is preferable to testing that requires frequent starts of a large compressor but you should still confirm that the compressor can't start when one or both of the switches are actuated by low flow.

A low temperature switch is normally provided in any evaporator used to chill water to prevent freezing of the water. Even though water flow is maintained through the evaporator, a flow imbalance created by foreign substances or associated with the piping connections to the evaporator can reduce flow sufficiently that ice could form in one or more tubes. The result is similar to what is shown in Figure 5-30. The thermowell for the switch is typically inserted into the refrigerant space to sense the refrigerant temperature which, being at saturation, should be consistent throughout the evaporator. During each annual inspection of an evaporator containing water the operation of the low temperature switch should be confirmed by removing the sensing element (normally a bulb in a thermowell at the bottom) and inserting that bulb in a glass of ice water. After a few minutes in that condition you should not be able to start

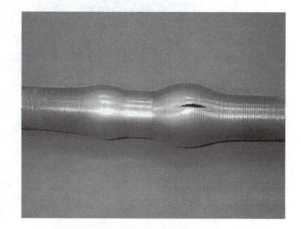

Figure 5-30. Chiller tube after freezing

the compressor. Don't forget to put the bulb back in its thermowell after testing it.

Note what appear to be grooves around the damaged chiller evaporator tube shown in Figure 5-30. They're not exactly fins but they do increase the surface available for heat transfer.

When all else fails, refrigerant systems are fitted with safety relief valves or rupture discs that will open and dump the refrigerant to atmosphere should the system pressure get too high threatening an explosive rupture of some part of the system. A safety relief will vent refrigerant until the pressure drops below the setting of the valve. A rupture disc, once activated, will remain open completely draining the system of refrigerant. With all modern refrigerants that are mostly heavier than air and, therefore capable of displacing the air in the equipment room a safety relief or a rupture disc should have connecting piping discharging to atmosphere high above the plant.

Heat pumps

In truth every refrigeration system is a heat pump. They all absorb heat at a lower temperature and pump it up to a higher temperature. However certain pieces of equipment are reversible and others are simply used to pump heat up. Perhaps you, a family member, a friend, or a neighbor have a heat pump serving their house. It's reversible so it can pump heat from the outside air into the house in the winter time as well as pumping heat out of the house into the outside air in the summer. There are also heat pumps that operate only in the heating mode. The most common of those is a pump that extracts heat

Figure 5-31. Residential heat pump

from the air or a water source and pumps that heat into a swimming pool. Systems such as the pool heater incorporate a simple refrigeration cycle.

Note that label of "pool heater." A piece of refrigeration equipment that's only used for heating is normally called a heater. The label of "heat pump" is primarily reserved for systems that can cool or heat, they're reversible. Reversible heat pumps operate to switch the condenser and evaporator operation. Figure 5-31 is a photograph of the most common one, a residential heat pump. The outdoor coil that forms almost three sides of the unit is a condenser in the summer and an evaporator in the winter. The enclosure also contains the compressor and the reversing valve. The solenoid operated reversing valve switches the suction and discharge between the evaporator and condenser to control the route of the refrigerant. It fails to the heating mode because, normally, loss of heat can result in more damage to a facility than loss of cooling. Heat pumps can utilize a single capillary connecting the evaporator and condenser to handle the reversed flow. The unit shown uses two TXVs and check valves to control liquid refrigerant flow.

Whenever outdoor air temperatures are less than about 42°F a heat pump will start forming ice on the outdoor (evaporator) coil and that ice can block air flow as well as insulate the coils to reduce heat transfer. The control systems for heat pumps that can be subjected to colder temperatures monitor the ice buildup by sensing air pressure drop across the coil or below freezing temperature at the coil surface to initiate a defrost cycle. The condenser fan is shut down, the reversing valve switches to cooling mode, and the compressor is operated to dump hot gas into the outdoor coil. For the comfort of occupants electric strip heaters in the indoor unit are powered to heat the air that is cooled during the defrost cycle.

Most residential heat pumps, and air conditioners are arranged the same so it's difficult to tell if the outdoor unit is a heat pump or simply a condensing unit for air conditioning only. When a residence has a gas or oil fired heating system a heat pump is only provided for operation at outdoor temperatures well above 42°F to produce heat more economically than firing fuel.

It's important to note that residential style units, whether heat pumps or simple condensing units, usually have a vertical discharge. While this dramatically reduces the potential for recirculating the cooling air it also provides a catch basin for pollen, leaves, and other debris. It's not easy but I typically shut down and disconnect my unit, temporarily remove the discharge grill, and clean out the bottom of the unit twice a year so the accumulation of debris doesn't block the bottom two or three rows of coil. I also inspect it about four times a year to catch any unusual accumulations.

Ground Source Heat Pumps

I would love to be able to afford a ground source heat pump for my house but my lot is so small that there's not enough room to do it. A ground source heat pump is also a big investment which I would gladly pay because it would provide a good return on the investment. The reason is rather simple. Knowing the refrigeration cycle you can see a significant difference between the typical air to air heat pump in a residence and a ground source heat pump. Instead of heating the house with refrigerant boiling from heat in air at temperatures as low as 5°F you heat it with refrigerant heated by the ground at an average temperature of 55°F. That defrost cycle isn't required. As for cooling, instead of condensing refrigerant with air temperatures of 90°F or more you're condensing it by dumping heat to the ground at that same 55°F. Lower temperature differentials with heat pumps reduce the pressure differential the compressor has to overcome with lower power consumption and electricity costs. Most residential ground source heat pumps are also used to heat or at least preheat the domestic hot water.

There are office buildings, schools, and other facilities larger than a residence that are using ground source heat pumps, taking advantage of the ability of the ground under their parking lots to absorb and give up heat to either the refrigerant or, more commonly, water circulated from the condenser/absorber of the heat pump through wells in the ground under the adjacent parking lots. The systems are operated and maintained just like any other refrigeration system with the added duties of monitoring and recording the ground water circulating pump, flow and temperatures.

Chillers

Hospitals, office buildings, retirement homes, schools, and similar buildings are commonly serviced with chillers for cooling and boilers for heating. Some hotels and motels are but many of those opt for what's the equivalent of a window unit to provide heating and cooling of the guest rooms. That's because the units serving empty rooms can be turned off to reduce operating cost when the hotel or motel isn't fully occupied.

Chillers are used to cool chilled water that is circulated to air handling units, convectors, beams and other heat transfer devices in buildings that cool and dehumidify the air. Centrifugal chillers, using a centrifugal compressor, are the most common but a chiller can use any type of compressor and there are modifications of the typical arrangement that permit air cooled condensers and operation with a mix of water and anti-freeze to permit production of cold water at temperatures as low as -25°F. They see applications like ice storage systems where the chilled water is cold enough to form the ice in the ice storage tanks. The normal chiller, however, produces chilled water at what has become a de facto standard of 42°F with a return temperature of 52°F. It's a standard because all manufacturer's make chillers with the same conditions to match users of the chilled water with the same conditions. Needless to say, you're always monitoring the temperature to ensure it's leaving the chilled water plant at a temperature of 42°F.

The typical centrifugal chiller, Figure 5-32, is designed to dump the heat to cooling tower water supplied at 85°F returning it to the cooling tower at 95°F. Again, these are adopted standards so that equipment of different manufacturers can be used in a system. A typical centrifugal chiller also operates with a vacuum in the evaporator, 15 to 16 inches Hg. Pressure in the condenser is typically 6 to 8 psig. The evaporator, commonly called a "chiller barrel" is a shell and tube heat exchanger with the chilled water flowing through the tubes and liquid refrigerant surrounding the tubes inside the shell. The condenser is similar with (normally) cooling tower water flowing through the tubes. Unlike most refrigerant systems the refrigerant is not superheated in the evaporator because all the heat is absorbed by boiling the liquid refrigerant. The evaporator is located below the inlet of the compressor and a demister in the top of the evaporator captures any droplets of liquid that are carried up from the surface of the boiling liquid

Figure 5-32. Centrifugal chiller with two stage compressor

combining those droplets until they're large enough to drop back into the liquid refrigerant pool. The refrigerant is superheated as it is worked on by the compressor and monitoring that superheat can detect problems with the compressor.

Despite the fact that the chiller barrel operates in a vacuum it's possible for the refrigerant in an idle chiller to reach the setting of the pressure relief device, which is typically a rupture disc which looks like a large version of the lid on one of your grandmother's canning jars. It's a convex sheet of metal that is bowed in toward the chiller barrel and, when the pressure reaches the design of the rupture disc it pops out (like the lid on that jar) but, in the process of doing that impinges on some sharp steel cutters that cut it so it peels back and dumps all the refrigerant. Temperatures above 120°F can create a saturation pressure that exceeds the typical rupture disc design of 15 psig. A full charge of refrigerant in a typical chiller costs more than $25,000 so, don't let those chillers get too warm. I've always questioned some locations for them, especially over boiler rooms.

Since the centrifugal chiller operates in a vacuum it's possible for ambient air to leak into it. You'll probably notice what looks like part of a window unit hanging around the condenser. Its purpose is to pull any air that leaks in off the top of the condenser and remove it, dumping it to atmosphere along with any moisture the air carried along.

Another device that can be applied to a centrifugal chiller is an oil separator which recovers oil that leaked into the evaporator and return it to the compressor sump. The oil can coat the tubes and block heat transfer so considerable effort is expended to ensure it doesn't leak into the evaporator.

The lubricating oil for a centrifugal chiller isn't circulated with the refrigerant like it is with other compressors. The oil is sealed at the bearings and circulated in its own system with a small portion of the chilled water used to cool the oil. If the seals leak so oil ends up in the evaporator it coats the tubes to reduce heat transfer and, therefore, the efficiency of the chiller. A separate electric oil pump is used to provide lubrication to the compressor bearings and it must be proven in operation before the compressor can be started. The oil sump is heated like other compressors and for the same reasons. The lubricating oil is typically sampled and tested regularly to determine the condition of the oil and the chiller itself. One concern you should have is properly labeling the oil samples because you do not want to mix up samples from several units and, like what almost happened once with one of our customers, a rebuild of the wrong chiller.

Sometimes people experienced with the need for superheat to control the flow through the throttling device insist that the gas leaving the evaporator of a chiller has to be superheated but that's not the case with centrifugal chillers. The throttling device in a centrifugal chiller is a float valve, Figure 5-22, and normally consists of two float valves which separate two different pressures. The system of two float valves and the chamber between them is commonly called an economizer because it captures some of the vapor that flashes as the pressure is dropped and uses it to cool the hermetic motor of the chiller, entering the compressor between two stages. That system saves the compressor pumping all of the gas from evaporator pressure to condenser pressure.

Being a large, expensive machine they are normally insured and the insurers typically require a breakdown inspection of a chiller every fifteen years. Examinations include electrical insulation testing, leak testing, and eddy current testing of the tubes. There's a hint here, the typical chiller is expected to operate for fifteen years without concern so it's a highly reliable piece of equipment.

Why chillers? A centrifugal chiller system can produce cooling for a fraction of the cost of the typical air to air cooling system. One chiller can produce chilled water at a lower first cost than self-contained units, all in a smaller space, for as little as half the operating cost. Centrifugal chillers are made with capacities, at the time of this writing, as high as 3,000 tons. That one would be a fraction of the size of 3,000 one ton window units stacked up.

Just like a boiler, the load for a chiller will change with outside air conditions. To accommodate reduced loads many chillers use inlet vanes that throttle the flow of the evaporated refrigerant to the compressor while causing it to swirl thereby reducing the horsepower requirements of the compressor (Figure 5-14). More modern chillers control the evaporator temperature by varying the speed of the compressor. When the vanes or reduced speed cannot prevent the evaporator pressure dropping to produce colder chilled water a low pressure switch shuts down the compressor. The low pressure switch has a differential setting which allows the evaporator pressure to rise over that range before restarting the compressor. A chiller cycling on and off is not much different than a boiler cycling and, whenever possible, a change to a smaller unit or shutting one down completely can reduce power consumption and demand charges.

Other considerations when operating centrifugal chillers include the monitoring of the temperature and corrosive/scaling condition of the cooling tower water and chilled water, the level of the liquid over the tubes in the evaporator, lubricating oil level and oil temperature.

A centrifugal compressor takes some time to come to a stop so a problem can occur if all power is lost to the plant. The result will be loss of lubricating oil pressure before the motor stops turning with wiping of the bearings. By wiping I mean the moving metal parts are not separated by the oil and rub stationary parts resulting in rubbing (wiping) metal off the bearings. Some centrifugal compressors have surge tanks for oil that provide enough for lubrication on a power failure but others require a UPS (uninterruptible power supply) to keep the oil pump running until the shaft stops rotating. Modern compressors with magnetic bearings will also require a UPS. The backup power in some facilities relies on startup of emergency generators before the UPS batteries lose their charge. A very few units might even have a hand pump that can be operated to keep the chiller lubricated until it stops turning. Be sure your SOPs and disaster plans include consideration for power loss to large chillers if you have them.

Figure 5-33. Absorption chiller

Absorption Chillers

Every time my lovely wife and I get ready for a trip I fire up the refrigerator in the RV and I do mean light a fire. Seriously? Yup! That refrigerator runs on propane. It boils a solution of ammonia and water to evaporate the ammonia which is then condensed in coils on the back of the refrigerator and flows into the refrigerator where it evaporates to absorb heat from the freezer and refrigerator sections. It uses the principle of partial pressures to achieve the cooling. The cooling is achieved because the liquid ammonia enters the evaporator sections that contain hydrogen gas which provides a sensation of lower pressure so the ammonia can evaporate. The evaporated ammonia is then absorbed in water cooled by another coil and the mixture returned to the furnace section to be boiled again. The refrigerator thermostat controls the firing of the gas. It's not an efficient cooling device but it works wherever I go as long as I keep propane in the tank and have a charged twelve volt battery to power the controls.

There's not much difference between that RV refrigerator and an absorption chiller in concept. Size, construction, only chilling water, and effi-

Figure 5-34. Absorption chiller rear view

ciency are the major differences. There are two systems for absorption refrigeration, water and ammonia like my RV refrigerator where ammonia is the refrigerant and a solution of lithium bromide and water where water is the refrigerant; the latter being more prevalent. The absorber is a large assembly of tubes, pumps and piping, all as shown in Figures 5-33 & 5-34. Instead of using gravity like the RV refrigerator the solution is pumped, spraying the solution on the coils for better heat transfer. Figure 5-35 is a diagram of a single stage absorption chiller.

Modern units are typically two-stage. The solution is pumped to the generator stage where heat is added to vaporize the refrigerant, either the ammonia or water. The source of heat can be a fire, steam, hot water, or any other fluid that's hot so absorption chillers can take advantage of heat that would otherwise be wasted. The refrigerant is then condensed by a condenser, normally served by cooling tower water but it could be ocean, lake, or river water. The concentrated sorbent (water or the lithium bromide solution) drains back to the absorber. The liquid refrigerant drains from the condenser through a series of orifices back into the absorber section, over the chilled water coils and into a pan connected to a pump that pumps it back up and through sprays over the chilled water coils. When the refrigerant evaporates to cool the chilled water the vapor is absorbed in the sorbent which is sprayed in a lower chamber to increase the rate of absorption.

The use of ammonia in an absorption chiller requires an additional feature not shown in the schematic. When evacuating the absorber to produce the low pressure required for the ammonia to evaporate at chilled water temperatures equipment is required to recover the ammonia rather than just dumping it into the chiller room. The potential hazards with ammonia are what compel most designers to use lithium bromide. While much safer to use, lithium bromide is a salt and under upset conditions it's possible to concentrate the solution so much in the generator section that the salt crystallizes and the generator section plugs up with the salt.

Just like the RV refrigerator an absorption chiller can be fired. After absorbing the vaporized refrigerant the sorbent is pumped through a generator which is just a boiler fired by gas or oil that boils off the refrigerant and concentrates the sorbent which is recycled back to the evaporator. They can also be operated similar to a heat pump where the sorbent is only heated and returned to make hot water instead of chilled water for a two-pipe air conditioning system. My service techs had the pleasure of starting up a fired absorber heater/chiller a few years ago. It was a single unit that did it all.

Absorption chillers have been in and out of favor depending on both the price of electricity and fossil fuels and advances in efficiency of centrifugal and absorp-

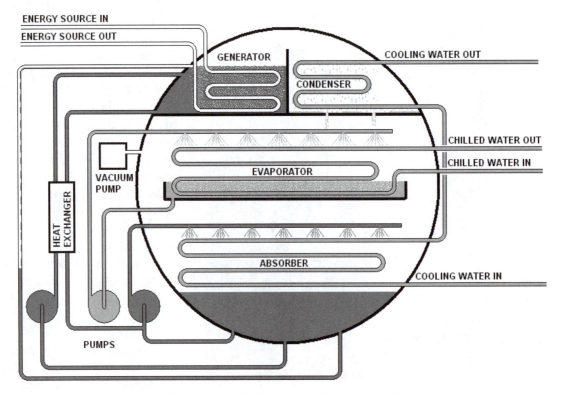

Figure 5-35. Absorption chiller schematic

tion chillers. Facilities that have both absorbers and centrifugal chillers can take advantage of the swings in the cost of energy to operate the chillers that are most economical. First there's the opportunity to reduce demand charges by operating an absorber during peak load periods then there's the constantly changing difference in energy prices. An absorber can be cheaper to operate at certain times of the day but starting and stopping one takes time and energy so there's clearly going to be a fair amount of calculating to determine how to operate since electricity prices can change every hour.

COOLING TOWERS

For refrigeration systems cooling towers are used to reduce the pressure differential a refrigerant compressor has to overcome in order to pump the refrigerant from the evaporator to the condenser. It's accomplished by cooling the water used in the water cooled condensers of refrigeration equipment, especially chillers. A cooling tower helps cool the water by evaporating some of it. A boiler operator knows that it takes about 1,000 Btu to evaporate a pound of water. In addition to reducing the cooling tower water temperature by 10°F the water is cooled by evaporation of some of the water. A good rule of thumb is the cooling tower will evaporate 2 gpm of water for every one million Btuh removed from the cooling water. The tower can cool the water almost to the

Figure 5-36. Blow thru cooling tower

wet bulb temperature of the atmospheric air (see psychometrics in the section on air conditioning) and that's seldom higher than 85°F. The cooling tower operates by distributing the water over a fill which has atmospheric air flowing through it, normally forced by a fan. The fan can be on the outlet of the tower, the common location, or force air into the tower where it's commonly referred to as a blow-thru cooling tower like Figure 5-36. Other terms for cooling towers are cross-flow and counter-flow; a blow-thru is commonly counter-flow where the air is pushed up through the tower as the water drops down through. In a cross flow tower the air enters the sides or ends crossing the flow of flowing water then rises to the fan inlet as in Figure 5-37.

Normally the water is delivered to perforated trays at the top of the cooling tower. Some use sprays but the perforated tray serves to distribute the water evenly over the top of the fill. Fill is the word used to describe the material in the tower that the water runs over and around as it drops from the inlet trays or sprays to the sump at the bottom of the cooling tower.

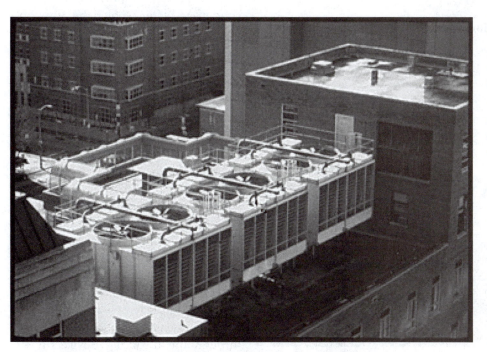

Figure 5-37. Cross flow cooling towers

The fill is designed to convert the falling water to a film along the surface of the fill and thin sheets or droplets of water between parts of the fill. The air is drawn through the fill to contact the air over that extended surface to achieve the heat transfer and sweep vaporized water out of the tower. The fill can consist of redwood slats in several alternating layers (Figure 5-38) but so many of them burned up when idled during the winter that it's an uncommon material in towers today, used primarily where the tower is always in service. There are many variations and designs of plastic fill used today.

Part of the plume can be drift. Drift identifies droplets of water that are swept off the fill by the air and leave the tower without being vaporized. Those droplets are usually much larger than the fog and drop out on adjacent structures. Excessive drift is an indication of air velocities in the tower that are higher than design. Sometimes it only occurs at high loads but it can be a problem at lower loads if the fill is distorted, damaged, covered with organic growth or otherwise altered to reduce or block air flow in some parts of the tower so the velocities have to be higher in other parts to compensate for it. Towers can contain baffles to redirect water splashing off the fill back into the tower and they can be bent or otherwise altered by high winds, falling tree limbs, and accidents to increase liquid water loss. Drift is simply a waste of water and should be addressed when it is detected. Consistent pools of water on adjacent roofs (evident in Figure 5-37) are indications of drift.

Cooling towers in electric power plants and large industrial plants can operate without a fan forcing the flow of the air. You're probably familiar with photographs of nuclear power plants and the tall hyperbolic cooling towers (Figure 5-39) which use the differential pressure produced by the difference in density of the atmospheric air and the air heated in the towers and containing a larger volume of lighter steam. The plume that we associate with cooling towers is actually droplets of water that condense as the exhaust of the cooling tower is cooled by the slightly cooler atmospheric air. A better name for that plume would be fog.

In an electric power plant colder condenser water, cooled by the cooling tower, increases the energy the turbine can produce. The water is cooled as much as possible, stopping in the winter slightly above 32°F to prevent freezing of the water. Cooling tower water used in industry for cooling of production equipment or product can also be allowed to vary in temperature but typically the lower limit is set higher to minimize thermal shock and ensure the temperature control valves on the cooling equipment maintain control. For chilled water systems a cooling tower is normally operated to maintain a leaving temperature of 85°F to match the design condition for most chillers. Maintaining the temperature maintains pressure to ensure adequate liquid flow through the throttling device. In circumstances of low loads, and where the

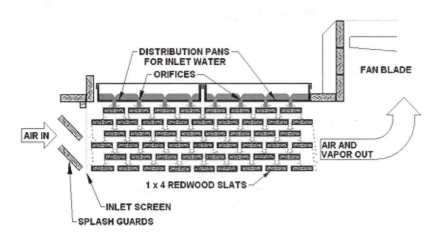

Figure 5-38. Cross section of redwood cooling tower

Figure 5-39. Hyperbolic CT in power plant

atmospheric temperature is colder, fan speed is adjusted to maintain a minimum temperature, unlike the typical water-cooled condenser where water flow is controlled. Variable speed drives on cooling tower water pumps can reduce power costs but are limited to ensure good distribution of the water in the cooling tower.

The water vaporized in a cooling tower has to be replaced with makeup. In addition, because the vapor doesn't carry off the solids that were dissolved in it, blowdown of the cooling tower is required to prevent scaling and deposits of mud on the fill. Small towers commonly have a float valve in the sump that adds makeup water to maintain the level. Larger towers can have level controls using an electric probe system that senses the water level to open and close a solenoid or motorized makeup water valve. Blowdown control can be manually set or automatic and except for cooling the water they work the same as a boiler's continuous blowdown. Any large system will also have meters on the makeup and blowdown water flows to produce a difference between the two that equals water lost to vaporization and drift so the owner doesn't pay to for sewage treatment of water that wasn't dumped down the sewer.

Just like a boiler the cooling tower water chemistry has to be monitored and maintained to prevent damage to the tower, the condenser, the pumps, and piping. In addition to scale and corrosion protection chemicals are added to control the growth of bacteria and algae in the cooling tower water.

AIR CONDITIONING

Normally we think of air conditioning as cooling the air in a space, a room, or building. More appropriately air conditioning should be considered to be the heating or cooling plus adding or removing moisture and removing airborne contaminants from the space to maintain conditions in that space that are comfortable for the occupants. That definition covers more than what we typically think of. In many zoos around the country there are spaces where the conditions are not necessarily comfortable for humans but are enjoyed by the occupants. With few exceptions air conditioning is accomplished by removing air from the space, altering its conditions, then returning it to the space where it is mixed with the air in the space to produce the comfortable conditions for the occupants.

Another term you should hear is HVAC, standing for heating, ventilating, and air conditioning. I would dissuade the use of the term if it were not for many fa-

cilities that only receive heating and ventilation. Since most people today expect complete air conditioning it's not necessarily an applicable label anymore. Air conditioning is the best overall label.

While they're not as common as they used to be the typical window unit (Figure 5-40), is a packaged air conditioning system that incorporates a complete refrigeration cycle within it to transfer heat from the room to the atmosphere outside the window, is still familiar to most of us. Air from the room (A) is drawn into the unit through a filter (B) that cleans the air, the air passes over an evaporator coil (C) with fins that removes the heat from the air and condenses some of the moisture in the air then a fan (D) blows the air back into the room (E) at a high velocity so that it mixes with the rest of the room air to produce the comfortable condition. To achieve all the requirements for air conditioning there's usually a small damper (F) in the barrier between the indoor and outdoor air that can be opened to admit outside air into the flow of room air to be conditioned to achieve ventilation. Condensate which is moisture removed from the room air drips (G) into a drain pan at the bottom of the unit and passes through a trap (H) formed in the bottom of the casing to the outdoor side of the unit. A slinging ring attached to the outside of the condenser fan (I) picks up the condensate and hurls it at the condenser (J). Heat from cooling the air and condensing the moisture is absorbed by the refrigerant with flow into the evaporator controlled by a metering device, in Figure 5-40 a capillary (K) and, for the typical window unit with R-134a into the evaporator (C) at a pressure of 35 psig which corresponds to a saturation temperature of 40°F. The boiling refrigerant absorbs the heat from the room air and condenses some of the moisture from the room air to cool it from a design room temperature of 75°F to approximately 55°F. Once all the refrigerant has evaporated (L) the gas is heated to a leaving temperature of 50 to 60°F (10 to 20 degrees of superheat). The refrigerant passes through the insulated barrier between the room and outdoors to the inlet of the compressor (M). The gas is compressed by the compressor to a pressure of 100 psig raising the temperature of the gas that enters the condenser to 200°F. The condenser (O) transfers the heat removed from the room along with the heat added by the compressor to outside air drawn into the outside casing by the condenser fan (I) through louvers in the side (P), passing over the condenser and out grills (Q) at the back and top of the unit. The grill at the top of the unit permits admission of rain water that can assist in cooling when the moisture contributed to the room air by the moisture from the rain penetrating the room increases the load on the unit. The first function of the condenser is to remove the superheat in the gas, reducing its temperature to the corresponding 123°F saturation temperature near (R). Then the refrigerant remains at the saturation temperature as it is condensed until all the gas is converted to liquid near (S). Once all the gas is converted to liquid the refrigerant is subcooled until it's about 105°F at the inlet of the capillary. As the pressure drops from the 100 psig in the condenser to 35 psig in the evaporator some of the liquid is converted to a gas to absorb the heat required to cool the 105°F liquid to the 40°F saturation temperature in the evaporator. To make that window unit a heat pump, simply requires a reversing of the refrigerant system, and it achieves all but one of the functions attributable to a proper air conditioning system.

> Clean the air, normally by filtering it.
> Cool the air when the temperature in the room is above set point
> Heat the air when the temperature in the room is below setpoint
> Remove moisture from the air when cooling it.
> Add moisture to the air when heating it.
> Incorporate outside air for ventilation.

Additional provisions can include pressurizing the room to prevent contamination of the room air from external sources or reducing the pressure in the room to prevent contamination of other spaces by contaminants that are in that room. The typical window unit cannot perform the latter nor either one of those with any degree of perfection and does not have means to add moisture to the air.

Introduction of outside air for ventilation may seem undesirable for facilities that lie in metropolitan areas with plenty of vehicle and industrial exhaust contaminating that outside air but it's essential to add sufficient air for ventilation to restore the oxygen content of a room where the oxygen is consumed by the occupants and replaced with carbon dioxide. You will not see many systems where there are enough plants within a space and enough sunlight to support photosynthesis to replace the carbon dioxide with fresh oxygen. To the best of my knowledge the only spaces that have systems to do that, independent of nature, are on board our Navy's submarines.

A well designed air conditioning sys-

tem will draw outside air from a location that is unlikely to be contaminated but you will still encounter some today that were designed when everyone thought the atmosphere was a giant sink that could absorb anything and remain fresh and clean. Be aware of potential problems with contaminated outside air and be prepared to do something about it, even if it's the poor, but essential, act of entering a regular comment about it in a log book.

The diagram of Figure 5-40 shows a feature common on small equipment, a service stub (U). That short length of tubing is provided to permit attachment of a tapping valve to permit connection of a gage set (Figure 5-3) to determine the unit suction pressure and add or remove refrigerant if necessary. A service stub is common on window units, residential refrigerators and freezers, wine coolers, etc. Whenever a tapping valve is used it should be removed by first crimping the tubing between the location of the tapping valve and the suction line then, after removing the valve, trimming the stub end and soldering it closed. If the valve is left in place it's likely it will vibrate loose followed by loss of all the refrigerant. A single connection is all that's required for equipment with a capillary for the throttling device because the manufacturer always provides tables for determining the proper charge of refrigerant based on load data and superheat or subcooling.

One of the keys to understanding the operation of an air conditioning system is learning what is comfortable for occupants and how to get the air in the space to meet those comfort requirements. In order to do that we need to understand Psychrometrics. Being able to use a psychrometric chart, Figure 5-41, is imperative.

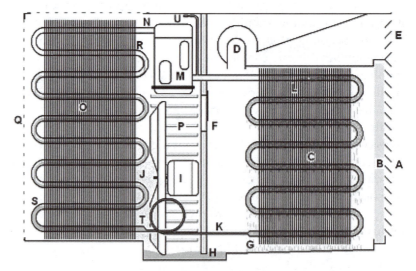

Figure 5-40. Window unit

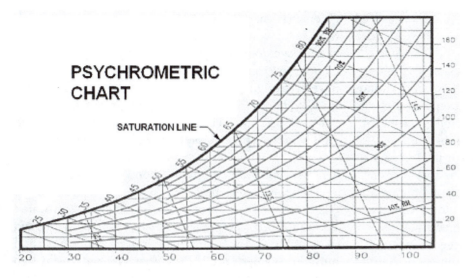

Figure 5-41. Psychrometric chart

Psychrometrics and Air Conditioning

Rather than produce separate chapters or sections on concepts and apparatus for air conditioning I've elected to describe them all together. So, we'll cover them in a manner consistent with air flowing through an AHU (air handling unit). There are other labels for air conditioning equipment which are mentioned as we go along but the generic term, apparatus, can be used to describe them all. Psychrometrics lets you read into the systems and understand what's going on as the air flows from the conditioned space through the ductwork and apparatus then back to the conditioned space.

A psychrometric chart can be more detailed than the one in Figure 5-41 and you should find that the salesmen that call on the plant will be more than willing to provide you with one that's encased in plastic so you can mark it up and erase it as often as you want. I still have one that was provided by Carrier™ back in 1969. I have no idea how many times I have plotted the conditions for an air conditioning system on that chart then copied it for a permanent record before erasing it but I bet the number is over one thousand. The salesmen also gave me pads of paper charts that I could use but, in today's world they're not commonly offered. The one shown here is much simpler because it doesn't contain all the values and lines that are on a complete psychrometric chart but it's more than adequate for the purposes of an operator. Normally you do not need to know the volume of the water and air mixture, the enthalpy, and other values used by engineers.

Why do you have to understand this? Because your concept of comfort is altered by being in a boiler room and there are times when you can't isolate the reason

that lady keeps calling and complaining about the conditions in her office without using the chart. Plotting the conditions for her office and the equipment should lead you to a conclusion as to why she's uncomfortable and what you might be able to do about it. The alternative is to receive constant calls to the boiler room and not being able to respond. I have noticed that the newer SCADA systems (System Control and Data Acquisition) frequently allow the operator to get a readout of the temperature in the space in question and the typical response is "I see that the temperature in that area is 77°F, ma'am so all I can do is suggest you put on a sweater because that's well within the comfort zone." It doesn't mean the humidity is under control and she will keep calling and complaining. So, become an expert in simple psychrometrics and you can reduce those calls.

There's also the matter of productivity. Many studies have shown an increase in productivity in a space when the conditions were changed. Further analysis also showed that it declined for a while until the conditions were changed again. It doesn't mean that you should be changing conditions, it simply means that people are more comfortable when they believe the guy operating the AC is concerned for their comfort. Ignore the psychology and simply give some attention to AC complaints and you'll be known as "that nice guy in the boiler room."

There are multiple scales on the pyschrometric chart so some explanation is necessary. The horizontal scale read at the bottom of the chart is identified as the dry bulb (DB) temperature. The vertical lines from that scale are representative of that temperature. The vertical scale on the right side of the chart in Figure 5-41 is in grains of water per pound of dry air (a grain being one seven thousandth of a pound). Your laminated chart will contain scales on the right side for pounds of moisture per pound of dry air as well. The horizontal lines are lines of constant moisture content, pounds or grains of water per pound of dry air. The scale on the curve of the chart is wet bulb temperature and the lines progressing down and to the right are lines of constant wet bulb Temperature. Curves within the chart that almost parallel the terminating curve are values of percent humidity. Lines sloping steeply from left to right along the width of the chart indicate cubic feet of air per pound of the air and water mixture but you will seldom use them as well as the scale on the chart for enthalpy. The chart will also indicate that it's based on a barometric pressure of 29.92 inches of mercury but deviations associated with changes in barometric pressure will not alter your understanding and use of the chart. Changes in barometric pressure are not significant unless you're maintaining the air conditioning of the visitor's shelter on Pike's Peak.

Your first task when you get your laminated psychrometric chart is to mark it up for the comfort zones, copy it and keep the copy handy. Because people typically change their clothes in response to outdoor conditions two comfort zones have been established by ASHRAE, the American Society of Heating, Refrigeration, and Air conditioning Engineers. A significant number of samples for both men and women were used to determine those comfort zones for situations where people are sedentary or slightly active and 80% indicated they were comfortable so you will not always be able to make everyone truly happy. When people are active lower temperatures and humidity are required. Keep it close to the original design conditions (found on system drawings) and you should keep a lot more than 80% happy. The winter comfort zone is outlined by dry bulb temperatures of 69° and 76°F at a moisture level of 31 grains per pound of dry air at the bottom and 68° and 74°F along a wet bulb of 64°F on top. Connect the left and right corners of the top and bottom lines to enclose the area and that's called the winter comfort zone. For the summer comfort zone stick with the 31 grains and dry bulb temperatures of 74° and 81°F at the bottom and plot the upper corners along the 68°F wet bulb line at 73° and 79°F dry bulb temperatures. Now, just because the wet bulb and dry bulb temperatures in that space indicate the room is in the comfort zone that's not an excuse to tell someone to get a sweater or take off their jacket, there are other factors that affect comfort and I'll cover them later.

You will also need a set of prints or drawing files on the computer for the air conditioning systems you're responsible for. They contain information that you can use to isolate problems when they come up. Plotting the values from the drawings for the equipment on your psychrometric chart during those rare periods when you have nothing to do and copying them can help you prepare to tackle a problem. When an addition to your facility is under consideration or construction it would help to find the design engineer during one of his visits and ask for copies of his charts for the systems and equipment to save you time.

In order to understand the function of a piece of air conditioning equipment you plot the values on a psy-chrometric chart. Comparing design and existing conditions helps you locate the source of problems. Learning about psychrometrics is best done by using the design values. You can do it before there's a problem and then have the design situation plotted out ready to use when a problem occurs. Now, follow the design data to produce one as described so you're familiar with the process and what you need to do to plot a system on the chart.

The Air Conditioned Space (Room)

Engineers use the generic word "room" to describe the air conditioned space. Room conditions are indicated on the Contract Drawings and are normally expressed in dry bulb and wet bulb temperature or dry bulb and humidity values. You'll learn that you only need two values for a given condition of air to locate its point on a psychrometric chart. Locate the intersection of the dry bulb and wet bulb or humidity values, mark it as a point with your trusty number 2 pencil then circle the point so it's easy to find again. Noting where the point is within the comfort zone may be a key to the design engineer's understanding of the use of the spaces served by the equipment. A normal design point for office buildings is 75°F and 50% RH (relative humidity). A home for senior citizens may have higher values because seniors move around less and tend to prefer warmer temperatures. If the space is a gym you might see cooler and dryer design conditions because the occupants are not sedentary, the same might be said for a plant's production line where workers are always busy. Indoor swimming pools provide some unique considerations. Sometimes the room conditions are specified for the manufactured product and not for the comfort of personnel; a good example would be a refrigerated space for food storage. On rare occasions a room's pressure, relative to atmospheric, is specified as well; we'll cover why in a bit.

Outdoor Conditions

Also on the drawings will be the design outdoor temperature conditions. That will be the condition of the air pulled in to provide ventilation. Plot that point and circle it then draw a straight line between the room conditions and the outdoor air conditions. The room air and the outside air are usually mixed in the mixing box of the apparatus before conditioning of the air is continued. The equipment inlet conditions listed on the schedule on the drawings should fall on that line when you plot and circle it. Why? Because when you mix two air streams the conditions for the mixture should always fall on a straight line drawn between the two points for conditions of the air streams on a psychrometric chart.

New regulations and design standards for building ventilation that are primarily concerned with the distribution of the ventilation air have led to the introduction of independent outside air supplies so don't be surprised if some of your equipment doesn't have a mixing box with connected outside air ductwork.

How much outside air is required for ventilation? A general rule of thumb used to be 5 cfm (cubic feet per minute) for each person in the space. New standards for ventilation produced by ASHRAE (the American Society of Heating, Refrigeration, and Air-conditioning Engineers) now require as much as 20 cfm per person. The design (essentially maximum) quantity for a given piece of air handling equipment is listed on the design drawings and you will find systems today that use a control interface with personnel access controls or the carbon dioxide content of the room air to determine actual values. As with night set-back temperature controls the amount of outside air should be reduced when the building isn't occupied or occupancy is very low. A round during night hours or weekends should include verification that the ventilation air is off or at minimums because it can cost a lot of money to heat or cool that air unnecessarily.

Economy Cooling

A feature of many air conditioning systems in the northern states is provision for economy cooling (not to be confused with another concept called free cooling). When outside temperatures are low enough air handling systems can be operated to take advantage of that colder outside air to cool spaces in the buildings instead of cooling the air with refrigeration equipment. Core spaces, in the middle of the building, will need cooling even in the winter because of the heat generated by people and lights alone. The savings in operating costs can be significant and an operator should ensure that the economy cooling is operating properly. When economy cooling is used the outside air ducts and dampers are larger so as much as 100% of the conditioned air flow can be drawn from outdoors. Typically the systems are fitted with a return air fan to draw the return air from the rooms and discharge it outdoors or into the mixing box. Room temperature controls are used to control the temperature of the supply air by mixing outside air and return air to produce the required temperature of supply air.

Economy cooling systems need to be watched to ensure the outdoor air is at minimum requirements during the summer. Only one of those large outdoor air dampers that hangs up open can prevent adequate cooling and really pump up the electric bill.

In the deep winter the moisture content of the conditioned air in an economy cooling system can get so low that the occupants encounter problems with static electricity. They get everything from shock to clothing sticking to them and hair looking … well, perhaps you've seen it. Means of adding moisture to the conditioned air to maintain a comfortable level of humidity are described later but some systems will have over-rides on the economy cooling to maintain desired humidity in the conditioned spaces and refrigeration or chilled water cooling as needed for temperature control.

Apparatus Inlet

If the inlet conditions don't fall on that line drawn between the outside air and room conditions it can be because the engineer has made allowances for heat gained between the room and the mixing box which can come from heated piping, lighting heat, or sun on the roof of the space, any one of which heats the room air that returns to the unit as return air flowing between the ceiling and the floor or roof above the space. It could also be heat gained in ducts run outdoors to the equipment. To represent that added heat gain in the return air erase the line from room air to outside air, plot the point for the return air entering the equipment and draw a line from the outside air conditions through the inlet air conditions down to the line for the same moisture level as the room air and draw a horizontal line that connects the room conditions with the end of the line you just drew. Unless it's an unusual air conditioning system, that should do it. When the air is simply heated its dry bulb temperature is increased without changes in moisture so the line is always horizontal and the percent humidity decreases. You should be able to confirm the condition by measuring the dry bulb temperature of the return air immediately before it enters the mixing box. If it doesn't match check wet bulb conditions because the return air must have picked up some moisture somehow. Any increase in moisture content between room air and return air entering the mixing box is potentially problematic, possible caused by steam, water, or outside air leaks into the return air stream.

Filters

The equipment airflow, now a mix of outside and return air, passes through filters as it leaves the mixing box. The filters are there to remove contaminants from the air like dust, pollen brought in with the outside air (or produced by plants in the conditioned space) and any other solid airborne contaminants that might plug up the cooling and heating coils or bother occupants.

Since there's only one filter on your home air conditioner it can also serve to remove bacteria, viruses, and even odors from the air depending on the grade of filter that you purchase. Where those contaminants are a concern with industrial and institutional equipment additional filters are normally installed at the outlet of the air handler to remove them.

Depending on the size of the facility the replacement of those filters can become one of the duties of the boiler plant operator. Arguments go both ways in this regard but I'm confident that it places the eyes, ears, and incomparable senses of the boiler operator next to the air handling equipment on a regular basis to detect problems that can become very expensive. So, if you're asked to replace filters be certain to use those senses to monitor the operation of the equipment. The argument that it takes you away from attending to the equipment in the boiler room has its merits but properly spaced intervals of absence from the boiler room should have little, if any, effect on the quality of monitoring the boiler room and you'll find time to fit it all in. Large installations with many air handlers are normally maintained by contractors. Checking up on those contractors by reviewing the status of the equipment with a regular walk-through or check of the SCADA system should also be a duty of the boiler plant operator.

One consideration of whoever is in charge of filters is the application of the right filter for the job. All too often the filters are not specified carefully or the installing contractor puts in the cheapest filter box and set of filters available. Don't throw out a filter salesman that walks into the boiler plant and wants to talk. There are modern filters that not only do a better job and last longer than conventional ones but ones that even cost less. When considering cost of replacement, place a value on your time or your contractor's time as well. Cheap filters that have to be replaced more frequently may cost less but when the combined cost of replacement labor and additional filters is considered more expensive filters can cost less.

Don't hesitate to consider the replacement of the filter frames. I'm certain the filter installation at 90 percent of houses can't compare to the one in mine. The typical provision for a filter is a hole cut in the side of the duct and some channels pop-riveted to the top and bottom of the duct as a guide. My filter box, and filters, (Figure 5-42) are designed to filter every cubic centimeter of air that goes to that air handling unit. When the salesmen come knocking on my door trying to sell me an air duct cleaning I just tell them what I have and ask if they will guarantee to show me dirt from my system or do the job for free. They all slink away quietly. A proper, well sealed filter system will dramatically reduce cleaning and other maintenance chores on the air handling equipment. It won't eliminate it but it can put the work off for years.

I've seen some real disasters at some facilities because filter replacement was ignored, abused, mishandled, or overdone. I'm a firm believer that one can tell how well a facility is maintained by simply looking at a few filter banks. Another typical indicator is the condition of replacement filters in storage. If they're laying on a shelf or in an open box without protection from the environment and collecting dust before they're ever placed in the filter box they can actually introduce dust and contaminants into the cooling and heating coils.

Cooling Coils

After the filters the air normally enters the cooling coil although the coil may have face and bypass dampers on it (Figure 5-43) that are used to control equipment outlet conditions. The air is not only cooled by the cooling coil but moisture is also removed from the air. What happens at that cooling coil is dependent on a lot of factors including the inlet temperature and humidity of the mixed return and ventilation air, number of people in the conditioned space, sunlight entering the conditioned space, outside air infiltrating the conditioned space through doors and leaks at windows, and a lot of things that occupants can do. The face and bypass dampers allow for varying supply air tempera-

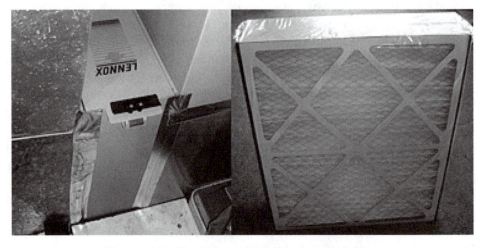

Figure 5-42. Ken's home air filter box and filter

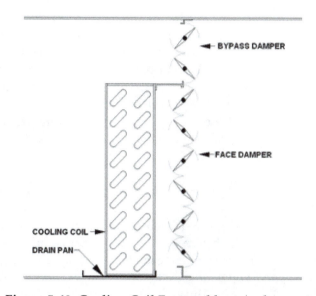

Figure 5-43. Cooling Coil Face and bypass dampers

tures to match the cooling load. Combined with reheat coils the face and bypass dampers also permit control of the moisture content of the supply air as well as the temperature.

Note that the damper blades do not operate in parallel. These opposed blade dampers provide a more linear position to airflow relationship for better control. The face and bypass dampers operate opposite each other, one closes as the other opens and the design provides for a pressure drop through the open bypass damper that equals that of the cooling coil and open face dampers so the air flow quantity isn't altered by the operation of these dampers.

In constant air flow systems two means are available for controlling the equipment outlet temperature when cooling, one is to cool all the air then to reheat the air with heating coils (not necessarily in the air handler) to maintain conditioned air temperature. To control the apparatus outlet temperature chilled water can be regulated by throttling the flow of water or bypassing some of it. Refrigeration temperatures can be adjusted with compressor controls and face and bypass dampers allow some of the air to pass through the apparatus without contacting the cooling coil and mix with the air passing through the cooling coil.

The cooling coil not only removes heat, it condenses moisture in the air to produce the equipment outlet conditions. Velocity of the air over the cooling coil is typically limited to 500 feet per minute so the droplets of condensed moisture drizzle down the fins to the drain pan instead of being carried off with the supply air. The condensed water dripping off the coils is collected in a drain pan connected to drain piping.

Blockage of the pan or drain piping by condensate contaminated with dust or other debris is a common problem with air conditioning equipment. Elimination of the condensate can also be a concern; in hospitals, similar medical facilities, and production plants the drain is piped to a sanitary sewer for obvious reasons. In some cases the drain piping can't slope down to discharge into the sewer or a suitable discharge point so small sumps and pumps are provided to lift the condensate into a suitable discharge area. The drains should be arranged in a manner that prevents sewer gases or other contamination from leaking into the air conditioning system and the normal solution for that is a 'P' trap like you find under a lavatory sink. A good remedy for those plugging up is to give them a fresh water rinse once a year. Regardless they should be checked quarterly to ensure water isn't flooding the drain pan.

Inspection and Access Doors

Checking the drain pan would include checking the condition of the cooling coil, inlet and outlet, to detect any undesirable accumulations on the coil or in the drain pan. In order to check them there has to be a means of looking into, and in large units, entering the space between the filters and the cooling coil. Inspection and access doors are normally provided for this purpose and, hopefully, access to them isn't blocked by the piping and wiring on the outside of the unit. The covers or doors for these openings range from a thin piece of sheet metal to a rugged, insulated door with latching handles and hinges. They should not be left open, half hanging, or loosely mounted because the air pressure in the apparatus is negative after the filters and air leaking in bypasses the filters.

If, as in so many cases, the openings are simply covered with a piece of sheet metal then it's good practice to replace the gasket that should be glued to the cover when it's accidentally damaged and to replace any screws that get lost. When replacing the cover care should be taken to simply draw up the cover until the gasket is seated then stop turning the screw. If you don't the metal cover will bow leaving gaps between the screw holes that allow unfiltered air into the AHU. Whenever the cover screws simply spin, because the sheet metal of the AHU is stripped, look for larger screws that have plastic knobs on them to replace those old screws. The larger screws will engage in the stripped holes and the plastic knobs make it possible to remove and replace them without tools while preventing application of too much torque that stripped the screw holes in the first place. Oh, by the way, get new screws that are just long enough to do the job so you don't waste time removing and replacing long ones.

Hinged access doors need to have those hinges lubricated. Hear a squeak when you're opening one? That's the AHU telling you it needs lubrication. You can also note that the hinges are worn down (cheap ones do that in a few years). You may have to replace them to get the door to fit properly but the first thing I always try is nylon washers. Remove the hinge pin, raise the door, slip the nylon washer in where the hinge has worn down, and replace the pin. I would prefer Teflon washers but they're hard to find.

Keep the access openings in condition so they seal out infiltration (air leaking in unfiltered) and you'll have less trouble with your AHU.

Back to the psychrometric chart, locate the point on the chart where the dry bulb and wet bulb temperatures at the outlet of the cooling coil intersect then mark and circle it. Then draw a line from the point of inlet conditions through the point of outlet conditions extended to the curve at the left side of the graph. The line between the inlet and outlet points indicates the changes in the air as it is cooled. The dry bulb temperature decreases and the moisture content decreases. Extending that line to a point where the line intersects the curve on the left marks the ADP (apparatus dew point) which is the temperature of the cooling coil, either the refrigerant in it or close to the chilled water entering temperature. The difference between the ADP and the air temperature leaving the cooling coil is due to the fact that some of the air passes through the coil without contacting the coil, actually bypassing it. Dividing the difference between the coil outlet dry bulb temperature and the ADP, then dividing the result by the difference between the coil inlet dry bulb temperature and the ADP produces a value called a bypass factor. Multiplied by 100 it's the percent of the air that didn't come in contact with the coil. A low bypass factor would be unusual and could be attributed to a replacement coil or change during construction where a coil with a lower bypass factor was installed. A high bypass factor, one higher than indicated by the original design drawings or more than 20%, can indicate leakage around the coil due to changes in the air handler's structure, corrosion and wasting away of metal baffles that

prevent bypassing, structural problems with the construction of the air handling unit or simply that the coil is dirty. Note that a dirty coil is detected by bypass factor, not by higher than normal outlet temperatures. A higher than design air pressure drop across the coil is also an indication. A dirty coil, like a dirty filter, changes the air flow which normally results in cooler unit outlet temperatures because the coil has less air to cool. These points are shown on the chart for cooling and reheat of Figure 5-44.

Ultraviolet Lights

When you encounter an apparatus with ultraviolet lights inside they're there to kill bacteria and other growths that occur on the cooling coil because its cool, moist condition is an excellent breeding ground for all those nasty things. They're high intensity lights that can injure your eyes which is why you should find signs on the access doors warning of exposure to them. You should be furnished with a set of UV filtering eyeglasses to wear when inspecting the lights because you have to open the access door and look in at them to see if they're working.

When inspecting the cooling and heating coils turn off the lights using the switch that should be nearby and, when ready to close the unit up, turn the lights back on and make sure they're all on before replacing the covers or closing the access doors. Check the upstream side of the bulbs while you're looking at the coils to detect any dust or debris that has impinged on them. Wipe that side with a white glove or cloth for best results. If you end up with a dirty glove or cloth you should look for problems with the filters.

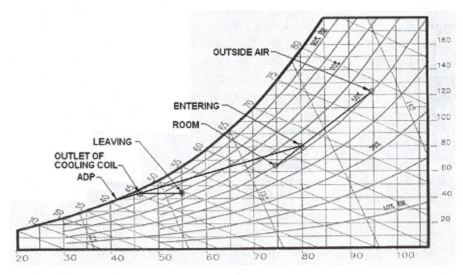

Figure 5-44. Psychrometric chart, reheat application

Heating Coil

Some apparatus, but not all, will have a heating coil after the cooling coil. When cooling of the air isn't required heating may be and the coil is there to heat the supply air leaving the apparatus. It becomes necessary in many installations because of the large, required, quantity of ventilating air adds so much cold outside air that heating it is necessary. Even systems with reheat coils (heating coils in the ductwork supplying different portions of the air conditioned space) will have a heating coil in the main apparatus because the cold air in the ductwork would promote condensation that could damage the insulation and ductwork. The clue here is don't shut down that heating coil in the winter, nor ignore it's lack of operation, thinking the reheat coils will take care of things because wet and falling ceiling tiles could be the last thing you remember of that particular facility. Heating can be performed with electric heating coils (very expensive to operate), steam or hot water heating coils, or a gas or oil fired furnace. Of course the furnace is not a heating coil but it accomplishes the same purpose and furnaces are discussed later.

When there is a heating coil it can be used for partial reheat or simply heating. The apparatus can be operated for humidity control of the conditioned space by cooling the air more than necessary for temperature control, in order to remove more moisture from the air, and reheat it to provide the desired supply air temperature.

When simply heating the line representing the changing condition of the air on the psychrometric chart is represented by a line drawn horizontally from the apparatus entering conditions to the apparatus outlet conditions. In reheat applications the line is drawn from the conditions at the outlet of the cooling coil to the apparatus outlet conditions, after the heating coil. A typical heating application is shown in Figure 5-45.

Humidifier

Whenever a considerable amount of outdoor air is used for ventilation it has to be heated when the outside air is cold and, because the heating increases the dry bulb temperature (as indicated by a horizontal line from left to right on the psychrometric chart) the humidity of that air drops. Unless moisture is added to the air the personnel in the conditioned space will experience electric shocks and other unpleasant conditions. Humidifiers add moisture to the air and, depending on their source of water, can also heat or cool the air.

The typical home humidifier consists of a high pressure spray that atomizes water to form fine droplets of it and inject them into the air stream coming off the heater. Others use a ceramic or fiber wick that's soaked with water and exposed to the air stream. Some of the heat in the air is used to evaporate the water. Being a boiler operator you know how much heat it takes to convert liquid water to a gas so you can see how the line on a psychrometric chart for that service would be drawn from the conditions at the outlet of the heating coil up and to the left, both adding moisture to the air and cooling it some.

Larger applications can use similar means of humidification but typically use steam humidifiers which not only add moisture but add heat to the air and the line on the psychrometric chart for that service would be drawn from the conditions at the outlet of the heating coil up and to the left, both adding moisture to the air and heating it some. They two methods are shown on the chart in figure 5-45. The dashed line shows that the air coming off the coil would have to be much hotter to vaporize the spray or wicked water because of the heat necessary to vaporize the water. The steam only adds about one degree to the air temperature while cooling the steam which is already vaporized.

One interesting element of the diagram on that chart is that the line for mixing the outside and return air indicates that, if more outside air was in the mix, the conditions would be to the left of the saturation curve. That's when the

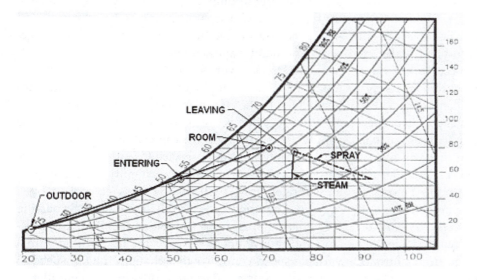

Figure 5-45. Psychrometric chart, heating and humidification

air would be supersaturated with moisture and droplets of water forming, producing what we refer to as fog. The condition could occur as the return and outside air mix resulting in wet spots in the mixing box or, worse yet, in the filters. One approach to eliminating the problem is a mixing box design that introduces sufficient turbulence to prevent droplets accumulating on the walls of the mixing box and ensuring good mixing before the combined air streams reach the filters. The other approach is to heat the outside air before it's mixed with the return air.

If a heating coil is used to boost the temperature of the outside air to prevent fogging at the apparatus inlet it requires close attention when steam or hot water is used as the heating medium. It's also true when large quantities of outside air are required for ventilation (typical for movie theaters and similar venues) then the cooling coil (if using chilled water) and the heating coil all require attention. Why? So they are protected from freezing. A poorly functioning steam trap isn't all that's required for a steam coil to freeze up. Sometimes the heating controls throttle a steam valve so much that the pressure is inadequate to push the condensate out of the coil and even so-called freeze proof coils will freeze; I had it happen. Hot water coils have also frozen when the flow to them is throttled enough that localized freezing in the coil is initiated and that leads to blockage of part of the coil and freezing damage. An idle cooling coil, unless drained for the winter, can easily be frozen if the outdoor air heating coil fails. Today's SCADA systems typically monitor apparatus temperatures to warn of freezing conditions and you should conduct an immediate inspection of the unit to determine if failure can be averted.

Furnaces

Gas fired and oil fired furnaces that heat the air directly can contribute to the discomfort and under extreme conditions injure the health of occupants. Even a small crack in a furnace section can introduce enough CO (carbon monoxide) to poison occupants. If you have operating carbon monoxide detectors check their operation regularly and don't disable one if it alarms repeatedly because it's probably telling you that you have a problem. Even a clean fire will produce CO under the right conditions. Unless shutdown introduces more serious hazards, such as freezing occupants, shut the unit down until the cause of the alarm is verified. Insist on having a portable analyzer if there are no fixed ones and use it to check all spaces supplied by air from a furnace and any complaint of uncomfortable conditions. Advise the Owner that all personnel should be evacuated from

an area that contains CO because any one of them can call a utility or fire company that will come to the site and, if they find CO in the building, shut off the fuel supply and order the place evacuated. CO kills and it shouldn't be treated lightly

Zone Dampers

When required to serve several zones (an independent group of spaces with different heating and cooling requirements) the AHU can be a push through design with the fan mounted between the filter housing and the coils and face and bypass dampers for each zone. Those dampers installed after the coils mix the outputs of the two coils to provide a supply air temperature to the respective zone controlled by that zone's thermostat. The zone dampers are like face and bypass dampers but simply direct air through either the heating coil, the cooling coil, or a mix of the two to provide the desired supply air temperature to a specific zone.

Fan Housing

The normal arrangement of an air conditioning apparatus is to have the fan in line after the stuff already covered making it what's called a draw-thru apparatus. If the fan or fans were located on the inlet it would be referred to as a blow-thru apparatus. An advantage of the fan at the outlet is that it is in conditioned air. Blow thru fans need regular attention to remove dust and dirt accumulating on the blades of the fan. That's a typical problem with return air fans as well.

The fan can be powered by a motor contained within the casing of the apparatus or a motor mounted outside the apparatus and connected by belt and pulley to a shaft passing through the fan housing and supporting the fan wheels with separate bearings. A motor mounted inside the housing requires more space plus room to access it for maintenance and is usually restricted to RTUs (covered later) or very large AHUs. The inefficiency of the motor (3 to 7% of power consumed by the motor) is heat added to the air but the fan simply working on the air adds heat that is normally considered a load on the supply air in the engineer's calculations.

The most common service for large units today is variable air volume where the quantity of conditioned air supplied to a particular zone is controlled by a VAV box (described later) so the amount of air passing through the apparatus is variable and, to limit duct pressure and reduce operating costs the fan is fitted with a variable speed controller that changes the fan speed to maintain a constant discharge pressure at the outlet of the apparatus or, when equipped, the HEPA filters.

HEPA Filters

A final element added to some systems where the air quality is critical are HEPA (high efficiency particulate arresting) filters and they're normally installed on the outlet of the air handling unit. Since it's on the pressure side it ensures there will be no infiltration of air after the HEPA filter. The pressure drop required to achieve the filtration would be added to the negative pressure in the apparatus casing which would make in-leakage of unconditioned air more of a problem along with requiring a stronger casing. I know I mentioned the synthetic fiber factory and HEPA filters were used there to prevent dust particles binding to the fiber during stretching and spinning because that would cause the fiber to break. HEPA filters can actually be so fine that they can remove odors. It's another thing that requires attention to the instruction manual.

Heat Exchangers, Heat and Energy Wheels

Applications that require very high percentages of outside air are frequently equipped with some means of exchanging some energy between the outside air and exhaust air. Heat pipes were mentioned earlier. Dilution of pollution of the conditioned space by manufacturing processes and concern for cross contamination of work are typical situations where 100% outside air is used. I'll resist commenting on the former but I did have the opportunity to design air conditioning for a forensics laboratory where contamination of samples was a concern and they used 100% outside air. My design incorporated a simple heat exchanger that kept the two airstreams isolated and ensure no cross-contamination.

Heat and Energy Wheels are what the name implies. The heat wheels contain a fine lattice of metal that absorbs heat from the hot air in summer and dumps it into the cooler exhaust air then absorbs heat from the exhaust air in winter and heats the cold outside air. Energy wheels, sometimes called enthalpy wheels, use a desiccant coating that allows the wheel to absorb and discharge moisture as well. The wheel is installed in a casing with outside and exhaust air inlet and outlet connections so as it rotates it picks up heat (and moisture) from one air stream and discharges it to the other. I believe you can tell that this application doesn't differentiate between heat and moisture and other things that can be in the air so it's possible that contamination from the conditioned space can be returned to the outside air drawn into the unit. A particular problem with the wheel units is maintaining the seals that separate the two air streams. Secondary problems include blockage due to filter leakage (yes, some of them require filters

on the exhaust air to protect the wheel) and they're frequently equipped with a number of pressure sensors to detect problems of leakage and blockage.

Complete Your Design Charts

The last line to draw on your psychrometric chart is from the equipment outlet condition to the room condition and that represents the heat and moisture added to the air as it enters the room and mixes with the room air to absorb the heat and moisture that's generated in the space. Once you have charts for all the design conditions you can use copies of them to compare to actual operation when problems arise. If you take actual measurements under full load conditions save those charts as well because they're "as-builts" not design situations and a better reference to use. The problem is that we seldom reach actual design conditions and have the time to take the readings at exactly that moment. Don't forget that you have at least two design conditions, summer and winter.

Now you have something like a steam cycle or refrigerant cycle only it's an air cycle. Label the chart with the equipment name or number, copy it and put it away for the day when you'll need it. Do a couple of others where conditions are different so you get a feel for what happens there and gradually build your library of the design and normal operating conditions for each piece of air conditioning equipment so you have the information you need at hand and fully understand what's supposed to happen.

Air Handling Units

The typical AHU consists of a mixing box where return and outside air are mixed before entering the filters, a filter box, face and bypass dampers (if included), an access space for maintenance of the cooling coil, cooling coil, another access space for both cooling and heating coils, heating coil, humidifier, and fan housing. Most AHUs are fitted with a fan driven by a shaft connecting to a sheave outside the unit with the motor, sheave, and belts. These are typically prefabricated in sections for field assembly but can also be shipped as an assembled unit that is simply rigged into place. Most AHUs use chilled water in the cooling coil and steam or hot water in the heating coil. Valves controlling the flow of the heating and cooling mediums have to be added along with diffusers, supply, return, and outside air ductwork to complete the installation. Openings for access to inspect and maintain the filters, fan and coils are provided in the casing. A drain pan is provided at the bottom of a cooling coil to collect and direct the condensate formed as moisture is removed from the air.

Ductwork

Every system other than window units and the thru-the-wall units at hotels and motels are fitted with ductwork to convey the conditioned supply air from the apparatus to the air conditioned spaces, return air from the conditioned spaces, and supply outside air to the inlet of the mixing box on the apparatus. As a general rule air leaks in ductwork were typically in the range of 20% to 30%. Jointing methods for sheet metal duct were developed for low cost installation and, except where specified otherwise, subject to sloppy installation that produced lots of leaks. Even insulated ducts had heavy leakage; I can recall looking into a ceiling space to see the insulation ballooning around the ducts because of the air leaking out of the ducts and pieces of duct tape flapping around in the breeze where the air came out of the insulation. Newer standards have reduced the amount of duct leakage but don't forget that it's significant in anything designed or built before 1973 when we had the first so-called energy crisis.

Ductwork is designed one of two ways, constant friction or static regain. The first is simply choosing duct sizes so the pressure drop is a constant (0.1 inches W.C. per 100 feet of duct) so the required discharge pressure at the supply air fan could be calculated readily by multiplying the length of duct (in feet) to the furthest register with corrections for elbows and other fittings by 0.001). Static regain design allowed for determination of the recovery of static pressure as the air velocity decreases providing additional force to convey the air to that furthest register. Regardless of the design the operator only has to know that a boost in static pressure will occur when the air slows down, provided that it slows down in a manner that's not produced by a damper or other restriction where the regain is considerably less than 100%. The example in Figure 5-46 shows how static pressure may be recovered from velocity regain in a duct expansion. Different configurations provide other values for the regain factor (R) but simply looking at a configuration and how abrupt changes are can give you an idea of how much of the potential regain can be recovered.

Don't confuse a lack of air flow to a conditioned space with blockage of ductwork. I've encountered several problems with duct failures, many including that flexible duct which is easy to crush.

Reheat Coils

When apparatus serves spaces containing many rooms, offices, or cubicles, it's difficult to ensure occupant comfort because the loads can change. Some of the

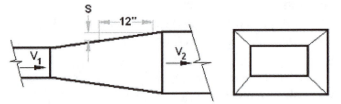

TABLE FOR DETERMINING THE VALUE OF 'R'						
$V_2 \div V_1$	SLOPE (S) INCHES PER FOOT (USE LARGEST)					
	1	2	3.25	4.5	7	10
0.20	0.83	0.74	0.68	0.62	0.52	0.45
0.40	0.89	0.83	0.78	0.74	0.68	0.64
0.60	0.93	0.87	0.84	0.82	0.79	0.77

VELOCITY PRESSURE $(P_V) = (V \div 4007)$

STATIC PRESSURE REGAIN $= R \times (P_{V2} - P_{V1})$

Figure 5-46. Duct velocity and regain calculation

area can be empty with a large number of people gathered in one part of the space to produce a concentrated load in one section. To handle situations where that can occur spaces served by a common piece of apparatus can be fitted with reheat coils that warm the air entering spaces where the cooling load is lower than normal (no people, sometimes no lights). Conference rooms and similar locations may also have time controls or occupancy sensors that actually reset the thermostatic control to allow the room to be cooler, or warmer, unless in use or scheduled to be in use.

Reheat coils are typically thermostatically controlled for the space in which they are located (which can be part of a number of spaces served by the same apparatus and reheat coil). The control system commonly uses a method of detecting the warmest of those spaces to control the temperature of the air leaving the apparatus and allows the remaining spaces to use their reheat coils to achieve comfort in them.

It seems strange to cool the air then heat it again but it's one way of ensuring occupant comfort. One chief operating engineer at a major college told me he could count on a considerable decrease in steam load when the chiller plant was shut down for maintenance. To eliminate the energy waste associated with reheat some new system designs, including VAV boxes, have been developed.

VAV Systems

Most modern systems have air handling units or rooftop units with a variable speed drive controlling the fan to supply conditioned air to VAV boxes. VAV stands for variable air volume. The space control thermostats are used to regulate the flow of conditioned air into the spaces to maintain the space dry bulb temperature. Varying the airflow, however, can create difficulties with space ventilation. Passing of small clouds that block sunlight to one or two rooms of a zone served by the same VAV box can produce detectable temperature swings in those rooms just like systems on reheat coils. A VAV box maintains temperature control in the room or zone by varying the quantity of conditioned air supplied to the room or zone. All the VAV boxes served by one piece of apparatus are supplied conditioned air to serve the largest load of the group so, theoretically, only one or two VAV box dampers will be wide open at any time. As long as the loads served by that air handling unit change uniformly no room supplied by a VAV box will have its damper throttled so much that the air registers cannot mix the conditioned and room air enough to prevent poor ventilation and drafts. Despite that many systems are now installed with separate ventilation air supplies.

Some VAV boxes are fan powered to provide the differential pressure required for proper distribution of conditioned air into the space. Electrical savings are realized when those fans are shut down during periods of no occupancy. The fans in the VAV boxes do not vary airflow because the airflow has to be maintained to provide proper mixing of supply and room air at the registers. So, the fan takes a suction on the inlet of the supply air after the VAV box damper and a return air connection from the conditioned space.

VAV boxes serving rooms or zones at the sides of the buildings, or under the roof, are typically fan powered and include a heating coil to absorb the load of heat losses to cold outside air. These are normally fitted with a filter in their return air connection to prevent accumulation of dust and airborne debris in the fins of the heating coil and on the fan (the conditioned air should be clean). Most of these maintain temperature control during the winter and on cold days by throttling the flow of the heating medium, usually hot water, through the heating coil. Since those filters are always in service monitoring of their condition until an optimal replacement period is established is an added predictive maintenance requirement for a new system. Other than situations where the air in the zone can be contaminated inconsistently during regular activity a filter replacement program can be established with longer periods between replacements than those recommended by the manufacturers.

Almost all VAV boxes are fitted with airflow sensors that normally permit connection of a manometer to produce a differential measurement that indicates airflow. This permits a quick comparison of actual airflow to the design airflow shown on the drawings. Before long I imagine the low-cost differential pressure sensors will permit you to check the airflow in any zone at a remote terminal. You'll need to use some judgment in monitoring the airflow because the loads served by the VAV box are seldom at design conditions. Most of the time however, you'll be looking at a problem during periods of maximum load where the readings will be meaningful.

A VAV box can be a simple device consisting of nothing more than an enclosure with a damper controlled by a modulating motor to open or close the damper depending on the temperature in the space served by the air passing through the VAV box. One VAV box can serve several offices along one exposure of one floor of an office building. Because of multiple exposures (sunlight being the major load in most air conditioning systems) corner offices typically require an independent VAV box. That's not always the case when the box serves all offices along one building wall and, since the higher up gets the corner office, the thermostat is in the corner office. When space utilization or heating or cooling load varies such as interior versus exterior exposures, office space versus conference room, operating room in a hospital compared to a patient's room, separate VAV boxes are normally provided. Changes in use of a space, installation of partitions and the like can make space temperature control impossible. So, despite the advantages of VAV boxes satisfying everyone isn't always possible.

The variable speed drive of the fan is controlled to maintain a constant pressure at one or more points in the distribution system. Cooling and fan motor horsepower are saved by these systems while simultaneously providing more precise cooling and heating in the conditioned spaces.

Air Registers and Diffusers

Registers, typically in walls and floors, and diffusers, usually devices installed in the ceilings of the rooms, are designed to mix conditioned supply air with room air to absorb most of the rising heat load when cooling. The intent of the design is to mix the air to within a couple of degrees of the room air temperature before it strikes a wall or occupants to produce sensations of draft. There is little need for concern with the delivery of heated air because very few occupants object to warmer temperatures of delivered air in the heating season.

Temperature controlled registers that vary the delivery of supply air by throttling the flow of air through the register are available and a handy solution to isolated problems. Since the register has to draw in room air to mix it with the supply air the temperature of the room can easily be detected at the register. Movement of the damper in the register is accomplished with a change in position of a bimetallic element in the room air stream. These registers are good for combating variations in cooling load in interior spaces where heating isn't required. They are only a trimming device, capable of making small corrections in the temperature of the room they're in. Sometimes they're a solution for that lady that's always cold in the summer.

Almost every design drawing of an air conditioning system will show air quantities at each register. When troubleshooting problems you can use a flow hood (Figure 5-47) that provides a reading of air flow through a diffuser or register and into the conditioned space to compare the existing condition with design. That will only tell you what the air flow is, and not necessarily give you a reason to start adjusting things like the damper on the register (if it has one). You can create more problems in other spaces by adjusting things to correct a problem in one space. Measure the actual air flow through a register or diffuser to compare it with the value on the design drawings and report problems with airflow that can be attended to by a TAB technician.

RTUs Roof-top Units

RTUs (roof top units) are applied in many commercial applications, schools, small office buildings, and other applications where the first cost of a facility is the primary consideration. A unit that combines all the equipment and pipe and tubing necessary to heat or cool spaces in the building are prefabricated and rigged to the roof of the building to connect to ducts that distribute cool and dehumidified or heated air to the spaces served by that unit. In addition to steam or hot water an RTU can be fitted with a natural gas or oil fired furnace for heating. Typical RTUs are shown in Figure 5-48.

Roof top units can also be fitted with a variable speed drive and temperature controls to serve variable air volume systems as described for AHUs.

I will not go easy on this and tell it like it is. I detest RTUs. They are a cheap substitute for a well designed building and air conditioning system and a pain in the you-know-where. I have a low

opinion of roof top units because so many of them are simply constructed to provide air-conditioning at the lowest possible first cost. I concur with a number of other engineers that say a typical rooftop unit will last 10 years whereas a typical air handling unit installed indoors will last more than 30 years. Self-contained refrigeration systems have twice the operating cost of systems using chilled water. However the one thing I've noted on practically every roof top unit I've seen the access opening covers are simply pieces of sheet metal along with the casing attached with a minimal number of screws and within a year or two are bent, lacking screws, and leaking outside air into the unit disrupting

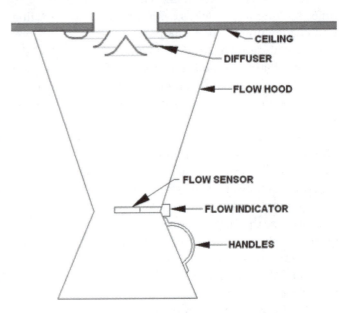

Figure 5-47. A flow hood

Figure 5-48. RTUs

its performance. I don't think that it's the maintenance crew's carelessness, is just cheap construction. I've seen them ready to fall through the roof and so corroded that I wonder what's preventing them collapsing. They seldom use more efficient chilled water sources. Normally they contain a complete refrigeration system with air cooled condensers and a DX coil. DX standing for direct expansion which is an acronym that separates them from separated systems that supply chilled water and means the air is cooled directly by the refrigerant. Their operating cost is easily twice that of a unit using chilled water. They frequently are direct fired for heating which would be more efficient than steam or hot water from a boiler if the burners and controls weren't so cheap that they are less efficient than a steam or hot water operation.

I have spent an evening at a meeting in a local school during the winter listening to the RTU and connecting ductwork creak, snap, and emit other sounds that challenge a box of Rice Krispies (but at several times the volume) while wondering if that night would be when the ductwork or the RTU comes crashing down because it has worked loose. Called for problems with a potential leak from an RTU I found it shifted on its curb (a curb, by the way, is a raised portion of the roof that supports one) until one corner of the RTU had slid over enough to expose the indoors to the outdoors. They're where I find most of the access panels lacking half their screws so unfiltered outside air leaks into them as well.

If you're stuck with operating and maintaining RTUs, may God bless you. All I can say about them is do your best and pay attention to those creaks, moans and snaps. If you can isolate them you may be able to eliminate them.

Air Conditioning Systems Control

The normal air conditioning apparatus control consists of a thermostat that starts and stops the fan, compressor and associated electrical equipment. That would be typical of our good old window units and the system in your home. While it's inexpensive at first cost the operating costs are considerably higher than equipment that can operate continuously and modulate to control space temperatures. When I had my heat pump replaced I had the new one fitted with a two speed scroll compressor. It means the fan runs more but the compressor shifts back and forth between half and full speed (hi and lo on the thermostat) to keep it running longer, sometimes constantly at high heating and cooling loads. Every time a compressor shuts down the refrigerant pressures balance out and the compressor has to reinstate

the differential before the cooling or heating gets started again. It's not the best (remember I mentioned I wanted ground source?) but my electric bills dropped to 38% of what they were.

Instead of controlling the unit with a thermostat in the space or sensing temperature in the return air duct the outlet temperature of air conditioning apparatus can be controlled by an independent system controller according to the lowest air temperature required at any outlet in the air distribution system. All other outlets in the system control the temperature of supply air to the space by a thermostat in that space through use of a reheat coil or a variable air volume box.

New regulations and design standards for building ventilation that are primarily concerned with the distribution of the ventilation air have led to the introduction of independent outside air supplies so don't be surprised if some of your equipment doesn't have a mixing box with connected outside air ductwork. Separate air handling units, frequently called ventilation air units, provide outside air distributed to the building through its own independent system of ducts and registers. Those units can be operated to deliver air at the room design conditions so the ventilation air does not have any effect on the operation of the heating and cooling equipment.

One of the major considerations for load of an air conditioning system is sun light. A well-designed system will account for variations in solar heat gain at different exposures (North, South, East, and West) through the use of separate AHUs, face and bypass dampers, and/or reheat coils for each exposure. Of course the bigwigs in any office building insist on having corner offices with double exposures and unless their space has independent temperature controls there will be periods when other occupants of the zone are uncomfortable.

The TAB Report

In any large installation the construction project usually calls for air balancing. Technicians will take readings and adjust air flows and temperatures, record them, and show the actual measured air flow at each apparatus, main distribution duct, VAV boxes, and registers. That document should also be kept handy so you can compare the setup conditions with actual.

Convectors

Convectors are usually floor mounted and usually below windows. They contain their own fan or fans to circulate room air over refrigerant or water coils cooling and/or heating the room air to adjust the room air for comfort. Many use the thermostat control to start and stop the fans while others will regulate the flow of the fluids through the coils. Ventilation is normally achieved by a small damper connecting to the outside so the fans will draw air in when they are operating. Refrigerant or chilled water coils will remove moisture from the room air by condensing it and the drain is typically run out the building wall but can be piped indoors. Convectors can be supplied with or without filters so filter replacement or regular cleaning of the unit is required. Convectors can also be two-pipe or four-pipe. Two pipe convectors use the same coil for heating and cooling, using heated or chilled water supplied by the central system. These require the operator monitor the weather, decide whether the building is to be heated or cooled and set up the system to deliver water at the required temperature. Modern convectors sense the temperature of the water supplied and automatically switch the thermostat response to control the flow of the water or the operation of the fans. Four-pipe convectors have separate cooling and heating coils and can heat or cool independent of the weather.

Achieving Comfort for Occupants

If someone is not comfortable you can count on repeated complaints and, in some cases, direct orders to do something about it. Unlike when they're home, the occupants are seldom able to walkover and adjust the thermostat because it's seldom that the entire occupancy is satisfied so the thermostat is locked up or concealed because some people produce excessive adjustments that result in additional heating or cooling expense. The first step is determining what the space conditions are and that involves using a sling psychrometer (Figure 5-50), a flow hood (Figure 5-47), a manometer set at a ten to one slope (Figure 2-3), pitot tube (Figure 5-49) an infrared thermometer (Figure 5-51) and a scientific calculator. I mention these early on because you might just be assigned to operate systems in a new installation and discover some of this equipment left behind by the installer or TAB (temperature and air balance) personnel. If so, you'll recognize it and put it away for your own future use. Much of the equipment and tools that still clutter the trunk of my car came from that source.

A combination of the manometer and pitot tube is used to determine air flow in a run of ductwork. It's great for checking for leakage in a long duct run. For best results the rest ports should be located twenty diameters downstream and four diameters upstream of any change in the duct (elbows, transitions, etc.). Oh, for square ducts use an equivalent round diameter deter-

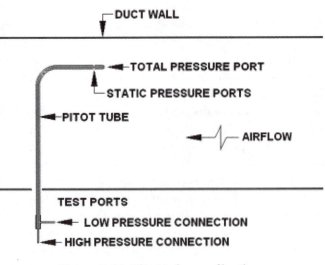

Figure 5-49. Pitot tube application

mined by a formula equal to 1.3 multiplied by the width times depth of the square duct multiplied and raised to the 0.425 power divided by the sum of width and depth raised to the 0.25 power. Width and depth are simply the two horizontal dimensions for vertical ducts. You should also be able to find a table for equivalent round duct sizes in one of the books available from the manufacturers of your air conditioning equipment. Test ports should be on two quarter points of round ducts and at least four on one side of a square duct. The more readings you take, the more accurate your calculated result. You connect the manometer to both ports on the pitot tube so you're getting velocity pressure, the difference between total pressure and static pressure. Velocity (in feet per minute) at each measurement is determined by multiplying the square root of the velocity pressure (in inches of water column) by 4006. (I use 4006 because there are competing groups that call it 4005 and 4007) With round ducts calculate the area of quarter points sampled in square feet and multiply by the velocity for the flow in that quarter for small ducts. For larger round ducts determine the area in segments and multiply by the calculated velocity in that segment. For square ducts it's much easier to simply select equal rectangles (that sometimes come out to be squares) centered on each measured point, and multiply that area by the measured velocity.

Even when the air in a space is within the comfort zone there may be other factors that affect the comfort of the occupants. These can be attributed to solar insolation, warm ceilings, cold floors, warm or cold walls and drafts; any one of which can result in more than 50% of the occupants being uncomfortable. Now you know why you need those instruments, you need to identify all the factors that have an effect on their comfort. It's not uncommon for complaints to disappear just because you were in the zone taking readings; people can get happy just knowing that you care. There are many studies that indicate changes in the room temperature settings above and below design at regular intervals increased occupant satisfaction with the air conditioning system.

Problems can be local or area wide and should be approached with checking the room or checking the equipment serving the area depending upon the complaints. Complaints from one room or zone require a different approach than those from all areas served by one air handling unit or rooftop unit.

Frequently comfort problems in a room or zone are due to the location of the thermostat. It's a common problem that I have encountered many times, especially when I was teaching seminars on boilers and air-conditioning and refrigeration in various hotels throughout the country. In some cases we would be in the room containing the thermostat but most of the time the thermostat was in another room in the zone. That was fine as long as the other rooms occupied with an equal number of people. All too often however, the room of the thermostat was empty with lights off and occasionally it had more external exposure than the room we were using. Although it gripped to me to do it I managed to reduce the problem by turning on the lights in the other room; that being a waste of energy, something I was teaching my students not to do. On a few occasions the other room containing the thermostat had a much larger crowd and my students are complaining of the cold. I managed to help that a little by blocking the returns. Yes, carefully tossed tissues help cover the return air grills blocking air circulation through our room to reduce the introduction of cold supply air. Those temperature controlled registers can help reduce those problems along with zones containing several small office spaces. Hotels, conference centers, and similar venues might reduce those problems with multiple thermostats and thermostat selection based on occupancy sensors. There's nothing much an operator can do to reduce the problem other than inform the person scheduling use of the rooms to make sure the first or only room in use is the one containing the thermostat. Altering the setting of the thermostat to balance the loads will normally result in more complaints

When there appears to a problem with air conditioning equipment serving an area you should collect the data on the system to see if it is operating as designed. Many factors can contribute to improper operation and you have to compare to determine what may have changed. Take dry bulb and wet bulb readings because the occupant discomfort can be due to humidity problems instead of temperature problems.

Action and Instrumentation

This point I'm certain many operators are saying "Wait a minute Ken, we are supposed to be operating the plant, not fixing it." However, I'm not telling you to fix anything, I'm merely suggesting you isolate the problem and determine what's necessary to fix it. Then, if it's something you can fix by operating controls or adjusting them and occasionally by fixing something that quit working the owner should be very pleased. You certainly don't want to be in the position where you hear the contractor tell the owner that it was a problem you could have fixed. Whether it was in the boiler room, the chiller room, or a mechanical equipment room, when that was the case I normally tried to tell the operator the solution to the problem and explaining how he or she could have fixed it so it wouldn't happen to them again. The only time I did tell the owner the operator could have fixed it was when the operator insisted it wasn't the operator's job or it was obvious the operator couldn't understand what was going on. I can only hope that after reading this book you will not be one of them.

Early in my years in this business I learned that listening got me further ahead than talking ever did. Of course that didn't stop me from putting my foot in my mouth at regular intervals. The sooner you learn to listen the better off you'll be. That's not to say that you have to believe everything you hear. I have yet to have an operator tell me what happened without adding his or her own opinion of what the problem was all about; most of the time that part of the discussion wasn't exactly right but their relating what happened was most telling. So, listen to what are facts, even from the dumbest observer you encounter and you'll be amazed at how it provides a good foundation for your analysis of what's wrong.

There are occasions when there's nothing wrong with the equipment, systems, or controls. I had one operator tell me that a certain lady was always complaining that she was cold and called the boiler room repeatedly. Each time he checked temperatures and humidity and had to inform her that he couldn't help her. Her calls stopped immediately after a visit when, during his examination of her office area, he received a cell phone call from his wife. I trust you can explain that one.

Hopefully all the foregoing explanation of equipment operation and things to check along with data collection and analysis will help you identify the source of problems with occupant comfort and resolve them or provide the owner with your analysis and suggestions for what a contractor would be needed to do.

Data Collection and Analysis

Using a psychrometer depends on what type you have. Call me old-fashioned but I prefer to use a good old sling psychrometer, the one shown in Figure 5-50. All I have to do is check that the wet bulb portion has water in the reservoir. You can use an electronic gadget but I suggest you make sure it's calibrated every time before you use it for serious work, especially if you've dropped it. Yes, I've dropped the sling, and I've tapped some furniture while using it several times, but I haven't broken it yet. And, since the gradations are on the thermometers it's always right if it isn't broke.

Progressively isolate the problem, determine if it's just one room, one zone, or affecting everything served by one piece of air conditioning equipment. Then check the obvious things: Is the equipment running? Does it have electric and control power? Has someone tampered with the thermostat or done something to block air flow?

In checking a problem with a room or zone collect and record room and supply air temperatures then use the flow hood (Figure 5-47) to measure the air flow through the registers. If the difference between room air and supply air temperature is equal to or exceeds design, air flow through the registers is equal to or exceeds design and supply air wet bulb temperature is equal to design or when cooling less, when heating more, then the system is working properly. That normally means the load in the room or zone is more than design and the equipment simply can't handle it.

Unless it's a new facility, which should be under warranty so the problem belongs to someone other than the owner, you should try to identify changes in load to see if they can be rectified. Look for any changes that increased the load and inform the owner of them.

To check a problem with an AHU or RTU collect and record the actual temperature and air flow data for that unit and mark a copy of the pyschrometric chart you prepared with that unit's design data to see what might be off.

Since the air carries the heat away, or delivers it to, the occupied spaces a lack of adequate air flow should be addressed first. Problems with unit air flow are commonly dirty filters followed by loose fan belts. For rooms and zones VAV box filters should be checked followed by isolating a loose splitter damper, control damper, dirty reheat coil, and failed or damaged ductwork. Note changes in the space that might have produced the problem; I recall one where a contractor replaced all the ceiling tile in an office area, including replacing the return air grills that allowed room air to return to the AHU. The project manager told me the contractor complained that he had to clip down all the new ceiling tiles because they kept popping up. Other unique problems included one where a contractor adding sprinkler piping had removed the flex connections between the ductwork and registers to get the pipes into the ceiling space. That was revealed by return air temperatures considerably different than room air temperature; the air was flowing, not into the room but bypassing it.

I could fill this book with many more stories, each a lesson in what can go wrong, but you'll have to apply what you understand the systems are supposed to do to isolate problems with them. I can't possibly tell you every possible scenario within the page limit of the book.

Figure 5-50. Sling Psychrometer

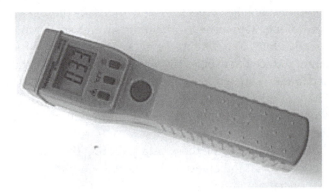

Figure 5-51. Infrared Thermometer

Chapter 6

Maintenance

Operating a system is not as simple as starting and stopping equipment and opening and closing valves. An operator not only operates, he ensures operability. That is the function of maintenance.

MAINTENANCE

You'll recall I said that maintenance of the boiler plant is an operator's responsibility. You can be called upon to do everything from sweeping the floor to rebuilding a turbine, the simplest job to one of the most complex, and everything in between. In a small plant with little equipment you might be expected to do it all yourself. As the size of the plant increases those duties will increasingly be performed by others but you still have a responsibility to make sure they don't interfere with the continuous safe operation of the boiler plant.

The purpose of maintenance is reliability and cost control. We ensure reliability of the equipment and systems in the boiler plant by limiting or preventing wear, vibration, erosion, corrosion, oxidation, and breakdown. Proper maintenance prevents failures of equipment that can result in significant repair costs. Maintenance includes many activities but the most important are monitoring and testing performed by the boiler operator.

There are many forms of maintenance and, contrary to many opinions, each one has its place. You choose which form of maintenance to use depending on the degree of reliability you want or can afford. Maintenance methods fall into three general categories, breakdown maintenance, preventive maintenance, and predictive maintenance. Despite what you may have heard, all three methods should be used to maintain your boiler plant. There are many items that you simply won't pay any attention to until they fail, then you'll replace them. That's breakdown maintenance and it applies to things like light bulbs, sump pumps, and other items that cost so little to replace and are so easy to obtain that any time spent maintaining them is a waste. Some, like light bulbs, only allow breakdown maintenance.

Maintenance requirements vary but should represent a cost relative to the potential loss. You wouldn't spend a considerable amount to check lubrication of a little cooling fan motor (normally they have permanent lubrication) when its replacement costs less than the labor to check it once; that's a situation where breakdown maintenance applies. On the other hand lubrication of a steam turbine can include testing the oil and operation of equipment that continuously cleans the oil because a failure would represent a significant cost.

A small 1/2 horsepower feed pump for a little heating boiler isn't eligible for much more than breakdown maintenance. A 2,000-horsepower feed pump for a super-critical boiler plant will have vibration and temperature sensors at every bearing, speed sensor, suction and discharge pressure and temperature sensors and probably its own flow meter.

Between those two extremes are all sorts of variations on monitoring and maintenance but most of them rely on the skill and dedication of you, the boiler operator. Each round of the boiler plant you will look and listen to the feed pump, noting its condition, look for signs of vibration or shaft leakage, possibly feel the motor and pump bearing housings to get a sense of their temperature; all that is predictive maintenance. When you add oil or grease to bearings you're performing preventive maintenance.

Breakdown maintenance has the advantage of low cost because we basically do nothing to prevent a failure. Preventive and predictive maintenance require an expenditure of effort and materials which represent an investment in reliability. There are varying degrees of effort expended in those activities depending on the cost of failure, the cost of maintenance, and the probability of failure.

The only caution here is to remember that some equipment becomes obsolete. It pays to think about the condition of something that would normally only deserve breakdown maintenance but could be irreplaceable and force a major expense if it isn't taken care of. An example would be a special bolt on a turbine speed control; the bolt might be easy to replace, if you could find one, but its loss would produce hours of turbine down time.

Preventive maintenance is performed on a regular schedule to, as the name implies, prevent damage to equipment or systems. Water treatment and lubrication are the two principle preventive maintenance activities in a boiler plant. Those activities prevent failures by maintaining conditions that do not allow corrosion, scale, or friction to occur. Proper operation of some systems can also be called preventive maintenance when they prevent erosion by ensuring velocities do not get too high.

Water treatment, properly performed, can prevent very expensive and catastrophic failure and the probability of such a failure if water treatment is avoided or ignored makes it the principle concern in all plants. It is so important that it deserves its own chapter in this book so we'll cover it later.

Predictive maintenance consists of monitoring, examinations and tests to reveal problems that will, if allowed to continue, result in failure. Annual inspections of steam boilers and less frequent inspections of other pieces of equipment are conducted to detect formation of scale, corrosion, vibration, wear, cracks, overheating and other problems that can be corrected to prevent eventual failure.

Of course there's that one instrument in the plant that is the best investment in predictive maintenance, the operator's ear. An operator can detect many problems indicating imminent failure and react to prevent the failure. An operator can detect changes in sound, vibration, temperature (by simply resting a hand on the equipment) that would require a considerable investment in test and monitoring equipment. Constant attendance by a boiler plant operator is one investment in predictive maintenance that helps ensure no surprises consisting of major equipment or system failures. It's normally the boiler operator that provides the principle maintenance of water treatment as well.

Since you're at the forefront of the maintenance program, and in many plants you're the one that will catch hell if it breaks down, having a sound maintenance program is an essential part of your job. Repeating what I said in the section on documentation, if your program isn't documented then you have no proof that you did everything that's prudent and reasonable to prevent a failure.

You may have changed the oil in that compressor the week before it failed but without a document indicating you did it... well, it will be very difficult to convince anyone you did. It's also very difficult to remember everything so a documented maintenance schedule serves as an excellent reminder of when something should be done. A schedule and a record of the work being done is the best evidence that you are doing your job and a failure will not reflect on your performance. If you've done a good job planning and executing the maintenance plan you shouldn't have any failures.

Every piece of equipment that requires preventive or predictive maintenance should have that maintenance scheduled. You have to generate the maintenance schedule for your plant because your plant is unique. The best place to start working on that schedule is the operating and maintenance manuals, doing what the manufacturer recommends until you get some track record to find what you have to add and what requirements you can extend beyond the recommendations.

Be certain you got everything because failing to maintain something can be hazardous. I was called in to investigate the third boiler explosion in as many months at one plant and found they had never bothered to replace the tubes in their ultraviolet flame scanners despite the manufacturer's recommending they be replaced annually. Three boilers had extensive damage all because nobody replaced some three dollar electronic tubes. By the way, those were "self-checking" flame scanners.

CLEANING

If there's any distinct impression you get when walking into a boiler plant for the first time it is the cleanliness of the plant, or lack thereof. I have customers with plants that contain flowers in the control room and you believe you could safely eat off the floor. There are others that are so dirty it's hard to see anything because the entire plant is black with soot. Which one do you think is better maintained?

Don't get me wrong, cleanliness isn't a sure sign of a quality plant. Lack of it, however, is almost always indicative of nothing but trouble. A boiler operator has the ability to make the difference in the appearance of the plant and it should be part of the preventive maintenance program. Many an operator claims he or she is too busy to sweep and mop floors, dust, etc. to keep the plant clean. They're usually the ones I can see holding down a chair for twenty minutes or more after I first enter the plant. I always had time to do some cleaning and you will too. Like any other activity it makes the shift seem shorter. You don't have to polish the brass like I did but the extent of work you do is up to you. Every time you leave the plant you should look around and ask yourself a simple question, "would I be proud to have anyone come into this plant and look at it?"

Certain cleaning functions are, by their very nature, considered to be part of the operating function.

That's because those devices are in operation and only experienced, knowledgeable individuals (like a boiler operator) should be allowed to touch them because improper action could shut down the plant. These include cleaning burners, operating soot blowers, and cleaning oil strainers to name a few.

Speaking of cleaning oil strainers… The typical duplex oil strainer (Figure 6-1) is one of those devices that is in service when cleaned. If you open the wrong side (you shouldn't because the handle is supposed to be over the side in service) the plant could be shut down. Another situation involves switching the strainer in service. It must be done carefully and slowly because it's always possible that the cover wasn't replaced properly and the strainer could leak.

One of those strainers involved my first lesson in reading instruction manuals. I had just joined a ship as Second Assistant Engineer and entered the boiler room to find the new fireman using a helper to change the strainer. You know what I mean by a "helper," a long piece of pipe stuck over the end of the handle. I chided him for doing that, advising that he could break the handle. After trying everything I had been taught about them I finally relented and helped him operate the helper to switch the strainer. On the next watch he reported it was even tighter than the day before. Noticing that the handle was bending precariously I told him to wait until I had time to look at the manual.

A visit to the chief engineer's office later that day produced the manual and revealed that there was a little jacking screw under the strainer that both lifted the plug valve so the strainer could be changed and tightened it back down. On the evening watch I looked under the strainer and, sure enough, there was that little jacking

Figure 6-1. Duplex oil strainer

screw. The fireman and I were both amazed that once we operated it the strainer handle could be turned with one finger.

There's one other thing I've learned about oil strainers. The day you decide that it isn't necessary to clean it because it's always clean when you open it… that's the day it will plug up.

INSTRUCTIONS AND SPECIFICATIONS

Read the manual first and every time before you perform any maintenance unless you know the book by heart. Then prepare a checklist that helps you make sure you follow the instructions. It's awfully easy to forget a step or get them out of sequence with component failure being a result. If you don't have the manual then contact the manufacturer to get one. They may charge an absolutely atrocious amount (you have to consider their cost in producing one copy compared to several hundred during the period they manufactured and sold your equipment) but even as much as $300 can save ten times that amount in damage to the equipment.

A checklist will help insure that all the steps are executed in the prescribed order and can save a lot of time. Just jumping in and doing it may seem faster until you have to tear it back down again because a part was left out or an adjustment wasn't made; it's even longer if you're documenting every step because there was a failure and the equipment is severely damaged.

You should check instructions despite your skill and knowledge. I recall one contractor that was adamant about the rotation of a fuel oil pump when I told him it was running backwards. He insisted I didn't know what I was talking about. When I persisted long enough he finally grabbed the instructions (which were still enclosed in the envelope wired to the lifting eye on the pump motor) yanked them open, flipped through the pages and prepared to point at the graphic while thrusting the paper in front of me. Almost as quickly he drew back and checked the diagram; he was wrong. He had created several days of delays, damaged the piping on the pumps, and possibly the pumps, simply because he refused to take a few minutes to look at the instructions.

Specifications define requirements and anything more complicated than a faucet or a toilet ballcock should be compared to the specifications to ensure you have the right type and grade of material. That includes things supposedly simple, like bolts and nuts. I have encountered several situations where the wrong bolts or nuts were used and a few of them were on my projects where, despite the drawings specifically listing the requirements, the steamfitters used the wrong bolts or nuts.

I'm very grateful none of those incidents had a result like using the wrong nuts on the Iwo Jima, a Navy aircraft carrier, in October of 1990 when ten people were killed because a valve bonnet blew off in a confined engine room.[5] A valve's bonnet is that portion of the valve that's removable without dismantling the attached piping to provide access to the valve's internals.

Something that sounds good or looks right isn't the answer. If you don't understand a specification or can't determine whether the material you have complies with it you should consult someone to ensure you have the right material.

Don't take the salesman's word for it because he can deny telling you after the catastrophe occurs so you end up holding the bag. Sometimes the mistake is immediately evident. I can still remember the look on a contractor's face when they started filling a piping system that took over a week for five men to install and water was spurting from the longitudinal seam of every piece of pipe. Nobody checked the material, it was all "untested" pipe; manufactured for structural use.

Sometimes you find out later, that's almost always the case when the material isn't capable of withstanding corrosive action of the liquids it contains. I can still remember the condition of a mild steel thermometer well we had knowingly installed in a stainless steel piping system because the owner wanted the system running and we didn't have time to get a replacement well. We got to replace the well with one of the right material a week later and discovered there wasn't much left of that mild steel. Had the plant run for a few more days the well would have corroded away, the thermometer would have blown out and highly corrosive liquid would have been spraying into the plant.

There's one other thing about materials that needs to be addressed. You may find that a modern material does a better job, something like graphite gaskets for cast iron boilers instead of rubber ones. Refer to the section on replacements that follows.

LOCK-OUT, TAG-OUT

First of all I want to say that I'm not one of those people that gripes about all the hassle associated with lock-out and tag-out regulations and requirements. I operated in the times before those regulations and have very vivid and unsettling recollections of incidents where peo-

ple were injured (including me) and others were killed because we didn't have those regulations. Follow them religiously, they are there to protect you and keep you alive. Second, it is the operator's responsibility to ensure all those regulations are followed and, more importantly, to be the person in charge of lock-out, tag-out.

Don't be too quick to allow that responsibility to reside in someone else, you'll regret it the day the contractor's crew closes and locks out the wrong valve (like on the plant's only water line) then go out to lunch! You're also the only one in the plant I would count on to know every valve that has to be closed to ensure a system or vessel is really isolated. Another problem is that the owner of a plant is responsible for the safety of the contractors because any hazard in the plant involves the property of the owner. If the boss says "let the contractor do it" you might point out to him that the contractor can do it wrong, sue the owner when someone's injured, and the contractor will win!

The regulations for lock-out, tag-out are in OSHA 29CFR part 1910. They are still changing and evolving so I don't intend to address them all here. You should obtain a copy of that document and be aware of updates. You'll have it to review every time you have to prepare a system for maintenance. Right now there are many methods for satisfying the requirements but one simple program shown to me by Ken Donithan of Total Boiler Control seems to be a really clean and simple approach that satisfies the requirements with a minimum of paperwork and a great degree of understanding. It's demonstrated in Figure 6-2 which was prepared for work on a steam boiler.

A diagram or schematic of the system is prepared and laminated with plastic to serve as the key element

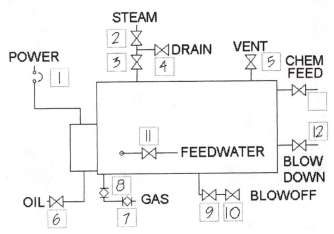

Figure 6-2. Lock-out/tag-out diagram

of the program. It's mounted on a stiff board and hung near the equipment while it's being maintained so it's easily seen and used. As each valve is closed or opened and locked the number of the lock is marked on the diagram with a non-permanent marker. A quick look at the diagram will tell you if all the valves and disconnects are set and locked. All the keys for those locks are placed in one box which has a lid secured by means of a latch that can accept multiple locks.

As each worker places their lock on that lock box his or her initials are added to the diagram so you can see who is in there (or left their lock on) during the progress of the maintenance job. When they leave they remove their lock and their initials. When all work is done and all workers' locks are removed you can remove the keys from the lock box and remove the locks that ensure the equipment or system was isolated, erasing the lock numbers as you go.

In some cases the job could have several operators removing locks and erasing the board as they are removed. This method ensures they're all off. Now the board can be put away for use on the next turnaround. You'll note it's simple and effective while not producing a lot of paper. The locks can have tags permanently attached but I think the number on the lock serves as the tag. The only time you may have to cut a lock is when some worker leaves a lock on the lock box and goes home. Of course, you have to make certain that's what he or she did.

It's always important to include venting, draining and purging of systems as part of your procedures of lock-out and tag out. That's very important when the system contains a hazardous substance, something corrosive or explosive. I've walked away from some locations when I've observed contractors starting work on pipes without making certain they're vented, drained and purged. I walked away so I wouldn't be injured if they opened a hot line. When dealing with certain substances additional requirements should be followed.

Don't say it's never happened. One of my crews cut open a hot line that was supposedly completely isolated. Caustic soda, if I recall correctly. The line penetrated several floors and the wrong valve got shut off at the lower level. I've also heard of several other incidents.

Any time a gas line is opened it should be vented and purged. If the gas is considered hazardous to the environment it should be purged through a flare or sorbent to prevent it escaping untreated. Flammable gases should be purged with inert gas. Usually that means a few bottles of nitrogen or carbon dioxide but large and long lines could be purged with inert gas from a special

generator. Once you're certain the flammable gas is out you follow up by purging the inert gas with air. Just using air is only acceptable for very small lines (less than 3 inches) because flammable mixtures could be produced in the piping and ignited. Keep in mind that inert gas not only prevents combustion, it doesn't contain any oxygen and you can't breathe in it.

We were installing gas burners in a plant that had a future gas line installed several years earlier. The gas line, a ten inch one, entered the plant through the west wall and was closed with a weld cap. I gave my foreman specific instructions to prepare a steel plug in case it was necessary and be ready to insert it in a hole drilled in the line. I also told him not to cut the line until I was there with a gas tester. Luckily an apprentice overheard me and suggested to the foreman that he should call me before taking a cutting torch to the pipe. The foreman relented and called so I went to the plant with the tester. He explained that he had talked to the gas company workmen, who had been there to check the meter location, and the piping was "dead." He finally allowed as to how I was just being safe and had the apprentice drill a one-eighth inch hole in the top of the pipe. The gas detector went nuts and it took a lot of pressure by the apprentice's thumb to stop the leak.

No, the foreman hadn't made up the plug either. We wandered around the plant looking for something until I finally found a piece of wood and used my pocket knife (which I'm never without) to make a plug that we used to seal the hole. The next day the gas company managed to seal off the pipe and we vented it for ages through that little hole.

What do you think would have happened if the apprentice had just started cutting with that torch? Safety is an attitude, acquire it. Lock-out tag-out, purging and environmental testing are things you should take for granted and insist upon happening before opening any equipment for maintenance.

That was only one situation involving that superintendent and I was never allowed to fire him. When I think back to the many times he created hazards or simply changed a job without approval, and got away with it, I don't wonder that I finally managed to get myself fired. Looking back at what happened later, I feel satisfied by the old adage "better safe than sorry."

LUBRICATION

Lubrication is probably the second most important element of preventive maintenance. On larger pieces of equipment drawing samples of the oil for testing is a predictive maintenance measure. It falls on the operator to ensure that every piece of moving equipment is properly lubricated. With the increased use of synthetic lubricants that portion of the job is becoming more complex. Synthetic oils can save thousands of dollars in power cost for operating large pieces of equipment. On the other hand, adding the wrong oil to a crankcase can result in an instantaneous breakdown of the equipment because the two oils are incompatible and one oil causes the other to break down. Keeping an up-to-date lubrication chart that covers everything in the plant is important. Paying some attention to proper lubrication schedules can save you time in the long run.

I've discovered that lubrication is one of the maintenance activities that is always a mixed bag. Most plants seem to have a program that consists of over-lubrication of some equipment and insufficient attention to the lubrication of other equipment. Many grease lubricated bearings need lubrication infrequently but are lubricated regularly simply because the program doesn't provide for a proper schedule; that results in unnecessary lubrication and over-lubrication of that equipment. If your program doesn't allow for lubrication schedules over periods as long as five years that will happen. Grease is not cheap nor is the labor that's required to move around the plant and lubricate equipment unnecessarily so developing a suitable program normally pays for itself.

Lubrication is a function of operating hours more than anything else so a program for scheduling it suggests installation of recording operating hours of the equipment to determine when lubrication is necessary. I'm in favor of installing operating hour meters on everything. Tracking when equipment is in service in a log book is another way to determine operating hours.

Frequency of operation is also a factor and equipment that is started and stopped frequently should be lubricated more often than those that run continuously because the constant heating and cooling of the bearing results in swell and shrinkage of the lubricant and can result in air and moisture mixing with it to degrade the lubricant and rust the bearing. Systems that are oil lubricated also have a requirement for replacing the oil at frequencies that are based on the greater of operating hours or time. Grease is replaced with each lubrication so there's no additional scheduling to replace it.

It's that replacing of grease that many operators fail to consider. I don't know how many times I've seen someone slap a grease gun onto a fitting and pump away with no thought or concern for where the grease

that was in the bearing is going. That frequently results in the bearing shaft seals failing because the grease forced them to upset (Figure 6-3) and additional grease is forced out around the shaft or into the equipment housing.

Combine that with the common over-lubrication associated with grease bearings and it promotes equipment failure because the grease eventually blocks cooling air flow passes within the equipment. Invariably there is a plug or cap that can be removed to provide a passage for the old grease and that opening should be provided before pressing new grease into the bearing. Don't forget to put the plug or cap back after the bearing is lubricated and, when the manufacturer recommends it, the equipment is operated to stabilize the volume of grease in the bearing.

Use of the proper grease is also important. I've observed some facilities simply use the highest grade of grease required to simplify their activities thinking that if they use the best in everything they won't have a problem. There are two problems with that thinking, first it's expensive because the high quality grease is very expensive and secondly that high priced grease may not work well in the bearings that can function with the less expensive material.

Grease requirements are a function of load on the bearing and speed so a grease designed for a high speed low load bearing will not adequately support the larger loads of a low speed bearing. A lubrication program that's designed to be simple or make life easy for personnel can result in shorter bearing and equipment life. So... give up on the concept that you can use one grade of grease and lubricate the bearings in accordance with the manufacturer's instructions or the recommendations of your lubrication specialist. Painting a circle around each fitting with special colors to denote the grease to be used and applying similar paint to the

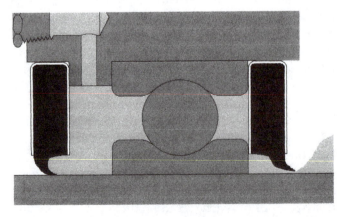

Figure 6-3. Grease seal upset by overpressure

barrel of the grease guns and tip will help to ensure the proper lubricant is utilized.

Another problem I see regularly is a failure to clean the grease fitting before attaching the grease gun. Use of a lint free rag to wipe off the fitting is recommended but it will not always remove the paint and other materials that manage to find their way onto grease fittings over time. If I had my way every grease fitting in the plant would be protected by a plastic cap that prevents anything getting on that fitting between lubrications. I would also still require the fitting be cleaned before attaching the grease gun. What if someone steps on the plastic cap or hits it with something and you find it off? If I had my way the grease fitting would be replaced before installing a new cap.

Eliminating contamination of the bearing with contaminated grease in the tip of the grease gun is also important. Always carry an additional lint free rag or small bucket to collect a small amount of grease from the gun before attaching it to the fitting. A quick shot into the rag or bucket will eliminate any dust or other debris that was picked up by the grease in the tip of the grease gun.

Sound like a lot more work? Perhaps you feel you aren't ready to go to all that trouble. The truth is that grease lubrication requirements are so infrequent that people I've convinced to establish a good grease lubrication program find they're doing half the work because they were lubricating their equipment too frequently. If you have a policy of greasing everything once a month, or more frequently, that's probably the case.

Oil, like grease, varies in its application and you must be certain you are using the proper oil for the equipment. A simple mistake involving oil can destroy a piece of equipment because one oil mixed with another can produce an incompatible mixture that loses all its lubricating properties. When that happens the mixture tends to split into a light fraction that is too thin to support the load and a sludge that settles to the bottom of the sump or plugs up the pump and filters.

Every piece of oil lubricated equipment should be marked to clearly indicate which oil is to be used in it. Refrigeration oils are addressed in Chapter 5. With all the changes in lubricants I wish someone would come up with some standard color coding (like those for refrigerants) to make it a little more difficult to use the wrong lubricant.

Preventing contamination of the oil in your equipment by adding contaminated oil is very easy. Oil interacts with its environment more readily than grease so you should always take every possible measure to protect oil in storage and en route from storage to the equipment. Many modern oils can absorb moisture and must be kept sealed until they are put to use. If your equipment contains an oil heater then the oil will probably absorb moisture right out of the air, contaminating itself if it isn't kept in sealed containers.

Oil, unlike grease, can be cleaned and rehabilitated while still in the machine. In addition to oil strainers and filters a lubricating system can contain water separators, magnetic separators, heaters and coolers to maintain the oil at its optimum operating temperature, and settling tanks to allow removal of solids and contaminants. The expensive oil is maintained by these systems to reduce the cost of regular replacements but it requires attention to maintenance of the oil systems.

If there isn't an oil maintenance system you may also have the option of an oil maintenance service, a company that will pick up and refine your used oil and give you credit toward the purchase of new oil. Regular testing of the oil in those systems is essential to ensuring proper system operation and maintenance of the lubricating quality. Normally the testing of oil (tribology) is performed by outside laboratories that have all the required equipment. The oil is tested for water, acidity, lubricating properties and microscopically. The examination by a skilled technician with a microscope can identify all the particles in the oil to reveal impending bearing failure or problems with gears or other parts of a machine.

Maintenance of oil lubricated equipment requires more attention than grease lubricated ones because the oil is exposed to the air in the plant. Grease systems are basically sealed so air doesn't contaminate them, that's why some grease lubricated bearings can go 40,000 hours, which is close to five years, without re-greasing. When equipment starts and stops it breathes because the oil and air heat up then cool off to change volume so air has to bleed out then is drawn in. The grease changes volume but it's normally such a small change that those seals expand and contract with it to prevent leakage of contaminants in or grease out.

Systems with oil temperature control will also breath with changes in load because the temperature of some of the oil increases and decreases depending on the load. Therefore equipment that is subjected to frequent stops and starts or varying loads requires more frequent checks of the oil than those that operate continuously. That's why you will frequently see an accumulation of oil around an oil sump vent, it's condensed vapors that were pushed out of the vent filter as the system breathes.

If you, or your boss, object to the accumulation

of oil around the vent you can try putting an extension pipe on it, raising the vent at least three or four feet. If you would like a more engineered design you can calculate the change in volume of the air and oil in the system then put on enough pipe to provide that volume. Overhead clearances may prevent extending the pipe at its connection size but that doesn't prohibit you from adding a couple of reducers and larger pipe to the extension to get the volume. The concept of this solution is to create a vertical settling space where the oil that would normally settle on something outside the vent settles in the piping to leave a volume of air substantially free of oil to flow out of the vent.

A simpler solution is to carry a rag with you and keep the area around the vent clean; observation of the oil around that vent can give you an indication of a change in the condition of the oil in the equipment so it may be a better way.

Oil has to be changed in any system that doesn't have its own conditioning equipment just like your car. Also, just like your car, there are rules of thumb that are wasteful. Most cars don't need an oil change every 3,000 miles but that rule of thumb is treated as inviolate. I change oil in my car every 7,500 miles unless I happen to do some driving on dirt roads or in similar dusty conditions when I think it prudent to change the oil right after that situation. No, it's not my idea, that's what the instruction manual says to do.

The instruction manual for the equipment will provide some guidance but you can judge the need for an oil change yourself by noting the condition of the oil. You don't have to be a tribologist to tell that the oil needs changing more frequently when you see distinct changes in color or particles in the oil before it's due to be changed. A problem with water supply to the cooling system that resulted in a significant rise in oil temperature should be followed immediately by an oil change or testing to see if it needs changing.

Other indications include presence of a whitish waxy substance that indicates water has contaminated the oil. The opposite isn't necessarily true however; just because the oil looks good you can't be assured that it's okay. If the cost of the oil and labor to replace it is not significant (less than $100 per year) then you might as well change it according to manufacturer's recommendations. If the cost is significant you should employ the services of a tribology lab to test the oil and make recommendations for changing it. I know of systems that have operated 100,000 hours without an oil change. A manufacturer's recommendations are normally based on the most severe use and the wise operator makes every effort to ensure the equipment isn't overloaded, or abused, so the oil can last longer.

Replacing organic oils with synthetic ones can reduce wear and power requirements for equipment. In addition, the synthetics last much longer than the organic oils. There are balancing factors in the additional cost of the synthetic oil and reduced power and maintenance costs. If you're changing large volumes of oil in equipment on a regular basis (less than annually) a hard look at synthetic replacements is recommended.

Oil lubrication systems require maintenance of more than the oil. Filters have to be changed along with the oil and more frequently in some systems. Coolers need to be cleaned on the water side to prevent fouling and maintain heat transfer. Temperature controls must be checked to ensure they're operating properly and maintaining the right temperatures. Centrifugal separators and the like have to be maintained according to manufacturer's instructions.

Anything that affects the temperature of a lubricating system is critical to continued safe and reliable operation. If a lubricant gets too hot it will break down and lose its lubricating properties to allow metal surfaces in the equipment to rub, gall, and scrape with failure occurring rapidly. That's why you're told to log an oil temperature that is always the same. The purpose is to notice when it suddenly does change so something can be done about it.

Cleanliness is the next important factor because clearances in bearings and gears are so small that a particle of dust that's almost invisible in the air can span the clearance to produce damage in the equipment. Any opening into a lubricating system should be fitted with a filter and systems should not be opened unless provisions have been taken to prevent dust and dirt getting into them. A little contamination of a lubricating system can result in total system failure costing thousand times more than the oil.

INSULATION

Insulation is one of those items that, for whatever reason, never gets the attention it deserves. It's not uncommon for me to be called to a plant for complaints of high fuel bills only to find that half the insulation has fallen off. You'll recall the story about rain load in the section on knowing your load; that was because of lack of adequate insulation. Burning fuel unnecessarily because the insulation isn't maintained is not what a wise operator does.

Any discussion about insulation raises the concern for asbestos bearing insulation contaminating the air in the plant. While many facilities have spent the fortune it costs to remove asbestos bearing insulation others have chosen to encapsulate it. If your plant is one of the latter then maintenance of that encapsulation has a priority. Damage to the cover can occur as a result of normal operating and maintenance activities, from vibration that occurs during normal operation or a plant upset. A tour to check the integrity of encapsulation should be performed on a monthly basis.

When it becomes necessary to gain access to something covered by Asbestos insulation you should notify your employer so he can have the insulation removed unless you have been trained to do it. The laws regarding asbestos bearing insulation do permit removal of small quantities without all the environmental controls required of a major material removal; and you could be trained to do it. If you are, follow the rules you were taught in the class. If not, and you think the contractor doing the removal is contaminating your air (lots of dust blowing around isn't to be accepted) scream and holler because once you've breathed it in it's yours for a lifetime. Once the work is complete make sure the asbestos that remains is encapsulated and don't forget to mention its removal, and who did it, in the boiler plant log.

Whenever insulation is removed for maintenance or repair make certain it's put back or replaced. I, if nobody else, will have a very low opinion of your maintenance practices if I come into the plant and find little bits of insulation missing here and there. Small areas tend to become bigger and, after a while, the whole system is bald. Not only is it a waste of energy, it's hazardous because you could be severely burned.

I was in one plant where I suggested the customer do something about his insulation for another safety reason. It had received no attention and was literally falling off the pipes. The hazard was associated with being hit on the head by falling insulation! Such instances aren't uncommon and they lead me to recommend you never accept an insulation job that consists of nothing but stapling up ASJ (All Service Jacket, that white paper like material with the flap that comes on most insulation) because it won't last. The staples eventually corrode and fail with the rest of what happens being most obvious.

At the very least piping insulation should be secured with minimum 20-gauge galvanized wire wrapped around it, twisted, and bent back against the insulation (to prevent the sharp ends catching or cutting anything or anyone) twice on each section. For longevity a light canvas wrap impregnated with a waterproof mastic will look better and will last even longer.

Outdoors and in areas where the insulation may be struck by people carrying objects such as ladders the corrugated aluminum jacket with aluminum straps and fasteners is necessary to provide long life. Long runs of hot piping pose a special problem, the pipe expands but the insulation doesn't expand anywhere near as much. And the jacket, particularly outside in cold weather, can shrink from its original length. When restoring insulation on long runs try to compress the existing insulation as much as possible without crushing it then compress the new material as much as possible when installing it; jackets should have a minimum overlap of three inches outdoors and the longitudinal seam should always be on the side of the piping lapped down to prevent rain entering the seam. On vertical runs of pipe make certain any jacketing is lapped to shed water. Do it indoors too because a leak can always spray water, or worse, all over the place.

Large flat surfaces require the installation of insulation studs, wire secured to the surface by stud welding or a special machine that shoots the wire into the surface. The studs hold insulation with special washers over the stud pressing the insulation against the equipment surface. An impregnated canvas covering or corrugated aluminum jacketing is necessary to protect the surface of that insulation. Any repair job should return the insulation to a like new condition using one of the methods I described.

What do you do if some insulation gets wet? If it got so wet that it collapsed it has to be replaced, otherwise let it dry. If it got wet while the pipe was out of service and the line contains steam or hot water you should warm the piping up very slowly or you may generate steam under the insulation that will blow it off.

Damaged or compressed insulation should be replaced as part of the annual clean up operation. Where the damage is repeated some consideration should be given to installation of better protection of the insulation, consider replacing or covering the jacket with heavy galvanized sheet metal thick enough to ward off the damage.

No, I don't want to hear the argument that it doesn't make any difference if the piping is only used during the heating season and it heats the building anyway. The heat lost through lack of insulation is almost never able to heat the space as intended. It's almost as weak an argument as the one that I'm always hearing which is "it's only a little bit." Little bits become lots when that attitude is taken. "We're out of that thickness" is another unacceptable argument; put something thicker on it! The energy lost in the month or more it takes someone to get around to ordering the right thickness will pay for the additional thickness.

Speaking of various thicknesses, it doesn't pay to maintain an inventory of multiple thicknesses, get pipe insulation in one inch increments, one, two, and three (if you need three inch) etc., and layer it for greater thicknesses. Limit your stock of one-inch thickness to pipes two inches and smaller. For flat and large diameter surface insulation all I would keep is two-inch thicknesses. Your inventory should also be limited to the insulated pipe diameters you actually have in the plant.

Be cautious with insulation on or near piping containing flammable liquids such as fuel oil. The insulation can absorb it like a wick to become a fire problem later. Insulation in the area of fuel oil pumps, strainers, burners and such other places that could be splashed by a leak should have full aluminum jacketing over a mastic impregnated covering to prevent a leak or splash soaking in.

Insulation for refrigerant piping, chilled water, and some ductwork has the added requirement that it be properly sealed. If there is an open path for water vapor to flow through the insulation to reach the cold surface of the equipment, piping, or duct it can condense there, soaking the insulation to reduce its effectiveness, corrode the metal, and occasionally harbor mold growth. Cellular glass and foam insulations are inherently vapor proof but they only work if their seams are sealed tight. When you observe damage to vapor-tight insulation try to take the system out of service for a while to allow any condensate to evaporate before restoring the vapor-tight seal.

Re-evaluate your insulation once in a while. The old rule that says it should be insulated if you can't hold your hand on it still applies. The only thing you should not add insulation to is any part of a boiler casing.

The wise operator maintains the insulation in his plant. The argument that the owner won't buy any insulation is easily covered. Explain to the owner that you're paid to be there anyway so the cost of material for repairing or even adding insulation is recovered in fuel cost in a couple of months. The owner might even consider boosting your salary a little with what is saved after that.

REFRACTORY

Refractory is unique material in one regard be-

cause no manufacturer will absolutely guarantee their material will remain intact. Materials exposed to the high temperatures of a furnace are also subject to components of the fuel that become very caustic or acidic at the high operating temperatures. Some components of fuels produce considerable damage with vanadium being particularly offensive.

Vanadium is common in many of the heavy fuel oils and has a particular means to damage refractory. Vanadium pentoxide is molten at flame temperatures and as low as 1200°F. It remains molten at the refractory walls and soaks into the refractory during boiler operation. When the burner shuts down the materials cool and the pentoxide solidifies. Being a metal oxide it shrinks at a different rate than the refractory. The difference in thermal expansion, where the pentoxide soaked layer shrinks more than the regular refractory, creates a shear plane between the two materials where they pull apart. The result is breaking off of a layer of the refractory from one quarter to two inches thick, a process we call spalling. The damage is very evident on inspection of the furnace because the pentoxide soaked layer has a glossy black appearance and is spotted with light tan areas where the pieces of refractory spalled off.

Yes, refractory does expand and contract with changes in temperature. It's nowhere near as much as it is for metal but it does grow and shrink and that must be accounted for. I've known operators to try repairing every crack that appears in the refractory in their boiler's furnace on each annual outage and, as a result, accelerate the damage.

I have a rule that says any crack that is smaller than a number 2 pencil, where you can't put a sharpened pencil in up to the yellow paint, should be left alone. Those are expansion cracks and will close up as the boiler heats up. Plugging larger cracks, as much as three-quarters of an inch, with hard refractory materials isn't recommended. Today we have access to ceramic fibers rated at temperatures as high as 3200°F that should be used to fill those cracks. The ceramic fibers shouldn't be packed into the crack to the extent that they're solid, leave it soft so there's room for the major pieces of material to expand into the crack.

In my days of operating we used asbestos for such repairs and you could encounter asbestos in joints and cracks of refractory in an older boiler. If you have good maintenance records you'll know what you're getting into but, lacking data, treat any fibrous material as asbestos until such time that it's proven it isn't.

One important location for providing thermal expansion is around the burner throat on oil and gas fired boilers, also pulverized coal burners. The throat material is usually rated for very high temperatures because the throat is closest to the fire and will be the hottest refractory in the furnace. Those of you firing gas know that the throat is glowing cherry red when the boiler is in operation. Actually it's always red hot, regardless of the fuel, you just can't see the glow with pulverized coal or oil fires because the bright fire lights up the furnace.

Throats are either made up of pieces of a pre-fired refractory material we call "tiles" or a plastic material. When we use the word "plastic" in discussions of refractory we mean a material that can be molded and shaped as desired until it is dried. Plastic refractory has the consistency of stiff clay and looks and feels like mud with lots of sand and fine gravel in it.

Either of the throat materials will expand considerably during boiler operation so there should always be some form of expansion joint around the throat. I've seen many installations of plastic refractory where the throat and burner wall were monolithic (all one big piece) and they do manage to stay intact for quite a while despite the differences in temperature; I just prefer separating them because a prepared joint provides a perimeter for expansion and eventually, a repair.

A problem we used to have, and one that I'm certain is still possible, is sagging of a plastic refractory wall which bears down on the burner throats to distort them. I still insist on a "bull ring," a circle of special pre-fired arch brick or tile around the burner throat that supports the wall and prevents it's weight bearing down on the throat tile. The bull ring should be designed to provide a half inch gap between the inside diameter of the bull ring and the throat tile which, today, would be packed lightly with ceramic fiber.

If you find yourself repairing your burner throat

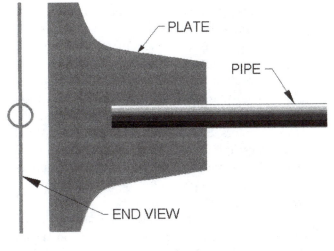

PLATE

PIPE

END VIEW

Figure 6-4. Throat sweep

again you might give serious consideration to rebuilding the entire thing to get that flexibility. Burner throat repair and replacement is best left to the experts, men and women skilled in installing the materials because it isn't easy to properly position throat tile so you get a perfect circle or shape a refractory throat in perfect form along the sweep.

Sweep? That's a special tool used to shape a burner throat out of plastic refractory. Normally it's a piece of flat steel plate welded to a pipe that fits into the oil burner guide pipe and cut to produce the form of the burner throat. (Figure 6-4) I had one on one ship that consisted of several pieces which, when assembled, formed the burner cone completely with four scraper bars and it was designed to spin into the packed plastic to produce a finished throat. I can also remember that a refractory crew in a foreign shipyard thought they didn't need that sweep to form the throats and I ended up going back into the boiler to replace their work shortly thereafter because they produced a completely different shape. If you have plastic throats make certain the installers use that throat sweep and use it properly.

If anyone tries to sell you a refractory "maintenance coating" kick them out of your plant. I may incur the wrath and ire of some manufacturers and salesmen that believe they're providing a valuable service but I don't care. So called maintenance coatings don't do squat as far as I'm concerned and I've never seen them do anything good, they're usually quite harmful. Those materials are, in some instances, nothing more than mud somebody dug up. Higher quality materials are seldom matched to the refractory in your boiler so their thermal expansion rates are matched. The result is that much of the spalling I've seen is just the maintenance coating breaking away. It also fills the small cracks that provided for expansion to create stress on the face of the refractory.

Another regular problem with those materials is they are applied carelessly. In many of the situations where I've been asked to help with problems with firing gas I've found the openings in the gas ring partially blocked with that so-called maintenance coating. Instead of spending money on that junk put it in the bank to pay for a complete replacement of the refractory some years in the future. If your refractory is suitable for the application there will not be any serious degradation unless you create it.

You shouldn't encounter all the problems I had with refractory because the materials and installation methods have improved considerably in the past forty years. If you do have a forty year or older boiler you may be seeing them but modern boilers with mostly water cooled walls will have very few refractory problems.

The one difficulty with modern boilers, especially the 'A' and 'O' type package boilers is retention of the refractory seal where tangent or finned tubes are offset or lacking fins next to the boiler drums. Those sections consist of very small pieces of refractory with very little to hold them in place and, for those particular boilers, the grip has to overcome gravity so their weight is a factor. The best way to repair those is to completely remove a section and replace it. You'll find that new material doesn't bond to old refractory at all. As the new material cures and dries it shrinks and simply pulls away from the old material.

Any refractory repair that isn't just for a short term should consist of complete replacement of a section with adequate provisions for expansion. That repair will last. Patches are exactly that and they don't last. Don't be afraid to improve on an installation either. If a repair is made because a furnace wall buckled into the furnace you should improve the anchoring as well as provide for thermal expansion. Either lack of anchoring or buckling due to thermal expansion was the cause of the failure so take measures to counter both problems.

Any temporary patch has to be anchored or it will be more temporary than you intended; falling out as soon as the boiler heats up. Since the repair material will shrink a little as it dries. It doesn't matter how hard you hammer on the wet plastic refractory material (or how thick any slurry of castable refractory is) it has to be anchored somehow. Castable, by the way, is a pow-

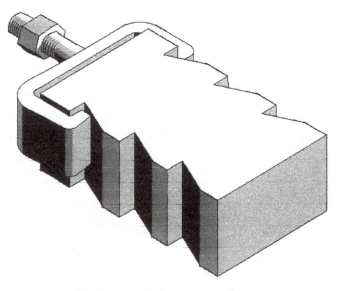

Figure 6-6. Refractory anchor

der that's mixed with water to form a very dense soupy mixture that can be poured into spaces surrounded by forms. Small areas, less than sixteen inches in diameter should be "keyed in" to the existing material. That's accomplished by undercutting the face of the existing material (Figure 6-5) so the patch is wedged between the edges of the existing material and the casing insulation.

Larger patches should be anchored by installing a refractory anchor (Figure 6-6) secured to the casing or brick setting so the patch is secured and will not tend to crack and buckle out as it's heated. Refractory anchors should be installed within 18 to 24 inches of each other if you don't have a successful wall to compare to.

Almost any refractory repair requires a "dry-out" as described in the chapter on new start-ups. If the repair consists of brick or tile laid up dry, a common arrangement for sealing the furnace access opening on many boilers, then there's no need for a dry out because there is no moisture imbedded in the refractory. Anything else will have to be dried out.

When the patch is made with plastic refractory the dry out will be accelerated if you provide vents in the material. You provide vents by poking the material with a small welding rod to produce small round holes about two-thirds of the thickness of the wet material on three to four inch centers. Steam forming in the material will then have an escape route. If the repair is due to vanadium pentoxide damage the venting isn't recommended because it will provide places for the oxide to soak into the refractory.

Some refractory materials are labeled as air drying, some are heat drying but most are combination air and heat drying. A heat drying material reacts to a small de-

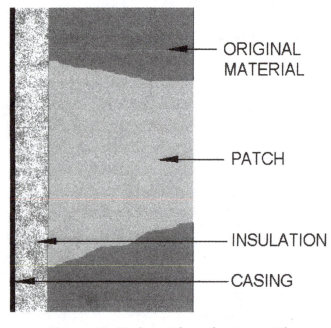

ORIGINAL
MATERIAL

PATCH

INSULATION

CASING

Figure 6-5. Undercut for refractory patch

gree with the water that's in it to create another chemical that helps bond it together. When using heat drying material it's important to avoid letting it air dry. You should fire up the boiler to apply the heat in accordance with manufacturer's instructions as soon as possible. The best option is to use a combination material and it's always important to treat all of them gently so the repair isn't destroyed in its first few hours of operation. Bring the boiler up to operating temperature as slowly as possible.

PACKING

A lot of modern designs and new materials are eliminating packing as I know it but it will be a long time before you won't encounter a pump, a valve, or other device with packing. Packing is material pressed into a space between a metal housing and a metal shaft to provide a seal to prevent or control leakage of water, steam, or another fluid.

I trust you noted that I used the words (or control leakage) because in many pumps that's very important. I've run into many a new operator or maintenance technician that was thoroughly convinced that the packing on a pump shouldn't leak and destroyed the pump by tightening the packing to stop the leak. Unless a small amount of fluid leaks along a constantly moving shaft to lubricate the shaft, and protect it from rubbing, the packing will cut into the shaft. If you ever see a pump shaft or sleeve reduced in diameter with gouges from the packing that's what happens.

Whether it's a pump, a valve, a control float, it really doesn't matter, there's a standard arrangement for installing packing. Many leaky valves I've seen consist of a repair where the installer simply wrapped packing around the shaft in a spiral, cut it off, jammed it in, and expected it to seal. That doesn't work. Packing should be arranged in cut segments that barely fit around the shaft stacked as shown in Figure 6-7. The stacking doesn't have to be precisely as shown, just alternate placing the open seams first 180 degrees out of phase then 90 degrees to produce a complex path for any leakage to follow.

It's actually better to have the packing rings cut a little short than a little long. If you have to jam the ends together to get the packing into the opening it will create a hard bump that can bear all the pressure placed on the packing gland so the rest of the packing ring isn't compressed and doesn't seal. If you jam ends when packing the gland on a gauge glass you've increased the odds

that the glass will break when you tighten the packing.

Packing of pumps usually includes a lantern ring (Figure 6-8) that has to be properly positioned in the packing gland. Always count the number of pieces of packing you take out from under one. The lantern ring provides a space for distribution of leakage into or out of the packing gland. When the packing is sealing the high pressure side of a pump the leakage into the space containing the lantern ring bleeds off to the pump suction, which is at a lower pressure. That recovers some of the fluid. The remaining packing, between the lantern ring and atmosphere is only exposed to suction pressure. For cooling and lubricating some flows between the packing and the shaft to the outside of the packing gland.

When the packing is on the suction side of a pump operating at pressures equal to or below atmospheric the lantern ring space is piped to the pump discharge or an intermediate pump stage. The purpose here is to provide lubrication of the packing and shaft plus sealing the pump to prevent air leaking into the fluid. That's important for condensate pumps to keep oxygen out of the condensate. Flow in that case is into the lantern ring space. It then splits with some flowing into the pump suction and the rest leaking out of the packing gland in the other direction.

Whenever you're re-packing a pump you should be aware that the gland could contain a lantern ring. I remember seeing one feed pump where the operators were not aware of the packing gland and had repeatedly pressed the packing down until the lantern ring was

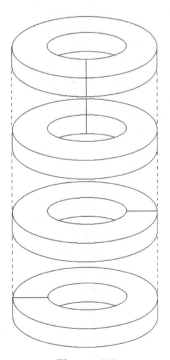
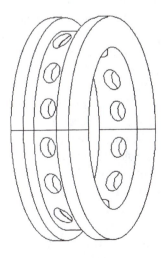

Figure 6-7.
Packing segment stack

Figure 6-8.
Lantern ring

pressed past the location of the bleed connection. They couldn't stop excessive leaking because the entire packing set was exposed to the high pressure water and the erosion along the shaft was getting worse.

The split in the lantern ring should always be set 90 degrees from the split in any pump casing to provide a clear indication that it's a lantern ring and not the bottom of the packing gland. If there's a piping connection at the packing gland I like to open it up so I can look into the gland while I'm re-packing it to make certain the lantern ring matches up to the opening. Sometimes you can get the count wrong when removing the packing because it comes out in pieces so it doesn't hurt to spend the extra time to make certain the lantern ring is positioned properly. Yes, I have had to take it back out to add or remove a piece of packing so the lantern ring is positioned properly.`

Packing of air actuators, compressors, etc., where there's no fluid for lubrication will have grease fittings or piped oil connections to apply grease or oil to lubricate them. These usually incorporate a lantern ring to distribute the lubrication. Those packing glands use the lubricant as part of the seal. It's important to follow the manufacturer's instructions with that packing because some have to be soaked in the oil or grease before installation in the packing gland while others have to be installed dry then "charged" with the lubricant before putting the equipment in service.

Valves and a few other pieces of equipment have very limited movement of the shaft through the packing so there is little need for extensive lubrication. In most cases the lubricant is part of the packing, typically graphite. There is no need for leakage of the fluid to lubricate the shaft. So, pumps and other devices with moving shafts should leak to a degree but valves and devices like a Keckley float controller shouldn't leak. The most important maintenance practice for those packing glands is to tighten the packing as soon as you see it leaking.

Every time you operate a valve check the packing gland afterward and tighten it immediately if you see a leak. Quick response to a leak can prevent the need to completely re-pack the valve. If that leak is allowed to continue it will cut through the packing, destroying it and making it impossible to seal by simply tightening the gland.

Since operators are the ones that open and close valves. And, since that's the only time the seal between packing and shaft is broken; there's no question that tightening valve packing is an operator's responsibility.

CONTROLS AND INSTRUMENTATION

Controls are the robots that do the boiler operator's bidding. Without them we would be very tired at the end of a shift because we would have to make every little adjustment that the controls make for us. Instruments are an extension of our eyes and ears to allow us to know what's going on in the process and it's important the information they give us is correct. It makes sense to maintain them so they keep doing their job. There's a separate section on the function and operation of controls and instruments in this book; this part is devoted only to their maintenance.

I'll go on several times in this book about how great the modern microprocessor based controls are; that's because they are. They make our jobs as operators so much easier than it was when I was operating boiler plants. They're almost maintenance free! You do have to make certain cooling is maintained by keeping dust and dirt out of the slots and vents of devices and panels and make sure they don't get wet but that's about it.

Speaking of getting wet, I've seen more control hardware lost to water leaking into panels than for any other cause. It never ceases to amaze me how we engineers manage to do such dumb things as lay out an entire control panel right under a shower room. It's also stupid to remove something from a panel and leave the opening for water to enter. I would sure like a nickel for every time I found a transmitter or control valve with the cover off because someone forgot or was too damn lazy to put it back. Even small conduit covers can admit water that can find its way into a control panel or device. The wise operator looks for such things on every round and does something to restore enclosure integrity when he spots a problem. He also carries a clean rag to dust off cooling vents.

Those of us that are still stuck with maintaining pneumatic controls know the most important thing to keep up is the air compressor, storage tank, filters and dryer. Makes sense doesn't it? If the compressor fails then the controls won't work. If the tank floods because we forget to drain condensate the controls get to try to work on water instead of air. If the filters get overloaded then the compressor won't work or the controls get to try to work on oil. The oil coalescing filter and dryer are there to ensure we have the clean dry air the controls manufacturer specified.

Without clean dry air all we can expect is control problem after control problem. Refer to the previous section on lubrication and make sure you always check the oil level in the compressor. Keep the fins on any air cooler, and the ones on the compressor head, clean so they reject heat the way they're supposed to. It's better to replace a coalescing filter a little early than to put it off until it's too late, once oil gets past that filter and into the system it will take what seems like forever to get rid of the oil problems.

If your pneumatic controls do get gummed up with oil you or a contractor will eventually have to clean them or replace them because the oil gets gummier as it dries and collects little particles of dust to really goo up the controls. If you simply ignore that problem you'll soon discover that efficient operation is impossible because the controls will always be hanging up. Hopefully I've put the fear into you and you will never fail to keep an eye on the oil removal system to ensure it's working.

What happens, however, when you inherit the problem? Say you just hired on in an old installation and discovered all the controls are spitting out oil, what do you do? The first thing I would do is try to convince the owner to replace the controls with microprocessor based hardware to eliminate all the problems with the old pneumatics. Failing that I would watch the systems for a while without changing anything. Some of the older pneumatic systems can work on oil or water; the old ratio totalizer seemed to be able to. I would hesitate to do anything about the oil getting into the system until I had a better understanding of how it affects everything. The expense of all the oil added to the compressor may help convince the owner to upgrade but that's not the reason to let it go on; fresh oil flowing through the instruments will flush them and limit gumming up.

Situations where the controls work anyway should probably be left alone, the only thing you can do is keep good records of the costs associated with the problem to give the owner a justification for replacing those controls. Switching a system by adding coalescing filters or other oil removal devices could result in system failure because the oil remaining in the instruments will gum them up.

Keeping the control devices clean, free of dust and dirt, oil and grease is the most important thing you can do. Electrical and electronic, including microprocessor based controls are subject to dirty power supplies as well. No, not real dirt, power with harmonics, spikes and all those other things that do dirt to electrical equipment.

Whenever a contractor tries to hook up a welding machine in the plant make sure a connection designated for welding machines is used. Be certain any welding lead is not run over or around control cabinets or conduit containing control wiring. If you test an emergency generator regularly you may find you need a UPS (uninterruptible power supply) on your controls to keep them from dropping out and doing stupid things (some set up by the logic designer) like restarting everything in manual. Actually I prefer a UPS on all electronic and microprocessor based control supplies because the UPS isolates the controls from the line and will protect the controls from surges and power line noise. It's like putting an oil-free compressor with a dryer on a pneumatic control air supply.

Today there's a lot of UPS systems designed for computers that can handle the normal control system load for a boiler. Putting one of those on your boiler control power supply will be well worth the little bit they cost.

Logging readings not only allows an evaluation of the continuing performance of the plant (see boiler logs) but also provides indications of instruments and controls losing calibration or operating inconsistently. Maintaining local instruments like pressure gauges and thermometers provides a reference for your control and instrument indications that can be used to identify problems and schedule control and instrument tune-ups.

You may be allowed to do the instrument calibration yourself. With the proper training, tools, and by carefully following the manufacturer's instruction manuals it's possible for an operator to maintain a calibration schedule during his normal shift. That is not only a big saving for the employer in contractor's costs it will help keep fuel and power costs down as well. I know, it sounds like I'm trying to keep an operator moving every minute of his shift with no time to rest… I am.

I did everything including polishing brass to make my shift seem to go faster. Just sitting there listening to the plant gets boring and makes the time pass slowly. Count your plant instruments, transmitters, controllers, etc. and multiply by four hours then compare the result to the 2,000 hours you normally spend in the plant (not counting overtime) and you'll see that it's not that big a deal. Many plants are manned around the clock so there's over 8,000 hours to share operating and maintenance time. You'll have at least three other people to take their share of the work load.

Tuning firing rate controls isn't always something an operator can do. There's a certain amount of skill and experience required to do it without blowing the boiler up. You can do it if you you've had hands on training under the watchful eye of an instructor and that instructor tells you that you have an aptitude for it and can do it. I'm not confident that I can put enough guidance in a few paragraphs of a book to guide someone through the process and refuse to let anyone tune a boiler until I've watched them do it. That's because I've discovered many an operator that just doesn't get it and can't tune a boiler without turning a screw the wrong way or too much to create a dangerous condition. If I'm not confident about someone I just taught in a class I'm sure not going to count on somebody that's only read this book.

If you choose to tune the controls of a boiler without hands on training I can't stop you but I will say that you're taking your life in your hands. One of my service technicians who just retired after thirty two years in the business was given the nickname "Boomer" for obvious reasons. He was present for two boiler explosions that I can remember and several heavy puffs plus had a plant burn down shortly after he left. All that despite his skill. In every incident that I investigated, and several I heard of, he wasn't the one that created the unsafe condition. A lot of them occurred due to operator action before or after his visit. Unless you have the training to add to your confidence, and the confidence of a qualified instructor, I would strongly recommend you let the experienced tune your boiler.

Pressure and draft gauges require maintenance to insure their readings are accurate and reliable. All pressure and draft gauges in the plant should be checked for calibration every five years. If the gauge is observed constantly swinging (the needle is moving constantly) or it is subjected to frequent bumps (like the discharge gauge on an on-off boiler feed pump) they should be checked more frequently. The sensing lines of the gauges require more attention than the gauge itself. Lines to gauges (provided the gauge is protected by a siphon) should be blown down at least once a year and that blow should be long and large enough to fully flush out the piping.

Draft gauges should be checked for zero every time the boiler is shut down. There is little pressure available to blow them; don't use compressed air because it has little effect and it's too easy to damage the gauges. Draft gauge lines are normally fitted with tees and crosses that permit cleaning them with a wire brush attached to special fiberglass extension rods; if they're dirty that's the way to clean them.

Another important annual operation is to ensure there's an air cushion in pressure sensing lines that are supposed to have them and no air in sensing lines that shouldn't have it. Air in a sensing line can act like an accumulator, compressing when pressure is applied to the system to take on liquid then expand when the system is shut down to push the liquid back out. That's not a good thing for something like an oil burner gauge because the oil that is pushed back out will allow continued firing of the burner when it isn't supposed to be.

With heavy fuel oil make sure the sensing lines are full of the separating fluid by pumping some through the sensing line during start-up after the annual inspection. Light fuel oil and other liquids that burn are best for this.

LIGHTING AND ELECTRICAL EQUIPMENT

Yes, in many plants you're also the one that has to change the light bulbs, so do it wisely. With modern lighting technology there's more choices in lighting and you should take advantage of them. Many of the modern lighting fixtures are energy efficient but will not pay for themselves in electrical savings because they cost so much more. So what! A fluorescent bulb has an average life of about 10,000 hours, five times that of an incandescent. All you have to think about is the value of your labor to replace one of those bulbs five times and the owner should be willing to pay the higher price.

Compact fluorescents, those curly bulbs, are becoming so common that their prices are dropping; so they will pay for themselves in energy savings in less than a year, on top of your labor savings. Typically you can replace a 60 watt bulb with a 17 watt fluorescent. Use that ratio to get an idea of the right size. LEDs are another story, very expensive but they have a life of about 100,000 hours (over ten years of continuous operation) so they're really invaluable for those applications where the reliability of the light is important. They take about one quarter of the power of an incandescent bulb for comparable illumination and even less in applications that are not involved with illumination so, with the extended life, are fantastic for applications like control panel indicating lights.

When I was designing and installing burner management panels I always made sure I had spare light bulbs because one would always blow. I insisted on testing every new system on a simulator in the shop before it went to the field. That way I caught all the little surprises before fuel went in the furnace. Almost always, after a couple of days of testing, one or more indicating lights would fail. Some of that problem was solved by going to transformer type lights but the best solution is those LED indicating lights.

When it comes to a question of what's happening because a light burnt out the reliability of LED lights overshadows all the arguments about the little bit extra they cost. I would rather buy new LED light assemblies than spare incandescent light bulbs.

Some operators are expected to perform normal checks and maintenance of electrical equipment in addition to maintaining the boiler plant. I don't expect you to pull wire or perform other functions that are appropriately performed by an electrician but… in many cases it won't get done if you don't do it. Changing light bulbs and performing the following maintenance functions can make you more valuable to your employer. It's also possible it will save you being called out in the middle of the night to start up the boilers after an electrical malfunction.

Contrary to popular beliefs, electrical systems require maintenance. You may think the systems in your house are so reliable you don't have to worry about them. I thought that way until I spent a cold Christmas Eve working on an outside receptacle (where you put the plug for Christmas lights and your electric hedge trimmer) to restore power and lighting in all the bathrooms in the house. One wire had come loose from the receptacle and all the power to the bathrooms was routed through it. The circuit breaker kept tripping because it was a ground fault interrupter and that complicated finding the problem. I don't expect you to fix a problem or even find one but some regular maintenance activities would have saved me freezing that night while relatives were using candles to go to the bathroom.

Those ground fault interruption devices, called GFCI for ground fault circuit interrupter, all have a test push-button on them. No, they're not there for the electrician to use, they're there for you to test the darn things on a regular basis. Instructions for the smaller units say to test them monthly. So, to protect yourself from shocks, and both you and your employer from a very expensive lawsuit, do it! Record the test in the log though. Don't use those little stickers that come with the breakers.

Insert a test light or some other device that is obviously using power to determine if the device passed the test for certain. When you're confident that everything powered by the circuit can be shut down, push the test button. The test light should go out and then come back on when you push the reset button or reset the circuit breaker.

GFCI circuit breakers trip without shifting the operating toggle all the way to the off position, just like a normal circuit breaker when it trips, so you have to turn it off and back on. The GFCI has current detection devices in them to compare the current going out the hot conductor and the current coming back on the other conductor; if the two currents don't match precisely it trips. Smaller GFCIs are also called personnel ground fault protectors because their real purpose is to prevent anyone that accidentally touches a hot electric wire or any conductor (metal, wire, copper pipe, whatever that will carry electricity) while in contact with a ground.

The concept of grounding needs some clarification. Grounds in electrical terms are conductors that are not supposed to carry electrical current but they can convey it to the ground, the dirt below you. A concern in any installation is the lack of grounding, where a conductor that's not supposed to carry electricity is not connected to the ground, it's ungrounded. The concern with ungrounded conductors is they can become hot by coming in contact with a hot conductor.

A hot conductor is anything in an electric circuit that is designed to carry electric current and there is a difference in voltage between it and ground. If you touch the ungrounded object and your feet are on the ground you can close an electrical circuit between the hot conductor and ground. Electricity will flow through you and, if the current range is right, it will kill you instantly. If it's low voltage (less than 600 volts above ground) it shouldn't kill you but it can cause everything from a mild shock to severe burns.

Personnel GFCIs will sense the fact that the current is going to ground (because of the difference between the currents in the two conductors) and trip before the current reaches a value that could give you a tickle. Regular testing of those devices helps to shift dust and debris that can settle in the mechanism and prevent its operation. Personnel GFCIs are very important in a boiler plant because you have a lot of grounds around you. All receptacles in a plant should be fitted with personnel GFCIs because everything around you is grounded (or should be) and if an electric tool or trouble light you're holding has its hot conductor short to something you're holding you want that device to prevent you getting shocked.

Larger GFCIs (in current carrying capability) are required because a current flowing through devices not intended to carry current can overheat them to the degree that they burn or explode. Look at the thickness of the metal in any large electrical panel compared to the size of the wiring supplying it. If the current were to suddenly start flowing from the wiring through that thin panel to ground it would damage the thin metal in that panel. Those devices should be tested regularly by an electrician and you should record it in the log.

Operating circuit breakers has the same effect as GFCIs, you help ensure they will function when necessary by keeping them loose. It's always a good idea to open the circuit breakers in addition to disconnects when servicing equipment so add them to your lock-out tag-out procedures.

Maintaining grounds is a constant problem in many plants and I always rely on the eyes and skill of operators to spot problems before they become serious. A common way to ensure a good electrical connection between steel building structures and the ground is installation of a grounding grid and bonding. A grounding grid is a pattern of copper rods laid out in the ground around and under a building to provide good contact with the earth, they are welded or mechanically attached to each other and to bonding jumpers that extend to the building structure.

Bonding is the process of installing jumpers connecting one piece of metal to another to ensure electrical current can flow from one to the other. If buildings were not grounded lightning could create thousands of volts of potential between the building and ground, let alone the static electricity differences in a building from a cloud passing over it. If you touched the building with your feet in contact with the ground, well... you would become the grounding conductor.

Look around at the bases of steel columns and you'll see an occasional wire run up through the concrete to an attachment on the steel, that's a bonding jumper. The connections can be mechanical or the metals can be fused using a thermite welding process. Thermite welding creates a puddle of hot molten metal that attaches itself to the steel and wiring. The bonding wires serve as the bonding jumpers because there's no guarantee that the anchor bolts, nuts, and column bases will maintain electrical continuity.

The problem with those connections is they are exposed and can be broken loose by any number of methods. Your effort should simply consist of noting any damage to one and repairing it or having it repaired immediately. Caution is advisable because there could be a voltage difference between the two so always make certain you have no voltage difference before attempting to restore a connection and be aware that any number of incidences in and around the facility could create a difference, including a cloud passing over.

I'm particularly concerned with grounds in and around boiler systems because we're dealing with so much steel and water, all good conductors of electricity (well water normally is) and lack of a ground invites problems with control operation. The deadly explosion of a boiler at the New York Telephone Company in 1963 was associated with ground paths bypassing some limit switches so the boiler continued to fire and build pressure until it exploded.

To ensure that can't happen again all control circuits must have one leg grounded and all final devices (control relays and fuel safety shut-off valves) have one side connected to the grounded conductor. (A grounded conductor is a wire for carrying current that is connected to ground at one point to ensure its electrical potential is the same as ground) Any ground that forms in the control circuit should produce a fault that will trip the fuse or circuit breaker. If that doesn't happen the ground should produce a short circuit between the fault and ground so there is no voltage across the associated relays or safety shut-off valves to keep them open.

Of course, if the conduit or other parts located where the wiring insulation fails is not grounded it not only becomes a point of high potential that can cause personnel injury. It's also a conductor that can bypass some of the limit switches on the boiler. To ensure there are no inadequately grounded metals around a boiler an annual check should be made of their resistance to ground. Using a simple multi-meter set at the lowest resistance setting and one very long test lead check the resistance between the grounded conductor in the burner management panel and every metal object (except wiring) on and around the boiler. The resistance should be less than 5 ohms everywhere. Usually you will find the resistance is less than one ohm with 0.3 to 0.5 being common. I chose 5 ohms because a little more resistance can produce enough potential to keep a small control relay energized.

Just like you check motors for overheating bearings, you should check out your electrical panels and switchgear for loose connections that generate heat. The wiring can loosen especially when the equipment is started and stopped frequently because the wire does heat up a little bit every time it runs and that results in expansion and contraction of the metal that can loosen the connections.

Loose connections are very common with aluminum wiring because aluminum has a larger coefficient of expansion than copper. During a normal round you just lay your hand on the front of each panel and compare what you feel to previous rounds. With large panels it's a good idea to sweep your hand over the front to note hot spots which are indicators of loose connections. If you detect one plan to shut down that equipment to correct the problem... before the equipment picks its own time to go down!

Prior to annual inspections you should perform a detailed examination for hot spots at connections, opening panels whenever possible and scanning all connections with an infra-red thermometer to find any hot spots. On a five-year interval you should open all peckerheads at motors to check the motor connections and open rear covers on motor control centers to check the bus bars, make that two years if they're aluminum. You don't even have to check connections in your home, shut down the circuit and tighten them, there aren't that many. You may find that regular annual tightening of aluminum conductors is required, my kitchen stove and heat pump have aluminum wiring and I check them annually.

High temperatures are the worst enemy of electrical systems. There is a rule of thumb that claims the life of electrical equipment is halved for every ten degree increase in temperature. It's important that you do what you can to limit the temperature of the electrical equipment you operate even if you don't maintain it. It's a simple matter of keeping cooling passages clean and unobstructed.

Don't let painters lay their drop clothes over operating pumps or electrical enclosures so they block the flow of cooling air. I've noticed a fresh coat of paint on and around electric devices that failed is very common. In one case the coat of paint actually froze a motor bearing on its shaft. Regular cleaning of vent screens, louvers, and the like will prevent blockages that could kill your equipment. Always use a vacuum to clean them, blowing air and brushing simply loosen the dirt and allow it to flow into the equipment, not keep it out. Use a damp rag for removing dust from the top of motors and electrical enclosures so you pick it up instead of brushing it off and into the vents.

Okay, somebody jumped on it. A wet rag! I don't want to get electrocuted! First of all, let me dispel one myth that's always perpetuated by Hollywood. If you're in a bathtub full of water and someone drops an electric appliance in the water you are not automatically electrocuted. You can only suffer harm if the current passes through you and the only way the current can do that is if you are in the circuit between the electrical appliance and the water which serves as a grounding conductor. You have to touch the electrical device and the current has to flow from you to the water to do you any harm.

The concern in bathrooms and kitchens is that the water is there, contacting drain piping, etc. and is a ground which you can contact at the same time as a hot conductor. Re-read the above on GFCIs; that's why all new bathrooms have to have them. Electrical enclosures and motor housings should be grounded, not hot, so a little scrubbing with a damp rag can't cause a problem. If you're using a soaking wet rag that's squeezing water out and into the electrical appliance to become a conductor between hot and ground you could get stung but a damp rag can't do that.

Transformers are frequently allowed to die for lack of maintenance and it's a shame that so many of them are neglected because they not only represent a significant repair or replacement cost; there's the matter of the downtime associated with their failure and the very large and very real additional cost of power that's wasted when the transformer is operating inefficiently. Whenever a transformer can be taken out of service you should use the opportunity to maintain it. Opening the enclosure and removing accumulated dust and dirt then inspecting it for apparent hot spots and tightening all the connections is the minimum you should do.

Samples of oil from oil filled transformers should be drawn and sent to a qualified testing lab at least every five years; the lab should provide you with sampling kits. Refer to the manufacturer's instructions because there are a variety and forms of transformers with different requirements. You also have to be careful with some real old transformers that may still contain PCBs, a known carcinogen.

During the operation of the transformers a regular cleaning of any external fins should be scheduled based on an observed difference between metal temperature and ambient air. Also make sure you maintain the ventilation equipment for any electrical enclosure, it's a lot easier to replace a hundred dollar exhaust fan than several thousand dollars worth of transformers. If you do no more than walk through the room containing a transformer while noting temperatures you will still improve their reliability.

Newer transformers can produce dramatic savings in energy cost because they're so much more efficient. Add to that the problem with many transformers operating at very low loads (where the losses are more significant) to be aware that replacements should be considered on a regular basis.

VOLTAGE AND CURRENT IMBALANCE

Changes in electrical loads in facilities can create voltage and current imbalance that can generate serious problems with alternating current motors and other equipment. One significant contribution to changes in facilities includes the conversion to more efficient lighting. If that's not done with some attention to the share each phase of a three phase power supply is altered you will start having strange situations with motors and other electrical devices, including lighting. Prior to each annual shutdown, if you have one, you should perform voltage balance checks and check all large motors for current imbalance.

Checking for voltage imbalance is relatively easy, is only applicable to three phase power, and only takes a voltmeter and some PE (Protective equipment including face shield, leather jacket and rubber gloves suitable for the voltage). The PE is required to protect you from accidentally contacting hot conductors and arc flashes. Read the phase to phase voltage for the three phases at each three phase motor starter. This requires opening the starter cabinet, which usually requires opening the disconnect, then restoring power by closing the disconnect with the door open. That usually requires manually over-riding the interlock that proves the door is shut. After recording the voltages, calculate the average voltage by adding the three values and dividing the result by three. Then calculate the voltage imbalance by dividing the maximum difference between the average voltage and each of the three phases by the average voltage and multiply by 100 to get a voltage imbalance in percent.

Do the same thing with the motor operating. Operating a motor with a voltage imbalance of more than 2 to 3% is not recommended. If voltage readings vary considerably between the motor off and motor running values the wiring to the motor should be checked. Voltage imbalance can be caused by a lack of balance of single phase loads in the facility or, rarely, a problem with the utility feed.

Current imbalance in a three phase motor can be caused by voltage imbalance (which should be corrected first) shorted motor windings, or a high resistance connection. For current imbalance you'll also need an amprobe to measure the current on each leg of the three phase motor. Current imbalance is calculated in the same manner as voltage imbalance and a motor should never be operated with a current imbalance of more than 10%. When you detect a current imbalance of measurable value then it's advisable to check to determine if the problem is in the source or the motor. To determine the source of the problem shut down the motor, isolate the power and rotate the leads then take another set of readings. Note that rotating the leads is not the same as switching the leads; to change a three phase motor's rotation you switch any two leads. Rotate the leads by connecting L1 to M2, L2 to M3, and L3 to M1 where L denotes the line and M denotes the motor lead. If the readings at the line match then the problem is in the power supply. If the readings at the line show different values for each phase the problem is with the motor.

EDDY CURRENT TESTING

Eddy current testing has become a standard practice for large chillers on their 15 year overhaul and is gaining acceptance as a method for checking the condition of boiler tubes. A technician (because the interpretation of the readings requires someone skilled with it) has a helper insert a probe connected to a cable into each tube and pass it through the tube while watching a monitor for indications. If you ask nicely the technician will show you a sample of a tube (that matches yours in size and material) with holes drilled in it and sections where the metal is cut to be thinner that must be detectable with the equipment at setup before beginning the testing of your tubes. The testing will not go well if your tubes are scaled up, especially if there's so much scale that the probe can't be pushed through the tube. This technology is relatively new and improving daily so you can expect to see it in the near future.

MISCELLANEOUS

As mentioned in the section above on electrical equipment, painting is a maintenance activity that can create problems. In many plants painting seems to be the only form of maintenance. If it's necessary to paint then make sure nameplates, gauge faces, and other items that shouldn't be painted are adequately masked before the painting process begins.

Keep in mind that multiple layers of paint are insulation and can shorten the life of electrical equipment. Paint can block tiny openings that are required for proper operation of self contained control valves and other equipment. Regular painting of screens and narrow louvers can reduce the free opening to reduce air flow with possible hazardous or damaging consequences.

I dislike inspecting a plant where I have to scrape several layers of paint off nameplates in order to get the information and I consider painting a poor excuse for maintenance. Instead of painting the plant, try cleaning it. Proper use of cleaners, soap and water can restore the condition of a plant at a lower cost and with less harm than painting. It will look good when it's done and some people will think you painted. As far as I'm concerned the only things that should need regular painting are floors and handrails because they are exposed to wear.

ASME CSD-1 and the NFPA 85 Codes are adopted by law in many states and contain requirements for maintenance. Factory Mutual and other insurance underwriters also have their own requirements for testing of fire and explosion prevention devices to ensure their reliability. Be certain to incorporate all the applicable requirements in your program. A recommended program of testing safety devices is included in this book but it may not contain every requirement you are legally or contractually required to perform. Keep in mind those requirements are only safety related and concentrate on devices that were found to contribute to significant failures and warranted investigation due to their cost or loss of life. A system that is as safe as some insurer's and code writers would like is not necessarily reliable because it can shut down more frequently.

Maintenance of stored fuel oil is one item many operators forget about because they're primarily firing gas. Checking the inventory to be certain the tanks aren't leaking and checking for water in the bottom of the tanks is critical to ensuring a reliable source of oil is available if it's needed. There are additives that can extend the life of fuel oil in storage and tests for the condition of the oil as well, check with your oil supplier.

I have to say it somewhere and this is the only place I could conveniently choose. Whenever you pull maintenance on a piece of equipment please, for the sake of yourself and others, please replace the belt or coupling guard. Don't just set it there either. You haven't seen what happens when a loose coupling guard vibrates around until it's caught by the coupling bolts and flung across the boiler room at someone. Always replace all the parts, especially protective guards.

REPLACEMENTS

I'm regularly called in to provide recommendations when the customer's management is upset with repeated failures in an aging boiler plant. A review normally results in a recommendation for a major replacement program because everything has been ignored and is so worn that it all needs replacement. Frequently it's due to the plant being operated in a manner that ensures everything wears out at the same time (see rotating boilers in the section on operating modes) a common practice that should be avoided.

Rotating equipment (fans and pumps) and similar devices where movement promotes wear, top the list of equipment that must be replaced on a regular basis. Motorized valves, pressure and temperature switches, pressure gauges and bi-metal thermometers all have moving parts that can wear, gall and fail so they need to be replaced at regular intervals. Those devices can last for years when their use is infrequent and they are subjected to a limited number of operating cycles or changes in condition.

Scheduling replacements is not a simple process. You have to have some reasonable degree of expectation when the device is going to fail so you are not wasting money by replacing them too frequently. That's one of the problems with a program that only considers preventive maintenance.

If you have scheduled operation of equipment that consists of an operating unit and a spare the first failure provides a basis for determining the life of the other. Of course if you operated them for equal periods of time the probability is the spare unit will fail... right now! By ensuring operating hours are proportional to the number of pieces of equipment you ensure some time to operate the remaining piece or pieces before they will fail. Scheduled replacement of spares that have failed shouldn't be questioned and you have a reasonable basis for establishing a deadline for the replacement. The concept here is breakdown maintenance and works well when you have one or two spares to deal with.

When you don't have spares the scheduling of replacement of devices is dependent on how critical it's continued operation is. If the decision is yours weigh the cost of the replacement of all the devices that have a greater than 50% probability of failing between now and the next maintenance period. Include the cost of labor to replace the devices and such contingent costs as disposal expense to establish a reasonable cost for replacement. The cost of a failure is dependent on the type of facility served by the boiler plant and can vary dramatically.

A hot water heater in a Boy Scout camp will have a minimal failure cost, they can use the time spent replacing the failed heater to train the scouts in providing their own hot water. On the other hand, failure of a hot water heater in a hospital borders on unacceptable because the lack of hot water prevents proper hygiene. The cost of canceled operations, bringing in food, and possibly relocating patients can all be reflected in the cost of failure of a steam boiler. Any production facility will normally have a high cost of failure because the costs could include damaged product and loss of sales that will destroy customer confidence; let alone the high cost of paying employees when they aren't making product and securing the facility then restoring it once the repairs are completed.

If you don't have a spare you should have a contingency plan in the event of a failure. Possibly you are operating a heating plant for an apartment complex that has only one heating boiler. In the event that boiler fails you have several options but lack of a plan will see you looking unprepared and could generate significant unnecessary costs. The wise operator will always have contingency plans for failure of each piece of equipment and service.

Service? Yes, you need to have a plan for the failure of every utility. Loss of electric power is a common occurrence and I'm always amazed at how some customers respond to it. They are always in a quandary when the generator fails to start or shuts down shortly after the electricity is lost. You need plans that include procedures in the event standby equipment fails, loss of the utility becomes long term, or conditions prevent delivery.

When replacing small parts and items make a concerted effort to ensure you're replacing something of equal quality. A big problem with valves is they cost less when furnished with reduced trim (a smaller opening). Motors with a service factor may be using it and a larger motor may be required. Modern technology has also provided better and lower cost alternatives, especially motors and controls, that should be considered when replacing parts and equipment.

Boiler Tube Cleaning—Replacement

One thing that is designed to be replaced is a boiler tube. They're designed to transfer heat rapidly so they are more likely to be coated with scale. They're thin, also for heat transfer, so they will corrode through first. There are means for cleaning scaled tubes so they don't have to be replaced but water side cleaning occasionally penetrates the tube so replacement is necessary.

Fire side cleaning can be performed by wire brushing the tubes of fire tube boilers. A modern piece of equipment (Figure 6-9A) that connects to a vacuum to collect the removed soot and a motor driven brush makes the job relatively easy and a lot cleaner than using a brush on a pole like I used to (Figure 6-9B). Without the machinery your spouse won't let you into the house until you've stripped and put all your sooty clothes in a bag.

Fire side cleaning of water tube boilers is normally accomplished with the boiler in operation using soot blowers. Note that soot blowers should be used only when the boiler is firing. During boiler operation the flue gas inside is essentially an inert gas. If soot blowers are operated with only the forced draft fan running you are creating an explosive mixture of dust and air with enough energy added by the steam to create a static spark. I've noticed a lot of new designs with soot blowers connected to a header instead of the respective boiler, that's wrong!

Of course soot blowers have to be intact and installed right to do a good cleaning job. You should be able to tell by the sound if they're working right. If the end of the soot blower has corroded or burnt off or the element is misaligned so the steam jets are hitting the tubes (a good way to cut through the tubes) you should be able to tell by the sound. When soot blowing doesn't do the job and fuel additives don't do the job then the boiler has to be cleaned with a high pressure water wash.

Figure 6-9A. Firetube cleaner

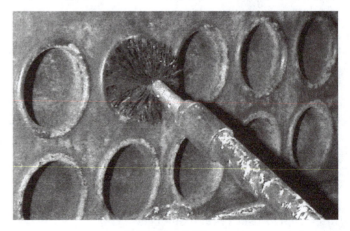

Figure 6-9B.

We did it occasionally on ships using boiler water. A heavy steel reinforced hose was connected to the blowdown of an operating boiler. A valve and homemade lance was attached to the other end and we proceeded to try to wash the soot accumulations from the boiler. The hot boiler water would help dissolve the deposits and the caustic solution would help neutralize the acidic soot. That's also a very dirty, and hot, job that shouldn't be necessary with proper firing, properly adjusted soot blowers, and fuel treatment.

Another boiler expert I know insists soot blowers are installed on boilers only to give operators time to learn how to operate the boiler. That's not true, but he usually gets a snicker when he says it.

There are three methods for cleaning water side scale from boiler tubes but none should be required under normal circumstances. If you have adequate pretreatment facilities and adequate boiler water chemical treatment you should never need tube cleaning. Turbining is the method I was introduced to when I started and is occasionally used as a general maintenance method in plants with very poor water pretreatment.

Turbining tubes is accomplished with a special water powered tool that rotates a set of small sharp gears around inside the tube (Figure 6-10). The water not only powers the tool but flushes the debris away. A tube cleaning turbine will remove most of the scale but leave small pieces unless you repeatedly run it up and down the tube until you've removed a lot of metal as well. They're not difficult to operate. It's just difficult to control the enthusiasm of young people that might remove half the tube metal. Of course they only work for removing waterside scale from inside water tubes. I should say from inside round water tubes. If you have a very old boiler you may find the tubes are closer to square where they're bent. Turbines will jam in them and you tend to poke holes in the flats of those squarish tubes. With modern water treatment methods, this picture should be the only time you see a tube turbine.

High pressure washers are used to remove scale from the water side of fire tube boilers. Operating with nozzle pressures as high as 40,000 psig they blow the scale away and sometimes take some metal as well. These are best handled by contractors experienced with their operation. The application usually requires a vacuum system and truck to remove the scale from the boiler as it's washed off and separate it from the wash water to allow recycling of the wash water.

The third method for scale removal is acid washing. An inhibited hydrochloric acid is used to eat the scale off the tubes. The application requires care and regular testing to ensure the acid is removing scale and not boiler metal. The acid solution is heated and circulated and the entire boiler has to be flooded so all the boiler metal is exposed to the acid. Any mistakes don't result in just tube replacement. This method is also best left to contractors with the equipment and skill necessary to do the task. They also haul off the spent acid and dissolved scale when they're done.

When cleaning fails, and so much energy is wasted by scale that something has to be done, plugging or replacement of the tubes is required. If you have a modern flexitube boiler then all you need is a wrench, special tool, and big hammer. They're designed to be replaced by individuals with a reasonable mechanical sense. Otherwise your boiler tubes are installed by rolling or a combination of rolling and welding, processes that require more skill.

When you have only one or two defective tubes it's usually easier and more frugal to plug them than to replace them. Some tubes can't be plugged because they serve purposes other than heat transfer. Tubes that form boiler walls or flue gas baffles can't be plugged because they will melt down or burn off without water cooling and allow heat and flue gases through.

For watertube boilers it's a little more than simply a matter of obtaining some machined steel plugs that fit into the ends of the tubes and inserting them. The first, and a very important, thing to do is to make sure you have located the leaking tube at both ends. Testing using rubber plugs and a water hose is recommended. To be certain the plugs don't blow out because steam is generated in the tube from water leakage you should drill or chisel a hole in the tube so any leakage is bled into the flue gas. You

Figure 6-10. Tube turbine

should also remove any scale from the end of the tube, making certain it is clean, round, and smooth so there's a good metal to metal fit between the plug and tube. Gently tap the plug into the tube, the water pressure will hold it and hammering excessively can distort the drum or header.

Plugging of fire tubes requires not only a plug but a means of holding them in because the water will leak to the fireside of the tube and apply pressure to the plugs. A piece of cold rolled steel rod longer than the tube and threaded at the ends is required along with nuts and plugs that are bored to accept the steel rod. At least with the rod you are certain you've got the right tube ends. The rod has to be large enough to overcome the force of the water pressure against the plug and produce enough force to seal the plug and the end of the tube. The tube end must be cleaned as described for watertube boilers. The plugs also have to have a means of sealing the space between the rod and the plug. An advantage of plugging a firetube boiler is you can tighten the plug while the boiler is under hydrostatic test to try to seal a leak. You can also plug a boiler while it's under water pressure but, for most operations, the plugging of a firetube boiler is so involved that it's much easier to just replace the tube.

A common repair for many water tube boilers involves replacing a section of the boiler tube. Frequently it's only a portion of the half of the tube that faces the furnace. When bulges or blisters form due to scale buildup, and sometimes rupture, the rest of the tube is still intact and the original thickness. The repair requires a skilled boilermaker welder. The tube is cut out around the failure, normally in an elliptical form, and a piece cut from another tube is inserted in its place with the edges of the original tube and patch butt welded. Since the tube walls are so thin (less than 1/8 inch) the weld is normally made by GTAW (gas shielded tungsten tip arc welding also called TIG) welding.

Entire sections of boiler tubes can be removed and replaced in a similar manner. The elliptical patch is used at either end so the welder can reach through the opening provided for it to reach the butt joint at the back of the tube. The welder has to work on the inside of the tube at the back because there's no room to get to it from the back. Once the back is welded the patch is set and welded to complete the repair. That method is referred to as using a "window weld."

Replacing a boiler tube is best done by a boilermaker who has the skill and experience necessary to do the job right but you can do it if you have the tools. If the tube is welded you should check with your insurance company or state boiler inspector to be certain you can re-weld them under the local law. Most states require all welded repairs be performed by an authorized contractor that is approved by the State or holds a National Board Certificate of Authorization to repair boilers, what we call an "R" stamp. It really is a stamp, the authorized company actually has a steel stamp that is used to mark the boiler when the welded repair is done.

The first step in replacing a tube is removing the old one. Whenever it's possible the tube should be cut off and removed, leaving the ends in the drum, header or tube sheet. Replacement of some water tubes in bent tube watertube boilers requires removal of other tubes to gain access to the tube that's to be removed. It's possible that you will have to remove several good tubes to remove a defective one.

Removing a tube from a firetube boiler is pretty much restricted to pulling it out of the hole it's installed in. If the tube is heavily scaled it may be necessary to remove it from the inside and that could require removal of several other tubes. A single tube replacement in a firetube boiler is seldom located where the tube can be removed via a handhole or manhole. The holes in the tubesheet of a firetube boiler are made a bit larger than the tube so slight accumulations of scale will slip through the hole. In some cases the scale is stripped from the tube as it is removed. In extreme situations it's necessary to split and collapse the entire tube to get it out.

Removing the tube requires crushing or cutting away the tube end where it is expanded into the drum, header or tube sheet. I've seen several boilers seriously damaged by inexperienced or careless contractors cutting the tubes with a torch and one case where a repeat repair was necessary in very few months because the contractor cut the tube sheet with a torch and put new tubes in without repairing the cuts. If your personnel or a contractor uses a torch to cut the tubes inspect every opening to ensure the tube holes are smooth and clean so a new tube will seat properly in the hole when it's expanded.

The best way to remove a tube end is to chisel it out, making certain you never touch the tube sheet, drum or header with the chisel. It eliminates the risk of cutting the inside of the tube hole but it takes longer and, quite frankly, takes more skill. By cutting a shallow (about half the tube thickness) groove through the tube where it's expanded you produce the same effect as flame cutting. After the tube is cut by driving it to the center you can collapse the tube into the middle, away from the tube hole, so the end or whole tube, can be removed.

Once you've removed the tube you should "dress up" the hole, removing any tube metal stuck to it and any corrosion that would accompany a leak or defective rolled joint. Careful use of a file and sandpaper should produce a smooth surface. The edges of the holes should also be smoothed over to eliminate any sharp edges that will cut the new tube. The tube ends should also be dressed up to remove any corrosion for a tight metal to metal fit.

The new tube is expanded with a roller (Figure 6-11a) to compress the outside of the tube against the inside of the tube hole to seal the joint. The roller in Figure 6-11a expands the end of the tube inside the boiler, flaring it. The roller in Figure 6-11b has a beading attachment which forces the metal end of the tube out and back against the tube sheet to form the ends shown in Figure 6-12b. As shown in Figures 6-12 of completed joints a water tube, Figure 6-12a is flared but a fire tube end is beaded; Figure 6-12b or restricted in protrusion to limit heating of the end of the tube. Typically the inlet of the first pass of a four-pass firetube boiler is welded (6-12c) to increase its ability to transfer heat to the water because the flue gases are much hotter in that first turn of a four pass boiler.

Once your tube replacement is complete the boiler should be subjected to a full one and one-half times maximum allowable working pressure hydrostatic test. Many contractors and most inspectors will accept an operating pressure test but why accept anything other than a test that proves the repair has returned the boiler to a like-new condition?

Refer to the section on hydrostatic testing a new boiler. Testing a repaired boiler is done the same way.

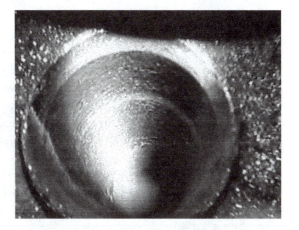

Figure 6-12a. Rolled tube—flared

Figure 6-12b. Rolled tubes—beaded

Figure 6-11a. Tube roller with flare　**Figure 6-11b. Tube roller beading attachment**

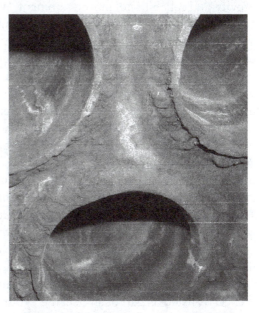

Figure 6-12c. Rolled tubes—welded

MAINTAINING EFFICIENCY

An important part of maintaining the plant is maintaining efficiency. Since the cost of fuel is the largest single expense in a boiler plant activity it's essential to prevent that cost getting out of control. Efficiency maintenance relies on two activities; monitoring to detect any changes and tune-ups when a problem arises. Monitoring is the boiler operator's responsibility; tune-ups are usually performed by outside contractors that have the necessary equipment and skills to perform that work. I would prefer to do my own tuning but there's nothing wrong in having an outside contractor do the work in small plants where the energy saved cannot justify the purchase and maintenance of the equipment required to tune up a boiler. An operator should know enough about tuning to ensure the contractor is doing a proper job and the sections on combustion and controls in this book are sufficient to impart that knowledge.

RECORDS

How do you remember when it's time to change the oil in your automobile? That sticker on the windshield or side of the door is a record that gives you that information. I don't know about you but I can never remember the mileage when I changed my oil last and that record is important because without it I may fail to change the oil until the engine lets me know I should have.

Schedules for maintenance are essential to ensure the longevity and reliability of most equipment. Whether you let it run until it breaks or perform significant PM (preventive maintenance/predictive maintenance) documentation is essential. For breakdown maintenance items it allows you to know about when you need to order a spare device because the operating one is scheduled to fail. More importantly, the documents tell you what to buy, what oil to use, what grease to use, etc., so you perform the maintenance in a manner that keeps the equipment and systems running.

Maintenance isn't complete until all the documents are properly filed away (see the chapter on documentation). To anyone investigating your plant after an incident a lack of maintenance records is an indication of a failure on your part to see to it that the work was done. You can say you did it, describe the day and what you did, but without that documentation you can't prove it. When a check is listed as part of an SOP then your entry into the log that you performed the procedure is docu-

mented proof you did it. Be careful, however, that it's done consistently or the entire log is questionable. Do what you say you will and say what you did consistently for the protection of your employer, your job, and the health and welfare of you and your fellow employees.

Welding

While I had the perception for years that a certified welder was always part artist, part perfectionist, and extremely skilled at the craft my later years in semi-retirement exposed me to another world that clearly revealed my perception to be wrong. I will limit my description of that other world to most federal contractors and trust that the few that really do comply with the applicable codes aren't insulted. Welding of boilers (including attached piping on high pressure boilers) and pressure vessels must be performed by a Company that is authorized by ASME to do that work. Welding on ductwork and other containments that operate at very low pressures are not covered by the ASME Code but all piping is. Piping for steam at pressures over 15 psig and hot water operating at pressures over 160 psig or 250°F must comply with the ASME B31.1 Power Piping Code. Piping for other services inside your facility must comply with the ASME B31.9 Building Services Piping Code. Both Codes require all welding to be performed in accordance with Section IX of the BPVC (ASME Boiler and Pressure Vessel Code) which describes the requirements for qualifying welding procedures and welding operators. The Piping Codes also contain requirements for examination of welds using methods and procedures covered by Section V of the BPVC. I can't recommend an operator read, let alone understand, the content of these codes but I will offer some advice on what to look for when a contractor, or even your own maintenance personnel, are welding piping in your plant. After reading the following you should be able to determine quickly if the welding is in accordance with Code.

First there's the requirement for materials. Inside a boiler plant every piece of pipe must conform to at least an ANSI (American National Standards Institute) specification and for all high pressure work over 125 psig MAWP must also comply with an ASME specification. The two specifications differ very little with the requirement that the piping must be supplied with mill test certificates when conforming to the ASME specifications. Once upon a time I had a salesman call on me and claim he could provide me with piping and all I had to do was tell him how it was to be labeled, clearly insinuating that he was prepared to falsely label the pipe so we could buy the product at his low prices. He was told thanks, but no thanks and I've been very conscious of what I was buying thereafter. As far as I'm concerned any pipe that goes into a boiler room should have a stencil running repeatedly along its length that begins "ANSI" and for B31.1 piping also "ASME" followed by a specification number and other data including a heat number which makes the material traceable to the piping manufacturer. Pipe fittings and valves must bear certain markings but certificates aren't required. They should, however, bear a marking that identifies the manufacturer and the appropriate ANSI specification.

As for the welder, the only way to determine the quality without requiring RT (radiation testing, frequently called X-ray) is visual examination by someone who knows what to look for. Here's what to look for:

Fit-up

The fit-up consists of matching the joint between two pieces of pipe or a pipe and fitting or weld end valve and attaching them with tack welds at no fewer than four points evenly spaced around the pipe. The two ends should be ground clean to gray metal (no rust, paint, oil or other coating) for at least one-half inch from the surfaces to be welded. The pipe should aligned such that any differences in the inside edge of each pipe is less than 1/16 inch (less for very high pressures and temperatures). The joint should also be prepared in a manner that conforms with the WPS (Welding Procedure Specification) which the welder should be able to show to you. Typically the ends of the pipe or fitting is ground to a bevel as shown in Figure 6-13. That's the standard preparation for an open butt welded joint where the two ends of the pipe are positioned so a cross section of the fit-up joint would look like that in Figure 6-14. Four tack welds are typically used and located at quarter points straddling the vertical and horizontal centerlines. A keyhole is formed at the end of each tack weld providing a starting point for the rest of the root pass. This arrangement permits the welder to fill in between the two ends to provide a complete weld like the cross-section diagram in Figure 6-17. A section through a weld wouldn't look like that because you shouldn't be able to see the black outlines shown in Figure 6-17. That's because the pipe and weld metal should be fused together so you can't tell them apart. The joint should be inspected by someone other than those preparing the joint and should incorporate use of the gage shown in Figure 6-15a to check for the appropriate bevel and internal alignment. Figure 6-15b shows use of the gage to check for internal misalignment and a proper bevel.

The tack welds should all be checked to be sure none of them are cracked; a light tap with a hammer will normally produce a ringing sound, any static is indicative of a crack that can't be seen without magnification or other testing. Some procedures require the tack welds be replaced as the root pass is made but the tacks are normally incorporated into the root pass.

Figure 6-13. Tee showing beveled end prep

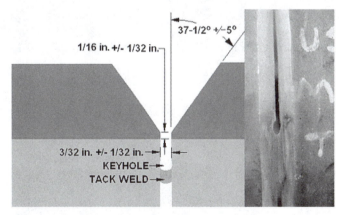

Figure 6-14. Cross section of fit-up

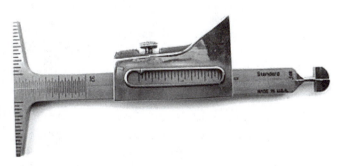

Figure 6-15a. Welding inspection gage

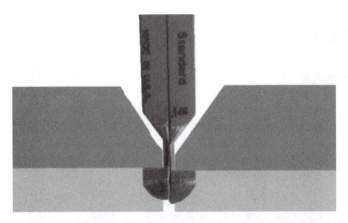

Figure 6-15b. Checking fit up with gage

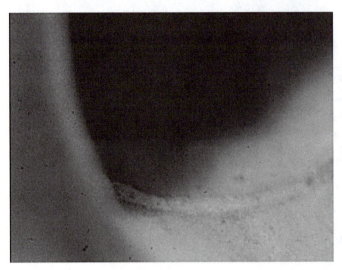

Figure 6-16. View inside of root weld

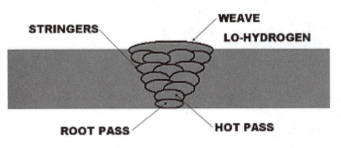

Figure 6-17. Weld cross section

Root Pass

The root pass indicated in Figure 6-17 is the first full circumferential weld and the inspector should check it in stages to ensure there is no excess internal reinforcement as shown in Figure 6-18. Welders normally refer to this as "grapes" and it can produce damaging eddy currents in the flowing fluid that will erode the piping

downstream of the joint where indicated. Normally internal reinforcement cannot exceed 1/16 inch but that decreases with higher pressures and temperatures. On the other hand incomplete penetration as shown in Figure 6-19 isn't permitted in excess of 1/32 inch. The root and following passes should have a surface texture similar to what's difficult to see in Figure 6-20, it is produced by the welder swinging the rod in overlapping circles. It's difficult to see because the photo is one of a filler pass, commonly called a stringer, using low hydrogen electrodes. It's also hard to see in the photo of the inside of a root pass in Figure 6-16. A reasonably smooth root weld inside the pipe and within the parameters described here is always a sign of a good welder.

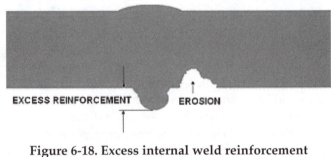

Figure 6-18. Excess internal weld reinforcement

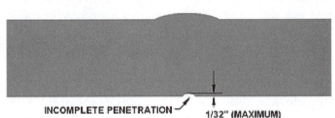

Figure 6-19. Incomplete penetration of a weld

Hot Pass

The hot pass follows the root pass and is commonly checked for cracking. Once the hot pass is complete, the stresses associated with the cold pipe metal and hot weld during fit-up are reduced by even heating. When welding carbon steel pipe most welders use a procedure calling for low hydrogen electrodes for everything except the root pass. Low hydrogen electrodes have to be kept hot after the container is opened to prevent the absorption of water and special ovens are usually used for that purpose. If you see those electrodes laying around and getting cold then used in a weld, that weld will have porosity (holes) in it. Low hydrogen electrodes produce a weld that flows and fills well with a slag (molten minerals that coat the electrodes that melt and form an airtight barrier over the weld to prevent oxygen from the air combining with the molten metal) that's readily removed.

Stringers

Additional passes that cover the hot pass can consist of several "strings" of weld to fill in the weld until nearly flush with the surface. Careful cleaning of the slag from each of these passes is necessary to ensure it isn't left in the edges which dramatically weakens the joint. The stringer in Figure 6-20 is as wide as they are typically made with additional ones running alongside each other.

Figure 6-21. Cover pass of completed weld

Figure 6-20. Stringers in progression of a pipe weld

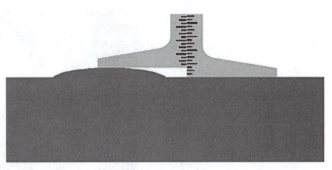

Figure 6-22. Check for excess external reinforcement

Cover Pass

This is normally called a "weave" pass because it provides a smoother look to the welded joint as shown in the photograph in Figure 6-21. On completion it's checked with the gage to measure the external reinforcement as shown in Figure 6-22. The external reinforcement shouldn't exceed 3/32 of an inch (less as design pressure and temperature increase). Also, any groves in the external joint shouldn't exceed 1/32 of an inch which is also called incomplete penetration.

Why did you have to go through all that? Because lately I've seen several jobs where the weld section looked like the section in Figure 6-23 and a few where failed joints, when sectioned, look like the one shown in the photograph of Figure 6-24. Even though a weld can look good on the outside like Figure 6-25a—when you can, take a look at the inside because it might look like Figure 6-25b. When someone makes a weld like this in your plant he or she is showing no concern for your life. I've heard all the welder's excuses like "It's just low

pressure steam" or "it's only low pressure gas" and I don't buy them. Joints like those shown can fail dramatically. Low pressure steam in an 8 inch pipe produces an axial force of 750 pounds which is more than enough to do some serious damage if it gets loose. In the typical high pressure boiler plant the force in the same 8 inch pipe is over three tons and in the typical utility plant at the throttle inlet it's thirty tons. I trust that explains why we have Codes and why you shouldn't allow welders in your plant to be cavalier about the welds they make.

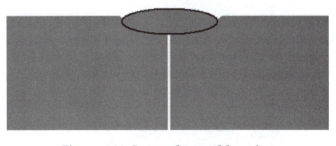

Figure 6-23. Incomplete weld section

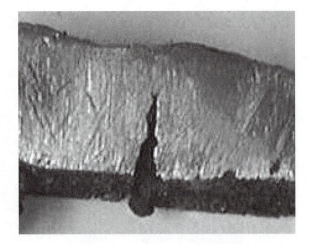

Figure 6-24. Section from failed incomplete weld

Figure 6-25a. Apparently completed weld

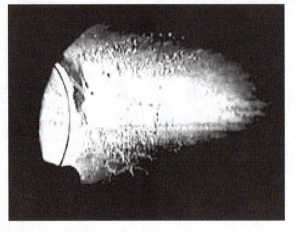

Figure 6-25b. View of inside of weld shown in 6-25a

Chapter 7

Consumables

Few people realize the value of consumables. The typical boiler plant consumes a million dollars worth in each year. Boiler operators can have a significant impact on their consumption.

I'm not talking about the illegal activities that can involve things as simple as rags or pallets. I will only say that operators that entered into those have, in my experience, always been caught and the punishment is severe.

There is significant trust placed in operating personnel to protect the income of their employer and, as a result, their fellow employees. The use, or abuse, of consumables is where the true value of operation is measured.

FUELS

The principle purpose of most boilers is to convert the chemical energy in a fuel to heat absorbed in water, steam, or another medium for use in the facility served by the boiler plant. (We can't forget that there are electric boilers). A wise operator should know as much as possible about the fuel he's burning both to get it done safely and to get the most out of that fuel. We'll cover the most common fuels first then touch on some of the others you might encounter. In the process you should get an understanding of what's required to burn any fuel so you're comfortable working with something that is unusual.

Oil, gas and coal are called "fossil" fuels because they are found in the ground where they were trapped as vegetable and animal matter hundreds of thousands of years ago. As they decayed they became the fuels we know of today. Wood, bagasse, corn and similar fuels, all produced from living plants are called "biomass" fuels.

The ultimate analysis of a fuel is a determination of the percentage of each element in a fuel. An element is a material that consists entirely of one kind of atom. The determination is made in a laboratory using standard procedures which are included in the appendix. An ultimate analysis will normally list the amount of Hydrogen, Carbon, Sulfur, Oxygen, and Nitrogen in the fuel along with any other element of significant quantity and, for

fuel oils and coal, water and ash. An analysis of fuel oil will also list "BS&W" which stands for bottom sediment and water (I'll admit I normally call it brown stuff and water except I abbreviate the second word a little).

The laboratory will usually include the higher heating value of the fuel as well. Results are typically listed as pounds of an element per pound of fuel, a value that is readily converted to percent by multiplying by 100. The values for the fuel are dependent on the fuel source and any treatment it endures before it is delivered to you to burn. When fuel gas is analyzed and you don't ask for an ultimate analysis you will be given a list of the gases in the fuel and their respective percentages by volume. Normally methane is listed as the primary constituent of natural gas with much smaller fractions of other gases.

It's a simple matter to convert a volumetric analysis (one that shows the percent by volume) to a gravimetric analysis (one that shows percent by weight) and to use those analysis. It's only essential for a boiler operator to know what the words mean and to be aware that the ratio of hydrogen to carbon in fuel will vary to affect boiler operation. The reason is clear when you do an efficiency calculation, see the chapter on efficiency.

Sulfur in fuel contributes a small amount to the energy released in combustion. The problem with sulfur is its products of combustion, sulfur dioxide (SO_2) and sulfur trioxide (SO_3) combine with the water in the flue gas and atmosphere to produce sulfurous (H_2SO_3) and sulfuric (H_2SO_4) acids. When surface temperatures in the boiler and ductwork are so low that the acid gas can condense the acids attack the metal and extreme damage due to corrosion is the result. The last half of the twentieth century saw a concerted effort to reduce the sulfur content of fuels to reduce the problems with acid rain caused by the burning of the sulfur in fuels.

Liquid water in fuel can create all sorts of problems. It absorbs a lot of heat from combustion to convert it to a vapor (the hydrogen in fuel burns to a vapor, not a liquid) and it creates corrosive conditions that can damage the fuel handling and storage system. Water in coal is a major problem in the winter because it will freeze to convert a pile of coal to one solid chunk that can't be fed to the boilers. Similarly it can freeze in gas or oil systems to block valves and regulators resulting in dangerous operating pressures.

When water separates from the oil in storage tanks it settles to the bottom. It will eventually accumulate until, all of a sudden, you find yourself trying to burn water. Water in fuel oil also provides a medium for corrosion of the fuel tank and piping. It's one of the reasons for leakage of underground storage tanks (USTs) with some serious consequences. Water can be emulsified (a process that mixes the fuel and water distributing water throughout the oil) but it can still produce corrosion and will always require the addition of latent heat to vaporize it in the furnace.

Small and controlled quantities of water emulsified in oil can help reduce soot formation which can improve heat transfer to the degree it compensates for the latent heat loss. When I was sailing for Moore Mc-Cormack Lines in the 1960's we were conducting an experiment with injecting small quantities of superheated steam into the fire to reduce sooting. I never did find out what the results of that were.

Water in fuel gas systems can be a considerable problem when the gas pressures are low because it can collect and produce blockages in the piping as well as promote corrosion. When you have wet fuel gas you'll have additional requirements for handling the liquids that settle in your piping because there can be liquid fuels as well as water. Water draining from a coal pile is highly corrosive and must be discharged to a sanitary sewer after it is neutralized.

The discussion in the chapter on combustion helps explain why firing conditions change when the fuel changes. Most of the time the air-fuel ratio is close enough to ignore the variations. When a service technician uses a portable analyzer to calculate combustion efficiency that analyzer contains a "typical" fuel analysis for the fuel and determines efficiency based on that typical analysis. I've always wondered if those analyzers are calibrated for the area because the carbon content of natural gas can vary from 20.3% to 23.5% between the east and west of the country. That amounts to a 15% variation in higher heating value of the fuel and it's one reason I refuse to believe the efficiency on one of those machine's printouts. It's only important that you know that the analysis can change and have an equal distrust of those electronic analyzers' efficiency indications.

In the Baltimore area we can experience changes in natural gas depending on the source of the gas in Pennsylvania, Texas or Louisiana and the blending of gases from those sources. We also have a chance to burn some of the LNG (liquefied natural gas) imported from North Africa which has an air-fuel ratio ten percent higher than domestic natural gas. LNG is compressed and cooled until it becomes a liquid; is loaded into tanks aboard ships built exclusively for the purpose; then transported across the Atlantic Ocean to special port facilities near Boston and Baltimore among others.

Ash in the fuel, whether it's coal, oil, or biomass can create problems with firing. The ash fusion point is the temperature at which the ash melts. If furnace conditions produce higher temperatures the ash will melt then solidify again when it cools, usually forming large accumulations of solidified ash that can block air or gas flow passages or grow in the upper portions of the furnace. They grow until they get too heavy to maintain their adhesion to the tubes or refractory and fall crashing to the bottom of the furnace doing damage to tubes, grates, etc. When firing fuels with a low (less than 1800°F) ash fusion temperature the operator has to monitor the furnace conditions inspecting it and recording draft readings to detect hardened ash accumulations early.

One of my projects included burning dust from a laminate sanding operation where portions of the ash had very low fusion temperatures. We operated that boiler at very high excess air just to keep the furnace temperatures down to prevent the ash melting and sticking to the tubes. Someone advised that customer they could save a lot by decreasing excess air (true in other situations) so they did; and ended up with huge globs of solid ash stuck to the furnace walls and tubes.

You should always know what the vanadium content of your fuel is because that material produces a lot of low melting point ash. Just last week I spent a Saturday evening crawling into a boiler to see the result of blockage due to low melting point ash. The customer's fuel oil only had about 30 ppm of vanadium in it but was enough to completely block up the first pass of the boiler with ash that took about two days to clean with a high pressure washer.

FUEL GASES

Natural gas is mostly methane (CH_4) with portions of other flammable gases, oxygen, carbon dioxide, and nitrogen. A typical volumetric analysis is 96.53% methane, 2.38% ethane, 0.18% propane, 0.02% iso-butane, 0.77% carbon dioxide, and 0.12% nitrogen. That's east coast gas. Gas constituents will vary depending on the well the gas came from. When a boiler is fired with oxygen trim controls to achieve very small quantities of excess air those controls accommodate the varying air-fuel requirements of the gas supply. Domestic natural gas has a higher heating value of approximately 23,165 Btu per pound, approximately 1,042 Btu per standard cubic foot. For combustion it requires 11.48 standard cubic feet of air per standard cubic foot of gas, 185 standard cubic feet per minute of air per million Btu per hour.

Liquefied petroleum gases (LPG) are primarily butane or propane with propane being the more common. They are transported as a liquid under pressure. They combine the clean burning properties of gas with the transportation properties of oil but at a premium in cost. In boiler plants where LPG is used it's normally as an alternate fuel for interruptible natural gas. Propane can be mixed with air in proper proportions to produce a blend that will fire in natural gas burners without adjustment of the burners.

Propane has a slightly higher heating value of approximately 21,523 Btu per pound, approximately 2,573 Btu per standard cubic foot and it requires 28.78 standard cubic feet of air per standard cubic foot of gas, 186.45 cubic feet of air per million Btu. You'll note that the air required per million Btuh is about the same for all gases. All but very large LPG installations will absorb enough heat at the tank to convert the liquid to a vapor. Large installations require a vaporizer, a heater fired by vapor off the tank that provides the energy to evaporate a liquid stream for use in the boilers. Propane will condense at normal atmospheric temperature (70°F) at 109 psig.

Butane will condense at 17 psig. On a very cold day butane will not vaporize and most installations require a vaporizer. Butane has a higher heating value of approximately 21,441 Btu per pound, approximately 3,392 Btu per standard cubic foot and it requires 37.57 standard cubic feet of air per standard cubic foot of gas, 184.64 cubic feet of air per million Btu.

You've undoubtedly heard a lot about hydrogen as a fuel lately because it's the principal fuel for fuel cells, those devices used on the space shuttle to generate electricity and water. By now you can probably envision one taking on hydrogen and oxygen to produce water and the energy generated comes out as mostly electricity. Fuel cells do produce some heat but that's considered a by-product in their application. I've only had one experience with burning hydrogen in a boiler. It was a waste gas from a chemical process and we burned it to recover the energy. You can imagine that at 61,000 Btu per pound it was a very hot fuel and burner construction and maintenance was very demanding. If I knew then what I know now I would have blended it with something before trying to burn it, either natural gas or lots of air to avoid the terribly high flame temperatures. This fuel is one where you better read the instruction manual and be aware that leaks are very hazardous.

Digester gas is actually natural gas, just very young natural gas. Like a young bourbon it has a kick, lots of things in it that make it less desirable than natural gas, which had thousands of years to cure in the ground. Digester gas is a by-product of waste water treatment where the water is enclosed in the digester and anaerobic bacteria (bugs that don't like air) literally eat the waste and generate methane and carbon dioxide in the process.

The principal difference between digester gas and natural gas from wells is the digester gas contains a lot more carbon dioxide and usually has some other materials in it that carry over with the gas as its generated. Some of the less desirable materials include water, hydrochloric acid, and solids. Some digester systems are fitted with filters to reduce the solids and separators to remove most of the water and acid before it gets to the boiler plant. The largest variable in digester gas is the amount of carbon dioxide. It's basically inert (the carbon and oxygen already combined) so it dilutes the methane content of the gas to reduce its heating value to numbers in the 250 to 800 Btu per standard cubic foot range, 25% to 80% of the energy normally found in natural gas.

Special considerations for firing digester gas include concern for blockage of valves (especially safety shut-offs), regulators, etc. All the piping should be fitted with drains, usually drain pots where the collected moisture, etc. can be captured for return to the digester. The piping also has to be arranged so it can be cleaned in the event of an upset in the digester which could send over considerable quantities of water and solids (another name for that "s" word) to plug things up. Piping materials may be constructed of stainless and other alloys to prevent corrosion by the acids in the system but pre-cleaning usually reduces the acids enough that normal steel can be used. When you do have steel piping it's advisable to check its thickness regularly and after any severe plant upset.

The large fractions of carbon dioxide can dilute a digester gas so much that it will not burn with a stable fire. Special burners are required to pass the larger gas volumes required to get the fuel value needed for the boiler capacity and many of them are fitted with standing pilots. Most of the applications I've worked on include real natural gas as a support fuel to maintain ignition of the digester gas and to make up any additional energy requirements. Both fuels are fired simultaneously and the controls have to be able to cope with that.

If you're firing digester gas you usually will have a responsibility to monitor the digester itself. A little training on how those anaerobic bugs work and you're a wastewater plant operator as well. You'll quickly learn that if you don't burn the gas in the boilers and allow it to escape to the atmosphere everyone in the neighborhood will be complaining about the odor. When a boiler plant can't burn all the digester gas or the boiler plant is temporarily shut down for maintenance the gas is usually burned off using a flare (Figure 7-1). You'll find yourself responsible for the flare too, but it's only a burner without a furnace and boiler around it so it isn't that difficult to handle.

Landfill gas is very much like digester gas. The anaerobic bacteria work on the garbage in the dump (a landfill is, after all, nothing more than a well maintained garbage dump) to generate the gas. There are some potential problems with landfill gas that are not encountered with digester gas. The carbon dioxide content can vary more (over extended periods of time) and air can leak in through breaks in the cover of the landfill. The gas will also vary in mix of fuel gases because the garbage in the landfill is not consistent.

Refineries produce a variety of gases with various blends which have different heating values and air fuel ratios. I remember the familiar sight of flares burning off those gases but problems with hydrocarbon emissions from those flares and the waste of energy combined with modern technology that allows us to burn them efficiently has reduced their numbers and use. Control systems that continuously measure the heating value and combustion air requirements of the gases can provide real time information to a control system on a boiler to burn those gases. Here again, you'll need to read the instruction manuals and will more than likely receive special training for operating a boiler burning those gases.

All too often gas is taken for granted. You just assume it will continue flowing out of the pipeline. The gas flow can stop if a line ruptures, a compressor station breaks down or has a fire or other emergency, or someone burning gas near you has a failure. We also have to stop burning gas when we're on an interruptible gas service. If we don't the owner will pay a serious fine for burning gas.

Some older plants had "gas holders" expandable tanks that used the tank weight to pressurize the gas in storage. You probably can recall seeing one on some city skyline in the past. Those gas holders provided a source of gas in case of an emergency. Utilities use mines where they compress the gas for storage and there's liquefied natural gas storage facilities in a few spots in the country. Regardless of all these provisions most of us have to be prepared for an interruption in the gas supply.

Being able to burn one of the LPG choices is one way to have a standby provision in the event the gas supply fails. LPG is expensive and a storage facility capable of providing any extensive operation of a boiler plant is very expensive so few plants use that option. Most of the time we use fuel oil as a backup to loss of our natural gas supply. Either LPG or fuel oil will be stored on site for interruptions to a natural gas supply regardless of the reason for the interruption.

FUEL OIL

Fuel oils are identified by ASTM specification D-396-62T which replaced the Pacific Specifications (now obsolete) that originally identify the oils by a grade number. Number 1 is basically kerosene and is seldom used in boilers. The common fuel oils are grades 2, 4, and 6. The term "grade" was dropped so now they're normally identified by the number alone.

Number 2 is called "light fuel oil" which is not as dense as the others. Light fuel oil is basically the same as diesel engine fuel. It has a typical heating value of 141,000 Btu per gallon, weighs about 7.2 pounds per gallon and has an air-fuel ratio requirement of 16.394 pounds of air per pound of fuel that is approximately equal to 218 cubic feet of air per gallon, 189 cubic feet of air per minute per million Btuh. It is relatively clean burning and has almost no ash. There is one common myth about Number 2 fuel oil, it is not a low sulfur oil. It

Figure 7-1. Flare

contains about the same amount of sulfur as low sulfur heavy oil.

To reduce sulfur oxide emissions, transportation and off road equipment is now required to burn ultra low sulfur diesel oil (which doesn't materially differ from No. 2 oil) with a sulfur content of less than 15 ppm. It stands to reason that restriction will shortly apply to heating fuel as well.

Grade 3 was dropped from consideration in 1948.[6] That's why nobody knows about it unless they're over 60.

Numbers 4 through 6 are referred to as "heavy fuel oil," they are dark in color, require some heating before they will burn and exhibit varying degrees of soot formation and other problems with burning. Numbers 5 and 6 require heating to reduce the viscosity of the fuel so it can be pumped. Number 6 fuel oil has to be heated so it will flow. I have a sample of it that I carry for seminars. It looks like a puddle of oil when it's resting on a table but I can pick it up and tap out a tune with it, it's that hard at room temperature. I then explain that it will flow like water if it is heated to about 200°F.

The viscosity (resistance to flowing) of these fuels varies considerably with temperature. The viscosity, not the temperature, has to be maintained at the value prescribed by the burner manufacturer and the operator has to set the oil temperature to achieve the required viscosity for proper atomization. The analysis of the fuel, provided by the fuel supplier, will indicate a viscosity at a standard temperature and charts or graphs furnished by the fuel supplier or the burner manufacturer must be used to determine the required temperature for burning. If you're burning a heavy fuel your fuel supplier should furnish you with temperature—viscosity charts and guidance in maintaining the proper viscosity.

That will give you a starting point. An oil burner is designed to atomize the oil at a specific viscosity, most of them at 200 SSU (Seconds Saybolt Universal). That simply means it takes 200 seconds for a 60 milliliter oil sample at 100°F to flow through an orifice in the Saybolt Viscometer. I like to vary the viscosity, by varying the temperature, a little each side of the specified value and see what it does for the boiler performance. It the performance improves or I seem to be getting cleaner combustion at that viscosity I'll change it a little more. Eventually I'll find the best viscosity for my burner and that's what I'll heat the oil to get. The result of that activity should be recorded in the maintenance log for that particular burner.

When we hear the term "heavy" applied to oil it can conjure up thoughts of extreme weights but the truth is that all oil is lighter than water. Heavy oils are just heavier than lighter oils. One other confusing factor is the use of "gravity" to define an oil. The API gravity of a fuel oil increases as the fuel gets lighter. API gravity is the ratio of a weight of oil of a specified volume compared to the weight of the same volume of water at the same temperature. To determine the specific gravity of an oil add 131.5 to the API gravity and divide the result into 141.5. Multiply that result by 62.4 to determine the pounds per cubic foot. An oil with an API gravity of 10 will have the same weight as water. Higher numbers are lighter than water.

Number 4 oil has a typical heating value of 146,000 Btu per gallon, weighs about 7.7 pounds per gallon and has an air-fuel ratio requirement of 14.01 pounds of air per pound of fuel. That is approximately equal to 108.2 cubic feet of air per gallon, 0.74 cubic feet per million Btu.

Number 6 oil has a typical heating value of 150,000 Btu per gallon, weighs about 8.21 pounds per gallon and has an air-fuel ratio requirement of 13.95 pounds of air per pound of fuel that is approximately equal to 114.6 cubic feet of air per gallon, 0.76 cubic feet per million Btu.

Pour point is one of the important values the operator should monitor when firing heavy fuel oils, especially Number 6. Before acid rain was recognized as a problem the pour point of fuel oils was fairly stable. When it became necessary to remove the 3 to 5% sulfur in the oil to reduce emissions the process changed the characteristics of the oils introducing a problem with elevated pour points. The Pour Point is the temperature at which the oil will start to flow. Oil in a storage tank that is allowed to cool below its pour point will not flow to the heater to be heated and pumped out of the tank. Heating the oil to a higher temperature ensures the oil will flow.

Desulferized fuels have a tendency to develop elevated pour points. Once the oil cools below its pour point and sets up it must be heated to a much higher temperature before it will flow again. Repeat the cooling and heating process enough times and the oil becomes a solid mass that will not flow and can't be pumped. The only solution to a gelled oil tank is to add chemicals and oil to dissolve the mass. Regrettably it can't be chopped up and burned as coal because once it gets in the furnace it will melt, becoming a liquid again at the high furnace temperatures.

Flash point is another property of fuel oils that should be watched. Those Pacific Specifications required Number 2 fuel oil have a flash point higher than 100°F. Heavier oils were listed for higher flash points, above 150°F. There are two methods for determining flash point, the common one being the open cup method where the oil is heated and a technician passes a standard match over the top of the cup containing the oil. When the oil is so hot that it generates enough flammable vapor to be ignited by the match the temperature of the oil is the flash point.

It's called flash point because the flame starts and extinguishes rapidly, flashing rather than continuing to burn. When you're burning oil with a low flash point any leak should be a concern. Temperatures in a boiler plant are frequently higher than 100°F, especially in the summer, and steam and hot water piping is so hot that they can generate flammable vapors if the oil leaks onto them. What about gas you say? Natural gas has a comparable flash point and it's around 500°F. When we started converting boiler plants to natural gas in the 1960's there were a number of concerned people expressing a common phrase "go gas—go boom!" But the truth is gas requires more energy to ignite than oil and it isn't as hazardous. Of the boiler explosions I've investigated the worse were always light oil fired.

In addition to the normal grades of fuel oil there are several sources of waste oils that can be burned in a boiler as fuel. A common one used in small installations is waste lubricating oil. If you are firing waste lubricating oils you're firing a very dangerous product because it can be tainted by gasoline. In one army base I visited the waste lube oil was from helicopters and it could contain a considerable fraction of jet fuel.

Usually waste oils are burned as a second fuel to limit the effect of their variable heating content and air requirements. Some systems use density meters to measure the waste oil flow to get a concept of air requirements and energy content according to its density. To date there isn't an economical means of obtaining instantaneous measurements of higher heating value and air requirements for waste oils.

Typical problems with waste oil firing include dirt and grit in the oil. There's also a concern for lead from bearings oxidizing in the furnace to produce high ground level concentrations of lead oxide around the plant.

Any grade of fuel oil is a hazardous waste if it escapes the normal containers and piping to leak into the ground or sewers. Of particular concern is any floor drain in the plant. The wise operator should know where the floor drains in the plant discharge. I remember years ago when we were converting a major university from coal to oil and a line leaking at the fuel oil pump and heater set ran away to a floor drain that discharged into a small creek right outside the boiler plant. No more than four or five gallons of oil escaped before the leak was discovered but the cost of cleaning up the

mess eliminated any profit we expected to make on the entire job. Today an oil leak can cost tens of thousands of dollars to clean up so you should always seek to keep any leak contained.

Oil can be supplied directly to the plant via a pipeline. In such cases you're relying on the supplier just like you would for natural gas. Most plants could not justify a pipeline directly from a supplier so they have fuel oil delivered by truck and have to store the fuel on the plant site. Storage doesn't have to be in tanks but potential hazards of leaks has eliminated use of open pits, old mines, and similar measures.

Tanks are generally one of three types, underground, above-ground horizontal, and above-ground vertical. Underground storage tanks are now labeled "UST's" for underground storage tanks and are a lot different than fifty years ago. Above-ground horizontal tanks are common for small plants and include the ones enclosed in concrete vaults for physical protection as well as fire safety. They're called horizontal because the tank is formed around a horizontal (parallel to the ground) centerline. Larger ones may exist but the typical horizontal tank is limited to around 90,000 gallons capacity. Vertical tanks are formed around a vertical centerline and can range in size from a few hundred to hundreds of thousands of gallons.

UST's became a hassle when it was discovered how many of them were leaking. From tanks at gasoline filling stations to those at every boiler plant more tanks were leaking than were intact. Much of it was due to a lack of understanding of how the tank and soil interacts as the fuel was added and removed. For years there was a standard procedure for installing an underground tank that consisted of pouring a concrete base then resting the tank in the concrete. Only after several years did we discover that the tanks changed shape, becoming more elliptical as they were filled and compressed the soil. The point between where the tank metal was trapped in concrete and bearing only on the soil provided a sharp corner that the tank was always bending around and that's where they cracked and leaked. There were other problems, mainly corrosion due to electrolytic action in the soil, that provoked leaks in those steel tanks.

The initial solution to the UST leakage problem was their replacement with fiberglass tanks properly installed so they could flex with the soil. It hasn't proven itself a wise decision. If you have a UST and it's to be replaced with another one, it should be fiberglass resin encased steel to get the best of both worlds. All installations since the early 1990's are required to have means for testing the tank and connected buried piping for leaks. Most of the piping is also installed inside conduit so a leak can be detected.

Any UST installed today has to be of a double shell construction, actually a tank within a tank. Monitoring of those tanks includes checking the "interstitial space" (the space between the two tanks) for signs of leakage. That's leakage either way, fuel oil out or ground water in. Any tanks of single shell design have to pass a "petro-tite" leak test regularly; those tests are performed by a licensed third party.

An operator's responsibility, when it comes to UST's is monitoring the existing tanks for leaks. That means that you keep track of the oil. You know how much you had, how much was delivered, how much was burned and, therefore, how much should be in storage. Storage equals previous quantity plus fuel delivered less fuel burned. Then you sound or stick the tanks to determine how much fuel is in them and compare that to your calculations. Some modern microprocessor based equipment is available that does all this for you, issuing an alarm when a leak is indicated. Regardless of that provision you should know if you have a leak of any significance.

Above-ground tanks aren't exempt from consideration. There have been many discoveries of leakage of Above-ground vertical tanks so monitoring them and testing them on a regular basis is necessary. Aboveground horizontal tanks are usually completely above the ground so a leak is apparent. That doesn't mean you shouldn't keep track of the fuel inventory. More than one Above-ground tank user has discovered mysterious disappearances of oil with no explanation. That's because some people know they can get away with burning No. 2 in their diesel vehicles if they're not too concerned for injector wear. Most of the heating oil that's not subjected to motor vehicle fuel taxes is now colored red and anyone caught with red fuel in their car or truck faces serious fines so that's not such a problem today.

One special purpose label we have is "day tank." That's a small fuel oil tank which is filled daily from the larger tanks in storage and used to supply the boilers. The initial purpose of a day tank was providing a supply of oil heated properly for pumping to the burners. It also eliminated double piping of oil suction and return to all the field tanks (the larger storage tanks). Oil in larger field tanks was allowed to be much cooler. A day tank requires means of filling it from field tanks and accepts the returned fuel oil from burners that aren't operating and oil relieved from the fuel pump discharge. The day tank could be heated to supply oil at burning temperature or just heated enough to flow properly through the

high pressure burner fuel oil supply pumps.

The oil is transferred from trucks to Above-ground tanks by fuel oil unloading pumps, "unloading pumps" for short. Those pumps are designed for high volume and low pressure to move the fuel from a typical delivery truck containing 8,000 gallons to the storage tanks. Oil transfer pumps are used to move the oil from one tank to another and from field tanks to day tanks. An installation with UST's may have neither of these because the truck can drop the oil into the underground tanks and fuel is drawn from the tanks by the burner pumps. In some cases fuel is drawn from storage tanks and transferred tank to tank using the burner pumps.

The pumps used to deliver the stored fuel to the boiler burners are the only ones called fuel oil pumps even though the others also pump oil. They are traditionally furnished in a package construction mounted on a steel base that supports the pumps and serves as a big drip pan underneath them to catch spills. When used for light fuel oil the pumps and a suction strainer are mounted on the base and we call that a "pump set" or "fuel oil pump set." Heavy oil fired installations include some heaters with the pumps to raise the temperature of the oil to a proper value for burning and another strainer with smaller openings in the screen to further clean the heated oil. The complete assembly with suction strainer, pumps, heaters, and discharge strainer is called a "pump and heater set."

What do we call oil pumps that are mounted on burners and fitted with a connecting shaft to the fan motor? I call them "wrong!" Try not to get stuck with them. At Power and Combustion we used to stock up on fan wheels before December because of those arrangements. We sold a lot of new fan wheels every time plants with those pumps had to switch over to oil.

Those burners are arranged so a short shaft with two coupling halves is inserted in the burner housing inside the fan wheel where a matching half coupling receives one end. The other end of the shaft engages a coupling half on the oil pump which is mounted on the outside of the fan housing with its shaft through a hole in the housing. You have to practice yoga or something to be able to get your hands in there and install that shaft properly and tighten the set screws that secure the coupling halves. Do it wrong and the shaft flies off when you start the burner with subsequent damage to the fan wheel. We, along with other burner representatives, made a lot of money on that design but I refused to sell the darn things. You want a pump set, not one of those monsters.

Heavy oil is not heated to a certain temperature so the oil is hot enough to burn. It's heated so it flows properly; viscosity giving us an indication of its ability to flow. Storage tanks should be heated only enough to get the oil to flow to the day tank or fuel oil heaters, anything hotter is just a waste of heat. That's because most storage tanks are not insulated. Heating the oil to the right viscosity for burning should happen just before it goes to the burners.

It's necessary to run some of that heavy oil through the piping of an idle boiler to keep it flowing. We call that recirculation and it's essential for oils that could become solid in the piping and prevent our starting the idle boiler. There's normally one globe valve in the piping that returns the oil to the pump suction or the tank (return oil piping) and that valve is throttled for several reasons. If we open it too far it can return more oil than the pump is delivering with a resulting drop in oil supply pressure. Carelessly open a recirculating valve too far and you can force a shutdown of the entire plant.

If you don't recirculate the oil enough the heat losses in the piping will lower the temperature until the oil is too cold when it gets to the burner. You need to open the valve enough to get the hot oil to the burner. On the other hand, the oil can return to the day tank to raise its temperature so high that the pumps can't create enough pressure and you're shutting the plant down again. That happens because more oil slips back through the pump as the viscosity increases and, therefore, less is forced out the piping to the heaters and burners.

In many plants the operators aren't trusted to do it right so the recirculating control valves (those globe valves in the return piping) are set and locked or the handwheels are removed so you can't mess with them. The best of both worlds is to throttle the recirculating control valve enough to keep oil flowing to the burners and back the return line with only one boiler (the one that you would start up if necessary, sometimes called the standby boiler) having enough recirculating flow to get the right temperature at the burner. That way flow is assured but you're not returning so much that the oil entering the pumps gets too hot.

Almost all fuel oil pumps are positive displacement pumps. Gear types and screw types for the most part, they're capable of raising the pressure of the oil considerably so it can be delivered to the burners at a pressure high enough for proper atomization. Since it's a positive displacement pump it'll deliver a relatively constant quantity of oil. The oil you don't burn is returned, sometimes to the pump suction, others to the fuel oil day tank or storage tank. To maintain pump discharge pressure and control the flow of oil to the tanks requires a relief valve, either pump mounted or piped onto the pump set. A relief valve, not a safety valve. Even when more precise pressure control is provided you normally have relief valves at the pump set.

The self contained relief valve has to experience a change in pressure to change the flow of oil. In order to be stable in operation a reasonable pressure droop of ten pounds minimum is required between conditions of no fire and full load on all boilers. A relief valve might return all the oil to the tank at a pressure of 180 psig and close off that port so all oil flows to the burners at 170 psig. If you try to install one with a smaller droop the flow and pressure will be unstable. It has to be that way so don't expect the pressure relief valves at the pump set to maintain a constant oil supply pressure.

It's the variation in supply pressure that makes for tiny variations in flow through the fuel oil flow control valves in certain burner systems so additional provisions for pressure control are usually provided in an oil system. Since the pump set is usually remote from the burners a second pressure adjustment is made closer to the burners by a back pressure regulator. It's a self contained control valve that maintains a more constant pressure on its inlet by dumping some of the supplied oil into the fuel oil return line. It's really a relief valve but normally has a much larger diaphragm so the swings in pressure are not as great as they are for the pump set relief valve. The two in combination produce a much lower droop.

For really precise oil supply pressure control two measures are used. One is a pressure regulator at each boiler. The regulator has a large degree of droop but since it's repeatable the pressure at any particular firing rate is the same regardless of oil supply pressure. The other is installation of a more elaborate back pressure control system, from pilot operated valves to a complete control loop with transmitter, PID controller, and control valve.

Heating of the heavy fuel oil on small systems may consist of a simple temperature actuated control valve but most of the systems use a temperature piloted pressure control valve. A valve that acts on temperature alone will allow large swings in oil temperature with swings in flow to the burner because the control valve doesn't know the oil flow has increased until the colder oil gets to it. By then the lower steam pressure has allowed the metal of the heat exchanger to cool as well so the temperature controller will have to overreact. A temperature piloted control valve simply uses the temperature of the oil to set a steam pressure to be maintained in the oil heater. When the oil flow increases it will use more steam to heat it and the pressure in the heater will start to drop. The pressure controller opens the valve to compensate for the pressure drop, maintaining the pressure and a more precise output temperature.

Newer strategies include viscosity control. An instrument is installed in the piping to sense the viscosity of the oil and there are several methods for analysis ranging from vibrating a heavy wire in the oil to trapping a sample and dropping a plunger through it. Whatever method of sensing is used, there's still the problem of response time so a viscosity controller should only be used to produce a set point for the steam pressure.

On many of the facilities that we converted to light oil in the 1980's we suggested the customer retain the fuel oil heaters. Testing at that time indicated a light oil burner would operate cleanly and more efficiently with a little better turndown if the oil was heated to 120°F. That was, of course, a temperature still below the flash point of oils supplied at that time. Now, with lower sulfur requirements some light oils have flash points just barely above 100°F.

The most difficult activity regarding fuel oil handling for operators today is keeping the system in operating condition when it's never used. Your SOPs should include regular operation of the system to ensure it's operational and a drill for switching to oil for each operator in the fall before the first winter interruption can be expected. I still remember one plant I was asked to investigate where none of the operators could switch over to oil.

COAL

Coals are commonly identified by their source, either by the area or state in which they were found or a particular mine. There are three distinct classifications of coal, Anthracite, Bituminous, and Peat which are principally related to the crushing strength of the coal with anthracite being the hardest. Other criteria includes size of the coal particles and characteristics that affect handling and burning. Coals that are fired on grates must be large enough that they don't fall through the holes in the grate and have a limited portion of fines that would fall through.

Coals that are pulverized to something close to talcum powder so they will burn in suspension (floating in the air very similar to an oil fire) are graded by how difficult it is to grind them. An operator in charge of a coal fired boiler plant should be aware of the specifications of the grate or burner manufacturer and boiler manufacturer and how variations in those specifications affect its firing.

Coal and oil require less air to burn than natural gas and LPG for the same heat output. That's because the gases have a higher hydrogen-carbon ratio, more hydrogen in the fuel produces more water which increases stack losses. Some hydrogen in the fuel is always desirable because it helps form coal gas which is a gas that burns far more readily than solid carbon. Also, higher moisture content in flue gas seems to improve boiler performance because water makes it possible for gas firing to clean up soot that collects on boiler surfaces when firing oil during a gas interruption. The accumulations of soot associated with firing oil and coal are due to fixed carbon that can't be readily converted to gas so it can burn.

Another form of coal that's being considered as a fuel is culm. That's the waste material removed from coal mines which contains some coal but is mixed with dirt. There are several huge piles of culm around the mines of this country. Some are big enough to supply a plant for several years. Modern fluidized bed boilers are capable of burning that material.

As a solid, coal requires different handling methods than oil or gas (both fluids which can be transported in pipelines). Every once in a while I'll run into another attempt to burn a coal and oil slurry, a mixture that handles like a fluid but contains particles of solid coal. Some utilities manage to burn it successfully but I haven't seen a successful operation in a small boiler plant.

Coal is usually delivered by truck or railroad car. In either case they can present a serious problem in the winter if rained or snowed on with subsequent freezing. Usually a plant with railroad supply will have a melting shed where the cars are heated to melt the ice so the coal will flow. The trucks or rail cars are dumped into a hopper where a conveyor picks up small quantities of it and lifts it to the bunkers or a storage pile.

Storage piles are simply piles of coal stored for burning. Unlike a fuel oil storage tank there's no enclosure so the coal is subject to degradation from weather. They also have a bad tendency to ignite spontaneously if left sitting too long. When it comes time to burn, or move, the coal another conveyor can do it or it might just be handled by you operating a little front loader. In either case the coal is eventually transferred to the burners.

Conveyors come in a wide variety of sizes and styles. Many use a belt, a wide fabric reinforced rubber or synthetic rubber riding on rollers that shape it into a trough that holds the coal. At some point in a belt conveyor system the belt is pulled taught by a roller that's adjustable and the belt makes a full 180 degree turn over the roller. Belt conveyors are used mainly in large plants where a constant movement of coal is required. I've never had the pleasure of working a coal fired plant with belt conveyors so about all I can tell you is to treat the belt with care. Sudden stopping and starting of large belt conveyors tend to break the belt.

One coal requires careful handling on conveyors. Coal from the Powder River Basin in Wyoming and Montana is used throughout the country in pulverized coal fired boiler plants because of its lower sulfur content. PRB coal, however, is very friable (it breaks up easily) and it generates dust readily. After several explosions in conveyor systems major rebuilds were required when handling PRB to ventilate transfer sheds and other areas thereby preventing flammable concentrations of airborne dust accumulation and to conveyors to reduce dust generation hot spots and sparking. Pay close attention to the special instructions you normally receive when a plant is switched to burn PRB coal.

The typical small coal fired plant will use a front loader to move coal from storage to a bucket elevator that lifts the coal from grade level to the bunkers. A bucket elevator can be a belt with small containers (buckets) attached to it or any number of unique arrangements of chains, connectors, and buckets that form a continuous and endless string of buckets to scoop up the coal and lift it to a higher level in the plant where it's dumped into the bunker. In some medium sized plants the bucket elevator will dump the coal onto a special belt conveyor that distributes the coal into the bunkers. The belt conveyor will have a special assembly consisting of a couple of rollers that flip the belt twice, all mounted on a set of rails so it can be moved along the length of the belt. When the coal gets to the assembly it's dumped as the belt turns at the first roller and is deflected past the second turn of the belt to fall into the bunker.

Another special conveyor for coal is a Redler conveyor. It consists of a continuous chain with metal paddles that ride inside a rectangular metal tube. The top of the tube is eliminated at the in-feed hopper (where the coal is dumped or falls from a storage pile) so the paddles can intercept the coal and start dragging it along. The tube is closed for lifting and transporting the coal horizontally past a series of gates. Each gate consists of a section of the tube where the bottom can be opened to allow the coal to fall out. The conveyor can then deliver coal to a large number of bunkers.

Okay, now it's time to explain what a bunker is. It's sort of like a day tank for coal. I can go in many boiler plants today and look up to see the bunkers are still there. That's even when the plant hasn't burned coal for several years. A bunker can be a concrete room (for all practical purposes) or the more common catenary form of hopper. The shape was developed to hold coal without a lot of reinforcing and structural members. Steel plates were made long and literally slung, somewhat like a hammock, from the building framing in the space above the burner fronts, what we call the firing aisle. The result was something like a half ellipse in shape hanging down above you with trap door openings that were used to release the coal for feeding to the burners.

To keep track of the coal and transport it from the bunker to individual burners many plants have weigh lorries. These are mounted on tracks with wheels similar to those on a railroad car so the lorry can be moved from under the bunker horizontally along the tracks to a position above the coal hopper of the boiler being fired. The lorry incorporates a hopper to hold the coal dropped from the bunker and its own drop gate to empty the lorry hopper into the boiler hopper. The hopper on the lorry is suspended from the wheels and arranged like a scale so the operator can weigh each load of coal. That way you can get an idea of your coal firing rate in pounds per hour and track how much coal you burned on a shift.

Weigh feeders which consist of a short (up to five feet long) belt conveyor with the belt assembly suspended so it's weight can be consistently monitored are used for coal fed to pulverizers and, in a form similar to the ones used to feed bunkers, hoppers for stokers. They provide an indication of the rate at which fuel is being fed to the boiler.

One rare (I've never seen one) but possible system to encounter is pulverized coal storage. It would consist of a bunker but be covered and incorporate additional safety measures because the fine powdered coal readily forms a combustible mixture when exposed to air and any agitation. I imagine I've never seen one because of the hazards associated with them; they're just plain rare.

When firing coal, whether on a stoker or pulverizer (see the section on burners) a continuous supply of coal to the hoppers or pulverizers is always a function of the operator. Usually in a coal fired plant we'll say "operators" because it takes more than one person to keep the coal moving and burning properly. You may also be operating processing equipment that changes the size of the coal particles or screens to actually separate out some of the coal to provide the size and form of fuel that's required for the burner.

Coal also requires handling after it's fired. A certain amount of ash and unburned fuel (frequently called LOI for loss on ignition) collects in the bottom of the furnace, on top of the grate, and in the dust collector at the boiler outlet. It has to be handled back out of the plant to be dumped in a landfill or used in cement operations.

OTHER SOLID FUELS

Biomass fuels can vary from firewood, the most common, to bedding which contains some unpleasant animal waste but still burns. There are many varieties of wood and a considerable variation in other vegetation that can be burned. There are more ways to burn those fuels than there are fuels and new methods of burning them are still being developed.

As with coal you have to prepare the fuel to conform to the specifications of the burner manufacturer so it will burn well. They have a higher hydrogen content so they tend to burn cleaner. The major problem with these fuels is their high moisture content, liquid water in the fuel cools the fire in the furnace and the vapor produces high latent energy loss up the stack. The fuel's lower cost normally compensates for that.

Wood can be fired in several forms, logs like on a campfire, chips as large as a playing card and about one-half inch thick down to sizes rivaling sawdust and various sizes of dust from sawing (where the dust is more like a chip, sometimes as big as one-quarter inch square) to sanding. Some of the finer and lighter materials can be burned almost entirely in suspension (floating in the air) in a flame that is similar to an oil fire. Most of the chip is burned on a grate although it's common to introduce the chips by tossing them in above the grate where the finer dust in the fuel is burned in suspension.

Some wood burners are dealing with raw wood which has a high moisture content and much of the energy in the fuel is used to vaporize that water. Others fire kiln dried wood which has less than 10% moisture and is an excellent fuel. The construction of the boiler and the grates are designed for the fuel to be burned and it's usually difficult to handle a different material. A boiler designed for dry fuel will probably fail to reach capacity when burning wet fuel and may not maintain ignition. A boiler designed for wet fuel will probably have problems of burning up grates due to the higher flame temperatures of the dry fuel.

A principal problem with wood firing is sand. When the fuel is cut, hauled, and prepared for firing a certain amount of dirt comes with it and sand can erode boiler tubes quickly as it's carried by the flue gas out of the furnace into the tube banks. Sander dust will always contain a certain amount of flint and other sharp sands that are very damaging to the boiler. When a boiler is de-

signed to fire wood that is sand contaminated the velocities through the tube banks are intentionally reduced to limit the erosive effects of the sand. Operators should also avoid any action that produces high gas velocities (too much excess air, over-firing) to reduce erosion damage.

Leaves are another potential source of boiler fuel that isn't used as much as it could be. A principal problem with leaves is they're only available at certain times of the year. Firing problems with leaves include an ash content greater than wood but the big one is that the fuel is tough to handle, can be messy if it gets wet, and can be contaminated with sand and dirt. There are some systems that convert dry loose leaves to compact fuel packages by extruding them.

Bagasse is sugar cane after all the sugar juice has been squeezed out. Since I've never spent any appreciable amount of time in the south where the cane is grown I have no experience with burning bagasse. I do know that the long stringy material is tough to handle and burn.

Other natural sources of biomass include hay, animal bedding (yes, it all burns), and corn cobs. Dried corn itself has been used for a fuel.

Waste paper, cardboard and similar materials that are contaminated, so they can't be recycled into more paper, are burned in trash burners but some major government and industrial facilities that process a lot of paper may have boilers fired by those fuels just so they destroy the material for security purposes. Corrugated cardboard is one of those fuels that's very dangerous to burn. That's because it comes with its own air supply within the corrugations. When a corrugated cardboard is fed into a hot furnace the heat will start boiling away the glues and wood to form gaseous hydrocarbons that mix with the air within the corrugations. When the mixture reaches its explosive range it explodes!

Hospital waste is normally burned in an incinerator with energy recovered by a waste heat boiler. The purpose of the separate incinerator is to ensure that all the material is exposed to the heat of the fire so all the diseases and pathogens in the waste is destroyed. My experience with these systems is that's not always the case. Unless the waste is mixed up so it's all exposed to the heat there will be unburned, and sometimes untreated, fuel in the waste. The waste heat boilers must be designed for high ash loads and capable of withstanding occasional acid attacks because of the acids produced while firing the waste.

Trash burners, large boilers burning tons of garbage are considered an air pollution hazard and many localities chose to landfill their garbage rather than burn it. Today the cost of landfill space and the offset of better flue gas cleaning systems has restored interest in trash burning plants. It's a unique boiler plant because you actually get paid for burning the fuel, a far cry from paying for fuel. The cost of operating the plant, personnel, and the continuous repairs required (like when you try to burn an entire engine block) are covered by the value of the steam produced and the payments for processing the trash. I say processing it because there's a considerable amount of ash left over, around 10%, that's usually returned to the county or city for placement in a landfill so what you're doing is reducing the volume of trash they have to deal with.

I've worked on boilers burning many other forms of solid fuels including such unique materials as laminate trimmings and plastic bags. Almost any organic material can be fired, the question is whether the source of the fuel is consistent in generation of quantity and quality and how much it costs to prepare, handle, and burn the fuel. If you have the opportunity to work in a waste fuel plant you should realize that the cheap fuel has a lot of heating value and should be treated as if it was as expensive as any purchased fuel. If you don't burn the waste fuel efficiently then any deficit has to be made up with purchased fuels.

Know your fuel, know what the fire looks like when it's burning normally and get real concerned when it isn't normal. Keep in mind that how the fuel is stored and handled on its way to the burner can have an effect on plant safety.

WATER

I consider a major problem with most Americans is their attitude about water. As a consumable water is not an unlimited resource. Despite a recent three year drought in the Northeast I find my friends and neighbors still acting as if there was an unlimited supply of potable water. Continued growth of the human population will constantly expand the demand for fresh water and, like it or not, we're dangerously close to conditions of real water shortage; first a shortage of drinking quality water.

A boiler plant has the potential to draw on, and waste, millions of gallons of water each year and some plants consume and waste those millions in months or even weeks. I consider it regrettable that we place such a low value on water. I hope that's beginning to change. In the Baltimore metropolitan area we're charged something in the range of three to four dollars per thousand

gallons of water consumed in a combined water and sewer charge. It should be interesting to note that the majority of that cost is for sewage treatment. I know a few localities where the rate is much higher, $15 per thousand and higher. A recent EPA estimate is $16.40. Wise operators will address those costs and recognize their contribution to the preservation of this invaluable natural resource.

Major utility plants are doing something about it because water represents a significant cost to them. Where possible they're using treated waste water (from sewage treatment plants) for make-up instead of fresh potable water. Despite the yuck factor there's no reason to question the quality of that water after proper pre-treatment (see the section on water treatment) and the boilers don't care what it may have been. PSEG's power plant in New Jersey saves 10 million gallons of precious drinkable water each month by using waste water as makeup, saving more than thirty thousand dollars a month in the process.[7] Sooner or later you will be working in a plant that does it too.

It's very important to understand what a gpm is worth. I've discovered that many operators have a change in understanding once they do the math themselves. What is a gallon per minute worth? First, it's a good idea to know that a minute doesn't give us a fair measure of the cost. There are 525,600 minutes in a year, more than half a million. A two gallon per minute leak that we allow to continue represents more than one million gallons wasted every year. At the low range of water costs a one gallon per minute leak costs $1,500, what some people consider minimal. I don't consider such waste minimal. When you consider the fact that half a million gallons of clear fresh water was converted to sewage that leak is very expensive.

I'm always objecting to something I see regularly, a boiler water sample cooler operating constantly. I know that it takes a few minutes to clear lines and tune up a sample cooler each time you draw a water sample and a little more time to close the valves when you're done. It's also easy to argue that the boiler water would be removed by blowdown anyway. However, the typical sample cooler uses about 12 gpm to cool a boiler water sample and leaving it running constantly wastes over six million gallons of water every year and costs at least $18,000 per year to convert good water to sewage. Don't do it.

Recycling the water in a boiler plant is becoming increasingly important. Some utilities are actually committed to zero discharge, where they don't put one gallon of waste water into the local municipal sewer or dump it otherwise. Part of that effort is to avoid the heavy cost of treating the plant's discharge of waste water which is highly concentrated with solids and chemicals compared to water that's simply wasted to a drain. It's an action I am glad to see. We should all understand that especially blowdown contains considerably more solids and chemicals than normal waste water so minimizing blowdown is important.

Another consideration is the draining of a boiler and refilling it with fresh water during every annual outage. As the cost of treating sewage continues to rise and concern for the treatment of very caustic waters grows there may come a time when dumping a boiler is restricted. If your plant doesn't have a connection to a sanitary sewer, and many don't, I strongly recommend you rent a tank trailer to store your boiler water while performing your annual inspections. That way you aren't discharging caustic water into the environment unnecessarily and you save on the cost of the chemicals it contained (although loss of sulfite is expected) and the cost of treating fresh makeup water.

I've also seen a fair number of operators use boiler bottom blowoff as a means of water level control. In the chapters on control I mention one probable reason that an operator feels compelled to do this but even if the controls malfunction there's no reason to consistently waste water and boiler chemicals in order to maintain boiler water level. If the level tends to rise it can be prevented by restricting feedwater flow.

It's also not sensible to use bottom blowoff as a means to reduce the dissolved solids content (TDS) of the boiler water instead of using continuous blowdown, what we sometimes call surface blowdown. Removal of the boiler water to limit the TDS is best done with the continuous blowdown because it removes the most concentrated water in the boiler, the water that's left right after the steam is separated. Bottom blowoff tends to be a blend of the boiler water and feedwater that just dropped to the bottom drum so it contains a much lower concentration of solids. None of the water or heat is recovered from blowoff; water and heat is recovered by a good blowdown heat recovery system.

What about that bottom blowoff? It's used to remove sediment, mud and sludge that either enter the boiler with the feedwater or created by the water treatment. With chelants and polymers the scale forming salts are sequestered or trapped in solution and removed in the continuous blowdown so blowoff isn't needed anywhere near as much as it used to be. The chemical salesman may give you a schedule for blowoff that wastes chemicals and his commission is based on the value of the chemicals you purchase. Especially after a switch to chelant or polymer treatment from 100 % phosphate (maintaining some phosphate as an indicator of the operation of the others isn't a bad idea) your chemical treatment representative should produce a blowoff schedule that is less frequent. If not, check it yourself. I suggest making certain that you perform a bottom blowoff before shutting down a boiler for inspection at a time before shutdown to match the scheduled interval for blowoff plus a shift or day. Use a shift if you're blowing off more frequently than every two or more days. Then, slowly drain the boiler after cooling and look at the bottom of the mud drum and headers right after you open the boiler to see if there's any accumulated sludge or mud. If none is there and the boiler had been operating normally (not in standby or other low load modes) before the shutdown, increase the interval between blowoffs so that it matches the interval you just established. After a period of time you'll come up with a small accumulation in the mud drum which means you just exceeded your limit and stick with the interval that you know works.

On steam systems blowdown heat recovery systems capture much of the heat and a little bit of the water that's dumped by continuous blowdown. The blowdown is dumped into a flash tank which operates at a pressure slightly above the deaerator pressure. Since the water is much hotter than the saturation temperature at that pressure some of the water flashes into steam. The steam is separated by some internals then flows to the deaerator where it replaces some of the boiler output that would need to be used to heat the feedwater. The remaining water then flows to the heat exchanger. Low pressure plants and small high pressure plants may not be able to justify the flash tanks so all the blowdown water flows to the heat exchanger. The heat exchanger transfers heat from the blowdown to the makeup water.

In low pressure plants the heat exchanger can be as simple as a barrel set above the boiler feed tank and arranged so the makeup water is fed into the barrel then overflows into the boiler feed tank. The blowdown is passed through a coil of tubing in the bottom of the tank then to the blowdown control valve (which can be manually set) or an automatic one. In a plant with multiple low pressure boilers each one could have its own coil. If the flow isn't throttled after the heat exchanger the boiler water will flash in the coil, making a lot of noise and eventually damaging the coil.

A heat exchanger in high pressure plants should be of high pressure construction and heat the makeup water before it goes to the deaerator. The control valve on the heat exchanger outlet is usually controlled by the level in the bottom of the flash tank. That way the heat exchanger is always flooded and the blowdown is not flashing into steam which can leave deposits that plug up the heat exchanger.

Blowdown heat recovery does save some energy and, with a flash tank, a little bit of water. The real savings, however, is in the water that would be used to cool the blowdown if you didn't have the heat recovery system. Blowdown dumped through the blowoff flash tank will dump some heat in the form of steam up the vent but the 212°F water has to be cooled to less than 140°F before it's dumped in the sewer and the typical practice is to use good city water for it. It will take a volume of cold water about equal to the blowdown to cool it before it's dumped in the sewer. You're wasting all that cold water if you aren't using the blowdown heat recovery system.

The best way to reduce water waste in a steam plant is to recover condensate and use it as boiler feedwater. There are many reasons for this in addition to saving water. Recovery of condensate recovers heat, eliminating the need to heat cold makeup water before it's fed to the boiler. Condensate is basically distilled water, converted to steam in the boiler and then condensed so it doesn't require all the pretreatment and chemical treatment needed for fresh makeup water. Recovery of condensate saves money that would be spent on additional fuel, boiler water treatment chemicals, and the additional water required for blowdown to remove the solids brought in by fresh city water.

All too frequently the only consideration for recovering condensate is the value of the heat. After evaluating several condensate recovery projects I can assure you the cost of heating the water is minimal. The cost of the water itself is more valuable than the heat and the cost of chemicals adds even more to it. Treat condensate as a valuable resource.

Recovery of condensate is the best way to minimize water waste from a boiler plant but there are times when recovery for use as boiler feedwater is undesirable. Wasting of condensate is not unusual in a chemical or petroleum facility because the potential for contamination of that condensate is so high. In some cases we recovered some of the energy from it using a heat exchanger but that doesn't preserve the water. Capability to monitor the water and filter it with carbon filters and other measures, including reverse osmosis, make it possible to recover and use condensate in those plants today.

In instances where the capital cost expenditure to recover condensate is so high that recovery can't be justified it's possible that the condensate can be used for other purposes, anything from makeup for cooling towers (I know it's hot, but it's also distilled water) to use as sanitary water (where it has to be cooled). In chemical and petroleum facilities there's considerable water used in scrubbers and condensate makes a great replacement for scrubber makeup. In other words, if it has to be wasted try wasting it in another system instead of fresh water.

Appropriate recovery of condensate is another matter. I've found that several plants allow considerable waste of high temperature condensate by collecting it in open systems where as much as 15% of the water and over 50% of the energy in that condensate is lost in flash steam. High pressure condensate should be recovered in a way that prevents flashing. The best way to recover high pressure condensate is to return it to the deaerator. Some of the condensate may flash off in the deaerator but it simply displaces some of the steam required for deaeration. There's no reason to be concerned for oxygen in high pressure condensate, but it's typically returned in a manner that allows some scrubbing of it to remove any oxygen that may be in it from start-up and other operations.

Having explained that condensate that would flash off should be recovered in a manner that uses that steam it only makes sense that any signs of steam escaping from a condensate tank vent line is a problem that requires an operator's attention. The normal reason we have steam leaking is leaking traps. Trap maintenance is very important in reducing water waste.

TREATMENT CHEMICALS

I'm always listening to plant chiefs complain about the price of water treatment chemicals. They aren't cheap and they sure aren't anything that you want to treat casually. In the normal plant they're about two percent of the total cost. The amount of chemicals we use are a function of the amount of makeup water entering the plant so preserving water is the first important step in minimizing the cost of water treatment chemicals. The following section deals with water treatment because it's definitely one of the most important things that boiler operators have to do. Considerations of the chemicals as consumables are addressed here.

The concentrated treatment chemicals are definitely hazardous waste if they escape their containers or treatment equipment. They are hazardous to handle and can cause severe burns. I know the attitude about how we can be perfectly competent in handling the material and shouldn't need the protective gear because we never make mistakes, right? Now that I know a few people that have been seriously injured handling treatment chemicals I can honestly say that the wise operator uses all the protective gear.

I regularly thank God that I'm not one of those hard heads that got hurt handling chemicals, there's nothing other than will and dumb luck that prevented it. You may feel you look stupid in the clunky rubber boots, silly rubber apron, klutzy rubber gloves and the face shield that steams up so it's hard to see what you're doing—but you're safe. Not wearing that outfit is taking a chance on living with a serious injury for the rest of your life; wear it.

Frequently people don't think of salt as a water treatment chemical. It is, and it's one of the cheapest and safest to handle so you want to make sure you make the best use of it first. Ensure the water softeners are regenerated with adequate brine concentrations and regenerate them before they're depleted to minimize consumption of phosphate or other scale treatments which are a lot more expensive than salt.

Take regular samples of the incoming makeup water to check for changes in hardness that will alter the capacity of the softeners and adjust the softener throughput accordingly. You don't want to be like one plant I visited for problems with their new boiler. Blisters at the bottom of the boiler's waterwall tubes were a

sound indication of high degrees of hardness in the water. When I asked about the regeneration of their softeners I was told they did it just like they always did, every Wednesday. It didn't seem to matter to them that the steam demand on the plant, and makeup, had tripled in the last three years. The softeners ran out of sodium ions on Monday.

Applying the chemicals in a uniform matter, consistent with the rate of boiler water makeup will minimize their use by making them most effective. Some systems, such as low pressure hot water heating systems, require very little treatment because the system is closed and losses of water are very limited so shot feeding of chemicals using a shot feeder (Figure 7-2) is capable of providing adequate treatment.

Those shot feeders do, however, often look much like a mess where it's evident that the chemicals were spilled and wasted as opposed to injected into the system. Proper use of a shot feeder requires closing the isolating valves and proving them closed by slowly cracking the vent valve while holding a bucket under the vent pipe to capture any discharge. It's possible for the shot feeder to accumulate some air or gas from the system so the contents could expand out dramatically when it is opened to atmosphere. It may require waiting several seconds or even minutes to allow pressure to bleed off slowly before it's relieved. If only liquid flows out that's an indication that one or both of the isolating valves are not shut. Be sure you wear the silly outfit because expanding gas can carry out slugs of water that could still contain concentrated chemicals and splash them on you.

Once the pressure is relieved the shot feeder should be drained by opening the drain valve with a bucket under it to capture the contents. If the contents are system water it's the best thing to use for mixing the new charge for chemicals. If the contents appear to be a concentrated mixture of chemicals it means the feeder didn't discharge its contents; in that case, close the drain, open the fill valve, pour it back in and return the feeder to service to get the chemicals where you want them, in the system. Be certain the drain is closed, checking it by adding a cup or two of fresh water, then open the fill valve and slowly pour in the new mixture of chemicals.

To charge the chemicals close the fill valve, close the vent valve down then crack it a little and crack the feeder outlet valve to fill the feeder pot. Hold a small container under the vent line to capture the first shot of water and close the vent valve as soon as the water appears. Finally open the feeder outlet valve and the feeder inlet valve to

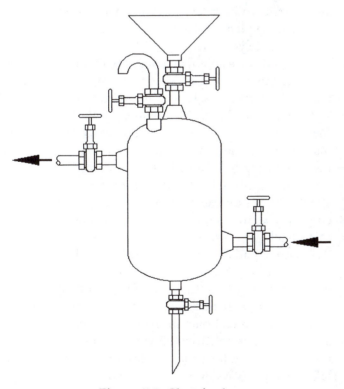

Figure 7-2. Shot feeder

discharge the contents to the system. When the feeder is flushed by a high differential pressure (a typical arrangement is from the system pump discharge to the same pump's suction) it's advisable to limit opening the feeder inlet valve to limit thermal shock from any cold contents of the feeder. It also prevents sending a slug of chemicals into the system instead of a solution of them.

Failure to vent a pot feeder is a common problem. Always flood it before putting it in service. If you don't then you stand the risk of having a compressed gas burp blow concentrated chemicals on you when you attempt to open it. It's also possible to send some air into the system to collect in some obscure spot and restrict system water flow.

To prevent loss of valuable sodium sulfite you should keep the containers tightly closed. A sulfite mix tank for a chemical feed pump should have a floating top or be otherwise sealed to limit atmospheric oxygen getting at the contents to consume the sulfite before it even gets into the system water.

Be careful mixing and handling caustic mixtures. I can still remember being so stupid as to try to use a piece of galvanized lagging as a funnel to add boil-out chemicals to a boiler drum. The gas and splashing from the reaction of that caustic solution and the galvanizing could have blinded me or caused serious burns. Aluminum and galvanized steel (actually the zinc in the galvanizing) react violently. I remember another incident where someone used a galvanized bucket to mix some caustic solution and it literally boiled out of the bucket to create a hazardous spill and almost burned the individual seriously.

MISCELLANEOUS

One consumable that a plant always seems to have troubles with is small tools. I can remember one chief that had a policy of buying seven of any new tool, one for the plant and one for each of the operators to take home. He explained that by doing so he eliminated his personnel stealing the tool and, since they all had one at home, ensured the extra one he bought would be at the plant when it was needed. Even though his policy seemed to work it didn't speak well for those operators and I thought it was actually berating them. They didn't seem to mind because they got new tools regularly but I would have considered it an insult.

I can remember more stories about stealing of small tools and how many people treated it as an acceptable practice, even implying a respect for the skill of some of the thieves. I have no respect for them and I have a problem with anyone that steals the owner's property. If you can't be trusted with a little tool that probably costs less than one hour's salary how can you be trusted with a plant that costs thousands of dollars a day to operate?

I tried to institute a policy of loaning tools. That way if an employee had a project going at home that required a particular tool he or she could borrow it, like taking a book from the library, and return it when finished with it. It included items that aren't easy to steal like scaffolding and tall ladders. I was very disappointed to discover that some people felt it was more manly to steal a tool than borrow it so the program didn't work very well. Good, wise, operators don't steal the owner's tools.

It would be nice if more of us treated other people's property with respect. Please join me in doing that and ask your boss if you can borrow a tool that you would like to use at home. I still have something I stole that I keep just to remind myself that I was very ashamed afterwards, it wasn't worth much at all and, had I been caught, it could have cost me my career. Stealing may seem heroic and being one of the guys (male or female) but it's still stealing and somewhere down the road you will be ashamed of it. Try taking pride in the contention that your plant has tools that have been there for years.

Another problem with small tools is breaking or damaging them. Wood chisels don't cut nails very well and electric drills make lousy hammers. I do hope you will treat the plant's tools with as much care and attention as you do your own.

Batteries are another commodity that is frequently converted to private use. Somehow people get the idea that their alarm clock is required to get them to work so the batteries should be provided by their employer. Rags are another commodity that can be abused. I once discussed this matter with a plant chief that had a $500 per month rag bill! Nobody was stealing them necessarily they just wasted the darn things. Wise operators should always treat every little thing supplied by their employer as the employer's property, not something that somehow reverts to their possession.

Paper pads, pencils, erasers, scotch tape, and making copies, it doesn't amount to much and many employers say to use those resources without concern because it costs them more to account for it than to let you take it. Since I had an expense account I always allowed enough to cover the value of things I used. If there's no policy for using the owner's property don't take it. Between Wal-Marts and Kinkos on almost every corner there's no reason to.

Chapter 8

Water Treatment

Water is, unquestionably, the most unique substance we will ever encounter. It's unique character is important to every form of life on earth. Controlling water's unique characteristics is one of the major occupations of the wise boiler operator.

WATER TREATMENT

There's more to it than H two O. Perhaps its because there are so many water treatment companies and salesmen that insist their product is the do all and end all that boiler operators tend to believe they can't do much about water treatment. The fact of the matter is that nobody can do a better job than a boiler operator that's been trained. Water treatment isn't a black art and it doesn't require a college degree to understand it. The only problem with it is you can't see what's going on in there and you have to accept certain statements about it as fact and base the rest of your decision making on them. Let's see if your understanding of water treatment isn't improved in these next few pages.

Water is called the "universal solvent" because it dissolves just about anything. It's such a great solvent that it even dissolves itself! I like to think it's because the water molecule is lopsided; it contains two hydrogen atoms with an atomic weight of one, so they're very light, and one atom of oxygen with an atomic weight of eight; the two hydrogen atoms hang around one side of the much larger oxygen atom. That lopsided condition results in a concentration of protons at the one side, where the hydrogen is hanging out, and nothing but electrons at the other side so the molecule of water has a magnetic polarity.

It's a reasonable explanation for why a microwave oven works. The microwaves are building up and then dumping a magnetic field in the food in the oven several times a second and the polarized water molecules keep twisting back and forth to align with the magnetic field. Other things, like plastics, don't have any polarity and aren't affected. All those water molecules twisting back and forth inside the food rubs the other molecules and heats everything up by friction.

That polarity of the water is what makes it such a good solvent. It has a negative charge on one side and a positive one on the other so it can pull other molecules apart. It pulls molecules of H_2O apart, converting them to hydrogen ions (H^+) and hydroxyl ions (OH^-). That's how water dissolves itself.

Every solid material dissolved in water is present as an ion. You'll note the little plus sign and little minus sign which indicate that the atoms have something like an electric charge on them, not unlike the static electricity charge you build up on a wool suit so everything sticks to it. It's what makes it possible for water to dissolve just about anything. In its pure form, where the only ions in water are the hydrogen and hydroxyl ones, water is hungry. It looks for things to dissolve and will dissolve them until it has dissolved enough to satisfy its appetite for ions.

Once it has dissolved a fair amount it isn't as aggressive, that's why it doesn't viciously attack the pipes, hot water heater, and other parts it contacts in our homes. Everything we do with water treatment is associated with what is dissolved in it, either ions or different types or gases.

One of the critical values in water treatment is the relative proportion of hydrogen ions in the water. Careful experiments have been developed to determine that there is one hydrogen ion in each million deciliters of pure water. That's 0.0000001 ions per deciliter. The normal range of hydrogen ion concentrations in water solutions runs from 0.01 ions per deciliter to 0.00000000000001 ions per deciliter. Since these numbers are a little cumbersome to work with someone decided to measure the hydrogen ion concentration according to the number of decimal places so the range of measurement is easily described as 2 to 14 (the number of zeros after the decimal place plus one) and the number labeled "pH."

It really does represent the number of hydrogen ions in solution; since it's the count of decimal places it gets smaller when there are more hydrogen ions. There are far more hydrogen ions in the solution when the pH is 2 than when the pH is 14. Whenever you deal with pH you have to keep in mind that a change in value is a change in decimal places, not a proportional change. If you add chemicals to water and increase its pH from 7 to 8 it will take ten times as much to increase it from 8 to 9 and one hundred times as much to raise the pH from 9 to 10.

The value of pH provides a measure of the acidity or alkalinity of water. When the pH is less than 7 it is called acidic and when the pH is greater than 7 it is called alkaline. Acidic water is very corrosive. Highly alkaline water is also very reactive, highly alkaline water will react violently with aluminum and generate some very toxic gases. Normal values of pH in a boiler plant are 7 to 8 for make-up and feed water, 10 to 11 for boiler water, and 5 to 8 for condensate. Water supply plants in the United States are required by law to maintain pH in the range of 7.6 to 8.5.

We measure all the other things that dissolve in water using a scale that is a lot simpler than pH. The standard units of measure are parts per million (ppm) which is a ratio, the number of pounds of material that would be dissolved in a million pounds of water. Some operators find it easier to think in terms of pounds per million pounds of water. Of course we don't have to have a million pounds of water to determine the ratio.

Some water treatment departments will measure the concentrations of ions in solution in terms of micrograms per deciliter. That value is very close to ppm so use it as such unless you're trying to do some critical evaluation of your water treatment facilities.

Occasionally you will see an analysis described as "ppm as $CaCO_3$" to describe a condition of water that includes a combination of materials dissolved. Since the materials have different weights, they are corrected so the analysis can be expressed as an equivalent to calcium carbonate ($CaCO_3$). If you should ever need to know the precise concentration of a substance dissolved in water there are tables of equivalents that give you a multiplier. You shouldn't have to be doing this though, it's best left to the water treatment specialists.

Most of the time we don't need to know how much of a chemical is in the water, only its proportion compared to the amount of water. Therefore parts per million is an easy way to measure the chemicals dissolved in the water. In those rare instances when we need to know how much is in the water it's a simple calculation. Find out how much water is in the boiler; (or whatever it is you're working with) if the value is gallons then multiply by 8.33 to convert to pounds; dividing the number of pounds of water by one million then multiplying by ppm tells you how many pounds of chemical is in the water. Normally this only comes up when you're charging a system, filling the boiler and in some cases the

piping with water that you want properly treated. An initial fill of a hot water boiler system can be calculated by estimating the total length of pipe, multiplying the pounds of water per foot from the table in Appendix D then adding that result to the number of pounds to fill the boiler. To establish the initial charge of sodium nitrite in the water (to achieve a content of 60 ppm) divide the weight of water by one million then multiply by 60. The result is the number of pounds of nitrite to put in the chemical feed pot, sodium nitrite is 63% nitrite so multiply by 1.58 to determine how much of the actual chemical to add then divide that result by the purity of what your chemical supplier provides.

We treat water for two principal reasons, to prevent corrosion and to prevent scale formation. The most common form of corrosion is destruction of metal by hydrogen ions but other chemicals dissolved in water can also attack the metal in our systems. Another form of corrosion is oxidation, where the oxygen in the air or water combines with the metal to form rust. A severe form of oxygen corrosion is oxygen pitting.

Scale formation coats the heat exchange surfaces of the boiler to act like a heat insulator. The scale being on the inner surface of the boiler separates the water and metal so the water can't cool the metal. When enough scale builds up the metal overheats and fails. The various water treatment processes serve to prevent corrosion and scale formation by pretreatment which changes the corrosive and scale forming properties of the water before it gets to the boiler and chemical treatment which changes the properties of the feedwater and boiler water.

WATER TESTING

Testing of water is required to learn what's in the water, what other people and other systems have done, and to check on the actions you have taken to maintain quality water for the system be it boiler, chiller, or cooling water. Most operators do water testing and I've seen variations in that activity ranging from something equal to hospital grade testing to something I can describe only as early cave man. Before you decide to skip this part ask yourself if you're absolutely certain you can't learn anything new about testing water.

The first requirement of water testing is to draw what us engineering types call a "representative sample." That means the sample of water you take to the test bench should be the same as the water in the system you took the sample from. If the sample is drawn from blowdown piping it must come from a section that's almost the same pressure as in the boiler. If it's drawn after the water pressure drops and some of it flashes to steam you have no assurance that your sample is representative. It could be the water left after the steam flashes off and contain higher concentrations of solutes (the stuff dissolved in it, including your treatment chemicals) or it could be condensed flash steam and contain almost none of the solutes. If you're trying to draw a sample off the blowoff piping or any other volume where the water is stagnant you're not getting a representative sample. The best point to draw a sample from is the continuous blowdown piping before the water passes through any orifice or throttling valve.

When I see someone put on chemical preparation gear and try grabbing a sample off the blowdown valve at the base of the water column I know it's not a representative sample. Samples of raw water, softened water, etc., can be collected by simply draining water from the systems and making certain the sampling piping is flushed so the sample is fresh and representative of the water flowing through the system. Samples of boiler feedwater, hydronic system water, boiler water, and most condensate require cooling to ensure you get a representative sample.

Sample coolers can be as simple as a large coil of copper tubing, to allow air cooling of a low pressure boiler water sample, to units designed for operating pressures up to 5,000 psi. You should read the instructions for your sample cooler and follow them, but when they're lacking, the following guidelines are suggested.

A sample cooler should be shut down except when it is used to draw a sample. To ensure there's no vacuum created in it to draw air in and corrode it and no way to over-pressure it through thermal expansion leaving it under pressure is recommended. That means its cooling water and sample outlet valves should be closed, right? Well, that's fine as long as it isn't leaking and you'll never know when it springs a leak under those conditions until the cooling water system is contaminated with boiler water. I like to connect a sample cooler so there's only a cooling water supply valve and the cooling water outlet is piped to form a loop up above the cooler, vented, then dropped to a drain. The loop keeps the cooler under static pressure so air can't get in and will allow for expansion of the water and even generated steam to escape if someone opens the sample line first. If the cooler leaks the boiler water will go to drain, not back into the cooling water system. There's no way some dummy can close an outlet valve that isn't there to force leaking boiler water into the cooling water system

or heat up the cooling water side of the cooler to blow it. It's also almost impossible to dilute a water sample with cooling water.

Close the water supply and sample outlet valves to shut down the cooler. When ready to draw a sample you first check the cooling water drain to be certain the cooler isn't leaking then open the cooling water supply. Once cooling water is flowing open the sample outlet valve to flush the sample piping and get a fresh sample up to the cooler. Boiler water and deaerated feedwater should start flashing at the outlet so you know you have a fresh sample. Throttle the sampling outlet valve until you get a reasonable flow of cooled water.

To ensure there's no vapor vented off your sample, or condensate from the air getting in, the sample should be cooled to the same temperature as the air in that area. A thermometer sensing the water temperature leaving the cooler (Figure 8-1) works well but it has to be able to take the maximum possible temperature of the sample, the temperature in the boiler. The thermometer is also suggested to be certain you don't burn your fingers when drawing the sample.

Once a sample is flowing you should rinse all the apparatus that will be contacting the sample so previous samples don't contaminate what you're analyzing. If you must draw a sample from a location away from the test bench always draw enough to rinse the testing

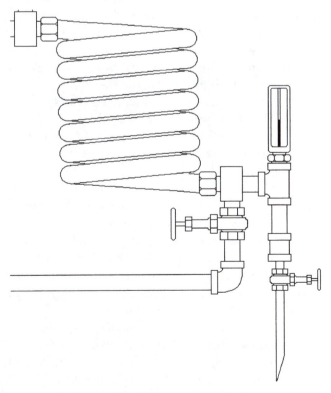

Figure 8-1. Water sample cooler

apparatus when you get back to the test bench. Note that the sample line in Figure 36 is shown long enough to submerge it in a sample bottle. That's necessary to provide a representative sample for testing sulfite content. Once the sample is exposed to air some of the sulfite will start reacting with the water in the air.

To minimize contamination of your water sample with air insert the sample line to the bottom of the sample bottle, leaving it submerged as the bottle fills, and allow the bottle to overflow for a couple of seconds to eliminate mixing of air with the sample and displace all the air from the sample bottle, flushing off the surface so you have a sample that wasn't in contact with air.

If you're drawing from a remote sampling point take another bottle for rinsing your apparatus. Unless you're testing the sample for sulfite immediately you should cap the flooded sample bottle. That's the right way to draw a sample even if you're not testing for sulfite. Always draw at least twice as much as you'll need, that small amount of sample is negligible compared to the cooling water you're wasting, see the section on water consumption.

I seldom find a water test bench closed up. Most of the time everything is setting out and the stand is well illuminated. Didn't anybody ever read the instructions for the test reagents that state they degrade when exposed to light? A good bench will be closed up and dark. Also, the extra reagents and other test chemicals will be stored in their shipping containers in a dark area that has a reasonably constant temperature. Stacking them on shelves leaning against the sheet metal outside wall that's cold in the winter nights and heated by the summer sun is not the right place to put them.

It's also a little dumb to order a ten-year supply of reagents (yes, I've found bottles with ten year old expiration dates on them setting in a plant's storage locker). It's a pain to order stuff at regular intervals but some of it has a short shelf life. You want to be confident of the results you get when testing your water so make sure you have fresh reagents. If the expiration date is before next week, throw it away and get new.

Most test stands I see are kept clean but I do remember one in a poultry processing plant that had… you got it, chicken droppings all over everything. Part of the cleanliness is associated with operating the test bench because some reagents can damage or discolor paint if they're spilled. The automatic filling burettes will spray reagent out a little hole in the back if you force too much reagent up. Those spatters on the back of the test cabinet are an indication of carelessness. If you do accidentally pump some out the discoloration won't

happen if you clean it up right away.

To make it easier to clean and limit breakage of glassware many plants have rubber mats under the test equipment. I regularly tell someone "you can get white rubber." The entire test stand should be white. It's a lot easier to see color changes and other things with a white background. I would like to have a picture of a test stand after regular use to hold up as a good example but I haven't seen one yet. I can't say too much because I know I never kept the ones I used that clean; now I know better.

If I'm watching an operator running water tests I can tell quickly if he's up to the task, even when they're nervous with me standing there watching them. They know what the results should be and add most of the reagent quickly to get to the point where it should be added drop by drop. That saves time in the process and has no effect on the outcome. Holding the sample container up so its lip is above the reagent spout prevents spilling but you can get awful tired if it takes too long to add the reagent until the color change is evident. Occasionally you'll overshoot. No big deal, just measure up another sample and do it over. That's one reason you drew a large sample to begin with.

Speaking of measuring; you do know you're supposed to measure to the meniscus right? That's the level inside the glass (Figure 8-2) not the line at the edge of the glass where the water tries to climb the sides. There's less than 99 milliliters in the cylinder of the figure, not 100. There's very little liquid in that edge so you don't want to read the level there.

Write it down as soon as you read it. Make it a habit. No matter how good you think you are at remembering numbers the time will come when you can't remember them long enough and you'll have to repeat the test to get the results right. Also, never assume you'll get

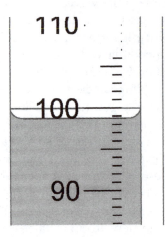

Figure 8-2. Meniscus

the exact same results. I remember one customer calling me up in a panic and requesting a boilermaker crew as soon as possible. It seems an operator decided that the two boilers always tested the same so he saved himself some time by copying the values for one to the log for the other. They rotated shifts each week and the next operator to come on that watch tested both boilers to find the one had very excessive levels of chealant. When the boilermakers pulled the baffles out of the drum they looked like Muenster cheese, full of holes. In another few days the boiler would have failed dramatically. Testing is one of the most important things you do and you shouldn't take it lightly.

Some operators are color blind. It's not a significant problem except for colorimetric testing and it's not something you need to be ashamed of. If you're color blind make sure the boss knows it and sees to it that the chemical consultant provides a test method that you can use accurately. Some operators also have vision problems and trouble reading the little numbers on the burettes. That's okay, there are magnifying glasses for that. It's better to admit you have trouble reading those little numbers (I do) than to guess at what you're reading and destroying a boiler.

If you're still using one of those testers that provides a conductivity reading for the water suggest purchasing a new one. Conductivity is measured in micro-mho where a mho is "ohm," the label for resistance to electricity, spelled backwards. What you end up doing with one of those meters is looking up the matching TDS level on a chart. It's a lot easier to have a meter that is simply labeled with values for TDS.

Oh, you're one of those guys or gals that's interested in operating boilers but doesn't know what TDS is. Mentioned earlier… it stands for total dissolved solids, a measure of the amount of solid material that's dissolved in the water. Those solids include what the water managed to dissolve as it hung around as droplets in a cloud, including gases from the atmosphere and fine particles of dust, what it picked up as a raindrop falling from that cloud, from the dirt and rocks it ran over going down the stream or river or as it trickled down through the earth to the well, and everything it managed to get from the piping until it entered the boiler plant plus the chemicals we added to it.

TDS is measured in ppm. Steam boiler water should have the highest value of TDS and condensate the lowest with makeup and boiler feedwater falling in between so it's a value that's useful in determining percentage makeup and condensate as well as providing values for blowdown control (described later). Anyway,

there's less of a chance of error if the tester reads directly in ppm instead of micro-mho.

You're luckier than I was. When I was testing for hardness we only had one method, soap. I'm sure you know that hard water causes problems in the laundry. It's because the ions that cause hardness, calcium, magnesium and iron have to be captured by the soap before it can foam. We call water "hard" when it's hard to get a foam with soap and soft when the soap lathers easily. You don't have to worry about lather factor and maintaining the soap solution.

Modern titration or colorimetric methods for hardness testing are much easier to use and provide a better determination of the amount of hardness ions in water than our obscure method with so many drops of standard soap solution.

Testing for acidity is a lot easier too. Now all you have to do is stick the instrument in the water sample and read the pH on the little screen on the instrument. We had complicated probes that were always a problem.

Testing for alkalinity hasn't really changed much from my day and still depends on titration testing with phenopthalein for partial alkalinity. Acid is added to neutralize all the OH⁻ ions from the caustic soda added to the water, half the alkalinity produced by carbonate dissolved in the water and one third of the alkalinity produced by phosphates dissolved in the water. The result is rather simple and straightforward, the water is either pink or it isn't. The color changes at a pH of approximately 8.3.

Testing for total alkalinity uses the same sample. Using methyl orange or methyl purple indicator you add more acid until the color changes. The acid removes the remaining half of the alkalinity due to dissolved carbonates and the other two thirds produced by dissolved phosphate. The color changes at a pH of approximately 4.3. Good results is another matter because the color change is very subjective. You add acid until the yellow turns pink or the green turns purple. I seemed to always get on ships that used methyl purple and had a lot of trouble deciding when green turned to purple.

Those tests can be problematic because you never know how much of what you're looking at is carbonate alkalinity and how much is phosphate alkalinity. We don't use sodium carbonate for water treatment anymore so you can count on most of it being due to the phosphate you added to the water. Some carbonate is dissolved in the makeup water with the amount varying depending on the location of your plant and your source of water. It's really not important how much of each is in there, only that you realize that changes in results of alkalinity testing can be due to the phosphate you added to the boiler water.

The main reason for looking for the difference between partial alkalinity and total alkalinity was the determination of how much scale treatment (carbonate or phosphate) was in the water. Keeping up the spread between partial and total alkalinity was, at one time, the only way to tell.

I always disliked the chloride test because it used silver nitrate solution which made your skin brown and I just never managed to keep from getting it on me. My hands were always blotchy from that stuff. Chloride tests are very handy however. The chloride ion doesn't really react with anything once it's in the water so chloride measurements provide an excellent means of determining the mixture of different waters. For example, you can figure out your percent makeup by testing the makeup water and the boiler feed water. The condensate should have zero chlorides in it (it is, after all, distilled water) so all you need do is divide the feedwater ppm by the makeup ppm and multiply by 100 to get percent makeup.

Of course that doesn't work when there's some leakage into the condensate at hot water heaters and the like—which is best caught by testing for chlorides. We used to use chlorides as a measure of dissolved solids on ships but that was a given since our major source of contamination was salty sea water.

In addition to checking for ratios of mixtures of water, chloride tests can indicate the performance of a dealkalizer, where chloride ions are exchanged for other anions (ions with a negative charge). It also allows a determination of the concentration of the boiler water, provided there's no carryover because they're concentrated as the steam leaves. Of course they're used to check for carryover because otherwise there's no reason for them to be in steam line condensate.

Despite getting brown finger spots you should make judicious use of the chloride test to answer your own questions about what's happening with your water.

As far as I know you still test for phosphates like I used to. Mixing some boiler water with an indicator and filling another sample tube with plain boiler water then comparing the color. Those color comparitor tests were always subjective, and I'm not color blind. Similar tests are available for chelants and I don't know of any for polymers.

Whatever ion you're looking for, or the test method used, you should read the instruction manual and carefully follow the instructions if you want to get reliable, repeatable results. When I say repeatable I mean that the guy on the second shift should get the same reading as the gal on the first shift and the third shift should concur. If everyone gets different results one or more of you are doing something wrong or the test is no good.

Any trainee should be allowed to test water with the operator repeating the test to see if the results are identical. If the results don't make sense there's always a possibility that you missed a step or upset the sample and the best thing to do is draw another sample and repeat the test to see if you get the same results. I discovered long ago that I had to ignore everyone when I was doing a water test or I had to put it down and walk away. If I stood there talking to someone I had a tendency to let the sample bottle tilt to dump a little and blow my results out of the water.

Sampling and testing water is the first step in a good treatment program. If you know how to measure the quality of the water and how to determine what is in it, both desirable and undesirable, you're that one step closer to ensuring the boiler plant remains intact. Keep in mind that carelessness and inattention to detail can result in major, sometimes catastrophic failure of a boiler, and you're the closest one to it.

PRETREATMENT

Pretreatment is the conditioning of water to prepare it for use in the boilers. It is less expensive and easier to alter the conditions of the water before it gets into the boiler because we can do it at lower pressures and temperatures. Only the more common pretreatment methods are described in this book. There are other resources, with the best being your water treatment supplier, for descriptions of other methods.

Filtering is the most common form of pretreatment but it's seldom done at the boiler plant. If you use well water you should filter it. City water is normally filtered by the city and is adequate for boiler makeup water. Filters vary from a simple cartridge filter to large sand filters that are tanks filled with sand that does the filtering. Sand filters are back-washed at regular intervals or when the pressure drop through them increases to a predetermined value. A back-wash removes the accumulated material by pumping filtered water through the filter in the opposite direction of normal flow. The water used to back wash is sent to a sewer as waste and can, at least in the first few minutes of back-washing, contain a large amount of solid material. Back-washing also serves to fluff up the sand so the water will flow through it at a lower pressure drop. Some other pretreatment equipment also does a certain amount of filtering.

The most common piece of pretreating equipment found in a boiler plant is a water softener. Softeners are just one of several types of ion exchange equipment. They're called softeners because they reduce the hardness of the water. A water is considered hard when it is difficult (hard) to make soap foam in the water. The original tests of a softener involved mixing a sample of the output water with a standard soap solution to see if it would foam. Water is soft when soap produces a foam readily.

The softener tanks contain resin that fills the tank one third to half full. The resin just lays in the tank so we call it a resin bed. The resin in a softener has an affinity for specific ions, (ions with a positive charge) principally (Na^+) sodium, (Mg^+) magnesium, and (Ca^+) calcium. The beads of resin are selfish little things, always wanting what they don't have. They tend to collect ions until they are in balance with the solution surrounding them. The purpose of the softener is swapping the magnesium and calcium ions in the makeup water with sodium ions, exchanging one for the other. The reason for the exchange is that calcium and magnesium form scale in the boiler and sodium doesn't. The resin traps some of the dirt and large particles in the water so it also acts as a filter.

Where do we get the sodium ions for the softener? From salt. Salt is sodium chloride (NaCl) a common and very cheap material. It's dissolved in water by forming sodium (Na^+) cations and Chlorine (Cl^-) anions. By using brine (concentrated salt solution) in the softener to remove hardness we reduce the amount of expensive chemicals that we have to use in the boiler. In very small plants with very little makeup water or where city water is fully softened or naturally soft a softener isn't justified but there aren't many situations like that. The smallest plant can benefit from a softener if it doesn't use a more exotic form of ion exchanger or reverse osmosis.

Operating modes of a softener include backwash, brine draw, fast rinse, slow rinse, and service. Backwashing removes dirt and "fluffs up" the resin. Water flow during a backwash is up through the bed. The space in the tank above the resin provides room for the resin to separate from the backwash water before the water leaves the tank. If the water flow rate is too high then resin will be flushed out of the softener so it's a good idea to look at the water draining during a backwash to spot resin loss. That's best done with a flashlight pointed into the water, the resin will cause the light to sparkle. You might notice an occasional piece of resin leave because small pieces of resin break off occasionally.

The backwash also flushes out most of the dirt in the water that was filtered out by the resin bed. Under unusual and upset conditions there can be a lot of dirt and mud collected by a softener so you should try to take a look at the backwash water near the end of the cycle to ensure it's clear. Sometimes storms, and at other times the water company crews flushing hydrants, can stir up mud and dirt to put a concentrated amount in the water for short periods.

After the backwash is complete brine is drawn into the softener. The brine solution is a high concentration of dissolved salt. Since salt is sodium chloride, brine is a solution of sodium and chlorine ions. The resin beads exchange ions to balance with the high concentration of brine in the softener, giving up the magnesium and calcium ions collected during the service mode and increasing the number of sodium ions they hold.

When the brine draw is complete the softener is rinsed to remove the spent brine and the calcium and magnesium salts removed from the resin. A fast rinse flows down through the bed to quickly displace most of it. A slow rinse then follows to completely remove all the brine. A salt elutrition test is run occasionally to ensure the softener is operating properly, absorbing most of the brine.

Those previous modes of operation were all part of the regeneration cycle which restores the softeners ability to remove calcium and magnesium ions from the water. They take from a few minutes up to two hours depending on the size and capacity of the softener. Most of the time the softener is in the service mode where makeup water enters at the top and, as it flows to the bottom, calcium and magnesium ions are exchanged for the sodium ions on the resin beads. Since they're oversaturated with sodium from the brine draw operation the resin beads readily give up those sodium ions when they can grab one of the calcium or magnesium ions from the water.

That explains those greedy little resin beads, they always grab what they don't have. The drop the calcium and magnesium when they're loaded up with sodium then readily toss the calcium and magnesium when the water around them is full of sodium for them to grab.

An important element of managing a water softener is knowing the hardness of the inlet water. A softener's capacity is normally listed in kilograins, thousands of grains. It depends on how much resin there is in the softener and how many sodium ions each particle of resin can exchange. Grains, by the way, are a measure of weight equal to one 7,000[th] of a pound. The amount of water your softener can soften depends on it's capacity and the hardness in the makeup water. Since resin eventually degrades (chlorine is rough on it), some of it breaks up and is washed out, and the hardness of makeup can vary, you have to check operation by testing the water.

A condensate polisher is almost identical to a water softener. The differences are mainly due to the high temperature of the condensate. The resin beads and mechanical parts of a polisher are designed to take the higher temperatures. The resin also has an affinity for iron (FE^{++}) in addition to calcium and magnesium to remove iron from the condensate. So, products of corrosion, dissolved iron oxides, get removed by condensate polishers. Operation of a polisher is very similar to a softener, using brine to regenerate.

Dealkalizers are also similar to softeners and are regenerated with salt. The principal difference is dealkalizers contain anion exchange resin, accumulating a concentration of chlorine ions on the resin beads instead of sodium. Their principal purpose is exchanging the chlorine ions to replace the bicarbonate ions in makeup water. Now you would think that salt water isn't the best thing to put in a boiler but we just explained that a combination of softener and dealkalizer do exactly that. The reason is that salt, unlike many other chemicals, will stay dissolved in water as the water is heated up. The calcium, magnesium, and iron will not; they'll drop out of solution as the water is heated to form scale. Some dealkalizers are also regenerated with a little caustic soda added to add hydroxyl ions for exchange instead of sodium. That helps to remove other anions while raising the pH of the water.

Demineralizers are combination ion exchange units that incorporate both cation and anion exchange resins. They can consist of trains of two tanks (one cation one anion) in series or a "mixed bed" that contains both resins in one tank. Demineralizers differ from other ion exchangers because they actually remove dissolved materials from the water. The cation resins are regenerated with an acid to build up a concentration of hydrogen ions on the beads. The anion resins are regenerated with caustic soda to build up a concentration of hydroxyl ions on their beads. As the makeup water flows through the demineralizer all the dissolved material is replace with hydrogen and hydroxyl ions which combine to form water. The result is an output that is pure water, better than distilled.

One of the most important things an operator can do to maintain ion exchange equipment is to prevent condensation on them. The constant formation of moisture with access to air accelerates corrosion of the equipment and piping. Usually good ventilation in the room containing the equipment is adequate but sometimes special coatings are required to act as insulators. Check the backwash water after any system maintenance to ensure the resin isn't washing out and when water temperatures drop. Colder water is more dense and can carry out resin that warmer water couldn't.

Another important thing to remember is the ion exchange process isn't perfect. A few ions manage to sneak through depending on the equipment design, loads, and how they are operated. Demineralizers are almost perfect ion exchange devices. Softeners reduce hardness to 2 to 5 ppm and dealkalizers are about 80% to 90% effective. All ion exchange devices have limited turndown and tend to "channel" at low flow rates where the low flow of water takes the easiest route through the resin to consume the ions there and allow leakage of untreated water. Know the limitations of your equipment.

A good rule of thumb is to maintain a velocity of less than 2.5 gpm per square foot of resin surface to prevent channeling. When water demand is low you're better off shutting off water flow through one or more of several ion exchangers to keep the flow rate up. To better understand this, imagine that small creek you played in as a child. When the water flow was filling the stream all the rocks were wet; when flow was down many of the rocks stuck out of the water and were dry on top; the water flowed by in little channels. When the flow is low in an ion exchanger some of the resin never sees water flow while the resin in the channels sees it all. The result is the resin in the channels is exhausted (all the ions it had to exchange are used up) while the rest isn't used much at all. Since we control ion exchange units by measuring the throughput of the water there's a good chance untreated water will be passing through before the quantity of water that could be treated by the exchanger passes through.

An important part of an ion exchange operation is cleaning and replacement of the resin bed. The normal backwash doesn't remove all the sediment and particles that get imbedded in the resin beads during operation. Chemical cleaning with a resin cleaner that's pumped into the idle exchanger then rinsed out is a normal function in many plants. A complete replacement of the resin every five years is common where the chlorine in the makeup is high.

Reverse osmosis (RO) is becoming more common as the cost of the membranes decreases. Rather than absorbing all the theory of osmosis, treat them as filters that will let water through but won't let the ions dissolved in the water get through. The pressure drop is high because the filter has very tiny holes in it and some of the water has to be used to constantly carry the dissolved stuff away (sort of like blowdown). The filter membranes, depending on their make, can be susceptible to heat or certain chemicals in the water, chlorine being one, so you may have to pretreat the water before it gets to the RO unit. Reverse osmosis performance varies as well, expect anything from 70% to 99% efficiency. Note that while they eliminate ions indiscriminately they don't get them all so boiler internal water treatment is still needed despite what the salesman says.

High quality RO requires wasting a considerable amount of the water to carry off the contaminants, nominally about 20% of the water fed to the unit. The purified water is called "permeate" because it penetrated the membranes, and is, therefore, about 80% of the makeup water supply. Lower waste rates usually accompany lower efficiency but some can be low efficiency with high waste rates.

This is one piece of equipment that requires reading the instruction manual immediately. The membranes can't be allowed to dry out. If they sit too long without water flow there's danger of microbiological (very little bugs) growth. You can't shut it down for the summer and walk away. Feeding with a biocide (bug killer) during idle periods is required. They require some chemical treatment at their inlet to prevent chlorine damage. Cleaning at intervals as frequent as every month is necessary to keep the capacity up.

Finally, the membrane cartridges have to be replaced about every five years. Current replacement cost is about $100 for every gallon per minute capacity.

Some water sources, especially those in the middle of the country, have a high concentration of bicarbonate ions. The bicarbonate produces two problems for boiler operation. In the boiler, where the water is heated, the bicarbonate breaks down to form carbon dioxide gas and hydroxyl ions. That raises the pH and alkalinity of the boiler water, frequently so much so that blowdown is based on alkalinity, not dissolved solids.

The carbon dioxide that evaporates in the boiler flows with the steam to the steam users where it is absorbed in the condensate that forms. Each molecule of carbon dioxide dissolved into the water produces a bicarbonate ion by combining with a hydroxyl ion. When it obtains the hydroxyl ion another molecule of water is dissolved to replace the hydroxyl ion and increase the number of hydrogen ions in the water. The result is condensate with a very low pH and corrosion of the piping and other parts of the condensate system.

The best approach for high bicarbonates today is to use a dealkalizer but other equipment was used, and is still used today, to remove the carbon dioxide before it ever gets to the boiler. These are caused decarbonators or degassifiers and consist of a tank, usually wooden or fiberglass, with wood slats or pieces of plastic stacked inside to form what we call "fill." Treated water is dumped into the top and trickles down over the fill while air is forced by a blower into the degassifier and up through the fill. The water has to be treated so the carbon dioxide gas will separate from the bicarbonate ion. In some plants the treatment simply consisted of adding acid, usually sulfuric, to the water to lower the pH so the bicarbonate ions would break down. The other pretreatment consists of running some or all of the water through a cation unit. The hydrogen ions exchanged for others lowers the pH of the water. In many demineralizers the cation and anion units are separated by a degassifier because the bicarbonate is broken down and removing it as carbon dioxide gas takes load off the anion units. The carbon dioxide, now a dissolved gas, is "stripped" from the water by the air flowing up through the degassifier so it can't recombine with a hydroxyl ion to form a bicarbonate ion again.

A dealkalizer is an ion exchanger regerated with salt, taking on chloride ions that are exchanged for the bicarbonate ions.

BOILER FEED TANKS AND DEAERATORS

Boiler feed tanks with heaters and deaerators are other common pieces of pretreating equipment. They have three principal functions, removing oxygen from the boiler feedwater, heating, and storing boiler feedwater. In the case of some deaerators the three functions are served by separate tanks, a deaerator and separate storage tank. Both systems remove air from the water but there are variations in equipment construction and differences in how much air is removed. Neither removes oxygen completely. A boiler feed tank can only remove oxygen to small values. Deaerators, operated properly, will remove oxygen to minimal amounts.

Removal of the oxygen is achieved by raising the temperature of the water. As the water temperature approaches the boiling point the amount of oxygen the water can hold decreases. Heating the water to 180°F reduces the maximum oxygen absorption to less than 2 ppm. Raising the temperature to boiling reduces that to 0.007 ppm. When the water is ready to boil every molecule of water is prepared to change to steam so the water has very little ability to hold dissolved oxygen.

The dissolved oxygen forms bubbles of gas in the water. Complete deaeration is not achieved until those bubbles are removed. It's getting the bubbles out that makes the difference in deaerators.

Boiler feed tanks come with two kinds of heaters. The water in the tank can be heated by a submerged heating coil or a sparge line. A sparge line simply injects steam directly into the tank. The steam heats the water, condensing and becoming part of the feedwater, while agitating the water. Agitation is important in that it helps remove the bubbles of oxygen from the water. Sparge lines are noisy and that should be considered when adopting a method of heating the water although I prefer the noise and lower oxygen content to a quiet steam trap that needs maintenance.

For all practical purposes boiler feed tanks simply provide a place for storage of boiler feedwater and to return condensate with some capability of oxygen removal provided occasionally. They're normally fitted with a float controlled makeup valve to admit makeup water to maintain a constant level in the tank. The cold makeup water, being more dense than the condensate, tends to simply drop to the bottom of the tank, mixing with the condensate as it enters the feed pump suction piping. Dripping or, better yet, spraying the makeup into the top of the tank will help reduce oxygen from it. Heaters and sparge lines seldom manage to effectively deaerate that water. Deaerators, on the other hand, are designed to remove air and the key is their operating pressure. Boiler plant deaerators are always operated so pressure will force any removed air out of them.

Deaerators are provided in five types, vacuum, flash, spray, scrubber, and tray. A vacuum deaerator is typically a vessel filled with packing and operated under a vacuum. The packing is not like pump or valve packing, it's like fill, loose pieces of ceramic or plastic materials stacked randomly that act sort of like splash blocks so a lot of the water surface is exposed as it tumbles down through the packing. Producing a sufficient vacuum in a vacuum deaerator will bring the water to a saturated condition. For example, pulling a vacuum of 29"Hg (inches of mercury) produces a condition where 79°F water will boil. As long as the water is warmer than the saturation temperature that matches the pressure inside the deaerator it will be at boiling and a little is actually vaporized. The air and noncondensable gases are removed from the deaerator by the vacuum pump or steam jet ejector, whichever is used. A steam jet ejector will normally discharge to a condenser that uses the remaining energy in the steam to preheat the water before it enters the deaerator. When a vacuum pump is used

provisions are made to heat the water and can include any type of heat.

Vacuum deaerators are not normally used in boiler plants because the water is heated to higher temperatures anyway. I thought I would explain vacuum deaerators because someone in the plant may be having trouble with one and might say "gee, the boiler operator should know about this thing."

By heating the water to a saturation temperature higher than 212°F the pressure in the deaerator will be above atmospheric and that higher pressure will push the air and noncondensables out to atmosphere. That's typical of all boiler plant deaerators. The variations in the four types depends on how difficult it is to get the bubbles of air and noncondensables out. Noncondensables are gases other than air that can be released by bringing the water to boiling. They include chlorine gas, ammonia, and others that aren't normally found in air but can be found, in very small quantities, in water.

Flash type deaerators use this concept to produce a pressure just slightly higher than atmosphere to remove the gases. The makeup water is heated in an external heat exchanger to a temperature higher than 212°F then passed through a spray valve into an open tank where some of the water flashes into steam. Since all the water is above the saturation temperature it cannot hold any oxygen so it should be removed with the flash steam which may, or may not, be recovered. There are a number of these devices in the field but (and I know I'm going to get some heat from manufacturers for this one) I don't think they're capable of doing a decent job and I don't recommend them.

The best choices for deaerators for boiler plants are spray, scrubber or tray types and which one depends upon the normal temperature difference between the makeup water and the boiler feedwater. They are all called DC heaters (for direct contact) because the water is heated by mixing steam with the water; the steam is condensed and becomes part of the feedwater in the process. Heating the water to saturation only removes the oxygen and gases from solution, it doesn't get the little bubbles of air and gases out of the water. To do that you need some agitation and how you get the agitation is determined by the temperature difference. All these deaerators have spray nozzles that serve to break up the water as it enters the deaerator. The purpose of the water spray nozzles is to break the water up into small droplets so they can be heated rapidly by the rising steam.

These deaerators also always have a vent condenser. A vent condenser can be an external heat exchanger or, as shown in the following figures, simply a length of tubing inside the water box above the water spray nozzles. The purpose of the vent condenser is to condense most of the steam that is carried out with the air and gases. The idea is to have only air and gases leaving the deaerator. Of course we always adjust the vent valve on a deaerator to produce a "wisp" of steam, just enough so we know all the air and gases are pushed out because a little steam is coming out with them. Throttling the vent valve too much will recover all the steam as condensate but can also trap air and gases in the top of the deaerator to prevent steam contacting the makeup as it enters through the sprays and prevent proper deaeration. Opening the vent valve too much is just wasting steam.

Operation of that vent valve is the key function of a boiler operator. The trouble is most operators solve any control problem by simply leaving the valve so far open that steam is blowing out dramatically. That's a considerable waste of energy and water. The wise operator keeps that vent adjusted so there's only that wisp of steam coming out.

I always dealt with spray type deaerators (Figure 8-3) aboard ship because the water from the condenser was relatively cold and only heated slightly in the air ejector condenser and turbine bleed heat exchangers so there's a considerable difference between make-up and feedwater temperature. If you're operating in a plant that generates a lot of power by condensing turbines (a utility) then a spray type deaerator may be all that's needed. The large difference in temperature requires a lot of auxiliary steam to heat the water and the steam can be directed into the spray section where it creates a violent mixing with the droplets of heated makeup water before it flows up to mix with and heat the water entering at the spray nozzles. It's the effect of all that steam rattling those water droplets around and breaking

them up further before they reach the outside and drop into the storage section that removes the bubbles of air and gases.

Many people get confused with the term "spray" because all these deaerators have water spray nozzles. Even I will use the terms "spray-scrubber" and "spray-tray" to describe scrubber and tray type deaerators to avoid that confusion. A spray scrubber uses a steam spray to provide the agitation to remove the bubbles of air and gas so there's no real reason to prefix the titles of the other two with the word spray.

Except for power generation plants where the makeup is primarily colder water from a condenser few plants can use a spray type scrubber. The combined condensate return and makeup water temperature is so high that steam requirements aren't enough to perform the agitation. When the temperature difference of the condensate and feedwater can be consistently more than about 50°F then there's enough difference for a scrubber type of deaerator (Figure 8-4) to work well. The flow of steam along with the water up through the baffles of the scrubber provides enough energy to separate the bubbles. Some of the energy is achieved using the difference in density of the water and steam.

When the temperature difference between blended makeup and feedwater is less than 50°F, always insist on a tray type deaerator. The trays (Figure 8-5) don't look like what you put your lunch on at the cafeteria, they're made up to produce hundreds or even thousands of little waterfalls. Distributing the water over the trays and producing thin little falls produces hundreds of square feet of exposed water surface for the bubbles to escape from. Some scrubbing of the falling water is achieved by the steam flowing up through the trays to the water sprays but most of the energy that's used to force the bubbles out of the water is provided by gravity. A tray

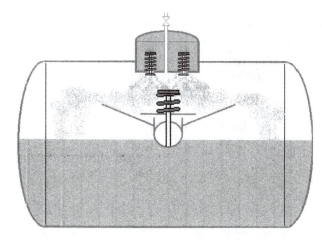

Figure 8-3. Spray type deaerator

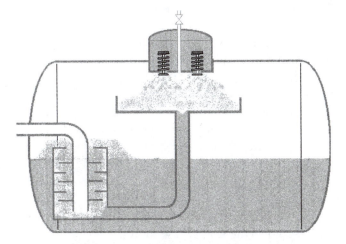

Figure 8-4. Scrubber type deaerator

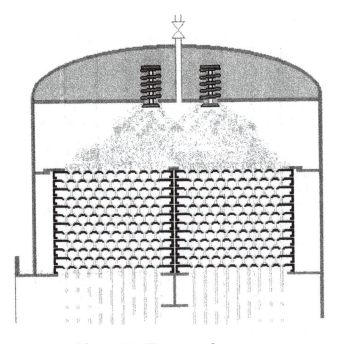

Figure 8-5. Tray type deaerator

type deaerator costs a lot more but when compared to the added cost of sulfite and blowdown over the operating life of the boiler plant the additional cost is justified.

I should mention that there's a scrubber type deaerator on the market that looks something like a combination of a vacuum and tray type, using packing instead of trays. I also have a concern for those pieces of equipment and will not recommend them because they tend to channel at reduced loads, where all the water goes down one path while the steam goes up another so it doesn't do its job. Vacuum type deaerators are designed to operate continuously at one load so they don't normally experience that problem.

Occasionally I'll see a deaerator that isn't operating properly because the pressure control for the steam is at the control valve or senses the pressure in the steam line going to the deaerator. A proper installation, regardless of type, senses and controls the pressure in the top of the storage tank after the scrubber or trays of the deaerator to eliminate the pressure drop through the scrubber, or trays, and connecting piping. If you have a deaerator problem, check where you're sensing pressure.

Why does a boiler operator have to know all this stuff about deaerators? So he won't screw them up! Modern tray type deaerators are normally furnished with tie-bolts to hold down the trays because some people managed to dislodge all the trays. They work properly only when the trays are all stacked properly and leveled so the water flows uniformly over the entire bank of trays to interact with the steam.

Imagine what happens someone shuts off the steam to a tray type deaerator. Colder makeup and condensate still enter through the sprays but now there's no steam to heat it; what little steam is left condenses almost immediately and a vacuum forms, right? Nope. Below the deaeration section is a storage tank full of water at the original steam temperature; it's going to start flashing off steam as the pressure falls so there is some steam provided for deaeration. Assuming the sudden flashing of all the feedwater doesn't produce so much cavitation in the feed pumps that they trip (turbine driven ones normally do) the feedwater in storage will boil as the colder makeup continues to enter the deaerator. Before they started bolting down the trays the only sign an operator had that something was wrong was some clanging as the flashing steam and water swelled up out of the storage section and lifted all the trays. Frequently the insulation on the deaerator prevented the noise reaching the operator's ear so the next thing he got to notice was all the sulfite in the boilers just disappeared. Of course, by the time an operator gets around to discovering the sulfite was wiped out because the deaerator's trays were all laying in the bottom of the storage tank, and not deaerating, a lot of oxygen had reached the boiler to corrode it.

You have to lower the operating pressure gradually until you get down to atmospheric conditions or you'll rattle a deaerator. A deaerator should also have a vacuum breaker, normally a check valve installed backwards connected to the steam space to admit air should you lose steam pressure.

I should also say that you can shut down the steam supplying a deaerator at full boiler load the odds are that check valve used for a vacuum breaker will not allow enough air in once a vacuum starts forming and the storage tank could be crushed by atmospheric pressure.

I've looked into the dearation section of many a tray type deaerator to see the trays all jumbled up. Other times they were stacked at different heights, indicating they shifted. One plant told me they had been that way for several years! Another problem that affects any unit is a water spray valve coming apart. When that happens you have the equivalent of a fire hose hitting the trays and no breakdown of the water initially so it isn't heated. When you have a feedwater temperature lower than the saturation temperature matching the steam pressure that's a good sign that you have a defective water spray valve, regardless of the deaerator design.

Except for vacuum deaerators the feedwater temperature has to be above 212°F (unless you're in Denver where it has to be above 203°F) or the deaerator isn't working. The saturation pressure has to be above atmospheric or there's no pressure to push the air and gases out. I've found at least three plants that were operating in the 180° to 190°F range and thought there system was working just fine. So did their sulfite salesman!

Deaeration, getting the oxygen out of the water before it gets to the boiler is principally done to reduce the cost of chemically treating the water to remove the oxygen. If the oxygen isn't removed it will create pits in the boiler metal, something that looks almost as if it was done with an electric drill. Oxygen pitting can destroy a boiler in short order so the sulfite is always added to remove the little bit of oxygen that slips past a deaerator even when it's working fine. In order for the sulfite to be effective and remove the oxygen that gets past the deaerator and before it gets to the boiler the sulfite should be added to the deaerator storage section.

Sulfite generates sulfate ions when it reacts with the oxygen in the water and, since sulfate salts form the hardest scale, you don't want to put in any more than absolutely necessary so maintaining proper operation of the deaerator is important.

If your plant happens to be one of those where the sulfite is added before the deaerator you should change that so it gives the deaerator something to do. The sulfite salesman won't be happy but your boss will be. I prefer to see the sulfite fed right below where the water drops from the deaerator (but below the low water line) so it can start doing its job immediately.

Since the first edition I've encountered a problem with corrosion of a deaerator associated with the injection of sulfite. Apparently sodium sulfite solution is corrosive when concentrated and heated and it will produce a pattern of corrosion on the bottom of the deaerator storage section that looked like it was sandblasted unevenly if the sulfite isn't distributed properly. Having stainless steel sulfite distribution piping in the storage tank with a long row of perforations to distribute the chemical just below the low level should prevent that kind of damage.

Hot water boiler plants don't normally have the experience of constant makeup. Many of them are treated with sodium nitrite. The nitrite ion converts to nitrate, absorbing oxygen in the process. It's only usable at the low pressure hot water heating temperatures.

BLOWDOWN

I do hope you know the difference between blowdown and blowoff. It's rather important from the standpoint of energy waste and water treatment. Read the portion on water in the section on consumables for further information on this subject. That section was concerned with wasting water, now we're going to talk about wasting some of it to maintain boiler water quality. Blowdown, and I do mean continuous blowdown or so-called surface blowdown on steam systems and low point blowdown on hydronic systems, is used to reduce the concentration of solids dissolved in the water.

Even if we have demineralizers or are using distilled water for makeup we will still get a growing concentration of solids in the water in the system. Some will come from corrosion of piping and other parts of the system by our condensate. Even in tight hydronic systems we'll get increasing solids from gradual dissolving of materials left in the system during construction and minor vapor leaks that aren't always apparent. In steam boilers all the solids remain in the boiler water, concentrating there as the water leaves the boiler as steam. If some of the solids carry over with water droplets in the steam they're returned in the high pressure condensate so the boiler is where all the solids end up.

The amount of solids and some liquids dissolved in water has an effect on the surface tension of the water. There are two sticky properties of fluids, cohesion and adhesion. Cohesion is a measure of how the material sticks to itself. Adhesion is a measure of how much the material sticks to something else. Water is high in both. You'll notice that water actually climbs the sides of a glass because it adheres to the glass. High cohesion is evident at the surface of water where it sticks to itself. When separated from a large body of water a small droplet becomes perfectly round because of the high cohesion at its surface, what we call surface tension.

The combination of adhesion and cohesion contributes to the capillary action of water. It will literally pull itself up into narrow spaces after adhering to the surrounding walls then reach out again. It's what makes water flow up those three hundred feet high redwood trees in California.

As the quantity of dissolved solids increases the physical characteristics of the water change, increasing the surface tension of the water until eventually the water starts to foam and carry over into the steam piping. While this is one way to get the solids out of the boiler it doesn't do the steam piping a lot of good. Increasing solids can also result in saturation of the water with solids in the risers so some of the dissolved materials drop out as scale.

We need a way to limit the concentration of solids in the boiler water to a value just below that point of carryover or scale formation and blowdown is it. By removing some of the boiler water from where it contains the highest concentration of solids we provide space for some makeup water that contains very little solids to enter the boiler and reduce the overall concentration of solids.

In a steam boiler that means removing the water right after it has separated from the steam in the steam drum. That's why the continuous blowdown piping is in the steam drum and the piping has the holes located where they are. That was a hint for those of you who didn't put the piping back the last time you removed internals for inspection because you figured it was just a waste of your time.

In hydronic systems the blowdown is usually drawn from the boiler at the same place as for steam boilers but you may want to check the system for places where the solids are more concentrated. Usually the return water will be more concentrated because the water shrunk as it cooled but contains the same weight of solids so blowing down return water will waste less of it.

I've also had some unusual encounters with multiple drum boilers, older sterling designs, where the solids managed to concentrate in one section of the boiler, not the one where the continuous blowdown connection was, with scale forming despite maintenance of low TDS at the point of blowdown. Regardless of the system, its operating pressure and temperature, and the quality of the makeup water you should be aware that someone could have made a careless decision regarding the location of a blowdown connection. Any time you experience scale formation or problems with carryover that isn't related to pressure fluctuations you should re-evaluate the location of your testing and continuous blowdown connections.

We determine how much to blow down by the TDS reading (described above in testing). The ABMA has set standards for proper levels of solids concentration for boilers according to operating pressure and your boiler's instruction manual may contain that table or specific recommendations for what levels of solids concentration to run at. Note, that's a recommendation, not an absolute value. You may find that your boilers can operate with a considerably higher level of solids without forming scale or carryover. It depends on many factors including boiler load. I always recommend a customer raise their settings for TDS levels gradually until some problem is detected or they get as high as 4,000 ppm either stopping at that value or backing down below the value where problems occurred.

They're also told to establish values for each boiler

load because they can operate with higher solids content at lower loads. Usually carryover is the limiting factor but scale formation can be so I also recommend raising the level at 50 ppm intervals doing so each year one month before the annual internal inspection while keeping a close watch on relative stack temperatures. Back off on any increase in stack temperature because it could indicate scale formation. Since blowing down wastes energy and water minimizing it is a wise operation; it's worthwhile to minimize blowdown.

We used to adjust the blowdown manually but modern technology has produced instruments and equipment that provide a reasonable degree of automatic blowdown control. There are systems that provide continuous blowdown as intended, with continuous measurement of TDS and modulating of a control valve to vary the rate of blowdown to maintain a maximum level but the more common systems are intermittent in operation.

The typical system incorporates a timer that opens the continuous blowdown control valve at fixed intervals. The valve then remains open until the TDS, measured at a probe in the blowdown piping, drops below the preset value. One potential problem with that type of blowdown control is introduction of a surge of high solids water fed to the boiler by opening up a previously shutdown system. The solids will be high in the boiler until the valve opens again. I would prefer a method where the automatic control has a high and low setting and blowdown is continuous through a manually set valve with the automatic valve opening to dump additional blowdown when the high point is reached and close when the low point is reached. It doesn't cost any more than the system with valve timing, constantly monitors solids, provides a continuous flow of water to any blowdown heat recovery system, and will react immediately when additional solids are introduced to the boiler. The only thing that's better is a modulating control valve but they're also rather expensive for small boilers.

Blowoff is designed to remove solids that settled out of the boiler water. The sources include solids from makeup water, rust and other solid particles returned with condensate, and the intentional production of sludge by chemical water treatment. It contributes to the reduction of dissolved solids but at a considerable expense in water and energy because bottom blowoff is not recovered in any way. Use continuous blowdown to remove dissolved solids concentration and limit bottom blowoff to its purpose of removing sediment which will vary depending on the quality of the makeup water, degree and type of chemical treatment. See the discussion on water as a consumable.

CHEMICAL TREATMENT

If there's any time for you to make a bad decision regarding reading this book it's right here. I know that many times chemical treatment of water is treated like a black art but hopefully you have had no trouble understanding any other part of this book and this section should be no exception.

I will admit that chemical treatment suppliers have, and will continue to, make it difficult to understand what their product is doing by using obscure names and numbers to label what are really common chemicals. The first rule in understanding your chemical treatment program is knowing what's in the container. There aren't that many compounds for water treatment and they do the same thing regardless of the name or number on the barrel so you can understand the purpose and function of the chemical if you know what's in it.

If your supplier will not tell you my suggestion is to go find one that will. Given the true title of the active chemical and the following paragraphs you should know enough to properly maintain chemical treatment of your plant's water.

I've said it before and I will repeat it; the only person that can effectively operate a water treatment program is the educated boiler operator. Those chemical treatment consultants that arrive at the plant every month or two have no idea what has transpired between visits. They can't possibly know that the boiler was shut down, drained and refilled, left sitting idle or operated continuously. They might if they bothered to look at the operator's log but I don't recall ever seeing one do that. They may not know that there was an upset in the level controls and the operator used the bottom blowoff to restore water level several times dramatically reducing the chemical levels. All too often I've noticed that a water treatment consultant has changed a program due to an upset condition with resulting over-treatment.

You and your fellow operators are able to communicate so you're aware of all the variables that affect the chemical concentrations in your boiler water and can make sound decisions about changes in the treatment program much better than a consultant. Knowing this, I trust you'll be able to explain the activities that changed the water content to your consultant so you get better service. Note that I didn't say get rid of that consultant.

It's like having a boiler inspector, always better to have some other, somewhat disinterested, party looking at the chemistry. It's really best if the consultant doesn't get to sell you the chemicals.

A water treatment program only has two goals, prevent corrosion and prevent scale formation, it's that simple. The causes of corrosion and scale formation have to be understood to prevent them and knowing how the chemicals prevent (or enhance) those conditions must be understood to maintain proper chemical water treatment. The process of obtaining representative water samples and properly testing them for chemical content has been covered and how to use that information to achieve the goals of the program is described in the paragraphs that follow.

Recording everything that happens, every test run and follow-up actions is important to understanding what's happening and the result of your actions. Don't limit the record to the space provided on the log supplied by the chemical treatment supplier. I've already mentioned a few incidents that can occur and alter water chemistry but there are many others and I'm counting on you knowing enough about it to determine when something has altered the chemistry and logging it in addition to correcting for it.

Your boiler or water system has boundaries and contains a certain volume of water. That volume or weight of water contents can be determined from manufacturer's instruction manuals and estimates using actual measurements and the data in the pipe tables in the appendix to calculate the volume and determine the weights. Once you have an initial volume or weight determined you know what the weight of the water in the system or boiler will be when it's cold, at 70°F where water weighs 62.27 pounds per cubic foot. Freezing (32°F) water weighs 62.4.

Once the system is up to operating temperature the weight will be lower and you may want to adjust your data for the effect of thermal expansion. Determine the ratio of cold to operating by dividing the specific volume of water at 70°F (0.016025) by the specific volume of water at the operating temperature using the data from the steam tables and multiplying it by the weight of the water when it's cold. That's the weight of the water in the system when it's operating. Move the decimal place of that result six places to the left or, if you're using a calculator then divide by one million, to know how many million pounds of water are in your system. Unless it's a very big system the number will be small but you will know how many million pounds of water you have so the results of chemical tests in parts per million will have some meaning and you can use it to estimate the effect of chemical additions. Don't forget about the complication with pH being steps of ten.

There are basically four sources for the chemicals that are in your boiler system's water, makeup, corrosion, leaks, and treatment. In order to effectively control your water treatment you need to be able to determine where the chemicals came from. The principal source is the makeup and it's a function of the quality of the water you get from the well, river, city water main or wherever it comes from. You have to test that water to know how much it's capable of adding to the chemical burden of your boiler water and how to treat it.

Testing that water for hardness provides an indication of the required frequency of regeneration of the water softeners. Tests for bicarbonates or TDS provide indications for other ion exchange equipment and bleed requirements for reverse osmosis systems. When you're using well or river water you may also need to test the water for suspended solids to determine the loading of water filters.

In the Baltimore metropolitan area we have a concern for the source of the city water. Most of the time our water is drawn from reservoirs filled by surface runoff in the northern part of the state but during periods of drought or when work is performed at the reservoirs the city switches from that source to the Susquehanna River. Some of the water in the Susquehanna has traveled from as far away as New York State and most of it's from Pennsylvania so it's spent a lot of time flowing over rocks and dissolving them. The TDS levels of the Susquehanna River are substantially higher than those of the reservoir water and adjustments in softener throughput are essential to make sure all the hardness is removed. Also, blowdown has to be increased to compensate for the heavier solids loading. Regular daily testing of that raw city water is essential because they don't always tell us when they make the switch.

Your softener's capacity is based on hardness removed so testing the hardness and recording the meter give you a clue. If the hardness of the makeup is 50 ppm and the softener is set to regenerate after 20,000 gallons you'll have to reset the meter for 10,000 gallons when the hardness increases to 100 ppm. Stick with the ratios to avoid all those kilograin calculations. As the resin deteriorates, which you detect by noting some hardness increase at the end of the softener run, you should adjust the meter setting accordingly.

Testing condensate can identify leaks into the system. A common source is steam heated service water heaters and that's always a concern because the water is not routed through the pretreatment equipment such as softeners. Condensate will also contain iron, copper, and other metals from corrosion of the steam and condensate piping. It's also possible to receive water contaminated by some operation in the facility. I've seen or heard of boilers filled with fuel oil, sand, salt, sugar, and milk to name a few. A boiler plant operator has to know a little bit about the facility served to be aware of the potential for such contamination and to watch out for it.

One odd one was a boiler contaminated with softener resin. It formed a hard, baked on coating over all the boiler tubes where the resin hit the tubes and melted on. The operators found one of the strainers in the softener had broken off allowing the resin to leave with the treated water.

Water that's passed through a piece of pretreatment equipment has to be tested to ensure the equipment is operating properly. Some of the tests are only significant at specific stages of the system operation. For example, testing of the output of a water softener near the end of the run is critical to make certain the resin has not deteriorated to the degree that hardness is bleeding through. Some tests have to be combined with analysis of chemical use; if you find yourself using more sulfite than normal it's an indication of problems with the deaerator.

Of course only you are aware of operations that affect that chemical use; I remember dismantling a deaerator to find nothing wrong based on a consultant's analysis. The consultant didn't know about a complete plant shutdown and draining and refilling of the boilers using a tank truck. Since the water was exposed to air the sulfite was consumed but all the other chemicals were recovered.

Reverse osmosis and blowdown reduces the concentration of ions in the boiler water but it doesn't eliminate them completely. Softeners and other ion exchange equipment, except hydrogen softeners and demineralizers, swap ions replacing those that produce difficulties with ones that are not as damaging.

They don't get every bad ion out. By maintaining a certain amount of special chemicals dissolved in the boiler water we provide for the final demise of the nasty ions and any oxygen that may have managed to sneak past all our pretreatment equipment. We say we have a "residual" of water treatment chemicals in the water. They reside there, waiting to pounce on any scale forming ions or oxygen that gets through before they can damage our boiler. Another reason we maintain a residual is that we can measure it. If it's there so we can find it with a chemical water test, we know it's there to do the job. For protection from corrosion due to oxygen in the water we normally maintain a residual of 30 ppm of sulfite. To stop hardness, a residual of 60 ppm of phosphate (less with chelant and polymer) is common.

There's one problem with sulfite use. When it's done the job the sulfite ion is a sulfate ion and sulfate ions can combine with calcium and magnesium to form the hardest, toughest scale there is. Low pressure hot water boiler systems and chilled water systems occasionally use sodium nitrite for oxygen removal. The mode of oxygen removal is the same as sulfite. Neither the nitrite nor the sulfite produce desirable elements in waste water so science is still looking for a better solution.

Chemicals can't reduce the solids content of the boiler water; they actually increase it as we add them. Most of our water treatment chemicals are sodium based, consisting of sodium and other molecules that dissolve in the water. The sodium ions tend to remain dissolved so they are not a problem. The other ions from the material are what we use to treat the problems of corrosion and scale formation. You don't test for sodium, you test for the ions that do something and TDS which is a measure of all the ions in the water.

PREVENTING CORROSION

There are two basic ways corrosion occurs in a boiler and an additional one for condensate systems and piping. As the number of hydrogen ions in water increases the pH gets lower and the free hydrogen ions attack the metal in the boiler, changing places with the iron molecules in the steel. Preventing this kind of corrosion is solved by adding hydroxyl ions to the water to combine with the hydrogen ions, making water molecules, so there are very few hydrogen ions and they can't attack the iron. The chemical normally added to boiler water to raise the pH (which means fewer hydrogen ions) is sodium hydroxide ($NaOH$).

It's easy to envision that chemical dissolving into sodium (Na^+) and hydroxyl (OH^-) ions in the water. Enough is added to keep the pH of the boiler water in the range of 10 to 12. Adding too much caustic soda will raise the pH so high that other problems, caustic embrittlement and caustic cracking, will occur.

In some localities the water is already caustic so additions of caustic soda are not required. Some of those actually require additional blowdown to prevent the pH going too high, usually allowing it to go as high as 12. When you have a problem with caustic water or high pH you have to be very careful of leaks in the boiler because evaporating water leaves a concentrated solution where the pH is way too high and severe damage to the boiler near the leak can result. The damage is said to be the result of caustic embrittlement.

The other cause of corrosion in a boiler is dissolved oxygen. We all know that oxygen in the air will combine with iron to form rust but the conditions in a boiler are different. Oxygen in a boiler will produce what we call "pitting." It looks as if some strange worm tried to eat a hole straight out through the metal or someone used a poor drill on it. Oxygen pitting is usually easy to identify because it happens where water is heated to free the oxygen from solution and the oxygen comes in contact with the metal.

Heating of boiler feed tanks and deaerators remove most of the oxygen but we need some chemical treatment to get the little bit that leaks through. If we don't have a heated feed tank or deaerator we'll need a lot of chemical to make certain the oxygen doesn't eat away our boilers. The standard chemical for steam plants and lots of hot water plants is sodium sulfite ($NaSO_3$) which dissolves to free sulfite ions to remove oxygen. It takes two sulfite ions to remove a molecule of oxygen gas ($2SO_3^- + O_2 => SO_4^-$) so it takes a while for two of them to get around to ganging up on that oxygen to remove it from the water. That's one reason we feed the sulfite back at the boiler feed tank or deaerator, so the sulfite has time to work.

Other reasons for feeding the sulfite there include protecting the feed system, storage tank, pumps, and piping along with any economizer we have on the boiler. I know I'm probably going to take some heat for this next one, but... Many chemical salesmen try to sell catalyzed sodium sulfite. It's supposed to have some special ingredient in it that makes it operate faster. I'm sorry, but I don't know of any chemical that will make ions move around in water any faster. The ions move around and the sulfite ions will contact the oxygen in proportion to the temperature of the water (molecules and ions move around faster as they're heated) and mixing of the water, not some additional chemical. Like the guy on TV says, "don't waste your money;" catalyzed sulfite will not necessarily do any better than regular sulfite and if you have the recommended installation of the feeder the regular stuff has lots of time to find and interact with those oxygen molecules.

What about that business with the condensate? I'm sure you've seen many a condensate line eaten up, usually by a groove at the bottom of the pipe, by carbonic acid. The question has to be how can the condensate have acid in it if it's distilled water? The problem is associated with carbon dioxide gas coming from bicarbonate ions in the water. We mentioned in testing for alkalinity that the methyl orange or methyl purple test showed either phosphate or carbonate alkalinity and it's the result of those ions. Bicarbonate ions (HCO_3^-) in the water break down when the water is heated in the boiler to form hydroxyl ions and carbon dioxide gas ($HCO_3^- => OH^- + CO_2$[8]) the gas leaves the boiler and travels with the steam.

Decarbonators and degassifiers mentioned earlier help remove the bicarbonate but, like other pretreatment processes, they don't always get it all.

When the condensate forms the carbon dioxide is dissolved in the condensate to return to bicarbonate, leaving a hydrogen ion in the process ($CO_2 + H_2O => H^+ + HCO_3^-$) It's those hydrogen ions that do the corroding of condensate lines after the carbon dioxide is dissolved again. The only effective treatment is to put something in the water to raise the pH (just like we did in the boiler) but it's not a simple matter of adding caustic soda. If we were to add caustic soda we would have to put it in at every little condensate trap in the system and then try to come up with a way of controlling it. We can't put it in the steam because it would be a dry chemical and plug up the steam lines.

Special chemicals called "amines" and cyclohexylamine in particular will flow with the steam as a vapor then dissolve in the condensate along with the carbon dioxide and act to raise the pH of the condensate to prevent the acidic corrosion. I can remember using "filming amines" which were supposed to coat the piping to protect it from corrosion at a lower cost than the "neutralizing amines" which raised the pH but most of those chemicals were discontinued because they cause cancer. Even cyclohexylamine is questionable for cancer so you should limit its use to what's necessary.

When I was sailing we were using another water treatment product called Hydrazine (N_2H_4) which combined with oxygen to produce water and gaseous nitrogen. It also formed ammonia which flowed with the steam to dissolve in the condensate and raise the pH of the condensate. While it still may be used in some plants the concern over it's caustic properties, potential as a carcinogen and generation of poisonous ammonia require special handling and operations so its use is not general.

So, preventing corrosion is simply a matter of maintaining the pH and removing oxygen. I should add that it's also keeping oxygen out but that's addressed in many other places in this book.

PREVENTING SCALE FORMATION

Now that we've taken care of the corrosion problems all that's left is preventing scale formation. Scale is the result of all the rocks that water dissolved as it traveled from the rain cloud to the makeup water piping in your plant. Once the water leaves the boiler as steam it leaves all those dissolved rocks behind. Frequently the water has so much dissolved in it that it isn't a matter of converting it to steam, all you have to do is heat it up to get scale formation. I recall one application where well water at 57°F formed scale in a heat exchanger that only raised the temperature 6°F. Water with that kind of scale forming property is going to plug up service water heaters with scale, let alone a boiler.

Softeners and other forms of pretreatment can reduce the amount of scale forming ions in the water by, with the exception of demineralizers, swapping them with ions that normally don't form scale (sodium). But that doesn't eliminate the potential for scale and some of the scale forming ions always manage to sneak past all that pretreatment. Chemicals are added to the water to either convert the scale forming salts to sludge or "sequester" (the word means to surround and isolate) them to prevent them forming a scale.

Both methods work fine as long as some water remains to hold the sludge or sequestered ions in solution. If all the water is boiled away to steam then the dissolved solids that remain will appear as scale no matter what we do. After all, when salt water is evaporated there will be crystalline salt left and it will be called scale if it's on the boiler tubes.

There are several chemicals that will combine with the magnesium and calcium ions that tend to form scale and convert them to a sludge. The idea is the sludge isn't going to stick to the heating surfaces of the boiler but will settle out in the mud drum (where it can be removed by bottom blowoff) to eliminate the scale forming salts from the water. Sources of treatment that accomplished this ranged from potato peels (a source of tannin which is the actual chemical) to the many blends of phosphates that are in use today.

An advantage of the sludge forming treatments is they combine with the salts to form a solid thereby reducing the TDS of the boiler water, they don't contribute to the dissolved solids content. Disadvantages of sludge forming treatments include problems handling the sludge and problems in certain boilers where there isn't enough room in the mud drum to reduce water velocity to the point where the sludge can settle out. If the sludge doesn't settle out it can be swept around by the water and eventually reach a concentration where, despite treatment, the sludge sticks to a heating surface and becomes scale.

If your boiler contains scale and tests of it indicate a high percentage of phosphate that's an indication you have that problem. Sludge handling problems include plugging of blowoff piping and valves, usually resolved by more frequent bottom blows. Problems with sludge remaining in suspension in the water is attacked with other chemicals called "sludge conditioners" that are designed to reduce the tendency of the sludge to stick and increase the density of the sludge so it will settle out.

The conventional system for treating boiler water is called "soda-phosphate" and now you know the derivation of the words. Caustic soda is added to raise pH and alkalinity and phosphate is added to remove scale forming salts by combining with them to produce removable sludge. The performance of the phosphates is dependent on the maintenance of alkalinity and to work best the pH should be maintained between 10.5 and 11.5.

To be certain that there's phosphate laying in wait for any calcium or magnesium ions that manage to find their way into a boiler we maintain a residual of 60 ppm of phosphate. Sometimes that is a little tricky to do. I recall one ship where the method of treatment was sodium hexa-meta-phosphate. I actually liked the treatment because the water was clear (many of the treatments produce a muddy looking water) but it had a bad habit of changing concentrations depending on boiler load.

I don't to this day know if it was the chemical or the boiler but the residual values would shoot up into the hundreds when we were in port (boiler loads were low) then drop to almost nothing when we got underway (full boiler load). The water treatment consultant the shipping company used told me it was "hidden phosphate" but never came up with a good explanation for why it did that. I learned to live with the high values in port and always checked it the minute we were under way.

In instances of other scale treatments phosphate is also used as an "indicator." By maintaining a residual level of phosphate in the boiler any failure of the other program is indicated by a reduction in the phosphate residual.

A better, and more complicated, method of controlling scale emerged in the late 1960's. The treatment is generally called "chelant" and it comes in many patented forms. Phosphates are used to remove scale forming salts from the water but chelants simply build a barrier around them that prevents their combining with other ions to form scale. Keeping the scale forming ions in suspension allows their removal in the continuous blowdown where the energy and some water are recovered thereby reducing the losses associated with bottom blowoff. Chelants also attack scale that's already formed, returning it to solution so it can be removed with the blowdown. Used properly a chelant treatment program can remove scale formed on a boiler as the result of an upset condition.

There are two hazards associated with the use of chelant treatment. First, if used to remove existing scale it has to be performed in a manner that doesn't result in fast removal of the scale. The chelant tends to break the bond of the scale to the iron first so any rapid attack on the scale will result in large pieces of scale releasing into the boiler water and transporting to points of restriction where it can plug tubes resulting in overheating and failure. Even when that extreme isn't reached it can produce so much loose scale accumulating in the bottom of the boiler that you'll have problems with blowoff valves and piping plugging up. The second hazard has to do with the fact that iron is related to magnesium and calcium and chelant insists on having something to sequester; if it runs out of calcium and magnesium then it will grab iron, the stuff the boiler is made of. That requires careful and closely controlled use of chelant.

To ensure the scale forming ions are sequestered before they get near the boiler heating surface chelant is normally introduced into the boiler feedwater. The typical means is to introduce it using a "quill" which is best described as a thermometer well with a hole drilled in the side at the tip. By injecting the chelant into the water through the quill into the center of the feed piping it will encounter the scale forming ions in the water before it reaches the iron in the pipe. The quill should always be installed upstream of a long straight run where it can uniformly mix. Any elbows, valves, or pipe fittings downstream of the injection point should be inspected one year after beginning treatment and at five year intervals thereafter to ensue they aren't corroded away by the chelant.

To ensure the chelant doesn't attack the iron it must be fed at the same rate that the scale forming salts enter the system. A chemical feed pump capable of varying the feed rate automatically is required to feed proportional to feedwater flow and testing of the hardness of feedwater before the chelant feed to make adjustments of the proportions of chemical feed to water flow must be made regularly. Testing of the water for any residual should be frequent when boiler loads or feedwater blends vary to ensure a residual doesn't build up that would result in attacking the boiler metal.

The typical commercial or industrial boiler plant today uses a combination of phosphate and chelant which is introduced into a boiler the same way as phosphate. The phosphate residual reacts with any ions entering the boiler and the chelant works on the scale that has formed in the boiler because the phosphate residual beat them to the ions.

Polymers are the new innovation in boiler water treatment today and, to be perfectly honest, I don't have enough experience with polymer treatment to address it. I do know that a boiler treated with polymer will have a thin gray coating on the steel parts when a boiler is opened for inspection and some of that coating breaks off like scale. Hopefully I'll learn enough to give you some guidelines for operating wisely using polymers in a later version of this book.

That's it! Hopefully not as complicated as you thought. Simply believe in ions, good ones and bad ones, and oxygen control to protect your boilers. If the treatment program you're using doesn't make sense to you keep asking the water treatment consultant to explain it until you get something you understand and can manage. If the consultant isn't interested in training you to do a good job tell the boss he had better get a different one.

Chapter 9

Strength of Materials

An understanding of the strength of the materials used in construction of the boiler plant is essential. No element of a plant is designed to operate anywhere close to its breaking point for reasons of safety and maintenance of that margin of safety protects the life of the operator and others.

STRENGTH OF MATERIALS

Lots of boiler operators are not like me. They've never broken anything. Are you one of them? I've broken everything from the standard lumber 2 by 4 to some rather expensive fiberglass piping and witnessed some serious destruction of everything from a bag of rags to pressure vessels and boilers. It's not uncommon to break things and there's nothing I can tell you that will ensure you never do.

I tend to argue that those operators that don't break anything manage to do so by not doing anything. If you are doing something it's important to understand a little bit about the strength of materials in order to make sound operating and maintenance decisions. That way you may break less than I did. I'll try to do that without all the technical engineering but still give you an adequate understanding of what's involved and what some of the buzzwords mean.

STRESS

Stress in materials is very much like pressure. We measure it in pounds per square inch and it's basically force (in pounds) applied over a surface area measured in square inches. We can determine the stress we apply to a material and, by testing, know how much it takes. Since most of the materials involved in a boiler plant are metal I'm going to use it to explain the application of stress and the strength of the material.

We'll start with tensile stress because it's the most common. A material is subjected to tension when you try to pull it apart. You expose a material to tensile stress on a regular basis, the material isn't steel, it's rubber and you call it a rubber band. It's a little hard to imagine yourself stretching a metal but you can do it; just clamp one end of a piece of lightweight wire in a vise, lead it out about twenty feet and grip it with a pair of pliers. Set a stepladder or something else next to the wire to get a reference point then pull on the wire and ease off. You'll discover that the wire is just like a rubber band, you can stretch it and it will shorten when you ease off your pulling on it. That's the elastic action where you apply a stress and the material resists it. You'll also notice if you pull a little too much that the wire suddenly gives and will not shorten to its original length when you ease off the pull; you just over-stressed it.

That operation is a simple form of the tensile tests that are performed on materials to determine their strength. In the case of the wire you pulled with a variable force that could be measured in pounds and you applied that load to the cross-sectional area of the wire which is the area of a circle with a diameter equal to the diameter of the wire. Since any wire we could stretch would be very thin the area is very small. The more cross sectional area of the material the more force you need to stretch it. You can easily stretch a rubber band but a rubber hose is another story.

Tensile tests on material use a sample a little larger than a piece of wire to get an average value. The typical tensile test specimen consists of a piece of metal about six inches long with the center three inches machined uniformly to a thickness of about one quarter of an inch and a width of three quarters of an inch to produce a cross sectional area of three sixteenths of an inch (1/4 times 3/4 equals 3/16) so the cross sectional area is 0.1875 square inches. The two ends of the specimen are clamped in a machine that pulls them apart. For standard metal testing the sample is also marked with a center punch about one inch from the center on each side so the machine can sense the location of the punch marks and measure very precisely the distance between them.

The stress-strain diagram (Figure 9-1) shows a common graph produced by the machine as the material is tested. The stress, which is the applied force per square inch of material, is indicated on the left of the diagram and the strain, which is the amount the material is stretched is indicated on the bottom. As the machine

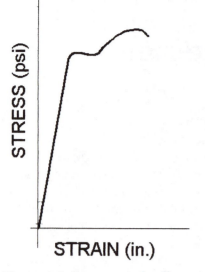

Figure 9-1. Stress - strain diagram

pulls on the material the force or pull on the material is recorded. That value is converted to stress by dividing by the cross sectional area of the specimen.

Modern machines allow the operator to enter the area on a keyboard so the machine also calculates the stress (pounds pull divided by the area in square inches) to imprint it on the diagram. The machine measures the change in distance between the two center punch marks to determine the strain.

The stress strain diagram shows what is normally called the proportional range where, from zero stress, the stress and strain are proportional. If the machine were stopped while the metal was in the proportional range and the force removed the metal would return to its original length. Metal in that range acts the same as the rubber band, always returning to its original shape. At the end of that straight line is the proportional limit where the metal's properties change and it will not return to its original size when the force is removed. It's the same situation when we were pulling on the wire.

Application of a little more force creates a stress where the metal simply stretches out without adding resistance (the slope of the line is horizontal). The point where that starts is called the yield point. When metal reaches its yield point it deforms. That action is similar to "cold working" the metal which hardens most steels making them stronger. I'm sure you've heard that cold worked metal is stronger than hot worked metal. The sudden cold working of the metal increases its strength and, despite the cross sectional area being reduced a tiny bit, it can handle more stress.

The metal continues to resist force but it stretches dramatically until the ultimate strength is reached,

where the stress doesn't go any higher. That's where the coupon is deforming so much that its cross sectional area is reducing so, even though the stress in the coupon increases, the force it can withstand decreases because the area is decreasing. Shortly after the ultimate strength is reached the material ruptures. If the coupon is not too deformed we can measure the cross sectional area at the rupture to determine the actual stress when it ruptured. That's how metal is tested and although you may never see it done this explanation should give you a better understanding of material strength and what us engineers are talking about.

Cast iron and similar materials, including concrete, that are not extremely strong in tension but very strong in compression are tested differently. The test method helps describe what compressive stress is all about. A metal sample is machined to prescribed dimensions over its entire length to form a test coupon. All those short round chunks of concrete you've seen laying around any construction site are test coupons that were poured. The coupon is placed in a machine with a firm bottom plate and pressure is applied to the top of the coupon. (Figure 9-2) The force applied by the machine is divided by the cross-sectional area of the coupon to determine the stress. Some materials, like cast iron and concrete, withstand considerable stress until they fail and they fail quickly when their yield strength is reached. They produce a failure that is closer to shear

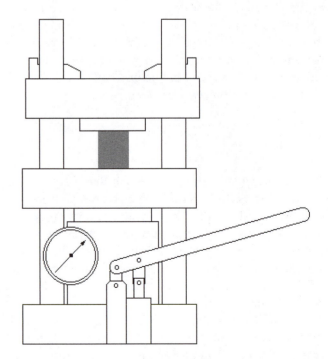

Figure 9-2. Compression stress coupon in machine

than compression because it goes across the coupon at an angle. Since most metals would swell (increasing the cross-sectional area and strength of the coupon) when their yield point is reached, the test is not run past the yield point. The slope of the curve is usually the same for metals under tensile stress so the compressive stress-strain diagram matches the tensile stress-strain diagram in the proportional range.

Shear stress, as it's name implies, is resistance to being cut and is considered primarily for fabrication activities where the material is cut by shears. Unlike tensile and compressive stress, where the force is applied through the cross-sectional area in tensile stress it is applied parallel to the cross-sectional area. It's seldom a consideration in boiler design. Mainly because you're not allowed to make a riveted boiler anymore. If you run into a situation that requires knowledge of shear stress you should be able to understand its function from the previous discussion.

Bending stress is not a special kind of stress, it's a function of compressive, tensile and shear strength. To describe how it relates I use an example that you can reproduce yourself. Take several pieces of 1 by 4 (that's lumber which is really about 3/4 of an inch thick by 3-1/2 inches wide) and stack them up on the floor between two bricks and stand on them. The result is something like that shown in Figure 9-3 because the layers of lumber can't support your weight. Note, however, that the lumber ends are not flush like they were when you laid them out. Gluing all the layers of lumber together (or even securing them to each other with several nails or screws) prevents the equivalent of shearing stress from occurring in the material and they will support your weight when you stand on them. The force of your weight is countered by tension on the bottom layers of the material and compression on the top layers with shear stress applied to the individual layers.

Once you've glued (or fastened) the layers together you might not notice that they still bend a little

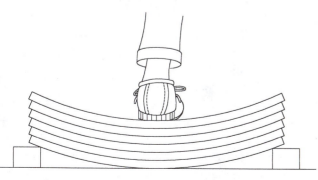

Figure 9-3. Layered board sample of bending stresses

when you stand on them but they hold you up. Just like the rubber band the material length changes when force is applied to it. The bottom layers get longer and the top layers get shorter to compensate for the applied force of your weight. Since the layer at the middle neither shortens or lengthens it doesn't do anything to counter the applied force. The stress in the material increases from zero at the center to maximum at the extreme outer fibers (engineer's word for edge) and that's why all the steel beams we see are made in the form of the letter I, by putting most of the material at the outer layer (where the maximum stress is) we get the strongest beam.

Now that you know about the actual measured strength of the material we can talk about "allowable" or "design" stress. For everything boiler and pressure vessel related those values are listed in the ASME Code in Section II which is called "Material Specifications." Section II is broken down into three parts. Part A is for ferrous (engineer's and scientist's word for iron) metals, Part B is for non-ferrous metals (like brass and copper), and Part C is for welding materials (welding rod). Those sections define the quality of a material and how it must be made and tested.

For the most part the Section II contents is identical to the material specifications prepared by ASTM (The American Society of Testing and Materials) and differs primarily in the certification requirements. A boiler or pressure vessel manufacturer has to buy material that is certified by the manufacturer to conform to the specifications in Section II.

Part D is called "Properties" and it lists the allowable stress for each of the metals described in the three other parts. If you were to look at Part D you would discover that the ASME has different values for allowable stress depending on the use of the material and the maximum or minimum operating temperature. Allowable stresses vary for use as boilers (BPVC Sections I and IV) and pressure vessels.

To relate to that yield strength determined by testing a coupon you could look at the minimum yield values for a material in the applicable Part (A, B, or C) and the allowable stress in Part D. Since you really don't want to pay ASME's price for those books it's not recommended. I can tell you that what you would find for ferrous metals, the allowable stress is one fifth to one fourth the yield. That means the boiler is constructed of metal that should not fail (by deforming) until the pressure is four or five times higher than the maximum allowable pressure. It's a safety factor of 4 or 5 and it's one thing that helps protect you from injury due to a material failure.

CYLINDERS UNDER INTERNAL PRESSURE

The basic calculations for determining the required thickness of a cylinder under internal pressure (like a boiler tube or drum or shell or piping) is best explained by looking at a cross section of the cylinder like that in Figure 9-4. The Figure shows half the cylinder with arrows beside where we imagine that we cut through the cylinder. When we're evaluating that view we make the section over a unit length of the cylinder, normally one inch. So imagine the dark line is a piece of metal that's one inch deep into the page. Any inch along the length of a cylinder would be the same so we can work with one inch and it applies to the whole length. The gray arrows show the direction of the forces that are applied.

The pressure is inside the cylinder trying to get out and pushing against the area that is equal to the inside diameter (I.D.) of the cylinder. The area equals the diameter because the width is unity (one inch). The pressure times the diameter equals the force produced by the internal pressure (p×d). We're applying a pressure, pounds per square inch, against an area measured in square inches so the overall force can be measured in pounds. That force has to be balanced and the balance is the force produced by the metal cylinder. If the force were not balanced the cylinder would rupture. The area of the metal in the cylinder is equal to twice the metal thickness so we can determine the stress in a known thickness of metal. Alternatively, we can calculate the minimum thickness of the metal for a given stress because the forces have to be equal.

The force from pressure equals the pressure (P) times the diameter (D) and it must be equaled by the force on the two thicknesses of metal (2T) and the stress in the metal (S) so the mathematical formula for a cylinder under pressure is P × D = 2T × S. Substitute known values for any three of the letters and you can calculate the fourth using simple algebra. If you don't know algebra then here's what you do for the four options:

- To determine the stress on the metal you multiply pressure times the diameter and divide that result by twice the metal thickness. $S = (P \times D)/(2 \times T)$

- To determine the minimum thickness of the metal you multiply pressure times the diameter, divide that result by the allowable stress and divide that result by two.
  $T = ((P \times D)/S)/2$

- To determine the maximum diameter for a cylinder of a given thickness at a selected operating pressure you multiply the thickness and the allowable stress, that result is multiplied by two and you finish by dividing by the pressure. $D = (T \times S \times 2)/P$

- To determine the maximum pressure for a cylinder of a given thickness, diameter, and material, you multiply the thickness and the allowable stress, that result is multiplied by two and you finish by dividing by the diameter. $P = (T \times S \times 2)/P$

The ASME Code isn't quite as simple and it's because the overall length of the material around the cylinder gets larger as the thickness increases. The code formula is:

$$T = (P \times D)/(2 \times S \times E + 2 \times Y \times P) + C[8]$$

to determine the thickness and

$$P = (2 \times S \times E) \times (T - C)/(D - 2 \times Y) \times (T - C)$$

to determine the maximum allowable pressure for a given thickness. There are values in addition to those in the more simple explanation above represented by C for corrosion allowance, E for a factor that depends on the method of welding (sometimes called weld efficiency) and Y which is a coefficient that depends on maximum operating temperature and the type of steel. These formulas are for power boilers. The ones for heating boilers and pressure vessels are a little different.

For your purposes the simple formulas should be fine. As long as you know there's a little difference between them and the actual code formulas it's okay.

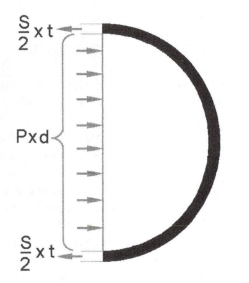

Figure 9-4. Cylinder analyzed for pressure stress

You aren't expected to design the boiler but I think you should have an understanding of how the design is determined and that's why I've subjected you to this math business.

Calculating the stresses and required thickness of a pipe or boiler shell is rather simple. Complications enter the equations when you have openings in the pipe or shell; for example, all the holes in a water-tube boiler drum. In those cases allowance for the holes is based on the required thickness of a cylinder without holes and how much metal has to be added where the cylinder is complete to make up for the holes in other locations. A steam drum where the tube holes are two inches in diameter and installed on 4-inch centers has to be about twice as thick as one without the holes.

Normally the code doesn't require any special consideration for an occasional opening for a connection smaller than two inch nominal pipe size. Larger openings may have enough extra material in the cylinder (because the standard steel plate thickness, greater than what was required by the code formulas, provided it). It may be included in the structure of the opening (like manhole rings) or a doubler (additional steel plate surrounding the opening) that's added to provide the required material. If you would like to know any more about boiler design and construction requirements I would suggest you take one of the courses provided by the National Board.

Cylinders under internal pressure are easy to understand and the calculations are rather simple once you get the gist of them. We have other situations that are complex, cylinders under external pressure is one. All the tubes of a firetube boiler and its furnace are cylinders with the pressure on the outside of the tube. I'm sure you know there's a difference in the amount of pressure a cylinder can withstand depending on whether it's on the inside or the outside.

CYLINDERS UNDER EXTERNAL PRESSURE

Even as a child we knew our soda straw would collapse if it got plugged with some ice cream in our shake and we continued to suck on it. To clear the plug we would blow on it. Sorry, for those of you that don't or can't remember, shakes used to be made with real milk and ice cream mixed by something like a blender.

You can get the same evaluation by blowing into or sucking on the top of a plastic soda or water bottle. The bottle could change shape while you are blowing on it if it isn't a cylinder (it becomes more cylindrical) but it has no trouble returning to its original shape and you can't rupture it unless you're a real blow hard or used compressed air. If you suck on it the results are much different. By removing the air you expose it to external pressure from the atmosphere and it collapses and, it's usually permanently deformed.

The stresses that are applied to anything exposed to external pressure produce both compressive and bending stresses and usually the bending stresses produce the failure. Cylinders and other parts exposed to external pressure and flat parts of vessels exposed to internal pressure are thicker than they would have to be for the same pressure applied internally or they are made with stiffening rings, bars, etc., to help them resist the bending forces. The corrugated steel furnace of a firetube boiler (called a Morrison tube after the man that determined it would be stronger) handles external pressure better than a simple cylindrical furnace of the same thickness and diameter because the corrugations stiffen the cylinder.

Calculating stresses becomes a lot more complex when you're making a valve, flange or other pressure retaining structure. Many standard arrangements have been developed and, in most cases, tested to failure to determine their strength. That was a much easier proposition in the days before computers and all their capabilities. We have standards based on a maximum operating pressure. The ones you'll normally encounter are 125, 150, 250, 300, 400, 600, 900, 1,500, 2,000 and 3,000.

An important thing to know about those standards is they have secondary ratings. Perhaps you've seen a valve with "500 WOG" cast onto it and wondered what it's about because the other side has "250" and you understand it to be a 250 psig valve, what's that other stuff about? The 500 WOG means the valve is also rated for operation at a maximum allowable pressure of 500 psig if it's used for water, oil, or gas at normal atmospheric temperatures. 250 psig steam is at 400°F and the valve's strength is less at that temperature.

An operator argued violently with me once about this, he was absolutely certain that I was endangering his life by allowing a boiler feed pump to operate at a 270 psig discharge pressure when it was fitted with 250 psig valves. A copy of the valve manufacturer's table of secondary ratings (that's what they're called) for the valve didn't convince him. You shouldn't question secondary ratings (he had taken a position and didn't want to back down from it) because any manufacturer's chart is based on standard tests and they're all alike, the secondary ratings are an American National Standard.

The secondary ratings allow for differences in the maximum temperature of the system and permit using less expensive, but perfectly fine, materials for some processes. You will always find 600 psi rated steel valves and flanges on feedwater piping for boilers with a maximum operating pressure of 600 psig even when the feedwater pressure is as high as 800 psig because the secondary ratings of 600 psi standard valves and flanges is 900 psig with 250°F feedwater. An abbreviated copy of secondary ratings is in the appendix.

PIPING FLEXIBILITY

Tension, compression, and bending stresses are all involved in determining the flexibility of boiler plant piping. I should explain that what we're talking about when we mention the words "piping flexibility" is the stresses in the piping and the stress and forces applied to boilers, pumps, turbines, building structures and other things the piping is connected to where those forces and stresses are produced by the thermal expansion or contraction of the piping.

I can still recall being asked to look at a problem in the warehouse section of a plant where a wall had been damaged. The wall was at the south end of a large warehouse, it was made out of concrete block and it had a very large hole in it right around a piece of insulated pipe. In the shipping area opposite the wall was a pile of broken concrete block. You could see the remnants of a thin steel plate that was welded around the pipe to seal the opening in the wall (required for a fire wall construction) still hanging on the pipe. According to the drawings a similar plate was in the north wall of the warehouse. In between those two plates was 84 feet of four inch steam piping; a straight 84 feet of pipe. Operating pressure was 150 psig and the pipe was installed when the outdoor temperature was about 70°F. Using the tables and procedures in the Appendix you can determine that the pipe would lengthen by about 2 inches when heated. Since the pipe had no place to go but straight south it broke the wall. Later a 'U' bend was installed in the pipe inside the warehouse and the wall repaired. Unlike the stiff straight piece of pipe the pipe with the U bend was sufficiently flexible that the pipe bent slightly and the seal plates and walls were able to withstand the forces applied to them.

If the concept of piping flexibility doesn't gel in your mind I suggest you do what I have done in the past to picture it, make a model of the piping out of a piece of coat hanger wire then grasp it at two points where it will be anchored (attached to something that doesn't give) then try to move your two hands toward each other to simulate the effect of the pipe expanding. You could also clamp it in two vises properly positioned and heat it up but that's a little more complicated.

Keep in mind that pipes get stiffer as they get larger, note the sag in different sizes of pipe when you pick them up in the middle; bigger is stiffer, smaller is more flexible. You can also relate to the fact that valves and other devices in the piping make it stiffer. When stiff piping is heated it tends to grow in length and diameter. Getting a little larger in diameter isn't much of a problem to handle but the added length is.

Sometimes the pipe can do the same thing that railroad tracks do, just spring sideways a little to convert the straight line of pipe to a shallow S. That doesn't cover much of a change in length and we're just lucky that railroad tracks don't get that hot. Other examples include roads. I can remember one hot summer when a lane of the Baltimore Beltway got so hot that the pavement buckled up at a joint producing the equivalent of a two foot high speed bump. Luckily I was driving in the other direction but I saw two cars hit it and they didn't fare well. The compression stress gets so high that any little change in cross section (the roadway joint) permit translation of some of that compression stress to bending stress and, in that event, the roadway bends.

I can also remember looking at two 16-inch HTHW lines in an underground tunnel where they made a 45-degree bend. The adjacent support for the piping had moved, shearing off its anchor bolts. The piping movement drove it back so far that some conduit behind it was overstressed in tension and split like an old paper soda straw to produce a gap over an inch wide exposing the wires inside. You have to respect the forces associated with thermal expansion.

Back to piping flexibility. Usually you don't notice any problems with it in a boiler plant because the designers are aware of it. That doesn't mean the designer did it right. There are times when the installing contractor changes the piping arrangement and it produces excessive stresses. If you fail to maintain joints in the piping or the piping supports you may have some problems with overstress.

I've seen welded steel elbows buckled because an adjacent packed type expansion joint (Figure 9-5) froze up. This form of expansion joint allows the pipe to expand into the space between the flanged connection and a bare end. They have to have anchors somewhere else to take the axial pressure forces of the pipe or the darn thing will come apart. If you have some of these joints be very certain that the anchors aren't corroded.

The forces produced by a packed expansion joint

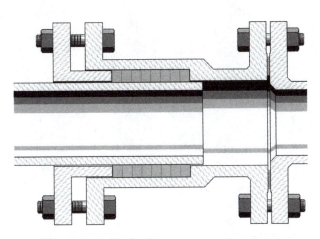

Figure 9-5. Packed type expansion joint

or bellows joint on the anchors is readily determined by multiplying the MAWP (maximum allowable working pressure) of the fluid in the system by the cross sectional area of the pipe (including the metal area) for packed joints like the one in Figure 9-5 or the bellows area of bellows joints. Perform the calculation where you have these joints and I'm certain you will be amazed at the forces that are produced. It's only one reason I do not recommend their use.

I've also encountered many a valve that leaked because the piping stresses applied to it were too high. It happens frequently where large stiff piping is reduced at a control valve making the valve and its flanged joint the weakest point in the piping and the one that bends or breaks.

During the first trip on the last ship I sailed I had a piece of gasket blow out of a flanged joint and bean me on the head. It was a good thing I was more than three feet away because 90 psig steam followed the piece of gasket. After I replaced the gasket I cut off one of the pipe supports that was obviously (at least to my engineer's mind) causing the stress at the gasket.

The largest problem with stiff piping is its effect on pumps, blowers, and turbines. Where piping is attached to boilers and pressure vessels the vessel wall is normally more flexible while remaining strong and can take the stress; the shell simply changes its shape a little. Pumps, blowers and turbines, on the other hand, go out of alignment when they are overstressed. A piping system that has very little stress in the piping can still overstress a pump or turbine to the degree that the impellers hit the casing and shaft seals are rapidly worn and fail.

Anytime a customer has a problem with a pump the first thing I look at is the connecting piping. I was called to resolve the problem on one job where the stress got so high that it broke the concrete pad under the pump away from the floor and moved it almost two inches. The maximum allowable forces on pump connections are described in API-610. When you look at those forces you'll notice that some are so low that enough pipe to get from where it's attached to the pump to overhead will, along with the weight of fluid it holds, weigh enough to exceed the standard's limits.

Over the years I've encountered many situations where the operators of a plant modified or had a contractor modify piping without careful analysis of the flexibility; and they suffered the consequences. I'm not talking about application of snubbers that are like shock absorbers and restrict the dynamic flexing of the piping (when it acts sort of like a tuning fork) they don't restrict the thermal growth. The wise operator realizes that the piping has to remain flexible and will not attach stuff to it or impair its movement to reduce its flexibility.

I hope this little discourse in strength of boiler plant materials gives you some guidance in operation. You should feel a little more comfortable with what you're dealing with and, at the same time, gain some respect for the pressures you're operating at. Look at some of the vessels you're operating and calculate the force by multiplying the area by the operating pressure then dividing that result by the area of the metal holding that force back.

Chapter 10

Plants and Equipment

It would take volumes of books to adequately describe all the variations of design and construction in boilers since Hero first produced steam under (a little) pressure. And that's only boilers, nothing to do with all the other plant equipment and systems. You may encounter a design that isn't described in this section. I encountered two true Sterling design boilers, an 1890 design constructed in 1952, in 2004. I've limited the descriptions in this section to what you would normally encounter. If you do encounter an older design it should be described in one of the references listed in the bibliography.

TYPES OF BOILER PLANTS

When you want to get a definition right you go to the source and, in the case of boilers, the source is ASME, the American Society of Mechanical Engineers which produced and maintains its Boiler and Pressure Vessel Code (BPVC). The code is the accepted rule for construction of boilers and pressure vessels in the United States, Canada, and much of the world. According to the code a boiler is a vessel in which a liquid is heated or a vapor is generated under pressure by application of heat from the products or combustion or another source. Vessel is the code word for an enclosed container under pressure.

Now let's get the meaning of pressure straight. You'll encounter a large number of people in your career that have their own idea of what is low pressure and high pressure then we all get to disagree on what we mean when we say medium pressure. The BPVC in its various documents defines high pressure and low pressure but never addresses the term medium pressure.

High pressure boilers are defined by ASME in the first document prepared to address the construction of boilers and pressure vessels which is now known as Section I of the BPVC and it's simply titled "Rules for Construction of Power Boilers." That is a roman numeral one, not a capital letter i. All sections of the code are numbered using roman numerals. Within section I a high pressure boiler is defined as a steam boiler that operates at a pressure higher than 15 psig or a hot water boiler that operates at a water temperature greater than 250°F or a pressure greater than 160 psig.

Low pressure boilers are defined by ASME in Section IV of the BPVC "Rules for Construction of Heating Boilers." It defines a low pressure boiler as a steam boiler that operates at a pressure no greater than 15 psig or a hot water boiler that operates at temperatures not greater than 250°F and pressures not exceeding 160 psig.

Now you can understand why there's so much confusion regarding medium pressure, there simply isn't any room for it! If the boiler makes steam it's low pressure until 15 psig and high pressure at any pressure higher than 15 psig. Hot water boilers aren't quite as clearly defined but the temperature is normally the clue, almost any hot water boiler operating at temperatures less than 250°F is a low pressure boiler.

I zipped through that discussion of high and low pressure without making note of some other defining labels. The titles of the code sections is one key. A high pressure boiler is also called a "power" boiler, low pressure boilers are called "heating" boilers and the definitions apply to those titles as well. A boiler plant that is only used for heating but operates at steam pressures above 15 psig or heats water to a temperature greater than 250°F is a high pressure plant with power boilers. A low pressure boiler could be used to power a steam engine to generate electricity but it is still called a low pressure boiler or heating boiler, the use has no bearing on the definition of the boiler.

A superheated steam boiler is any boiler that raises the temperature of the steam above saturation temperature. It's possible that low pressure steam could be superheated but virtually all superheated steam boilers are power boilers. On rare occasions you will encounter a separately fired superheater which is also a power boiler by definition in Section I of the code.

One other definition that isn't clearly defined in the code but is commonly used is "High Temperature Hot Water" abbreviated HTHW. When we talk about these plants we typically say the initials rather than the words. An HTHW boiler is simply a power or high pressure boiler that heats water rather than generating steam.

Since we've adopted the label of HTHW any low pressure hot water heating boiler plant is simply called a "hot water" plant with the understanding that it complies with the code definition of a low pressure hot water heating boiler. With water heating plants labeled as such we understand a low pressure or high pressure label to mean a steam generating plant. Don't ever be afraid to ask what somebody means. Requirements for licensing of operators frequently depends on whether a boiler is a power boiler or heating boiler so you want to get it right.

BOILERS

Boilers do not have to have a burner. All of these types can generate hot water or steam by absorbing heat from another fluid. That other fluid can be steam and create steam or hot water, it can be HTHW and generate steam, or it can be a hot liquid or gas from some chemical process that is hot enough to do the job. I imagine I worked on one of the largest low pressure steam boilers that was ever built in the late 1960's and it generated steam by oxidizing a liquid. The heat source was a large volume of oil which air was forced through to oxidize the liquid similar to combustion but at a low temperature and nowhere near complete combustion. Twenty-four feet in diameter and ninety feet tall with thousands of square feet of heating surface it made about 25,000 pounds per hour.

Other projects included a hot water boiler using 500°F air from a steelmaking operation rated at 100 million Btuh. Operating that type of equipment to get the most steam out of it is wise because you save on fuel that would have to be used to generate that steam. These boilers can be constructed as unfired pressure vessels in accordance with Section VIII of the ASME Code, "Rules for Construction of Pressure Vessels."

Boilers that are fired must be built to Section I or Section IV but their construction is limited to materials that can handle the high rates of heat transfer required for direct fired equipment. Boilers using waste heat can require materials of construction that can't handle direct firing but are essential to prevent corrosion in the waste heat application. In simpler words, a fired boiler can't be built in stainless steel, an unfired boiler can be.

Since there's a fixed relationship between pressure and temperature for steam and water, pressure has to increase. When we need to heat product or other materials to high temperatures the pressures can get very high. To obtain temperatures greater than about 500°F, which

would require steam or water pressure over 666 psig another fluid is used. There are several liquids, mostly hydrocarbons, that can be heated to temperatures as high as 1,000°F without operating at such high pressures. The liquids are identified by the trade name given by their manufacturer and include Dowtherm™ and Paracymene™ as the more common names. They are supplied in different materials according to the temperatures required. The common label for boilers that heat these liquids is "hot oil" so we call them hot oil boilers. The Appendix contains tables, similar to steam tables, for the more common of those hot oils.

Some of those liquids can be vaporized just like converting water to steam. A common name for them could be oil vaporizers but it's far more common for the label to use the trade name of the fluid and add the word vaporizer so you'll normally hear them called Dowtherm vaporizers, but there's no strict rule. Since all these plants operate at temperatures higher than 250°F they require power boilers built in accordance with Section I. You could be operating one of these boilers in addition to the steam plant because steam is usually required to quench the fire in the event the hot oil leaks into the furnace to feed the fire.

Equipment that heats water in an open container or very small one is not a boiler. Your teapot doesn't have to be constructed in accordance with the code because it's so small. The hot water heater in your home isn't considered a boiler unless it holds more than 120 gallons. Another limit on the size of a boiler is an internal diameter of 6 inches or less. The exceptions found in the code are occasionally stretched to create boilers that, by definition, are not.

Fired air heaters are not boilers unless the air is under pressure. Any application that heats air, or any other gas for that matter, that doesn't contain the heated fluid in an enclosed vessel is normally called a furnace. If the fluid is air or another gas and it's under pressure then it does meet the definition of a boiler.

There are many boilers unique to their respective industry. You may encounter asphalt heaters, flux heaters (a raw material that becomes asphalt), many forms of waste heat boilers and equipment like recovery boilers (used in the paper industry) which convert product by burning it. I've chosen to limit this book to the more common types of boilers so you can acquire a basic understanding of them. The principles discussed here will allow you to understand those unique boilers which, by virtue of their uniqueness, are best understood by reading the operating and maintenance instruction manuals for them. This section contains general descriptions of the basic elements of a boiler plant to provide a basic understanding of the systems and equipment. Hopefully an operator can append this information with the contents of the instruction manuals to develop a full working knowledge of his or her boiler plant.

HEAT TRANSFER IN BOILERS

An understanding of heat transfer is a fundamental requirement for a boiler operator because a lack of understanding of heat transfer can result in the operator's death; it's that simple. The energy transferred in a little 100-horsepower boiler is about eight times the amount it takes to power an automobile at sixty miles per hour. Screw up to get that energy going in the wrong direction and you're inviting an accident that can only be compared to eight or more cars running you over, all at the same instant!

There are three ways that heat is transferred, conduction, radiation, and convection and all three means occur in a boiler. Conductive heat transfer is the flow of heat through a substance molecule by molecule. A molecule is the smallest piece of a substance that we can get without destroying it's identity. The heat is absorbed by one molecule which passes it onto the next and so on. The best example I know of conductive heat transfer that you can readily understand is toasting marshmallows. Of course toasting marshmallows is done best over a campfire and I love campfires; if you haven't had these experiences go out and do it so you can learn to love them too.

You should remember the time when you and some friends were toasting marshmallows and you got stuck with the fork without the wooden handle. As your marshmallow was toasting you could feel the metal get hot in your hand. The metal over the fire was heated and that heat was conducted up the metal of the fork to your hand. You should also remember those cold nights at the campfire when the front of you was hot and your back was cold so you stood up and turned so the heat from the fire would warm your back. You were using the radiant means of heat transfer.

The sun is another good example, when you're laying there on the beach you are soaking up heat from the sun. It's almost 93 million miles away with mostly space (nothing) between it and us but the heat is getting here. Radiant heat transfer is the flow of heat energy by light waves that can penetrate empty space and the air above us but is absorbed by solid and liquid in its path.

The last means, convective heat transfer, uses a transport to get the heat from one spot to another. In your home the furnace or boiler heats air or water which is then moved (blown or pumped) to the room you're in and heats the air in the room which then heats you. There are two types of convection heating, natural and forced. Forced convection is the result of a fan, pump or blower forcing the movement of the fluid over a heated surface where it picks up heat then on to another surface where it gives up that heat.

If you're sitting in a house with a radiator next to the wall that radiator is heating the air around it and the air gets lighter (less dense) as it expands from the heating so it rises up in the room like a lighter than air balloon. When it reaches the ceiling it starts to cool because it's giving up heat to the ceiling and it's pushed aside by hot air following it. When the air reaches a cooler outside wall it gives up more heat, shrinks to become denser, and drops to the floor then travels back to the radiator. That's natural convection heat transfer. All these methods of heat transfer occur in a boiler.

The modern boiler with its water cooled walls absorbs about 60% of the heat from the burning of the fuel using radiant energy. That heat travels in the form of light waves from the glowing hot fire directly to the boiler tubes, in water tube boilers, or furnace tube in fire tube boilers. The reason so much heat is transferred is due to the low resistance to the radiant heat flow from the fire to the tubes. Though not quite as hot as the sun a fire is an awful lot closer so there's a lot of heat flowing there. You can feel the radiant heat of a fire if you can open up an observation port to look in. Once it hits the fire side of the tube the heat is transferred by conduction to the water side of the tube and by convection to form hot water and steam.

Conductive heat transfer to the boiler water and steam is limited to the flow through the boiler metal itself. The steel parts of a boiler are selected for their ability to transfer heat with as little temperature difference as possible. The outside of a water cooled tube is no more than 60 or 70 degrees hotter than the inside both because the heat is passed through the tube easily and because the heat is drawn off the tube by the water and steam rapidly.

Other parts of a boiler count on poor conductive heat transfer to protect them from the heat of the fire. Refractory material not only can withstand high temperatures it's a poor conductor of heat. When it's backed up with some insulation the outer surface of the boiler's metal casing is less than 140°F which is the maximum temperature that should be allowed. (anything hotter will give someone a serious burn in a matter of seconds,

140°F is that temperature where you can just barely hold your hand on it for a few seconds)

Now is a good time to point out that heat flows from points of higher temperature to points of lower temperature. If there is no difference, there will be no heat flow. The converse is almost true, if there isn't any heat flow there can't be any temperature difference. If we were to put a layer of insulation with a super high resistance to heat flow on the outside of the boiler the refractory, insulation, and casing would get almost as hot as the inside of the furnace. That's why you never add insulation to a boiler casing that's not water cooled because it will overheat.

If the boiler tubes are coated with fireside deposits they will get hotter and reflect heat back to the fire to reduce heat transfer to the water and steam. If the boiler tubes are coated with scale on the water side then the tube wall will get very hot because the scale acts like insulation to block the flow of heat from the tube metal to the water.

Other mechanisms are involved when the scale on the water side accumulates and it provides an early indication of potential failure. If the metal gets too hot it will lose its strength and begin to bulge under the force of the boiler pressure. Usually found on the top of fire tubes and in the bottom of water tubes where exposed to the furnace, bulges are evidence of excessive water side scale formation.

When the tube metal bulges the hard scale is released, breaking away from the metal that's stretched to form the bulge. Once the scale is broken away the metal is exposed to water again, cooling it to stop the growth of the bulge. Repeated incidents of bulge formation can occur with some of the metal stretched until it is very thin and its chemical composition changes so the surface becomes rough oxidized metal, something we call a blister.

Sometimes the bulges or blisters can be left in place if the processes that promoted scale formation are eliminated but blisters should eventually be replaced because the metal is thinner than permitted by code. Slight bulges, where the tube metal is not distended or deformed beyond its own thickness, can be left in place. See Code repairs for replacing bulges and blisters.

Changes in heat conductivity of materials in the path of conductive heat transfer can create conditions that are inconsistent with the original boiler design to result in failure. Hopefully you will operate and maintain your boiler in a manner that doesn't interfere with the design heat flow.

As for the radiant energy that hits the refractory wall, it's reflected right back to the flame or is reflected off toward some of the heat transfer surface. Actually you could argue that very little heat is transferred because the face of the wall and the fire are at almost the same temperature, but the truth is it's radiated back almost as fast as it's received.

Everything radiates energy, we radiate energy. If you can recall a time when you sat with your back to a window in the winter time you'll realize you radiate heat energy. The heat radiating from you goes right out the window into the cold making your back feel colder than when it faces a wall and most of the heat from you is radiated back from the wall. It's also the reason you feel cooler when you go into a parking garage. Even in the heat of summer those floors and walls are colder than you are (because they lost their heat overnight) and they absorb more radiant energy than they emit so you feel cooler. Okay, there are rare times when, after several warm days, you enter a parking lot on a cool evening and feel the heat radiating out of the concrete.

You'll discover that your boiler loads are a little higher on clear nights because of the black sky effect. Heat radiates from the earth and everything else right out into space on a clear night so it takes more heat to keep the buildings warm. On a cloudy night the clouds act like a mirror reflecting the radiant heat back toward us so we're warmer. An important factor in radiant heat transfer is the emissivity of a substance. It has more to do with the color and finish of a surface than the actual material of construction. White and mirrored objects have a higher emissivity than black and rough surfaces so they tend to emit more radiant energy than the black and rough surface even though they're at the same temperature. Keeping those white rubber roofs clean in the summer and letting them get dirty in the winter will actually help maintain desirable building temperatures.

As the flue gases leave the furnace they carry the remaining heat into what we call the convection section of the boiler. That's where convective heat transfer takes place so it's reasonable to call it the convection section. When we're dealing with water tube boilers it's also called the convection bank. (a bank being a group of boiler tubes that serve a common purpose) Heat transfer in the convection section is driven by much lower temperature differences, (typically the flue gas leaves the furnace at less than 1800°F. 1400°F to 1600°F is a normal range, which is almost half of the 3200°F plus flame temperature. The temperature difference drops to a typical leaving differential of 75°F to 150°F so we need a lot more heat transfer surface in the convection section of a boiler to get rid of the 40% that wasn't transferred by radiant energy in the furnace. Okay, there was some convective heat transfer in the furnace but it was minimal compared to the radiant heat transfer and, no, there shouldn't be any measurable flame to boiler conductive heat transfer in the furnace because the steel can't handle those flame temperatures if the flame touches the tubes.

I had better mention flame impingement right now because that's when we have conductive heat transfer from the flame to the boiler tubes. It's also called flame gouging because the tube metal is melted and swept away when flame impingement really happens. You must have seen what happens when someone heats metal to cut it with a cutting torch, that's flame impingement. If you have flame impingement you can see the damage during an internal inspection.

The truth is that we seldom have flame impingement problems in a boiler despite many people arguing that they have it. I have only seen a couple of incidents of true flame impingement in my forty-five years in the business so I refuse to believe anyone's claim of it until I've examined the boiler. It doesn't happen because the flame is cooled so much by radiant heat transfer that it's normally quenched (below ignition temperature) before it gets to the tube.

When I can look into the furnace and see the flame bouncing off the tubes or furnace wall just like you would see water bouncing when a wall is sprayed with a water hose that appears to be flame impingement. Even then you can examine the boiler and find no damage at all on the tubes.

Bulges and blisters (mentioned earlier) are not due to flame impingement, they're due to scale formation. If the flame seems to be rolling along the tubes or passing along them so close that they must be touching we call it "brushing" the tubes and it doesn't do any damage.

The same thing that helps prevent true damage from flame impingement also makes it difficult to transfer heat by convection. The molecules of air and flue gas that are in contact with the tubes stick to the tube and each other to form what we call a "film." It's a very thin layer of gas that acts like insulation separating the hot flue gases from the tubes. In the course of heat flow from the flue gases to the water and steam it contributes the most resistance to heat flow. That film is mainly what protects the tubes in a furnace from the hot flue gases in the fire. Otherwise the metal temperature would be so high that it would melt. The typical boiler steel will melt around 2800°F and it begins to weaken at temperatures above 650°F. (It actually gets a little stronger as it is heated up to 650°F.)

A film forms on most gas to metal or liquid to metal surfaces to resist heat transfer. Water really sticks to other surfaces. Its adhesion is greater than its cohesion as evidenced by the meniscus (see water analysis) and I'm sure you've noticed that water clings to surfaces so the concept of a film is not difficult to envision. To improve convective heat transfer the fluid flowing past the heat transfer surface is made turbulent (all mixed up and swirling around) to sweep against that film and transfer the heat from the fluid through the film to the metal. As velocities in a boiler drop, a point is reached where the flue gases can't disturb the film, it gets thicker, and the heat transfer drops off dramatically.

When flow is so low that the flue gases simply meander along, like congested traffic where the vehicles in the middle can't get to the sides of the road, a lot of the gas leaves without contacting the tubes. It can't give up its heat so it's hotter, carrying that valuable energy out of the boiler and up the stack.

Something unique happens to that film on the water side when we're making steam so heat transfer from metal to boiling water is a lot greater than heat transfer to water or steam. If you think about it, it's easy to understand. I mentioned it earlier in the chapter on water, steam, and energy. When heat is transferred from the tube to the water to make steam a bubble of steam forms and it grows to several times the volume of the water it came from (in the typical heating boiler operating at 10 psig the steam expands to 981 times the volume of the water) so there's a dramatic movement of the steam and water interface. The steam bubble then breaks away from the metal (steam is nowhere near as cohesive as water) and water rushes in to fill the void. All that activity makes steam generation much easier than simply heating water or superheating steam and it requires less heat transfer surface to get the heat through. Similarly when getting heat from steam the steam forms condensate at almost one thousandth of the volume and more steam rushes in to fill that void while the condensate drizzles down the heat transfer surface effectively scrubbing it clean.

The range of heat transmittance (U) for steam condensers is 50 to 200 Btuh-ft^2-°F (British thermal units per hour per square foot per degree Fahrenheit) compared to water to water heaters at 25 to 60 Btuh-ft^2-°F, and superheaters have values of 2.6 to 6 Btuh-ft^2-°F[9]. Also see the comparison of E.D.R. in the Chapter 1. No wonder steam is an excellent heat transfer medium.

CIRCULATION

In addition to heat transfer a boiler operator has to have a sound understanding of the circulation of steam and water in a boiler to operate it without damaging it. If circulation is interrupted for more than a few seconds all the water will boil away in areas of high heat transfer and, only able to heat the steam, metal temperatures will shoot up and the boiler will fail.

To be certain you understand what boiler water circulation is and how it works I'll use some simple examples and develop them to the more complex provisions. If you've never watched a pot of water at what we call a rolling boil on the stove take a break and go do it; you'll waste a little energy but the lesson is worth it. Those of you who already have can read on.

Notice how rapidly the steam bubbles and water moved in that pot? At a nominal one atmosphere, where water boils at 212°F the volume of steam is 1,603 times greater than the volume of the same weight of water so the weight of the steam is about six ten thousandths of the weight of an equal volume of water. Try to push a balloon full of air down into a bucket of water to get an idea of the force created by the difference in density.

If you manage you'll get your feet wet because the water in the bucket will be displaced by the balloon and come splashing out. The steam forming in that pot of boiling water would blow all the water out of the pot if it were not for the fact that it rises to the surface of the water and breaks out so rapidly. The steam bubbles have to move fast to get out of the water without displacing it completely. If you get the pot boiling too fast the level will rise and the water will spill over the top anyway. That's despite the fact that some of it is converting to steam so there's always less water in the pot than when you started.

Watching the pot you can see that the water is circulating, water and steam bubbles rise up, the steam separates and goes into the air, and the water that came up with the steam returns to the bottom of the pot, usually in the middle but not always and not consistently. Being much heavier than the steam the water manages to find its way down with a force comparable to the one that you had to use to get the balloon down in the water. It will tend to go where the velocity of rising steam bubbles and water is lowest.

The water in a boiler has to move around, or circulate, just like it does in the pot on the stove in order to let the steam out of the boiler. Enough water has to flow with the steam to carry the solids dissolved in the remaining water and keep them dissolved or they will drop out on the heat exchange surfaces to form scale. Luckily water is highly cohesive (it sticks to itself) and tries to hold itself together around those steam bubbles so there are many pounds of water circulating up to the

water surface along with each pound of steam that's formed.

Recall that in the boiling pot of water you saw lots of round bubbles? In among all of them was a lot of water. A sphere (bubble) occupies 52.36% of a cube that would have sides equal to the diameter of the sphere so even if every steam bubble was touching another one only slightly more than half of the volume of the rising steam and water mixture would be steam. In our pot on the stove the steam occupies 26.8 cubic feet per pound and water occupies 0.01672 cubic feet per pound (see steam tables, in the appendix.) If the volume of the pot was one cubic foot we could calculate the weights of steam and water if all the bubbles were touching each other. The steam would weigh 0.01954 pounds (0.5236 $ft^3 \div$ 26.8 ft^3/lb) and the water would weigh 28.498 pounds ({1-.5235}$ft^3 \div$ 0.01672 ft^3/lb). The weight ratio of water to steam would be 1,458 pounds of water per pound of steam (28.498 ÷ 0.01954).

I won't apologize for the math, it's just adding subtracting and dividing and I believe it's necessary because without supporting math most operators refuse to believe that the rate the water circulates inside the boiler is hundreds of times greater than the rate of steam flowing out the nozzle. The ratio gets smaller as pressures increase, if you would like to know what the ratio would be for your operating pressure all you have to do is substitute the volumetric values for your operating pressure from the steam tables into those formulas. Of course you have to admit that the bubbles aren't touching each other so there's a lot more water flowing around than this calculation would indicate.

Now that you have a good mental picture of the water and steam rising in a pot on the stove let's translate that to the inside of a boiler. A firetube boiler might have a pattern like that of Figure 10-1. It's more complicated than that because the amount of heat transfer changes from the front of the boiler to the rear. In the typical scotch marine boiler the water rises around the furnace over the entire length and drops at the sides to varying degrees and considerably against the front tube sheet.

Water tube boilers have circulation patterns that vary considerably with the boiler design and the firing rate. The typical example shown for circulation in a water tube boiler is that shown in Figure 10-2. The water and steam rises in the tubes that receive the greatest amount of heat because more steam bubbles are in that water. Water along with a little steam that is generated drops in the tubes that receive less heat.

The tubes where water and steam flow up toward the steam drum are called "risers" and the ones where

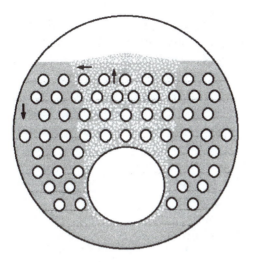

Figure 10-1. Steam flow pattern in firetube boiler

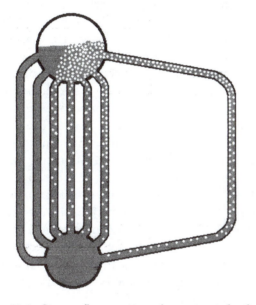

Figure 10-2. Steam flow pattern in water tube boiler

the water drops are called "downcomers." It stands to reason that all the tubes that face the furnace of a boiler must be risers. Remember that 60% of heat transferred by radiation? When the boiler is operating at low loads only a few of the tubes, those along the sides of the boiler that are heated on one side only (and don't face the furnace) will be downcomers. As the boiler load increases even the downcomers will have some steam bubbles forming in them because they're absorbing heat and more tubes will have to become downcomers in order to move all the water that has to circulate in the boiler. Some tubes will always be risers, some will always be downcomers, but many of them switch back and forth.

Some water tube boiler designs encountered problems with the translation from risers to downcomers.

The water flow tended to be so low in those tubes that scale formed in them, you might still run into one of those boilers and be told that there are certain steaming rates you want to avoid to prevent scaling problems in portions of the boiler. A number of designs were modified to include "unheated downcomers," tubes or pipes installed between the top and bottom drums (or headers) on the boiler to provide an unheated path for the water to circulate through.

We actually added some unheated downcomers to a boiler in an effort to correct a problem with overheating of the boiler's roof tubes despite the fact that I didn't agree with the solution. Sometimes unheated downcomers aren't obvious, they're buried in a tube bank where flue gas can't get at them.

Okay, some wise guy is asking "what does this have to do with hot water boilers?" The truth is that there is some steam generation to force circulation in most hot water boilers; there has to be. Maybe there isn't at low loads but the differences in density of heated water are not enough to produce the rapid flow of water needed to carry the heat away from the heat transfer surfaces. The steam that's generated condenses again when the bubbles separate from the heat transfer surface and find their way to colder (by a few degrees) water in the boiler.

There are some hot water boilers, HTHW generators for example, that are designed to force the water along and absorb the heat fast enough to prevent steam formation but I'm willing to bet that you would find steam bubbles forming and collapsing in any conventional hot water boiler. If you watched that pot on the stove while the water was heating up you probably noticed signs of movement which was due to differences in density of water heated at the bottom and the colder water on top (cooled some by the air) and along the sides. You also should have noticed that bubbles formed on the bottom of the pan and lifted off then disappeared before reaching the surface. I'm certain that must happen in most hot water boilers.

Keep in mind that circulation is absolutely necessary to prevent scale formation and blocking of tubes to the degree they overheat and fail. If bottom blows aren't adequately removing the accumulating sludge in a boiler the normal circulation can sweep some of that sludge into some risers with almost instantaneous failure a certainty.

Growth of scale on tubes will restrict flow in the boiler and accelerate the scale formation as a result. If you have scale in your boiler its demise is only a question of timing. Loose drum internals that will break loose when exposed to the rapid movement of water and steam can block flow resulting in loss of circulation and boiler failure so don't let those broken bolts and supports go, get them fixed.

One of the ships I sailed had a special baffle in the top of the side waterwall header. The tubes sloped up from the front of the boiler to the back between two headers. The purpose of the baffle was to scoop some of the descending water into the top rows of tubes. It was discovered that the velocity of the water coming down the downcomer to the header was so great the water shot past the inlet of the top rows and they were starved for water.

I doubt if you'll encounter a boiler with water circulation baffles but if you do find some strange looking piece of metal bolted in a boiler don't remove it. If you're like the engineers on that ship some time before I sailed her and find the piece loose in the bottom of the boiler, go looking for where it should be and put it back. They didn't and the top waterwall tube failed on the next ocean crossing after they found the baffle and left it laying on a workbench.

There are watertube boilers that are designed in a manner that doesn't permit natural circulation. They're equipped with boiler water circulation pumps that are designed to operate at the high pressure and temperature of a steam boiler and produce the differential that's required for the boiler to operate properly. You'll also find small boiler water circulating pumps on hot water boilers that serve a different purpose. That's explained in the chapter on operation.

BOILER CONSTRUCTION

The construction of a boiler can be attributed to many things but the principle ones are code compliance and cost. The manufacturer has to build a boiler that complies with the applicable section of the ASME BPCV but the key to building a boiler is to make the cheapest one that will do the job. Low price can be as simple as first cost but should be based on life cycle cost where the selected boiler should provide the required steam or hot water with the lowest combined price, installation, fuel and maintenance cost over its expected life.

There is always an ongoing effort to design a better boiler and it has resulted in many changes during my lifetime so you can expect to see more changes in boiler construction in the future. There are many books that show the extent of construction variations so I'll only touch on this subject to give you an idea of the develop-

ment of the designs and why they're made that way.

Not only is a teapot a simple boiler, it's representative of many of the earliest designs of boilers. They were nothing more than an enclosed pressure vessel full of water suspended above a fire with some piping leading off to the user of the steam. Some, like the early Roman baths, were even simpler, separating the fire from the water by a simple row of mud bricks, the earliest refractory.

Any fired boiler has some refractory in it so it's appropriate to explain what it is. It's material that can withstand the heat right next to a fire. Looking like cement or regular brick it contains chemicals to bind it that will not melt under normal furnace conditions. There are very few that can stand to be right next to a fire and none can tolerate the highest possible flame temperatures. Refractory materials come in different grades based principally on the temperature they can reach without melting or failing. They range from 1200°F stuff on the low end to 3200°F material. Normally the higher grade materials are used closest to the fire and lower grades are used where the temperature will be lower. Upsets in flame shape, openings in baffle walls and other problems in a furnace can direct hot burning gases against refractory that can't tolerate the higher temperature resulting in early, and sometimes quick, failure of the boiler.

There are basically three types of refractory, brick or tile, plastic, and castable. Brick or tile are preformed and fired at the factory. A burner throat is normally made up of tile. Plastic is moldable, usually applied by positioning chunks of it then beating it into position with a hammer. Castable is mixed and poured into forms like cement.

In any large wall of refractory special "anchors" are furnished with steel or alloy material that penetrates any back-up insulation and attaches to the setting, casing or buckstays for support. Some anchors are made up of a combination of metal and a piece of tile (Figure 5-6) to provide better attachment to the refractory. Setting is the name used for a boiler and furnace enclosure that consists of brick stacked up like walls to enclose the boiler and furnace.

Casing is the name we use to describe the outside of the boiler enclosure when it's typically made up of steel plate. It's not the same as Lagging. Lagging can range from steel plate to painted canvas but is normally thin sheet metal covers used to protect insulation applied to a boiler. Buckstays are structural steel components that stiffen the casing of a boiler or provide attachments for panels of water tubes.

There was a time when all boilers were enclosed in

a setting or casing, insulation and refractory. The typical form was a box. and could consist of a mixture of materials. Boilers were constructed with bottom support, top and intermediate support. Top supported boilers require inverted thinking because they grow down as the boiler heats up. Intermediate supported units grow both ways. Top supported boilers required an external structural steel frame to hang from; sometimes they are made part of the building and other times they're independent of the building.

Yes, boilers grow. There's a list of materials and the amount their length changes in the appendix. Since a boiler is made mostly of steel it will grow around 0.6% for each one degree change in temperature. The steel in a boiler will always be very close to the temperature of the steam or water (saturation condition for a steam boiler, average temperature for hot water). So, if the boiler is supported at the top, basically hanging from the structural steel, it will grow down. If it's supported at the bottom it will grow up. We don't attach the boiler to the building structure, the tendency of boilers to grow as they are heated prevents it. There are times when you'll find some platforms supported off the boiler steel; be aware that they will move!

Today there are three basic types of boiler construction, cast iron, firetube and watertube. Cast iron forms produce spaces for water, the fire, and products of combustion. A firetube boiler contains the fire and products of combustion inside the tubes and the water and steam is outside the tube. A watertube boiler has the fluids on the other side, tubes surround the water and the fire and flue gas is on the outside of the tubes. There are also tubeless boilers (which I would classify as firetube) that, like the whistling teapot on your stove, are small and inefficient but are so cheap to build they are more than adequate for some small operations.

Cast Iron and Tubeless Boilers

Cast iron boilers are made up of cast pressure parts bolted together or connected by piping. There are arrangements of castings that form a furnace as part of the boiler (Figure 10-3) and others that require additional setting (Figure 10-4) and lagging. Cast iron boilers are restricted to heating boiler service, the maximum pressure rating being 60 psig.

The corrosion resistance of cast iron makes the cast iron boiler very durable. I've seen many of them in hot water service for more than fifty years. Their largest problem is that durability, they get ignored and they fail.

The tubeless boiler (Figure 10-5) uses the outside of its shell as part of the heat exchange surface. The flue

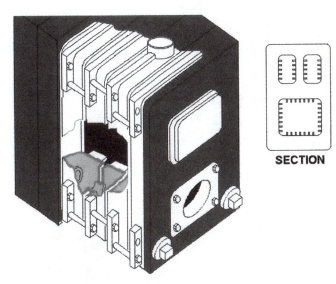

Figure 10-3. Cast iron boiler, integral furnace

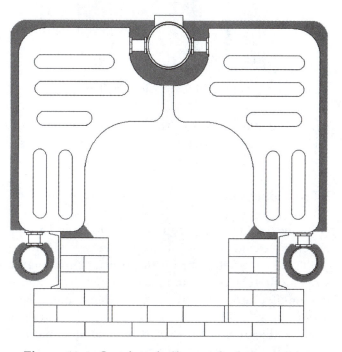

Figure 10-4. Cast iron boiler, pork chop sections

gases exit the furnace through a nozzle that connects the furnace and shell then makes a couple of passes along the shell between fins formed by welding steel flat bar to the shell before exiting the stack. One manufacturer adds another pass around a boiler feed tank attached to the boiler shell and forming part of the assembly.

I think of them as crab shack boilers because so many of them, mostly made by Columbia Boiler Company (here in Columbia, Maryland), are sold to restaurants and other facilities for the sole purpose of steaming crabs. Since the crabs are exposed to the steam there's no condensate return and these boilers don't last very

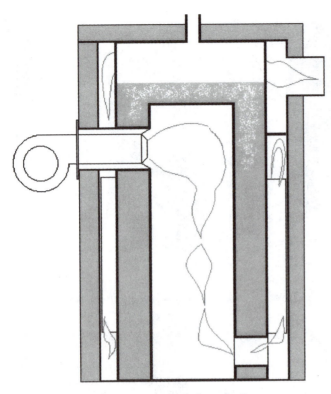

Figure 10-5. Tubeless boiler

long using 100% makeup. Their low price and vertical construction allows relatively inexpensive replacement.

FIRETUBE BOILERS

The firetube boiler requires a "shell" to enclose the water and steam to complete the pressure vessel portion of the boiler and that shell is the principal limit on the size of a firetube boiler. To understand why the shell is the limiting factor we have to understand some basics about strength of materials and how we determine the required thickness of the shell, tubes, and other parts of a boiler. If you skipped the chapter on strength of materials you may have trouble understanding this.

You should have noticed that the required thickness of the shell of a boiler or a boiler tube is a function of the radius. As the tubes get larger the thickness has to increase to hold the same pressure. Since the outer shell of a firetube boiler is very large it has to be quite thick. Thicker materials require more elaborate construction practices in addition to more weight so the price of a boiler increases proportional to its diameter with sudden large steps in price associated with different construction rules depending on the thickness and temperature.

A big break point for high pressure boilers come at

1/2 inch thick and 650°F. The increasing thickness has imposed a normal limit on firetube boilers of 250 psig MAWP (maximum allowable working pressure). It's possible to get a firetube boiler for a higher pressure but it's not a common one. The other practical limit on the size of a firetube boiler is its diameter. Anything larger than 8 feet 6 inches in diameter will require special permits for transporting it on our nation's highways. Shipping a firetube boiler without trim and panels on the sides (but with insulation and lagging) and without special roadway permits and escort vehicles limits the diameter to eight feet.

To allow shipment with control panels mounted the normal firetube boiler is limited to shell diameters of seven feet. There's also a limit on length which is around twenty feet (to fit inside a low boy trailer) but longer units are made. Since you need twice the length of the boiler to permit replacing the tubes a twelve foot boiler would require twenty-four feet of space and that's the nominal distance between building columns in average construction. Many are built backed up to roll-up doors so the tubes can be pulled outdoors.

All those factors place a reasonable limit on firetube boilers at about 500 horsepower for a normal unit rated five square feet of heating surface per boiler horsepower, 600 horsepower if all the trim is removed or the boiler is rated at four square feet of heating surface per boiler horsepower, and about 800 boiler horsepower if roadway problems are not too expensive and the customer can handle a permit load or delivery by rail. That

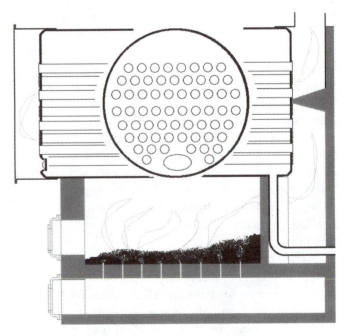

Figure 10-6. HRT boiler

doesn't mean a firetube boiler can't be larger, I saw a 1400 horsepower firetube boiler a couple of years ago. It was a monster some ten feet in diameter and almost forty feet long; I would love to know how they kept the tubes in it from sagging. The lower cost of manufacturing firetube boilers has also increased the manufacturer's offering to 1,000 boiler horsepower. Sometimes they do it by simply increasing the size of a burner on a 800 horsepower boiler.

Firetube boilers come in several configurations and arrangements. Basically they are cylindrical in shape (Figure 10-6) and are further defined by position and modifications to the general form. The arrangement in Figure 10-6 is typical of an HRT boiler (the letters stand for Horizontal Return Tubular) which is an early design of boiler that has survived to modern times. Return in the label indicates the flue gasses flow down some of the boiler tubes from one end to the other then return through the remaining tubes.

A cross section is shown in the middle of the figure that shows the tubes, how they're arranged to permit the baffle at the rear and location of an access door for scraping off the bottom. Typically the shell of the boiler is extended at the end where the gas makes the turn to form a "turning box" which is closed by large cast iron doors (Figure 10-7). The doors could be at the front or rear of the boiler depending on how it's constructed relative to the furnace.

Most of these boilers were assembled without welding. The joints in the shell, the tubesheet to shell joint, and piping connections were all made using rivets. See a later paragraph about riveted boilers. The furnace

is typically a brick walled enclosure constructed below the boiler. Many were built with the brick serving as a base to support the boiler. Few of those remain because a furnace explosion which dislodges the bricks would result in the boiler collapsing into the furnace. More modern HRT boilers are constructed with steel bases that support the boiler or a steel frame straddling the boiler and supporting it with suspension rods.

A constant problem with HRT boilers is maintenance of protection for the bottom blowoff piping. In many cases that pipe drops vertically through one end of the furnace and has to be protected by refractory because it would absorb so much heat that steam couldn't escape it fast enough to allow water in. They go dry, overheat, and rupture.

The other concern with HRT boilers is the bottom where radiant heat from the furnace is absorbed by the shell. Any accumulation of mud in the bottom of the boiler tends to prevent cooling of the shell with resultant failure. The only service one of these boilers is purchased for today is in firing solid fuel, normally small biomass applications because those applications require a large furnace and have low radiant energy emissions compared to oil and gas fired boilers.

Take the standard form of firetube boiler and turn it on its end to get a vertical firetube boiler. These are seldom used for steam service because the top tube sheet is exposed to steam instead of water and the tubesheet to tube joints are exposed to considerable heat. They are

Figure 10-7. Cast doors on HRT boiler

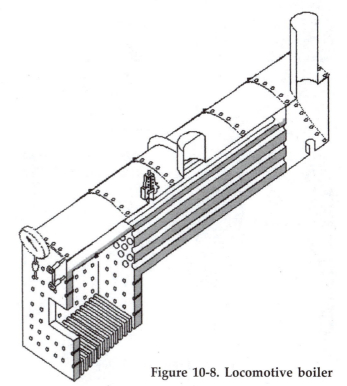

Figure 10-8. Locomotive boiler

commonly used for service water heating (Figure 4-9) and may find occasional use for hydronic heating and in waste heat service.

A locomotive boiler (Figure 10-8) is a good example of a firetube boiler modified to provide some water cooling of the furnace. The increased cost of the boiler to create a water jacket around the furnace was justified for locomotive service because the steel and water were considerably lighter than the refractory that would be required while providing more heating surface to make the locomotive more powerful. Staybolts are used to hold the flat surfaces against the internal pressure and their failure was one reason many of these boilers are no longer around.

The techniques developed in the railroad industry were translated to stationary boilers to create the fire-box boiler (Figure 10-9). The firebox boiler was the first potential "package" boiler because it only required construction of an insulated base in the field with all other parts assembled in the factory. A partial form of the boiler was also built to provide comparable performance at lower construction and shipping costs by requiring construction of part of the furnace as a brickwork base then setting the boiler on top of that base. It included some of the cast iron boilers shown previously. You may hear the terms "low set" and "high set" referring to these boilers. A high set firebox boiler incorporated all the furnace so the burner was set high in the firebox. A low set firebox boiler normally requires the burner be installed in the brickwork base.

Finally there is the construction that is typical of all our modern fire tube boilers. We call them scotch-marine although you probably won't find one on a ship and there's no proof that they were a Scottish design. This construction incorporates the insertion of a large

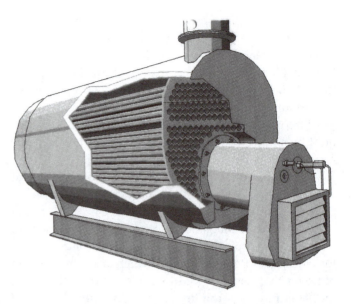

Figure 10-10. Scotch Marine boiler

furnace tube in the boiler (Figure 10-10) eliminating the requirements for an external furnace and providing a furnace that is almost completely water cooled.

Many of the original boilers of this design, the ones that were used on ships, were coal fired and required multiple furnaces to provide enough furnace volume and grate surface. The furnace tube diameters range from two feet to four feet and are welded to the tube sheets. The tube sheet to shell joint is also welded. The scotch marine design comes in two general arrangements, the most common is a dry back design where the turning chambers at either end of the boiler are formed by an extension of the shell and/or a door that forms the turning chambers. In either case both ends of the boiler are fitted with doors to gain access to the tube ends.

The doors can be full size, covering the entire end of the boiler or they can be multiple with separate doors providing access to various portions of the tube ends and furnace. In almost every case the door covering the end of the boiler and furnace tube is refractory lined because the temperatures of flue gas leaving the furnace can be over 1200°F. Some doors contain integral baffles (Figure 10-11) to divert the flow of flue gas back into other tubes in the boilers. The baffle arrangement varies with the boiler design principally to separate the passes. The wet back arrangement (Figure 10-12) is a more efficient boiler with less refractory to maintain but the higher cost and limited tube removal (front only) has resulted in a decline of its use.

Something common to firebox and wet back boilers and possible to find on others is a fusible plug. It's shown in Figure 10-12, where it belongs, at the top of the

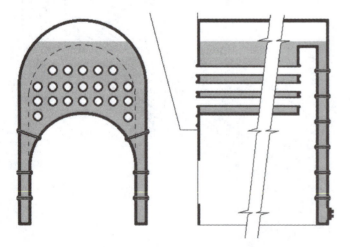

Figure 10-9. Firebox boiler

Figure 10-11. Baffled rear door of four pass firetube boiler

turning box in the middle of the flat top which is a surface that would be exposed to high temperatures if the water level dropped enough that the only thing to cool it was steam. The plug is filled with a low melting point metal so it would (theoretically) put the fire out with steam if the top of the turning box was overheated. Of course if it had to work the boiler had to be shut down, cooled down, drained and the plug replaced before returning to operation. I still regret not keeping four or five that were in Power and Combustion's stock before they were thrown out because nobody used them anymore. They would be good for parlor games; when you play what's this?

The locomotive boiler (Figure 10-8) is a basic single pass design. The flue gases enter the boiler proper and flow through all the tubes to the outlet of the boiler. The HRT design provided improved heat transfer by pro-

viding two passes, the flue gases are turned and return down a portion of the tubes on their way to the stack.

Note that a pass consists of a path for flue gas to travel from one extreme end of the flue gas containing parts of the boiler to another. Neither of these designs required a baffle to direct the flow of flue gas. Scotch marine designs can have two, three, or four passes. A two pass scotch marine boiler requires no baffles other than means to separate the burner from the returning flue gas. Three pass scotch marine construction requires one baffle in the rear of the boiler to separate the first and second pass turning box from the third pass outlet while four pass boilers require a baffle there plus one at the front to separate the second and third pass turning box from the fourth pass outlet (Figure 10-13).

Four pass firetube boilers have a construction unique to them, the tubes at the inlet of the second past are normally welded to the tube sheet. That's because the flue gases in the first to second pass turning box are much hotter in those boilers and the welding provides a better course for heat to pass from the metal to the water to prevent overheating those tube ends (however, see why they fail for a discussion of problems with four pass boilers).

Since I mentioned that the tubes are connected differently in four pass boilers I should also explain how they are normally connected. Whether firetube or watertube, the normal means of connecting the tubes in the boiler is by rolling. It's a mechanical method of attachment that is strong, watertight, and reliable but also relatively easy to break so the tubes can be removed. Refer to the section on maintenance for a description of installing a tube by rolling.

The furnace tube is normally connected by welding to the tube sheets. That's because it is large and thick

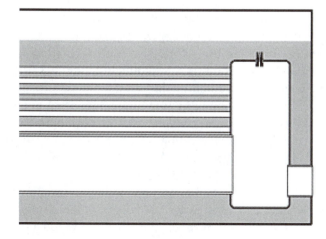

Figure 10-12. Wet back scotch marine boiler

Figure 10-13. Front baffle of four pass boiler

so it is difficult, if not impossible, to install it by rolling. Also, I wouldn't want to be the guy that has to pick up that tube roller.

Sometimes furnace tubes are called Morrison tubes, and it's done without distinction. Some furnace tubes are not Morrison tubes; they're the ones that are basically a simple cylinder. Morrison is the guy that realized the furnace tube could be made thinner and still withstand the external pressure without collapsing if it was corrugated (Figure 10-14). If you look closely at Figure 10-11 you can see that boiler has a Morrison tube. Now you know the difference, if it's corrugated it's a Morrison tube and if it's not it's just a furnace tube.

The section through a firetube boiler in Figure 10-14 also reveals another important element of their construction, staybolts. The tube sheet isn't supported by the boiler tubes in the top of the boiler (what we call the steam space) so staybolts are required to keep that portion of the tube sheet from buckling out. Part of a boiler internal inspection is checking the fillet welds attaching the staybolts to the top of the boiler shell, and the staybolts themselves, for corrosion. The staybolts normally penetrate the tube sheet and their welds should be checked on the outside as well as the inside.

There's another classification of firetube boiler that you may encounter. They're called "oil field boilers" and they're designed for that application. Boilers used in oil fields get little care, normally run on raw water with little condensate return and don't get the quality treatment provided by a wise boiler operator so they're designed for the abuse. They have thicker shells, thicker tubes, and lower heat transfer rates.

There are many advantages to a scotch marine fire-

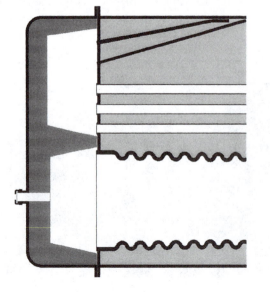

Figure 10-14. Morrison tube

tube boiler which includes simplicity in design. They're relatively easy to clean completely on the fire side, once you get those heavy doors off. They can be packaged in most of the sizes, they contain minimal refractory. Tube replacement is less expensive because all the tubes are straight. They also hold a larger volume of water compared to a watertube boiler so they absorb load swings a little better.

WATERTUBE BOILERS

Whether tubes are straight or bent is probably the first distinguishing characteristic for a multitude of designs of watertube boilers. I started operating straight tube boilers and learned later that there was such a thing as a bent tube boiler. Actually the last boiler I operated while in the merchant marine was a straight tube boiler and in the process of rebuilding and retrofitting boilers with Power and Combustion in the 1980's we designed a new burner installation and furnace modifications for a straight tube boiler that had a riveted drum.

You may never see a riveted boiler outside of a museum because they are no longer built and many have failed, never to fire again. Most state laws require replacement of any riveted boiler that has a failure after a certain age and those laws have effectively eliminated riveted boilers. When I mention a riveted boiler the normal response is a question, "how did they keep them from leaking?" The answer is caulking, not the goo in a tube type you're thinking of. To caulk a joint in a riveted boiler you used a special chisel and a good heavy hammer to deform the metal at the joint working the two together.

Blacksmiths still weld metal by heating the material until it's soft then beating two pieces together to form one piece. Most of the time we managed to seal the joints in a boiler by caulking them cold. The real problem with riveted boilers wasn't leaks, it was cracks forming between the rivets. The crack formation was eventually identified as a byproduct of tiny leaks that left water concentrated in the metal to metal joint and caustic corrosion cracking (see water treatment). A lack of skilled riveters and caulkers and the development of gas and electric arc welding, which formed a stronger and cheaper joint, produced the change from riveted boiler construction to welded construction.

Just like firetube boilers need a shell to contain the water and steam most watertube boilers require drums or headers to close off the ends of the tubes, provide a path for the water and steam to flow into and out of the tubes, and provide a place for steam and water to separate.

I've never come across a distinctive definition that differentiates drums and headers but I know drums are big and headers are small and I differentiate them by whether or not I can get inside one with the exception of the steam drum which, to me, is always the pressure vessel part where the steam and water are separated. That rule doesn't always work when it comes to what we call a mud drum which is the lowest drum in a boiler and has connecting piping for blowoff so the mud can be removed from the boiler. I can't say it's the lowest point because there are boilers where the mud drum is several feet higher than the lowest header. Those low headers have to be blown down because mud will collect in them but they require special attention to prevent problems with circulation during the process. Anyway, drums close off the ends of tubes and it's the tube and drum arrangement that further defines a watertube boiler. There were a few firetube / watertube combinations created over the years but I'm not aware of any that are left; I did help tear a couple of them out.

Occasionally you may see a boiler with two nameplates, one will be for the boiler proper, and one will be for the water walls. As boilers got larger the area of furnace walls increased to the point that they represented a considerable waste of heat. Fuel was so inexpensive then that it wasn't the primary consideration, keeping the boiler room cool and limiting the cost of refractory was.

Another problem was refractory walls were getting so high they couldn't be self supporting and expensive structural steel was required to hold them up. To solve many of those problems boiler manufacturers started making water walls which are rows of tubes that help protect the refractory or actually replace it. The waterwalls on large utility boilers actually occupy more space than the boiler itself. Most of them are tangent tube walls (described later) and constructed in "panels" that are subsequently welded together to form waterwalls, some over two hundred feet tall.

Waterwalls consist of tubes that may be bent to connect to a steam or mud drum or connect to a header that is connected to one of the drums with more tubes. Despite the two nameplate labeling (which was abandoned shortly after it was started) the waterwalls and boiler are all parts of the same pressure vessel.

The first boiler I worked on was a cross drum sectional header boiler (Figure 10-15) where all the tubes were straight; which made it a straight tube boiler. I doubt if you'll ever see one, let alone operate with one but it's a good one for explaining some of the unique characteristics and requirements of watertube boilers. Note first that this is a three pass boiler. The flue gases traverse the furnace from the burners to the rear but

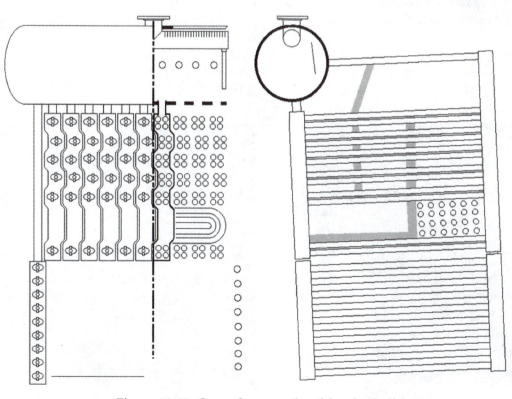

Figure 10-15. Cross drum sectional header boiler

that's not counted as a pass. The gases turn up at the back of the boiler and pass up through the superheater and boiler tubes until they reach the top (first pass) then drop down through the middle of the tubes (second pass) and finally up through the tubes at the front of the boiler and out the stack. The baffles are made out of refractory and include tile laid on top of the screen tubes to form the bottom of the second and third passes.

The bottom two rows of tubes are called screen tubes because they form a screen that blocks the radiant energy from the superheater (more on superheaters later). They also protect the baffle. The sectional header part of this boiler involved the forged square headers shown in the detail which were connected to the steam drum and bottom header by tube nipples (short lengths of tube) and contained handholes on the side to gain access to the tube ends so they could be rolled. The headers were forged in a semi-square shape to provide a uniform surface for rolling the tubes. Drums are normally of sufficient diameter that there is no problem rolling a tube in them.

To gain access to the tube ends to roll them and for other parts the drums have manholes, usually a 12-inch by 16-inch oval opening. Handholes are simple openings in the drum or header that are closed by a cast cover (Figure 10-16) which is inserted inside the boiler and bears on the inner surface of the shell, drum, or header usually against a gasket so the internal pressure of the boiler helps hold the cover in place. To keep them in place when the boiler is not under pressure the bolt, nut and dog are applied. Key caps (Figure 10-16A) are similar but tapered cast plugs that wedged into the header or drum openings to form a metal to metal fit.

A special "puller" was required to seat the key caps so they wouldn't leak as the boiler was filled.

That old sectional header boiler provides a simple look at the complex conditions surrounding circulation in watertube boilers. Water separated from the steam and boiler feedwater mixes in the steam drum (a common arrangement) then drops down the front headers (which are exposed to the coolest flue gas) and rises up the sloped tubes going from the front of the boiler to the rear. In those tubes the water is heated to the point of saturation and starts boiling, changing from water to steam. The steam forms small bubbles in the water, displacing the heavier water and reducing the density of the steam and water mixture as it travels along the tube.

By the time the mixture reaches the rear headers it is significantly lighter than the water so the weight of the water in the front header is just like a piston pushing down to force the water and steam mixture up the rear headers and back the return tubes to the steam drum. There's only a little difference in pressure between the water in the front header and the mixture at the rear header, perhaps half the height of the boiler (inches water column) but that's enough to force the water and steam to flow around with the flow rate of the steam and water mixture through the top tubes at least five times the rate of the steam going out the nozzle, perhaps more. In the case of this boiler all tubes are risers, the front headers are downcomers.

Another form of straight tube boiler was the box header boiler which used fabricated boxes containing stud bolts (see discussion for firebox boilers) and handholes opposite the tube ends in an arrangement very similar to the sectional header boiler. The straight tube

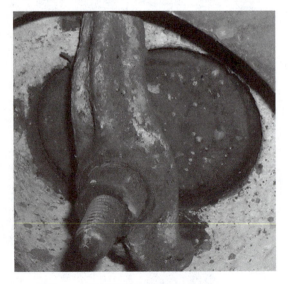

Figure 10-16. Handhole and cover

Figure 10-16A. Key caps

boiler with its headers limited boiler size (it was difficult to support the tubes as they got longer) and included multiple sources for leaks (all those handholes) so, in 1890 a man named Sterling came up with a better concept for constructing boilers to eliminate a lot of those problems, he decided to use bent tubes. There are particular designs of boilers (Figure 10-17) that are identified as Sterling boilers but for all practical purposes all bent tube boilers are identified as Sterling. Bent tubes added flexibility to the design of boilers to permit hundreds of designs and arrangements.

The evolution of bent tube water tube boilers consisted of many arrangements of the sterling design, so many that it would take an entire book to cover all the variations so I have no intentions of trying to describe them all. Keys to sounding intelligent about them include how they're supported (top, bottom or intermediate), drum position relative to movement of the fire (cross drum if the fire moves perpendicular to the centerline of the drum; we don't say anything if it isn't) and the pressure on the flue gas side (forced draft, balanced draft, or induced draft).

Firetube and package watertube type boilers are mostly forced draft design, the hot water heater in your basement is most likely induced draft, and most forms of Sterling boilers built today are balanced draft design.

A forced draft boiler has a fan blowing air into it and the pressure produced by that fan is used to force the air and products of combustion all the way through the setting and out the stack. An induced draft boiler can use stack effect to produce the differential pressure necessary to get the air and flue gases through the setting or the boiler can be fitted with an induced draft fan that creates a negative pressure at the boiler outlet and forces the flue gases up the stack. Induced draft methods basically create a lower pressure at the outlet of the boiler so atmospheric air pressure can force the air and gases into

the burner, furnace, etc..

The first boilers were primarily induced draft designs because motors and fans were more expensive to buy and run than building a tall stack. The stack effect is also a lot more reliable but you seldom see a tall stack erected today because it's considered an eyesore, not the indication of prosperity that was welcomed in the 1930's and 1940's. If you do see a tall stack going up its purpose is to disperse pollutants, not to create a draft for induced draft boiler operation.

As industry flourished the cost of fans and electricity dropped and the pressure drop across the boiler heating surfaces increased to the point that a stack alone was not sufficient and induced draft fans were developed to save on the cost of a tall stack and low pressure drop boiler. Almost all of those boilers were coal fired and had brick settings so use of forced draft fans was not desirable because pressure would force the flue gases out little cracks in the setting into the boiler room.

As boilers got larger the low furnace pressures required to draw the combustion air into the boiler and mix it with the fuel also increased admission of tramp air to lower the boiler efficiency. Tramp air leaks in after the burners. On large units it required additional structure to overcome the force of atmospheric pressure on the furnace wall. To reduce the low furnace pressures balanced draft boilers were developed where the induced draft fan, or stack, produces a slightly negative pressure in the furnace and provides the force to move the flue gases out of the boiler while a forced draft fan delivers the combustion air to the furnace.

Modern fossil fuel fired electric power generating boilers are all balanced draft and have significant pressure drops on the flue gas side to overcome draft losses in the environmental controls as well as the heat transfer elements. Some operate with induced draft fans capable of generating over fifty inches of water column differential, so much that they could, if conditions were not controlled, implode the boiler. They create so much differential that atmospheric pressure would push in the casing around the furnace of the boiler because it isn't designed to operate with that large a differential. Should controls on those boilers fail we will get an "implosion" the furnace walls collapse in.

I think that's enough on Sterling design water tube boilers, most of you will be operating other types.

Package Watertube Boilers

An interest in other watertube boiler designs can be satisfied by looking up a copy of Steam[10] but most of the watertube boilers you encounter today, except for a

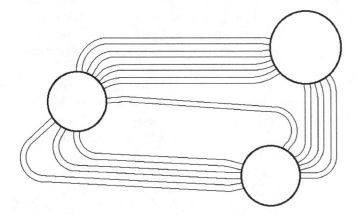

Figure 10-17. Sterling boiler

few rare Sterling designs, will be what we loosely term "package types" that come in one of four basic arrangements, A, D, O or Flexitube. These designs provide the current optimum in cost and performance, some better than others, and represent the heart of the packaged watertube boiler industry. A good understanding of their construction and operation will serve you well in developing an understanding of any other watertube boiler you come upon.

The A type (Figure 10-18) was originally developed by the Saginaw Boiler Works in Michigan and subsequently purchased by Combustion Engineering. Other manufacturers produce comparable designs. The A shape is attributed to the single steam drum at the center top and the two mud drums, commonly called headers, at the bottom. They require a second blow down line and more soot blowers but provided features like a water cooled furnace from one end to the other and balanced construction which makes them easy to transport as package boilers.

The tubes inside that form the furnace have alternating shapes. One will drop from the steam drum around the furnace and down into the bottom header while the next tube turns above the bottom header and crosses the bottom of the furnace to enter the side of the opposite bottom header. Shifting the tube arrangement by one sets up the crossing pattern with a tangent tube wall construction (Figure 10-19) in most of the roof and sides of the furnace. The furnace floor (the tubes at the bottom) have a maximum spacing of one tube width.

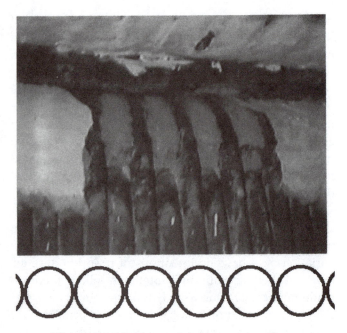

Figure 10-19. Tangent tube construction

Normally the bottom tubes are covered with refractory tile to limit heat absorption on the top of the tube. The tangent tube walls and installation of sealing refractory in the "crotch" under the steam drum close the furnace so all the flame and flue gases are restricted to the center of the boiler. Four to eight rows of tubes from the back of the boiler are installed without the drop to the bottom header forming tube gaps that allow the flue gases to turn and proceed down the convection bank tubes back toward the front of the boiler.

Most of these boilers have the flue gas outlet at the top front but some were made with the convection bank terminated part way down the boiler to create a larger furnace. In that case the side wall tubes are also the furnace wall tubes. One serious problem with the A type boiler is the crotch refractory falls out on occasion forcing an outage of the boiler because a lot of capacity is lost and there is concern for damage to the steam drum. They're also a pain to maintain because all the trim is above the burner and fans and ductwork connected to the burner at that point makes access to the front drum manhole almost impossible.

The front wall of all these boiler designs is normally a simple 13-1/2 inch thickness consisting of 9 inches of plastic refractory over 4-1/2 inches of insulating brick with a 1/4- or 3/8-inch thick steel front wall plate. There are variations in thickness and materials of construction including use of ceramic wool, insulation instead of brick and precast fired tile instead of the plastic refractory but all perform the basic function of closing the front wall. A few, very few, use additional tubes bent to

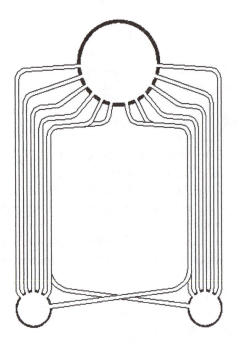

Figure 10-18. "A" type boiler

spread over the front wall to help protect the refractory.

The rear wall, on the other hand, is usually fitted with bent tubes spread out to cover it. The wall is typically much lighter in construction than the front wall, an allowance partially provided by the tubes and distance from the heat of the flame. Frequently the rear wall is called the target wall because the flame is shooting straight at it and the tubes against the rear wall are called target tubes. The tubes form a framework of steel that helps to hold the rear wall in place, especially during shipment of the boiler and that's a major consideration in the wall thickness.

The O type boiler (Figure 10-20) is similar to the A while eliminating one header by providing a drum in the bottom center just like the top. The headers required many handholes for rolling the tubes in an A type boiler so the single drum eliminated that expense but produced a boiler with a smaller furnace cross section.

The single bottom drum saved one longitudinal weld as well. All the longitudinal welds in modern boilers are X-rayed making them more expensive to form. The O type boiler is only manufactured by Erie City Iron Works of Erie, Pennsylvania (now Zurn), and is the only boiler I know of where the feedwater line enters the bottom drum. Some of the same difficulties experienced with the A boiler are associated with the O design. This boiler is not a good candidate for firing solid fuels or heavy fuel oil because it's almost impossible to remove the soot and ash from the bottom of the boiler. It does work well on gas.

The predominant design is the D type (Figure 10-21) which has only one drawback and that's the problem with transporting and supporting something with most of the weight on one side. The D tubes extend out of the drum to form the roof of the furnace, drop to form the furnace side wall, and return under the furnace to the mud drum. It has one convection bank of tubes centered between the drums to limit sootblower requirements. This construction makes it possible for the flue gas to leave the boiler via the front or side. A more detailed diagram (Figure 10-22) will help you identify some of the standard features of this construction.

There are many modifications to this design with different manufacturers featuring different details. D type boilers are also manufactured in semi-shop fabricated form where the furnace portion is shipped as an independent assembly from the convection bank with the two drums. Another arrangement is the D tubes and casing are shipped loose for installation in the field. These may still be referred to as "packaged" boilers despite final field assembly. Shipping the furnace or its components separately allow for larger capacity boilers without the restraints of shipping clearances and still retaining most of the advantages of a package boiler.

Unlike the scotch marine firetube and other smaller boilers "package" doesn't clearly describe the assembly for water tube boilers. A package boiler can be shipped without any burner or connecting piping. Almost any package water tube boiler with a capacity over 25,000 pph is not ready for connecting pipe and wire and start-

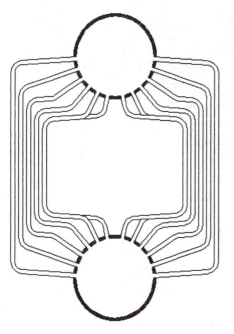

Figure 10-20. "O" type boiler

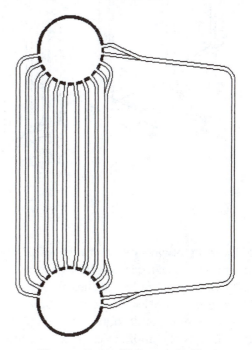

Figure 10-21. "D" type boiler

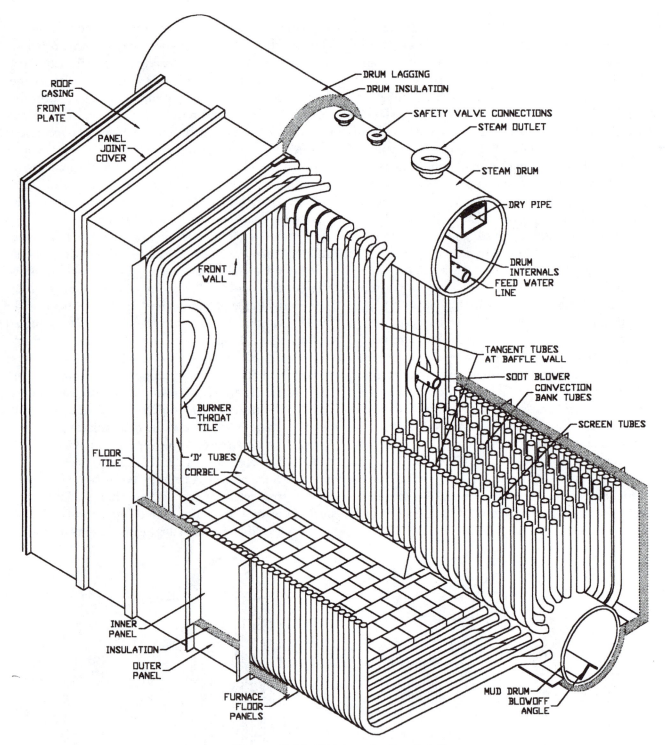

Figure 10-22. "D" type boiler details

ing up, there are always different degrees of assembly. When specifying a package water tube boiler an engineer has to explain very carefully what he calls a package.

There are also a lot of package boilers setting around that were not built in a factory, they were field erected. Problems of shipping clearances where a bridge or tunnel near a plant prevented delivery of a factory packaged boiler or clearances into a building where the owner wanted the boiler installed resulted in field erection of those boilers. In the middle 1960's boilermakers working for Power and Combustion felt they were in a contest to see if they could field erect more Combustion Engineering package boilers than Combustion built in the factory. I don't know if that was a close competition but I do know a lot were field erected. During my time with Power and Combustion I think we field erected half of the package boilers we installed.

The boiler in Figure 10-22 has tangent tube walls at the side of the furnace, side of the convection bank, and the baffle wall between the furnace and convection bank (except for the short section of screen tubes). Other manufacturers provide finned tube walls (Figure 10-23) where bars are welded between the tubes to form a heat absorbing fin and eliminate the special bending of alternate tubes near the drum which is required to get a tangent tube wall.

Babcock and Wilcox provide an integral finned tube (Figure 10-24) which provides the equivalent of a tangent tube construction without the need to weld the tubes. The finned tube provides a gas tight envelope around the furnace (with the exception of a gap where the tubes enter the drum) tangent and integral fin tubes are easier to replace.

Combustion Engineering produced several boilers with swaged tubes to simplify construction of the boiler, each D tube, outer wall tube and baffle tube was swaged (mechanically formed to reduce the diameter, Figure 10-25) from four inch to two inch so the tangent tubes could be installed in one row of holes. CE also built several boilers where the D tubes are made progressively shorter, top and bottom, so the rear wall of the boiler could be formed of tangent tubes.

In looking at the construction of the A, O and D type boilers you get the impression that they are only two pass boilers. Many of them are, with flue gas traveling down the furnace to the back then back to the front and out. A lot of D type boilers are not simple two pass design because they're fitted with baffles consisting of steel plates set between the tubes near the outlet of the boiler. Those baffles redirect the horizontal flow of the

flue gas to an up and down flow path to introduce additional passes, usually making them a four pass design when the switching of directions is accounted for. The boilers without baffling have higher velocities through the screen tubes and the initial portion of the convection bank with attendant higher pressure drop on the gas side and higher furnace pressures to provide a balance of heat transfer comparable to a multi-pass boiler.

Notice that I said most water tube boilers require drums or headers, a boiler that consists of continuous

Figure 10-24. Integral fin wall construction

Figure 10-25. Swagged tube

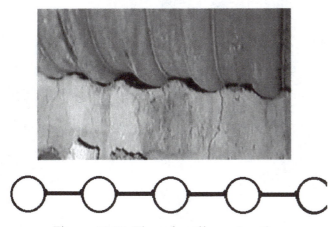

Figure 10-23. Finned wall construction

tube doesn't. Many hot oil heaters and some steam and hot water boilers consist of one coil of tube or two coils to produce a furnace and convection pass. A boiler consisting of one continuous tube or several tubes connected in parallel are called once-through boilers. If they generate steam the water used is ultra pure or some water leaves the boiler with the steam and is separated from it to remove the solids and impurities. Such boilers have no controllable steam and water line so other means are necessary to ensure they aren't dry fired. Some are fitted with temperature sensors that can identify conditions by superheat. One uses the coil of tube itself, when the tube gets hotter than saturation temperature its thermal expansion trips a limit switch. Should you encounter one of those boilers in your plant the best thing to do, once again I say it, is to read the instruction manual.

New in my time is the "flexitube" boiler (Figure 10-26 being one example) which has taken advantage of the bent tube construction to produce a boiler that is lighter, easy to repair, easy to field erect, and highly efficient. The only disadvantage of these boilers is their very low water content. Tubes in these boilers are bent to very small radii to achieve the form that allows them to use the tubes as baffles and produce a five pass boiler. In order to comply with code restrictions on bending of tubes (which makes the wall at the outside of the bend thinner) they are constructed using 3/4- or 1-inch tubes compared to the typical water tube boiler that is principally 2-inch tubes.

An additional feature of the flexitube design includes a new way of connecting the tubes to the drums or headers; that construction is shown in Figure 10-27. The ferrule is a forged tapered plug bored to accept the tube and the tube is rolled into the ferrule instead of into the drum. They can also be welded together. To install the tube the ferrule is driven into a correspondingly tapered hole punched or drilled and reamed into the drum or header. Precise machining of the ferrule and drum provides a tight fit and the dog is used to clamp it in position for added security.

I haven't seen this method used on high pressure boilers but it makes field erection of low pressure boilers much simpler. There are some questions about the long term operation of these boilers because thermal cycling could loosen the ferrules and movement could wipe out the ceramic fiber insulation used to seal the ends of the passes but when weighted against

the ease of removing and replacing tubes those questions are a little moot. There is a question in my mind as to whether higher efficiency, ease of repair, and other price advantages can compensate for lower reliability that may be associated with these units because they have a wider range of thermal cycling under normal operation due to the small volumes of water.

I have discovered that there are problems with the field erection of flexitube boilers because I served as an expert witness in an arbitration case where a contractor had installed the tubes improperly. While it's practically impossible to mis-align the tubes on the sides where the length of tube fixes their position it is possible to mis-align the tubes where they form the baffles that separate the passes. That's what the contractor did and the leakage of flue gases from the furnace into the second pass before combustion was completed resulted in very noisy operation and regular explosions.

Superheaters

Most commercial and industrial boilers produce saturated steam only. Superheaters associated with electric power generation and driving large equipment will

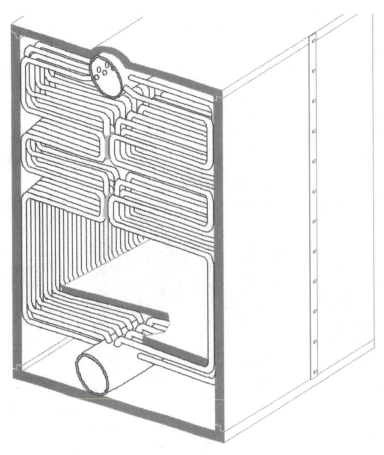

Figure 10-26. Flexitube boiler

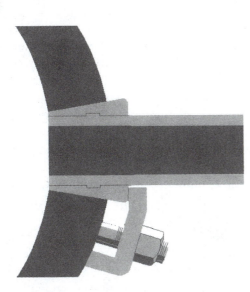

Figure 10-27. Flexitube tube to driving joint

become more prevalent after the writing of this book because the deregulation of electricity has finally forced utilities to become more efficient so more distributed generating systems will be built. Your boiler plant will eventually become a power generator as well as a steam generator unless it's a very small boiler plant or has a very inconsistent load.

Since steam can only be superheated when there is no water left around to evaporate, any superheated steam boiler takes steam at the boiler outlet to super-heat. The steam flows through a connecting pipe to a header where it's distributed through a number of par-allel tubes exposed to the furnace (radiant superheater) or flue gases after they pass through the screen tubes (convection superheater). There the steam temperature is increased as it absorbs heat from the flue gas.

Since the heat transfer rate is not as efficient as boiling water the steam velocity is rather high in the superheater to ensure turbulent flow for the best pos-sible cooling of the metal tubes. The full load pressure drop in a superheater is typically 10 psi because it takes a lot of pressure drop to create the turbulence for good heat transfer. The thin gas film that makes a conven-tional boiler tube much cooler than midway between the furnace gas and boiling water temperature when boiling water is repeated on the inside of a superheater so the tube metal in a superheater is considerably hotter. The superheater materials of construction are designed for those higher temperatures. Many of them use tube metal that is not as malleable (easy to mold or bend) as normal boiler tubes; in fact they're so brittle that they can't be rolled. A short piece of malleable tube that's rolled into the header is frequently provided as a

stub end welded to the more brittle superheater tubes. Those stub ends are protected from the heat by baffles or refractory coatings.

To prevent problems with water depositing in them many superheaters are designed to drain com-pletely by installing the headers at the bottom with the tubes extending up from the headers. We call them "drainable" superheaters. Boilers in most utility plants are of a construction that doesn't drain, the tubes hang down from the headers into the furnace or flue gas pas-sages and they're called "pendant" type superheaters.

Some superheaters are separately fired. Boilers on ships of the Navy usually have two furnaces, one before the superheater and one after it so the superheat temperature can be controlled. In shoreside applications there's frequently a requirement for small quantities of superheated steam so a separately fired superheater is installed to boost the temperature of that steam.

Large power generating boilers can also have re-heaters. They're the same as a superheater in construc-tion but steam passing through a reheater will be steam that has passed through part of the steam turbine after leaving the boiler outlet. To ensure the steam remains superheated in the lower pressure stages of the turbine it is reheated in the reheater. Construction is about the same as a superheater.

Boilers with superheaters will always have a safety valve at the outlet of the superheater and a valved vent line to atmosphere for ensuring flow through the su-perheater during startup and upset conditions. Another pressure gauge and a thermometer are also standard trim items.

Steam Drum Internals

All that steam and water entering the drum needs to be separated so the steam can go out the steam nozzle and the water can drop down the downcomers. To aid in separating the steam and water parts are installed in the steam drum. Everything that's installed inside the boiler is described as "internals" and that includes steam and water separating devices. Most steam drum internals are something like the details shown in Figure 10-28. Baffles deflect the steam and water mixture entering the drum to prevent water splashing up to the outlet. They spread the water and steam out over the surface so it can separate by gravity (heavier water falls, lighter steam rises).

The steam then has to go up over the top of the dry pipe and down through the holes in it to get inside the dry pipe which is connected by a tee to the boiler steam outlet. Small pipes connected to each end of the dry pipe

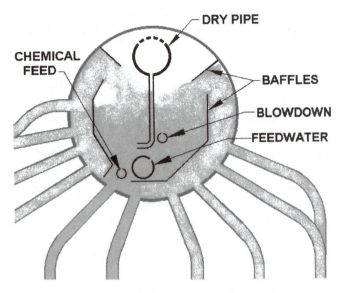

Figure 10-28. Steam drum internals

extend into the water to drain any water that does carry over into the dry pipe and settles out before leaving via the steam nozzle.

The other common form of steam and water separation device at the steam outlet is a chevron separator (Figure 10-29) which provides a tortuous path for the steam to travel on its way to the outlet with several changes in direction that tend to throw entrained water droplets against the chevron elements where they accumulate then drain by sliding down the surface of the chevron to the bottom forming large drops that fall off. Some modern boilers will have more complex baffling arrangements for separating the steam and water but a dry pipe or chevron separator usually do the job.

The baffles are bolted to steel bars welded to the side of the drum to support them and keep them in position during operation. Since they have to be removed to allow for each internal boiler inspection they're frequently broken. They should be replaced when broken because the movement of the water is so violent the lack of one connection could allow a baffle to break away and disrupt circulation to cause a boiler failure.

Another common internal for a steam drum is the boiler feed line. To prevent thermal shock the boiler feedwater piping enters the drum through a special arrangement (Figure 10-30) that diminishes thermal stresses on the thick steam drum by isolating it from the feedwater (which may be considerably colder than the steam and water mixture). The feed pipe extends into the drum, sometimes going the full length, and is capped off at the end. Holes are drilled in the feed pipe, normally in the top, to distribute the feedwater over the length of the pipe. At least one hole is drilled in the bottom of the feed pipe to ensure it will drain. Occasionally there are baffles added to the boiler to further distribute the feedwater and there are always supports for the pipe attached to the drum and the pipe to prevent it moving. A flanged, threaded, or slip joint is provided just inside the drum penetration so the feed pipe can be removed to gain access to the tube ends.

In addition to that boiler feed pipe drum internals commonly include a chemical feed line and a continuous (surface) blowdown line which are installed similar to the feed piping. The continuous blowdown line doesn't require the tempering fitting used for feedwater but a chemical feed line normally does. They are located in the drum in positions best suited for their purpose. The chemical feed is installed so the chemicals can mix as thoroughly as possible with the water before it starts its trip down the downcomers.

The continuous blowdown piping is located near the surface but not so close that it would draw off steam. You want it as close as possible to the water that just separated from the steam because it will contain the highest concentration of solids.

Occasionally a mud drum will have one internal, an angle set in the bottom to spread out the flow of water when blowing off the boiler. There are more elaborate boiler internals but most of the time these are

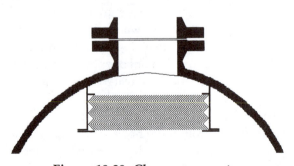

Figure 10-29. Chevron separator

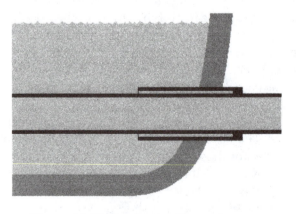

Figure 10-30. Feedwater line entrance

all you will encounter. A review of any drawings and instruction manuals for something different along with the basics of steam generation in this book should let you figure out what they're there for and how they affect the boiler's performance.

TRIM

Just as we hang decorations on our Christmas trees our boilers have a multitude of objects hanging on them; that's why it's called "trim." There is no concrete definition for what is included in boiler trim. I choose to say it includes all devices normally attached to the boiler including anything within the jurisdiction of the ASME Boiler and Pressure Vessel Construction Code and anything that isn't attached to something else.

Since the code for construction of power boilers usually extends to the far side of the second steam valve from the boiler I consider those valves and connecting piping part of the boiler trim. Others seem to include the blowoff and feedwater piping and valves but not the steam piping and valves. The discussion that follows is based on my definition. Some manufacturers provide covers or enclosures around all or part of the trim to change the appearance of their product but most of the trim is always there and some of it is essential.

Safety Valves

First, I'll point out that the correct title for safety valves is "safety relief valves" not to be confused with "relief valves" or safety shutoff valves. I'll continue to use "safety valve" because all us boiler operators know that we mean the safety relief valves.

Safety valves are the most important part of our boiler trim. They're the final defense against a real disaster, a boiler explosion. A safety valve may look simple but it's the most refined device in the world. The ASME Code contains extensive requirements for construction, testing, certification and labeling them. A safety valve manufacturer has to be qualified to use one or more of the various stamps ASME issues that authorize the manufacturer to make those valves. There are also rules and procedures for repairing safety valves.

Our safety valves have to have a nameplate or stamp on them that includes the appropriate ASME Code Symbol Stamp for the application. The stamps (see appendix) identify valves that have met all the requirements of the code. Notice that they're application specific, you shouldn't use a safety valve for a pressure vessel (UV stamped) for a boiler. Your valve doesn't have a

label or stamping but you think its okay? The only thing I can say to you is that's not a lot different than driving a car without any brakes! The ASME valve is an assurance that the valve will work when it has to, to operate without it is foolhardy, not the actions of a wise operator.

The valve nameplate should also bear the set pressure and capacity of the valve. The valve has to be large enough to dump all the steam (or heat) the boiler can generate or the maximum fluid input to a pressure vessel. I recall visiting a church to look at the burner on their boiler and noticed they had installed piping with reducing fittings on the two inch safety valve connection and reduced it to a little 3/4-inch safety valve. I shuddered, then turned to the deacon who was escorting me and said "you must want your congregation to go to heaven all at once." Never replace a valve with less capacity than the valve you have.

You should also never add piping between the safety valve and the boiler and under no circumstances should you install a valve or a blind between the safety and the boiler. There are times, when testing the boiler and for other maintenance activities, that you will install a blank or plug in place of the safety valves but never operate the boiler without them. Safety valves must be installed with their stems vertical so adding an elbow to turn the valve so the boiler will fit under some obstruction is unacceptable.

Steam safety valves have a special arrangement in their construction that makes the valve open completely. Sometimes operators call them "pop valves" because they pop open. When the valve is closed the disc of the valve is exposed to the pressure in the boiler over the area that's inside the seat as shown in Figure 10-31. As soon as the valve starts to open the pressure in the boiler is exposed to the full surface area of the disc (the larger circle) so there's more force on the valve and it pops open. The pressure has to drop to a value lower than the set pressure of the valve before it will close; we call the difference "blowdown" (which has nothing to do with boiler blowdown). When you operate too close to the set pressure of the safety valves you'll have to drop your operating pressure to get the valve to reseat.

Service water heaters (for domestic hot water heating) have an added feature on their valves. They're called PTVs for (pressure, temperature relief valves) and they're essential for preventing the explosion of a service water heater. The hot water heater in your house has one. The problem with domestic water heaters is the pressure isn't provided by the source of heat. A typical valve setting is 125 psig so it won't lift to dump water with the normal variations in the water supply pressure.

Figure 10-31. Safety valve seat exposed to pressure

About the only time a PTV will operate on pressure is when the water is trapped by a check valve or backflow preventer (see service water heating) and the pressure is increased by the water expanding as it is heated.

If the controls fail to shut down the burner or electric element or steam supplying a service water heater the pressure usually doesn't increase because the pressure is dependent on the water supply. Expansion of the heated water simply pushes the cold water back down the line out of the heater. Herein lies the problem, when the heat continues the water eventually gets so hot that it starts to turn to steam. The steam takes up a lot more room than the water and pushes the hot water back the cold water line until the heating element or the bottom of the heater is exposed to steam instead of water. Now, the steam picks up some heat as it is superheated but it can't provide all the cooling that evaporating water does so the temperature of the heating element or the bottom of the heater rises until they get so hot that they fail.

Luckily for those of us that have electric hot water heaters the element shorts out or burns open to stop adding heat. If you have a piece of fired equipment the outcome is not so pleasant, the weakened surface of the heater ruptures. The steam expands and the hot

water flashes to form more steam resulting in an explosion. Hot water heaters commonly rocket their way up through as many floors as are above them and have flattened many houses.

The temperature element of a PTV is a small cylindrical tube that extends from the inlet of the valve. The valve must be installed so that element will be immersed in the hot water. Mounting the valve on connecting piping will not work because the element isn't exposed to the heat. Since the element must be in contact with the heated water PTVs can be installed horizontally and, when labeled for it, even upside down. Don't make the mistake of one contractor in Oklahoma who decided the PTVs were installed wrong (the stems weren't vertical) so he went out to the local hardware and got some street ells (piping elbows with male thread at one end and a female thread at the other) to add and turn the PTVs. The worker assigned the job of changing the valves had a problem with the little pencil like things hanging out of the bottom of the valves (they prevented installation on the street ell) so he broke them off. Lacking the thermal element the PTVs didn't work when other controls failed and the heaters exploded. Six children and one adult were killed and forty-two others were injured. It was an 80 gallon water heater.

Boilers larger than 100 horsepower must have two safety valves, that's a code requirement. Also, boilers with superheaters have to have a safety valve at the outlet of the superheater which is set lower than the safety valves on the steam drum. It's essential that the superheater safety valve opens first to maintain a flow of steam through the superheater to prevent it overheating.

In addition to monthly and annual testing of safety valves (see normal operating procedures) you may be required to send the safety valves out to be replaced or rebuilt. That's normally a requirement of the insurance company that doesn't want their inspectors to spend time observing the pop testing of safety valves. It's less expensive to simply replace a small valve but valve prices increase with size and set pressure to where you would want to have them rebuilt at a much lower cost. A contractor that rebuilds safety valves should have ASME or National Board authorization to do that work.

You'll also want to replace a valve or send the valve out for rebuilding if it starts weeping or leaking. The steam condensing on the spring and stem will accelerate rusting in the topworks of the safety valve which can prevent it operating. Continuously operating a boiler with a leaking safety valve is hazardous.

When I encounter leaking safety valves I always check the vent piping immediately. In my experience it's the most common reason for a safety valve leaking. The boiler always grows (normally it expands upward) as it heats up. The conventional high pressure package boiler will grow at least three eighths of an inch from cold to operating pressure and a little more before reaching set pressure. Unless the vent piping allows the safety valve to move up with the boiler a considerable amount of stress is applied to the valve to spring the vent piping and that stress can deform the valve so it leaks. To prevent any stress on the safety valve we normally install a drip pan ell (Figure 10-32) which allows the safety valve to move with the boiler without any restraint.

When the boiler is installed pipefitters commonly stack nuts or washers under the vent pipe in the drip pan to provide the required gap between vent pipe and drip pan. One plant I visited had all their safety valves leaking and I found washers stacked in the drip pans. When I asked the operators why they were there they replied that the contractor put them in so they always made sure they put them back. After they removed the washers their problems with leaking safety valves disappeared.

Buildings do settle as they age and there are times when the structure (which supports the vent pipe) shifts independently of the boiler and its foundation will change the relative position of the safety valve discharge stub and the vent pipe. The settling can shift the structure so the vent pipe is not centered around the stub but pressing against it for another way to stress the safety valve. Annually, preferably right before doing your annual pop tests, check that the vent pipe is centered

Figure 10-32. Drip pan ell

around the stub and there's a 1-1/2-inch gap between the vent pipe and drip pan.

Water Column

In the list of trim the water column and gauge glass comes right after the safety valves in order of importance. The water column is a surge chamber that provides a stable water level independent of the splashing and bubbling inside the boiler so the level in the attached gauge glass is a true representation of the water level in the boiler. The water column is usually fitted with other trim items like a low water cutoff or cutoff and pump controller combination. It can incorporate probes for remote water level indications. Usually the controlling and high steam pressure switches are mounted on the piping connecting the water column to the boiler.

There was a time when the code required petcocks on the column to provide a means of checking the water level if the gauge glass was damaged or its indication questioned. Many manufacturers still provide them and they're always a good idea for the original reason. One problem with petcocks was some operators had the attitude that they would check their water level using the petcocks and shut off the gauge glass so it wouldn't blow. I'm sure you won't be that stupid.

Some operators will argue that you can't tell if there's water or steam there so the petcocks are useless. That's not true, you can tell. If there is steam at the level of the petcock then a second after you open it you will not be able to see anything between the end of the petcock discharge and the cloud of condensate that forms, steam is invisible. If you want to argue that statement then maybe you can explain to me why you don't see anything in the top of the gauge glass. If there's water there you will see it coming out of the petcock.

A water column is always equipped with a drain valve. That permits blowing down the column to ensure the connections between the boiler and water column are open. Refer to checking the low water cutout in the chapter on normal operation to learn more about blowing down water columns.

Water columns can be separated from the boiler by valves, provided they are rising stem gate valves. You'll notice that they're seldom valved off. If they are you should make it a habit of ensuring the valves are open (stems are sticking up) and keep in mind that the discs can come off the stem of a gate valve. The only time those valves should be closed is when the boiler is shut down to allow maintenance of gauge glasses and other water column parts while the boiler is still hot or under pressure. Don't be like one laundry I encountered a few years ago where the procedure was to close the valves every time the boiler was shut down. It's no wonder that they had dry fired the boilers so frequently that they had to replace all the boilers in the plant and that was only since they were all replaced six years earlier.

Piping connecting the water column and boiler must be installed so it can be inspected and cleaned. That normally results in the installation of crosses in the piping. I always insist on the opposite end of those crosses being closed with nipples and pipe caps. It provides two possible joints that will break so you can gain access to inspect the piping and it's a lot easier to remove a pipe nipple than a pipe plug. In a plant with a boiler damaged by dry firing, and after several hours of effort to remove the plugs, we found the piping hadn't been inspected for years because the operators couldn't get the plugs out. No matter how good you think your water treatment is there is a potential for those pipes to plug and you must inspect them annually.

Another important consideration with the piping is connections. Nothing more than operating pressure switches should be connected to the water column piping. In one plant I found someone had decided to connect the atomizing steam line to the column piping. All they had to do was remove the pipe cap and hook up to it! The pressure drop of the steam flowing from the inside of the drum to the cross immediately outside it was twelve inches of water column when the atomizing steam was on. Luckily the boiler had a separately piped low water cutoff because the level at the gauge glass and water column read a false twelve inches higher than it actually was in the boiler.

You should never accept a leak in that water column piping for the same reason. Any small flow of steam out a valve packing or leaking pipe joint can change the indicated level of the water.

Another important factor with the column piping is it must be installed so it stays in position relative to the boiler. Any maintenance activity that involves removing the water column or part of its piping should be preceded by measuring the height of the column relative to the steam drum or above the boiler room floor so you can confirm its proper reinstallation later. Most columns will have a mark in the casting that's the normal water line. You can use it as a reference.

Gauge Glass

The gauge glass is normally mounted on the water column and can be isolated with special shutoff valves. The valves are designed to shut off in about one quarter

turn and are fitted with a T type handle so they can be closed by pulling a chain hanging from the ends of the handle. For more effective shutoff a chain link or small triangle shaped piece of metal is attached to the bottom valve handle and connected to balance the force of the pull chain between the two valve handles (Figure 78) for a positive shutoff.

The purpose of that valve arrangement is to permit an operator to close them when (not if) the gauge glass breaks. On any ship I worked on I added a little style to those chains by making two different tabs for the ends of the chains, one that was a miniature copy of a stop sign and one looking like a yellow yield sign. The stop sign shape had "shut" instead of "stop" and the yield

Figure 10-33. Gage glass shutoff chains

sign shape had "open" instead of "yield" painted on it. I also made sure that, even with the valves open, the shut tab hung a little lower than the open tab so it was easy to grab and pull when the glass broke. A couple of trips under the spray of hot water from a broken glass trying to grab the right chain to close the valve would convince you that my arrangement will pay in the long run. The idea is you get to be prepared so you won't have to make several passes at those chains.

Locomotive boilers and a few others are fitted with gauge glasses independent of water columns. They normally have a liquid line that penetrates the boiler with a few holes in it to restrict surging flow so the glass level is stable.

Gauge glasses come in many forms but they all perform the same function. The water level inside the boiler is repeated in the gauge glass. The water in the glass is usually very clear because it's all condensate. Steam is constantly condensing in the water column, connecting piping and the gauge glass then draining to the bottom of the column and gauge glass and returning to the boiler through the connecting piping. Occasionally when the water level is fluctuating so the boiler water is surging into and out of the water column it will mix with that condensate and any color in the boiler water will appear in the bottom of the glass.

Aboard a ship the entire boiler moved so the water was always swinging in and out of the gauge glass. During a storm the determination of water level got very interesting. My last ship had steam drums that ran thwart ships (that's left and right as you face the bow or stern) and we determined drum level during a storm by the timing between the level rising above the top of the glass then coming back into view compared to the time it spent out of the bottom of the glass.

Since the steam and water are both clear it's difficult to identify the actual water level in some gauge glasses. The most common one is a simple glass tube like the one in Figure 10-33, twelve to twenty inches tall with some paint applied along one side. The paint is applied to form a thin red line along the length of the glass with a wider white line applied over that and it's the minimum you should have.

I have been in plants where someone decided to save a few bucks and buy plain glass tubes instead of the red line tubes. In one they also bought a new boiler because the operators made a wrong decision about water level. With a plain glass you can't tell if it's full of water or completely empty when the level is beyond the limits of the glass. The red line glass utilizes the natural diffraction of light through steam and water

to help you determine what's water and what's steam. When the level is within the limits of the glass and you position yourself opposite the side with the red line you will see the narrow red line above the water level but it will appear to be much wider below the water level. It works because the light is bent at the intersection of the glass and water but it isn't at the intersection of glass and steam. One important consideration is to install the glass with the lines painted on it away from your normal position when viewing it. I saw one job where the operators thought the light shined through the lines somehow and put all the glasses in backwards; you couldn't see the water level.

Tubular glasses should be fitted with an additional glass enclosure, usually wire reinforced, to protect personnel in case the glass breaks. I don't think it's necessary when the glass is ten feet in the air where you can't get close enough to it to be hurt by it breaking but the glasses are used on vessels where you can be right beside them, and those should be guarded. On my first ship I blew down a gauge glass on an evaporator to flush it out so I could see the level and the glass cracked from the thermal shock. I had to bend over to reach the drain valve and my eyes were about three inches from the bare glass. I had a burn across my left forehead, the bridge of my nose, and my right cheek; if I had been a few centimeters to the right or left I probably would have lost an eye. It's another reason for red line glasses, I might have been able to see the level through the dirt and not blown the glass down.

Tubular glasses can't handle pressure above 150 psig so higher pressure boilers have other products that permit viewing the water level. Prismatic gauge glasses are heavy steel frames with a groove cut in them to form a tube between the steam and water connections and a special glass bolted to one side. The glass is thick and narrow to eliminate the stress associated with the difference in temperature between the water and air sides.

A tubular glass tends to expand more on the inside where it's hot and the colder outside of the glass restrains that expansion resulting in stress that will eventually result in the outer layer cracking, being pulled apart by the tensile stress. Since the prismatic glass is narrow the stress is minimized. The glass to steel frame joint is sealed by a gasket and the glass is pressed against the gasket and frame by dogs which are held against the glass and frame by bolts (Figure 10-34). The notches cast into the glass that produce the sawtooth appearance use the diffraction principle to differentiate between water and steam. Part of the installation of a prismatic glass requires a light shining on it from the side so it illuminates the notches, they appear bright, almost white. Diffraction in the water shifts the light below the water line so that portion of the glass containing water looks dark.

When pressures get higher than 250 psig the glass can't withstand the heat of the steam so flat glass is used with a thin sheet of mica (a mineral that forms natural transparent sheets) installed between the gasket and glass.

As pressures increase the problems with differential expansion prevent use of full length glass so the gauge glass is converted to several small round flat glasses stacked one over the other on a steel frame. These have areas between each glass where the level isn't visible. To allow differentiation of water and steam the gauge glass is doubled up with another round flat glass behind installed at a slight angle to the other one. Lights shine through red and green lenses and through the gauge glass. Diffraction in this case determines which color you see, red if the glass contains steam and green if it's under water. You should make it a point to carefully read and make sure you understand the manufacturer's instructions for your gauge glass, it will pay you by reducing the number of times you have to change it.

A problem with gauge glasses that I've seen recently is regular packing leaks. Read the section on pumps to get some guidance on how to install packing properly so you won't have leaks right after you packed them. Another technological advance is graphite tape which can be wrapped around the glass to form a packing ring that will do an excellent job of sealing a gauge glass.

Low Water Cutoff

Frequently integral with the water column, occasionally (on hot water boilers) built into the boiler, and regularly mounted as an external device, a low water

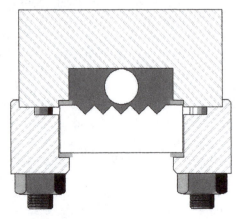

Figure 10-34. Section through refractive gage glass

cutoff is the primary protective device to save the boiler in the event the water level goes too low. The cutoff must be installed in a manner that keeps it in position relative to the boiler so thermal expansion doesn't shift it relative to the lowest safe water level. A low water cutoff normally has a mark in the casting that indicates its operating point. That level has to be higher than the lowest safe operating level established by the boiler manufacturer.

If there's no indication of that level in the documentation or on the boiler the bottom of the gauge glass is a good place to set it. Since I've discovered cutoffs installed at different levels on identical boilers (they had replaced the piping but the contractor wasn't concerned with matching construction) and cutoffs lowered by maintenance personnel (the darn things kept shutting the boiler down) you should be aware that a low water cutoff can be installed improperly. Any cutoff located significantly lower than the bottom of the gauge glass is a potential problem.

Any steam boiler should have two low water cutoffs (see why they fail at the end of the book) and they should be piped to independent connections on the boiler. That way if one connection gets plugged the other cutoff can still work. Many are installed with a common steam connection because they're less likely to plug with two water leg connections (I have, however, seen a couple of steam lines to low water cutoffs plugged). If there are valves located between the low water cutoffs and the boiler they should be full ported valves to reduce the potential for plugging and they must be rising stem type or quarter turn valves that indicate their position at a glance.

The drain valve for any low water cutoff should always be a globe valve. Gate valves and quarter turn valves do not throttle flow adequately to permit the operator to drop the water level slowly.

The cutoff must be installed in such a manner that it will drain back into the boiler. A major university lost a brand new field erected boiler because the erector installed the cutoff in a trap (Figure 10-35). Notice that the figure shows tees and crosses in the piping closed with nipples and caps, that's so you can gain access to the piping to inspect it and, if necessary, clean it.

If your boiler has low water cutoffs at the front and rear of the boiler don't be surprised if they are not at the same level. Since the fire is concentrated in the front of the boiler a slope in the surface of the water in the boiler from front to rear is not unusual. Depending on the distribution of the flue gas and tube arrangement the level in the back of the boiler can be higher or lower than the

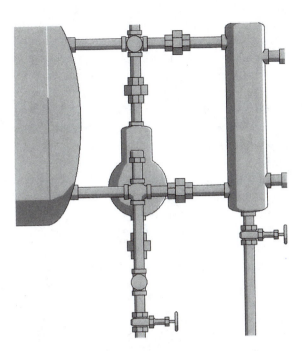

Figure 10-35. LWCO piped into a trap

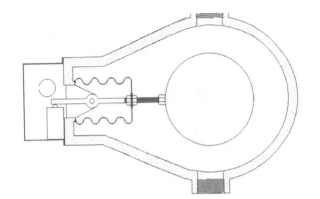

Figure 10-36. Bellows on float switch

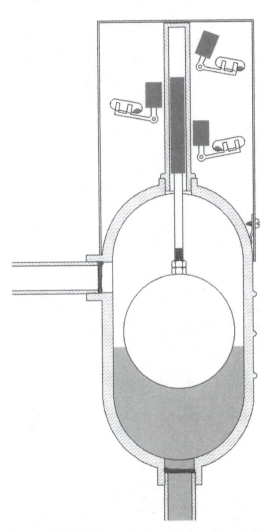

Figure 10-37. Magnet actuated level switch

front and there are some boilers where the level in the back shifts relative to the front with load changes.

Float actuated cutoffs require some means of sealing the part which connects the float rod to the electrical switches to prevent steam or water leaking into the portion of the switch that contains the electrical contacts. The most common method of sealing is to use a bellows (Figure 10-36) which allows the float shaft to transmit the float motion to the switches. The bellows provides a water (and steam) proof seal which is flexible to allow movement. Another common method is to use a magnetic coupling where a magnet connected to the float shaft is followed by external magnets connected to the switches (Figure 10-37). They work well in very clean environments. Another method is to transfer float motion using a transverse shaft (Figure 10-38) with packing but these are prone to leakage.

I should also mention that I've seen each type fail. Any one can fail if the float is banged around by improper testing or fluctuating water level to create a crack so the float fills and sinks. That's a fail-safe mode because the boiler should shut down. The problem with that happening is I've seen two of them where the operators simply bypassed it to keep on running. I've seen the bellows so coated with scale that it couldn't drop to the cutoff level and holders for the magnetic sensing switches slip down (they're usually clamped to the tube) until they were set too low.

If someone wonders why the cutoff is listed here after the water column and gauge glass it's because the operator watching the level in that gauge glass is more reliable than the low water cutoff. If low water cutoffs were as reliable as we would like them to be we would have almost 30% fewer boiler failures. Recall the low water cutoff testing in operations and read the comments in why they fail later in this book.

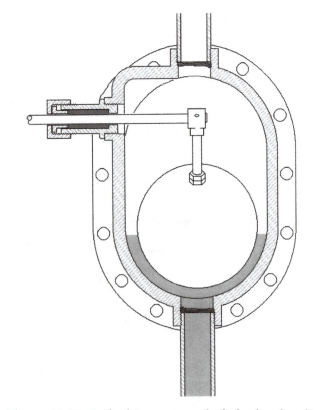

Figure 10-38. Packed transverse shaft for level switch

Pressure Gauge

A pressure gauge is a required piece of trim on a boiler. It's obvious that you need a pressure gauge to ensure that the controls are doing what they're supposed to but I've seen plants where the gauges were missing and the operator's didn't seem to miss them until I started asking them questions. A plant without a pressure gauge on the boiler is bound to have a lot of other problems and you always wonder exactly how safe it is to be there when you run into such a simple deficiency.

The pressure gauge is required by code and its size and scale are also subject to requirements of the code. A favorite violation in many plants is replacing the gauge with a much smaller one. The owner thought to save some money but ended up buying two gauges because a smaller one doesn't meet the code requirements. There's no specific size required by code but the interpretation of the code requirement that the gauge be "easily readable" is interpreted to mean nothing smaller than what the manufacturer installed originally. For low pressure boilers the size is dictated by the travel of the pointer which must be at least 3 inches for pressure swings from 0 to 30 psi. Manufacturers do not put on larger gauges to make the boiler look better, they put that large gauge on because the National Board Inspector monitoring that boiler's construction considered it as small as it could be

and be easily readable.

A pressure gauge is normally selected so at normal operating pressure the needle on the gauge is pointing straight up. That makes it easy for the operator to determine if the controls are operating properly. The normal hydrostatic test pressure for a boiler is 1-1/2 times the maximum allowable working pressure so the gauge must always have a scale range that extends to that value.

Hot water boilers must also have a thermometer that indicates the highest temperature in the boiler and code rules for size and scale should also apply to them.

The piping connecting the pressure gauge to the boiler can't have any other connections except a drain connection that's open to the atmosphere and an extra valved connection for the inspector to attach a gauge. A valve in the piping must be a quarter turn valve with handle in line with the piping when the valve is open on low pressure boilers and a rising stem valve locked open during operation on high pressure boilers.

Code requirements do not include provision of crosses and tees in the piping to permit cleaning it but I strongly recommend you have them because I have encountered several instances where the boiler's pressure gauge piping was plugged with mud that managed to get into the sensing line over the years. The piping should be opened and inspected at the connection to the boiler every year. The rest of the piping should be inspected when there is any reason to believe it may contain some sludge or mud. The piping should also be blown down every year right after bringing it up to pressure and before picking up the load. The piping should include a siphon or pigtail, a curl of pipe or tubing, to ensure water is trapped between the gauge and the boiler to ensure the heat of the steam never gets to the working parts of the gauge. Sometimes the gauge is connected to a section of piping that traps water for that purpose. Refer to the section on controls and instrumentation for other important considerations for application of pressure gauges.

On many larger boilers the gage can be considerably lower than the steam drum so it's visible at the normal operating level. Those gages have to be calibrated for the installation because they have several feet of condensate standing in the connecting piping and the head of that water adds to the gage pressure. If the gage is twenty-three feet below the drum connection it will read 10 psi higher unless it is adjusted to compensate for that static head. Don't be like one boilermaker I had that tried to return a gage because it was reading below zero when he took it out of the box.

Pressure and Temperature Limit Switches

A boiler that vaporizes a fluid should always have a high pressure switch to stop the burner or isolate the source of heat in the event the pressure in the boiler gets too high. If the boiler simply heats a fluid it should have a high temperature switch. In some instances fluid heating boilers are served by a common expansion tank which can be isolated from the boiler so a high pressure switch is also provided.

A hot water or fluid system boiler can also have a low pressure switch to prevent operation when the system pressure is so low that vapor would form out in the system (at the high point) to block liquid flow and possibly cause the equivalent of water hammer to damage the piping or heat utilization equipment.

All high limit switches require a manual reset. In some jurisdictions this is interpreted as a switch which will not close, once it's opened by a high pressure or temperature, until the operating personnel push a button located on the switch. I prefer systems that require the operator reset it at the control panel or boiler front and will argue with anyone that insists they be put on the switch. In most cases that switch is above the reach of a boiler operator and it's seldom mounted where the operator can conveniently get at it to push that reset button. I always suggest the proponents of reset switches picture themselves alone in the plant at two in the morning trying to climb up to the switch to push the button. Remember that first priority?

In addition to the high limit switch a boiler can have a pressure or temperature control switch which provides for on / off control of the boiler. These switches are all directly connected to ensure they sense the actual pressure or temperature in the boiler. Location of temperature switches is important, see the discussion on boiler construction. Pressure switches will not have any valves separating them from the boiler unless they're rising stem valves and are locked in the open position when the boiler is operating normally; a provision on a boiler whose continued operation is critical.

Pressure switch sensing piping can plug up just like the pressure gauge sensing lines although a pressure switch is normally mounted close to the boiler connection (it doesn't have to be seen all the time like a gauge). It's always important that the pressure switch has a siphon or piping arrangement which traps and holds some liquid in the switch and immediate connecting piping to protect the switch from the high temperature of the vapor.

Another concern with pressure switches that use mercury switches is the mounting of the switch. If mounted on a siphon the switch can be tilted as pressure builds in the boiler because the siphon tends to straighten just like the Bourdon tube in a pressure gauge. That would alter the switch operating point. If you have a mercury bulb switch mounted on a siphon make sure the travel of the mercury switch is perpendicular to the siphon.

Temperature limit switches are normally installed with a thermal bulb penetrating the boiler and the switch assembly right on the end of the bulb. When they're mounted separately to keep the wiring and switch isolated from the high temperatures in the boiler it's common for the assembly to include a capillary between the bulb and the switch bellows or diaphragm so the fluid expands in the bulb as it's heated and some of the fluid is pressed into the capillary which displaces fluid in the capillary, pushing it into the bellows or diaphragm chamber to expand it and actuate the switch.

If the capillary is crimped by bending or by physical abuse then the flat ends of the bulb tend to bulge out, making more room for the expanding fluid because the crimping of the capillary restricts the movement of the fluid in the assembly. The bulging of the flat ends of the bulb can act like a spring, maintaining pressure on the fluid so it eventually bleeds through the small restriction and acts on the switch. After the boiler cools the bulging is restored first and it may even reverse, caving in at the end as the liquid shrinks to produce a pressure differential that forces the fluid back through the small restriction and the switch resets.

The liquid slowly bleeding through the restriction results in the switch operating after a delay. Any significant delay in the response of a limit or operating temperature control is probably due to damage to the capillary and the only solution is to replace the switch. If the restriction is due to a repeated situation (like a plant where the operator's climbed to reach a valve and repeatedly stepped on a capillary draped over a support) the capillary can be shut off entirely and the switch won't work.

Since the switches are normally mounted on the boiler or the steam drum it's not at all unusual for them to be located where they don't get regular attention. The heat that radiates from the boiler and leaks through the casing or lagging creates swirling air currents around the boiler and its trim. The air currents can be warm one minute and cold the next so the air around pressure and temperature switches promotes breathing of the switch housing to suck dust into the housing. Dust settles inside the housing and can eventually block its operation. That's assuming the cover is on the housing; I would love to have a nickel for every limit switch I found with a cover dangling or removed. They're usually quite full of dust. Yes, you have to clean them.

Valves, Steam

The boiler codes don't have any requirements for steam valves for low pressure boilers but you might want to follow the discussion on high pressure boiler piping because the reasons for valve arrangements could apply to your low pressure plant. When a boiler plant has more than one boiler and they're connected to a common header two valves have to be installed on the steam outlet of each boiler with a manhole; and, the piping between them has to be fitted with a free blow drain valve.

The primary purpose of that arrangement is to protect anyone that's inside the boiler by providing a vented section of piping between the two valves to isolate them from steam generated by other boilers, or high temperature hot water. Despite that strict requirement I've encountered a few plants without the second stop valve and many without a free blow drain. Sometimes the conditions of the requirement result in a failure to provide comparable protection.

What exactly is a "free blow drain"? It's actually a combination of piping and a valve or valves that connects a piece of process piping to atmosphere with a constant decrease in elevation of the piping so it drains. In normal parlance the label refers to the valve that's in that combination. The purpose is to provide an uninterrupted path from the process piping to atmosphere at a low point where any liquid in the process will drain out and, because it's open to atmosphere, any fluid in the process piping will also flow out until there is no pressure (or at least an insignificant pressure) in the process piping. A free blow drain ensures the process line is not at pressure and doesn't contain any process fluids under pressure. A free blow drain should be provided in all piping systems that contain a process fluid that can be safely drained without contaminating the atmosphere specifically to allow personnel to confirm lack of fluid under pressure in that piping.

I can recall watching some boilermakers working inside a boiler removing tubes with steam blowing out the leaking packing gland of the valve mounted on top of the boiler. There wasn't any blank between the valve and boiler either. Needless to say that was in the days before lock-out tag-out. If your boilers have manholes you should use the double valve and drain provision for safely working in them; it doesn't matter whether they're high pressure or low, steam is deadly at any pressure above atmospheric and can be dangerous at any pressure and temperature.

No, the code doesn't require a non-return valve on all high pressure boilers. Non-returns are recommended in multiple boiler plants but are not required. Don't tell your boss that if a new plant or new boiler is under consideration though, they cost more and someone that's only interested in first cost will try to save a few bucks by using regular valves. Non return valves just make operation of a multiple boiler plant a lot easier for the operator so the investment in a more expensive valve saves in operating headaches.

A less tangible reason is they prevent high thermal variations in the boilers (when operators don't isolate the boiler and there's cold water under the hot upper blanket of steam) and flooding (as the steam condensate accumulates in a cold boiler) which can result in early equipment failure. Since a non-return valve acts as a combination globe valve and lift check valve it's treated by operators as an automatic shutoff valve for idle boilers (the check function isolates the boiler) that automatically opens when the boiler starts making steam. With non-return valves the operator has to make a trip to the top of the boiler only for isolating it for annual internal inspection. It is important to use the free blow drain to remove any condensate from the piping between the two isolating valves to prevent a slug of condensate rushing down the piping with the first flow of steam.

Wait a minute, I didn't say to use a non-return valve on a low pressure boiler. The piston type disc in a non-return valve is heavy and it takes a lot to lift it so you can expect to see a two to ten psi pressure drop across a non-return valve. Since low pressure boilers typically operate at around ten psi a non-return could prevent any steam getting to the facility. That doesn't mean you can't get the same performance by installing a low pressure drop check valve on the boiler outlet. Just remember that it has to have a low pressure drop. If you intend to use the check to prevent steam entering and condensing in an idle low pressure boiler then it should be soft seated. When you add that soft seated check valve to the steam outlet also add another one as a vacuum breaker to a branch off the vent connection so a vacuum won't crush the boiler.

When you don't have a non-return or check valve in the steam piping the valves have to be operated with each startup and shutdown of the boiler so access to those valves should be as simple and convenient as possible. Either chainwheel operators or properly located platforms with safe ladders should be provided so the operator can get at them. Operation of steam valves is scheduled by the boiler and the load more than anything else so the operators have to get at them quickly.

Valves, Feedwater

On low pressure boilers the code requires a stop valve and check valve on the boiler feed piping but the pipe itself isn't controlled by the code. The Code has specific requirements for the feedwater piping on a high pressure boiler out to the second stop valve and also requires a check valve. That arrangement normally means the bypass valve and isolating valve for the feedwater control valve are both within the limits of the boiler external piping.

The shutoff valve is there to allow you to isolate the boiler from the feedwater supply when it's shut down and, more importantly, isolate the feedwater system from pressure in the boiler. The check valve is there to help prevent draining water from the boiler in the event there's a failure of the feedwater supply and, more importantly, preventing boiler water leaking out to produce a steam explosion if there's a failure of the piping. Unlike the steam valves and piping there is no code requirement for a free blow drain connected between the two isolating valves on a high pressure feedwater line; there should be, and for the same reasons.

Valves, Blowdown and Blowoff

The valve for manual control of continuous blowdown (surface blowoff) should be provided with an indicator that shows the position of the valve so an operator can restore a particular valve position. Some valves are fitted with indicators and tapered throttling guides as part of the disc so the flow rate through the valve is proportional to the indicated valve position. The ability to closely regulate the flow of blowdown (independent of automatic blowdown controls) permits the operator to closely control the concentration of solids in the boiler.

It isn't essential and not required by the code but I would strongly recommend installation of a check valve between the continuous blowdown control valve and the boiler. Should you forget to close the continuous blowdown valve it will prevent water from another boiler entering the idle boiler. It's also like using a nonreturn valve, if you chose to you can rely on the check valve so you don't have to close the blowdown control valve (and reset it later) for short outages.

I've seen many an automatic blowdown control system isolated because the blowdown control valve failed. On most small boilers these are quarter turn motor actuated valves or solenoid valves which aren't designed to handle flashing steam. There's supposed to be an orifice or manually adjusted throttling valve to take the pressure drop located in the piping close to, and farther from, the boiler than that automatic valve. If not, or the orifice is removed or the throttling valve opened wide the automatic valve will most certainly fail.

Bottom blowoff valves come in a variety of forms but the most important part of their construction is they don't have any pockets or cavities that sludge can settle into and plug up. That's the idea anyway. I can only remember one time where I was tearing valves off to unplug a line and that's because someone had installed the valves backwards so all the mud settled on top of the discs preventing opening the valves. There are two things that are stressed by these points, use proper valves (ones designed for bottom blowoff applications) and make sure you installed them in the right direction. See the section on normal operation for operation of blowoff valves.

The code for high pressure boilers requires two valves for bottom blowoff designed for the service. The piping connecting them and the boiler must be at least schedule 80 (extra heavy) of materials certified to comply with ASME Codes and all welded. Piping inside the second valve must be fabricated by a manufacturer or contractor certified by ASME to do that work (See the section on ASME Code construction). All other blowoff and blowdown connections only require one valve and the code piping requirements are limited to the portion between valve and boiler. You may find two valves in other lines because the owner or contractor elected to have an accessible valve to use with the code required valve and piping located close to the boiler thereby limiting the extent of the code piping.

Low pressure boilers and some high pressure boilers are equipped with quick opening valves, a valve that works something like a gate valve but has a steel plate with a hole in it that is positioned in line with the pipe so there's no way for mud to plug the valve. The code permits one of the two valves required on high pressure boilers to be a quick opening valve. There are rules for operating those valves (see normal operation) and they should be installed in a manner that makes it easy to use them.

The seatless blowoff valve (Figure 10-39) is a commonly used bottom blowoff valve and one that I have seen operated improperly more than any other valve (see normal operation) but it is easily repaired. Unlike other types of valves it doesn't require skill or special tools to repair or even adjust to restore its shutoff capability. The manufacturer's instruction manual is very important reading before working on these valves.

Figure 10-39. Seatless blowoff valve

Boiler External Piping

The extent of the jurisdiction of the code for construction of power boilers extends to the far side of the second shutoff valve from the boiler on steam, blowoff, and feedwater piping. The jurisdiction extends to the far side of the first shutoff valve on all other connecting piping. All boiler external piping must be made of materials certified to comply with ASME Codes and all welded piping must be fabricated by a manufacturer or contractor certified by ASME to do that work

A piece of welded piping should be stamped or fitted with a securely attached nameplate containing the stamping required by the ASME Code. The stamping should include either the "S," "A," or "PP" Stamp. You may also find the National Board "R" stamp which indicates a contractor has repaired the piping. Be aware that the boiler inspector could look for those stampings at any time and they better be there or you will not be allowed to operate the boiler until a complying section of piping is installed.

I recall one incident where an owner moved the entire boiler plant to make room for a new baseball stadium and the contractor's personnel threw away the boiler feed piping thinking they could replace it easily when they reached the new site. We made a fair amount of money installing new piping (replacing what the contractor had installed and at the contractor's expense) after the job was ready for inspection and the boiler inspector couldn't find the stampings. The work was accelerated because everything else was ready to make steam. Luckily the contractor did move the steam and

blowoff piping.

Threaded pipe and fittings can be assembled by anyone and you can replace damaged piping yourself provided you use the materials required by the code. Replacing flanges and fittings is usually simple because they are marked and all you have to do is find materials with matching marks. Pipe, on the other hand, can vary from code quality to what we call "untested" pipe with some different grades in between and you usually won't find any markings on the damaged pipe to get a clue as to what material is required.

There are many different grades of quality of material that can be provided in compliance with the ASME Code and there are many ASME material specifications for material that isn't satisfactory for boiler external piping. Your insurance inspector should be able to tell you what material to specify when replacing boiler external piping. When you buy it you should request Mill Test Certificates and check the stamping (grade and heat number) on the pipe against the data on the certificate. Record in your maintenance log that the material and paperwork match and return both if they don't.

If the pipe is welded you can only repair or replace it if you are certified to do so by the ASME or the National Board. Unless you work for an employer that maintains several boilers with a need for regular repair of boiler external piping it doesn't pay to obtain that certification. It's much less expensive to locate a contractor that is certified and have them do the work for an occasional repair.

Adding a connection to boiler external piping can only be done by a certified contractor and many an owner has had to employ a certified contractor to remove and replace connections that were installed by the operators, the facility's maintenance personnel or an unqualified contractor. For a short period in history we built a lot of sections of boiler external piping around Baltimore to replace piping installed by unauthorized contractors. That was usually at the contractor's expense because the installation didn't pass initial inspection. Sometimes, however, the owner had to pay the bill.

There's probably no difference in the quality of the work but unless the contractor is qualified you will never know. How would the inspector know if you had added a connection? Well, all you have to do is look for the ASME P-4 and any National Board R-1 forms you have. They describe the initial construction and all repairs. If a connection is not described on those forms it isn't in compliance. You should keep all those forms, which are actually certifications, for the boiler external piping along with the forms for the boiler itself.

HEAT TRAPS

There's a general use of the term heat trap to refer to anything that is added to a boiler to absorb heat remaining in the flue gas. They normally return that heat directly to the boiler. Conventional heat traps are economizers and air preheaters. Condensing heat exchangers can be used as either an economizer or air preheater but are commonly used to heat water for other purposes.

Economizers

An economizer traps heat by transferring energy from the flue gas to the boiler feedwater so that heat doesn't leave the boiler. Economizers are only found in high pressure steam plants. They don't work on most low pressure or any of the HTHW plants, and some high pressure plants can't benefit from the addition of an economizer. An economizer can work in a low pressure steam plant that has no condensate returns because the feedwater temperature would be much lower than steam temperature. If you have a low pressure plant with little condensate return such that the feedwater temperature (before heating in a feed tank) would be around 100°F lower than the steam temperature then an economizer could be used to trap some of the energy lost up the stack but we would probably call it a CHX for reasons that will become evident.

When the boiler feedwater is colder than the steam and water in the boiler, it can extract more heat from the flue gas. Fluids colder than what's in the boiler can also be used in an economizer to recover the heat. An economizer on a high pressure boiler plant makes it as efficient as low pressure boilers because the feedwater supplied to the economizer inlet is about the same temperature as steam and water in a low pressure boiler. It's important to be certain the feedwater flows through the economizer in the opposite direction of the flue gas so it sees hotter flue gas as it heats up and the coldest water is exposed to the gas just before it leaves the economizer. Economizers can heat feedwater to a higher temperature than the flue gas leaving the economizer because of the counterflow arrangement.

At low loads there are some concerns with economizer operation which can restrict the turndown capability of the boiler. When the economizer is mounted in the stack or on top of the boiler the water has to flow down through the economizer. The natural tendency of heated water is to rise up through colder water because it's lighter (the thermal-siphoning effect) so water flow through the economizer can become unstable at low loads. Combine that with the fact that the heating surface doesn't change so heat transfer improves at lower loads and you have an opportunity for generating steam in the economizer. Generating steam in the economizer will promote scaling of the water sides of the economizer and potential damage from water hammer as flows change.

When the feedwater control valve is between the economizer and boiler the probability of steaming is reduced because the economizer operates at a higher pressure but the control valve will take a beating as some of the water flashes to steam as it goes through it. The feedwater piping in the boiler drum will also be exposed to water hammering and erosion from the flashing steam. There are such things as steaming economizers but they're designed to do it; a normal economizer is not designed to generate steam at any load.

If you have wide variations in load the economizer of each boiler should be fitted with a return line to dump some of the feedwater back to the deaerator. By adjusting a globe valve in that line you can control the outlet temperature at low loads.

I always provide bypass and isolating valves because there's no reason to limit boiler operation to include the economizer. If the economizer has problems draining it and bypassing it will not damage it because the flue gas temperatures will not be hot enough to hurt it.

An economizer is typically constructed of tubes just like boiler tubes with those tubes rolled or welded into headers. The tubes can be bare but are usually fitted with fins to increase the heat transfer surface (Figure 10-40) There are two standard arrangements of construction, square, where the tubes are straight and connected to each other by bends, and circular where the tubes form a coil between the two headers. The circular econo-

Figure 10-40. Finned tube economizer

mizers are less expensive initially but almost impossible to repair.

Since economizers can be subjected to corrosive conditions more frequently than the boiler the materials of construction may be special to resist corrosion. Combustion Engineering developed cast iron muffs, cast pieces that look like finned tubes pressed over the steel tubes, which provided a corrosion resistant covering for excellent performance on coal and heavy fuel oil fired boilers because the iron conducted heat well in addition to resisting corrosion. Modern metallurgy has created materials that permit construction of economizers that can withstand very corrosive conditions permitting closer operation to the flue gas acid dewpoints without concern for serious damage.

In addition to corrosion economizers can have problems with soot accumulation, occasional plugging with unburned fuel, and unique situations (Figure 10-41) with waste fuels. The tube in the top of the picture is the soot blower, you can see the coated fins in the bottom of the picture. Soot and dirt manage to build up between the fins of finned economizers to almost completely block heat transfer. Even if there's no reason to believe you will have problems of blockage an economizer should be supplied with a means to clean it or provisions to install them in the future. The common in-service method of cleaning is using soot blowers but they are ineffective when the deposits forms a hard gelatinous mass so there should also be means to gain access to the economizer to clean it with water washing.

Some economizer applications (like the one in Figure 10-41) require regular cleaning, a tough and dirty job for the boiler operators. The savings in fuel makes the effort worth it.

Gas fired operations produce flue gas with very low acid dewpoints so you can operate a deaerator supplying economizers or low pressure boilers at lower feedwater temperatures when firing gas than when firing oil or other fuels with higher carbon and sulfur content. If gas is the primary fuel you can adjust (slowly) deaerator pressures to raise the feedwater temperature when firing fuels with a higher acid dewpoint and lower it for firing gas.

An alternative to that, found in plants with auxiliary turbines designed for low exhaust pressures is to use a steam heated feedwater heater between the deaerator and economizers to raise the feedwater temperature to the point that corrosion will not occur when firing high sulfur fuels. Power generating plants normally use feedwater heaters to condense some of the turbine steam and raise the feedwater temperature. It raises the feedwater

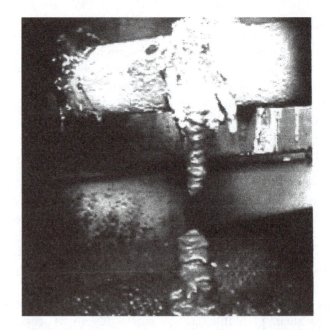

Figure 10-41. Plugged economizer, firing waste fuel

temperature to reduce potential for corrosion in the economizer and reduces the required size of later stages of the turbine for overall cost savings.

The best economizer arrangement (and also the most expensive) is where the flue gases flow down through the economizer. For counterflow the feedwater flows up through and that prevents problems with stratification and thermal siphoning. A turning box under the economizer can also serve as a drain pan for wash water used to clean the economizer. However, those installations introduce a hazard when the boiler is idle because any natural gas or other fuel vapors which are lighter than air and get into the setting can accumulate because the boiler and economizer arrangement forms a trap to hold them. I prefer to install an access door in the top of the ductwork between the economizer and boiler to vent it prior to entering the setting for inspection or maintenance.

I hate to call an economizer a heat trap because it's so much more than that. In addition to capturing heat that would be lost it provides additional heating surface for transferring the energy that's in the fuel into the steam and water. Adding an economizer to an existing high pressure boiler installation can also increase the capacity of the plant because heat that was absorbed through the boiler surface to raise the feedwater temperature is now used to generate more steam. I've seen capacities increased by as much as 8%. Of course the fan has to be replaced to overcome the pressure drop through the economizer or that added capacity is lost.

Economizers require some attention when starting

a boiler and at low loads (to avoid steaming you have to keep enough water going through it). If you have a feedwater recirculating loop you would use that to maintain temperatures, otherwise during startups you should add water to the boiler more frequently to provide some consistency to cooling of the economizer and you may even have to accelerate blowdown to provide enough water flow (that's a typical operation with HRSGs with integral economizers). That little bit of extra work is worth the savings in fuel over the operating life of the equipment.

Air Preheaters

Use some caution with this term. Normally we mean a heat trap when we use the term preheater but some manufacturers will call a steam coil installed in an air supply an air preheater because it does do what the name implies. Within the trade such devices are called "steam air heaters" to differentiate them from our traditional heat traps. An air preheater uses energy left in the flue gas that leaves the boiler to preheat the combustion air. This makes the air preheater the only true heat trap because it does trap the heat without adding any surface to the boiler. The way the air preheater increases the efficiency of the boiler is by raising the temperature of the combustion air using the stack heat instead of fuel. There's also a slight increase in heat flow through the boiler heating surface due to higher furnace temperatures.

An advantage of air heaters is higher temperature differentials. Instead of using 200°F plus feedwater to cool the flue gas you use combustion air entering at 80°F. There's potential for lower flue gas outlet temperatures which means higher boiler efficiency but corrosion of metal parts of the preheaters and ductwork to and including the stack must be given consideration.

There are two basic designs of air preheater, tubular air preheaters which consist of a box and tube heat exchanger to transfer heat from flue gas to combustion air and regenerative air preheaters.

Tubular air preheaters are normally arranged with the flue gas passing through the center of the tubes and combustion air surrounding the tubes. Corrosion during startup and low load operation is eliminated by bypassing the air around the heat exchanger so the flue gases keep it hot. Modulating the bypass damper to allow partial air flow doesn't work very well because the cold surfaces where the air first enters will promote condensation anyway. The bypass should be either open or closed.

Regenerative air preheaters use a rotating element to transfer the heat. A shaft rotates an assembly of "baskets" from the air side to the flue gas side and back. The metal baskets absorb heat from the flue gas then give it up to the combustion air. The major manufacturer of regenerative air heaters makes a "lungstrom" (for it's designer) air preheater (Figure 10-42) in a plant right near where I grew up, Wellsville, New York. The regenerative air heater occupies less space than a tubular heat exchanger and can prevent corrosive conditions by simply stopping the rotation.

To accommodate varying combustion air supply temperatures air preheaters are frequently fitted with steam air heaters to prevent acid condensation. There's a loss of efficiency associated with the steam use but it's recovered in added energy from the flue gas which couldn't be absorbed without damage to the preheater. Some of those heaters are adequate to permit startup and low load operation without starting and stopping or bypassing the air heater.

Air heaters are a little easier to operate than economizers since you can leave them off line until the boiler is carrying a load then close the bypass damper or start the rotor motor to put them into service. By simply not running the rotor motor during boiler warm-up the flue gas side is heated to prevent corrosion. The rotor should be run while purging the boiler to ensure all the baskets are purged. There are small pie piece shaped sections that are sealed between the gas and air sides while the rotor is stopped. Regenerative air heaters require additional maintenance because of the moving parts and seals to separate the flue gas and air sections but performance is usually more consistent than tubular air heaters. They can be cleaned in service whereas tubular air heaters are usually bypassed for water washing or require a full boiler shutdown to clean them.

You may never encounter them but I should men-

Figure 10-42. Lungstrom air preheater

tion that there were several attempts to use heat pipes for air preheaters in the 1980's, usually without success. At least I don't recall a successful application. They should work but there are times when technology isn't applied correctly and it results in failures. Heat pipes contain a refrigerant (not the typical ones used for freezers and air conditioners but something capable of operation at high temperatures) and employ a simple concept. The pipes are sloped so the liquid refrigerant flows down the pipe to the end exposed to flue gas where the hot flue gases heat the refrigerant until it boils. The gaseous refrigerant then flows up the sloped pipe to the end exposed to the colder combustion air where the refrigerant condenses, giving up its heat to the combustion air. The tubes are sealed in a casing where they penetrate a wall separating the hot flue gas and cold combustion air. If you encounter a working one I would love to have some information on it.

Condensing Heat Exchangers

A condensing heat exchanger (CHX for short) could be an economizer or an air preheater as well as other devices. What makes a CHX a CHX is the use of materials of construction that are corrosive resistant, allowing the heat exchanger to operate at temperatures below the acid dewpoint. Condensation of acidic flue gas components is expected and accounted for.

There's a definite difference between a CHX and the other heat traps because the others aren't designed to recover the latent heat in the flue gas. When the hydrogen in the fuel burns to form water it normally leaves the boiler as steam. With natural gas firing the energy that could be recovered by condensing that steam amounts to about 11% of the total heat input. A CHX is designed to condense as much of that water as possible to recover an additional 970 Btu per pound of water condensed.

The additional heat that can be recovered by a CHX helps pay for the exotic materials of construction but many of the materials that can withstand the corrosive acids can't tolerate the high temperature of the flue gas. Metals that can handle both haven't been proven as of the writing of this book but they may be. In the next ten years so condensing air preheaters and other CHX options will become standard boiler plant devices. Right now they aren't because of many unsuccessful applications and, to be perfectly honest, operators not understanding the benefits of them and how to operate them properly.

The current common application is a CHX used for preheating boiler water makeup and service water independent of the boilers. Flue gas is drawn from the boiler stacks by an induced draft fan downstream of the CHX. By using city water you're running high temperature differentials (city water is normally between 40 and 70°F) so the poor heat transfer capability of the corrosion resisting materials is overcome. The typical applications right now use high grades of stainless steel for gas fired applications and Teflon coated copper for more acidic flue gases.

To withstand the corrosive properties of the flue gas after passing through the CHX the exhaust ductwork is constructed of corrosion resistant materials, usually FRP (fiberglass reinforced plastic piping). Those materials are not common to boiler plants despite the fact they've been used in some cooling tower operations for several years now. They're not difficult to deal with in operation or maintenance, they just require learning about them. It's best to read the instruction manuals for the materials your plant may have because there are considerable variations in capability and handling.

The condensate from a CHX has a low pH because the condensed water will absorb the CO_2 and SO_2 in the flue gas to form acids. The drain piping should be FRP to a point where the condensate can be neutralized. Mixing the acid condensate of a CHX with the caustic blowdown from the boiler can produce a mixture that may meet the local jurisdiction's requirements for sanitary sewage. If it doesn't you'll have to add caustic soda to neutralize the mix before it is dumped to the sewer.

A final consideration for heat traps is they don't have to be used on boilers or trap the heat from boilers. I've had some very successful projects that saved customers a lot of money by using these devices to recover heat lost up the stack from process operations. What is an economizer, for all practical purposes, sits in a steel mill recovering an average 75 million Btuh (120 million peak) and all it's doing is preheating boiler plant makeup water. Many a boiler plant can save a fair amount of energy in the winter if normal building exhaust can be trapped and used as combustion air. In those cases the building and its occupants preheat the air.

BURNERS

Most boilers get their heat for the hot water or steam from the combustion of fuel which requires a burner. There are some devices for combustion that aren't called burners, including stokers but all of them combine the fuel with air to form a combustible mixture so the air and fuel react to produce combustion products and heat. The purpose of the burner is to control the mixing of fuel and air so the combustion occurs smoothly and uniformly within the furnace of the boiler. There are several components of a burner and variations in construction that are designed to assist in this function and I'll try to explain most of them. First I want to explain some of the important aspects of combustion that a burner design has to address.

The burner has to control the mixing of the fuel and air in a manner that ensures complete and stable combustion. Stability of combustion requires the burner produce a fuel rich mixture right at the upper explosive limit where the burning begins and that mixing point has to be stable as described in the chapter on combustion. If the burner fails to produce a stable ignition point the flame front will shift around in the furnace producing pulsations that disturb the process and make it worse.

The quality of the burner is indicated by noise, as the quality of mixing gets worse the noise gets louder and some burners are so bad that flame spurts out any open inspection port. Resolving that mixing problem is not a simple matter, it's a combination of engineering and art with many solutions achieved solely by trial and error. It's not uncommon for a service technician to try several combinations of burner tips, diffusers, and burner adjustments to resolve an unstable ignition problem, sometimes taking days or weeks.

During startups many owners and operators get frustrated with a new boiler because the startup takes so much time to resolve an unstable combustion problem and, despite the fact that the problem is solved, will never trust the boiler as much as they would if the problem never occurred. It's not uncommon and it's not something that's predictable so if it happens don't blame the manufacturer and take a position that the boiler will always be a lemon. Unless the owner accepts something less than reliable operation out of a new boiler it will always be more reliable than an older one.

It's the nature of burners that a deviation in any one part can produce several conditions inconsistent with good combustion all of which can be due to several things. Many times an operator unwittingly does something that alters a burner's performance without being aware of it and the owner pays the price in higher fuel costs for long periods before the deviation is discovered and corrected. Understanding what the many adjustments on a burner do is one way of preventing such things happening.

Air Supply and Distribution

The burner is normally fitted with some means of controlling the amount of air supplied to the fire. The means can vary from a simple single bladed damper to variable inlet vanes on the fan inlet and can include a VSD (variable speed drive) on the fan motor. To provide stable combustion the dampers or VSD have to control the air flow without sticking or flopping around which produces variations in air flow.

The dampers also have to control the flow without producing distortions in the flow of air to the burner. If the dampers tend to shift air to one side of the burner inlet (or the fan inlet) it can shift the point of ignition to one side of the burner and that can produce instability. Sometimes obstructions around fan inlets can produce unusual swirls that are carried through the burner.

Installation of boilers that position building columns, pipes, racks of conduit and similar obstructions within seven diameters of the fan inlet should be avoided but sometimes you're stuck with one. There are partial to total solutions to air distribution problems caused by such things when it's impossible to move the boiler. Of course setting portable equipment, storage, and other things in front of a fan inlet can also cause problems with burner operation; so don't do it.

The devices controlling the flow of air must present it at the burner throat in a manner that ensures mixing of the air and fuel to produce a mix in the flammable range where the heat of the furnace will ignite it. To establish that ignition point where it's desired in the burner there's always a primary air adjustment. It can be sectional dampers in a stoker, position of multiple burner registers, adjustment of cylindrical tubes in the burner that vary air flow and, the most common, positioning of a diffuser.

A diffuser (Figure 10-43) contains slots or vanes that restrict air flow. Since the flow through the diffuser

Figure 10-43. Burner diffuser

is restricted the fuel-air mix there will be richer in fuel than the mixture passing around the diffuser so the ignition point is usually aligned with the diffuser and it can be altered by changing the position of the diffuser. On a typical gas or oil fired burner the diffuser normally has two positions, one for firing gas and one for firing oil. The reason for the different positions has nothing to do with the diffuser itself and everything to do with where the fuel enters the burner. In the typical burner oil is admitted in the center at an oil nozzle and gas is admitted through a gas ring or spuds on the outside of the burner. The diffuser positions must be switched to control the primary air ratio for each fuel. When that's the case, a semi-permanent marking should be applied to the adjustment for each fuel position so an operator knows the diffuser is properly located. Paint a ring around the diffuser guide pipe with arrows pointing to the point where it enters the burner and label them for each fuel.

An inexperienced operator positioned a diffuser improperly on a boiler in south Baltimore in 1999. The pipe wasn't marked but he knew it was pushed in for firing oil so he pushed the diffuser guide pipe all the way in, as far as it would go. The burner failed to light several times until enough oil had accumulated in the furnace to feed the explosion!

An induced draft oil or gas fired boiler doesn't have a forced draft fan and doesn't need any provisions to supply the air to the burner but will still need means of controlling the distribution of air. Single burner boilers are typically fitted with a screen or perforated plate to provide uniform flow of air to the burner. Burners on multiple burner units are typically fitted with a register, a set of bent damper blades that form a circle around the burner inlet (Figure 10-44). Some are independently set with a locking bolt or screw on each blade while others are fitted with linkage attached so the blades turn uniformly and the flow of air to each burner can be adjusted while maintaining an even flow of air around the burner.

Burner registers will not only serve as a means to throttle air flow they also deflect the air to create a swirl in the air. That produces additional turbulence for better mixing. Sometimes two registers are employed, one to supply air around the outside of the fire and one for air supplied to the center, around the diffuser. When they are used, dual registers typically produce swirls in the opposite direction for better mixing. Another function of the burner registers and diffuser is flame shaping. Modern package boilers have very small furnaces and older sterling boilers have short furnaces. To prevent flame impingement on the furnace walls the burner register and diffuser position combinations help shape

Figure 10-44. Burner register

the flame. On some boilers the registers are modulated along with the air and fuel controls to alter the shape of the flame for different loads.

You probably won't see a burner register throttled down for better mixing today because we've learned that rapid mixing makes for quick burning, hotter fires and more NO_x production. The register burner is being replaced by the axial flow burner which is designed to minimize turbulence but ensure even distribution of air to the flame front. The original concept of the axial flow burner was developed in England in concert with the Royal Air Force to improve performance of multiple burner boilers at the English Air Force Bases and included creation of a venturi throat for each burner that not only improved air flow distribution but also provided a means of measuring the air flow at each burner to allow final tuning of air distribution through them (Figure 10-45).

One advantage of the venturi is it creates a large static pressure to velocity pressure conversion at the burner inlet, most of which is recovered in the diverging section. The velocity conversion tends to balance the air flow through each burner to improve air distribution on multiple burner boilers. Control of air flow, including shutting off burners on axial flow units is achieved by a damper that forms a sliding sleeve at the inlet of the venturi. Most low NO_x burners applied to single burner boilers can't benefit from the venturi design so other means are used to improve air distribution.

Large single burner and multiple burner boilers

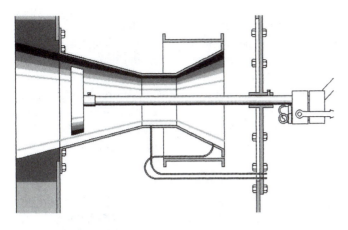

Figure 10-45. Venturi burner with flow sensing ports

normally have one air supply with the air flowing to the burner distributed within a windbox. The windbox receives the air from the forced draft fan and provides sufficient space around the burners for the air to be distributed evenly. At least that was supposed to be the idea. Several installations I've seen in past years didn't really provide adequate air distribution in the windbox, especially between burners, so the fires were not truly uniform.

A windbox has to be large enough to distribute the air and that's always larger than big enough to fit the burner. A burner manufacturer's dimensional tolerances for a burner are based solely on construction clearances so the minimum distance from the center of a burner to the inside of a windbox as listed by the manufacturer is just enough to prevent the register blades hitting the windbox. Let's face it, if the blades are just clearing the inside of the windbox there's no room for the air to get between them. If you're stuck with one of those poorly designed windboxes you'll know it because you can't get stack gas oxygen content down without generating a lot of carbon monoxide.

If you have air distribution problems you can try adding shrouds. Shrouds were developed to resolve the problem with limited room for registers within a burner windbox that would fit on the front of small package boilers. They consist of a cylinder of perforated plate, around 50% open area, larger than the open register (of course) but weighted for each application to achieve the most uniform distribution. The shroud is usually a couple of inches wider than the burner register.

I found shrouds beneficial in knocking down the concentrated discharge of windbox mounted fans. They proved to be more reliable than the methods I originally used, structural angles across the windbox and turned so the heel pointed at the fan discharge. In most applications I started with several sizes of angle cut to length

and set them in the windbox temporarily until I got good air flow distribution.

In many single burner systems we found proper placement of one or more 4 by 4 angles mounted near the windbox air inlet created sufficient turbulence and deflected the high velocity air from the fan enough to achieve good air distribution. Cut long enough to be a press fit they will hold position while testing for air velocity at the burner and can be moved to find the optimum position. One centered on the burner when the duct entering is centered, or at the point where the entering air velocity is highest, usually does most of the work because it prevents the direct blast of the air striking the shroud or register. It has to be far enough from the register so some air can eddy behind it or you'll lose a lot of air immediately behind it. Once you've got the best distribution you can manage, be certain to weld them in because they'll fall out when the windbox and boiler heats up in normal operation and they make a lot of noise when they blow into the register.

To check for uniform air distribution through a single burner (or each burner) measure it. First do all the lock-out, tag-out required for access into the boiler but provide a means for operating the forced draft fan. Leave all the normal burner components in place except for a center-fired oil gun. Hang a manometer against the tubes in the furnace and connect flexible tubing at one end to a copper tube about three feet long that you can point at the burner. Take some paper on a clipboard and pen into the furnace with you to record your measurements.

Be certain to wear good safety glasses because the breeze can do all sorts of strange things including blow your own hair around so roughly that it jabs you in the eye. With the fan in operation point the tube directly at the burner while holding it so the end is about flush with the face of the furnace wall and the tube is horizontal to get each reading. Begin with air flow consistent with low fire and record the total pressure read on the manometer at each point on the burner. I like to use clock positions (1 through 12) as a basis because everyone understands where the measurement was taken and the twelve readings provide reasonable resolution of the velocities around the burner.

Since the point flush with the furnace wall and the open tube on the manometer are both exposed to the furnace the static pressure is the same at both points and you're reading velocity pressure. Take readings at increasing air flow rates in steps of about twenty percent until you get to the top end or the velocities get so high that you can't stand up to hold the tube up to the burner or, in the case of a fire tube boiler, you're blown out of the furnace tube.

The actual velocity is reasonably estimated by multiplying the square root of the differential reading by 4005. That's done on any calculator by typing in the value of the differential (example, 0.08 for that many inches of water column) pressing the square root key (√) to get the square root then × (for multiply by), 4005 and the equals key to get the velocity in feet per minute (1132 in the example). There may be some argument about how much variation is permitted in the air flow around a burner but I would try to do something to cure any deviation that exceeded ten percent. I take the sum of the readings (add them up) then divide by twelve to get the average then multiply that result by 0.9 and 1.1 to get upper and lower limits. If any of the other readings are outside those limits I try ways to improve the air distribution in the burner including baffles, like the angles already mentioned, then proceeding to shrouds. Usually corrections made at low fire do not alter air flow at higher firing rates so correct the low fire variances first and repeat tests to determine their effect at higher flow rates.

That's a lot of work and is all after the fact but it doesn't cost as much as what they do for large utility boilers. Determining the best design of air distribution is such an art that utility boiler manufacturers will make models of the system and test them for proper air flow. They'll repeat that process to get it right before they build the boiler. It's much easier for them to spend all the time on a model than to try to solve distribution problems on twenty four or more burners in the field.

A large number of burners were built for staged combustion in the last half of the 20th century. Some of those burners incorporated secondary air ports (openings in the refractory front wall around the circumference of the burner) with adjustment of the air flow to them consisting of a piece of angle or other steel form surrounding the burner. I've noted that most of those provisions for adjusting that air flow are so flexible that they don't provide uniform air flow around the burner. Some are so limber they actually flop around in the air flow. If I had to set one of those burners up today I would wait until proximate requirements are established then measure the flows at the ports, adjust the flexible steel to equalize the flow through them then tack weld the adjustment in position at each port.

The mixture of fuel and air has to be heated to ignition temperature before it will start burning so the burner has to provide means to heat the incoming air and fuel. The normal and best means of heating the mix is application of a refractory throat. The radiant heat of the fire is reflected back into the entering fuel and air to heat them

to the ignition temperature before they reach the proper mixture so we have stable combustion. The throat is also part of the insulating portion of the burner that protects the boiler front and burner housing from the heat of the furnace. There is a considerable variation in temperatures across that refractory during operation. Any large cracks, spalling, or shifting of pieces of throat tile can distort air flow at the burner to produce unstable combustion.

Gas Burners

A gas burner can be premix or post-mix. While most boiler burners are post-mix, where the gas and oil mix after they enter the furnace, premix burners are available. Many operators think of a premix burner as hazardous, after all we make a combustible mixture outside the furnace! Many operators that moved from firing process equipment to the boiler plant are comfortable with premix burners because they have a lot of experience with them. As long as the mixture isn't heated above the ignition temperature it can't burn and premixing permits a low cost arrangement of multiple burners which is frequently necessary for good heat distribution in processing equipment. There aren't many boiler applications with premix burners so I won't spend any more time on them than this. Your understanding of combustion and the instruction manual should be all you need to operate a premix burner.

Of the post-mix gas burners there are two choices which are normally identified as atmospheric burners or power burners. Atmospheric burners do not normally have fans or blowers to deliver the combustion air to the burner and seldom have induced draft fans. Lacking the power of the fan to introduce and help mix the fuel and air, atmospheric burners use some of the gas pressure for that process.

The typical atmospheric burner has a "jet" which is a nozzle the gas flows through on its way into the burner and that jet acts like an inducer to draw primary air in with it. The gas and primary air mixture is then distributed through the burner head (Figure 10-46) or flame runners (Figure 10-47) into the furnace. Secondary air is delivered by natural draft and mixes with the primary air—gas mixture as it burns. The several forms of flame runners shown in Figure 10-47 all seem to work well with no discernable difference in performance. Cracks between the holes and holes in the bottom, usually caused by rust, can produce distorted, inefficient, and dangerous fires.

Some gas furnaces can be subjected to very corrosive conditions between heating seasons so it's always a good idea to check an atmospheric burner right before the heating season and clean it if necessary. I've seen them with large pieces of scale from the heat exchanger laying on top of the runners and, on one occasion, removed the runners and held them up to drain about a cup of rust from the inside of each tube!

If your home has one of those gas furnaces you also want to check the furnace sections for cracks and open seams. If there's a way for the products of combustion to get into the heated air side of the furnace it will most likely contain considerable quantities of poisonous carbon monoxide. The price of a furnace isn't worth the risk of dying so you should replace any rusty, mis-shaped or cracked furnace.

On atmospheric burners primary air adjustment is accomplished by moving a sleeve (Figure 10-48) or rotating a shutter (Figure 10-49) thereby changing the opening for primary air. The gas nozzle (D in Figure 10-48) converts much of the static pressure in the gas to velocity pressure. The high velocity gas shoots into the distribution header (B) drawing primary air along with it through the opening that's adjusted by the sleeve (E). The mixture then flows into the flame runners (A) and out the ports where heat from a spark or adjacent fire provides ignition energy to start the combustion. Always remember that additional air, secondary air, is required to complete the combustion and enters through openings like the one at (F).

The primary air—gas mixture is adjusted to produce a stable flame over the head or flame runners by adjusting the sleeve or shutter. Either one has a locking screw to ensure the piece stays where it was adjusted. The flame should burn clean and stable just above the distribution ports. Lighting these burners can be interesting at times, especially during initial startup, because the pilot only lights one to four ports on the burner head or runner and the rest of the burner is ignited by flame at the adjacent port.

Some atmospheric burners provide a degree of modulation and turndown by cutting out some of the jets or controlling groups of jets with individual shutoff valves and may be augmented by matching combustion

Figure 10-46. Gas burner head

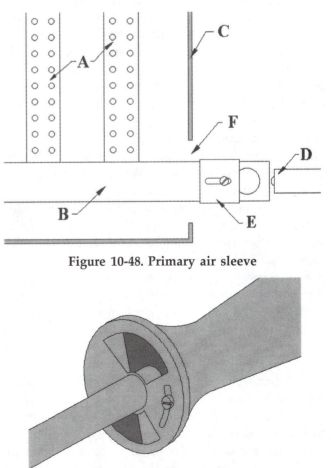

Figure 10-48. Primary air sleeve

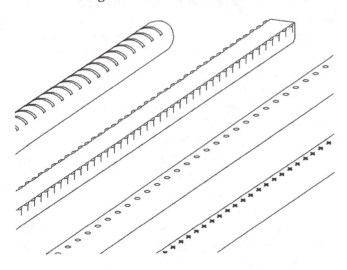

Figure 10-47. Flame runners

Figure 10-49. Primary air shutter

air blowers. I don't like them because it's very difficult to balance the fuel distribution to get them to burn cleanly and efficiently. The few I've encountered can't seem to fire without a considerable amount of CO.

Atmospheric burners are only used on small boilers, hot air furnaces, and service water heaters for the most part because they are normally fixed fired and have very little control of overall excess air. I've seen large boilers, as big as 150 horsepower, with atmospheric burners and have shuddered at the thought of what it costs to operate them. If they serve a constant load for which they're well matched then there's some sense in their application but in any service with a varying load the off cycle losses are very large.

Controlling those losses with dampers that shut off air flow through the boiler when it's not firing can provide significant reductions in those losses. The dampers have to be proved open before the boiler fires. Modern high efficiency heating equipment with pulse combustion or power burners should replace most of that equipment in the next few years once gas prices rise again. I've been able to show a boiler with a power burner can pay for itself over an atmospheric fired unit in less than a year. Any medium to large boiler should have a power burner and unless one isn't available, it should be modulating.

Fuel gas can be introduced into a power burner via a gas ring, spuds or a gas gun. A gas ring is a piece of pipe, fabricated steel or a casting surrounding the burner right at the boiler front plate with holes drilled in it to distribute the gas evenly around the outside diameter of the throat. Some gas rings are fitted with spuds while other burners have spuds at the end of pipes which deliver the gas from the front of the burner or a gas ring located outside the front of the burner.

Spuds are high temperature metal nozzles drilled with holes to admit the gas into the passing air stream. A gas gun consists of a pipe central to the burner with a closed end drilled to admit the gas very similar to an oil burner. Some gas guns consist of two concentric pipes that permit insertion of an oil gun down the center of them. The arrangement, distribution and mix of holes drilled in gas rings, spuds and guns varies with manufacturer and in many instances is customized during startup to achieve smooth stable combustion.

Retaining data on the drilling of your gas burner is essential because your information may be the only accurate copy around; it's not unusual for a manufacturer to fail to update the records for changes made by the service technician. One element of your annual boiler inspection should be checking the diameter of the holes in the gas ring, nozzle, spuds or combination thereof with matching drill bits. That's very important to do before closing a burner when refractory work is done because there's a tendency of masons to leave smears of refractory on and in the openings of gas rings.

The gas ring is usually bolted to the boiler front plate, that thick piece of steel that seals the front of the boiler, provides support or is integrated with support of the refractory front wall. The front plate supports the burner throat to keep it centered, and provides means of attaching the burner or windbox. If the gas ring or the boiler front plate is distorted then air leakage around the gas ring at different points can produce very unstable firing conditions. The condition of the gas ring and clearances (if any) between the gas ring and boiler front plate should also be checked annually.

If you find a warped front plate or other problems with irregular air spaces around a gas ring you can plug them with ceramic fiber rope. Always put the rope on the windbox side and be certain it's large enough it won't blow through. It's very embarrassing to have someone ask you what that thing is fluttering in the fire.

Gas rings can fail. Failure of adjustments of firing rate, like linkage slipping, and other contributions can produce situations where the heat of the fire is shifted into the throat where it can overheat the gas ring to create cracks in it. Any cracked gas ring should be replaced before the boiler fires gas.

There was a time when we attempted to deliver the fuel gas into the flame in the furnace as uniformly as possible to ensure complete mixing and permit low excess air firing. The discovery of NO_x as a problem has resulted in irregular gas delivery schemes, usually using spuds (Figure 10-50) installed with pairs facing each other to produce alternating fuel rich and lean zones in the burner. See the section on emissions for an explanation of why.

The fuel gas piping has to penetrate the windbox or burner to deliver the gas to the gas ring. There was a time when we used a flanged connection on the gas ring to permit disassembly but it also placed a potential point for leakage of gas inside the burner windbox. There it could light off producing heat in a windbox that wasn't designed to absorb that heat. There are also many variations in design of packing glands and other means of sealing the gas piping where it penetrates the windbox. If you ever have problems with air leakage at the gas line entrance the best solution is welding it to the windbox. Normally the windbox is flat and flexible enough at the gas line entrance that thermal expansion is accommodated by the windbox flexing. If you have problems with leaking piping joints inside the windbox

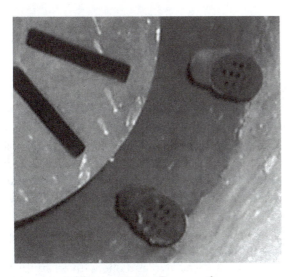

Figure 10-50. Gas spuds

and the gas ring isn't cast iron I would cut the flanged joint out and weld the piping. Gas free it before welding!

There are few options for the operator when it comes to gas rings, there's nothing to adjust. All the adjustments for fuel-air mixing have to be made by altering the combustion air flow. There are problems you have to watch out for. Gas rings can crack due to thermal shock, warping of the front plate, and improper repairs. The drilled openings for the gas can be blocked by dirt accumulation, careless application of refractory materials (a common one), and dirt when the burner port is used for furnace access. The ring can come loose from the boiler front plate because the mounting bolts vibrate loose. Any change in the appearance of a gas fire should be followed on the next shutdown by a careful examination of the gas ring.

There are a considerable number of gases that are fired in boilers in addition to natural gas (see FUEL GASES in Chapter 7). In many cases they are burned only because the alternative is to waste them to atmosphere and several are considered a source of pollution. The petroleum gases can, for the most part, be burned in equipment identical to natural gas burners with adjustments in nozzle size or fuel supply pressure to compensate for the difference in the heating value and air to fuel ratio of the gases. Others, such as digester gas, can contain a large percentage of non-flammable gases and require special burners that can accommodate the larger volumes of fuel gas required to satisfy the heat input requirements. Digester gas (from sewage treatment plants) and landfill gas (tapped off a landfill) are noted for containing hydrochloric acid so the piping and burner has to be capable of handling the corrosive material. As with everything else I've said before, take the

time to seriously review the instruction manual so you understand how those burners are to be operated.

Oil Burners

Fuel oil is introduced into a burner using a burner tip which is normally mounted on the end of what we call an oil gun (Figure 10-51). The design and arrangement of the tip and gun is dependent on the type of atomizing system. Pressure atomizing burners have one or more tips on the end of a pipe positioned in the burner at the point where the oil has to be injected to develop the air/fuel mix. Pressure differential, air atomizing and steam atomizing burners need two pipes, one to convey the oil to the tip and another to supply the air or steam or return the oil from the tip. Traditionally the two pipes are concentric with the oil supply down the center pipe and the annular space between the two providing the passage for air, steam or return oil but (like the one in the figure) some manufacturers provide two separate pipes running side by side.

The tip introduces the oil into the furnace in a way that makes it possible for the oil and air to mix and burn. As I sit here writing this the news on television is showing where the Iraqis have created large pits of oil and set them afire. The smoke released from those pits is clear evidence that you have to do more to produce a clean fire. To ensure the oil and air mix and burn completely a fuel oil burner tip provides a means for "atomizing" the oil. Atomization is breaking the oil up into tiny droplets (not as small as an atom but small enough) so the air can mix in between all the droplets for complete burning. If the oil isn't atomized it will not burn well. In some cases it won't burn at all.

Don't accidentally leave the tip off an oil burner and try to start it that way. I know one apartment house boiler operator that did that; the burner didn't light the first few times he tried it. After several tries he had

Figure 10-51. Oil gun

dumped enough oil in the furnace that the lighter portions, which evaporated, produced a flammable mixture that the ignitor managed to light! The resulting explosion didn't destroy the apartment building but it did manage to destroy the boiler.

Atomization is accomplished in different ways; all of them work. The principle difference between them is the degree of turndown they can accomplish. Pressure atomizing burners produce a fine spray pattern of oil just like you do when you use a water hose to wash your car and pull the trigger on the sprayer just enough to produce that fine conical mist of water. The quality of the atomization varies with the pressure drop across the burner tip. Many burner tips will have internal channels that divert the flow of oil (Figure 10-52) so the oil accelerates as it approaches the central chamber and produces a whirling motion in the oil. As the oil flows out the tip that spinning motion forces the oil to swirl out by centrifugal force and that causes the oil to tear apart into tiny droplets.

Figure 10-52. Burner oil tip showing swirler pattern

A similar principal was applied to a burner that isn't legal to use anymore. Rotary cup burners used a brass cup mounted on the end of a pipe. The pipe and its cup are mounted on the shaft of the blower of the burner and centered in the burner throat. The oil enters the rotating pipe through a flexible connection and literally drizzles into the cup. The cup's rotary motion whirls the oil around the inside of the cup until it reaches the top where it shears off into the combustion air stream. You can simulate the operation by swirling water in a cup with sloped sides. You'll notice the water doesn't leave the cup in a fine spray because you'll get pretty wet. The poor atomization of the water demonstrates the reason the rotary cup burner is no longer legal.

Steam and air atomizing burners use one of two methods to atomize the fuel. Some of the burners introduce the oil into a jet of steam or air that cuts into and breaks the stream of oil up into tiny droplets then transports them into the furnace. Most, however, simply mix the oil and steam or air in a chamber in the tip. When that mixture leaves the burner tip holes the gas (steam or air) expands rapidly breaking the oil up into tiny droplets. Since the energy for atomization is provided by the air or steam turndown of these burners is typically

about 5 to 1, some as high as 10 to 1.

The typical pressure atomizing burner is limited in turndown capability to about 2 to 1. Once the flow is reduced to half (the pressure drop through the tip is one fourth) the velocities are so low that the oil doesn't break up well. Oil return atomizers were produced as a solution to that problem. The full load flow of oil is delivered to the burner tip and flows through those slots to produce the spinning that breaks up the oil. To reduce the firing rate some of the oil is returned from the tip to the suction of the fuel oil pumps. The turndown is generally a function of the differential pressure where the turndown is equal to delivered oil pressure divided by one hundred. A system firing oil supplied at 300 psig will have a 3 to 1 turndown and one with oil supplied at 800 psig will have an 8 to 1 turndown. The practical limit for those burners seems to be 1200 psig because the price of pumps, pumping, maintenance, and all the pressure containing components of the oil system get very high.

With large boilers additional turndown is accomplished by using multiple burners. Half the burners operating at maximum oil supply pressure will produce half the boiler load. One fourth will produce one fourth, etc.

If the boiler is limited to one to four burners then other means of achieving additional turndown may be required. The typical solution is different sizes of burner tips. A smaller tip will pass a reduced amount of oil at the same pressures. The important thing to remember is all the burners in a boiler should have the same size tips installed so fuel delivery is uniform.

I skipped by an obvious question. What does a boiler plant with steam atomizing burners do to get started? There are two solutions, one is to use a small pressure atomizing tip to produce enough steam to get going. The other is to use compressed air temporarily. Temporary use means exactly that, a burner designed for steam atomizing will consume considerable amounts of compressed air at a high cost in electric power to generate it. A small control air compressor could be overloaded and damaged attempting to maintain fuel atomization.

As with the gas burners oil guns have seen modifications in recent years to produce alternating zones of fuel rich areas in the flame for NOx reduction so irregular drilling of a burner tip (Figure 10-53) is now common.

The gun in many cases is nothing more than a piece of pipe connecting the fuel delivered to the burner on out to the tip. Many burner guns can be removed by breaking the connections outside the burner and pulling it out for cleaning and maintenance. There are guns which disconnect at a union (Figure 10-54) or simply break at tubing fittings and guns with elaborate quick-connect capabilities, with many variations in between. Most arrangements are specific to a particular manufacturer but the common arrangement is a yoke coupling (Figure 10-55) which is used by many manufacturers. A yoke with a set screw (Figure 10-56) clamps the two together.

With a yoke coupling the gun has openings for oil and any atomizing medium that match with holes in the yoke coupling. To ensure alignment of the openings in the gun and yoke there are usually ferrules (short smaller pieces of pipe or tubing) set in the yoke holes (Figure 10-56). The gun holes pass over them. The ferrules are removable because they can be damaged, by pressing or throwing the gun against them, so the holes in the gun won't fit over them. The installation also includes some provision for sealing the joint of gun and yoke.

Sometimes it's using a softer material like brass, normally on the gun, that will deform under the pressure of the set screw to seal the joint. Frequently it's a gasket; most commonly a thin layer of copper surrounding a fiber material that will conform to variations in the two surfaces to seal the joint.

Being careful when inserting an oil gun prevents damage to the ferrules which can prevent proper fit-up of the yoke and gun. I know it's difficult when you're swinging around a five foot oil gun to insert it gently that last half inch but it's a skill all operators of oil fired boilers have to develop. If you do get the urge to slam the gun into the guide pipe then twist the gun so the joints don't match up and you don't bang up the ferrules and gaskets.

Figure 10-53. Irregular drilling in oil tip

Figure 10-55. Oil gun yoke coupling

Figure 10-54. Oil gun quick assembly union connected

Figure 10-56. Yoke coupling clamp and set screw

Most manufacturers' instructions state that you should replace the gaskets every time you change the burner. If you saw what they charged for those little gaskets you would get the same impression that I have, there's more than preventing leaks on their mind! You should keep a set of gaskets handy to replace them when what you're using fails or you can tell they're damaged but there's no reason to replace them every time you change out an oil gun. I've fired boilers with brass grips on the oil guns that mated up with a steel yoke where there's no gasket and they don't leak unless you get some dirt or grit in the joint. If they can last several hundred gun changes the copper wrapped gaskets should too. I can recall changing guns every shift and seldom changing gaskets.

A skilled operator can remove one oil burner and install a fresh one in a matter of a few seconds; however, if the boiler has a single burner the speed of the operator is not much of a consideration because the burner has to be shut down to remove the oil gun. To avoid the shutdown of the boiler along with the processes of purging, low fire positioning, and trials for ignition some burner manufacturers will provide single burners with the ability to accept two oil guns while others provide as many as four.

The typical two gun arrangement is designed to insert a temporary oil gun, transfer the fire to that gun then transfer back to the main oil gun. The fire may be lopsided or have voids when firing with the temporary gun. Other arrangements use guns with special tips that produce a uniform flame pattern when all the oil guns are in position and operating. Changing out the oil burner so it can be cleaned is accomplished by switching guns one at a time. During the period that one of those oil guns is removed there is a definite gap in the flame pattern. While changing guns the operator should increase the air to fuel ratio so the variations in fuel delivery do not produce fuel rich conditions in some portion of the furnace.

Of course the reason we have oil guns is the tip gets dirty so they have to be removed for cleaning. Spare oil guns and tips are provided so you have a clean one ready to put back in the burner to permit continued firing. Frequently I was asked "How often do we have to change out the oil guns for cleaning?" The answer is always "as soon as they get dirty." I know that's a flip response but there are no hard and fast rules for cleaning burners, it depends on the oil, contaminants in the oil, the firing rates, and the condition of the burner itself. We have to change guns and clean the tip of carbon before it builds up enough to start hampering atomization. You'll

be able to determine how long that is for your particular boiler, burner and load combination.

I was in one plant that claimed they only changed their heavy oil burners once a month. One look into the furnace explained that. They had the atomizing steam running so high that the flame didn't start until it was about eight inches from the tip. The fire was just barely stable. I didn't analyze the situation to see how unstable the fire could get on load changes nor how much it cost for all that extra atomizing steam.

I should explain that it really isn't all carbon, the accumulation of unburned fuel that has been heated to drive off much of the lighter fractions and leave mostly carbon is called "carbon" by boiler operators. Carbon is a common problem when firing oil. It is less of a problem when firing light oils. There are many reasons for carbon buildup on burner tips, burner throats, and the floor and walls of a furnace when firing oil.

The most common reason for carbon buildup is poor atomization. That can be produced by dirty oil that plugs burners or ties up the oil like glue so it won't atomize. Other reasons are using tips too large for the load, worn tips, loose tips and tips and other burner internals assembled improperly. One of our service engineers solved a poor atomization problem on a burner by assembling it improperly. Nobody could get a decent fire out of the burner but he did almost immediately by putting a couple of parts inside the mixer of the steam atomizing burner in the wrong order.

Steam and air atomizing burners can also suffer from condensate in the air or steam, the wrong pressure, and blockage of the atomizing medium piping. A problem we encountered regularly with differential controls was a significant variation in the differential at the burner tip due to pressure drop in the oil or steam piping. Usually the problem involved lighting the burner at low loads where the differential was so high the fuel air mixture was always lean because the atomizing medium broke it up too much. The solution to that problem is adding an orifice nipple (steel bar simulating a piece of pipe with hole drilled through it) which allows adjustment of the differential at low fire to get stable firing. As the load increases the nipple introduces a pressure drop in the oil that increases the differential at the burner tip as load increases.

I do know that many operators create their own problems when it comes to cleaning oil burners; they damage the tip so it gets dirty faster. Every hole drilled in a burner tip comes from the factory fresh and sharp, with a pure 90° angle between the edges of each hole and the face of the tip. That's so there is a sharp separation of the oil stream as it leaves the tip. Operators that get frustrated with the brass tools and wire brushes then resort to steel tools and wire brushes to round off those sharp edges so the oil stream doesn't make a sharp break with the tip; some of the oil tends to follow the curved edges created by abrading the tip with steel tools and that oil forms carbon very quickly. Save yourself a lot of trouble and stick with the brass tools.

Coal Stokers

Don't skip this part too quickly. We have a very limited supply of natural gas and oil in the world but, at our present rate of consumption, over one thousand years of coal. Despite the many undesirable features of coal firing it's the one fuel that will always be available in the future.

There are many options for introducing coal into a furnace and "coal burner" and "stoker" provide some differentiation. Stokers handle coal as a solid material. Coal firing can be as simple as a grate in the bottom of a furnace with openings for the air and an individual opening a door in the side of the boiler to introduce the coal with a shovel. It can be as complex as a multi-tiered tangentially fired furnace with over-fire air ports and re-burners.

I've seen a few of the first type in small plants throughout the country and only photographs and drawings of the latter. I don't expect many of the operators reading this book to be working in an electrical utility plant which is about the only place you will find the latter. Since utility plants normally have good training of their operators on those large and complex boilers I'll leave that to them.

Stokers come in a variety of forms and have basically been reduced to under-feed, traveling grate, and over-feed types. The difference in these is how the coal is introduced to the fire. An under-feed stoker pushes the coal up into the furnace from below the grate. The coal is removed from storage or a hopper by a screw conveyor (Figure 10-57) or ram (Figure 10-58) which pushes the coal along through the "retort" and against the pile in the bottom of the furnace.

As the coal is pushed up it is mixed with air entering via the tuyeres (C in Figure 10-59), pipes, tubes or slots in the grate that admit the air into the furnace. The mixture is ignited by coal already burning above the grate. The coal air mixture partially burns on the grate and completes burning of hydrocarbons vaporized by the heat of the furnace in the space immediately above the grate. Air at the tuyeres and most active portion of the grate is considered primary air and is controlled by dampers supplying the air to the primary air zone (B).

As the hydrocarbons and sulfur in the coal are consumed the remaining ash is pushed to the edge or sides of the grate where it can be removed by hand or dumped (D) for removal by hand or screw conveyor. For final burnout and handling high loads temporarily a controlled flow of air is supplied to the dump grate zone at (E) which must be reduced dramatically to permit

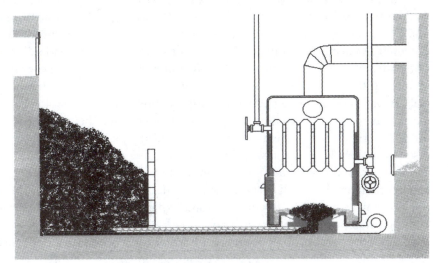

Figure 10-57. Coal screw conveyor

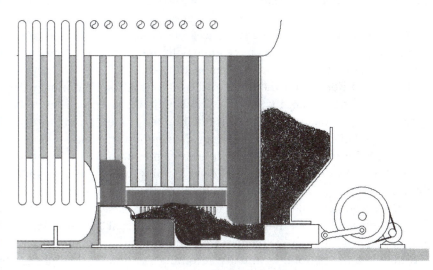

Figure 10-58. Underfeed stoker ram

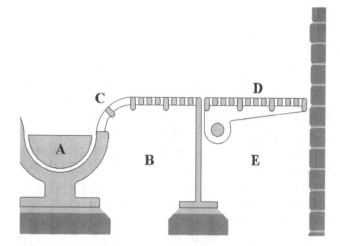

Figure 10-59. Dump grate

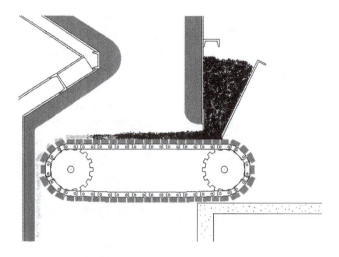

Figure 10-60. Traveling grate

removing ashes from that chamber manually.

Under-feed boilers with screw feeders like the one in Figure 10-57 are still found in homes in Pennsylvania, Ohio, and other coal states. Ram fed boilers can be powered by steam to eliminate the need for electricity. They are also available in sizes up to 100 million Btu by increasing the number of coal feed locations in a "multiple-retort" stoker. Some people might be surprised to learn that most of our nation's capitol was heated by those boilers up until the early 1990's. Under-feed stokers are capable of burning a wide range of coals and sizes. The common specification limits fines and particles smaller than one half inch because the fines sift through the equipment and tend to compress and expand preventing proper operation of the feeder.

Traveling grate stokers burn coal particles in the range of one eighth to three quarters of an inch in size. The grate (Figure 10-60) is a continuous belt of steel chain mounted between shafts spaced ten to sixteen feet with lengths up to twenty feet. The steel is protected from the heat of the furnace by pieces of refractory which form an external layer on the grate with openings around each piece to admit the combustion air. Coal is stored in a hopper on the front of the boiler and is dragged into the furnace by the grate. The depth of coal over the bed is adjusted by a plate in the hopper at the front of the boiler. Proper control of air distribution in the zones below the grate and the ignition arch maintain combustion. As the coal burns down the flaming particles under the ignition arch are blown up by the flow of combustion air and follows the flow of air and gas diverted by the arch so they land on the entering coal to ignite it. That way the coal burns from the top of the bed down to the bottom, eventually becoming ignition particles. Ash left over drops off the end of the grate as it

makes the turn back toward the front of the boiler.

Over-feed stokers have a grate just like the traveling grate stoker. The difference is the way the fuel is introduced. Frequently an over-feed stoker is called a "spreader" stoker because the fuel is, to a degree, spread over the grate. Over-feed stokers are further classified by the height of the feeders above the grate. 'Low set' stokers will have feeders injecting the coal in the neighborhood of three to five feet above the grate while 'High set' stokers can be as much as eighty feet above the grate. The grate on over-feed stokers typically runs in the opposite direction of spreader stokers, delivering the ash to the front end of the boiler. The coal feeders come in a variety of forms, from plates connected to eccentrics on a shaft that toss the coal dropped on them into the furnace to rotating blades and rotary feeders with air blown into the feeder to transfer the fuel into the furnace. Over-feed stokers are designed to fire fine coal, from dust size particles to pieces under one quarter inch. The fines are burned in suspension over the grate and the heavy particles drop to the grate to complete burning.

Operation of stoker fired boilers normally requires more manpower than oil or gas fired boiler plants. The coal has to be received, moved to storage, and moved from storage to the "bunkers" that supply the coal to the stoker. The considerable amount of ash has to be removed from the boiler, moved to storage and loaded into transports for final disposal. Occasional "dressing" of the fire is required to maintain uniform combustion over the bed of coal and to remove "clinkers" which are accumulations of carbon and ash that harden into solid deposits on the grate. Lighting a stoker fired boiler is accomplished by building a wood fire on the grate then introducing coal to be ignited by the wood. Cleaning the plant of coal dust and equipment that accumulates the fines is an ongoing task. All those activities require more personnel. The lower cost of coal justifies the added cost of personnel to handle it.

Coal can tend to "cake" before entering the stoker. The large pieces of compressed, usually wet coal will not burn completely in the furnace unless it is broken up. Preventing caking is accomplished in the handling and preparation of the coal. Keeping the coal dry by unloading cars or trucks before it rains or snows and limiting exposure of the fuel to water will reduce caking.

Clinkers is the name we give to chunks of unburned coal and ash that form in the furnace. Those large particles can jam stokers and ash handling equipment. They're usually formed when you get low ash fusion temperature coal or coal with a lot of dirt and other materials in it that melt at the normal furnace temperatures. They can also form when you get a hot spot in the furnace that is higher than the ash fusion temperature (see fuels). When they form, clinkers have to be broken up to prevent them forming a blockage in the fire that reduces output and increases temperatures in other areas of the grate. The operator has to watch the coal bed and use special tools with one end inserted into the furnace to break up the clinkers.

Another operation that operators perform with coal stokers is "dressing" the fire. Despite all provisions the coal never distributes perfectly evenly over the grate. Dressing the bed (the layer of coal on the grate) is accomplished with tools like those used for clinkers to move the coal around until the bed depth is uniform and burning evenly.

Breaking clinkers and dressing a coal fire are activities that require on the job training and experience to do it well. I'll have to admit I could never do it well but I have observed several operators that, in my opinion, were artists when it came to dressing a fire.

Coal burners

Coal burners are principally designed to burn the fuel in suspension so it has to be pulverized before it's delivered to the burner. The bottom of a furnace fitted with pulverized coal burners will have means to remove the ash that drops out of the fire but much of the ash is transported through the boiler to be removed by dust collectors on the boiler outlet. Pulverizers form an integral part of most coal burners. There are (or were, I'm not sure there are any) plants that burned pulverized coal from storage but most plants have an integral pulverizer that grinds the coal to fine powder and mixes it with primary air to produce a fuel rich stream of air

and coal fed to the burners. The coal cannot be simply ground down. It has to be dried as well because it does contain water and the grind would become muddy without drying it. To dry the coal the pulverizers are supplied with preheated combustion air from an air preheater or, in the case of some small plants, steam heated air.

One type of equipment that pulverizes the coal is a ball mill. It consists of a large drum mounted with its axis on the horizontal and filled with cast iron balls. The trunions (extensions at the center of the heads of the drum which serve as a shaft) are hollow so air and coal can be fed into one end and the pulverized mixture leaves the other. As the drum rotates the balls are lifted and dropped on the coal to crush it. The finely ground particles are carried out with the heated air.

Bowl mills consist of a bowl spinning on a vertical shaft with rollers inside that roll around on the inside of the bowl crushing the coal that's dumped into the bowl. Some use balls instead of rollers. The crushed coal is carried away by heated air directed up around the bowl.

Hammer mills use something comparable to several metal hammers that swing freely on a shaft connection. The metal hammers pound on an accumulation of coal to break it into fines that are carried away by the air.

Attrition mills are something like a combination of fan and grinder with pins on the circumference of the fan wheel that strike the coal particles to crush them. The attrition mills have stricter sizing requirements for feed than the others and mill capabilities vary with construction and manufacturer.

The fans or blowers that transport the coal and air mixture to the burners are called primary air fans or exhausters with the latter term reserved for those that move the coal laden air. Most installations use exhausters to limit potential leakage of powdered coal into the plant. The fuel and air mixture exits the mill into the exhauster inlet which discharges the mix under pressure to the burners. In smaller equipment the pulverizer and exhauster are all in the same housing.

What's probably the most important part of a pulverizer—burner combination is the classifier. It's normally a static device (no moving parts) that separates large particles from the stream of coal dust and air heading to the burners and returns those particles to the mill for further grinding. The normal requirements for pulverized coal leaving a classifier are at least 85% of the coal through a 200 mesh sieve and no more than 5% over a 5 mesh sieve. Finally, the mixture of coal and primary air has to be fuel rich to provide a stable point for ignition of the fuel at the exit of the coal nozzle.

The pulverized coal burner can be as simple as a pipe from the mill or exhauster pointed into the furnace to a cast assembly with orifices, guide vanes, and other features that further mix the fuel and primary air and distribute it into the fire in the furnace. Over time the coal flow can erode some of the more important parts of the burner to destroy baffles, etc., that produce the mix and, more importantly, provide that fuel-rich concentration that's needed to get the fire started and stabilized.

Some utility boilers are equipped with cyclone furnaces which use a pulverized coal with less size restriction than conventional pulverized fuel burners. The cyclone is a water cooled, refractory lined cylinder mounted horizontally at the side of the boiler. The coal and air is fired at very high heat release rates within the cyclone with temperatures so high that all the ash is melted and removed as a liquid. The flue gases exit the cyclone furnace into the boiler furnace at temperatures around 3000 degrees. The primary purpose of the cyclone furnace is reduced size of the boiler.

Modern versions of coal burning boilers are fluidized bed boilers and circulating fluidized bed boilers (CFBs) where the entire furnace or the whole boiler is part of the burner. The coal is introduced as solid particles into a bed that's fluidized by the combustion air and flue gases passing up through it. Fluidizing is accomplished by distributing the air into the bed of coal over a broad area using special nozzles through refractory under the bed. The solid particles seem to boil just like water in a pot as the air flows up between them. In the case of a circulating fluidized bed the smaller particles are carried out of the furnace with the flue gas to be captured and returned after they flow through the boiler.

In addition to the coal the bed is fed finely ground limestone that reacts with and absorbs the sulfur dioxide. The reacted limestone and gas leaves the boiler as part of the ash instead of emissions in the flue gas. Circulating fluidized bed boilers actually allow considerable carryover of the bed into the initial passes of the boiler to prolong contact time of limestone and sulfur dioxide plus increased fuel—air mixing. Cyclone separators act like classifiers to remove the coal and limestone particles from the flue gases and return them to the furnace.

Coal firing requires consideration of the time it takes fuel and air to mix and burn. A stoker fired boiler will hold the coal for several minutes while the heat breaks each particle down, evaporating the lighter fractions of the fuel then converting the carbon. The furnace must be larger to hold the inventory of fuel. The fuel for a coal burner has to be pulverized because the particles have little time to burn in the furnace.

One important element of coal firing is very low air flow purges. Operators used to wide open damper full air flow purges for oil and gas should be aware that you can blow the boiler up if you do that with coal. There can be accumulations of coal or coal and ash in the boiler which a full air flow purge would lift and stir to form a combustible (make that explosive) mixture. A purge should be conducted at low air flows to prevent that happening. A high flow of products of combustion can stir that stuff up and move it without hazard because the flue gases are inert, they don't contain any air to mix with the fuel.

While I've had time to visit a few fluidized bed boiler plants and review descriptions of CFBs I haven't had an opportunity to spend enough time with them to identify any tricks the operator should know about them. Once again your best guidance is the instruction manual.

Wood Burners

Wood burners vary from a campfire to burners firing sander dust. On the one extreme we have large pieces of wood which require long retention times in the furnace and on the other we have wood so finely ground that it burns faster than fuel oil. There are a considerable number of different boiler, burner, and grate designs for burning wood, wood waste, and similar fuels.

Wood requires some special consideration if it's 'green' or 'wet' because the moisture absorbs a considerable amount of heat and is capable of quenching the fire to the point it goes out. Dry wood from lumber operations (kiln dried) planing, sawing (of dried wood) trimming, and sanding burns readily and must be handled with care because it can easily produce an explosive atmosphere during air conveying or handling operations that mix the fuel with air. Most wood burning boilers serve industries that process that wood for such things as pine chemicals and furniture.

The fine materials, fine sawdust (some sawdust can be chips as large as one half inch square) and sander dust are typically fed to a burner similar to a pulverized coal burner where the material is burned in suspension like fuel oil. The furnace is usually also fitted with a grate, normally water cooled because there is no layer of fuel to protect the grate from the heat of the furnace.

Larger materials are usually burned in a high set spreader stoker which allows for burning of the fine particles in suspension and the heavier pieces on the grate. A special consideration when firing wood is contamination with denser solids. Material cut specifically for firing can contain sand, rocks and dirt. Sander dust can contain some of the abrasive material from the sanders. Those heavier and denser solids can seriously erode the fuel handling equipment, burners, grates, and the boiler tubes. Although some people don't consider it wood fuel paper plants burn large quantities of bark that's stripped from the wood used to make pulp for paper. Bark is usually burned over a high set overfeed stoker where the bark is introduced as much as sixty feet above the grate.

Wood and wood and paper product manufacturers have an opportunity to convert waste to fuel but in many cases it's less expensive to landfill the waste and burn gas or oil as a fuel. Environmental restrictions also limit its use. The increased cost of fuel and landfilling may change that in coming years. There are also innovations in wood burning systems including fluidized bed firing but those technologies will only be applied and proven as decisions to burn wood and wood waste increase.

Wood firing problems and how an operator can respond to them vary with the fuel. There is usually less ash than with coal firing but the ash can be finer and plug up equipment more. Keeping the systems clean, and dressing the fire of stoker fired boilers are principal activities.

One thing that wood fired boilers have is plenty of air. Much of the air that's used for combustion comes with the wood. If you think about logs you put on your fireplace, wood stove, or campfire, you'll recall the wood is full of little air spaces, especially if it's dry. There's so much air available that a pile of wood in a corner that looks like it's all burnt out ash can have glowing embers underneath. They can also be there after several hours or even days. Always treat any accumulation of anything in a wood fired installation as a potential source of flame. Stir it up and mix it with a little air and you could have an explosive mixture, same as with coal.

I've been involved with four major projects to burn wood and wood waste. They have all had their technical and operational problems on startup but all are operating and reducing the amount of wood and wood waste going to waste and landfills. Some processing of wood wastes have produced specialized wood fuels (pellets) and included material like leaves.

Biomass Burners

Wood is one of the many fuels identified as "biomass," a label applied to materials that are grown or produced from vegetation. Biomass is commonly restricted to mean a solid fuel and the burner may actually be a device that processes that fuel to convert it to a gas or liquid fuel that's then burned. Other biomass fuels

are modified to produce a fuel in a different form. I'm sure you're most familiar with one, charcoal. You also use one regularly; the ethanol added to your gasoline is made from corn.

Unlike dried corn kernels much of biomass has to be processed to prepare it for burning in conventional fuel burning equipment. A typical example is wood pellets but the equipment can also take larger particles and reduce them to dust (or nearly dust) such that the fuel can burn in suspension in a furnace like pulverized coal. A fairly common reduction creates a fuel labeled RDF (for refuse derived fuel). RDF can be a fairly fine material that's typically blended with pulverized coal for firing in combination with pulverized coal. Another form of RDF is a reduction to something less than ½ inch in dimension for firing in a spreader-stoker.

I don't know of any at the moment that generate a liquid fuel but there's applications in pilot plant stages that could do that. Producing a liquid fuel that's easily transported and burned like our standard fuel oils has been attempted many times but no single method has stood out as economically viable. Normally a conversion process performs a reaction known as pyrolysis where heat is generated by adding oxygen (in air or pure) to burn some of the fuel with the remainder vaporized into combustible gases, principally carbon monoxide.

With biomass fuels the instruction manual may not tell you all that would be good to know about burning that fuel. That's because advances in the technology are made regularly. Looking through your equipment manufacturer's web site on a regular basis may make you aware of changes and improvements in operation that have been developed. Check it out and you may find a user's group that you can join online that shares problems, improvements and solutions.

PUMPS

Pumps are used to move all of the liquids around a boiler plant and there is a diversity of designs and arrangements for pumping that provides many options. When engineers use the word 'application' it means what the equipment is used for; applications include feedwater pumping, condensate pumping, fuel oil pumping, sewage pumping, etc. Over the years the applications of pumps to boiler plants has singled out a particular pumping method and pump construction for each service. As a result you'll seldom find any deviations in the type of pump used for a particular fluid service.

High pressure feedwater and condensate system

pumps are usually centrifugal. Low pressure feedwater and small volume condensate pumps are usually turbine type pumps. Fuel oil is moved with positive displacement progressive cavity pumps of the screw and gear types. There are other options but their use is not as common. Technological advances could alter one or more of these general rules in the future. If only someone could come up with something better than a centrifugal pump we could see dramatic reductions in electric power consumption because many of the centrifugals run at efficiencies less than 50%.

Pumps handle liquids, incompressible fluids, and they're an essential part of the boiler plant. Modern pumps have become so reliable that operators tend to ignore them until something fails. I've been in many a plant where the pumps have been there operating for so long that the manufacturer's name that was formed in the casting of the pump had corroded until you couldn't read it. When asked, the operators couldn't produce an instruction manual or anything else that would identify the make and model of that pump.

Now that's confidence, it will last forever so we don't have to know where to get one to replace it! Dream on. Pumps don't last forever and their capacity and differential capability declines as they age. Their efficiency also declines with age and pumps that are so old you can't read the nameplate may be using twice as much electricity as they did when they were new or, more likely, only pumping half of what they could originally. Monitoring the performance of your pumps is a wise thing to do.

Pumps are usually oversized too. I frequently discover that boiler feed pumps are selected so any one of them can run the plant at full capacity (all boilers on) which we know doesn't make sense when at least one boiler is usually a spare. Then, to compound stupidity, the engineer specified three or four pumps of the same size. It's virtually impossible for an operator to select a pump that matches the load when they're all too damn big!

In many instances replacing a boiler feed pump with one that will just barely handle a spring or fall load will save enough electric power to pay for the pump in one summer. When you have more than two pumps the capacity of each should be such that it takes all of them, less one, to carry the peak load. With three pumps they should each handle half the peak load. With four pumps they should each handle one third of the peak load. Five pumps should be sized at one quarter the peak load, etc. Since boiler feed pumps have to be capable of delivering water to the boiler when the safety valves are blowing

(a code requirement) they're slightly oversized anyway because capacity picks up as the differential is lowered to operating conditions.

Just because the pumps can be oversized don't ignore the possibility that an operator can compound the problem by making logical decisions. I was in one plant with three boilers and four feed pumps. If two boilers were on line the operator ran two pumps. During the winter when there were three boilers on line... you got it, three pumps were running. It made no difference what the boiler load was, run a boiler and run a pump.

A quick look at the instruction manual revealed that any one pump could supply three boilers. Savings of electricity by only running one pump the year round was well over $50,000.00. I think it's now obvious that proper choices in the operation of pumps and monitoring their performance as well as maintaining them can make a significant difference in the cost of operating a plant and can also justify a wise operator's salary. Fair warning, however, simple numbers don't always work.

With rare exceptions pumps are powered by electric motors or steam turbines. We say the motor or turbine 'drives' the pump so we call them 'drivers.' They all serve to rotate or extend and contract the shaft of the pump. The energy is transmitted through a metal shaft that connects the driver to the pump. The rotating parts of a pump can be mounted directly on the driver's shaft or they can be mounted on their own shaft. When the pump has it own shaft it is also fitted with bearings to maintain alignment of the shaft in the casing of the pump. Regardless of operation, rotating or extending and retracting the shaft moves and the design of the pump must allow it to move without allowing the liquid to leak out of the pump.

When I was operating and maintaining pumps we had to allow some of the liquid to leak. That's because all we had to keep the liquid from leaking out of the pump in large quantities was packing. (Also see packing under maintenance. Packing seals the space along the shaft where it penetrates the casing to limit leakage. Some leakage through the packing is essential to lubricate the packing to shaft joint. If the packing is tightened enough to stop or reduce the leakage too much then the packing and shaft rub with deterioration of each.

As a matter of fact, it was so common for us to screw up a shaft with the packing that manufacturers started making rotating pumps with shaft sleeves to help with that problem. The sleeve was like a pipe or tube that slipped over the shaft and was either clamped with other parts or threaded onto a matching thread on the shaft so it was removable. That way, when we ran the pump with the packing dry and tore up the shaft sleeve all we had to do was replace it, not the entire shaft. I've been in a few plants where annual replacement of shafts and shaft sleeves was common because the operators consistently tightened the packing too much.

If you don't know how much leakage is necessary, try measuring the temperature of what leaks out and compare it with the temperature of the liquid inside the pump, it shouldn't rise more than 5 degrees. That doesn't work for boiler feed pumps because the liquid in the pump would flash. Usually we look for a tiny stream flowing out of the packing as a rule. By tiny stream I mean something no larger than a pencil lead. Over time you'll learn how much you can squeeze down on packing and learn not to go too far because you'll end up rebuilding the pump. The pump will usually tell you when it's too tight because it will wear the sleeve or shaft until it gets enough flow. In that case, listen to what the pump is telling you and allow that leakage.

A pump construction ensures the packing, or seals described below, are not subjected to discharge pressures unnecessarily. By designing pumps with the packing or shaft seal at the lowest possible pressure point in the pump wear and tear on them is reduced. Some pumps still have a seal or packing exposed to the highest pressure and construction is modified to reduce the effect of the high pressure. Packed pumps will have lantern rings (Figure 5-8) which permit bleeding off of the high pressure leakage to the suction of the pump with the rest of the packing exposed only to the suction pressure.

In the case of condensate pumps and others that can operate with pressures below atmospheric, pressure on the suction side supplied by a connection from the discharge provides fluid to seal and cool the packing.

Modern rotating pumps are commonly supplied with a shaft 'seal.' It's a special construction with very hard materials consisting of rotating and stationary parts that provide the liquid seal. Those materials are machined to very close tolerances so there are only a few hundred thousandths of an inch separate them when operating but the two materials do not touch because a minute amount of liquid separates them. In many cases the liquid forms a vapor between the two wearing surfaces and the vapor becomes the lubricant with no leakage evident at the seal.

Most of them require some flushing of the seal to keep it cool enough to operate properly, using a small line from the pump discharge to the seal to provide flushing liquid. When there's an opportunity for the liquid to contain small particles of rust or other solids

that could damage the seal the flushing liquid is passed through a strainer to remove those solids providing clean flushing liquid and extending the life of the shaft seal.

Sometimes the flushing liquid is improperly applied. If you open a pump to find erosion around the area where the flushing liquid is admitted check with the manufacturer. I've run into more than one pump that left the factory without a required orifice in the seal flushing piping.

The seal materials have to be able to handle the temperatures under those vapor forming conditions. Some shaft seals require coolers to lower the temperature of the liquid. HTHW circulating pumps, for example, have strainers and coolers on the flushing liquid. Maintaining the coolers and strainers is an important factor in keeping those pumps in operation. Care is required to keep the temperature of the liquid within an acceptable range because it can also get too cool producing thermal shock where it mixes with the fluid in the pump.

Alignment

When the pump and its driver are riding on separate bearings the two shafts are connected with a coupling. Rotating shafts are equipped with flexible couplings which allow the two shafts to be centered in their own bearings. Shafts that extend and retract can be connected with rigid clamped couplings or a coupling containing a bearing that allows one shaft to swing like that for a chemical feed pump.

Proper alignment of couplings is essential for long pump life. If the alignment is poor the coupling will apply alternating forces to the shaft, constantly bending it back and forth until it finally breaks if the bearings or packing don't fail first. I will not go into the alignment of a reciprocating shaft pump because you should not have to do it. If you do have to work on a straight recip make it a point to follow the directions in the instruction manual carefully. The following discussion on aligning rotating pumps should give you all the clues you need to know about what has to be done; you will still need the manual to see how to do it right.

The process of aligning a pump and driver begins with determining the differences between operating and cold conditions. A boiler feed pump, for example, will heat up when the pump is in operation so its shaft can be higher when it's operating. A pump and turbine combination can have different changes in shaft position. You normally do not need to correct for operating temperature on most pump and turbine combinations because both will be centerline supported. That means the pump and/or turbine are constructed with supporting feet that connect to the pump or driver near the centerline of the shaft. The temperature of the feet will not change much in operation so the shaft position will be the same whether the pump is hot or cold.

When the pump or driver is not centerline supported you should calculate the amount of growth or relative growth, given the operating temperatures and material of the casing and use that value in rough alignment then check the equipment when it's up to operating temperature.

Alignment should be performed in a particular order. Correct vertical angular alignment (Figure 10-61) first; vertical height (Figure 10-62) second, horizontal angular alignment third and horizontal alignment last. Those last two steps are done the same as the first two but they don't require shimming.

You'll need shim stock of varying thicknesses. Commonly that's thin sheets of brass (preferably) or steel in varying thicknesses. Normally you'll need some materials in 10, 5, 2, and 1 mil thicknesses. (A mil being one thousandth of an inch) but occasionally thicker pieces are required. Of course this assumes that the pump was reasonably aligned in the factory or before you started on it to begin with. Sometimes it takes some major pieces to rough in before you can start dealing with the thinner pieces.

Shims should be prepared as shown in Figure 10-63 so they can be slipped under the supports of the driver (normally) and around its anchor bolts. It's important to make the slot at least a sixteenth larger than the anchor bolt and to be careful with their installation so they don't interfere with bolting. When aligning pump and turbine it's sometimes easier to align the pump to the turbine. An electric motor does not have any connecting piping so it's easier to move the motor to achieve alignment.

If you are trying to align a pump to resolve some wear or other problems that indicate misalignment but don't find any problems with cold or hot alignment be aware that a pump casing can be deformed by application of piping expansion stress at the pump nozzles. If that's the case aligning the pump again isn't going to solve the problem.

The base the pump and driver are mounted on also have to be firm. If the base can flex it will allow vibrating misalignment which usually results in coupling or bearing failure in a short period of time. I remember being asked to look at a pair of condensate booster pumps, fairly large ones, because their couplings were constantly failing. The owner wanted me to recommend a coupling that wouldn't fail. I told him I didn't think

any coupling would work until he filled the base with grout as specified by the manufacturer. The base was suspended above the housekeeping pad by about an inch, held up only at the four anchor bolts in the corners of the base. A quick setup of a long ruler over a pivot next to the base showed how much it deflected when I simply put my foot on it. The base has to be solid and not bend before you start worrying about alignment.

There are many different methods of pump alignment and which one you use is dependent on the speed, power requirements, and size of the pump. As speeds, power, and size increase the precision of alignment becomes more important. That doesn't mean that the smaller pumps should not be carefully aligned, only that

Figure 10-61. Angular coupling alignment

Figure 10-62. Coupling offset alignment

Figure 10-63. Shims

the cost of the pump may be so low that the cost of precision alignment is higher and you can afford to replace the pump more often than you can afford to precisely align it.

The typical small pump is fitted with a coupling consisting of two metal halves with a rubber insert (Figure 10-64) The common method for aligning these pumps is to place a small metal ruler along the side of the coupling as shown in the earlier figures and adjusting until the rule shows the two coupling halves to be in line. Holding the rule as shown and holding a light behind it is the best way to see any gaps between the rule and the coupling halves. Turn the shafts 1/4 turn and repeat the reading three times when you get close to the end because this doesn't correct for couplings that are bored off center or where rough surfaces produce errors.

Figure 10-64. Small pump coupling

To determine how much angular adjustment is required you have to compare the length of the coupling half to the spacing between the motor mounts. You either eyeball the distance or slip varying thicknesses of shim stock in the gap between coupling half and ruler as shown in Figure 10-65 then calculate the required adjustment by the ratio of coupling half length to driver mount distance for vertical angular adjustments.

To correct the 2 mil difference over the coupling half as shown in the figure where the coupling is 1-1/2 inches long and the driver mounts are separated by 6 inches you'll need to adjust the shims at one end of the motor mount by 8 mils (2 * 6 ÷ 1.5). Be careful when you discover a vertical angular misalignment, it can mean that some of the shims got knocked out from only one foot of the driver.

Sometimes one mount is loosened and the shims are shaken out, I can recall finding loose shims in and under bases many times. When starting with a previously aligned pump it's always a good idea to loosen all the anchor bolts of the pump and driver and see if either rocks in any direction. Correct any rocking first or you could distort the pump or driver frame which is worse than misalignment for the bearings. It could even crack the motor housing.

Once you've resolved any vertical angular misalignment all the driver mounts should be level and further adjustments involve adding or removing the same amount of shim stock under each of the feet. Be careful when performing the vertical center alignment because you can add or remove different thicknesses of shims.

The best thing to do is use a micrometer (Figure 10-66) to measure the shims to be certain you're alter-

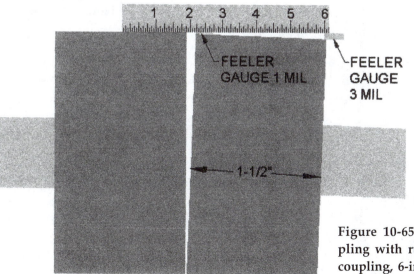

Figure 10-65. Aligning small coupling with ruler (shown 1-1/2 inch coupling, 6-inch motor mount)

ing each foot the same. If you don't have a micrometer use the ruler and light to compare the pieces of shim stock. Before you work on horizontal alignment check the vertical with the driver bolted down on the shims. Sometimes the shims can compress a little more or less to alter the alignment.

Once you've got vertical alignment down the jobs simpler because you don't have to mess with the shim stock. It is, however, hard to retain angular position horizontally while you're trying to correct centerline alignment. I always preferred the light hammer method. Once I got the pump close I used a small hammer to tap the feet. Once you get used to it you'll discover about how hard you have to tap to get a movement of one mil. Tapping both feet on one side consistently will shift the driver the same amount to retain angular displacement.

For better precision in aligning a pump and driver... Okay, I'll relent, I should say aligning a coupling

Figure 10-66. Measuring shim stock with micrometer

Figure 10-67. Dial micrometer

because that's what we always say. You really aren't aligning the coupling, you're aligning the shafts of the pump and driver but we still say we're aligning the coupling. Anyway, better is done with a dial micrometer (Figure 10-67) which eliminates problems with poorly machined couplings and provides hard readings instead of eyeballing it. You determine the error by clamping mounting bars furnished with the micrometer to hold it relative to one coupling half while the micrometer stub (sticking out at the bottom left of the figure) rests against the half coupling attached to the other shaft, zeroing the micrometer, then rotating the shafts to take a reading 180 degrees from the original one.

Zeroing the micrometer is accomplished by simply grabbing the dial and twisting it until the zero is centered under the needle. In this case you use twice the distance from the center of the shaft to the contact point of the micrometer instead of the length of the coupling to determine the ratio. Usually the ratio is close to one, making life a little easier, just use a shim matching the reading.

There are more precise methods using laser equipment and computers but that's best handled by a contractor that specializes in alignment. You have to align a lot of pumps in order to justify the cost of a laser alignment system.

NPSH

It stands for 'net positive suction head' and despite it being one of those terms that we engineers use it's absolutely essential that an operator understand what it is and how it relates to the operation of pumps. In many a discussion we'll use the term to mean one of two things without clarifying it and in other cases we'll clarify that NPSHR is the 'required' suction head and NPSHA is the 'available' suction head. Now let's get down to what they are.

Suction head is the pressure at the inlet of the pump produced by two things, the height of the liquid above (below) the centerline of the pump and any pressure acting on the surface of that liquid. When the pump is running the suction head has to account for the pressure drop in the suction piping so it will be a little lower when the pump is running. It will also decrease as the flow increases.

Why is NPSH important? When the available head isn't adequate the liquid in the pump will begin to boil, small bubbles of gas will form in the suction. If enough of them form the pump will be 'vapor bound' and can't pump any liquid. Once a pocket of vapor forms the pump contains compressible gas, not incompressible liquid. The pump parts either spin in the vapor producing no pressure or the vapor will constantly compress and expand. The net result is the pump stopped pumping. In some cases this will cause a surge of discharged liquid back into the pump which then gets pumped out again and that liquid surging back and forth damages the pump.

Sudden formation of vapor in a pump driven by a steam turbine will result in rapid over-speeding of the pump and turbine. Occasionally that happens so fast that the turbine over-speed trip can't respond before the turbine blades start flying out of the casing!

When the bubbles start forming they will collapse later when the pump increases the pressure in the liquid. In centrifugal and turbine pumps the result is bubbles forming then collapsing and the liquid rushing in, to fill the voids as bubbles collapse, hammer away on the parts of the pump. We call that 'cavitation' and it's evident by a small to fair amount of noise that you can hear. It's also evident when you dismantle the pump. You will see heavy wear consisting of lots of tiny indentations where the bubbles collapsed. To prevent pump damage you have to be sure you have adequate NPSH.

The NPSHA is the difference between the suction head and the vapor pressure of the liquid. To get away from the math let's assume a pump submerged to its centerline in a tank of boiling water at sea level. Since the water level is right at the inlet the suction head is zero gage, 15 psia. Since the water is boiling the vapor pressure is 15 psia and the NPSHA is zero. By submerging the pump in the tank so its centerline is four feet below the surface and there's no suction piping to produce friction, we increase the NPSHA to four feet.

I should also explain what happens when the water is colder. Lets assume our pump is in a tank of condensate at 162°F. The vapor pressure at that temperature (check the steam tables, is 5 psia. Subtract from 15 psia to get an additional 10 psi of pressure that the suction can drop before the water boils. Checking the head tables we find that the 10 psi converts to about 23 feet and we can add that to the four feet the pump is submerged to get an NPSHA of 27 feet.

To help explain how a centrifugal pump can lift water out of a lake once it's flooded, the NPSH of water at 60 °F is minus 14.5 psig equal to 33.5 feet. A pump can lift 60° water that far before it will start boiling. Of course the pump can't pump the air out so you'll have to install a foot valve in the lake and fill the piping and pump casing with water to get it started. Once it's started it will pump the water.

Now, back to the two additional labels. The NPSHR

is specified by the pump manufacturer for the design operating condition and is usually shown on the pump curves. It's the required NPSH for that pump at the rate of flow. Some of the requirement is a function of how much the liquid has to accelerate at the inlet of the pump impeller because some of the static pressure of the suction head has to be converted to velocity pressure to get the liquid into the impeller. The NPSHA is what's available, the actual NPSH at the inlet of the pump. That value always has to be higher than the NPSHR.

Operating a pump when the level in a tank it's taking suction on is too low can result in serious damage to the pump. Allowing a pump to continue operating when the suction head is inadequate doesn't make sense. If the tank is almost dry there's nothing there for the pump to move anyway, shut the pump down to prevent it being damaged. Remember priority number three?

I'm not talking about short term conditions here because I know we occasionally run a pump to the point of losing liquid. Stripping a fuel oil tank before cleaning is one example. In that case you should be prepared to stop the pump the instant it loses suction so you limit the potential for damage. I've cleaned all the metal shavings out of many a fuel oil strainer after somebody let a pump run for several minutes after the tank went dry. Then I helped rebuild the pump.

You'll notice on the pump curves that the NPSHR increases as the flow through the pump increases. Throttling the discharge of a pump to reduce the flow will also reduce the NPSH required and can stop a pump cavitating. Although this is occasionally required under unique operating conditions it shouldn't be the normal case. If you have to operate the pump at the lower suction head then you would do well to have the impeller turned down to reduce its capacity and horsepower requirement.

Cut it down! What's that about? It's a way of making a pump fit its application better. It can't always be done. However, in many cases it's something that should have been done and wasn't. If someone simply orders a new impeller giving the manufacturer nothing but the pump model number they could very easily get a full size impeller, not one that was trimmed for the application. You can tell what your impeller diameter should be by the pump curve that came with the original instructions.

Pump Curves

Pump curves provide answers to a lot of questions about our pumps. If you feel compelled to throw out a lot of unnecessary paper never include pump curves in that group. How much liquid can be pumped under varying differential pressures is the most important line on a pump curve. Any pump curve will normally have several of those depending on different construction and operating conditions. As stated above, the NPSH (NPSHR understood) will be shown when it's important. The pump will also have horsepower lines or efficiency lines or both. Either the horsepower or the efficiency will permit calculation of the other value because there's a standard formula for hydraulic horsepower.

The flow-differential curve is the first one to look for. In many cases they will be the darkest lines on the paper. The normal form of a curve lists the differential on the left side of the curve and the flow on the bottom. Differential is typically listed in feet, meaning head, and you have to convert that value to psi to see how much pressure boost you can get out of the pump. Some curves will show psi because the pump isn't affected much by density. The rate of flow is normally listed in gallons per minute but don't be surprised to see gallons per hour or hundreds of gallons per minute. If there's no label you should be able to safely assume gpm.

Now I know that you're going to find most curves will have several lines. There are several lines because the pump can pump more than one type of liquid and some have variations in construction. The typical centrifugal pump, where the curves are almost always for cold water, will have different lines for the choices of impeller diameters. Normally the curve is marked with the design point so you can see what diameter impeller was installed in your pump, otherwise you'll have to look elsewhere in the manual to find out what size impeller you have.

Once you've identified the line you can tell what the differential pressure will be for a given pumping volume. Sometimes it's valuable for determining how much you're pumping based on the difference in pressure. Other curves will address characteristics of the liquid. Fuel oil pumps, for example, will have a number of lines on the curve for different viscosities of the oil.

Unless specifically stated to the contrary a pump curve is supplied to show the flow and differential characteristics pumping cold water at 32°F and a density of 62.4 pounds per cubic foot. That provides a basis for determining the differential pressure at other fluid densities. Since we seldom pump ice water you have to adjust the head characteristic of a pump curve to determine the actual differential pressure which will always be lower than what the curve indicates. This gains some importance with water at high temperatures and is important for things like boiler feed pumps.

Boiler feedwater at 227°F (a common temperature) is not as dense as ice water, it only weighs about 59.4 pounds per cubic foot and while pumping that lighter feedwater the pump will only produce 95.3% of the discharge pressure that's produced when pumping ice water, enough to be significant when operating at high boiler pressures. It's also important to note that centrifugal pumps are volumetric machines, they pump so many gallons, not so many pounds so the 95.3% should also be applied to any calculation that converts the gallons per minute to pounds per hour.

The horsepower or efficiency lines are primarily used by engineers in selecting pumps, trying to buy the one with the lowest operating cost. At least that's what it should be. You, on the other hand, can use those curves to get an idea of the best mix of pumps for a given operation or to provide answers to problems with the pump. You may have a choice of running one or two pumps and decide that running one should be more efficient. While that's a logical decision it isn't always the case. Running one large pump far out on its curve could be less efficient than running two smaller pumps because they're operating at a better efficiency.

I always tell this story to make an important point regarding pump efficiency. I was asked to look at a problem with boiler feed pumps at a major laundry in Washington, DC. The owner complained that he was replacing the pumps every six months. They didn't sound too bad but it was obvious that they were cavitating during normal operation. A look at the pump curves and installation revealed inadequate suction head was the problem. I searched catalogs for alternates and submitted a recommendation for purchasing different pumps for two reasons. One was the NPSHR of the recommended pumps was less than what was available. The other reason was the new pumps did the job at 3.5 horsepower and the existing pumps took 7.2 horsepower. Yes, there is that big a variation in pump efficiencies. The savings in motor horsepower was worth $1,480.00 per year. The owner balked at my recommendation because the new pumps cost twice as much apiece, $2,500.00 more than the ones he had. To this day I don't know what happened because I was never called back to the site. If he had installed those expensive pumps he would be avoiding the $6,480.00 he had been spending for horsepower and replacement pumps every year. If you have any opportunity to choose a pump be conscious of power requirements in addition to NPSH.

Another point to consider in using pump curves is the occasional use of a pump for a purpose other than originally intended. You can use the curves to see if the pump will work and make certain you don't overload the motor.

I did say there's a standard formula for pump horsepower. There is, it's called Hydraulic Horsepower, is also called theoretical horsepower, and it can be calculated by multiplying the flow in gallons per minute by the head in feet and dividing by 3960. If the liquid isn't water at 8.33 pounds per gallon, multiply by the specific gravity of the liquid. Note that it's theoretical horsepower. Divide by the pump efficiency to get brake horsepower, the amount the driver has to produce. If you don't know the efficiency use 33% (multiply the theoretical horsepower by 3) to be safe.

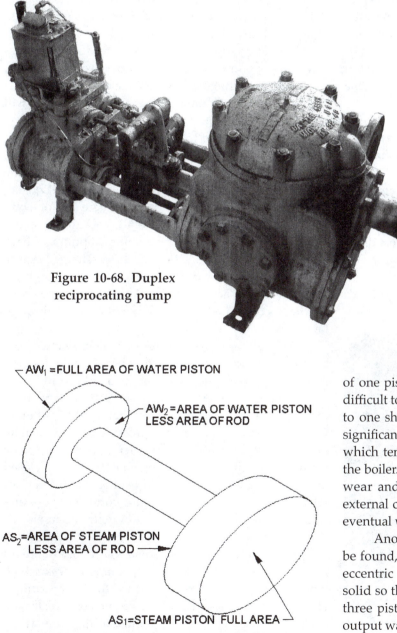

Figure 10-68. Duplex reciprocating pump

Figure 10-69. Areas of pistons for pump pressure

Reciprocating Pumps

Many boiler plant applications were predominantly served by reciprocating piston pumps until the middle of the 20th century when multi-stage centrifugal pumps displaced them. For that matter most of the liquids in the plant were moved by the standard duplex reciprocating pump (Figure 10-68) which was the mainstay of the power plant at the beginning of that century. The pump, powered by steam from the boiler, was capable of producing very high pressures and, despite the reciprocating operation, produced a reasonably constant output.

The pressure differential of the pumped liquid is determined by the difference between the steam supply and exhaust pressures and the ratio of the cylinder areas. The maximum pressure that could be produced, an important consideration for selecting valves and piping materials, is the area of the face of the steam piston less the area of the connecting rod times the difference in steam supply and exhaust pressures divided by the area of the fluid piston less the area of the connecting rod (Figure 10-69).

There were, and still are, single piston pumps consisting of one steam cylinder and one fluid cylinder but it was difficult to adjust them so they would operate continuously, occasionally hanging up at one end of the stroke or another. Most of those were larger pumps used for fuel oil and ballast (water) transfer aboard the ships. The duplex pump practically eliminated problems with the pumps hanging up because the stroking of one piston tripped the valve to reverse the other. It's difficult to see in the photograph but the linkage attached to one shaft operates the control valve for the other. A significant problem with these pumps was the lubrication which tended to get into the condensate and then into the boiler. They also had a lot of sliding parts that would wear and required constant maintenance. Internal or external check valves also slammed open and shut with eventual wear and breakage.

Another form of reciprocating pump that can still be found, principally in boiler feed use is a three piston eccentric cranked motor driven pump. The pistons are solid so they only pumped in one direction. Each of the three pistons operated off a different crank arm so the output was a little more uniform. The balance of pistons and a heavy counterweight on the shaft helped reduce the motor load from the imbalanced forces. A feature of the pump is control of the valves to vary capacity. The suction valve is held open on the discharge stroke (pushing the liquid back into the suction) for varying degrees of rotation to vary the amount of water pumped. If you think it an antiquated way of doing things I can only say that the first nuclear merchant ship, the Savannah, had one of those pumps for boiler feed.

The only reciprocating piston pump you'll normally find in a modern boiler plant is a chemical feed pump. Usually the piston is pumping a hydraulic fluid that transfers energy to the liquid being pumped using a diaphragm (Figure 10-70).

The capacity of a reciprocating pump is easy to determine. It's equal to the area of the piston times the length of stroke times the revolutions per minute if it's single acting. If it's double acting, where the liquid is admitted to and pushed out from both sides of the piston it's twice that much less the cross sectional area of the shaft times the length of stroke times rpm.

Reciprocating pumps are positive displacement pumps. That's a term we engineers use to mean that plastic, wood, metal, or whatever the pump is made of displaces (moves into the space that was occupied by) the liquid to move it through the pump. The steam powered duplex pump had some balancing features because the pressure on the liquid couldn't exceed the difference between the steam supply and exhaust pressures times the ratio of the areas of the pistons. That pump would simply stop if the pressure on the liquid got too high. Motor driven pumps seldom simply stop, they produce very high pressures because the motor's torque increas-

Figure 10-70. Piston chemical feed pump

es as it slows down. Usually the motor starter will trip but there are many reports where the pump or piping ruptured when someone accidentally started a pump without opening all the valves in the system.

To prevent damage of that nature and motor starters tripping or motors burning up a relief valve should always be installed at the discharge of a positive displacement pump. If it's reasonable to believe the flow through the relief valve will always be of short duration then the relief valve can dump the liquid back into the pump suction piping. It's always possible that the pump could be operated for some time pumping the same liquid and all the power will be diverted to heating up that liquid so it's better, whenever possible, to route that liquid back to a tank or sump where there is a larger mass of liquid to absorb the heat.

A final note is appropriate before discussing specific types of pumps. Any of them can be run backwards. Some, like centrifugals, can appear to operate, just not as well as with proper rotation. Gear and screw pumps will tend to pump the liquid in the opposite direction.

I'm reminded of the time I was asked to look at a fuel oil pumping installation that, for whatever reason, couldn't produce more than 30 psig. After arriving at the plant and introducing myself I looked at the oil system with special attention to the pumps. There was one odd provision, at least odd in my mind, because the check valves were on the suction side of the oil pumps. Check valves are normally mounted on the discharge because they will stop flow back to the pump if you stop it for something like packing failing. A pump with the check on the suction side will not prevent leakage through failed packing.

As I followed the lines to the boilers I noticed the back pressure regulator in the overhead piping, found a ladder to climb up and looked at the regulator and valving. I wanted to be certain the bypass valve had not been left open but didn't tell my escort that. When I got up to where I could see I noticed the pressure gage at the inlet of the regulator read 30 psig. After I made certain the manual isolating valves were open, I opened and closed the bypass valve. The gage still read 30 psig! When we returned to the pumps I asked the escort to start one and was informed that he couldn't do it without an electrician and there were no electricians on the job that day.

Dumfounded and concerned that I couldn't learn much more without operating the pumps I was asking him what happened when they tried to run the pumps. He claimed they made a lot of noise and some oil leaked out of the check valve as he pointed at the seam at the bonnet of the valve. When I told him that the pump was obviously running backwards he became very belligerent, telling me I didn't know what I was talking about, that he checked the rotation himself, and that couldn't possibly be the problem; besides, how could I know because I hadn't seen the pump running.

If you're now a wise operator I'm sure you already know how I knew. He proceeded to show me how he had determined the rotation, using a logic appropriate to a centrifugal pump, which was wrong for the crescent gear pump we were looking at. I suggested several times in the discussion that he would discover I was right if he looked at the instruction manual. Finally, showing signs of rage, he stepped over to the center pump, yanked away the envelope that was still wired to the motor lifting eye (as at all three pumps), extracted the instructions, flipped through the pages until he found a graphic, and pointing at the graphic approached me saying "see, right here it shows…" He suddenly stopped, turned to look at the pump again, shrugged and said "I'll get it changed tomorrow" then walked off. There's many a lesson in this story but making sure the rotation is correct on your pump is the one you should get right now.

Centrifugal Pumps

Our most common pump, the centrifugal, acquired that position for reasons other than efficiency and energy costs. In my experience it is also misapplied more than any other pump. The range of efficiency of pumps in service, again in my experience, runs from 30% to 70%. Now that has to be one serious variation, a pump at 30% efficiency will use 2.33 times as much energy as a pump operating at 70% efficiency. I told the story in the section on NPSH that reflects the differences that can exist in pump performance and it's primarily with centrifugal pumps. Why is there such a variation? Because engineers, contractors, and or owners all either ignore those significant differences or are so intent on getting the lowest first cost that they'll choose a pump that will chew up all the difference in first cost in comparison to an efficient pump in less than a year or two.

I'm an energy engineer and I've evaluated pumps for power costs for years but I also seem to be a voice crying in the wilderness because I keep finding them and continually encounter people that will purchase that cheaper pump anyway because the power cost isn't their problem. I've covered the matter in the previous paragraphs and I hope you learn to apply this knowledge of pump power requirements to operate your plant wisely even though, in normal situations, you have been given inefficient junk to operate.

If a centrifugal pump was installed in connecting piping with no valves and stopped the liquid would flow right back through the pump because there are no suction or discharge valves to block that flow. Some operators have a problem understanding how the pump even works. Centrifugal pumps simply grab the liquid and throw it. The impeller flings the liquid into the volute of the pump (Figure 10-71) where the velocity pressure is partially converted to static pressure and delivered to the discharge. To get an idea of its operation go to the kitchen, fill a pot or bowl half full of water, and start stirring it with a spoon. Stir the water in one direction (a pump only runs in one direction) and you'll notice that the level of the water in the bowl will vary from low in the middle to highest at the outside of the bowl. That difference in level is the head of your bowl pump at shutoff.

Stir faster and the head goes higher and when you spin it fast enough the water starts coming out of the bowl. Setting it under the spigot to add water and stirring it fast enough so the water spills out at the same rate you're adding water and you have a simple version of a centrifugal pump. Note what happens when you do various things with the spoon and you'll have a pretty good understanding of how a centrifugal pump operates.

A centrifugal pump does not move a fixed volume of liquid like a reciprocating pump. The amount of liquid moved varies with the differential. The flow of water pumped from a tank will vary with changes in the height of water in the tank or the discharge pressure at the outlet of the pump. If you open the spigot on the sink up so more water flows in and don't change the rate you're stirring it you will see more water flowing even though you aren't doing any more work. It might

Figure 10-71. Centrifugal pump impeller and volute

help to realize that a centrifugal pump simply boosts the pressure a certain amount and that boost is related to the flow of water through the pump. You can stop stirring the water in your bowl and it will still overflow once it has filled. You can also vary the difference (head) you're creating by changing the speed at which you stir it.

Back from playing in the kitchen sink? Good. I trust you now understand that there is no such thing as a limit on the flow through a centrifugal pump; the highest possible flow is much more than the design value and the minimum is zero. Without check valves in the discharge piping a higher external differential pressure than the pump can handle will result in flow backwards through the pump. The actual flow rate is dependent on the performance of the pump itself and the difference in pressure between suction and discharge.

Oh there's a design point, a flow and differential that the engineer calculated for selecting the pump and that's usually indicated in the manual and on the pump curve. What you, as an operator, have to deal with is the actual flowing conditions. The odds that the actual conditions are precisely the same as the design conditions are between slim and none.

A feature of centrifugal pumps that's frequently forgotten is the use of wear rings (Figure 10-72) The space between the casing and the eye of the impeller is all that separates the suction and discharge pressure zones of the pump so some water has to bleed back through that space because they can't rub. As the pump is used small particles in the liquid and the liquid itself can erode the material on either side of that gap and provision of wear rings makes it possible to restore a pump to a like-new condition by simply replacing the wear rings. The casing wear rings, right one hanging loose in the photo, are keyed to set in the casing and not rotate. The impeller wear ring is heated then inserted onto the end of the impeller where it shrinks on for a tight fit.

No, a strainer in the suction piping (standard requirement for most pumps) does not remove the small particles that erode the wear rings; the strainer does remove pieces that would jam between them. Usually a pump with wear rings will also have a shaft sleeve. I should mention that you should be cautious when replacing wear rings and anytime you reassemble a split case pump because the outer wear ring can be distorted when the two halves of the pump casing bear down on it. Always make sure the pump rotates by hand as you're drawing up on the bolts that hold the two casing halves together. Also, don't install a thicker gasket on a pump simply because you don't have the right thickness on hand, that will create gaps between the outer wear

Figure 10-72. Wear rings

ring and casing where erosion can cause problems. Too thin a gasket will normally bind the pump up.

You'll find a lot of variety in centrifugal pumps depending on their application. The pressure differential they can produce depends on the density of the liquid being pumped and the speed of the tips of the vanes in the impeller. To make a pump operate at a higher differential pressure with the same liquid the diameter of the impeller is increased. Once the impeller's maximum diameter is reached a faster motor is used. As the impeller diameter and speed increases the stress on the metal gets higher so there are practical limits on the pressure boost.

If a larger differential pressure is required the pump is supplied with additional impellers. We call them 'multi-stage' pumps. The pressure is increased a little in each impeller which, along with its volute and share of the casing constitutes a stage. That way high pressures can be developed without making pumps of very large diameter.

Since the eye (inlet of an impeller) is exposed to suction pressure at that stage and the rest is exposed to the discharge pressure of that stage there's a difference in axial forces on the stage (Figure 10-73). In single stage pumps holes are drilled through the back plate of the impeller and a second set of wear rings added to balance the pressure. (Figure 10-74) In multi-stage pumps the stages are reversed on the shaft (Figure 10-75) so the imbalance of one stage is opposed by the imbalance of another. Some pumps with vertical shafts are designed so the axial thrust helps offset the weight of the shaft and impeller. Despite the best design, there's no guarantee the pump will not see some axial forces so one end or the other is always fitted with a thrust bearing. If the pump is cantilevered off a single bearing it's also the thrust bearing. As pumps wear the direction of thrust can change so one excellent measure for pump condition is the axial position of the shaft when you can get at it to measure it. Taking initial measurements of how much a shaft shifts along its axis (using a dial micrometer) before it's ever operated provides baseline measurements for bearing wear. Take them anyway if the pump is in good shape then compare them every year or two to check for wear.

Your first clue of potential operating problems with a pump is the shape of the curve. If the curve has a negative slope at all times you should not have any operating problems with it under most circumstances. Slope is a value equal to the change in differential divided by the change in flow at any point on a curve, indicated by a line tangent to the curve at the point you're looking at. If the differential is always decreasing the pump is easy to handle. A lot of pump curves have a positive slope as the flow approaches zero. The curve will have a hump in it where the slope is zero (differential doesn't change) at the top. The curve will have a positive slope (differential decreasing) to the left of the hump where flows are lower.

Anytime you're operating at a point close to or to the left of that hump the pump's operation may be un-

stable. It may be unstable because, for one set differential across the pump, you have two possible flow rates. If the system somehow maintains a constant differential for those two flows the pump will not align with one or the other, switching back and forth between the two points. When a pump does that we call it 'surging' and it's usually accompanied by a lot of fluid noise in the pump and system to inform you it's going on. Multistage pumps can oscillate along the axis of the shaft

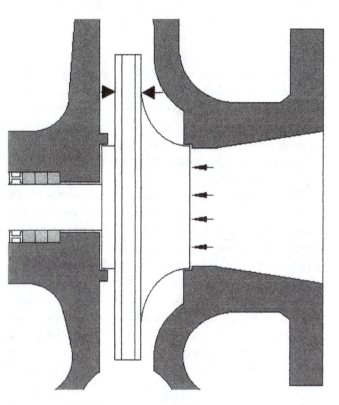

Figure 10-74. Back pressure with wear rings on centrifugal pump

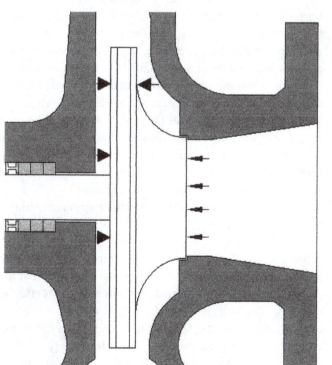

Figure 10-73. Axial forces on centrifugal pump

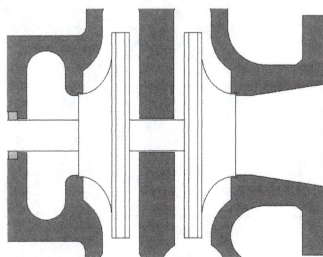

Figure 10-75. Opposing stages of centrifugal pump

when surging and that's another thing to look for when monitoring the operation of a centrifugal pump.

Someone is bound to say they have a pump with that curve shape and don't have a problem with it. I know there's many a situation where the hump in the curve is no problem. That's because the change in flow normally produces a change in pressure drop through the system. You'll remember in the chapter on flow where we found the change in pressure drop is proportional to the square of the change in flow. With that knowledge and some actual operating conditions you can spot the system flow curve on a pump curve to see when the problem of surging will occur.

First you look at the difference in pressure when there's nothing flowing, a piece of data that's not always easy to measure. Then note differences in pressure in the system to find the loss due to flow at some point. Draw a system curve on the pump curve by starting with the difference in pressure when nothing's flowing then add the pressure drop for corresponding flows to continue it.

The curve in Figure 10-76 is a sample of a boiler feed pump curve with a couple of system curves plotted on it. The system curve 'A' is for a normal plant. The system curve 'B' is for a condition with very low system pressure drop between pump and boiler, one with a feedwater control valve that's wide open for some reason. You'll note that there's no one flow rate where the slope of system curve 'A' and the slope of the pump curve are close to each other. The slopes of the pump curve and system curve B are very similar and that's where things get unstable because a change in flow that

increases the pressure drop in the system also rides up the pump curve to increase the pump differential by the same amount.

The rule of these curves is that the operating point is where the system curve and the pump curve intersect. It's the only point where both the pump and system have the same characteristics. If, however, one or the other didn't change then the flow through the system would be constant and we couldn't control the water flow. A control valve somewhere in the system or the differential at zero flow (the point where the system curves intersect the zero flow line) has to change to vary the flow. Picture the system curves being shifted up and down by the operation of the flow control valve and you'll notice how a curve like the one labeled B can hit two points on the pump curve.

If you have a problem with a surging pump this should be a clue to you on how to handle it; simply increase system resistance when operating at the lower loads by throttling a valve someplace. Alternatively, open a bypass line to recirculate fluid so the flow through the pump is beyond the hump of the curve where the slope is negative.

Recirculation of some fluid is typically recommended for centrifugal pumps that can be operated during periods of system flow stoppage to prevent overheating the pump or the fluid. If system flow is stopped the water simply churns in the pump, soaking up all the motor horsepower that is used by the pump in that condition (all inefficiencies) to raise the temperature of the pump and fluid. If the fluid can take the high temperatures it's possible that the heat will distort the pump or weaken the pump shaft until it springs off center or starts rubbing moving parts on stationary ones, and fails dramatically. If the pump can take the heat the next problem is the vapor pressure of the liquid in the pump. Once the temperature exceeds what matches the vapor pressure of the liquid then the liquid will start vaporizing, creating cavitation first, then flooding the pump with vapor.

Operating under shut-off can happen regularly with boiler feed pumps so you'll frequently find a recirculating line on a centrifugal feed pump that returns some water to the deaerator or boiler feed tank. On most jobs the line has an orifice between the connection at the pump discharge and an isolating valve on the recirculation line. The orifice is sized to bleed enough water off the pump to limit the temperature rise when the pump is operating in system shutoff conditions. If another orifice is installed in the piping before the deaerator or feed tank (included in the sizing to prevent pump and liquid overheating) there's an added advantage to these

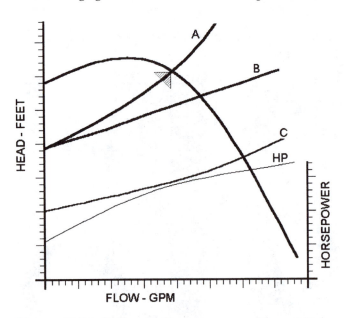

Figure 10-76. Boiler feed pump curve (A and B (no hump, hump, show horsepower)

systems because you can use the recirculating line of an idle pump to bleed some liquid back through it and keep it hot so it's ready to operate the moment it's started.

On the other hand, that's no small amount of water! If the engineer didn't include that flow with the design capacity of the pump you might find yourself short of pump capacity at high loads. However I've only encountered that problem once because the pumps are normally oversized. Engineers usually oversize pumps, including the recirculating flow before applying a safety factor. What that flow represents is a lot of electrical energy to replace steam energy. The power used to pump that liquid heats it up but electric power to do that costs a lot more than the fuel.

This is one place where an operator can reduce power costs. As long as the loads are such that the feedwater valves should always be open, shut off the recirculating line. The pump will back up on the curve, producing a little higher feedwater pressure, and the horsepower consumption will decrease. Open the valve when loads are low and periods of shutoff are possible. You won't save electricity then, but you will save on demand because forced draft fans and other equipment are at lower loads when you reinstate the recirculating feedwater pumping load.

I should mention that there are feedwater systems that recirculate large quantities of water to maintain a constant feedwater pressure or constant differential between feedwater and steam pressure. Sometimes it's nothing more than an engineer's concept of what should be done because the feedwater pressure gets too high as flow is reduced when the pumps have very steep curves. On the other hand the pressure regulation is there to reduce pressure drop across the feedwater control valves because they either can't shut off at the higher differentials or they throttle so much that the valves wear dramatically. I say change the damn feed valves and save electricity but not everyone agrees with me. Another solution is installing variable speed drives but the economics aren't always there.

When starting a centrifugal pump it's common practice to open the suction valve, start the pump, then open the discharge valve. The reason is the pump can't draw any more horsepower than what's used at shutoff during startup, reducing the load on the motor.

I recall one time when a discharge check valve had failed to close on a pump but we needed the pump in operation. When the pump driver stopped the fluid simply flowed backwards through the pump and tended to rotate it backwards. The additional motor load required to reverse the rotation before starting to pump resulted in heavy starting current for too long and the starter tripping.

When the pump is operating under system startup conditions you may have to leave the discharge valve throttled (I know they're normally gate valves and shouldn't be throttled) until system pressure builds. Not all pumps are furnished with non-overloading motors. If a boiler feed pump is running when the boiler pressure is way below normal and the feedwater valve runs wide open it's possible for the motor to overload. Look at that curve in Figure 10-76, you'll notice that the horsepower at the design operating flow (indicated by the little triangle) is less than the maximum. Draw a vertical line at the design flow (point of the triangle) and a horizontal line from where it intersects the horsepower curve to the right to read the pump horsepower at that design condition. It's always possible to pick a motor smaller than the maximum horsepower of the pump. Even though you pump to a higher pressure with curve A you can't pump as much volume as you can with curve B so horsepower is less.

If, however, the pressure in the boiler drops so the system curve is 'C' then the flow can increase considerably and the motor horsepower requirement for the pump at that point so much greater it could overload the motor. If you have a pump with a limited horsepower motor you have to take action to prevent it running out on the curve when the boiler pressure is low. Normal practice is throttling a valve down. Don't count on the throttling of the feedwater valve, it could suddenly go wide open.

There are hundreds of variations in pump construction because of the many different applications. The shape of the vanes in the impellers can vary from highly efficient backward curved to radial depending on desired efficiency weighed against the solids in the liquid they pump. They can be rubber lined for such purposes as pumping a slurry of limestone or ash. They can be "canned" where the rotor of the motor is sealed in an enclosure with the pump to prevent the leakage of hazardous liquids. The most common arrangement is the horizontal split case pump (Figure 10-77) but the ANSI pump (so called because the National Standard establishes fixed mounting dimensions so all manufacturer's pumps are interchangeable) is gaining popularity. They're end suction pumps that require the piping be disconnected to get to the pump for maintenance.

Turbine Pumps

When I say turbine pumps some people get the impression of a centrifugal pump powered by a steam

Figure 10-77. Horizontal split case pump

Figure 10-78. Turbine pump impeller

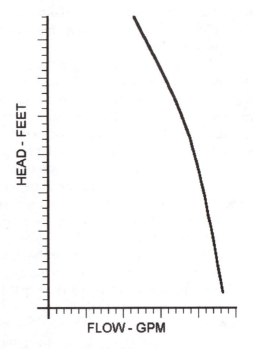

Figure 10-79. Turbine pump curve

turbine. That's not the case. A turbine pump is a type of pump and although they exhibit some characteristics comparable to a centrifugal pump they differ. The turbine pump grabs the liquid on the outer diameter of the impeller, spins it around inside the pump and heaves it out the discharge. A turbine pump impeller looks like the one in Figure 10-78 with little slots all around the outside. The fins formed by those slots is what grabs the liquid and whirls it around inside the pump casing until it gets to the discharge.

Turbine pumps can produce very high differential pressures because they act more like a positive displacement pump than a centrifugal. The typical turbine pump curve (if you got to see one that showed all conditions from zero flow) looks like a centrifugal pump curve but most of the curves you get look almost like a straight line with a steep negative slope (Figure 10-79). Since they operate more like a positive displacement pump you should treat them like one. Don't start a turbine pump with the discharge valve closed.

Turbine pumps are commonly used as boiler feed pumps, especially on low pressure steam boilers. Their steep curves permit them to handle the significant variations in boiler pressure without any effect on pump capacity. I've run into many a plant with centrifugal pumps that also have curves so steep that their flow isn't altered significantly by changes in boiler operating pressure.

I don't care for centrifugal feed pumps in heating plants because they can't handle the pressure variations. Take the typical heating boiler plant. Both centrifugal and turbine pumps can be obtained to produce a design flow of about 31 gpm (15,500 pph) at the normal boiler operating pressure of 12 psig (31.7 feet). The density of water for this example is assumed to be 54.55 pounds per cubic feet, 175°F water, which means the head relationship is 2.64 feet per psi. There is a big difference in their operation as the pressure changes. They're selected for when the boiler runs up to the limit of the safety valves (15 psig or 39.5 feet).

My concern with using centrifugal pumps is that any external pressure effects can result in total loss of water delivery (Figure 10-80). The curve as shown will deliver water to the boiler but changes in such things as level in the deaerator or feedwater tank can prevent delivery. The values of head used on this curve assume that the pressure drop through the piping is negligible and the level in the boiler is the same as the level in the feedwater tank. From this curve it's apparent that a drop in level at the feedwater tank of a couple of feet will increase the head requirement for the pump to the point that the centrifugal can't deliver any water until the boiler water level or pressure drops enough to produce a differential the pump can overcome.

On the other hand, any drop in boiler pressure will be accelerated by a centrifugal pump with a relatively flat curve. If the boiler pressure drops to 8 psi, a typical

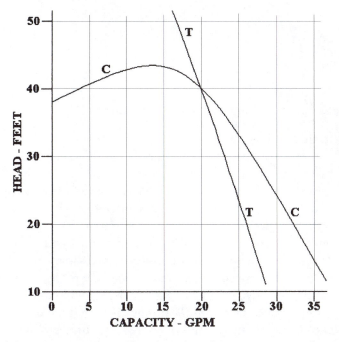

Figure 10-80. Centrifugal and turbine pumps on low pressure boilers

occurrence with a heavy load, the turbine pump output will only increase a little bit but the centrifugal pump will increase its delivery over twice as much. That additional water consumes more of the boiler's heat input leaving less to make steam so the pressure drops further. I think this shows that a poor choice in boiler feed pump selection on low pressure boilers can produce serious headaches for the boiler operator. Replacing those centrifugals with turbine pumps can reduce the swinging pressure problems encountered in some plants and eliminate others because the pumps create the problem.

Screw and Gear Pumps

Screw and gear pumps are used principally for fuel and lubricating oils. They can be more efficient at moving liquids with viscosities higher than water than other types of pumps and are capable of producing high differential pressures in a small package. Since they're positive displacement pumps one running at 3500 rpm can be half the size of one running at 1750 rpm to pump the same amount of oil.

Screw and gear pumps are positive displacement pumps and work pretty much alike. The pumps use two or more machined rotors that mesh closely together and produce a moving cavity as they rotate with each other. The cavity opens at the suction end and is sealed as the rotors turn then the cavity travels to the discharge end of the pump to deliver the liquid at the discharge pressure. The liquid serves to lubricate the rotors to prevent them rubbing each other or the pump casing. The ends of the rotors are enlarged to increase bearing surfaces to balance the axial forces or shaft bearings take the thrust.

Some liquid is squeezed between rotors and casing in the opposite direction of the moving cavity, the amount depending on pump construction and wear. Smaller pumps usually have one rotor or gear that is driven and the rest of the rotating parts are driven by it in turn. Larger pumps and pumps that produce high differentials or pump very low viscosity liquids can have external gearing so each rotating element is driven. That reduces the amount of force that has to be transferred through the thin film of liquid between rotating parts, replacing it with the lubrication of the external gears.

The quality of internal lubrication is dependent on differential pressure and pump speed. If the liquid is very viscous it will maintain a stronger liquid film between metal parts to prevent them rubbing. As the viscosity decreases the film gets thinner and will break to allow the metal parts to touch. The fluid bleeding back through the spaces between the metal parts is what provides lubrication. The differential pressure between each adjoining cavity pushes the fluid through so it wedges its way between the part. If pressure differentials are considerably lower than design there may not be sufficient differential to force the lubrication of the pump. If the pump speed is too low it won't generate that wedge effect as well so other factors like the viscosity of the liquid have to aid in lubrication.

The typical pump used for pumping heavy fuel oil will not effectively pump light fuel oil and may even fail if used to pump light fuel oil. Some people argue that a heavy oil pump is worn by the ash and sediment in the oil so the gaps between rotors and casing have increased. However, the truth of the matter is the pump's design and speed were established for heavy oil and don't work well on light oil. The lower the viscosity the faster the pump has to run.

Figures 10-81, 10-82, and 10-83 are the typical forms of screw and gear pumps used in boiler plants. A common gear pump consists of two gears in a casing. Usually one is driven and the other is an 'idler.' We use the term idler to imply it doesn't transmit power to anything else, not that it's lazy. The teeth of the driven gear engage in the teeth of the idler and they counter-rotate. Let's start with the gear pump in Figure 10-81. The liquid enters the pump where the gear teeth are disengaging, is trapped within the cavities formed between the teeth and casing and is carried to the discharge side of the pump where it is forced out as the two gears engage, filling the cavity the liquid was in with a tooth of the other gear. A sectional view in the other direction would not reveal much. The sides of the gears are flat and just clear flat sides of the pump casing. The view in Figure 10-81 shows all that's relative to the operation of the pump. Volumetric capacity of the pump is affected by the size, length of the gear teeth, and speed of rotation.

The crescent gear pump in Figure 10-82 simply traps the liquid between the gears and the crescent shaped piece of the housing. The inlet and outlet ports are outlined. Either of these pumps will pump the fluid in either direction.

The design capacity of a gear pump can be determined by calculating the area of the space between the casing and the root of the gear teeth, then multiplying that by the radius at the center of the teeth, the percent of the rotation where the liquid is trapped and the rpm times two to account for each side. The actual capacity will always be less because some of the liquid has to leak back past the teeth and the ends of the gears to lubricate the pump. In many of these pumps the spacing between the casing and ends of the gears is adjustable making them suitable for different viscosity fluids

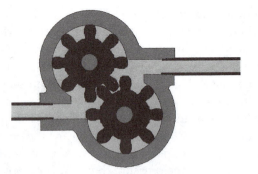

Figure 10-81. Gear pump

Figure 10-82. Crescent gear pump

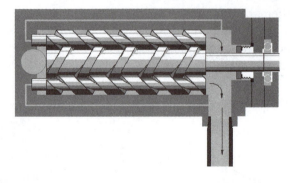

Figure 10-83. Screw pump

through adjustment.

The cavity in a screw pump (Figure 10-83) is formed by the intersection of the rotors and closed by the casing housing the rotors. The pump shown is supplied with two idler rotors that increases it's capacity without an appreciable change in size. A smaller pump can be had with only two rotors. Liquid enters the pump at one end of the rotors, fills a cavity that opens as the grooves in the rotors separate, is trapped between the casing and rotors as the grooves engage, then travels along the rotors to the discharge end of the pump. That movement and the difference between suction and discharge pressures produces an axial thrust on the pump that has to be opposed by the bearing of the driven rotor and the fluid film between rotors plus the end of the idler rotor bearing against the casing. Some manufacturers use an enlarged end on the rotors to increase bearing surface. Other schemes include balancing lines between suction and discharge ends applied to balance pressure forces. Another scheme is opposite hand ends of the rotors so they draw liquid from both ends and discharge in the middle to balance the hydraulic pressures almost completely.

Screw and gear pumps do not do a very good job of pumping compressible fluids. An oil pump can easily get air bound where there is a sufficient volume of air or vapors at the discharge and inlet to expand and contract as each cavity between rotors is opened and closed, thus preventing any flow through the pump. The air also leaks back just like the oil. It's not uncommon for the pump to generate a loud audible roar when air or vapor is trapped in it because the air or vapor doesn't do a very good job of lubricating the pump and it forms bubbles in the oil as it leaks back.

Operating a screw or gear pump with a vapor trap for any extended period of time will ensure complete breakdown of the film of lubricating oil on the rotating parts with subsequent damage to the pump as the metal parts start rubbing. It's necessary to vent them to eliminate compressing air or vapors in the pump that will prevent liquid entering. Properly vented the pump will move air to eliminate it from the suction piping.

When starting a dry pump (filled with air) it's important to ensure that lubricant film is maintained. Making certain some of the piping is full of oil that will be drawn into the pump is important to limit wear. The oil is also a sealing film that helps the pump trap the air in its cavities and push it through. The best way to do it is to fill the suction strainer with oil, shutting down at regular intervals and repeating the process until all the air or vapor is pumped out. Once you have a suction line full of liquid the pump will work.

Pump Control

There was a time when the only control we had over the operation of a pump was to turn it on or turn it off. That's still a common means of controlling the pumping of liquids, used almost exclusively for feeding low pressure boilers and returning condensate but modern technology has expanded our abilities and the wise operator should know how to utilize those methods. Note that this involves controlling the pump to control the flow. Before variable speed pump control we typically controlled the fluid flow to maintain operating parameters. Many times that meant recirculating the liquid that was pumped, wasting the energy we used to get the liquid up to pressure, but a necessary means of controlling the flow.

When we're dealing with on-off pump control there are opportunities to improve that method of control to reduce energy costs and wear and tear on the pump. An attitude of limiting the number of starts by extending run time is one you should adopt. I'm not talking about recirculating liquid to keep the pump running, that saves starts but also wastes energy. It is an option you should consider if the pump has extremely short off cycles where it may run for ten minutes then shut down for five to ten seconds. If that's the case then recirculating some liquid to keep it running past those short off cycles will save on pump and motor wear and reduce wear and tear on the starter as well.

Every time the pump is started the entire assembly is subjected to stresses above and beyond the normal operating conditions. Motor current is five to ten times normal operating current when the pump is started and those high currents produce rapid heating of the motor windings with attendant thermal stresses and also high magnetic forces that can dislodge the windings. The run then stop and run then stop operation is also rough on bearings, both in the motor and in the pump, because the bearings will heat up and cool down with some breathing that can increase the probability of air mixing into the grease or oil to corrode them (see Chapter 6, Lubrication). The pump always experiences pressure spikes when starting because the liquid in the connecting piping has to be accelerated from its stationary position and the check valve has to be lifted. By reducing starts you'll reduce the strain on the equipment to extend its life.

Reduce starts by stretching the pump's on-off settings as far as you can. Let the level get a little lower and run the pump until it's a little higher by adjusting the level controller. There are limits to this, including allowing the level in a boiler to get so low that a little upset results in operation of the low water cutoff. All that really does is tell you where the low limit is.

I doubt if you can take advantage of that to get something replaced. I ran into one operator that was starting and stopping a pump repeatedly keeping his hands on the enclosure with his two thumbs on the start and stop buttons. When asked about it he responded "I'm trying to blow this damn thing up so they'll give me one that works." On a repeat visit to the plant I noticed a new identical pump had replaced it. The operator was still grumbling because he got the same make and model pump back so it still didn't do what he wanted it to do.

As for controlling the flow through a pump automatically with modulating capability, it isn't done consistently. The only pumps that can provide modulating control are chemical feed pumps which use a reciprocating hydraulic pump acting on a diaphragm with an adjustment of the stroke of the reciprocating section. That's what the knob is for on the pump in Figure 10-70.

The centrifugal pump, which serves the majority of applications, is self aligning so the flow through it is determined by the system. By throttling the flow in the system at some point, preferably after the pump discharge, the differential pressure required to force liquid through the system increases and the flow through the pump decreases as it follows the differential up the pump curve. Applications with some centrifugals and most screw and gear pumps normally incorporate recirculation control where the flow through the pump doesn't change and a portion of that flow is diverted back to the pump suction to achieve a final control of delivery pressure.

Advances in motor speed control have made some pump control projects possible that were not possible before. They are limited; varying the speed of a centrifugal pump with a relatively flat performance curve doesn't produce much of a savings in horsepower above what the pump automatically provides. A pump with a nearly flat curve will supply a reasonably constant differential pressure automatically so there is no need for control. If the pump has a very steep curve varying the speed will save power costs and allow differential pressure control. If the differential required for varying flow also varies with load then some potential savings by controlling pump speed is possible, even with pumps with nearly flat curves.

Application of variable speed drives on pumps has resulted in significant operating cost savings because (1) a pump is typically oversized like every other piece of equipment, (2) systems seldom operate at the design capacity, and (3) it permits tuning the pump to operate most efficiently under different load conditions. Don't be surprised, however, if you encounter a pump with a VSD control that's simply operating at it's highest speed. It's not unusual for the system to have problems that were not anticipated by the designer or other operators that simply don't understand the application and, therefore, solve problems by eliminating the automatic control and running the pump full out.

I have one customer that manages to control the drum level in each boiler by controlling the speed of the respective boiler feed pump and it's successful despite many engineers telling me it can't be done. I would have said the same but came to understand how it works by observing that customer's system operation. Note that

this requires an independent feedwater pump and supply line for each boiler and will not work with a pump with a relatively flat curve. When the boiler is operating at low loads, recirculating some of the feedwater back to the deaerator, especially on boilers with economizers, will eliminate problems with surging of the pump and flashing of the feedwater in the economizer.

A more common means of boiler feed pump speed control consists of maintaining a constant header pressure. That can be enhanced, and should be, by actually maintaining a differential between the feedwater pressure and the steam pressure. It's invaluable for good control during plant startup and in systems with varying steam pressure. The control system should also incorporate a set point adjustment relative to load to account for the pressure drop in the feedwater piping and in the steam piping before the pressure sensing point.

Where I've seen systems that simply failed to work the major problem was with piping friction losses and most of those were in chilled water distribution systems. Application to systems with three way bypass valves on cooling equipment have to be changed to simple two way valves to produce a change in demand that can be sensed. The three way valves simply redirected flow so there was no way to reduce system flow. Problems also appear when significant differences between actual operating conditions and design conditions are introduced by such things as buildings shutting down or varying significantly in load. A common significant shift in load is an auditorium or gym where cooling is required for the high concentration of people and/or activity on a cool or cold night when the other buildings don't require cooling. That's usually solved with application of a number of differential pressure transmitters in the system, artfully applied so the pump speed controls can operate to maintain a differential at the lowest measured differential in the system.

Another potential problem is a pump was selected carefully for the design condition and the efficiency of the pump drops off dramatically as speed is reduced. The best one being to operate equipment that matches the load and stretching out operating cycles.

In those cases I tend to look for a replacement pump because it isn't hard to find one that wasn't the lowest first cost pump but, in terms of life cycle costing, the most economical pump for that application. If it doesn't seem to be working, or creates problems occasionally, seek out some help with the application. Otherwise, do what makes sense to achieve an optimum economical operating state for the pump.

FANS AND BLOWERS

Fans and blowers are used to move gases (compressible fluids) around a boiler plant. In many cases I will use the terms "rotating equipment" or "fluid handling equipment" to include pumps, fans, blowers and compressors without regard to the fluid or the from of the equipment because they all do the same thing, move a certain volume of a fluid and add energy to it to permit it to flow through the rest of the system. For every design of pump there is a comparable design of fan, blower or compressor. Be sure to look through what I've written on pumps; it will improve your understanding of fans and blowers.

Differences in the equipment are related primarily to the different densities, temperatures, and viscosities of the fluids the equipment handles and the effect the equipment has on the fluid. Fans and blowers are used to move compressible fluids, basically gases, not compress them. That's what makes fans and blowers differ from compressors which we'll cover a little later.

Even though they aren't designed to compress a gas fans and blowers do manage to compress the fluid slightly. In most cases we ignore the compressive effects because the density of the fluid does not change significantly. As the differential pressure of a fan or blower increases compression becomes more significant. There's a very gray line between blowers and compressors with no clear definition of when, specifically, one becomes the other. A fan, on the other hand, is almost never capable of compression.

The difference is principally intent. If we intend to compress the gas it's a compressor, if we don't it's a fan or blower. As for whether a particular piece of centrifugal equipment is a fan or a blower, that's also a gray area. A centrifugal pump can pump a gas; it doesn't produce much differential but it can do it. If you look at any centrifugal pump, fan, or blower their construction is pretty much the same and the dynamics that allows them to move fluid is the same.

These 'centrifugal devices' will all perform according to their performance curve regardless of the fluid that passes through them. The differential pressure they produce is directly related to the tip speed of the impeller and the density of the fluid because the impeller vanes throw the fluid and the pressure produced is related to the weight of the fluid flowing at a velocity related to the tip speed. You could take a centrifugal pump curve and realizing the differential head of the pump is feet of fluid, convert to determine the inches of water differential pressure it would produce while pumping air. A fan curve could be used to calculate the differential it would produce if pumping water. The problem is the denser water would produce so much load on the fan that it would break or the motor overload before it actually pumped any water.

So, a lot of the rules for pumps apply just as well to fans and blowers. Oh, there are differences, we're not as particular about some air leaking out of a fan so there's seldom any kind of shaft seal and, because the density of the fluid is so low, fans and blowers can get a lot larger than pumps in order to handle enough volume to deliver the pounds of air or other gases that have to be moved. The typical application of a fan or blower also doesn't involve raising the pressure of the fluid to move it into a reservoir at a higher pressure; the differential pressure in a system at zero flow is typically zero for a fan or blower because the pressure at the far ends of the system are the same. The system curve always starts at zero differential at zero flow. When it doesn't, the device is a compressor.

Propeller Fans

This prompts a question, why aren't there many propeller pumps in a boiler plant? If you ever bought gas for a day of running around on the water in a motor boat you would know why; they're not that efficient. Propeller fans have a niche in the world because a propeller can move air effectively as long as it doesn't have to produce any significant differential pressure. If you haven't installed some ceiling fans in your home to take advantage of the cooling effect they produce by simply moving air in the summer, you should.

The blades of a propeller fan simply push the air along and add some spin to it (Figure 10-84). Housings around the propeller can redirect the flow to eliminate

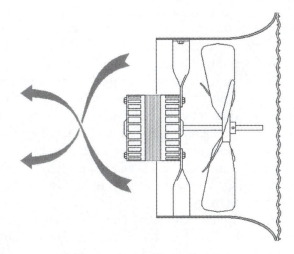

Figure 10-84. Propeller fan

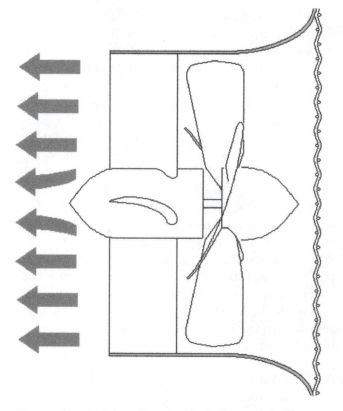

Figure 10-85. Propeller fan housing with flow redirected

some spin and make them more efficient (Figure 10-85). Propeller fans are primarily limited to ventilation services in a boiler plant although they were used in the middle of the last century for forced draft and induced draft service when differentials were low.

Some key things to know about propeller fans include the fact that they readily overload their motors if the system doesn't produce the design resistance. I remember visiting a job site in a synthetic fiber plant where a contractor had several propeller fans simply sitting on the floor and running with temporary wiring. We were informed that they were testing the fans because the motor on one of them failed and now they're finding more of them are failing. Luckily I was wise enough to pull one of the instruction manuals out of the envelope attached to a lifting eye and read enough to learn the fans were designed for a two inch differential. The drawings showed the installation would produce that but the fans sitting on the floor just blowing air were operating with no appreciable differential. I suggested they quit testing them immediately because they were destroying them by overloading them.

In turn I was lectured by one of the contractor's engineers that they couldn't possibly be overloading because fan horsepower is equal to the capacity in cfm

times the differential in inches divided by 6356 and, since the differential was zero, the horsepower requirements as they sat there on the floor should be negligible. Since I had the instruction manual in hand and it clearly stated that the fan had to be installed and the differential has to be at least 80% of the design value he agreed to have an electrician check the motor current. The motor current was three times nameplate rating and that's why they kept burning them up.

It was later, when I examined one of my engineering books, that I discovered the reason for the problem. The differential pressure in the horsepower formula is total pressure, a combination of static and velocity pressure differences. Those fans had no static difference but the velocity pressure was there and a lot higher because, without the static resistance, the fan could force more air through to produce a higher velocity and, therefore, a higher velocity pressure. The increased flow and velocity pressure added up to produce the high horsepower that overloaded the motor. This is a lesson for testing any electrical device, that contractor had simply wired the fans to a welding connection in the plant. No starter, no overload devices, is it any wonder he was burning up motors?

Fans, like pumps, have a theoretical horsepower. From the story I just told you know that it's the total pressure across the fan that has to be used. The formula is cfm times total pressure divided by 6356. If all you can measure is the differential you can calculate the velocity pressure. Divide the cfm by the area of the fan discharge to get velocity then look up the velocity pressure. Add velocity and static pressure differentials to get total pressure. Don't have a table? The velocity is the capacity in cfm divided by the area of the outlet. Divide the velocity by 4005 and multiply the result by itself to get velocity pressure. Add it to the static to get total pressure.

Many fans and blowers are belt driven. The use of belts will allow an engineer to pick a fan for optimum speed for a given application because any speed can be established by the proper mix of motor speed and size of sheave (those pulleys the belts run on). In some cases one of the sheaves is adjustable to permit field adjustment of the speed. All these features are very valuable for HVAC equipment where the flow is constant and the fan can be tuned to achieve the precise required flow without chewing up added energy with dampers. They're not as valuable in a boiler application where the air flow is varied.

Another advantage of belts is they can slip on startup to reduce the startup load on the motor, something to let go until you've checked the instruction manual. Belts are typically provided to the degree that one belt

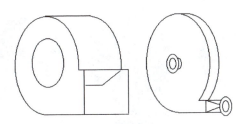

Figure 10-86. Centrifugal fan shape as opposed to blower

can break and the rest can still carry normal loads. The problem is that the one belt that breaks usually gets tangled with the others with complete failure. I don't like belt driven fans and blowers and believe that there are a sufficient number of choices of fans at standard motor speeds to use direct drive fans on boilers. With the growth of variable speed drives where we can run a fan at any speed we choose we don't need belts. I'm definitely opposed to belts because they're a maintenance item and produce unnecessary radial loads on fan shafts and bearings.

Centrifugal Fans and Blowers

The obvious question is, "what's the difference?" The answer is, I'm not entirely certain. I tend to look at a centrifugal fan or blower and call it one or the other depending on the relationship of width and diameter. When one is as wide, or wider, than the center to scroll distance at the discharge I call it a fan. When it's obviously narrow I call it a blower. So the two shapes in Figure 10-86 are fan on the left and blower on the right.

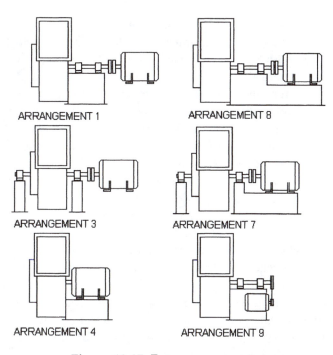

Figure 10-87. Fan arrangements

In more general terms, blowers produce significantly higher differential pressures than fans. Neither of those rules works every time and I'll call something a blower when the people in the plant call it a fan and vice versa. There are few times that happens so the two rationalizations I've developed usually work. One other label you'll run into is the term "exhauster." When most of the pressure drop in the system is incurred before the fan inlet they tend to be given that label. Primary air fans on pulverizers are commonly called exhausters.

Centrifugal fans are used in so many applications that standards have been developed to describe their construction. The different 'arrangements' which relate to bearings and motor connections are defined in Figure 10-87. The motors for arrangement 1 and 3 fans aren't left hanging in the air, the graphic only indicates that the fan manufacturer is not expected to provide anything to support the motor.

Discharge locations are shown in Figure 10-88. These are based on viewing the fan or blower as if you were sitting on the motor or the fan's sheve. You'll also note that the rotation can be determined by simply looking at a fan's discharge position. Strangely enough I've encountered fans operating with the wrong rotation, some for several years. Centrifugal devices will work with either rotation, only

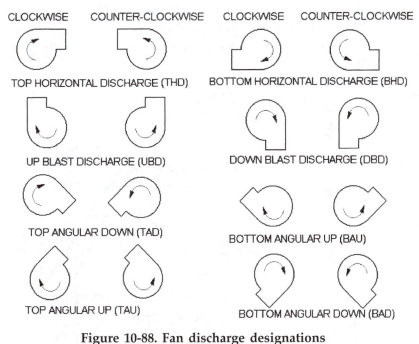

Figure 10-88. Fan discharge designations

difference, is one way works better.

You'll also encounter some definitions on width and number of inlets. I'm sure you have seen single inlet fans where air enters one side, but there are also double inlet fans where air enters both sides. They are defined by simple abbreviations with SWSI (Single Width, Single Inlet) being the most common and DWDI (Double Width, Double Inlet) where the air can enter both sides used in many applications from little convectors (those fan powered heating and cooling units mounted under windows in many buildings) to large forced draft fans. Please don't ask me to explain the width business, I just look at the fan and decide whether to call it single or double based on the ratio of wheel diameter to width and that's all.

Instead of calling the primary rotating element an 'impeller' we call it a 'wheel.' The term scroll is applied to the casing because the radius increases from the cutoff to the discharge. A casing is still a casing and many other labels are consistent with what we use for pumps. The cutoff is the portion of the scroll that's closest to the outside diameter of the wheel. It's where the swirling fluid in the fan is cut off so it heads out the discharge instead of riding around with the fan wheel. The inlet bell, is that specially formed section that connects the fan inlet to the inside diameter of the wheel. Makes sense, because when you take it out and set it on the floor it does look like the bottom of a big old church bell. Small fans won't have an inlet bell, only a hole in the casing that faces the wheel.

There are some additional gadgets that are not found on pumps because fans usually don't have seals or packing glands, although they are used on occasion. We have 'heat slingers' that are like little fan wheels located on the shaft outside the fan to draw cooling air over the bearings and protect them from hot gases and the heat that conducts along the fan shaft. Instead of strainers a fan will be protected by 'inlet screens' which keep sticks and stones out but not dust.

Dust is, therefore, something an operator has to keep in mind. Keep it in mind for two reasons; because it can damage the fan or hinder its performance and that dust can be converted from large harmless sizes to much finer particles that are injurious to human health after they pass through a boiler.

A certain amount of dust will be struck by the blades on the fan wheel and trapped there, accumulating until they form a rather thick layer if they aren't cleaned. The accumulation will tend to reduce the fan capacity. The bigger problem, however, is that once it reaches a certain level it will suddenly start breaking off. Losing a fair sized accumulation of dust on one

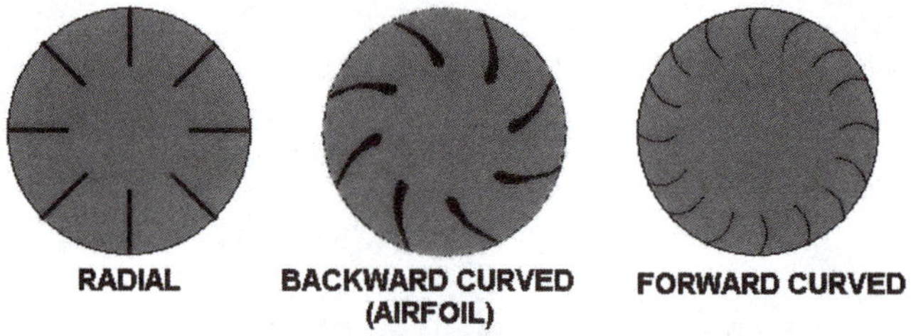

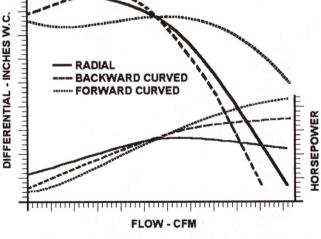

Figure 10-89. Different shapes of fan blades

blade will generate an imbalance in the fan wheel that adds load to the fan bearings, a variable shock load. If that's allowed to happen you can have everything from shaft distortion to where the fan wheel hits the inlet bell, cutoff or casing. You should clean every fan, or have it cleaned, during the annual inspection. Some forced draft fans or what's below them in the ductwork can't tolerate a water wash so you'll have to limit cleaning to brushing and vacuuming. Be sure to do the inside of the scroll too because the dust is thrown at it.

Centrifugal fans and blowers are used more than any other device for moving air. In order to accommodate a variety of applications they are also supplied in a significant variety of configurations. Three principle variations involve the shape of the vanes or blades in the fan. A fan is called 'backward curved,' 'forward curved,' or 'radial' depending on the shape of the blades as shown in Figure 10-89. These three shapes produce significantly different fan curves as shown in Figure 10-90.

Most applications in a boiler plant use backward curved or radial bladed fans because they are more efficient for the operating condition when backward curved and, in the case of radial blades, do not accumulate solids on the blades in operation. Radial bladed fans are

Figure 10-90. Fan curves, BC, Radial, FC

used almost exclusively for application as induced draft fans and primary air fans for coal pulverizers. You'll discover that most air conditioning and ventilation systems use forward curved fans because they are more efficient at delivering large volumes of air at low differential pressures. It's important to note that forward curved fans have a very stretched curve and it's not at all uncommon for the motors on those fans to be overloaded if nothing restricts the air flow.

Blowers will have radial blades or backward curved blades depending on the application and can experience the same problems with surging that was discussed with centrifugal pumps. That surging will also occur for the same reasons. It's seldom encountered in fan and blower applications but is frequently encountered when compression is involved.

An important thing to remember about these fans is that they're centrifugal devices, the differential pressure that's produced by them is a function of the flow through the fan and the density of the gas flowing through. When the gas is colder a fan will produce a higher differential pressure (in terms of inches of water) and, because it's moving denser air, more pounds of gas. When a fan is handling gases at higher temperatures they will not produce as high a differential and move fewer pounds of gas. In many cases the motor on an induced draft fan is not big enough to handle cold air because the power requirement is significantly higher when pumping cold air. That's why you have to be careful when starting a boiler with an induced draft fan to ensure you do not overload it. Once it's pumping hot flue gas the load on it drops off.

That's also justification for not getting excited when a boiler can't produce full load in the summer time. If an F.D. (Forced Draft) fan is installed to collect the heated air in the top of the boiler plant it will not pump as much air in the summer, when the temperatures are about 125°F to 130°F but it will in the winter when those temperatures are 50° to 60° lower. It is important to realize that the fan is moving less air (pounds of it) in the summer to reduce excess air because you might be running fuel rich. If the boiler is summer tuned then you'll find that excess air is higher in the winter because the air is denser than when the boiler was tuned.

Rotary Blowers

Rotary blowers don't resemble fans, the same construction is used for compressors and the main reason for rotary blowers is to produce high differentials that are necessary for material transport systems. Probably the only time you'll see a rotary blower in a boiler plant

is when it's used to provide air for ash or coal transport systems which require some rather high differential pressures. See the following discussion on rotary compressors for more information that would apply to blowers.

Fan and Blower Control

Control of the flow of gases in systems with fans and blowers is typically achieved using devices we call dampers that are a leaky version of valves. Sometimes the system uses valves or their equivalent when leakage is not acceptable. Dampers are not the best method for controlling air flow because they are typically made to be inexpensive and there isn't a linear relationship (see controls) between the damper position and air flow. Opposed blade dampers (Figure 10-91)[11] provide a better relationship than parallel bladed dampers. The different

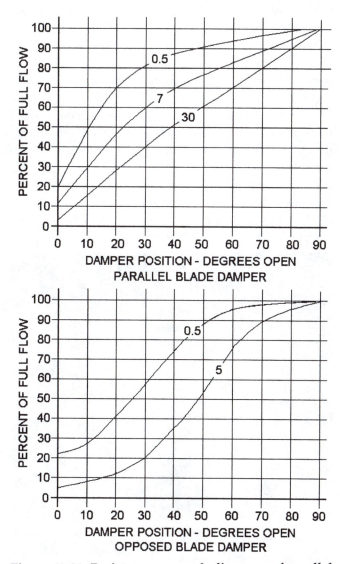

Figure 10-91. Resistance curves & diagrams of parallel and opposed blade dampers

curves relate to the damper's wide open pressure drop divided by the maximum system differential pressure.

In the most common fan application that requires air flow control, forced draft fans, variable inlet vanes are typically used to reduce fan horsepower requirements. Variable inlet vanes (VIVs, Figure 10-92) on the inlet of a forced draft fan not only act as dampers but also put a swirl on the air as it enters the fan. By turning the vanes in a way that puts a twist on the air entering the fan the air is rotated in the direction of fan wheel rotation. The inlet vanes reduce fan motor horsepower because they swirl the air so the fan doesn't have to. The reduction of fan motor horsepower attributable to VIVs is indicated in the curve in Figure 10-93. Note that the air has to be turned in the direction of fan rotation, if you manage to reverse the vane positions when replacing that assembly the horsepower could be much higher, so much that the motor will overload. VIVs are fine for boilers operating with a maximum four to one turndown but they usually leak enough air when closed that they're not adequate for higher turndowns. Some applications use a discharge damper in addition to the VIVs to extend turndown.

Today we have VSDs (variable speed drives) sometimes called VFDs (variable frequency drives) that permit an almost infinite control of fan speed and, therefore, the air or gas flow. I installed my first ones in 1989 on the forced draft and induced draft fans of a three-fuel boiler, and have been in love with them since. On that job I included braking resistors but discovered we can really run a boiler without them. I knew the resis-

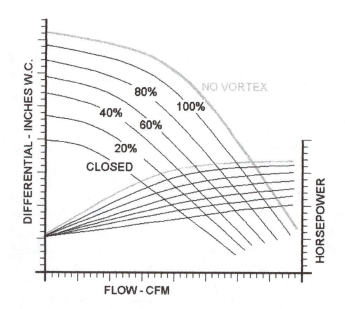

Figure 10-93. Fan curve, effect of variable inlet vanes

tors worked because they crackled and popped as they heated up and the only time they came into service was during setup when the controls were hunting a little trying to establish a fan speed.

When we started that plant up we discovered we could have put in a power feeder half the size necessary to operate two across-the-line started fans. When the boiler was at low fire the combination of 50 horsepower forced draft fan and 125 horsepower induced draft fan along with all the controls and lights pulled a total of 5 amps! That has to be compared to a full load motor rating of 218 amps. Any installation I design will have a VSD on the fan and a positive shutoff damper that's closed when the boiler is shut down to limit off cycle losses and rapid cooling of refractory by cold air.

Ejectors and Injectors

You've probably used a water hose to sweep down a floor at one time or another so you know the principle of ejectors and injectors by observation. The force of the fast moving water is capable of pushing a lot of additional water along. What happens is the high velocity is converted to pressure that pushes the rest of the water. When the motive fluid (the one going through at high velocity) is steam or air it has less mass to contribute to the pressure but it's traveling at a much higher velocity so it can do almost as much work. We occasionally refer to these devices as jet pumps.

Ejectors are used to produce lower pressures at their inlet (suction) by pushing a fluid along. The common use of an ejector is to produce a vacuum by pumping air, and sometimes water, out of a closed system.

Figure 10-92. Variable inlet vanes

They're commonly used to produce a vacuum in a condenser. Another common use is to remove condensate and rain water from underground vaults containing steam piping. An ejector with a float actuated steam shutoff valve is the least expensive means of automatically clearing water from underground piping vaults and they're quite reliable.

When ejectors are combined, or staged, as for a condenser ejector (Figure 10-94) they can produce an almost pure vacuum. The steam to the jets (C) entrains the air drawn from the condenser at (A) accelerating it through the venturi (B) to the first stage condenser (D) where the steam is condensed by the condensate pumped up from the condenser (J). Another jet draws the air from the first stage, accelerating it to a higher pressure through the venturi at (E) then into the second stage condenser (F). After the steam from the second ejector is condensed the air is vented into the boiler room a (G). The condensed steam drains from the second stage condenser through a liquid trap (H) into the first stage condenser. The liquid trap separates the two different pressures, the second stage being around atmospheric and the first stage being something in the range of 8 to 20 inches of mercury vacuum. The combined condensate in the first stage drains to the condenser through another liquid trap. A steam powered ejector can also lift water out of a vault even when it's hot and flashing because it will pump the flash steam.

Injectors are the same device but used to produce higher pressures at the discharge. You will normally see an injector on a coal fired boiler (Figure 10-95) to provide an emergency means of feeding water to the boiler in the event power is lost to the boiler feed pumps. Yes, you can use the boiler's steam to generate a higher feedwater pressure to feed the same boiler. The heat energy of the steam is converted by the injector to mechanical energy to pump the water.

Ejectors and injectors have limited use because they use a considerable amount of energy compared to pumps, blowers, and compressors and are only suitable for moving small volumes of fluid. Their use is limited to operations where there is little flow (condenser vacuum), or small flows and/or no electricity available.

I call an ejector or injector that doesn't boost the pressure or create a vacuum an eductor because it simply teaches the fluid where to go. They basically move water and are principally used to mix two fluids.

Compressors

Compressors are, of course, used to compress compressible fluids, mostly what we call air and gases. It's possible to compress a liquid a little but most compressors will simply break if you try to compress a liquid with them. That sounds like a simple and straightforward statement but I know a few operators that have tried to compress water or lubricating oil with devastating results.

Compression is simply packing more pounds of a fluid into a certain volume. A simple example is pushing fluid into a container. Since none of the fluid leaves the container and we keep putting more in each pound of fluid we add has to share the space with what's already there and there are simply more pounds per cubic foot every second we continue compressing the fluid into that space. When I say fluid I can mean a liquid or a gas, both flow. The distinction for gases and liquids is that liquids aren't what we would call compressible.

For most compressor operations there is some fluid

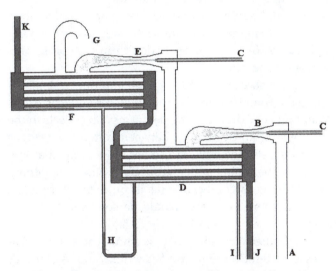

Figure 10-94. Dual jet ejectors for a condenser

Figure 10-95. Feedwater injector

leaving the container as we press more in but the two flows do not have to match. A control air compressor may run five minutes to fill a compressed air storage tank with enough air to supply the system after that tank for a half hour or longer. That should help explain why most compressor operations are on-off. The fluid stored under high pressure will expand to produce flow for the system, the fluid flows out of the container as it is used and fewer and fewer pounds remain in the container. The container or storage tank serves as a reservoir for the fluid required by the system and the compressor refills the reservoir when the fluid level drops to a preset value.

Specific compressor operations require special consideration because the fluid being compressed may contain other fluids or contaminants that interfere with or require consideration in the process. When compressing air we also pack in the moisture that's in the air, the humidity. Since we're packing molecules of air into smaller and tighter spaces the water vapor in that air is subjected to higher pressures so it condenses to form liquid water. Since compressors don't run well on liquids we have to remove that water.

We also don't want the water in our system because the combination of air and water is very corrosive. Water must be drained where it forms and collects in the compressed air system, in between compressor stages and in the storage tank. It also has to be drained at low points in the piping system, especially where the piping goes through a colder area (as in outdoors during the winter) where the water would be condensed by heat loss.

Despite what some people think, coolers on compressors aren't there to condense the water. As long as the water remains a vapor it acts just like the air and does little harm to the compressed air system. The coolers are required because the compression is not efficient. Some of the energy that's used by the compressor does the work to compress the fluid. The inefficiency of the compressor is associated with simply heating the fluid and since there is little mass in the fluid the temperature of the fluid increases dramatically.

Now is probably the best time to say that there's a simple formula for compression that says $P_1 \times V_1 \div T_1 = P_2 \times V_2 \div T_2$ which means that the pressure (P), volume (V), and temperature (T) are all related before and after compression. Pressure times volume divided by temperature at one condition for a gas will be equal to the pressure times temperature divided by the volume at another condition. If we double the pressure and the temperature remains the same then the volume has to be half as much. It's important to note that the pressure and

temperature have to be absolute values, add 15 to gage pressure to get absolute pressure and add 460 to temperature to get absolute temperature. For volume you could use cubic feet or cubic inches, it doesn't matter. Comparable metric units work just as well because it's the relationship, not the units that is determined; all that counts is using the same units on both sides of the equation. To eliminate any consideration of algebra, here are the solutions for each factor in the equation. To learn the second condition of any one of them perform the math on the right of the equals sign

$$P_2 = P_1 \times V_1 \times T_2 \div T_1 \div V_2$$

$$V_2 = P_1 \times V_1 \times T_2 \div T_1 \div P_2$$

$$T_2 = P_2 \times V_2 \times T_1 \div P_1 \div V_1$$

Sometimes this formula is referred to as the ideal gas law. It would be ideal if it worked perfectly but it doesn't. For what we have to deal with as operators it's more than adequate. It not only applies to compression but any change in the pressure, volume, or temperature of gases. It's most accurate with common diatomic gases, O_2, N_2, etc.

Since you know that it's the inefficiency of the compressor that produces the heat you can understand why you burned your arm on the piping or that compressor head the last time you got too close. It's a good thing to measure to monitor the health of your compressor system. The temperature will vary with load so you have to relate the temperature you're reading with one at a similar load at an earlier time to identify any pending problems.

There's a lot of confusion associated with compressor application that I want to make sure you don't get involved in. Unless otherwise indicated the capacity of a compressor is always described in scfm (standard cubic feet per minute) equal to air at 70°F and one atmosphere. Sometimes it's called 'atmospheric cubic feet per minute' and abbreviated acfm which many engineers, including me, interpret as 'actual cubic feet per minute' with unpleasant consequences.

If I'm looking at an application, such as air atomizing for a burner, and the burner manufacturer's table indicates I need 30 cubic feet per minute of air at 80 psig I'll call that 30 acfm. It's actually 190 scfm or a compressor salesman's atmospheric acfm ($30 \times 95 \div 15 = 190$). I know of several occasions where that confusion has resulted in attempts to change steam atomizing burners to air atomizing because the engineer didn't realize the compressor people don't understand anything but scfm. Of course I only made that stupid error once!

Normally all we deal with in a boiler plant is compressing air. It has its problems but it isn't as critical a process as compressing oxygen where the hydrocarbons from your fingerprint on one part can catch on fire in the compressor and do damage. Be aware of the hazards associated with any fluid you're compressing. Best way, of course, is to read the instruction manual. Just because the fluid is flammable or hazardous it's not something you should shy away from, with proper training and sensible operation you should be able to handle any compressed fluid.

We regularly use gas compressors, used to boost the pressure of natural gas high enough to fire in our boilers. The key to their use is that the gas is all gas; it's so fuel rich that it can't burn inside the unit. All you have to be concerned with is any leak that might form a flammable mixture and accumulate somewhere.

A unique feature of compressors that is not associated with other fluid handling equipment is the function of 'unloading.' Unloading a compressor consists of bypasses, valves held open, or other methods built into the machinery that prevents compression occurring but does not require stopping the compressor. It's not efficient operation because the compressor isn't doing anything but moving its parts around but the wear and tear of full blown starts and stops is eliminated to make life easier on the compressor and driver. Some equipment even has staged unloading where part of the compressor is actually working while the other part or parts are unloaded. The original purpose of unloading had nothing to do with continuing compressor operation, it still serves that purpose today; it allows the driver to bring the compressor up to speed before it starts compressing fluid. Even the smallest compressors have that feature.

Almost every boiler plant has a reciprocating compressor to produce compressed air for controls and actuators. That will probably be the case for a few more years until microprocessor based controls and electrically powered actuators are fully developed to eliminate both the compressor and all the compressed air distribution piping. You can pick any other system in the plant and you won't find one that is more inefficient than the compressed air. We compress air to 80 to 120 psig then use most of it at 18 to 30 psig.

I don't understand why I can't convince plants to install little compressors to produce air at about 25 psig and distribute that to all the controls then leave the other one to serve actuators that need it and provide atomizing medium for emergencies. Replacing pneumatic controls with microprocessor based controls in some plants has eliminated a lot of the waste but there's still more to do.

A wise operator can realize the opportunities for cost savings by locating and repairing leaks in air systems and eliminating wasteful use of compressed air. Waste can account for about 60% to 80% of the consumption of compressed air.

Reciprocating Compressors

Just like reciprocating pumps reciprocating compressors use a piston that changes the volume of a chamber to move the fluid. Intake valves are required to open as the piston moves down the chamber, increasing its volume, so the air can enter the chamber. They close as soon as the flow stops. Unlike a reciprocating pump the fluid doesn't start to leave the chamber as the piston moves up to reduce the volume, the fluid is compressed in the chamber instead. Not until the pressure is higher in the chamber than in the discharge piping connecting the compressor to its storage tank will the fluid begin to leave the chamber. When the piston reaches the end of its stroke there's no difference in pressure so the discharge valves close. As the piston moves down the chamber to increase its volume the fluid expands until the pressure in the chamber is lower than the pressure at the inlet. Then the fluid will flow into the chamber until the piston reaches the end of its stoke. The progression is depicted in Figure 10-96.

The typical air compressor valve looks something like a metal popsicle stick. For those of you who have never enjoyed a popsicle on a hot summer day, the valve looks something like the tongue depressor the doctor

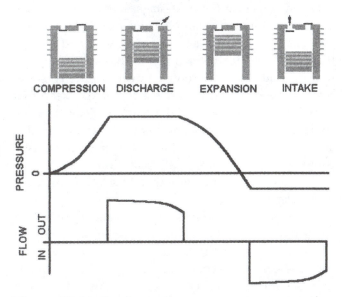

Figure 10-96. Reciprocating compressor operating stages

uses when he asks you to say "ah."

It's far more complicated than the typical liquid (incompressible fluid) pump which fills and discharges. You can't calculate the volume of the stroke and determine the capacity of the compressor because a good portion of the stroke is devoted to recompressing the fluid that expanded after the discharge valve closed. It should be obvious to you that the less fluid in the compressor at the end of its discharge stroke the less that will be there to expand and get in the way of more fluid coming in. That's why compressors are built differently.

The piston and chamber are designed for minimum clearances at the end of the stroke. There's very little room devoted to passages between the chamber and the discharge valves. It's all those close clearances that create the problem when a little liquid gets into a compressor, it will pass out through the discharge valves but it makes a lot of noise doing it and the hammering usually results in compressor damage. That's why it's so important to remove any liquid that forms between compressor stages.

Staging in compressors is similar to staging in pumps, you let one part of the compressor do part of the job and another finish it (two stage) although more stages are common. That little compressor you bought at the hardware store and keep in your garage is probably a two stage compressor, not two cylinders each doing the same amount of work. Two, three and four stage compressors are all common, some with multiple intercoolers.

Compressors are fitted with 'intercoolers' which are heat exchangers used to cool the compressed air between stages so the next stage doesn't get too hot. You'll probably notice an intercooler buried under the belt guard of your control air compressor and the fact that the sheave for the compressor has spokes formed like fan blades to force room air over the intercooler to remove the heat. Now there's a cue, if you keep that screen and all the fins on that surface clean then the compressor will run more efficiently. Please be sensible about it though; I told that to one operator who used compressed air to blow it all clean every day. Yes, he used more energy to clean it than he saved by keeping it that clean.

Usually the compression is such that a control air compressor will not condense any water out of the air in the intercooler so there are no drains on it. Larger compressors will be cooled to the degree that water has to be separated, collected, and drained from the outlet of each intercooler.

We used to count on the operator to open drain valves to remove moisture collected in the compressor and storage tank. Then, to give the operator time for other duties, we tried installing drain traps on them that would automatically drain the water off. We quickly learned that we couldn't count on those drain traps entirely so the operator still had to check them regularly.

Most systems today are equipped with timed drain traps, solenoid valves connected to a timer that opens them at preset intervals to drain the liquid. From what I've seen of them they drain some liquid and a lot of air, another waste. The problem is we don't know what the demand on the compressor is so we set the timers for the worst (full load) condition. Occasionally the compressor will be shut down or run unloaded between drain valve cycles so the only thing it's going to drain is air. They're reliable but waste a lot more air than a wise operator. As time goes by there will probably be a better device invented, but until then…

Reciprocating compressors are designed to start unloaded. The typical scheme is use of lube oil pressure where a small oil pump eventually builds up pressure as the compressor is started and that pressure is used to overcome the force of springs that hold the compressor's inlet valves open. During normal operation that same oil pressure can be bled off to the crankcase to allow the springs to hold the inlet valves open for unloading. In compressors with multiple cylinders it's possible to unload one set of valves while leaving others in operation to adjust the capacity of the compressor. That form of unloading is normally accomplished with a pressure switch that switches valves in the oil circuits although it can be done with an electric switch and solenoid valves. Staged unloading is common in refrigeration compressors. (See Chapter 5.)

You have to be aware of that unloading scheme if you proceed to adjust anything in the system. I knew one operator that thought he would save money by lowering the compressed air pressure. He lowered the setting of a pressure control switch but was dumfounded to see that the compressor would run longer. He had simply reset the unloading setting so the compressor always ran with half the cylinders unloaded. Someone else lowered the setting of the on-off pressure control below the unloading value of a compressor with hydraulic unloading and couldn't understand why the motor burnt up because the compressor was constantly starting and stopping. The partial unloader or unloaders must operate within the span of the on-off control switch. If the unloading settings are not in the operating range of the compressor they won't work.

Oil almost always requires attention in a reciprocating compressor. There are small compressors that use diaphragms instead of pistons to compress the air and others with synthetic rings that can operate without oil (oil-free compressors) but most of the ones you'll find in a boiler plant use oil. If you've never checked the oil in a reciprocating compressor before take this one small piece of advice; always wait until it has just shut down before checking the oil. If you just walk up to it and remove the cap on the oil reservoir it's bound to start and blow oil all over the front of you! Of course that's advice from the experienced.

Oil is required to lubricate the moving parts of a compressor and except for oil free units, serves to seal the space between piston and cylinder so the air can be compressed. (By the way, you still have to keep oil in some oil-free compressors, it's only the air that has no oil in it)

Since the oil is coating the cylinder walls, is scraped by the piston rings, and exposed to those parts heated by the inefficiency, some of it is vaporized and some droplets form to leave the compressor with the air. As compressors age they tend to load the air with oil more than when they were new. Your system should have an oil separator to remove that oil so it doesn't contaminate instruments, controls, and tools that use the air. At least that's true most of the time, some systems are only used for tools and the oil helps lubricate them. In that case the oil should be a non-hazardous type that doesn't form poisonous aerosols where it leaves the tool. In addition you could have an oil coalescing filter which absorbs the oil. For the sake of your controls, please watch that coalescing filter and change it when it's not quite saturated. Also make certain the separator is working to reduce the oil loading on the filter.

Other Types of Compressors

Centrifugal compressors were touted as the latest thing about forty years ago but they quickly faded away because the tip speeds had to be so very high to develop the necessary pressure. The compressor required large speed increasing gears to get that high tip speed and the stresses on the metals at those high speeds made them vulnerable to all sorts of problems. A reciprocating compressor, which runs at relatively low speeds, could take a small drop of water coming off the previous stage, a high speed whirling impeller couldn't. I still think a steam turbine driven centrifugal could be developed that would be efficient and reliable but nobody has built one that anyone would buy.

Screw compressors function about the same as a screw pump. The important difference is the screw is machined so the cavity becomes smaller as you move along the shaft. An added feature in the compressor world is a slide that bleeds air back to the suction to reduce capacity. Screw compressors are used extensively in the construction industry, that's what most of those little trailers towed behind the contractor's truck are. They also need lubrication because the oil is what seals the cavities and keeps the metal parts from rubbing each other. Since most construction tools need lubrication there's no problem with what's carried over with the air. A screw compressor in a plant is usually followed by an oil separator and coalescing filter to provide the specified 'clean and dry air' for boiler plant controls and actuators.

Some rotary compressors are very similar to gear pumps (Figure 10-97). They simply move air along with little concern for the fact that air rushes into the cavity as it opens to compress the air before it starts flowing out. Vane type rotary compressors (Figure 10-98) use the eccentrically positioned core to produce a cavity that changes volume to compress the air as the chamber rotates around the shaft. These compressors must be

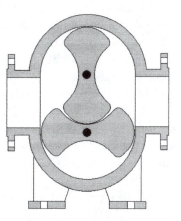

Figure 10-97. Lobe type rotary compressor

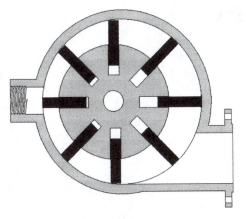

Figure 10-98. Vane type rotary compressor

lubricated and are typically used for low values of compression, producing air pressures in the range of thirty to fifty psig.

I used one rotary compressor as a gas booster on a job in the 1980's and was hoping to get capacity control later by converting the drive to variable speed. That never worked out because the oil lubrication would be lost if the compressor was slowed down.

I have to elaborate a little on gas boosters here because they are frequently found in a boiler plant. They are either the rotary or centrifugal type and can't be turned down significantly so the gas has to be recirculated through the booster to reduce output to match the requirements of the boiler's burner when it's modulating. If the boiler shuts down the booster must also shut down. During certain periods of boiler operation the booster must run to produce pressure while the burner isn't firing (to prove fuel pressure available) so full recirculating mode exists for a period of time. To prevent overheating the gas as it continues to recirculate in the booster some means is required to cool it. An air cooled heat exchanger is recommended. Water cooled heat exchangers can waste a lot of good water and need so much that it all can't be used for makeup. If you do use a water cooled heat exchanger, and it's using city water, allow the water to get up to at least 140°F before discharging it so you waste as little as possible and use whatever you can for makeup.

Don't run a booster you don't need either. I visited one plant where the booster was running but the service supply pressure was more than adequate. I suggested they try operating with the booster shut down and by-passed. They did, and it worked fine. I'm told they went through the winter without needing the booster. I would love to have all the money they saved on electricity with that one suggestion.

COGENERATION

There's no question that the de-regulation of electricity has changed the scene of electric power generation. Without their monopolistic position utilities have been compelled to produce power more efficiently. The ability of the ordinary steam boiler plant to convert fossil fuel energy at an 80% efficiency has allowed many facilities to incorporate power generation, what we call cogeneration (the generation of both heat and electric power), into conventional plants with a lower overall operating cost.

The new buzzword is 'distributed generation'

where electric power is generated by many smaller operations. When a boiler plant passes all the steam it generates through a steam turbine to produce electric power with a generator then uses all the steam in the facility the thermal efficiency is much higher than a power company that normally runs at 40% (60% of the energy they consume in fuel goes up the stack and out the cooling tower). Understanding steam turbines and their operation is going to become more important for the wise boiler operator.

The principal reason most plants have not generated power is the utility's standby charges. The utilities argued, with a certain degree of justification, that they had to provide generating capacity to replace any generator someone else owned to ensure an adequate power supply. In other words, they needed additional capacity to replace the power normally produced in one of their customer's plants in the event the customer's generator failed. The charge was almost always large enough that the customer abandoned any thoughts of power generation. Despite that and other disincentives some of my customers are cogenerators. With deregulation a lot more are going to be.

It really isn't a new thing. Cogeneration was the way of the world early in the twentieth century. Power companies had not strung lines everywhere and new facilities didn't have a source of reliable power so they generated their own. Buried deep in the bowels, and sometimes under the concrete, of many old industrial and institutional facilities throughout the country are old cogeneration plants which generated power with steam engines and used the exhaust to supply the process. Many industrial museums are popping up today with the remnants of many of those old plants as showcases. A visit to one is always worthwhile. I know because I learn something new with every one I visit.

So, unless you have a plant where your load is very small or very inconsistent you're going to be exposed to a change that involves cogeneration sometime in the near future. It's simply energy sense and, since the utilities don't have a monopoly on power generation any more, it's also economic sense. Of course economics doesn't always make sense. I know many a boiler operator that complained about the accounting practices of their employer and I've seen many examples of fiscal stupidity that converted an obvious economic advantage to a loss.

I remember a project many years ago where I had the opportunity to install a new boiler and back pressure turbine generator in a plant that was already a cogenerator but discovered when the economics of the project were evaluated that I couldn't justify the power generation. Further investigation revealed the reason: the facility's accountants charged all electrical maintenance in the plant to plant generated power. That produced a cost of generated power, on the books, that was only 80% of the cost of purchased power despite the fact the actual cost was only about 25%. The accountants ruled and the project reverted to a new saturated steam boiler operating at the turbine exhaust pressure. The operators in the plant were not happy, nor was I, but fiscal stupidity won that argument because I didn't have enough gray hairs at the time to get anyone to believe me. I know an operator can't do anything directly about such stupidity but maintaining good quality records of power generation and its cost would have allowed me to beat that argument back then. You got it! The old document or disaster rule repeats itself.

The key to cogeneration is to use the energy that's left over after generating electric power. One company is promoting tri-generation where the plant produces power, heat, and chilled water for refrigeration or air conditioning. The heat of the generator exhaust is used in absorption chillers in the summertime to produce chilled water. That allows plants that only need heat in the winter to become cogenerators (or, if you will, tri-generators) although they can't do much in the spring and fall. Of course that depends on your electrical contracts and fuel costs. In some cases it pays to generate electricity and waste some of the exhaust heat that you can't use in order to avoid standby charges (although they will give them a different name) and related expenses. You may also be expected to operate the generator to minimize demand charges.

Somewhere in this book I have suggested operating practices to minimize demand charges but operating a generator to minimize them, when possible, can also be the responsibility of a boiler plant operator. With cogeneration you can produce additional power, even if it isn't efficient to do so, to reduce a peak load and lower those demand charges. The degree you go to is dependent on the length of time a peak load is endured and the inefficiency associated with producing that extra power. If the peak is substantial and only occurs during a short period of time (like half an hour a week) it may pay to dump steam to atmosphere, as mentioned earlier, just to eliminate that peak. You have to look at the cost to generate the power for that period of time and how much you save overall on demand charges. Now you know why operation of an emergency generator can be a small cogeneration activity like what I explained under reducing demand charges.

When capable of tri-generation you can identify and develop SOPs for spring and fall operation to balance wasted energy, by heating and cooling at the same time, to optimize your operating cost by generating more electricity and reducing demand when you normally don't have the loads at the generator exhaust. It requires knowledge of the electrical contract and how to manipulate it and good records on power generation and system loads. Many of you will gladly allow an engineer or consultant to help you develop the program for such operations because it does get complicated. In time you'll probably find that it will take a computer to guide you in the decision making process because electricity costs will vary hourly. There are already situations where the cost of electricity varies with each hour of the day. An example was one in North Carolina where electricity cost as little as 2¢ per kW at night and 26¢ in the early afternoon with hourly variations in between. You're limited with controlling power usage to avoid the higher costs but cogeneration gives you the ability to really save your employer some money on power.

I do hope that any plant that allows a computer to do the controlling also has an operator to make certain the computer is doing what it's supposed to.

There are several options for generating power with exhaust heat to be used for steam, hot water, service water and absorption chillers. They include steam turbines and engines that have a long history in that service; turbines require substantially less maintenance and operator attention than engines.

Generating steam or hot water with exhaust from diesel generators, including those fired on natural gas, also have a long history but, like engines, the generators require a considerable amount of maintenance. Modern engines have improved on that maintenance requirement to the degree that they are being used. Modern devices include gas turbines and fuel cells. Let's discuss them just a little so you know what they're like. Once again, the instruction manual and other documentation is necessary for you to learn all that's required for operating them because they aren't a common element of today's boiler plant.

Steam Engines and Turbines

Take a good look at the photograph of Figure 10-99, it's a steam engine driven air compressor and it's probably one of the few that are still operating today. Specific problems with steam engines have almost eliminated their use today. Lubrication oil getting into the boilers has just about been eliminated with provision of better materials that can seal the piston and shaft of a

Figure 10-99. Steam powered air compressor

steam engine but the need for skilled workers to maintain and rebuild them and the high initial cost and cost of maintenance has pretty much priced them out of existence. It's not that they cost more than a motor over their lifetime, it's just that they cost more to begin with and their maintenance isn't understood. Direct conversion of steam to power should more than cover the maintenance costs but most owners are not willing to invest in these very efficient devices. Let's see if another revision of this book reflects change in that attitude.

Steam turbines have also seen a decline in use, mostly because electric power has become so reliable and there is such a low demand for steam turbines that they are only found in medium to large boiler plants which are willing to invest in them. There has been no argument for having turbines to ensure continued operation in the event of an electric power interruption because of power reliability and the ability to operate a generator to run motors so you don't need dual drives. I can remember many installations that had boiler fans and feed pumps with dual drives, a motor on one end of the piece of equipment and a turbine on the other. Auxiliary turbines, (see earlier discussion) which exhaust steam to the deaerator are just about the only surviving application.

Steam turbines convert the heat energy in steam to mechanical energy. It's a simple matter of passing the steam through a nozzle from a higher pressure to a

lower pressure in a manner that converts the static pressure in the steam to velocity pressure. Once the steam is moving at a high velocity the mechanical energy is in the steam and the turbine has to transfer the energy in the steam to rotation of the shaft. Turbines use two methods to transfer the energy to the rotating shaft, either impulse or reaction. An impulse turbine works the same as a pinwheel, when you blow on a pinwheel the air strikes the surface of the pinwheel, giving up it's velocity pressure to the blades of the pinwheel.

A reaction turbine works more like a loose garden hose. I'm sure you have turned the water on a garden hose at some occasion when the nozzle was open and the hose twisted around splashing all over the place. The hose reacted to the water spraying out the nozzle. Figure 10-100 shows the two types of turbine stages graphically and includes what is called a velocity compounded stage where the back splash from an impulse stage is re-directed to a second set of turbine blades to increase the performance. All steam turbines have a least one impulse stage because the steam is initially supplied through a set of stationary nozzles in the turbine casing. Moving blocks are shown in black, nozzles and satationary blades in dark gray, and steam flow in light gray.

Multiple stages just like pumps? Yes, by dropping the pressure in stages we get better efficiency. Utility

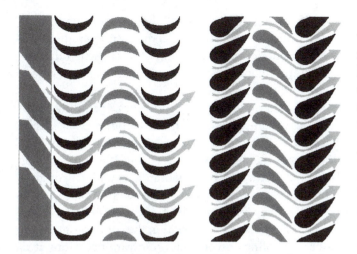

Figure 10-100. Impulse and reaction stages of turbines

style turbines for power generation have twenty or more stages in the high pressure turbine and another 18 or more in the low pressure turbine(s). As the steam pressure drops from stage to stage it expands. If you check the volume of a pound of steam in the steam tables you'll notice that the volume of steam increases rather dramatically. The manufacturer of the turbine either has to make the latter stages of the turbine much larger or provide for bleeding off some of the steam.

It's typical for a large power generating turbine to have at least two bleeds. High pressure bleed steam is regularly used for feedwater heating between the deaerator and the economizer or boiler. Some high pressure bleed steam is at a pressure high enough to be used as the steam supply for auxiliary turbines. Intermediate pressure steam can be used in the deaerator or for other purposes. Low pressure bleed steam can be used for plant services such as building heating and reheating condensate after it leaves the condensers. See the utility cycle (Figure 1-8).

Steam turbines, and engines, extract energy from the steam without condensing it. That's very important because the turbine would be severely damaged by droplets of condensate hammering the turbine blades. The energy that's removed from the steam to generate power is only enough to reduce the superheat.

Despite what some people think, a turbine that runs on saturated steam is only extracting superheat. The steam contains the same amount of energy after it passes through the first nozzles of a turbine as it did at the inlet of the turbine. Since the pressure in the steam is lower the steam has to be superheated. As long as the turbine doesn't draw too much energy from the steam it will still be superheated (a little bit) at the outlet of the turbine.

Of course to really generate power we superheat the steam in the boiler. That allows us to use a more efficient turbine that extracts more energy from the steam. In large power generation equipment the steam is piped off the turbine and back to the boiler to be reheated before continuing its trip through the turbine. The reason we use reheat is the superheat necessary to prevent condensation on a full path through the turbine would be so high that the superheater tubes and pipes would melt, we simply don't have metal that could take those high temperatures. By reheating we can boost the temperature at an intermediate stage in the turbine to about the same temperature as the steam at the turbine inlet without requiring more exotic metals in the superheater and piping. The heat exchange surface in the boiler that does this is called the reheater and it requires special consideration in the startup and operation of a boiler that's equipped with one.

The turbine arrangement that will probably become more common with the development of 'distributed generation' is the 'topping turbine' or 'high back pressure turbine' that will generate electric power. All the generated steam passes through the turbine for use in the facility. The steam will be produced at high pressure (600 to 900 psig being the most common) and superheated, then dropped through the power generation turbine to generate power while dropping to pressures you're operating at now, the level required in the facility served by the boiler plant.

The few things you need to know about turbines is that their lubrication is critical and the torque of a turbine is at maximum when it's not rotating and decreases as speed increases. Most power generation turbines have pressure lubrication; the oil is supplied to the bearings under pressure. The oil feed can be from a pump directly, in some cases one attached to the turbine, or from a head tank set well above the turbine which is constantly refilled from the sump by pumps. You may recall being in a power plant and seeing a viewport in some piping where you can see oil splashing through; that's the overflow from a head tank. As long as you see oil spilling down that overflow you know there's lubricating pressure for the turbine. If you don't see it you have a short period of time in which to get that turbine shut down.

As for the torque business; you don't want to damage the turbine. Spinning open a steam valve on an idle turbine inlet can result in so much torque at the first stages that the plate holding the turbine blades or the shaft can be bent enough to cause the blades to hit with serious damage. The rapid acceleration of the turbine from zero speed can result in serious over-speed conditions. Just crack the steam valve to any turbine, it takes very little steam to get it moving. The marine turbines I used to operate had a bunch of heavy gears, a fifty ton propeller, and long shaft holding it back so we gave it a bit of a blast to get it started, opening the valve a quarter turn or so, then quickly throttled back as it started moving.

There's still considerable force on the turbine blades when a turbine is operating under load. If a power generator trips off the line, instantly stopping any generation of power, the turbine is bound to over-speed because there's nothing to stop it taking off. That's why they all have over-speed trips and some even have hydraulic brakes to limit the speed. Don't skip that very important function of testing the over-speed trip when you start up a turbine. If it doesn't work you could be watching turbine blades flying out of the casing and all over the place.

Power generating steam turbines are usually large pieces of equipment with very thick parts. The shaft of a turbine can be several inches thick so it's important to prevent quick temperature changes during the operation of those turbines. During startup the turbine should be brought up to speed and loaded gradually. Always, and I do mean always, look for the instruction manual to see what the manufacturer recommends for startup then confirm that the facility's SOP's comply with those guidelines (they normally allow for slower heating) and stick with those rules. Luckily I never brought a turbine up too fast but I do know what one sounds like when it starts rubbing due to uneven temperature growth; that's a sound you don't want to hear, ever.

The large shafts and heavy metal of steam turbine casings also retain heat. If the turbine is stopped it should be given a bump to force a few rotations at regular intervals to prevent the shaft buckling because the temperatures are higher in the top of the casing than in the bottom. When we were at a stop during maneuvers to get into or out of a docking space we would give the main shaft turbine a little steam to rotate it once or twice at regular intervals. It should be noted that the propeller only moved a couple of degrees during that operation.

When a power steam turbine (or a gas turbine) is shut down it is normally connected to a "jacking gear" A small electric motor with (typically) a screw drive is engaged to rotate the turbine at a low speed (normally less than one rpm) to provide for a balanced cool down. The jacking gear is usually operated until the moment before steam is admitted to the inlet of the turbine throttle valve and some form of interlock is provided to prevent opening those steam valves until the jacking gear is disengaged. Our typical shipboard interlock was

a sign "Jacking Gear engaged" hanging on the throttle valves. Modern equipment has more serious interlocks to prevent accidental admission of steam when the turbine shaft is connected to the jacking gear (remember where torque is greatest).

All electric utility steam turbines, including those in nuclear plants, are condensing turbines. That means that at least some of the steam passes through the turbine to a condenser. The water from the condenser is then pumped up to the deaerator, usually through a number of heat exchangers. To condense the steam all the heat of vaporization (the latent heat) has to be removed. That heat is transferred in the condenser to river water or cooling tower water. On rare occasions it is dumped to air through air cooled condensers.

For maximum power generation the condenser must operate under a vacuum so non-condensible gases and any air that might leak in must be removed from the condenser. That's typically done with a steam jet ejector but may also be accomplished with motor driven vacuum pumps. The steam jet ejector (Figure 10-94) is usually two or more stages to pull as much vacuum as possible. The steam used to eject the air is then condensed in a heat exchanger using condensate. The actual vacuum achievable is dependent on the temperature of the cooling water or air but 27 inches of mercury is a typical value to shoot for. At that pressure the steam will condense at 92°F.

Any condensing turbine requires special provisions to seal the shaft of the vacuum pressure stages to prevent drawing air into the turbine. That's usually accomplished with a special regulator that supplies steam from a high pressure bleed or a reducing station to keep pressure on the shaft seals. The regulator also dumps excess steam leaking from high pressure seals into the condenser during operation.

Maintaining a vacuum by providing adequate cooling water or air and keeping non-condensible gases and air out of the condenser is imperative for best power generation. A boiler plant that's converted to generate power in addition to heat, as opposed to one that generates power as well as heat, may have a condensing turbine although most of the steam will be used in the plant. Various schemes including bleeding steam and separating stages with piping and control valves are used to maintain pressure of the steam supply to the facility while the steam to the turbine is controlled for power generation. One such scheme is a goggle plate inside the turbine with slots that are opened and closed to control flow to the lower stages right after the facility steam is drawn off.

GAS TURBINES, ENGINES AND HRSGs

This is just a small amount of information about the other types of cogeneration plants you may encounter. Once again, it's the manufacturer's instructions that are going to be most valuable in developing your operating knowledge of these plants. I have to admit I've never operated or even studied one of the land based plants that cogenerate, all I've done is visit some of them. Each is a little unique so once again the instruction manuals provide critical guidance.

Just because the fuel isn't burned in a conventional boiler furnace doesn't mean a boiler operator can't handle it. The combustion chemistry doesn't change, all the formulas stay the same, we're still simply burning hydrocarbons to release heat. Gas turbines and gas engines burn the fuel and some of the thermal energy is extracted from it to generate power. They're not a lot, if any, more efficient than a utility steam plant so there's heat left over to make steam. There are many engine generator plants with waste heat boilers in this country that have been operating for more than thirty years and gas turbines aren't as new as some people think. In many cases all that's new is a way of putting equipment together and a HRSG is a prime example of that.

Unless you're a rare individual you have general knowledge of how your automobile engine works. If you're like me you also know that modern technology has limited working on it to someone with computers that can talk to the computer in the car but that doesn't mean the combustion process is any different. Most electric power generating engines work the same, using either the Otto or Diesel cycle to convert energy in fuel to output power at the engine shaft which drives the generator. Otto is the guy that came up with the four cycle engine, the scheme of intake, compression, ignition and exhaust. A diesel engine can be two cycle or four cycle but most are four cycle with the only difference being Otto used a spark to ignite the fuel. Fuel is injected into a diesel engine right before ignition and ignites spontaneously in the hot compressed air.

The water jacket surrounding the engine's cylinders don't absorb much heat compared to a boiler so there isn't much energy recovered by the water jacket. I'll admit I've heard some of them called cogenerators but I really don't consider them as such. Some plants use the heat of the jacket for building heat and other purposes but most of the energy remains in the exhaust gases. Cogeneration plants using engines have a waste heat boiler that recovers the energy in the engine exhaust gases. The boiler or boilers are commonly manifolded to two or more engines so steam generation can be maintained. Frequently there's an auxiliary burner installed somewhere to provide additional heat or fire the boiler when the engines are shut down.

The auxiliary burner in those applications shouldn't be a conventional boiler burner. I saw such an application a few years ago with the conventional burner's inlet fitted with another fan to produce the static pressure that matches the engine exhaust pressures and overcomes the pressure drop through the boiler, economizer, and stack. That little burner looked somewhat ridiculous with the fan blowing into it and I'm not certain it won't blow apart under the pressures it is subjected to.

The principle difference in engines and turbines, as far as combustion is concerned, is that engines are typically fired fuel rich to keep them cool and turbines are fired air rich. There's a catalytic converter on your car because the engine would get too hot if the fuel was burned completely. By running the engine fuel rich the combustion products are much cooler, you don't get that extra 10,000 Btu from complete burning of the carbon. The catalytic converter combines the exhaust with the air added by the air pump to burn off the carbon monoxide later, after the energy produced by the initial combustion has been converted to mechanical power to drive the car. The catalyst simply provides a source of certain ignition of the lower temperature exhaust gas and air mixture to ensure more complete combustion. It raises the gas temperature in the process to waste the heat out the tailpipe. It isn't perfect or complete because a little CO manages to sneak by a converter but it does a pretty good job. You will find some engine cogeneration plants operating with catalytic converters, more for NO_x reduction than CO reduction.

Gas turbines, on the other hand, used to use lots of excess air to absorb the heat of combustion and lower the operating temperatures so the turbine blades don't melt. The high volumes of excess air make it difficult to get complete combustion but providing cooling water to spinning turbine blades is virtually impossible. New techniques and construction are changing the form of gas turbines to allow lower excess air. Once scheme now used is to bleed air or steam through holes in the leading edge of the first row of turbine blades to create a film of cooling air or steam flowing back over the blades. Designs of power generating gas turbines are evolving rapidly so you will have to read to keep up with what's happening with them.

Gas turbines are not a new thing. I said that a bit ago didn't I? Anyway, gas turbines are at least sixty years old. Every jet plane flying is powered by gas turbines. The first gas turbine powered ship, the Admiral Callahan, was powered by two airplane jet engines which exhausted to another turbine that drove the ship's propellers. Gas turbine plants that use that concept are now called 'aero-derivative.' The growing need for improved efficiency, fostered by the deregulation of electricity, has seen improvements in common shaft gas turbines (basically a jet engine with a shaft sticking out to drive the generator).

A gas turbine consists of three basic parts. A compressor, burner(s), and turbine. The compressor draws in atmospheric air and compresses it before supplying it to the burner. The burner mixes the fuel with the compressed air and ignites it. The parts containing the burner are protected from the heat of the burning fuel by baffles cooled by some of the compressed air. The products of combustion and cooling air mix to provide a cooler product before entering the turbine. The turbine, a reaction type, converts the heat energy to shaft power to drive the compressor (a large portion of the turbine load) and a load. I have to say load because there are some gas turbine driven pieces of industrial equipment; but most of the time they're used to power electricity generators.

It's the gas turbine and HRSG combination that form the plants we now call 'combined cycle' power plants. The HRSG (Heat Recovery Steam Generator) could be modestly called a waste heat boiler but is much more than that. It consists of a combination of all elements of a modern boiler plant in a carefully matched and packaged combination designed for maximum efficiency. With combined cycles utilities have been able to increase their efficiency to almost 50%! I should point out that it's a LEL efficiency so they're still nowhere near the performance of the common heating plant. The basic arrangement of a combined cycle plant is a gas turbine followed by a HRSG which generates steam to power a steam turbine with both turbines generating electric power.

What exactly is a HRSG? Why is it different than a waste heat boiler? It's because it is more than just a waste heat boiler. The typical HRSG is a combination of things with the most common arrangement being a connecting duct for the turbine exhaust with an integral duct burner, superheater, reheater, high pressure boiler, economizer, low pressure boiler for deaerator steam which flows to the integral deaerator, and low pressure economizer. The HRSG is designed to squeeze as much heat as possible at each section then follow with a lower temperature boiler or heat exchanger that can absorb some of the heat that's left. As the flue gases cool in their path through the HRSG they pass several "pinch points" where the flue gas temperature approaches the saturation temperature of the boiler or inlet temperature of the heat exchanger.

Many of the duct burners simply introduce fuel because the gas turbine at the inlet operates with very high excess air (300% to 400%) so there's plenty of air for the fuel. Some duct burners have air for the ignitors only and some have none, an unusual concept to some of us old boiler operators.

Microturbines

Microturbines are very small gas turbine generators with some unique differences. Most generators are limited to a speed of 3,600 rpm so they can generate 60 Hertz electricity. In Europe the speed limit is 3,000 to generate 50 Hertz. A gas turbine is more efficient at higher speeds. Microturbines generate direct current then invert the output with solid state electronics to produce alternating current. That way there's no link between speed and frequency so the turbine can be operated at the most efficient speed for the power it's generating.

The manufacturers of these small independent power plants, some no larger than a typical desk set up on one end, provide limited information about them. I've seen them sitting in a plant and making a little noise (they're surprisingly quiet) while generating power but that's the limit of my experience with them. Some of you may grow to learn a lot more when you have to try to operate them.

Microturbines are an assembly line product with common sizes being 30 kW and 60 kW. The largest I'm aware of is 250 kW. They also produce hot exhaust which can be directed to a waste heat boiler but many are used as emergency generators with no waste heat recovery.

Fuel Cells

This is a product I haven't seen in operation. It's relatively new and I know of several plants that use them but have no experience with them at all. I do know a little that I'll share with you because, if you know anything about combustion, you'll discover that they're given more credit than they really deserve.

Fuel cells do generate electricity without burning the fuel. That doesn't mean they run cool. Some of them operate at very high temperatures. The concept is one of hydrogen and oxygen combining to make water by a sort of reverse electrolysis. If you had chemistry in high school one of the things you did, at least I did, was bury two electrodes in water (the electrolyte), pass a direct current through them and the water, then watch gas bubbles form at each electrode and rise into an inverted test tube. One contained oxygen, the other contained hydrogen; the process broke the water down into its two basic parts. A fuel cell does the opposite, using reaction of hydrogen gas and oxygen to produce direct current electricity and water.

Fuel cells became the mainstay of electric power in the space program because they generated a lot of power with very little weight and produced water that could be consumed by the crew or jettisoned without degrading the environment. Their relatively low operating temperatures and lots of careful development produced a highly reliable electricity generator. When used in earthbound applications the direct current produced has to be inverted to alternating current. They're used principally in plants where highly reliable backup electric power is required. The important thing to note is that they are designed for, and work well with, pure hydrogen.

Since there are no hydrogen pipelines or storage tanks out there a conventional hydrogen—oxygen fuel cell is not the sort of thing that someone is ready to invest in. There's considerable hype around the development of fuel cells for automobiles as clean burning and that may result in some domestic supply of hydrogen but not enough to power any large systems. The typical earth based fuel cell installations currently burn a common hydrocarbon fuel such as natural gas with some modification.

The modification of a fuel cell to burn hydrocarbons incorporates a 'reformer' which modifies the fuel to produce pure hydrogen. As I understand the cryptic descriptions available, the reformer produces heat, generating steam. The steam is then exposed to the fuel in a catalyst where the hydrogen in the water is released as the carbon combines with the oxygen to form carbon dioxide. That way some of the energy produced by the carbon is used to create more hydrogen. The source of the oxygen is air. That means that the exhaust of a fuel cell contains carbon dioxide and water just like a normal boiler.

So, when you see those articles and advertisements that say a fuel cell produces less carbon dioxide than a boiler you should treat everything the author says as a potential lie because there's no alternative. Any hydrocarbon has to produce carbon dioxide and water; to claim it doesn't is blowing smoke. The only alternatives are to produce carbon monoxide, something we don't want to do, or leave pure carbon. A fuel cell suppos-

edly does neither unless it's fueled with pure hydrogen which has to be a very expensive fuel. Most of the fuel cells do operate at temperatures low enough that they don't produce any NO_x and there's no way for particulate matter to get through the liquid electrolyte.

As for carbon monoxide and volatile organics we'll have to assume they can't get through the electrolyte either although a reformer could dump them out with the carbon dioxide under upset conditions. I would be a lot happier about the future of fuel cells if someone would admit that they could go wrong and produce most of the other criterial pollutants except possibly NOx and sulfur oxides.

Sulfur oxides aren't considered because the sulfur would poison most of the electrolytes used in fuel cells to stop their operation in short order. Fuel cell applications require special fuel pretreatment to remove any sulfur. It's also highly probable that a fuel cell will require good air filters and some air pretreatment to limit the effects of the normal allotment of particulate and nasty gases that can be in the air around industrial sites. How much air cleaning and fuel preparation will be determined to extend the operating life of the fuel cell as we gain experience with them.

As I understand it right now a fuel cell has to be dismantled and rebuilt on at least a five-year schedule. It's the sort of operation the manufacturer insists on doing, probably to retain secrecy regarding their methods of construction and other details. The schedule may be based on experience with the degradation of the electrolyte from the problems we're already aware of, contaminated fuel, particulate and stray gases in the air, etc. so programs of life extension based on chemical treatment or reconditioning of the electrolyte may be able to extend that operating period in the future.

The actual operating temperature of a fuel cell depends on which electrolyte is used and can vary from very low (about 350°F) to high (about 1200°F) so the temperature of the exhaust can vary considerably and the possible uses of the exhaust heat will vary as well. If fuel cells reach the potential that many people try to give them the exhaust will only be good for heating service water.

Currently fuel cell applications are limited to sites where reliable electricity supply is all important and, by installing several fuel cells, an owner can be reasonably certain the power will never be interrupted. You may be called on to monitor fuel cell operation and, once again, the important thing to do is read that instruction manual.

Chapter 11

Controls

There was a time when the operator was the only controlling influence on the operation of a boiler plant. Today controls are an extension of the operator's eyes, ears, hands and feet that help the operator keep the plant running in a safe and efficient manner. There's no question that less manpower is needed in a plant equipped with controls but, unlike a human, they can't always let someone know when they aren't working correctly.

The modern boiler plant operator uses the latest controls made available to manage the ever increasing cost of operating a boiler plant. With these tools the operator can easily manage to reduce operating costs by as much as 15%.

CONTROLS

"Controls! What controls?" The fireman walked away shaking his head and muttering something about those dumb young engineers fresh out of the academy. I would discover later that what I thought were controls were considered a simple obstacle to walk around on that ship. They were Bailey Standard Line controls and the honest truth was that none of them were working. It didn't deter me because I had seen them working on ships when I was a cadet three years earlier. After a couple of weeks of tightening connections, adjusting settings, and replacing most of the diaphragms that sense air and flue gas pressures I managed to show that fireman that the controls could maintain the steam pressure as well as he could and all he had to do was sit back and watch them do it.

I believe the attitude that controls would put everyone out of work are gone and most operators consider them just another tool that only they have the skill to use. The last plant I was in where the controls were not allowed to do their job now operates automatically, without any operator in attendance. Because the controls do a better job? Nope, because the operators simply refused to allow them to work at all. A concern for keeping their jobs led to them losing all of them. I won't argue the fact that people have been replaced by technology, but I do know that it's better to embrace it than fight it. Controls are simply one of the things that help you do a better job and you should know how to use them.

There are two basic types of control, on-off and modulating. On-off control isn't as simple as you might think and I'll cover some of the unique conditions you should be aware of within the discussion of specific applications which follow. The following few paragraphs address the general elements of modulating controls which a boiler plant of any reasonable size will have.

If you recall the section on flow you know that controls change the rate of flow in order to maintain a desirable condition such as pressure, level or temperature. Despite the fact that we can't really control pressure, level or temperature we identify a control loop using those parameters. Just keep in mind the fact that you aren't controlling anything but flow.

Just like an engineer, right? Using big words like parameter! Parameter is one of those words that we use to mean several things and it's not that complicated a concept. We use it because controls aren't selective; the controller doesn't know if it's controlling to maintain a pressure or a level, that information isn't even important to the controller. We say parameter and we mean level, pressure, temperature, pH, or anything else we choose to maintain with a controller. The controller does the same thing regardless so we give it one generic name, parameter.

There are a number of words used by the control designer and technician that you should know. Why? Because they only know about controls and use words specific to their controls that don't differentiate among the hundreds or even thousands of different systems that can use those controls. A controller can be used for so many different applications that assigning names that are independent of the process being controlled was essential. Once you know what they are and what they represent you'll have a better understanding of controls.

Despite a tendency of some people to award a level of intelligence to controls they are really quite limited. They don't know what they're controlling nor what parameters they are maintaining, they respond to control signals and produce control signals. The signals are air pressure, fluid pressure, electrical voltage, electrical current, or a bunch of electrical charges in a tiny microchip that we relate to as ones and zeros. If they don't know any more than that we shouldn't have any problem understanding them. Understanding controls isn't that difficult, our controls can be used in any application, not just boiler plants. The really wise boiler operator will be able to relate to how the controls work with the boiler and its auxiliaries.

Let's start with parameter, it's a quantity, value, or constant whose value varies with the circumstances of the system. The controller doesn't know what the parameter is and it doesn't care. It can be pressure, temperature, level, count, pH, oxygen content in percent, differential pressure, a flow of any fluid, a weight, etc. The controller basically deals with parameters that are called inputs and they are used to create an output, or outputs. Inputs are assigned names that indicate what they are in relation to the controller with the two most important ones being process variable and setpoint.

The process variable is a value representing the measurement of whatever it is you are trying to maintain. If it's a pressure controller it's the pressure. If it's a level controller it's the level. It's the control system's representation of the actual value of the parameter you're trying to control.

The setpoint is a value representing what you want the process variable to be. If you want the boiler pressure to hold at 100 psig you adjust the setpoint until the parameter represents 100 psig. When properly applied the controller will indicate it is set at 100 psig and you don't even have to know what the actual value is. Setpoints are not always set by you; a setpoint can be the output of another controller.

We normally describe a setpoint as being "local" when you can adjust it and "remote" when it's the output of another controller. Note that the terms don't relate to what you would consider as local and remote. If you have to leave the boiler plant and go around the main building to the shed under the water tower to adjust the setpoint of the tower's water level controller it's still a local setpoint even though it's remote from the boiler plant. If it was a pneumatic setpoint you could install a little regulator and tubing in the boiler plant (local) and extend the tubing out to the shed under the water tower to produce a... well, I'm sorry to say it but it's still a local setpoint. As you'll see later, a remote setpoint can come from a controller right beside the next one in the control panel so don't confuse local and remote with location.

Now we can define a loop. We use the term loop to describe parts of a control system because each control loop is like a circle; there's no end to it. The parameter

we're trying to control (process variable) is sensed by the controller which compares that value to the setpoint then adjusts its output accordingly. The change in output produces a change in the process variable and the controller compares that new value to the setpoint to change its output again. A control loop contains a controller, a device to measure the process variable, a source of setpoint, an output device that controls the flow and anything else that changes the value of the process variable or the setpoint.

A loop can be as simple as a level controller consisting of controller with internal setpoint adjustment, a level transmitter and control valve to similar devices in combination with a large number of computers located in different parts of the plant. The practical limit of a loop is at the devices that affect the process variable and any one of those devices can be part of another control loop.

There is always a control range. The values the controllers use have an upper and lower limit. The range of transmitters has to be established to permit reasonable control and allow for the normal variations in the measured parameter. A range is selected by the applications engineer (the gal or guy that selects and specifies the controls to be used on a job) to ensure the system will control properly. What's the big deal? It's a question of accuracy and stability.

If you are operating a low pressure steam plant, then a transmitter producing a signal in the range of 0 to 30 psig can maintain a pressure of 10 psig plus or minus 0.15 psi because the transmitter (which typically has an accuracy of $\pm 1/2\%$) will produce a signal that accurate. On a plant operating at 3,000 psig a 0 to 4,000 psi transmitter would be accurate to $\pm$ 20 psi and that wouldn't necessarily be considered accurate control. So the engineer might use a transmitter that works in the range of 2,500 to 3,500 psig to get a transmitter accurate within 5 psi.

Control signals also have a range and each system normally uses the same signal range for all the devices in the system. There are many standard ranges of control signals with the most common ones being 3-15 psig (pneumatic), 0 to 5 volts (electric, electronic), 4 to 20 milliamps (electronic). Several other signal ranges were used and it's not uncommon to encounter a mix of these ranges within systems that are a mix of old and new instruments and controllers. Other signal ranges you may encounter are 0 to 30 psig, 0 to 60 psig, and 3 to 27 psig pneumatic, 0 to 10 volt, -5 to + 5 volt, 0 to 12 volt, and 0 to 24 volt values on electrical and electronic systems. There are others but their use is industry specific and very limited.

The control signal range is representative of the value of the measured parameter, the process variable. You can measure the control signal and, knowing the range of the transmitter, determine the actual value of the process variable. A simple example would be a loop to maintain 200 psig after a pressure control valve where the transmitter range is 0 to 300 psig and the control signal is a 0 to 30 psig air pressure. You know the control signal value for the setpoint has to be 20 psig (or equal to it) and the actual value can be determined by multiplying the control signal by 10. If we wanted a remote indication of the pressure we could extend the transmitter output with 1/4 inch copper tubing to a 0 to 30 psig pressure gauge and add a zero to each number on the gauge face. The tubing and lower pressure gauge would be considerably cheaper than running steel steam piping to the remote location with a high pressure gauge; the first demonstration of why we use instrumentation, it saves money.

Oops, I just used another big word. Instrumentation, as I understand it, consists of devices that could be used in control loops but they don't do any controlling. All they do is provide indications of the value of the process including parameters such as pressure and temperature and quantities like pounds, gallons, or cubic feet. We use the term controls and instrumentation to describe a complete system that not only maintains the desired parameters but provides outputs that tell you how it's doing and what's been done.

Before I jump off the subject of control (and instrument) signals I have to mention the concept of live zero, why we have it, and how to deal with it. When we engineers say "live zero" we mean something that isn't zero; ...oh well, so much for simple explanations. Live zero control signals are those for which the control signal value that represents zero is more than zero; like in a 3 to 15 psig or 4 to 20 milliamp control range. The 3 psi or 4 milliamps represent zero.

The main reason for a live zero is you can be sure of it. Our pressure transmitter in the previous paragraph can be set at zero output with zero pressure applied to it but we can't be certain that it will come off that zero properly; there may be slack in linkage or stiffness in the bellows that has to be overcome. With a live zero we can see that the signal value is right where it's supposed to be with zero pressure at the process connection and can adjust the output while watching the signal approach the live zero from either direction. It's darn hard to get a minus pressure or minus electrical current reading and live zero solves that problem.

That's probably more than you want to know about control labels but you will find that your understanding of them will help you get answers to inquiries about other control systems. Talk the talk and everyone thinks you're an expert. Understand the talk, and you are. Now lets talk actual controls and what they do.

A common application is a simple level controller and I'll use that to give you an example of control methods. We'll begin with a simple float control valve (Figure 11-1) which maintains the level (our process variable) by controlling the flow of water leaving the tank. If you're at all familiar with these float valves you know the level has to vary. When there's no flow out of the tank the valve has to shut off. Conversely, when water is drawn out of the tank at a high rate the valve has to open fully. In order to change the position of the valve the level in the tank has to change. When water use is low the level is higher and the highest level is at shut-off. The level has to drop for the valve to open fully. The level cannot be maintained at one precise point because the level has to change in order to control the flow.

The required change in process variable to achieve control is called "droop" and it's the difference between the value of the process variable at no flow and the process variable at maximum flow. The float controller is comparable to other "self-contained" devices that maintain desired pressures, temperatures, and other parameters; they work fine when the flows are low and the deviation in process variable is acceptable.

There are other factors that prevent all controls being as simple as a float control valve. The pressure of the water supply can be so great, or the flow so great that the float control valve simply will not work. If the pressure differential gets high enough it will force the

Figure 11-1. Float control valve

valve open regardless of the position of the float. The system in Figure 11-1 is obviously operating with very little pressure drop across the control valve. That's one of the few I've seen without a wire or cable led down to operating level so the operator can give it a yank to get it operating again.

You could calculate the maximum supply pressure for a float valve controlling water supply to a tank. Calculate the volume of the float and multiply by the density of the water in the tank (62.4 pounds per cubic foot for cold water) and the equivalent length of the float arm from the pivot to the center of the float. That's the maximum torque the float could impose on the valve because at that point the float is sinking. Divide the torque by the length of the pivot arm on the valve (from the pivot to the center of the valve disc) to determine the maximum force on the valve, then divide that force by the area of the valve disc that is exposed to the difference between supply pressure and the pressure in the tank. The result of your calculation is in pounds per square inch and that's the maximum pressure difference for the float valve. If the drain leads to another tank at atmospheric pressure the result is the maximum pressure (in psig) in the tank, the most the valve can handle.

If the flow is high the valve opening has to be large enough to handle the large flow and that requires the valve disc to be larger. Using the same procedure I just described you can see that eventually the disc will get so large that the water will force it open at very low pressures. You could use a larger float but there are limits to float size imposed by the largest float chamber or, for floats in tanks, the tank opening. That's why you'll occasionally see floats that are cylinders, able to fit in the hole in the tank but long enough to provide enough displacement volume to operate the control. Another problem with larger floats is they will collapse when exposed to high pressures inside an enclosed tank such as a boiler.

You could increase the length of the float arm to increase the torque but there's limits to that imposed by the inside of the tank and the increased droop. Now you probably realize why simple float valves are seldom used to control water level in a boiler. Small residential boilers are frequently fitted with one but it has a minimal water capacity and is limited to low pressure boilers.

A modulating controller that maintains a tank water level (on off control is described later) can be compared to that simple float controller. We can use a float operated valve to produce the control. It can work just like the float valve but control a much smaller volume of water with a very small valve so it can handle the

high differential. You'll probably never see anything exactly like this type of controller (Figure 11-2 which is a valve filling a bucket over the valve with an opposing spring) but it allows me to show you some concepts of control. The valve controls flow to a bucket on top of the main control valve. When the water level drops the float valve increases the water flow to the bucket to fill it. The heavier bucket overcomes the weight of the spring and closes the drain valve.

The drain hole in the bucket lets water out, somewhat essential because without it the main valve would close and never open until the water evaporated out of the bucket or you removed it. Control is achieved by changing the level of the water in the bucket; it fills to close the valve and drains to open it. Notice the differences between this system and the simple float control valve; an external source is used to power the system (weight of the water in this case) and the transmitter and main control valve are separate with no dramatic restrictions on the distance between them, another advantage of control systems.

There's another notable difference in this control system, the float valve that's used as the controller isn't the same as the typical float valve because it works backwards. Notice that the flow of water through it decreases with level, just the opposite of the simple float valve. It happens because the pivot point is on the other side of the valve. It was necessary to make the control system work and it reveals one of the control concepts you have to get used to, there are direct acting controllers and reverse acting controllers. A direct acting controller

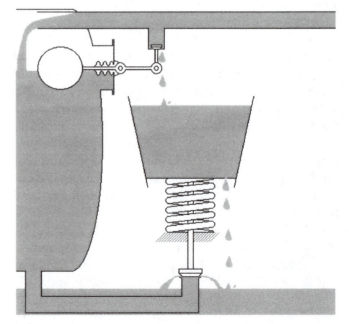

Figure 11-2. Bucket valve control

increases it's output as process variable increases; a reverse acting controller reduces its output as the process variable increases.

Controllers like the one we just described are seldom found today because there are a few problems with water; it's corrosive and contains solids that can eventually plug up the control orifices. In the prior example dust from the atmosphere could get into the bucket and close the drain hole to prevent the valve opening. We used to have hydraulic controls (which used oil instead of water in closed systems) but their expense and problems with corrosion and leaking resulted in their having a short period of acceptability. They were replaced by pneumatic controls which survived several years before they were outstripped in price and function by microprocessor based electronic controls, the current choice as of the writing of this book. Electrical and electronic controls saw some use and a share of the control market along with pneumatic controls as well.

I lived through the era of sophisticated pneumatic control. It provided more accurate control at lower cost than earlier mechanical and hydraulic systems. We're now living in the era of microprocessor based control. Who knows what will follow?

The system just described consisted of controller and control valve and is not consistent with modern control systems because the controller measured the process variable directly. A typical control system will have a transmitter which produces a control signal proportional to the value of the measured variable, a separate controller and a final element (control valve). We could relate the level of the water in the tank to the level in the bucket but that will change as the drain hole plugs or erodes and is also affected by the pressure drop through the valve and other factors. We could convert our float valve controller to a transmitter by drilling a hole in the outlet piping to let the water drain there and use the bucket as a reservoir. Installing a pressure gage on the piping feeding the bucket provides an indication of the output of our transmitter. The problem is our pressure transmitter can't produce a control signal that's precisely proportional to the level in the tank. A variation in the water supply pressure, wear in the valve and drain orifice and friction in the valve packing will all combine to generate changes in the signal that produce errors.

A desire for accuracy and, more importantly, repeatability resulted in the development of precision transmitters by introducing feedback. Feedback is the output and we use it to test or correct our output. In the case of a transmitter it's used to ensure the output is really proportional to what we're measuring, what we call the process value. Let's modify our float valve and use compressed air instead of water. There are two advantages to using air over water, one is it has very little weight so the weight of the air doesn't alter our signal value when the signal is piped up or down two or three floors in the building.

More importantly nobody complains when it leaks out. Let's face it, people would complain about our water powered transmitter constantly pissing water out but they don't even notice the air. The air leakage, normally undetected, proved to be a considerably costly part of control systems.

We also change the valve and float arrangement so the float arm compresses a spring and the spring force is opposed by a bellows that contains our output pressure so we get a transmitter that looks like the one in Figure 11-3. We have moved the orifice from the bucket to the air supply and created another one consisting of a nozzle. The nozzle discharging against a baffle becomes our valve (less expensive than a valve) and the valve moves with the float arm because we want the output to accurately represent the level in the tank. Flow through the valve doesn't change based on position of the float, it responds to differences between the position of the float and the balance of forces of the spring and the bellows.

This construction is typical of most pneumatic transmitters. As the level increases the nozzle is moved away from the baffle so more air bleeds out at the nozzle. The pressure in the output bellows decreases so the spring pushes down on the float and up on the baffle,

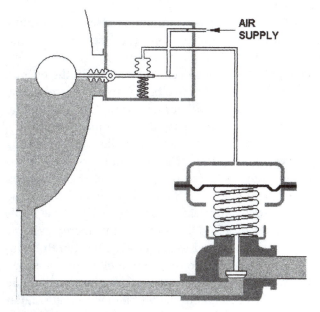

Figure 11-3. Pneumatic level transmitter and control valve

following the nozzle. When the level falls the nozzle is pressed against the baffle so the pressure increases and the bellows compresses the spring to push the baffle down. The transmitter uses a pressure balance principle where the output pressure of the transmitter is fed back (feedback) to restore the balance of the device, in this case the relative position of the nozzle and baffle.

This transmitter is reverse acting, the output increases as the level drops. In the figure the valve is draining the tank so it drops the level. The same transmitter can be used for direct control of a makeup water valve supplying a boiler feed tank because we could change the valve internals. The increasing air pressure would push the valve open.

The system shown is using the transmitter as a controller, and it would work, but it's seldom done that way for several reasons, price and power predominating. By switching to compressed air we were able to make a much simpler valve in our transmitter/controller and make it much smaller, lowering the cost of it dramatically. The reduction in size reduced air consumption a lot too so it costs less to operate. The small transmitter cannot, however, move lots of air so it would take a very long time for it to pass enough compressed air to increase the pressure in the diaphragm casing of a large pneumatic control valve. If the transmitter was used as a controller there would be a considerable lag in operation because it would have to pass all the air for the control valve in addition to filling the feedback bellows and connecting tubing. The very limited output of transmitters prevents them being used as controllers for those reasons.

Our simple transmitter would also have a droop, although not as noticeable as other methods, because the distance between the nozzle and baffle would have to change to raise or lower the pressure in the output. That produces a difference between the control signal and float position. Another important factor in the design of the transmitter also allows for increased droop. That's because the designer had to allow for something to go wrong (like loss of air pressure) so the baffle is usually a flexible piece of spring steel that can bend without breaking when the level is low and there's no air pressure to compress the spring and keep the baffle to nozzle position. As the control signal increases some of the pressure is used to bend the baffle slightly to introduce more droop. To reduce that effect on the transmitter and save on even more air the designers made the nozzle even smaller. The problem with that smaller nozzle was it could handle even less air and any leak in the tubing connecting the transmitter to other devices

would introduce an error, the output would be lower than it should be.

To eliminate the problem of leakage loading down a transmitter designers added boosters to their transmitters. The reduced size of the nozzle and baffle assembly and savings in compressed air consumption allowed them to reduce the cost enough to justify adding the booster which is a simple device. A booster installed in the transmitter eliminates any problems with tubing leakage loss because the nozzle air only feeds the feedback bellows and the booster diaphragm (Figure 11-4). The large area of the diaphragm provides ample force to position the output valve so the transmitter can pass enough air to compensate for small tubing leaks without a degradation in the value of the signal. It also allowed an operator to detect a leak by comparing a gauge connected to the output bellows and the tubing at another instrument. As designs of transmitters improved the nozzles got even smaller and, in some cases, a booster is used to feed the feedback bellows.

Believe it or not, you now know enough to understand almost any kind of pneumatic control device. That's because the bleed and feed and pressure balance principles we covered are basically what is used in all pneumatic devices. I'll continue using pneumatics for a while to show you the other concepts of controls.

Before we leave our level transmitter I do want to cover displacement transmitters. You're bound to run into a displacement transmitter some day because they do resolve some of the problems with floats. If you haven't had the pleasure of working on a toilet fill valve in your lifetime (highly unlikely for someone with an operator's skills) or even if you have, please go into the bathroom and lift the tank cover to check out the internals. Unless you have a modern pressure assisted toilet or a flushometer there will be a float valve there to control the water filling the tank. Gently push down on the float and continue pushing it until it is completely under the water noticing the force required, then dry your hands and come back to the book. I'm sure you

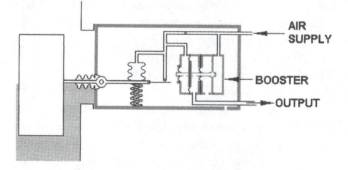

Figure 11-4. Booster for pneumatic transmitter

noticed that the force required to push the float down increased with depth. If you didn't notice, go back and do it again. The additional force is equal to the difference between the weight of air in the float and the weight of water it displaces, the buoyancy principle. Displacement transmitters balance the force on the float with a force produced by a feedback bellows.

Pressure transmitters use the same principles of force balance to produce an output by using another bellows or a diaphragm sensing the pressure in the process and balancing that with an output feedback. Different pressures are accommodated by changing the size of the process bellows or diaphragm. Pressure transmitters would be very expensive if a special bellows had to be made for each pressure range so they are made adjustable within standard ranges by allowing adjustment of a pivot between two beams connected to the bellows and feedback.

Temperature transmitters work the same, they just need a way to get motion or force proportional to the temperature then convert it to a signal. Bimetallic sensors use the movement or force produced by the difference in thermal expansion of two metals, fluid filled transmitters use the thermal expansion of the liquid to produce movement and gas filled transmitters use the increase in pressure proportional to temperature.

Electronic pressure and differential transmitters sense process values using the same techniques as described for pneumatic transmitters converting a force or movement to a voltage or current and generating a feedback force using an electromagnet. Temperature transmitters use a resistance to electric current where the resistor's resistance varies with temperature. Another means of measuring temperature that has been around for years is a thermocouple. Two wires of different materials connected at their ends will produce an electric voltage when the two ends are subjected to different temperatures. Note that the reference temperature (one end of the two wires) has to be stable to get a reliable signal proportional to temperature at the other end. Digital transmitters use similar methods then convert the analog signal to a digital one.

Gee, we got this far in the discussion of transmitters without mentioning the word "analog." That wasn't hard because, for all practical purposes, all pneumatic, voltage and current signals are analog signals. The signal represents (is analogous to) a process value, you can take a measurement of the signal and can determine the process value from the value of the signal. That's all an analog signal is, a value that represents another one.

What makes digital signals different? They change rapidly, commonly from a negative voltage to a positive voltage so there's no way you can put a meter on the signal terminals and measure it. The value of a digital signal is a function of the number of changes in value and the time between each change, so complex that it requires a computer to read it. Why are they better? Because the actual value is not important. Any significant resistance in the signal wiring for a voltage signal, like a loose terminal, can alter the signal to produce an error. Digital signals represent zeros and ones where a zero is considered anything between plus five and plus fifteen volts and a one is considered anything between a minus five and minus fifteen volts. That considerable range of voltage minimizes errors and the additional features of digital signal transmission provide more accuracy and reliability than analog signal transmission. All that and the lower cost, both hardware and installation, of digital controllers have made them the controls of choice, replacing all other types of control.

My understanding of control operation is based on pneumatic controls so I'm going to continue using them as examples in describing concepts. You may never see a pneumatic control system but the concepts work with any type of controller and a pneumatic understanding will help you comprehend them. I'll even use a controller that's no longer available (like most pneumatics), a Hagan Ratio Totalizer as shown in Figure 11-5.

The totalizer has four diaphragm chambers but they could also be fitted with bellows. The totalizer was designed to provide universal use by adapting it. The output chamber and A input chambers are secured to the base of the transmitter. Sliding in the middle are clamps that connect to the base and the beam. The beam floats in the middle of the assembly, attached to the diaphragms and the beam clamp. A very thin piece of spring steel connects the two clamps to form the pivot of the controller. The two clamps can be loosened and slide along the base and beam to positions right and left of center. The valve floats in the output chamber and

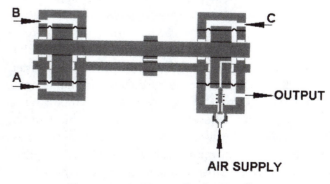

Figure 11-5. Hagan Ratio Totalizer

will open to admit air if the beam is rotated clockwise or close off the air supply and stop while the beam continues to rotate counterclockwise. Further counterclockwise rotation will open the bleed end of the valve to dump air to atmosphere.

Let's start with proportional control. That's where the output of a controller is proportional to the difference between the process value and the setpoint. Assume we're using the level transmitter covered earlier to produce the process value so, in this case, our controller will be used for level control. We'll also assume the level control valve is reverse acting so an increase in controller output will close the valve. When the water level in our tank increases the control signal decreases. To make the system work any increase in process value should result in a increase in output to close the valve. Now we can look at the ratio totalizer to see how to connect the process variable. The output bellows pushes up on the right side of the beam so any increase in output will tend to rotate the beam around the pivot in a counterclockwise direction. It's a pressure balance system so the process variable has to create a tendency to rotate the beam in the opposite direction to balance the forces.

If the beam tends to rotate clockwise more air is supplied to the output and output bellows to counter that rotation. If it tends to rotate counterclockwise the vent valve opens to decrease the output. Connecting the signal from the transmitter to the bottom bellows (A) does the job. Now a change in the level will produce a change in the output of the controller to open or close the control valve. As shown the controller acts pretty much like a signal booster because it produces a change in output that precisely matches a change in input. As the level transmitter output changes from minimum to maximum the controller output produces the same value because the bellows areas are identical. It works pretty much like our float controller, requiring the level change over the full range of the float to position the control valve between open and closed.

The whole reason for using a control system is to improve on the operation we get with a float control valve so let's see what we can do with it. We can reduce the change in level by moving the pivot on the controller closer to the output end. It works just like a teeter totter. Let's adjust the controller so the distance from the center of the process input to the pivot is twice the distance from the output bellows to the pivot (two thirds of the beam length). Now, if the level varies to produce a 1 psi change in the transmitter output the controller output has to change by 2 psi to maintain the force balance in the controller. There is a proportional difference in the change of the signals where the output has to change twice as much as the input. That's the concept of proportional control and in this case the controller has a gain of two which means the output has to change twice as much as the input. Now the controller will run the water valve from closed to open with half the change in the output of the level transmitter, between 25% and 75% of the signal range.

We could increase the gain until there was very little change in process value to produce a full stroke of the water control valve so the water level would not vary much. If we did that it wouldn't work too well because any little ripple in the water level would produce a dramatic change in the valve position and we would have a lot of valve wear. We would also have controller "noise" where the output is jumping around with little relationship to the actual level in the tank.

Conversely we could reduce the gain to something less than one which would create another problem; the water valve would never close. It might work during normal plant operation but when the plant is shut down the controller output couldn't increase enough to close the valve. Too much gain produces a lot of noise and erratic operation while too little gain can result in failure to operate at the extremes of load.

One problem with this controller arrangement is we have no way to adjust the setpoint. For all practical purposes the setpoint is the center of the transmitter's position. In order to have an adjustable setpoint we use the B bellows of the controller and supply it with a control signal that is adjustable. The setpoint signal in this case is produced by a simple air pressure regulator. By connecting the regulator to the bellows opposite the one sensing the signal from the transmitter we have developed a setpoint controller.

Now the output of the controller is proportional to the difference, what we call the error, between the setpoint and the transmitted level signal. Instead of acting only on the pressure from the float transmitter the action is dependent on the difference between the setpoint and the process variable. The setpoint pressure acts on the diaphragm at B pushing down on the right end of the beam opposite the process variable signal coming in at A. The force tending to rotate the beam is equal to the difference between the two pressures times the area of one diaphragm.

All modern controllers operate on the error, not the actual signal value. Now changes in output are proportional to changes in the error, not changes in the level. An important part of this to understand is you can introduce an error by changing the setpoint. We'll need to set the gain to much more than two in this case or the output may not change enough to fully stroke the valve.

Creating a setpoint controller allows us to use something more than the level control range for the transmitter so we can use the transmitter for instrumentation as well as control. We can put a long arm on the float and produce an output signal proportional to almost the full height of the tank so we can tell where the level is even when it's not in the control range. For example, our level transmitter could be set to indicate levels from zero to 60 inches in the tank. We select our control range and adjust the controller gain accordingly. If we want the level to control within ten inches we set the gain of the controller to six. If we establish a setpoint at fifty inches the control valve will be fully closed when the level reaches 55 inches and closed at 45 inches. We can also adjust the setpoint to anywhere from five inches to 55. We have to reserve half the control range to have control, that's why the setpoint can't be anywhere within the range of the transmitter when we're using proportional control. If we raise the setpoint to, say 58 inches, then we will not be able to stroke the output valve completely.

By now you're asking how good is a controller that needs ten inches to work when it comes to controlling the water level in a boiler. If we had to use the system described we would have to set the gain to sixty in order to keep the level within an inch of setpoint. There's two answers to that question, first we normally use a maximum of twenty inches for the range of the boiler water level transmitter (even if the boiler is over a hundred feet high) and we'll add reset to our controller. There are practical limits to an instrument's range when it is used for control but reset control is a refinement that can only be described as beautiful; it makes the setpoint realistic.

To convert our controller to a reset controller we add (Figure 11-6) some tubing, a needle valve, and a small volume chamber. It's these reset accessories that make our controller a reset controller. The controller has now acquired dynamic properties. The only time a reset controller will be in balance is when the setpoint and the process variable are precisely the same and the output has stopped changing. With our proportional controller the system could be stable with the level holding at a value below or above the setpoint. Now the left side of the controller is in balance only when the process variable and setpoint are precisely the same, when the error is zero. Even then the controller output can be changing, when the pressures inside the output and reset bellows are different.

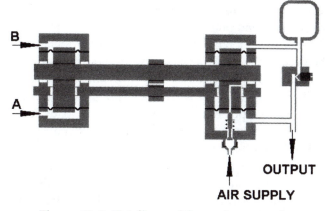

Figure 11-6. Totalizer with reset accessories

If this looks like an unmanageable concept don't quit yet. Reset control does some great things and after we get through this discussion you should be able to appreciate what it does.

Operation of reset control is difficult to comprehend and I've discovered many technicians have an inappropriate perception of it because they think in terms of speed, not response to an error. The operation of the ratio totalizer provides a basis of understanding because the dynamic effects are apparent. Let's start with a steady state condition where the pressure in the output bellows matches the pressure in the reset bellows and the error is zero. Assume the process variable drops a little so an error is generated. The proportional function of the controller responds immediately, changing the output an amount equal to the error times the gain. Also assume the error isn't corrected immediately by the output change and holds. Since there is now a difference between the output and reset bellows control air bleeds through the needle valve to (or from) the volume chamber and reset bellows. Since the error persists the output will have to continue to change to balance the error. If the error continued to exist the output would continue to change until it reached its practical limit (zero psig or supply pressure).

That doesn't happen often, usually the controller action results in the process variable returning to the setpoint. That's the beauty of reset control, it always works to return the process variable to the setpoint, not some value that's offset by the proportional value. It's real control. You can see that the only time the controller isn't changing its output is when the process variable and setpoint are exactly the same and the pressure in the output and reset bellows equal each other. You can also see that the controller can be in balance with any pressure at the output. The output signal can be anything from zero to supply pressure balanced by the same pressure in the reset bellows and the controller will be satisfied as long as the setpoint and process variable pressures are the same. Unlike proportional control we don't need a deviation in the process variable to get the required output.

It's reset control that keeps the boiler level right at the center of the glass while changing the feedwater control valve position from closed at low loads to almost wide open at high loads. Its also reset control that makes it possible to keep the steam header pressure at 120 psig whether we're at low fire or high fire and even when we're running five boilers instead of one. It's reset control that allows us to run air/fuel ratios so tight that oxygen in the flue gas can be held at one half percent.

Tuning a reset controller is nowhere near as easy as tuning a proportional controller but the additional feature of the controller (you have P + I, proportional plus integral) allows you more flexibility in matching the process. Aw shucks, another fancy word! Integral is a mathematical term that sort of means accumulating the average value. It's not important to understand mathematical integrals, only that it's another name for reset.

Tuning consists of changing the gain (proportional control) and reset (integral control) until the combination provides a response to an upset in process conditions where the process variable returns to setpoint within a short period of time and with only a little overshoot in response to the initial error. You've probably seen the curve in another book, where the error is plotted versus time, it starts as a big error with a rapid change in process variable quickly approaching the setpoint, overshooting it a bit, then turning back toward the setpoint and falling in line with it. It's a pretty picture but making it do that in the real world can be damn difficult at times.

On several occasions I've run into a reset controller that had all reset blocked out (like completely closing the needle valve) because an operator didn't understand reset control adjustments. Keep in mind that a simple proportional controller requires an error to do its job and you'll find that attempts to minimize that error result in some pretty wild swings in the output of the controller; what we call instability. Those swings are primarily associated with the fact that the process doesn't respond instantly to changes in the controller output. It can take a few hundredths of a second to several seconds before the full effect of a change in controller output is apparent by looking at the process variable.

You tune a reset controller to deal with those delays. The controller will have two adjustments, gain and integral. Gain is the proportional part, the output is the error times gain. The output changes when the error changes. Integral is the reset adjustment and it repeats the error multiplied by the value of the integral. Notice that the reset effect is the error repeated; an integral adjustment is normally marked to indicate repeats per minute, meaning that is how many times the error will be repeated in one minute. That doesn't mean the controller only repeats the error for a minute either; it continues repeating the error every minute. It also doesn't repeat it at the end of a minute. If the integral is set at 60 repeats per minute it will increase or decrease the output by value equal to the error every second.

A proper combination of gain (proportional control) and integral (reset control) will make the process return to the setpoint quickly and smoothly. Now that you understand the way the controller operates you should have a better idea of which adjustment to use and which direction to turn it, a big step in tuning a controller.

Adjusting the gain of a ratio totalizer was a lot more complicated than adjusting it on a modern controller. You had to release two set screws that held the pivot spring to the base and the beam then slide the pivot spring assembly to a new location and tighten the screws. While you did that the output was always jumping all over as you handled the parts and turned the set screws so you didn't have any idea of what the results would be until you got your hands off it. Gain on modern controllers can be adjusted without affecting the output except for the difference in the gain (times the error). Increase the gain and the output changes more for a given error. Just to make sure you understand it, the error is the difference between setpoint and process variable, what you want and what you've got.

Adjusting the integral of a ratio totalizer didn't upset the operation so much because adjusting the needle valve didn't have the effect that grabbing the beam to adjust gain did. It was more like a modern controller. If you opened up the needle valve you increased the repeats per minute because the air could flow through (adding air to or bleeding air from the volume chamber and reset bellows) faster. Closing down on the valve decreased the repeats per minute. Closing the needle valve entirely made for a pretty sloppy controller because the proportional part had to compensate for changing the compression of the air in the reset bellows and volume chamber as well.

As for tuning the controller, you adjust the gain or reset to balance the system response. If a change in controller output produces an almost instantaneous change in process variable then most of the control function can be left to proportional control. If, however, the process responds sluggishly to a change in controller output then the integral adjustment is more critical. Watching what happens when you introduce an error will give you a good idea of how to adjust the controller; a set-point controller allows you to do just that.

When starting up a new system I adjust the controller using some initial adjustments that are the average for comparable systems then switch it to automatic to see what happens. Quick swings in the process variable indicate instability and I would reduce gain immediately if they happen. Then I sneak up the gain until it starts getting a little unstable and back off some to eliminate the instability. If the process doesn't change due to external influences I introduce an error to see what happens. Actually it's much easier to work with an error you create because you can intentionally make it a value that you can relate to, something simple like 1% or 5% or 10%.

How do you create an error? You just change the setpoint, swinging it your selected difference from the process variable. If the process overshoots the setpoint considerably then reduce gain. If it seems to take forever for the process to return to setpoint then increase integral. If the process returns to setpoint while swinging back and forth either side of the setpoint then reduce integral. If the process slowly returns to setpoint then increase integral until the process overshoots the setpoint a little once.

Changes in one adjustment normally require an opposite change in the other when you get close to the desired characteristic of the controller (that curve where the process overshoots setpoint once then swings in to match it); an increase in gain will probably require a decrease in reset and vice versa.

Figure 11-7 is my rendition of that popular graphic you see in all the instructions for tuning a controller. Hopefully the previous discussion makes it meaningful to you now. It used to be rather difficult to get a graphic output on a recorder or other instrument that you could compare to something like Figure 11-7. Today you can use a recorder and speed up the chart or simply adjust the parameters for a trend screen.

Trend? Yup. That's the term used for recording today only it's not a pen on paper that leaves a permanent record. It's a graphic produced on a computer screen that draws a line between points of recorded values relative to time. It looks just like the old pen on paper chart, it's just done digitally on a video screen.

Now I have to say that it's not always that simple. Some systems are set so the data are only recorded every

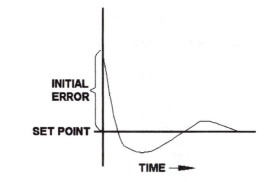

Figure 11-7. Controller error versus time, tuning guide

five seconds to every minute. In that case any graphic you're looking at can completely miss the swings you generated by changing the setpoint. Be very aware of that potential limit on electronic data.

When it comes to tuning controllers there is no substitute for practice to gain experience. If you decide to get some practice with a functioning controller you should record the gain and integral adjustments before you start changing them so you can restore the controller to its original settings. If it doesn't seem to work as well when you're done playing keep in mind that hysteresis can have an effect so restore the original settings by approaching them from the opposite direction.

Hysteresis! Yup, another one of those fancy words. It has to do with friction in mechanical systems but it can occur in almost any situation. The best way I know of to explain hysteresis is to relate to the operation of a pneumatic control valve without a positioner. The control valve in Figure 11-8 consists of a chamber over a rubber diaphragm where the control pressure can push down on the valve stem and a spring that pushes up to resist the pressure forces. Without hysteresis the position of the valve would be precisely proportional to the control pressure. The push down would be a force equal to the area of the diaphragm in square inches multiplied by the control pressure in pounds per square inch. For a 3 to 15 psig control signal and a 50 square inch diaphragm the force would vary from 150 pounds at zero control signal (3 psi × 50 sq. in. = 150 pounds) to 750 pounds at 100% control signal. The spring would be compressed to balance the 150 pound force when the valve is closed and have a spring constant equal to 600 pounds divided by the stroke of the valve. If the valve stroked 1-1/2 inches the spring constant would be 400 pounds per inch. (This is a typical valve so you can see why you can't move it)

Now to get to the hysteresis part; the valve packing is tight on the valve stem to keep it from leaking and that tight packing tends to hold the valve stem

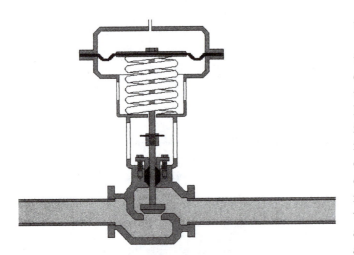

Figure 11-8. Simple pneumatic control valve diagram

where it is. The friction always acts in opposition to the travel of the stem so it will push against the diaphragm force when the valve is closing and oppose the spring when the valve is opening. It produces a difference in valve position for a given control signal depending on whether the valve is opening or closing. The graph in Figure 11-9 is a typical hysteresis curve and it applies to the valve just described.

Mechanical hysteresis isn't the only thing that creates a difference in position of a control valve operating on a control signal directly. There is a difference in the amount of air the controller must pass depending on the valve position because the volume of the diaphragm chamber increases and decreases with valve position to

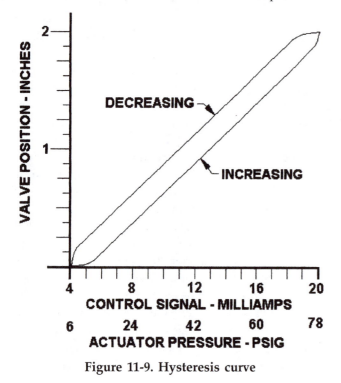

Figure 11-9. Hysteresis curve

upset the performance of our controller.

There's also the problems associated with the controlled fluid as well. When the valve is closed the difference in valve inlet and outlet pressures act on the area of the valve opening, adding another force to the valve stem. If the valve is a boiler control valve it can work perfectly fine when the boiler is operating but leak when the boiler is shut down because the pressure drop across the valve disc is so great that it overcomes the forces produced by control pressure. All these factors can be overcome by making sure the combination of diaphragm area and valve chamber pressure will keep the valve shut. Adding a positioner also helps because it can operate with higher actuator pressures using a separate air supply and match the valve position to the control signal.

A valve positioner is just another controller. It controls valve position by comparing the actual position (as a process variable) to the control signal (remote setpoint). The control signal becomes a remote setpoint because it is produced elsewhere and it's also a variable setpoint because it changes. A rather simple positioner is shown in Figure 11-10. The remote setpoint is the pneumatic signal coming to the positioner. The process variable is developed by the spring compressed by linkage attached to the valve stem; as the valve opens it compresses the spring.

Changes in the control signal change the force on the diaphragm so the spring is compressed or allowed to expand and that changes the position of the valve to divert air into or out of the diaphragm. The valve position is changed so the compression of the spring matches the control signal to return the valve to its center position. The pressure in the diaphragm is like the output of a reset controller, it's whatever it has to be to do the job. A positioner can also use a supply pressure higher than the control signal range to overcome high differential pressure on a valve and the friction of some packing that you tightened a little too much.

As far as I'm concerned, any control valve in a boiler plant should be equipped with a positioner. Today, with electronic control signals, the positioner has to adjust the air pressure to match an electronic signal. One simple positioner uses two solenoid valves, one to add air, one to bleed it off.

I think it's a good time to talk about reset windup because reset controllers and positioners did, and some may still, have that characteristic. Also these valve positioners can experience windup. The feedwater control valve mentioned earlier is a good example; we put a positioner on the valve and the pressure in the diaphragm of the valve actuator ran up while the

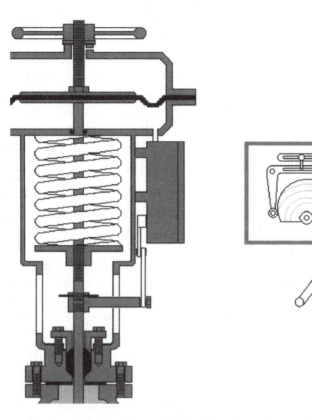

Figure 11-10. Simple valve positioner

boiler cooled because it had to overcome the pressure of the feedwater trying to open the valve. While the boiler came up to pressure the actuator pressure didn't change. (Why should it? It was happy because the valve position matched the control signal). When the boiler starts making steam and the water level drops the level controller will raise the control signal a little indicating the valve should open a little. In normal operation the valve would respond rather quickly but this first time, after a shutdown, it won't. It's because the positioner has to bleed off all that air that was compressed into the diaphragm chamber to raise the pressure enough to keep the valve closed against the high feedwater differential (that's now gone). That may explain why you've been surprised at the lag in response when operating a valve manually. I know I was, more than once, because I thought I had opened the valve a little and got no response so I raised it a little more and the next thing I knew the water level was at the top of the glass because once it did start to open, it opened!

The original pneumatic controllers did the same sort of thing and introduction of live zero made it happen at both ends of the control signal range. A generic 3 to 15 psig controller could wind up to an output equal to the standard 18 psig supply pressure or wind down to zero output. When that happens we're out of control,

the controller has done all it could to restore the process variable to setpoint. Unless that's an intentional condition (it could be) the output will eventually get the process variable going in the right direction and it will return to the setpoint. It won't stop there though, it will continue right on past because the output hasn't changed. With a reset controller it can't because the output is in windup, or wind down.

Once the error is in the opposite direction the output will start to change. During the period when the controller is building up from zero or dropping down from supply pressure the controlled device, a valve for example, doesn't respond because it only responds to signals within the control range. The result is always a long delay (seconds, not hours—although sometimes it seems like hours) before the output gets back into the normal control range and, as a result, the process variable is swinging all over the place. That's the effect of reset windup. You won't see most modern controllers doing it because the control manufacturers have designed the controls to eliminate it. You may still encounter it with valve positioners and damper actuators. If you do run into it, at least you will know why it happens.

Reset windup is not the only problem that I had to deal with and you probably will not. At some point in time you will hear the terms "procedureless" and "bumpless" applied to controllers. On the off chance that you will get to work in an antique boiler plant I think it's a good idea for you to get an understanding of why those features were added so you can deal with the situation.

Early pneumatic control systems that included hardware like the ratio totalizer had separate manual/ auto stations which were flush mounted on the panel and gave people the option of controlling the process by hand. When it was considered necessary to give people the option of changing the setpoint the station also included that adjustment. Figure 11-11 shows what one of those stations looked like.

The setpoint adjustment was nothing more than a pressure regulator with the adjustment knob penetrating the front of the station. The setpoint was indicated on a pressure gauge mounted above the adjustment knob. The output of the controller was indicated on another pressure gauge and another pressure regulator produced the manual output signal. The valve handle in the middle of the station was used to switch from automatic to manual and back to automatic; however, it wasn't as simple as just turning that pointed knob.

If you simply twisted the valve knob from the automatic to manual position and the automatic and manual pressures weren't the same you got a "bump;" the output would jump from the output produced by the controller to the setting of the manual station. To transfer from auto to manual without a bump the valve knob had intermediate positions at 1/4 turn for adjusting the outputs to match them up. When transferring from auto to manual the gage was switched to show the

Figure 11-11. Early pneumatic H/A station

manual signal and you got to adjust it until it matched the automatic output before turning the knob another 1/4 turn to manual. When transferring from manual to auto it showed the automatic output so you could bias it to match the manual output before switching to auto. (I'll get to explaining bias in a bit) As you can see, you had to perform a signal matching procedure during the transfer from hand to auto and vice versa or you got a bump.

Those old stations worked pretty well as long as they didn't leak much and you were quick at adjusting the signal for the transfer. In a way I was sorry to see them go because I could adjust my manual outputs where I wanted the controls to be if I switched to manual and I knew they would go there.

You may also run into controllers with a balance indicator. It consists of a clear plastic tube that's visible through a slot in the front plate of the controller and contains a small ball that fits inside the tube with very little clearance. One side of the tube is connected to the manual output and the other to the automatic. When you're ready to switch from one to the other you adjust the manual output until the ball floats to the middle then throw the auto/manual selector switch over.

As pneumatic controls improved the manufacturers included additional little controllers inside their devices so the automatic signal automatically followed the manual output and the manual output was automatically adjusted to match automatic to permit rapid and "procedureless-bumpless" transfer between manual and automatic operation. Electronic controls had similar procedures that were replaced by add-ins. Similar functions are understood to be included in modern controllers.

Now for bias, it's a control engineer's term for add or subtract. It is done a lot in controllers but most of the time you don't see it. It became an integral part of the manual/auto stations so you could line up auto and manual signals and it was done in one control regulator where the output of the regulator was a combination of the controller output and pressure that opposed a spring. The manual adjustment loaded the spring and the assembly looked something like Figure 11-12. When the control designers noticed that we operators used that spring adjustment to produce a difference in the output of two manual/auto stations using the same control signal (like on coal pulverizers where we could bias the primary air and coal feed) they simply manufactured another faceplate with that regulator on it and called it a bias station.

As far as I'm concerned I've given you enough

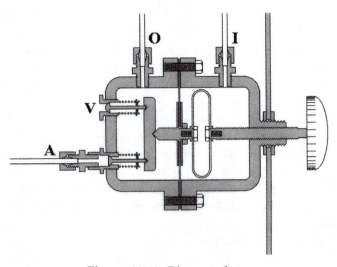

Figure 11-12. Bias regulator

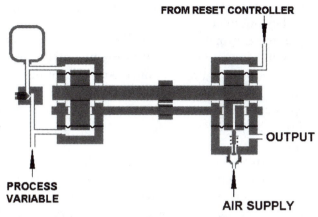

Figure 11-13. Ratio totalizer set up for rate control

about controls. If you've read this far you can handle most of the control problems you'll encounter. If you can relate the buzzwords proportional, integral, and bias you have the control world pretty down pat. Now I know someone is going to say "What about derivative control? Isn't that what the 'D' in 'PID' stands for?" Yup, that's derivative and you don't use it very often. I'm going to explain it but you'll use it sparingly; when it's needed in some unique application you can use it. There are a few other buzzwords that you need to know about and I'll explain them as we go.

Derivative control is 'Rate' in my parlance and it is a helpful control feature in systems where the process is upset quickly and erratically by external influences all the time. When there is no relationship between what an output controls and something that upsets the process a derivative control is almost a necessity. Let's take our ratio totalizer and convert it to a rate controller. It will look like the diagram in Figure 11-13. With no change in the process variable the output of the controller is equal to the output of the reset controller. The rate control occurs with changes in the process variable. If the process variable changes slowly it will have very little effect on the output because the control air will bleed through the needle valve fast enough that the pressure in the two bellows will remain about the same. If the process variable changes quickly the air can't bleed through fast enough so the difference between what it was and what it is produces an increase or decrease in the output on top of the reset controller signal.

Some manufacturers called the device a "pre-acting" controller because it changed the final output based on the action of the process variable. You can see how the output of this device would change according to the

rate of change of the process variable. When the process variable stops changing the output of the derivative element is always zero. It's called a derivative controller because the output is proportional to the rate of change of the input. Adjustments are in minutes per repeat.

To get an idea of where rate control can help, consider a system that maintains level in a small tank but the tank has an extra drain valve that's manually controlled. If the level is normal and there's no flow out of the tank the reset controller will wind down to shut off the water feed valve. Now someone opens the manual drain wide. With reset windup or even a stable reset control situation the level suddenly starts falling and the reset controller can't respond fast enough to keep the level from dropping quickly. The rate control senses the rapid change and forces the output valve open quickly.

Another control buzzword is "cascade" and it's used to identify the use of the output of one controller as an input to another. Cascade controls are useful when the output of one process feeds into another. Changes in the first process, which are made by the output of one controller, create proportional changes in the second process and you can reduce the impact on the control of the second process by using the output of the controller for the first process as an input for the controller of the second process. Okay, I know it sounds complicated, just read it again slowly and you'll get it. Typically drum level control and furnace pressure control on a boiler contain cascade control loops.

There are a lot of buzzwords that are specific to an industry that I don't have room to describe. You should be able to decipher what they mean by looking at the control schematics. A few words on reading schematics will be helpful before we get into control processes that are specific to boiler plants.

Control schematics are diagrams on paper that represent the elements of a control system for a process and how they are interconnected. Anything more elaborate than a simple proportional control system will normally have a diagram to show how the system is interconnected and help you figure out how it works. Control schematics or diagrams are not the same as PID diagrams (see documentation). The PID shows the process itself and where the transmitters and control elements are in the process.

Control schematics show how the transmitters and control elements are linked in a control system to control the process. The better control schematics will show the transmitters across the top of the drawing and the controlled elements (like valves and dampers) at the bottom so you have a flow from inputs to outputs going down the drawing. Lines on the drawing show the flow of information and may or may not indicate how that information is transmitted. They may be process signals like 3 to 15 psig pneumatic or 4 to 20 milliamps of current but can also be digital or something as unique as light (as used in fiber-optics); it isn't essential to know how the signal is transmitted to understand the system operation; only when you have to fix it.

Figure 11-14 is a simple single loop control schematic that I can use to explain some of the features of control diagrams. Diagrams of other systems later in this book will improve your understanding of them. This control loop is our level control system, described earlier, presented in control schematic symbology. These diagrams use symbols comparable to those standardized by the Instrument Society of American (ISA) and

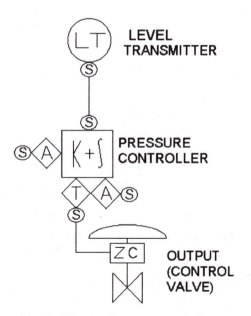

Figure 11-14. Single loop control diagram, SAMA symbols

the Scientific Apparatus Manufacturer's Association (SAMA). The level transmitter (LT) produces the process variable signal that is fed to the controller. The line to the controller represents the level signal traveling from the transmitter to the controller by whatever means the control system employs.

The setpoint of the controller (desired water level) is produced at the controller and is represented by the capital letter A in the diamond on the side. That symbol represents an analog output manually generated. The K in the box implies proportional control (Engineers commonly use the letter K to represent a constant value and the gain of a proportional controller is a constant value that we multiply the error by) the funny looking symbol after the plus sign is the integral symbol. I can hear someone saying "WHY?" We've been talking about PID and now he throws in K+∫, why? Because we engineers use those symbols on schematics to represent the functions to confuse other engineers and you—I'm kidding, that's not true, it's a holdover from earlier works and you may run into it so I want you to know it. On ISA drawings the collection of symbols in the center is replaced by one circle with PID in it to represent the controller. This method of diagramming shows more detail so I choose to use it.

The diamond with the T in it at the output of the controller is the symbol for a transfer switch (like for hand to automatic) and the analog output diamond next to it represents the manual output generator. When I make these drawings I include the little circles with an S in them to indicate the control signal value can be observed on a gauge or other visible output so personnel can see its value. You'll notice I showed it at the level transmitter; that's nice to have when the transmitter is a long distance or several floors above or below the control panel. It's one reason I like these symbols, I can show that I want to be able to see that signal value.

This controller should let me see the process variable, setpoint, manual output setting and controller output. It could use some switching device with only one display so it shows only one at a time. You'll notice that the controller output isn't shown at the valve positioner; it should be. Engineers use the letter Z to denote position a lot so a ZC is defined here as a valve positioner, more clearly understood as a position controller.

Figure 11-15 shows the same loop in the simpler symbol method, the ISA methodology. You can see that there's a lot of detail missing that you need other documents to clarify but the control function is the same. One distinctive clarification is the line through the center of the symbol (or the lack of it). The line through the symbol indicates it's panel mounted so the controller is mounted in a control panel. Sometimes double lines near the top and bottom of the drawing distinguish a separation between panel and field. The two figures could be modified to eliminate the lines representing logic flow by clustering symbols together. The controller in Figure 11-14 could be shown alongside the valve positioner which would indicate that all those functions were in a controller and positioner mounted in one enclosure on the control valve. The PID controller and transmitter symbols could be put together to indicate the controller and transmitter are in one package.

Now that we've covered control concepts and control diagrams let's look at some control systems used in boiler plants.

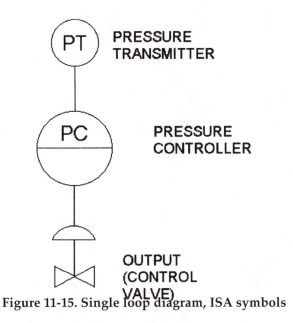

Figure 11-15. Single loop diagram, ISA symbols

SELF CONTAINED CONTROLS

You can be impressed by photographs of control rooms with long curved panels containing hundreds of knobs and buttons under a row of monitors that reveal graphic displays of the boiler systems but you should also be impressed by some of the self contained control devices including some that have been around for years and proven themselves to be so reliable and economical that they may not be replaced by the end of the twenty-first century.

A good example is the control valve on a little residential gas fired hot water heater. In one little box it is a burner management system, pressure controller and temperature controller. Simple versions use a

thermostat to monitor the gas pilot for burner safety. A bulb placed in the furnace over the pilot fire contains a liquid which evaporates to generate an internal pressure conducted through a length of very small tubing to compress a spring opposite a bellows in the valve body. As long as the pilot burns the bellows holds a latch which holds up a disc in the valve body to admit gas to the pilot. When you light one of these you hold in a button that opens the valve to admit gas to the pilot. Once the heat of the pilot generates enough pressure in the thermostat assembly it holds the pilot valve open and you can release the button.

When you release the button it allows a valve to open that admits pilot gas pressure to the main valve control. The temperature of the water in the heater is sensed by a bulb inserted into the side of the heater. That bulb can contain another liquid that expands to compress a spring or linkage that uses the difference in thermal expansion of metals for a mechanical movement relative to the temperature of the water in the tank. When you adjust the temperature knob on the side of the control valve you change the relative position of the linkage or spring and another small valve to select the desired starting temperature. When the water cools a switching valve opens to admit pilot gas pressure to a diaphragm which opens the main valve. The main valve admits gas that is ignited by the pilot and heats the water. During that operation the main valve also functions as a pressure regulator to maintain a constant gas pressure to the burner. When the temperature rises a set amount above the starting temperature the switching valve closes the pilot gas supply to the main valve diaphragm and drains the gas over the diaphragm to shut off the fire. The main valve closes until another operating cycle starts.

Modern self contained hot water heater valves do not operate on continuous pilot to save a little energy. They also eliminate the old problem we always had of pilots blowing out. They include a piezoelectric starter that uses pilot gas flow to power a generator that creates a spark to light the pilot as the water temperature drops. The main valves have double seated valve discs to ensure safe operation. Look at the instruction manual for one of them to gain some appreciation of how something that appears to be so simple is rather sophisticated. I think I would prefer the older type, however. If the electricity goes off I still want my hot water heater to work. Despite all modern miracles I still get a lot of power outages.

Some self contained control valves are simple and effective so they don't have to be sophisticated. A gas pressure regulator like the one in Figure 11-16 controls the flow of gas to maintain a constant outlet pressure. The position of the valve assembly is determined by the compression of the spring by the diaphragm. When the pressure at the outlet drops the force on the diaphragm is less so the spring pushes the valve further open. When the outlet pressure goes up the spring is compressed to close the valve.

It contains an internal sensing tube that points downstream allowing the velocity of the gas to produce a small venturi effect at the end of the tube to effectively reduce the pressure in the diaphragm chamber as the flow increases. That helps open the valve at higher flows and reduce the droop. These valves have a limited operating range as far as pressure drop is concerned because the spring has to have a low coefficient so the valve can stroke completely; it doesn't have the strength to open the valve if, during shutdown when the valve is closed, a high differential pressure between inlet and outlet develops. If you run into a regulator that locks up after a no flow situation then the differential pressure across the valve is too high. You solve the problem temporarily by shutting off the supply to the inlet and bleeding off the pressure upstream of the regulator.

Valves that lock up regularly need a larger diaphragm or should be replaced with a an internal lever actuated or pilot operated valve. Internal lever actuated valves use mechanical linkage to convert a longer mo-

Figure 11-16. Gas pressure regulator

tion at the spring and diaphragm to a shorter motion at the valve disc to allow higher pressure drops across the valve. The typical house regulator contains an internal lever.

When self contained diaphragm actuated regulators are used for natural gas the venting of the spring chamber requires special attention. If the diaphragm leaks the vent of the spring chamber must bleed off the gas or the spring will open the valve fully to raise the outlet pressure to unsafe levels. The gas bleeding out of that vent must be conveyed to a safe location (outside the building) to prevent flammable mixtures forming near the valve or displacing air to asphyxiate someone in the room (someone dies because there's no air to breath).

Occasionally the valve is fitted with an internal pressure relief valve which will drain gas to that vent in the event the outlet pressure gets too high (from thermal expansion or main valve leaking) so the vent piping location and size is very important. It's also important that vent lines for regulators on other boilers, and especially piping from the intermediate vent valves are not connected to the regulator vent lines.

I remember a time when one of our steamfitters was replacing a diaphragm on a regulator while the adjacent boiler was running. That was back in the 1960's when many men had long hair and there were few restrictions on smoking. He suddenly found his long hair on fire because his cigarette had ignited the gas leaking back down the vent line; gas fed from the regulator on the adjacent boiler which was also leaking.

Temperature control valves can use a probe mounted on the valve and penetrating the vessel or piping where it can sense the temperature used to control flow through the valve (like the one on the hot water heater) but that controls the location of the valve which can require extra piping or create other problems with installation or maintenance. Using a probe connected to a bellows by a capillary allows the control valve and temperature sensor to be located separately.

The capillary is a very small diameter tubing permanently connected to the bellows and probe assemblies. These consist of closed systems which are made up for a particular temperature range and valve actuating power. The contents of the system can be a liquid or a gas. Liquid systems are somewhat restrictive because the liquid expands and contracts with changes in temperature and develops high pressures quickly if the expansion is restricted. Gas filled systems change pressure with variations in temperature and many of them contain mostly liquid that evaporates when heated to produce the pressure in the bellows.

Any of these systems rely on minimal changes in temperature at the capillary and bellows which interferes with control based on the temperature at the probe. The capillaries are also very narrow to minimize the amount of fluid they contain and the effect of heating or cooling them. Those small capillaries are easily pinched to block the transmission of pressure from the probe to the bellows or nicked, cracked, or cut to drain the fluid and eliminate control.

Simple diaphragm operated valves and internal lever actuated valves have their limits when it comes to handling large pressure drops, large flow rates, or when low droop is desired. Pilot operated self contained control valves do a great job of handling those conditions. A pilot operated valve is basically a duplex valve where the pilot controls the pressure by controlling the main valve.

The pilot valve is like a regular pressure regulator but its output is fed to the diaphragm chamber of the main valve. (Figure 11-17) When the pressure at the outlet drops the pilot feeds fluid into the main valve diaphragm chamber to compress the main valve spring and open the valve further to match the flow out of the system and restore the outlet pressure. The pilot cannot close the main valve, it can only close off its flow. In order to close the main valve the diaphragm has a line connecting it downstream with an orifice in it so the fluid in the diaphragm chamber bleeds out to allow the valve to close. During normal operation the balance between pilot fluid flow and the flow through the orifice

Figure 11-17. Piloted gas pressure regulator

holds the valve in position. These valves have a droop but it is so small that you don't notice it. They require a minimum difference in inlet and outlet pressures and actually work a little better as the pressure difference increases because the main valve operation is determined by the difference between inlet and outlet pressure.

A self contained main flow control valve can be piloted by a small float valve, temperature element, or other devices to achieve control by using the difference between inlet and outlet pressure of the controlled fluid. Some important considerations for this control are filtering or installation of a strainer on the small stream of fluid used for control so it doesn't plug up the pilot valve or the orifice that bleeds the fluid downstream.

The flow for the pilot is so low that many piloted gas pressure regulators do not have a vent line. There's a small orifice in the spring chamber that can bleed off enough gas to allow the valve to work when the diaphragm is leaking slightly but restricts the flow to limit gas entering the adjacent atmosphere; it's called a restrictor. It's important to be sure you don't block the restrictor with paint; I've solved regulator problems many times by removing the paint from the little hole in the restrictor.

CONTROL LINEARITY

A wise operator will understand what I mean by linearity and how important it is after reading this section. Regrettably there are a lot of control technicians that don't understand it and throw on more and more control features to correct the problems created by a non-linear output. It's really a rather simple concept when you think about it. A control loop is linear when any change in controller output produces a proportional change in the process fluid flow.

Remember that all we can control is flow so we should expect a ten percent change in a controller output signal to produce a ten percent change in flow in the controlled system and it should be consistent throughout the control range. If we have 20% flow with a zero output of the controller (typical for a boiler with 5 to 1 turndown) then we should expect the flow to change 0.8% for every 1% of control signal change. If you were to plot a graph to compare control signal with flow it should produce something close to a straight line.

Why is linearity important? The system's response to errors produces an output to correct that error; if the output produces a different change in flow at various loads then the controller will overshoot at some loads and lag at others. Remember the joking comment "It always works fine when the serviceman is here?" The primary reason why that is true so often is the serviceman is always there when the loads are the same as they were when he tuned the controller. If you run into that situation you should insist the serviceman show up when the system is acting up; you can predict your loads and you should be able to relate load and control problems. If, however, the technician tunes the system for those loads it probably won't work well at the loads where she originally tuned it. If the system is linear those problems won't occur.

To understand why linearity is difficult to achieve let's discuss a typical forced draft fan actuator. The fan can be equipped with a discharge damper or variable inlet vanes, it doesn't matter, and you will find if you measure and plot it that the relationship of damper rotation and air flow looks something like the curve in Figure 11-18, hysteresis ignored. The flow at zero damper rotation is typical of leakage through a control damper.

At high loads the air flow doesn't change significantly but at low and in the middle it does and there's a big difference between that curve and the straight line (which represents a linear flow characteristic). The modulating motor or other actuator that drives the damper can't provide a linear response to controller output unless something compensates for that variation in flow relative to damper position. Adjusting the mechanical linkage connecting the damper and its actuator can eliminate some of the non-linearity to produce a curve similar to the dotted line which is the desired characteristic (linear). Pneumatic, hydraulic, and electric actuators with positioners can be fitted with cams to produce

an excellent linear relationship between control signal (controller output) and flow.

One problem I've noticed with microprocessor based controllers is technicians tend to avoid the rather laborious process of cutting a cam on a positioner by simply programming a function generator in the controller. That function generator produces an output that is a function of the controller output so the result is linear control. It works fine when the controls are in automatic but it ain't worth a damn when you're trying to operate a boiler on hand.

I insist that the linearity be established at the final drive (damper actuator, fuel control valve, etc.) so the response is consistent when operating on manual control. It's a lot nicer knowing you'll get a five percent increase in firing rate if you adjust the fuel and air controller outputs by five percent than tweaking each controller and watching the output changes to see what happens.

Liniarity adjustments are described in the chapter on maintenance.

STEAM PRESSURE MAINTENANCE

Somewhere back in this book I said you can't control pressure. That's true and there is no reason to believe you can. You can maintain steam pressure by controlling the flow of steam from a higher pressure source into a system at a lower pressure or you can control the operation of a boiler that generates steam. We use the steam pressure as the process variable to indicate how much steam is required and control the pressure reducing valve or boilers accordingly. The control loop for a pressure reducing valve is identical to the control loop we just looked at schematically for level maintenance, the difference is we're using pressure as the process variable instead of level. Controlling boilers to maintain steam pressure is accomplished in a variety of ways and we'll try to cover them all.

Regardless of the operating control method all boilers have on-off controls. The boiler in a house and most hot water heaters use on-off as the only method of control. As sophistication and complexity of systems grow on-off controlling seldom, if ever, happens; but it is always there. On-off control is normally achieved with one pressure sensing electrical switch that opens contacts to stop boiler operation and closes contacts to enable boiler operation, a pressure control switch. What do you mean "Ok, what's next?" There's a lot more to that pressure switch than the light switch on the wall. This book is about operating wisely and the

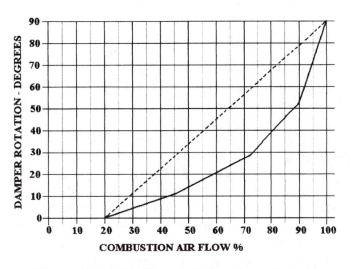

Figure 11-18. Non linear air flow from damper

wise operator should know that he can improve the quality of his operation by adjusting that switch. It has two adjustments; one is the pressure at which it opens its contacts as the pressure increases to stop operation. The other adjustment is the differential which is the difference between the contact opening pressure and the pressure when the contacts will close again. Well, to be honest the setting could be the pressure it closes at and the differential determines when it opens; there are both types.

Set (stop) pressure less differential equals start pressure. Many operators think they should set the differential as low as possible so the pressure won't swing as much. The result is an increased cycling of the boiler and lower efficiency (see cycling efficiency). To get the best performance out of your boiler you should establish the widest possible operating range.

For a simple on-off boiler operation your operating range is the differential setting of that switch and the differential should be set as large as you can tolerate. You should find that you can set it larger in the summer than you can in the winter. The boiler will not start as often. It will run longer on each operation but that reduces the frequency of starts so there's less of them for higher overall operating efficiency and less wear and tear.

The next obvious question is "how do you know how low you can go?" You need enough pressure so all the heating equipment in the facility your boiler is serving operates properly. Frequently it's the one that's the longest piping distance from the boiler but sometimes it's equipment at a shorter piping distance but the pressure drop to that one is higher or it is not as oversized as everything else. The best way to determine it is to gradually drop the lowest pressure (increase switch differential) until someone complains then raise it a bit.

If you can wander the facility you can read pressure gauges and find it. Unless the equipment operates at full capacity summer and winter and it has its own steam piping from the boiler plant you can do the same thing in the summer. Summer loads are usually lower than winter loads so piping pressure drops are less and steam demand on the equipment is less so you can drop the pressure a little more at the boiler.

A typical heating plant with a switch setting of 12 psig can usually operate well in the summer with pressures lower than the maximum differential adjustment of the switch. I've seen plants that operate as low as 2 psig; however, they had to install a special switch arrangement to get that spread. There's a caution here that is covered more later. Don't allow the starting pressure to go so low that the boiler will modulate above that setting.

Before we get off the subject of setting the pressure control switch there's a question of where we set the stop pressure, the main setting of the pressure control switch. Many operators are instructed to set it as low as possible because that makes the steam and water temperature lower to cool flue gases more and reduce stack losses. I will contradict that theory because the small savings in lower stack temperature will be lost with more frequent cycling of the boiler. Set the switch as high as it can go and still prevent operation of the high steam pressure switch (see burner management) and it's for two reasons. One, the larger the spread the longer the run time for a boiler when it's cycling and two, the more room you have for continuous operation.

Now we can talk about modulating controls and the most common of those is a simple electrical proportional control system. A pressuretrol (trademarked name of Honeywell) connects to the steam space in the boiler and consists of a diaphragm or bellows connected to mechanical linkage that adjusts the position of a wiper on a coil of wire. The coil has a constant electrical voltage across it supplied by a transformer. Voltage at any point on the coil is proportional to the position along the coil because the wire has a constant resistance. A matching coil is provided in the modulating motor that changes the firing rate of the boiler.

The wipers are not exactly like an automobile windshield wiper but they operate similarly, swinging so they touch the coil at any point from one end to the other. A schematic of the system is shown in Figure 11-19. When the steam pressure changes it moves the wiper along the coil in the pressuretrol. The voltage between ground and the wiper in the pressuretrol will change which produces an electrical current through the wiper to the balancing relay and the wiper on the coil in the modulating motor. The balancing relay is upset by the current when the two coils don't match so it makes one of the electrical contacts which drives the modulating motor.

The direction of the motor is determined by the voltage imbalance so it runs in a direction that moves its wiper until it is at the same position as the wiper in the pressuretrol. When the two wipers are in the same position the voltage is the same and no current will flow through the balancing relay so it centers to stop the motor operation. The system rotates the modulating motor proportional to steam pressure so it is basically a proportional controller.

The pressuretrol has two settings just like the pressure control switch. One establishes the center of

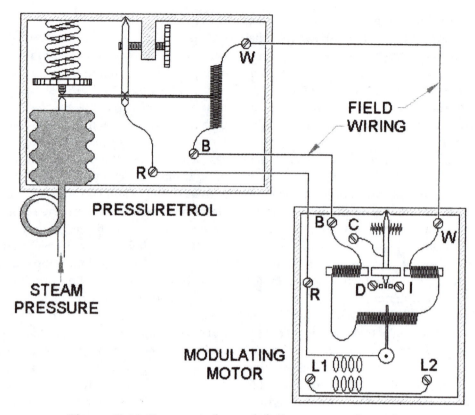

Figure 11-19. Pressuretrol—modulating motor schematic

the operating range (the steam pressure that will center the wiper in the middle of the coil) and the other is the differential which is the change in steam pressure necessary to drive the wiper from one end of the coil to the other. You tune it like any proportional controller, reducing the differential until the operation becomes erratic then increase it until it operates smoothly.

The setting of the center of the operating range of a pressuretrol should always be such that the entire operating range is below the start pressure of the operating pressure switch. How far below? Enough so the steam pressure after the boiler has started and purged at a load equal to low fire is slightly higher than the top of the operating range of the pressuretrol.

When a boiler is cycling on and off the steam requirement is less than the boiler produces at low fire. At those loads the boiler shouldn't be modulating because that increases the input during the firing cycle to shorten it and increase cycles. (See cycling efficiency)

If all you have is an operating pressure switch it's manufactured switch differential is your operating range. When you also have modulating controls your operating range is from the stop setting of the pressure control switch to the pressure that generates the maximum firing rate.

After you have tuned your modulating control

to the minimum differential for smooth operation you adjust the differential of the pressure control switch and the setting of the pressuretrol to establish an operating range as depicted in Figure 11-20. In many plants you will find that you can allow some of the differential of the pressuretrol to fall below the minimum operating pressure because the boiler doesn't have to modulate to high fire to handle the maximum summer load.

Heating boiler plants may have more than one boiler and a need to control operations where two or more boilers are required to serve the needs of the facil-

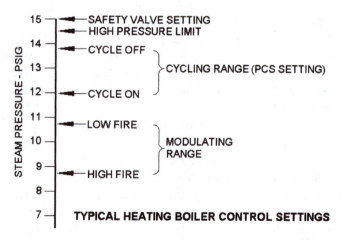

Figure 11-20. Range of control, modulating and on-off

ity. That requires a system that can stop and start each boiler as needed and may include modulating controls that fire the boilers at different rates. Several methods using complex arrangements of linkage, modulating motors that operate off another pressure control switch connected to the common steam header and powered by a shaft which in turn controlled switches and multiple electric coils like the one in a pressuretrol were provided and you may run into one of them.

Okay, I can't explain it in one sentence and that was part of their problem. The principles described for a single burner control apply to them but they required a lot of maintenance and are, for the most part, replaced by modern digital controllers that simulate their functions. If you have one of those, read that instruction manual several times and ensure yourself that you have an understanding of what it's supposed to do before you start making adjustments. Then watch what happens when you make adjustments because they may not do what you understood they should do. As of the writing of this book there is no national standard applied to those devices so their descriptions, labels, and settings vary considerably. Another problem is the people that write their instructions and label the panels may have completely different or erroneous perceptions of what gain, reset, and other control terms are. They will do a good job of controlling your boilers if you buy the right one and apply it properly.

If you have multiple boilers, need more than one in operation to serve all the loads, can't be there to make a decision regarding when to start or stop another boiler, and your boss won't put up the cash to buy one of those modern controllers you'll have to make do with the equipment you have. There's no reason to change modulating control settings on the boilers unless you have limited turndown (two to one or less) or you have adjusted turndown to the degree that you're very inefficient at low fire.

That, by the way, is a normal thing to do. Boiler operators don't normally like to see a boiler shutting down regularly. Creating load and other unwise operations are not the proper way to deal with it though.

You achieve multiple boiler control by setting your operating pressure switches within the range you would use for one boiler. It's easier to talk in terms of start and stop pressures where the stop pressure is the setting of the boiler pressure control switch and the start pressure is the switch setting less the differential. Figure 11-21 shows the start and stop settings for three boilers to achieve automatic control. The difference between stop settings has to be enough that the residual energy in a

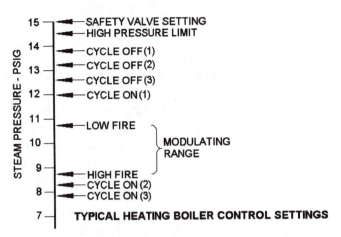

Figure 11-21. Settings for automatic three boiler control

boiler you just shut down will not generate so much steam that the pressure rise associated with that steam generation trips another boiler.

The difference in start (cycle on) settings has to be sufficient to allow for the pressure drop that will occur while the boiler just started is purging and lighting off. Figure 11-21 also shows how you can change the modulating range. You'll notice that the setup requires a considerable swing in pressure to satisfy all the criteria. If you want to change the order in which the boilers operate you have to change all the switch and pressuretrol settings, a lot of work. So it's much better to have one of those digital controllers to do it all.

Many plants have lead-lag controls as part of the package for controlling their boilers. The adjustment of settings in Figure 11-21 provides a form of lead-lag control because it varies the number of operating boilers, allowing Boiler 1 to carry the load until it can't then bringing on Boiler 2 and finally Boiler 3 to handle the maximum loads. All the boilers modulate together. Lead-lag controllers were designed to accomplish it in a slightly different manner. They would run the first boiler up to high fire and leave it there after starting the second boiler which would modulate to carry the load until the load exceed the capacity of two boilers when the third would start. Those controllers resolved a problem with the scheme of Figure 11-21 which provided different responses to load changes depending on how many boilers were on line. The lead-lag controller always had only one boiler responding to load changes, the others were either on at high fire, or off. One controller was actually capable of controlling as many as ten boilers.

High pressure boiler plants can operate with the same simple modulating control we just reviewed as long as there is no problem with the swinging pressure. Once a plant is large enough that someone installs a steam flow recorder that will change. Normally steam flow recorders require a constant pressure for accuracy (see boiler plant instrumentation).

When we want to keep producing steam at the same pressure we need an integral controller. In the early days of controls one of those controllers was an expensive item so we chose to use one to control all the boilers in a plant and called it the plant master pressure controller. It sensed the pressure in the common steam header so it wouldn't be affected by shutting a boiler down and it was close enough to the steam flow elements that it maintained a reasonably constant pressure at them for accuracy in recording. Those rules still apply today but lower costs for instruments and controls have made it possible to have a controller at every boiler if desired.

A plant master pressure controller produces an output signal that is used by each set of boiler controls to adjust their firing rate so they produce steam to satisfy the requirements of the facility while maintaining the steam pressure in the header at the setpoint. We'll be discussing the several types of boiler firing rate control systems later but they all change the flow of steam out of their boiler proportional to the change in the plant master signal.

You should always tune your plant master with the normal number of boilers on automatic. Most plants with multiple boilers in service and a plant master run one boiler on automatic and the rest on hand so the plant master will operate properly regardless of the number of boilers in operation. Note that this works only when all the boilers are the same size. If that's the case and you try operating two boilers on automatic you'll find the pressure will jump around a bit when there's a change in the steam load.

Under those conditions the two boilers change their steam output but the master controller expected the change in output to alter the steam flow the same as the output of one boiler. With two boilers in automatic the response to a controller action is doubled. If you run two boilers on automatic most of the time and one on occasion it's better to tune the master for two boiler operation and live with the slower response when one boiler is on. Plants with boilers of different sizes will also see a different response out of the plant master.

Prior to the days of digital controls it wasn't practical to deal with that situation in the controls, the plant operator had to adjust the tuning of the master controller if the operation was erratic. Some plants added derivative control to help account for it. I created a number of complicated logic systems that adjusted the gain of the master controller according to the number of boilers

on line in automatic.

Modern digital controllers can use digital (on or off) inputs to determine which boilers are in automatic and calculate what the response will be to a controller action so a good digital control system shouldn't be affected by the number of boilers in automatic or what size they are. The need for that degree of control sophistication isn't enough to justify a full explanation in this book. If you're constantly changing the number or combined capacity of boilers operating on automatic control and find the response of the master is never that good, there is a solution for it.

FLUID TEMPERATURE MAINTENANCE

Controls for heating fluids require special consideration that's not necessary for pressure controls. The largest single problem is making sure that the device that senses the temperature you are using as a process variable is representative of the heat flow you are really controlling. Always be aware that the sensor may be shielded by such things as air trapped above the fluid or scale or other material coating the sensor so it can't detect the temperature properly. It may be necessary to locate the sensor where it can't detect changes in temperature when flow is interrupted; additional sensors and controls (like a flow switch) may be necessary to prevent hazardous operation under those circumstances.

This chapter is dedicated to boiler plant controls, particularly hot water boilers for hydronic heating and similar applications. The control of boilers for service water heating (domestic hot water) is described in the chapter on water heating.

Most hot water boilers are supplied with a proportional control similar to that described for steam boilers. The only difference is the temperature control switch and modulating controller sense boiler water temperature, not pressure. In many hydronic systems the quantity of water in the boiler is large enough that it can operate much like a steam boiler, using temperature control instead of pressure. Simply convert the pressure values in the previous figures to the corresponding steam saturation temperature from the steam table in Apprndix A and you have it.

The decisions for setting the start, stop and modulating range for fluid temperature control are based on several considerations. The fluid has to be hot enough when it reaches the using equipment to transfer all the heat required. The fluid cannot be so cold that acids in the boiler flue gas condense on the boiler surfaces and corrode them. A normal low limit for natural gas is 170°F, fuel oils used to cause corrosion at all temperatures below the maximum operating temperature for heating boilers (250°F) so operation at 240°F was recommended. If you fire oil most of the time you should ask your supplier for the normal acid dewpoint temperature of the oil and try to keep your water temperature above that. The lower the start temperature the less loss due to cycling so review the section on steam pressure maintenance to get an understanding of how to set proportional fluid temperature controls. Also review the discussion on thermal shock.

For multiple boiler systems and large facilities the setting of hot water controllers is a little different because the pressure maintained in a steam boiler pushes the heat out to the facility; in fluid systems the heat is transferred by other means. There are basically two methods for transferring the heat and both rely on moving the heated fluid out of the boiler to the heat using equipment and returning the fluid, after it has given up some of that heat, to the boiler to pick up more heat.

The simplest method is gravity and it relies on the difference in density of the fluid as it is heated. Most fluids expand when heated. They take up more space. The density of the fluid (number of pounds per cubic foot) decreases. The hotter fluid tends to float up in any pool of colder fluid just like a block of wood floats to the top of water because it is lighter than the water. A boiler system with a proper piping arrangement can use this to force the heated fluid in a boiler to flow up through the pipes to radiators on the upper floors because the fluid cooled in the radiators fills the return lines to the bottom of the boiler. The colder water is heavier than the lighter, hotter water producing a thermal siphon. We call it natural circulation.

Only simple small systems use natural circulation. Even most small residential systems use an electric pump to circulate the water. The pump can produce far more force to circulate the water than the thermal siphon effect so pipes can be smaller and the system costs less to install. If you're buying a hot water system for your house you may want to think about that; the initial cost savings achieved by installing the pump is rather quickly eaten up by the cost of electricity to run that pump. A system designed for pumping won't work well on gravity when you need heat, the power is out, and you try burning some wood in your furnace. Some increase in initial cost may save a considerable amount on electric bills and ensure the ability to get heat if the pump or power fails.

Large hydronic heating systems for schools, of-fice buildings, etc. simply can't justify a system without pumps so they all include pumps to move the fluid around between boiler and heat user.

Unlike steam boilers where load is balanced by the flow of steam from its source of generation to the load, hot water boilers cannot function with a plant master that controls the firing rate of all the boilers. Some systems have a master temperature controller but it doesn't control the firing rate of each boiler; more on that in a minute. There have been attempts to produce common control by operating the boilers in series (water flows through one boiler then the next and so on) but I have yet to see one that works well.

When fluid heating systems become so large that the volume of fluid in the boiler is a small part of the entire system, control of the water temperature becomes difficult. Another factor is the volume of water in the boiler; fire tube boilers contain a large volume of water and can have long residence times (how long the water stays in the boiler) but water tube boilers can hold so little water that it's replaced every few seconds.

Boilers like that (with short residence time) can have a problem because the temperature of the water at the sensor is not the same as the average temperature of the water in the boiler. Temperature maintenance of those units can get erratic so another control method is required. The controls are very typical of high temperature hot water boilers (HTHW) which operate at temperatures over 250°F and pressures over 160 psig.

For boilers heating water the method is easy to understand, you're adding Btus to the water so the energy required is equal to the pounds of water going through the boiler and the temperature difference. The actual heating load is determined by multiplying the pounds of water flowing through the boiler by the difference in inlet and outlet temperatures. For other fluids all you need do is multiply by the average specific heat of the liquid. Control logic that performs that calculation provides a very responsive control because any change of inlet temperature or fluid flow rate produces a change in the control signal, increasing or decreasing the firing rate of the boiler to compensate. Since multipliers were a problem in early controls most plants relied on a constant fluid flow so only the temperature difference was needed to develop the control logic.

These systems cannot operate on that logic alone because there's no way to correct for changes in boiler efficiency or small errors in flow and temperature measurement that would produce an imbalance between the actual load and the firing rate. A temperature controller is used in these systems to provide a means of correcting for those differences. The typical HTHW boiler load control system is shown in the schematic in Figure 11-22. Refer to the following discussion on two element boiler water level control, for an explanation of this particular type of control loop.

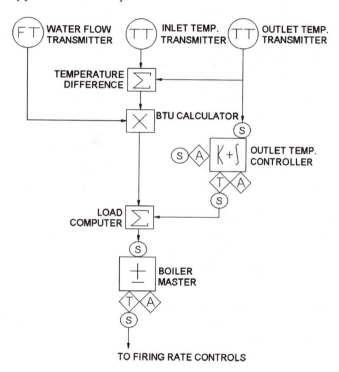

Figure 11-22. HTHW boiler control

FLUID LEVEL MAINTENANCE

There are several locations where water level must be maintained in a boiler plant but the most important is the level in the boiler itself. The method of control varies significantly depending on the size and complexity of the plant. The simplest is a float controlled valve and the most complex (and expensive) is a three element boiler drum level control loop. Each has its place, its advantages, and its problems. I'll try to give you all those in the following paragraphs.

A float controller for level control is common in boiler water feed tanks, condensate tanks, make-up tanks and other source tanks of water for the boiler plant. They are found in boiler feed service only on residential and small commercial boilers. We've covered their operation in our earlier general discussion of controls. They're not found on large or high pressure boilers because the float would have to be very large to produce the force needed to operate the control valve while operating on a very small change in level.

As they get larger they have to get stronger to

prevent crushing them so they get heavier and the float chamber has to be thicker (see strength of materials) so they become uneconomical. They do work fine for open tanks at small flow rates. One place where float controls have problems that you can relate to is in brine tanks used with water softeners. The salt tends to crystalize on the float and surrounding materials, usually a still pipe (a pipe placed around a float to prevent swinging operation due to wave action) so it can be trapped in the brine crystals and fail to operate.

A float that only has to open and close an electrical contact can be quite small by comparison to something that has to open and close a valve so we have many systems controlled by float operated switches. The switch can energize a solenoid valve to open it and admit fluid to the tank or boiler. All the energy required to operate the valve is provided by the electricity (and in many cases the fluid itself, see pilot operated valves in the general discussion on controls) so a small float can control any volume of fluid at any difference in pressure. The float still requires a change in level to function and only provides on-off control of the fluid flow but that's satisfactory in many situations. The switch can also be set to power a valve as the level rises to provide a system that allows fluid to flow on controller failure.

The typical heating steam boiler and small commercial and industrial boilers use float controls that start and stop the boiler feed pumps to control feedwater flow for maintenance of the water level instead of controlling a valve. These systems solve some of the problems with valve control by preventing operation of the feed pump when the control valve shuts off, a situation that would overheat the pump. It also eliminates feedwater control valves as a maintenance item.

Each boiler has to have its own pump for this control method to work and operation of standby pumps is complicated because the electrical control has to be switched along with pump isolation valves. It is a simple and inexpensive method for level control and works well in many applications. However, it can't be used with economizers and the higher electrical demand (pump and motor are normally sized at twice the boiler capacity) can create higher electrical power costs.

If you're operating a boiler with very little reserve capacity like most water-tube boilers, you have an economizer, or you can't tolerate the swings in load associated with feed pump on-off control a variable feed level control is required; one that modulates the feedwater flow control valve to maintain the level.

If a boiler has little reserve in it the cold feedwater

rushing in at twice the boiler capacity can, for a short period of time, consume so much of the heat to simply heat up the feedwater that some of the steam in the boiler is condensed so the water level drops suddenly every time the pump runs (see shrink and swell discussed later). Sometimes it's enough to trip the low water cutoff. Considerable differences in boiler level is required for them to operate without false trips. Many of the new flexitube boilers are equipped with two level controls, one set for controlling level when the boiler is off and another for when the boiler is firing, set at a higher level.

If the boiler has an economizer the continuous

Figure 11-23. Thermo-mechanical boiler level control

Figure 11-24. Thermo-hydraulic boiler level control

flow of water is required to prevent generating steam in the economizer. The feed pump on-off operation produces a significant change in output of a boiler, especially at low loads, that can cause bumps in the whole steam system. Anything larger than a small commercial boiler operation should have a better method of water level control.

There are two unique self contained control systems that you should be aware of. They were used only on boilers, and can still be found in many locations. One is a thermo-mechanical system; the other is thermo-hydraulic. The key to these controls is that prefix, "thermal" which indicates that we use temperature to detect level and power the control valve. The thermo-mechanical systems (Figure 11-23) are manufactured by Copes-Vulcan. The thermo-hydraulic systems (Figure 11-24) are manufactured by Bailey and Swartout among others. Both systems use the difference in heat transfer rates between steam condensing and simple water heating.

They incorporate a tube connecting ends to the water space and steam space in the boiler. The water level in the boiler is repeated in the tubing so the tubing above the water level is exposed to steam and the tubing below the water level is exposed to boiler water (actually it's mostly condensate from the steam condensing in the tube). Since steam condensing transfers heat much faster than hot water the portion of the tube that is exposed to steam is hotter. Both systems arrange connecting piping so the tube is at an angle, the slope of the thermo-mechanical tube being much shallower than the thermo-hydraulic, to provide additional tube length and (as a result) heat exchange surface for better control.

Since the heat transfer is much higher for steam condensing the lower the level of the boiler water the hotter the tube. The heat transfer from the finned water jacket of the thermo-hydraulic controller or from the tube of the thermo-mechanical controller to the surrounding air is increased slightly because of the hotter water jacket or tube. The expansion of the tube, or the water in the jacket, is converted to movement of the valve; opening it as the tube or jacket gets hotter.

The thermo-mechanical system uses a short pivot at the end of the tube which consists of a lever point at the end of the tube and a pivot attached to the two steel channels on either side of the tube. The lever connected to the control valve moves as much as six inches from its end with a very small change in the length of the sensing tube. As the tube expands the lever is pulled down by the weight to open the valve.

The expanding water in the jacket of the hydraulic version acts on a diaphragm (Swartout) or bellows (Bailey) on the control valve, opening it. As the water rises in the tube as a result of adding water the tube or water jacket shrinks. The shrinking tube pulls the valve closed on the mechanical system. A spring pushing against the bellows or diaphragm of the hydraulic system closes the valve as the water in the jacket shrinks. Both systems will stabilize to maintain a constant water level but they do not respond rapidly to level changes and always open the valve fully as the boiler cools down so you have to manually close off the water and manually control level on boilers equipped with these systems until the boiler is at operating pressure.

The Copes-Vulcan system (by the way, we've always called them Copes valves, failing to give Vulcan any credit) has another system with a feature to aid in response to changes in load. The control valve is fitted with a diaphragm connected to the feedwater valve with sensing lines to the steam header at either side of an orifice. Increasing steam flow produces a higher pressure drop across the orifice which produces a higher differential pressure on the valve diaphragm to force it further open. The lever of the thermo-mechanical tube is fitted with a chain extension that runs over a sprocket on the valve to the weight. The sprocket is connected to the valve stem like a rack and pinion to aid or restrict the diaphragm action for final water level control. This provides something comparable to two-element control, which I'll get to.

Experience and modern controls and instruments have convinced me that I would never want to use one of those thermo-hydraulic or therm-mechanical control valves again. I tell people that have them not to buy spares and replace them when they need repair. They are not the easiest things to work with, they don't control the level when the boiler is cold and they're relatively expensive. Now that level transmitters and controllers are so inexpensive the cost of those older designs can't justify their existence. They were fantastic controls years ago but new controls can do so much more.

Shrink and Swell

A simple single loop control system like the one covered at the beginning of this section will satisfy the requirements of most heating boilers and commercial and industrial loads with fairly constant steam demands. If, however, the steam requirements change significantly the control will actually operate in the wrong direction due to shrink and swell. Shrink and swell are terms we use to describe what happens when the boiler load changes and feedwater addition changes.

When the boiler is generating steam some of the volume below the water surface has to consist of steam bubbles. The amount that is bubbles depends on the load, the volume of the boiler below the water line in proportion to the capacity, the surface area of the water line, and the operating pressure. Many boilers, mostly fire-tube boilers, contain so much water in proportion to steaming capacity that the percentage of volume occupied by steam is small and the shrink and swell are not noticeable.

On the other hand, a low pressure water-tube boiler is most likely to show the most dramatic change because the steam density is low (volume of steam per pound is high). When a sudden increase in load occurs the steam pressure in the boiler drops and the steam bubbles in the boiler water expand. Also a small percentage of the water flashes to steam adding to the number of bubbles. The result is an increase in the water level which we call "swell" because the water level increases with no water being added to the boiler. A single element level control will react to the swell by closing down on the feedwater valve, the opposite of what is needed because more water is required for the larger load. Closing of the feedwater valve reduces the heat requirement for raising the temperature of the feedwater so more heat is used to make steam (and more bubbles) simply make the water swell more.

When the opposite occurs and the load decreases suddenly, pressure increases, the bubbles are compressed, the water in the boiler is not up to the new saturated condition so some of the steam condenses to heat it up. The water in the boiler shrinks and the level drops. A single element control senses the drop in water level and opens the control valve to increase the flow of feedwater. The additional feedwater requires heat to warm it to saturation condition so some more of the steam is condensed to collapse more bubbles. Increasing the water flow is not required because the steam flow decreased, but the shrinking water level makes the control open the feedwater valve more.

Two Element Control

To reduce the impact of shrink and swell a water system that doesn't enhance the effect of it is required. Two and three element systems actually counter some of the effect by adding water when the level is swelling up to quench bubbles which reduces the swell. Conversely they reduce the addition of colder feedwater when the level is shrinking.

I mentioned single element control operation. Single element feedwater controls have a single process variable for control, water level. I've already spent a lot of time discussing them. Two element controls use another process variable (that isn't maintained) and that is steam flow. Since the steam flow is not controlled as part of the feedwater system it is usually treated as a remote signal. The third variable for a three element control is feedwater flow. The two and three element systems act to maintain the balance of steam and feedwater flow with adjustments for level.

Both two and three element systems actually control the flow of water to match the flow of steam. It's a given that every pound of steam that leaves a boiler must be replaced by a pound of feedwater so that's a logical way to do it.

These systems require a control element called a signal summer which combines two or more control signals. The term "summer" is used instead of "adder" because a summer can subtract signals as well as add them. When mathematicians and control engineers use the word "sum" they mean to add up all the values and some of them can be negative. The ratio totalizer described earlier can be used as a signal summer. One input signal can be applied to the bellows opposite the output (port A in Figure 11-5) so the output equals that signal plus another signal be applied to port C of the totalizer for adding or port B for subtracting.

We could introduce a gain on the A and B values by adjusting the pivot. We could also add a spring to the assembly so we could introduce a fixed bias (spring force) at either end of the ratio totalizer. The mathematical equivalent of the summer output would be input C plus input A minus input B plus or minus a bias provided by a spring at their end plus or minus the bias provided by a spring at the output end. The output equals $(I_A - I_B \pm K_B) \times G + I_C \pm K_C)$ where the suffix identifies the port indicated on Figure 11-5, the letter 'I' refers to input, 'K' represents a spring attached to the pivot arm at that port and 'G' is the gain.

That's the basic concept of a summer but most microprocessor based controllers allow you to include the summer function inside the controller to eliminate the need for additional hardware; that's why we can make a two element controller out of a single element one by simply wiring the steam flow signal to the drum level controller. Actually, in many systems and any future system it is simply a matter of telling the controller to get the steam flow signal because all the controllers have access to all the signals in a system.

The two and three element systems control the feedwater valve in proportion to steam flow with an adjustment for drum level. A two element feedwater control system is shown in Figure 11-25. Two element control is very common today because any boiler that needs the control is large enough to justify steam flow metering for monitoring the boiler demand and performance. Since the steam flow meter is there it's simply a matter of adding, at most a little wiring, and normally just a few software instructions (for microprocessor based controls) to make a two element system out of a single element system. If the boiler has pneumatic controls another device (summer) is required to create a two element control and another hand automatic station may be necessary.

As steam flow increases the output to the feedwater valve increases. Provided the valve is selected or its positioner is set to provide a linear output the valve position for each value of steam flow will produce a feedwater flow that matches the steam flow. You can always tell if a two element system is set up properly by noting the output of the level controller at different boiler loads when the level and steam flow are relatively stable. The output of the level controller shouldn't change and should be about 50%.

Why 50%? I've encountered several systems where the operators were always fooling with the level on a

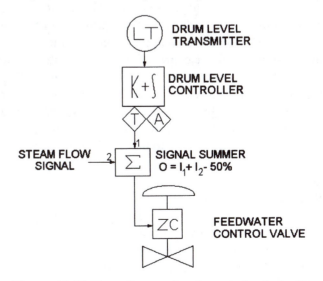

Figure 11-25. Two element level control schematic

boiler because the controls were not set up for that 50%. The setup of a boiler feedwater control system usually ignores the blowdown (which requires a little more feedwater) so the valve position is set to handle the correct amount of feedwater for the normal steam flow and blowdown. The technician setting up the system looks at the schematic and realizes that the level controller can add to the steam flow signal to increase flow and raise the water level but he doesn't think about what has to be done to lower the level if it gets too high.

A 50% bias opposed to the steam signal and balanced by the 50% output of the level controller has a net zero effect on the steam flow to valve position relationship but allows the level controller to modify the relationship up and down by 50%. Without that bias the level controller output is around zero and it can't do anything to lower the water level if there is a slight upset in operation (like lower steam pressure that will allow more water to flow into the boiler) that results in a high water level.

Three element control accomplishes the same thing as the two element control but it measures the feedwater flow as a process variable and the feedwater flow controller adjusts the feedwater control valve until the feedwater flow matches the steam flow plus or minus any adjustment from the water level controller. Three element systems are necessary whenever feedwater pressures are frequently changed or affected by heavy load swings so the linearity of the control valve can't be maintained.

Of course there's going to be a problem with a two element system that's set up right if the steam flow signal is lost. The controller won't be able to open the valve more than 50%. I don't consider this a big deal because I don't think you should be operating a boiler with a two element control in single element mode unless it's at very low loads. Modern digital controls sort of solve that problem because their instructions include switching to single element control that can fully stroke the valve when the steam flow signal is lost or low. Failing that and circumstances require you to operate a boiler without a steam flow signal you can simply find the summer and adjust the bias from a -50% to zero and the valve will respond directly to the level controller signal.

BURNER MANAGEMENT

Before I cover the control of fuel and air to produce a flame and add heat to the boiler I'm compelled to cover the burner management system. For years we called them flame safeguard systems because the prin-

ciple purpose of the system is to make sure it is safe to light a fire and to continue operation of the boiler. The burner management does two things; it "supervises" the operation to ensure all operating parameters are within "limits" and it supervises or performs and supervises the procedures required to place a boiler and its burner (or burners) in operation.

It can also provide a controlled shutdown of the boiler and its burner(s). The bulk of the controls manufactured to serve the purpose of burner management are supplied by one of two manufacturers, Fireye (now a division of Allen Bradley) and Honeywell. Their devices are competitively priced and do an excellent job of burner management for small and medium sized single burner boilers. The controller becomes a system when it is connected to a burner's controls, level, pressure, and temperature limit switches and a flame scanner or flame rod.

Accurate detection of a fire is the most important function of the burner management system. It has to ensure there is no fire when it shouldn't be there, as well as ensure a fire is present when it's supposed to be, and respond accordingly. Detectors are either flame rods, infra-red sensing, ultra-violet sensing, or more modern multiple frequency sensing units. The detectors, with the exception of flame rods, are all called scanners.

The basis of flame rod operation is that a fuel and air mixture does not conduct electricity but a flame does. The rod has to be positioned where it is in the flame, a pilot flame on large burners. The flame also has to have a grounding electrode that touches the flame so electricity can be conducted from one electrode to the other. The grounding electrode is normally connected electrically to the metal parts of the burner. It can also be another flame rod positioned at another point in the flame. The flame rod itself has to be constructed of a material that will not melt or oxidize in the fire and the insulation separating it from the metal parts of the burner have to be capable of withstanding the high temperatures. Normally the rod is made of high chrome steel and the insulators are ceramic material. The portion of the burner management system that identifies the presence of a fire has to produce a voltage adequate to push a detectable current through the flame and sense that current and distinguish between no flame, false signals (like the rod shorting out on some metal in the burner) and a flame.

The flame scanner is a sensor that detects a fire in the burner by absorbing some of the light energy emitted by the fire. Scanners may detect infra-red light or ultra-violet light, any frequency of light in between, or a combination. Some are called "self-checking" but that label can be inappropriate. I'll call them self-checking when they contain a device that blocks the light from the sensor at intervals and the detector circuit has to sense a no flame signal during that interval. Some scanners simply block the light at regular intervals so the detector circuit can determine a flame is present because the signal from the scanner is constantly swinging to produce an alternating current. A constant signal from the scanner indicates no flame or scanner failure including failure of the self checker. Self checking of scanners shouldn't be confused with self checking of the flame detector circuitry which is different.

Someone will also come up with a device and call it self checking by virtue of some software scheme. I spent several hours fooling with some smoke indicators years ago and found they tested well but didn't indicate the smoke I could see by looking at the top of the stack. Their scheme for calibrating a black stack consisted of shorting out a terminal on the sensor, not turning off the light at the other end of the stack. When I finally decided the system wasn't turning the light off and put a blank in front of it the indicator happily showed a clear stack! If the scanner doesn't block the sensor's view of the fire it isn't a self-checking scanner.

A self-checking circuit simply confirms a flame isn't detected when it shouldn't be. Of course I've discovered several systems that were installed and connected in such a manner that the self checking functions of the circuit were not allowed to work. If, on a burner that's turned off, your scanner doesn't detect a flame on a candle or cigarette lighter that you hold in front of it (you have to remove it from its mount) and it doesn't alarm and lock out as a result, yours is one of those. Since many of them were only found after a boiler explosion I urge you to perform that test. If the system doesn't lock out it isn't safe.

The typical BMS (burner management system) provides for automatic operation of the burner, performing all the steps described in the section on boiler start-up. When a pressure control switch closes the BMS should first determine there is no flame in the burner. Provided operating limits like low water cutoff, high steam pressure and low fuel pressure are all satisfied (contacts closed in a series circuit) it closes an output contact to start the burner fan or fans. When air flow is proven by closing contacts on an air flow switch the firing rate control system is instructed to increase damper position to high fire. Some systems may include provisions to start an oil pump and prove it operating as well. The open damper or a purge air flow switch senses purge air flow to close a contact for another input to the

burner management system. The system then waits for the prescribed period of time for a purge.

Some are set with fixed timing but modern units have provisions for setting the purge time to comply with the code requirements. The controller supervises the purge by requiring the damper open or purge air flow switch contacts remain closed during the purge period. Some will simply restart the purge timing if the input is interrupted while others will stop the start up. Once the purge timing is complete the contacts for high fire are opened and another set close to instruct the firing rate controls to go to a low fire position for ignition. When the firing rate controls are at low fire they close a low fire position (or ignition permissive where low fire is lower) switch contacts to provide an input to the burner management system.

During all of that portion of the start-up sequence the scanner should be looking for a flame. If it sees one the system should lock out. The reasons can be anything from defective scanners to oil dripping out of a gun and lighting to glowing hot refractory from the previous firing. On more than one occasion an operator figured out that he could prevent the lockout by pulling out the scanner and covering it with his glove during the purge and low fire positioning. Needless to say, he didn't have any accurate flame sensing and eventually an explosion occurred. If that scanner thinks it sees a flame where there isn't one it's not safe to operate that boiler.

With low fire position proven the controller closes a contact to energize an electric spark in the ignitor and another contact to energize the ignitor gas shut-off valve(s) (if the burner is equipped with an ignitor). The controller then waits ten seconds to see if the valves open to admit gas that is lit by the spark to create an ignitor flame. If the flame isn't detected in that time it stops operation and energizes an alarm horn. If the flame is detected it closes another contact to energize the main fuel valves. Some also de-energize the electric spark. At a prescribed time after the main fuel valves are open it de-energizes the ignitor gas valves. If a flame remains detected the controller opens the low fire contact and closes an automatic contact to permit automatic operation of the modulating controls to control the firing rate.

How it did it in the early days was clumsily and with a lot of errors. The timing was all controlled by a timer motor powering a shaft with several epoxy impregnated fiber discs on it that served as cams. Each cam had a flat metal spring riding on it and that spring made contact with another one where the cam was notched. The program was initiated when the boiler's pressure control switch closed to energize the motor to start rotating the cam assembly. As the cams rotated a change in the diameter of one would drop its spring to close the motor circuit and another closed in the modulating motor circuit to drive the controls to high fire. Another contact then opened to stop the cam drive motor. Once the modulating motor got to high fire it closed its high fire interlock contact which bypassed the open cam contact and restarted the cam drive motor beginning the purge timing.

I won't explain the whole operation, you'll figure the rest out if you happen to get in a plant that's old enough to still have one (I doubt it) and you'll want to replace it anyway. Those cam contacts were always a problem because the springs would stretch and get weak and the contacts on them would get dirty so they wouldn't complete a circuit (a boiler room wasn't that clean in the good old days) and lower priced microprocessor units have replaced most of them. You can't access the program or fix the microprocessor based units so you simply replace them; it saves you fooling with those springs and cleaning those contacts about once a month.

One last comment on the old cam operated controllers; there was always a dial connected to the cam shaft. The dial looked something like the bottom of an aluminum can that was cut off and it had numbers on it that corresponded to the timing of the motor so it could be checked. There was also a large black dot on the dial that indicated where the timer was to stop for normal firing operation. The beginning of the cycle, which was also where the timer stopped when the boiler shut down for any reason was always marked with a zero. The explanation of the cam operated switches and timing does help explain some of the instructions for the microprocessor based equipment because they were written to help us old farts relate to what was going on in that new black box.

One of the most important elements of the burner management control is the checking provisions. The burner management controller had to include at least two relays, a power relay and a flame relay. The power relay could only be energized via a normally closed contact on the flame relay (proving the flame relay was de-energized) and it closed a normally open contact to bypass the flame relay contact to continue operation. Once the power relay was up the flame relay closed its contacts to power the main fuel valves and stop the timer motor.

Any indication of a flame when there isn't supposed to be one can mean the burner will continue to operate when there is no flame present, generating an explosive atmosphere in the boiler. Any component

failure in the burner management system should also act to safely shut down the boiler or, if it's failure does not present an immediate danger, prevent a subsequent start-up of the burner.

Of the many boiler explosions I've investigated only two were found to occur during operation, the rest occurred on start-up and problems with the system arrangement (design) or an alteration of the design could be attributed to the explosion. Unlike airplane accidents where the reason is regularly attributed to pilot error I don't find operator error to be a primary reason for a boiler explosion. Many times the operator is present and doing something but that doesn't mean the system operated flawlessly, usually the system prevented proper operation.

Those of us that design the burner management systems have a directive to make the system "damn fool proof and moron approved" so it's supposed to be virtually impossible for operators to create an explosive condition unless they intentionally defeat limits and interlocks. Don't get me wrong, I've seen many a bypassed interlock or limit switch. Why was it bypassed? Because the damn burner wouldn't work if it wasn't!

Speaking of those limits and interlocks reminds me of the many ways they can fail to do what they're supposed to, mostly because of improper design or application. Every other facility I visit for the first time is usually set up with minimum air flow limits and purge air flow switches that, quite frankly, don't work. It's because they don't sense flow, they only sense pressure.

I'm sure you've seen many burner assemblies where the air flow switches are air pressure switches with one side connected to a burner windbox. Any burner windbox, however, normally has burners with air registers that can shut off the flow. Even if they don't a blockage in the boiler will ensure pressure to actuate those switches. I've seen many an installation where the operators closed the burner registers to produce enough pressure in the windbox to trip the purge air flow switch. Needless to say, if the registers are closed there is no way to get purge air flow. I prefer air flow switch installations that measure the air flow, normally simply sensing the pressure at the fan inlet as shown later for adding air flow metering. See initial boiler start-up for more clues on proper operation of burner management controls.

The key actions for a wise operator when it comes to burner management is 1) know what they're supposed to do, 2) shut the boiler down when they don't do it, 3) Report inconsistencies in operation and regular interruptions in operation, 4) don't change switch or position settings without permission. That last one is a real key because many states have adopted the ASME and NFPA standards that relate to burner management and both standards are very exacting about the requirements for changes in burner management systems.

No discussion of a burner management system should be left without mentioning the important concept of fail-safe design. Every element of the system should be arranged so it's failure will not compromise the safety of the boiler operation. Each wire, relay, pressure switch, etc., should be evaluated for failure modes and analyzed for what will happen if the device fails. Only when every evaluation indicates the result will be safe should the system be considered fail-safe.

Fail-safe concepts should be applied to all controls and applied in a sensible manner. Too many designers view fail-safe solutions as only resulting in a complete burner shutdown. That's not necessarily the safest thing to do because, while that burner is operating, most of the furnace and boiler is full of inert gas. There are many other examples where a shutdown is not necessarily the safest solution to a failure.

There are always arguments as to what is safe as well. Is it better to have a feedwater valve fail open, so the boiler will not run dry? Most of the time we have the valve fail closed because there is no safety to prevent water flying down the steam lines and hammering them apart but we should expect the low water cutoff to safely shutdown the boiler.

If you're replacing a component of a control system, and it's operation isn't exactly the same as the piece you're replacing, consider what will happen if it fails. Much thought has gone into deciding if a particular component will fail in the safest manner and replacing it with one of another action could reduce the safety and/ or reliability of your plant.

FIRING RATE CONTROL—GENERAL

Firing rate controls regulate the flow of fuel and combustion air to the burner to produce a flame and heat input that satisfies the demand for heat at the boiler outlet. We'll also call them combustion controls. These are independent of the steam pressure controls on any system except a simple jackshaft system. Typically we don't talk of combustion controls or firing rate control with a jackshaft system.

The heat input is primarily a function of the amount of fuel flowing to the fire; control of air is also required to produce the heat input. In the chapter on fuels we discussed the importance of maintaining an optimum air to fuel ratio. Part of the job of firing rate controls is to maintain an air to fuel ratio that is adequate for safe and efficient operation of the burner and boiler. There are different control schemes for controlling the fuel and air, to maintain the air to fuel ratio, and their ability to do the job varies with system cost and complexity.

The choice of control system for your boilers will depend primarily on the size of the boilers. Size of the boilers implies a certain annual fuel consumption and the increasing cost of more refined controls has to be weighed against the savings that can be produced by improving the controls for better control of air to fuel ratio. There's also the question of maintaining a certain steam or vapor pressure or a boiler outlet temperature that may, or may not, be critical to the facility served by the boiler plant. If the pressure or temperature is critical the controls will be more refined.

I have seen boiler plants where there were no pressure controls. In one the operators increased the firing rate when the pressure got down to around 5 psig and backed it down when the pressure got up to 90 psig. They raised that low point in the winter to 40 psig because anything less produced complaints in the college.

That's an extremely clumsy operation that could cause a considerable number of problems both for the operators and the equipment but they managed to keep the facility happy with that performance and that's all they cared about. Not very wise was it?

Swinging pressures will vary blowdown rates, increase the opportunities for carryover, and if not caught at the right time, result in boiler shutdown or lifting of safety valves which do reflect on the performance of the operators. The changes in temperature are adequate to define the operation as cycling and the standard boiler is constructed for a life of 7,000 cycles; swinging operation shortens boiler life. My perception of that operation provoked words like careless, lazy, and inconsiderate to name some of the printable ones. The boilers were equipped with firing rate controls but they were either inoperable due to no maintenance or not used for reasons I can't begin to understand. If the temperature swings you are inviting problems with thermal stress.

A low pressure steam plant can swing from a low of 8 psig to a high of 12 psig with a temperature swing of 9 degrees; to me that's the limit. Higher pressure plants have thicker boiler parts and swings of more than 4 or 5 degrees can cause problems with thermal stress in them so normal pressure swings should be held to less than 10 pounds.

Another common trick when maintenance is lacking is to operate with the fan damper wide open. That way there's always enough air to burn the fuel, right? Actually that's wrong because at lower firing rates the high excess air quenches the fire to produce combustibles, primarily carbon monoxide, and sometimes unburned fuel products that are carcinogenic. Such careless operation is not only lacking concern for the cost of fuel but is potentially hazardous to the health of the operators as well as everyone within a one or two mile radius of the boiler. Now that you're a wiser operator you will not, I hope, poison yourself and other people by failing to have adequate control of your air to fuel ratio.

Following are descriptions of the five most common methods of modulating a boiler's firing rate followed by four possible enhancements to some systems. They run from the simplest to the most refined and complex. You shouldn't be disappointed if you don't have the Cadillac nor be disgruntled because you have to deal with a complex system. They're selected to provide optimum performance when they're working right. It's your job to ensure they're working right, keep them working right as much as possible and report it when they aren't doing what they're supposed to so they can be fixed by qualified technicians when you can't handle it.

There's enough information in this book for you to make adjustments and correct problems in any of these systems but that doesn't guarantee that you can relate the indications you see to the right source of the problem. If you're confident you can fix something let the chief know and get permission to fix it, otherwise let one of the contract technicians do it. If they do something wrong their insurance company will pay the bill, not your employer's. Let's discuss these systems and we'll see where you stand.

A simple on-off boiler doesn't have a firing rate control system as far as I'm concerned and the first two simple systems aren't a lot better. They do, however, change fuel and air flow rates so they have to be considered.

FIRING RATE CONTROL—LOW FIRE START

A low fire start control system only regulates the input of fuel and air to the furnace during the ignition period. The system limits fuel input for ignition then allows it to increase to the maximum firing rate which is maintained for the rest of the burner operating time. The controls for gas typically consist of a two position fuel safety shut-off valve with a rack and pinion on its shaft connected to linkage that controls the position of

the fan damper. The valve opens to a preset position during the main flame trial for ignition and the linkage limits opening of the fan damper to another preset position. Once a flame is established and the ignitor is shut down the valve opens the rest of the way and the fan damper opens with it.

For oil burners the typical setup is a small hydraulic cylinder sensing the oil pressure at the burner. Two oil shut-off solenoids are used to produce the two different oil flows or a solenoid is powered to bypass a manually set throttling valve for full fire. The cylinder contains a spring and it moves the damper according to the burner oil pressure, low then high.

There is little advantage to a low fire start control system. Primarily all it does is permit the use of a cheaper ignitor that would blow out if exposed to full load combustion air flow. As far as I'm concerned you should have a high-low firing rate control if you're considering low fire start; there isn't enough difference in price that would prevent recovering the added cost of the modulating system in one or two heating seasons.

Adjusting low fire start controls is not easy and the manufacturer's instructions should be followed to the letter. You have to establish a suitable air to fuel ratio at the full load and ignition positions and ensure that the air to fuel ratio doesn't go too far awry as the controls swing from low fire to high fire. The process requires a thorough understanding of geometry to arrange the linkage so the ratio is maintained.

FIRING RATE CONTROL—HIGH-LOW

High-low firing rate control is similar to the low fire start system (described above) except the controls can switch between the low (ignition) position and high firing position to vary the heat input to the boiler. Another pressure control switch is added to the boiler to control the positioning between high and low. Of course if it's expected to work it has to be set lower than the setting of the on-off pressure control switch to prevent pressure or temperature swings above the high-low switch settings shutting the boiler down. Setting of that pressure switch and the on-off pressure switch can be varied with the season as described for the on-off pressure switch and electric positioning control.

Maintenance of a suitable air to fuel ratio during load swings is more important with the high-low system than the low fire start because the linkage has to maintain the ratio as the firing rate drops to low fire as well as when it increases to high fire and the burner may be frequently swinging from one to the other.

The only reasonable way is to watch the fire as the control swings from high to low. You don't want it smoking and you don't want it where it's about to blow out. Preferably it will be something close to a normal clean fire as it changes. Again, the process requires a thorough understanding of geometry to arrange the linkage so a reasonable ratio is maintained.

FIRING RATE CONTROL—BURNER CUTOUT

Certain gas fired appliances incorporate this method of controlling heat input and it's not the same as having a multiple burner boiler. The application consists of installing multiple shut-off valves (not safety shut-offs necessarily) between the main safety shut-off valves and parts of the burner. Oil burner cutout controls can shut down one or more burner nozzles leaving the rest to continue supplying oil. Gas burner cutout controls typically shut down the gas to one or groups of flame runners.

Sometimes the combustion air is not changed (very inefficient operation) while several means of changing the air flow are available including adjusting a damper, closing a valve in the air supply branch to the portion of the burner that's shut down, stopping a fan dedicated to that portion of the burner, or changing the fan speed.

I've only seen burner cutout systems on inexpensive equipment and, to be perfectly honest, I haven't seen a one that I like. All of them are difficult if not impossible to adjust to achieve optimum combustion for each stage of operation. In my judgment the people that buy such inexpensive equipment pay for it several times over in added fuel cost and maintenance headaches for the life of that equipment.

The last system I saw was touted as a real breakthrough by the manufacturer but neither his technicians nor two of my best could get it to operate with less than 5% excess oxygen, about 30% excess air, without generating excessive levels of CO and never got the CO down to levels that a conventional burner could provide.

FIRING RATE CONTROL—JACKSHAFT

This is the most common method for firing rate control if you go by the number of boilers equipped with modulating controls. The modulating motor described in the section on steam pressure control or another form of actuator responding to a device that is attempting to maintain the pressure or temperature at the boiler outlet is connected to a shaft (A in Figure 11-26) by mechanical linkage. The shaft is supported on the boiler by two or more bearings (B).

As the modulating motor shaft (C) rotates, or the actuator (not shown) changes position, the linkage (D) rotates the shaft. Some burners may not have a single central jackshaft, especially with small burners the linkage may simply connect one device to the next, but most burners will have one. In Figure 11-26 the gas valve (not shown) is driven by a cam (E) which pushes on linkage (F) and the burner register is controlled by another link (G). Notice that the linkage that controls the air, moving either a damper or register, is directly connected to the shaft without any adjustable cam.

The jackshaft is connected by additional linkage to the fuel valves, Figure 11-27 shows the extension of the shaft (A), an end bearing (B) and the cam (H) that directly positions the fuel oil flow control valve. On this particular boiler the cam for the gas valve is used to change the stroke of the linkage (Figure 11-28) for gas. Figure 11-29 shows another arrangement controlling a damper for air flow.

Figure 11-26. Jackshaft

The controls are all linked to the one common shaft so fuel and air flow controlling devices are all positioned together. Some people will call this system mechanical parallel positioning but I call them jackshaft systems.

To maintain a pressure or fluid temperature the modulating motor aligns its potentiometer with the pressuretrol or temperature controller as described earlier. The movement of the motor changes the position of the fuel flow control valve to increase or decrease the quantity of fuel entering the burner and, therefore, the heat released in the furnace and transferred to the fluid and vapor inside the pressure vessel. This is commonly a proportional control. You should be able to mount a lever on the jackshaft and a scale at the end of the lever

Figure 11-27. Link to oil valve

marked with the corresponding pressure or temperature. Occasionally you will find a reset controller powering an actuator to position a jackshaft.

The first step in setting up controls with a jackshaft is to establish linearity of air flow. That's all you have to do to get linear control because the fuel will be adjusted to match air flow. With simple linkage like that shown in Figures 11-26 and 11-29 establishing linearity can be very difficult but it's an exercise that's essential to get consistent control. I'll cover it in more detail in a bit.

After establishing linearity, tuning consists of positioning the controls at each screw on the fuel valve (Figure 11-27 or 11-28) then adjusting the screw to increase or decrease actual fuel flow at that position until the desired air to fuel ratio is established.

That process should be repeated at each screw although some technicians will do every other one or every third one then adjust the ones in between to provide a smooth transition from screw to screw. Sometimes the screws are not evident, they're concealed beneath a cover (Figure 11-30) to provide some tamper resistance. The series of screws form a cam that the roller on the fuel control valve shaft rides on as the jackshaft rotates. With some difficulty you can usually position yourself where you can see the shape of that cam.

I've seen a number of those cams adjusted in a manner that they look more like a woman's figure than a smooth cam. Look at yours to see if it's a smooth transition from low fire to high fire. If it isn't then the system is probably non-linear. You may have trouble

Figure 11-28. Link to gas valve

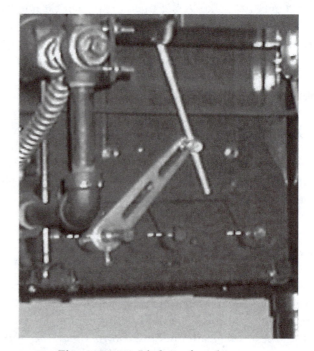

Figure 11-29. Link to fan damper

Figure 11-30. Linkage control valve with covered adjustments

finding a technician that's even capable of understanding linearity, let alone adjusting the actuating motor and fan damper linkage to produce it. Some technicians will tell you it's impossible to establish linearity on a small boiler but that's because they don't know how to do it. In a few pages I'll tell you how.

A jackshaft system provides simple, highly reliable control but its performance is affected by external conditions and devices. Wise boiler operators need to be aware of how they can alter the air to fuel ratio independent of the jackshaft controls and maintain their plant accordingly.

The flow that is the most susceptible to external influences is combustion air flow. I've often walked up to the door of a boiler plant and banged my nose because the door didn't budge when I pulled on the handle (I pulled myself into the door or wall) once I put enough pull on it the door opened and I found myself blown into the plant by the air flow. It's no wonder the operators were having problems with the boiler smoking, they had closed all the windows, doors and operable louvers (many also boarded up the fixed louvers) so the air had only cracks and seams to get through on its way to the inlet of the forced draft fan.

Such conditions also aggravate the situation because the soot formed on the fire sides of the boiler from the smoke act to restrict the flow of flue gases through the boiler to block off the air flow even more. A wise

operator knows his combustion air comes from the outdoors and makes sure the sources of that air flow are not blocked by leaves, snow, and other forms of debris. There are some offsetting conditions because a fan will deliver more pounds of cold air than hot air (see the section on centrifugal fans) so the air to the burner actually increases as the boiler room gets colder. It tends to offset the additional friction as the operators start closing everything to keep warm; but it can't do it all.

I have a problem with the typical approach of tuning up the boilers in the summer when there aren't a lot of no heat calls so the technicians have plenty of time. I would rather pay the technician overtime to tune my boiler in the winter when the doors are shut and the air is cold. That's when the boiler is burning the most fuel and I want the most efficient operation. Any jackshaft controlled boiler should be tuned in cold weather with all doors, windows, etc., adjusted to winter positions.

Some of that increased flow of colder air is required later in the winter when the gas or oil gets colder. There isn't a significant difference in the volume of oil as it cools and change in flow is not as measurable as it is with gas. Colder gas is more dense and the boiler will burn more gas at each setting of the control valve. The colder air doesn't necessarily compensate for it.

There are also variations in fuel and air flow associated with changes in atmospheric pressure because the pressure of fuel after a pressure regulator is equal to the sum of spring force and atmospheric pressure in the pressure reducing valve. The fuel gas pressure can vary a fair amount depending on where the regulator vent is. Pressure is higher when the vent is on the side the wind is hitting. A pressure below normal atmospheric is often produced on the downwind side of a building.

Wind forces can also affect the difference between the air inlets to the building and the stack to alter combustion air flow. Air density also varies slightly with atmospheric pressure. All these variations in temperature, wind, atmospheric pressure, and human generated interferences require all burner adjustments have a cushion of excess air to absorb those variations. We'll accept a little loss in efficiency to ensure we don't operate fuel rich so we generate carbon monoxide and other hazardous and poisonous gases.

A typical jackshaft system is adjusted for about 15% excess air at high loads, producing a flue gas with 3% oxygen remaining, to ensure the boiler will always operate without going fuel rich. In testing it can probably fire at 1/2 to 1% excess air without combustibles. Almost any boiler will require some increase in excess air below 50% firing rate because the drop in velocity through the burner reduces mixing of air and fuel. As the lower firing rates are approached the excess air may go as high as 100% and, due to damper leakage, can go even higher.

The principle concern with the jackshaft control system itself is linkage slipping. It's not uncommon for one of the linkage connections to come loose. I know one plant that chose to solve that problem by welding all the linkage only to discover that the heat from the welding distorted the linkage and they had to replace it to restore the adjustments.

Other tricks including drilling the links and inserting tapered pins didn't work either; they weakened the shaft and linkage which subsequently broke. The best solution to loose connections on jackshaft linkage was provided by technicians at the Louisiana Army Ammunition Depot outside Shrevesport. They stopped at the auto supply store every fall to buy a different colored can of automotive spray paint and, after making their adjustments, sprayed all the connections with that paint. Any change in position was immediately apparent because the paint was cracked or a different color was showing.

That doesn't mean they won't slip, only that you'll know it if they do. Judicious use of lock-tight or, preferably, star washers to prevent them coming loose is also a wise thing to do.

A less common problem, but one you have to be aware of, is that linkage rods can be bent to change their length and the relative position of the controls. That won't be evident with the paint trick described above. Some arrangements make this a difficult situation to spot because the rods are bent to begin with so they can clear some obstruction on the burner. If you have any of those rods the best thing to do is mark their angle on a cardboard template and keep it for reference.

Another problem, typical with firetube boilers, is the linkage gets disconnected when the boiler is opened for inspection or cleaning. The wise operator scratches match marks at all the connections before breaking them to open the boiler. That way the linkage can be put back (almost) precisely where it was. Fresh paint after matching the scratches will restore confidence in the settings too.

ESTABLISHING LINEARITY

Okay, it's not a word in the dictionary, that's because I created it to describe what's necessary with any system to operate properly at all loads. If systems are not linear you will experience problems with performance at different loads. Loss of linearity associated with repairs, rebuilding, and even other maintenance functions can suddenly create different operating characteristics for any system. To understand why let's use a principle that's discussed in the Chapter on controls and the graph in Figure 11-31.

You have to accept that the only thing you control in a boiler plant, refrigeration plant, and any process is flow. When a system load changes the flows must change for it to be in balance and it's important that a change in flow is proportional to a change in other flows or controlling parameters. The most desirable relationship between flow and any control or instrument signal that's used to control that flow is described by a straight line drawn on a graph between zero and 100% of the flow and controlling parameter. It was made apparent to me on a job years ago and I've been preaching the gospel of linearity ever since.

At that time we were converting two boilers from firing oil to gas / oil firing at the Central Avenue Heating Plant in Baltimore. One boiler had been finished and we were waiting for a two week check-out period to complete on the first boiler before starting work on the second. Repeated calls from the plant regarding slow control response were resolved by a control adjustment that produced more calls regarding erratic and swinging operation. Adjusting the controls to respond properly at one load would produce slow or swinging operation at another load. Investigation revealed the problem was due to the control of the boiler air flow which had a characteristic similar to curve A in Figure 11-31. At low loads a change in control signal (in response to a change in steam pressure) resulted in a significant change in air flow. At high loads a large change in control signal was required to achieve a similar change in air flow and, as analysis of the curve will reveal, inadequate change in air flow compared to lower loads. Steam pressure controls respond to a change in steam pressure by increasing or decreasing the control signal output in a consistent manner and the controls don't know that the controlled flow (air flow in this example) produces a varying change in flow. When the controlled flow changes vary considerably depending on load the controls don't know it and can't respond properly to it. The best performance will always be achieved when a change in controller output produces a consistently relative change in controlled flow; a 10% change in controller output should produce a 10% change in the controlled flow.

Adjustment of the linkage achieved performance similar to curve C which is close enough to eliminate most of the variation in performance and that resolved

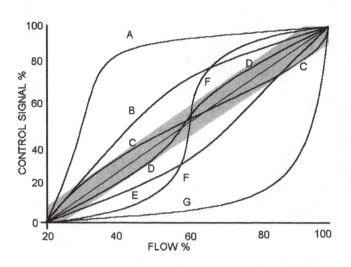

Figure 11-31. Flow characteristic curves

the problem for a while. Some months later the plant personnel pulled maintenance on the fan for that boiler and restored the original linkage positions which restored all the problems. So, whenever working on a device that controls a flow be certain that you maintain the linearity of that device. You may also find signs that one of your existing systems is not linear (note the curve of the cam in Figure 11-32) and resolve its non-linearity during a maintenance outing. Be prepared, however, to have changes in related systems resolved after correcting linearity. If the air flow of the boiler was corrected to be linear, or close to it, then the fuel control cam shown in Figure 11-32 has to be adjusted by a boiler technician (actually the technician should have established linearity of air flow first so the cam wouldn't look like it does).

Modern digital controllers are frequently adjusted to achieve linearity. The problem with them is they only

Figure 11-32. Non linear cam

correct it in automatic. You may encounter a system that responds to manual adjustments of the control output unsatisfactorily but the system works well in automatic. I insist that every flow control device provide a linear response whether the control is in automatic or manual to ensure when the signal is changed for a respective output to a flow control device it doesn't matter whether it's in automatic or manual. Imagine trying to operate a feedwater control valve in manual that has a flow characteristic similar to curve F in Figure 11-31. There's no reason for an operator to have to check a chart or curve to decide how much change in flow will occur with a given change in a controller output in manual. That requires the final element that actually changes the flow produce the linear characteristic.

When working on anything that controls flow the first thing to think about is retaining linearity. It's not at all uncommon for an operator or maintenance man to wonder why linkage is arranged as it is and change it so it "looks right." Adjusting the position of linkage on a boiler can alter the fuel air relationship. This has happened many times with different degrees of disastrous consequences. Before dismantling any linkage, including ones that aren't on boilers, always mark it and restore its original position after maintenance work is done. The correct position for linkage is seldom uniform as shown in Figure11-33a so position on each shaft, length of shaft arms (from center of shaft to center of hole for connecting rod) and length of each connecting rod should be recorded on a sketch along with labeling of the driving and driven shaft that will be clear to another worker replacing the linkage all before dismantling it.

On occasion the linkage will have an unusual configuration, Figure 11-33b is one example. It's where the driving shaft on the left has no measurable effect on the driven shaft on the right for the last few degrees of rotation. I trust you can see that within the dashed lines the rotation of the driven shaft will not change measurably while the driving shaft rotates for almost 15 degrees. This would be consistent with a setup where the forced draft fan damper, attached to the driven shaft, is closed as much as possible and the linkage has to permit more rotation of the driven shaft to reduce the fuel flow rate controlled by another set of linkage connected to the same driven shaft. So, don't expect this to be wrong if you observe it. If the link arm on the driven shaft is nearly aligned with the connecting rod there is a danger of it flipping and trying to rotate opposite of the intended direction. When such an arrangement is used it's advisable to have a stop mounted that will not permit the link arm and connecting rod flipping.

The rotation of the driven shaft will vary when the two link arms are not parallel when the link of the driving shaft is perpendicular to a line drawn through the center of the two shafts. This feature can result in nonlinearity or be used to produce it. Note that with the linkage shown in Figure 11-33c, counterclockwise rotation of the driving shaft on the left will produce a varying rotation of the driven shaft on the right. As the driving shaft begins to rotate counterclockwise the movement of its linkage connection is almost perpendicular to the connecting rod so there will be very little movement of the driven shaft. As the two shafts continue to rotate the rotation of the driven shaft will accelerate until the driving link is perpendicular to a line between the two shafts then begin to decelerate. The graphic in the figure shows the relationship of the two links as they rotate. Note that the driving link could rotate clockwise but the intent here is for it to rotate counterclockwise from the initial position shown. The driving shaft link cannot rotate more than 140 degrees in either direction because the driven shaft link and the connecting rod would be aligned. If the driving shaft rotates further it will not be able to return to the original position.

With most linkage arrangements there will be one link arm that has multiple holes in it for connecting the connecting rod at different distances from the attached shaft as shown in Figure 11-33d. Attaching the connecting rod at one of the holes that's further from the shaft than that of the link on the other shaft will result in more rotation of the other shaft than the one with the adjustable link. Conversely, attaching the rod at one of the holes closer to the shaft will produce less rotation in the other shaft. Combining that variation with those shown in Figure 11-33c should reveal why it is best left to the technicians to adjust linkage to achieve linearity.

Simply checking for linearity requires some care and understanding. You should recall the discussion of the lawn sprinkler in the first chapter and remember that flow is proportional to the square of pressure drop.

There are two graphs in the appendix that can be

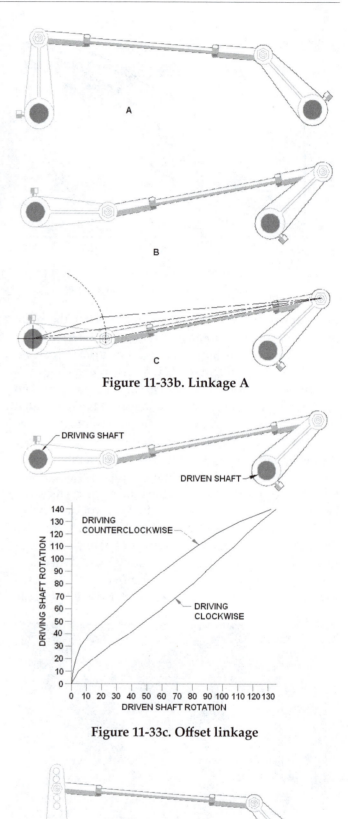

Figure 11-33b. Linkage A

Figure 11-33c. Offset linkage

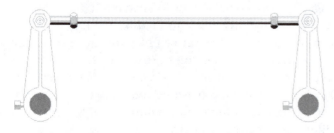

Figure 11-33a. Parallel linkage

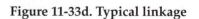

Figure 11-33d. Typical linkage

used to relate pressure drop and flow to get linear air flow characteristics. The easiest one to use is the square root graph paper in Appendix H. Measuring the pressure drop between furnace or burner housing and stack with a manometer and while the fan is running (no fire) will provide all the information necessary for establishing linear air flow.

Setting your manometer on a slope (Figure 2-3) will allow you to measure the pressure drop in hundredths of an inch. Extend tubing from the manometer, connecting the bottom end of it, into a hole in the stack. Be certain the end of your tubing is not pointing towards or away from the direction of air flow so you avoid getting any velocity pressure reading. Extend another piece of tubing through the observation port of the burner and connect it to the top end of the manometer. The end inside the burner has to be positioned to avoid velocity pressure as well; it's best to put a 90 degree bend in the end so the end is perpendicular to the flow.

If you have air flow measurement then you could use plain graph paper and simply record the air flow. This exercise is useful, however, when there is any reason to question your air flow measurement. Compare the flow indicated at the differential using the graph in appendix G.

To ensure the boiler will not fire while you're working on this it's best to remove the burner management chassis. On small boilers it may be necessary to jumper the fan starter to get it running independent of the burner management system. Once you get the fan running locate the terminals that drive the modulating motor so you can jumper them to control the position or, with other control systems, simply put it in manual so you can stroke the damper. Run the controls up to high fire to get the maximum air flow and record the pressure differential on the manometer.

If you're working with a jackshaft system you should operate the modulating motor to lower air flow, stopping when you're even with each screw and recording the air differential. With more sophisticated controls set the air flow controller output at maximum then decrease it and read the differential at 10% intervals (90%, 80%, 70%, etc.) Once you have differential pressure readings for all the flow values you can draw up your graph.

Make a copy of the graph in Appendix H and sit down with it and a calculator. Write "air flow—%/100" on the bottom of the graph and "differential—%/100" on the left side. The chart values are 0 to 1 so the "%/100" indicates that the range of your data is from zero to one hundred percent. If you had ten cam positions or used

the percentage scale of your air controllers output then all you have to do is use the scale on the bottom of the chart, remembering that each value indicated should have a zero after it and one is one hundred. If you have the typical cam with twelve positions then 100% is 12 and 1 is zero adjustment with a span of 11. For each cam position (1 through 12), subtract one from it then divide by 11. Note the result on the calculator, locate it on the bottom of the graph, draw a vertical line on the graph and write the cam position under it.

For each corresponding differential pressure reading, divide the reading by the maximum measured differential. Locate that value on the vertical scale of the graph and draw a light line horizontally until it intersects the corresponding cam position or controller output line and make a big dot there. Once you've applied the ten or twelve dots draw a line connecting them. The line should always extend to the upper right corner where both values are 100%. Don't be surprised if a line from zero and zero isn't appropriate, the lowest position or controller output is at low fire and the air flow at that point should be anywhere from 10 to 25% and the differential would be between 0.001 and 0.06.

Now what? Hey, if the line is straight, or nearly so, you're done. If, however, the line is anything but straight (like the curves A, B, E, F or G in Figure 11-31) you had better adjust that linkage to get a more linear system. You want something that falls in that narrow gray band on Figure 11-31 for good control.

If you're dealing with a jackshaft you'll have to change the position of the linkage. When possible, restore the original settings by the manufacturer, they should be linear. Otherwise, opt for changes that make sense to you then take some more readings to see how you did, repeating the process until you get something linear.

For the best world, a damper actuator with a positioner, the data you just collected will allow you to produce a new positioner cam. Linear control should produce a straight line from low fire to 100% so simply draw a straight line from the low fire point to the upper right corner. Draw horizontal lines through your data points until they intersect that line. The height of the existing cam at the data point is the height you need for the new cam at the position coinciding with the straight line.

Achieving linearity with a valve positioner or an actuator for dampers and other shaft operated devices was simply a matter of plotting the control signal, process flow, and height of the positioner or actuator cam along with its rotation. Then, take the blank cam and mark off the desired height at each angle of rotation, the desired height being determined by graphing the cam rotation, height, and flow then using that data to note the required height. The data can be collected with whatever cam the manufacturer furnished.

Before collecting cam data for linearity the stroke of the valve or actuator should be established to provide 120% of design flow. That extra flow provides an allowance for the system to catch up. Some valve manufacturers select their valves to produce 100% of the specified flow rate at 70% of the valve opening which means the valve could pass 143% of design flow when full open; I believe that's too much and is one reason I always select the trim of a control valve when I order it. Once the stroke is determined, adjust the linkage of the positioner so it rotates the cam 100% of the design rotation at that stroke. Keep in mind, however, that the valve stroke can't be shortened excessively, it will have problems at the bottom end and may jump on and off the seat, because of the Bernoulli Effect, making control sloppy. It's better to have the trim replaced to something smaller if the stroke is reduced more than 30%.

The record of the valve characteristic before and after linearizing should look something like figure 11-34a which is a sample alignment record sheet. The cam rotation relates to the valve stroke and degrees are simply values for scale.

The graph of actual measurements (normally on the back of the record sheet) and marking of the blank cam for a sample alignment are shown in Figures 11-34b and 11-34c. Yes, I made mistakes in the math and that's why the cam graph is plotted so I can see something that doesn't look

right and go back and check it.

I'm sometimes asked why I bother with the columns for zero percent signal. The answer is simple, for some applications there has to be flow through the valve or damper, like minimum flow of fuel and air for a boiler. This positioner was set up for a feedwater control valve and it has to shut for zero flow. The flow was read by the feedwater flow meter and divided by the maximum flow to plot that curve in percent. The height of the new cam positions was calculated using the desired flow divided by the measured flow and multiplied by the original height (in percent divided by 100). The heights are shown plotted on the blank cam of Figure 11-34c. A French curve (drafting instrument) is used to draw a smooth curve on the cam as a guide for cutting it. This may seem like a lot of work but it provides an accurate flow through the valve proportional to the signal so a change in the control signal always produces a desired

POSITIONER ALIGNMENT RECORD

VALVE / ACTUATOR SERVICE - **BOILER 3 FEEDWATER CONTROL VALVE**

VALVE / ACTUATOR AND POSITIONER TAG NO. - **1087**

POSITIONER MFR. - **FISHER**

MODEL NO. - **667/503E**   SERIAL NO. - **E503-347801**

INSTALLED MEASUREMENTS: CAM NO. - **2**

CONTROL SIGNAL	4MA										20MA
SIGNAL PERCENT	0	10%	20%	30%	40%	50%	60%	70%	80%	90%	100%
CAM ROTATION	0	30	60	90	120	150	170	190	210	240	270
CAM HEIGHT	0	10%	20%	30%	40%	50%	60%	70%	80%	90%	100%
PERCENT FLOW	0	26%	51.5%	68%	82%	94%	102%		108%	118%	120%
MEASURED FLOW	0	6,440	15,800	20,860	25,150	28,830	30,070		33,120	36,190	36,800

CAM CUT DATA AND FLOW VERIFICATION

CONTROL SIGNAL	4MA										20MA
SIGNAL PERCENT	0	10%	20%	30%	40%	50%	60%	70%	80%	90%	100%
CAM ROTATION	0	30	60	90	120	150	170	190	210	240	270
CAM HEIGHT	0	6.3%	10.3%	17.6%	26%	35.4%	49%		61.9%	88.5%	100%
PERCENT FLOW	0	11.1%	22.2%	33.3%	44.5%	55.6%	66.7%		77.4%	88.8%	100%
MEASURED FLOW	0	4,090	8,180	12,270	16,360	20,450	24,530		28,500	32,700	36,800

TECHNICIAN: **KEN HESELTON**   BADGE NO. _____   DATE: **11** / **08** / **88**

NOTES: **NONE OF THE STANDARD CAMS PRODUCED LINEAR FLOW.**

Figure 11-34a. Record of linearizing a positioner

flow through the valve (precise provided the feedwater header and steam pressures are near design). The controller will not see an unexpected response and control operation will be smooth. The added benefit is that you can always expect a proportional change in flow when operating in hand.

START-UP CONTROL

The only type of start-up control that I believe I didn't create in a system is one that's called "low fire hold." The control consists of an extra pressure or temperature switch that opens contacts to prevent automatic modulation of a burner when the pressure of temperature of the boiler is below the switch setting. Once the temperature, and temperature can be used even on steam boilers, or pressure rises to a level higher than the switch setting the automatic controls can operate.

What usually follows is a modulating control run-

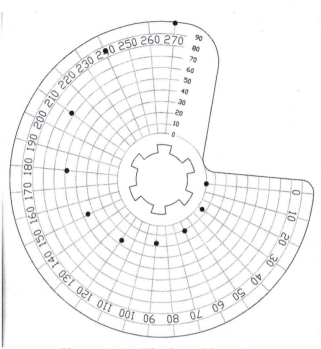

Figure 11-34c. Blank positioner cam

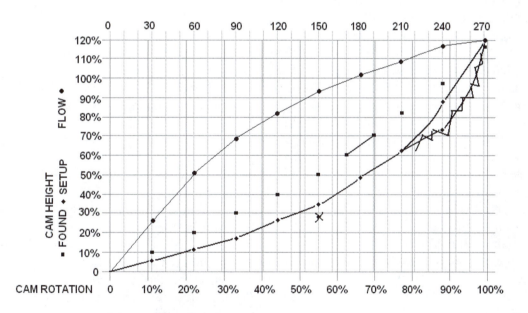

BOILER FULL LOAD FLOW = 30,000 PPH X 120% = 36,000 PPH
STROKE ADJUSTED TO 1.12" BEFORE CAM READINGS   100% - 30,6?0
FULL LOAD FLOW AT 100% STROKE = 36,800 PPH, 2.2% OVER,
  -30° - 6440/306.7 = 26%   60 - 15,800/306.7 = 51.5%   90 - 20,860/306.7 = 68%
120 - 25,150/306.7 = 82%   150 - 28,830 = 94%   180 - 30,070 = 102%
210 - 33,120 = 108%   240 - 36,190 = 118%   270 - 36,800 = 120%

CAM CUT CALCULATION
                                                150
  30° 4,089 │8,178│12,267│16,356│ 20,445 │24,534│28,623│32.712│36,800
  6.3% │10.3%│17.6%│26% │ 35.4% │49% │61.9%│88.5%│100%
                                                 ▪(65%)

Figure 11-34b. Graphs for positioner

ning on up to high fire. Now, supposedly, the boiler is warm enough that it won't experience any thermal shock or excessive thermal stresses in the process, but I'm never certain of that. I've never had occasion to even think about a better way to do that before writing this book. It's probably because I've never been required to design any for the smaller boilers. Now that I've thought about it, I would do something a little different.

All of the large boiler systems where I designed the controls, and they included an automatic start-up provision, we used a ramping provision. Once the pressure exceeded the setting of the low fire hold switch vent valves were automatically closed and the ramping system put in service. It simply allowed a very slow increase in the firing rate and prevented a more rapid increase until it had completely ramped out.

The first ones were applied on pneumatic control systems and consisted of a low signal selector, three-way solenoid valve, and a volume chamber with a metering valve. The solenoid valve dumped the contents of the volume chamber to atmosphere while the boiler was off and applied supply air, usually at around 18 psig, to the chamber via the metering valve when the boiler had started and the low fire hold switch released. The pressure in the chamber was piped to the low signal selector along with the boiler master output or the plant master output. The low signal selector then fed either the fuel and air controls or the boiler master. Which one depended on the plant master operation. If all the boilers were operated off the plant master then the output of the ramping control was fed to the boiler master. Otherwise it went to the fuel and air controls so the operators could set a firing rate manually at the boiler master.

Once the low fire hold was released the air bleeding into the volume chamber slowly increased the firing rate. Where we could we set that ramp rate as slow as possible, sometimes to get it out to two hours to reach high fire (and that's still faster than some boiler manufacturers specified) we would have to use a larger volume chamber. Once the firing rate exceeded the plant master or boiler master output it was no longer the low signal and the boiler was operating on automatic.

The ramping control actually allowed a boiler to automatically come on line, generating steam and picking up load without upsetting the operation of the other boilers. I should mention that this control feature wasn't all that was required. Steam traps to drain the boiler steam headers and many other features were required for fully automatic control. Oh yes, some of those boiler plants were actually unattended, no licensed operator

present most of the time. That wasn't my choice, however, I only designed the systems.

If a sudden increase in plant load required the boiler firing rate to increase once it was on automatic the low signal selector would not allow the firing rate to increase any faster than the ramp rate. Once the pressure in the volume chamber bled up to supply pressure the boiler operated automatically as if it wasn't there, the automatic signals were always the low signals.

Another nice feature of the system was it drained off pressure from the volume chamber as slowly as it filled it up. If the boiler tripped for some reason, then started back up, the ramping simply swung back up but allowed the already hot boiler to fire at higher rates.

With modern digital controls the same feature can be added. It can be augmented to provide a ramp rate from a cold start and a different ramp rate from a restart. All it takes is some additions to the software. It's one of the times when computers are wonderful.

FIRING RATE CONTROL— PARALLEL POSITIONING

Parallel positioning controls perform about the same as a jackshaft control system because they simply establish the position of the fan damper and fuel valve. Large boilers and boilers with air heaters or fans located away from the windbox can't easily use a jackshaft type of control because the weight of the linkage becomes a problem. A parallel positioning system allows the fan and fuel valve to be located convenient for the construction and for other reasons.

The most commonly used parallel positioning system is an electric positioning system which uses potentiometers like the modulating motor controls to compare the position of the fan damper and fuel valve actuators and adjust them to match the position of a boiler master. Plants with parallel positioning control usually have a plant master for steam pressure control which actuates a bunch of potentiometers that are matched in position by the boiler masters or fuel valve controllers on each boiler plus or minus any bias introduced by adding resistance in the potentiometer loop.

Some advantages of parallel positioning controls include the ability to run on a plant master and bias boilers, it doesn't constrain the location of fuel valves and fan, it permits isolation of gas and oil control valves on two fuel boilers so both aren't operating to influence operation of another boiler firing a different fuel, and they permit maintenance on the valve and actuator for the alternate fuel. It also permits independent operation of the fuel and air controls so a boiler operating on hand can be trimmed (adjusted by the operators) to reduce excess air.

A principle disadvantage of parallel positioning controls is variations in response of the actuators, especially for pneumatic actuators, which can produce temporary upsets in air to fuel ratio during load changes. Unlike the jackshaft system there is nothing to prevent the fuel valve actuator from moving faster than the fan damper actuator or vice versa. They also have all the disadvantages of the jackshaft system with one provision to help offset the problems with maintaining air to fuel ratio, adjustment of the air to fuel ratio.

By adjusting the resistance in the loop of a system where the air flow actuator follows the fuel flow actuator the relative position of the two actuators can be varied to provide an excess air adjustment. The adjustment is a bias type adjustment in most systems but it does permit running the air to fuel ratio a little tighter if the operating personnel choose to. It also allows the operators to compensate for soot accumulation in the boiler, something that's not easy to do with jackshaft controls. To help overcome the problems with the independent actuators the controls can be enhanced to include a leading actuator provision so the fan damper actuator follows the fuel on a decrease in load and the fuel actuator follows the fan damper actuator on an increase in load.

In addition to problems with response jackshaft and other modes of parallel control can suffer a degradation due to hysteresis associated with wear in linkage connecting the devices used to position dampers and valves. To overcome the problems with response and hysteresis manufacturers have developed systems utilizing synchronous actuators which are, for the most part, mounted directly on the damper or valve shaft. The synchronous motors are basically electric actuators with an integral positioner that responds to a current signal or digital instructions from a controller. I have seen these devices operating smoothly and erratically but don't have enough experience with them at the time of preparing this second edition to judge them one way or the other. I think we all know that a mechanical device that's constantly changing position back and forth isn't exactly working right so don't accept one that's doing it.

For all practical purposes a parallel positioning control system is set the same way as a jackshaft control. You need some way to set the fuel valve to match the air. Normally a parallel positioning system will have cams just like a jackshaft system.

FIRING RATE CONTROL—ADD AIR METERING

The full title of this control logic is parallel positioning with air metering. The next evolution in control systems after parallel positioning was to add air flow metering. Since air flow is influenced by so many factors it makes sense to measure the air flow and control it. The measured air flow provides feedback to the control system so the air flow controller can adjust the fan damper actuator to produce a repeatable air flow. Instead of simply positioning the fan damper the control system adjusts the damper until the air flow signal matches the plant master position signal.

The decision to measure air flow started the still standing arguments about where it should be measured. The type of control system and boiler has some effect on the choice and you should be aware of all the variations. Air metering with measurement across the burner windbox to furnace on multiple burner boilers made it possible to compensate for the number of burners in operation. Because each burner throat is one orifice in the flow path; changing the number of burners doesn't change the differential measured at the air flow transmitter when the register on one closes, but it does change the air flow.

Measuring the air flow using a differential across the boiler itself is measuring the flue gas flow, not just air flow, but that's not a significant variation since the air is 93 to 94% of the flue gas. The problem with using the boiler is sooting can change the differential relative to air flow and other problems like refractory seals breaking up can also alter that differential without the operator being aware of it.

The best measurement is in a suitable metering run or venturi between the forced draft fan and burner windbox but most boilers don't have enough room there for any kind of precision flow measurement. I've always preferred using the inlet of the forced draft fan to measure air flow, provided it's a single inlet fan. When a boiler is large enough to justify a double inlet fan then a good metering element between fan and burner windbox is justified as well. Many operators don't understand fan inlet metering and some even manage to screw it up so I want to explain it well enough that you'll understand it.

Within the boiler room there is air movement but it's very slow except for right at the inlet of the forced draft fan or its silencer. In order for the air to accelerate from something very close to zero velocity in the boiler room to the speed it needs to get through the fan inlet there has to be a difference in pressure between the two locations. The fan creates a lower pressure at the fan inlet by removing the air that enters it and it's that void created by the fan removing the air that the room air rushes into.

The boiler room itself is nothing more than a big pipe that the combustion air flows through; the fan inlet is just like an orifice. By measuring the static pressure at the orifice and subtracting it from the pressure in the room we get the velocity pressure which tells us how fast the air is flowing into the inlet of the fan. There's one thing rather nice about this flow measurement, there's no orifice coefficient because there's no measurable friction applied to the airstream between the boiler room and the fan inlet.

My standard arrangement for this measurement of air flow is shown in Figure 11-35 and requires: a ring of half inch tubing forming a circle equal to two thirds of the diameter of the inlet; the holes in the ring drilled just a little past center to minimize plugging with dust from the air; the ring mounted outside any screen or other obstruction in the fan inlet that could get dirty to vary the signal; mounting of the transmitter at least five fan inlet diameters from the inlet of the fan and independent of any obstructions that would produce air velocity near the high pressure sensing port of the transmitter; a drop leg to prevent dirt entering the high pressure connection of the transmitter; mounting of the transmitter above the ring so there's no way condensate can form and collect in the transmitter and sensing piping to block the signal. Any condensate that does form will run out the holes in

Figure 11-35. Fan inlet flow measuring ring

the sensing ring.

There's only one caveat with this method of air flow measurement. You have to be certain there's no way for the air you're measuring to go anywhere but to the burner. I've encountered more than one embarrassing situation where this method measured the air flow but it didn't all get to the burner. It won't work if the there's air leakage, branch ducts, or the like between fan and burner.

The original systems were a little lax in producing a true flow signal. Recall that the pressure drop we measure is proportional to the square of the flow? Some simply used the differential signal and counted on the screw cam type fuel flow control valve for setting the fuel air ratio. Others provided a flow signal somewhat related to actual air flow but still counted on the adjustment of a cam type fuel valve. The problem was developing an output proportional to flow from a differential pressure signal. Some of the original air flow transmitters used cams, others used combinations of springs, and others used the stretching of the diaphragm used to sense the differential.

They all gave way to differential pressure transmitters with panel mounted square root extractors until microprocessor based transmitters were developed. If you ever have an opportunity to visit a museum that displays controls and devices you'll quickly appreciate the many tricks used to determine the square root of a signal. Modern microprocessor based instruments either calculate the square root right in the transmitter so the output is directly proportional to flow or the square root is calculated after the differential pressure signal is input to the controller.

An air metering addition to a parallel positioning controller allows tighter control of air to fuel ratio and should permit operation at less than 15% excess air, in the range of 2-1/2 to 3% oxygen in the flue gas and less on single burner boilers.

FIRING RATE CONTROL— INFERENTIAL METERING

I mentioned the fact that extracting the square root to convert a differential pressure signal to a flow signal was a little difficult. Early inferential metering systems simply avoided the problem by comparing the differential signals. After all, if it's proportional to the square root for air flow it must be for oil flow or gas flow so just match up the differential signals, right?

Well, it does work, there are some differences between orifice coefficients and other factors that had to be taken into account and most control systems had provisions (adjustable cams) to compensate for it so inferential control provided many of the features of metered control without the expense (and difficulty) of square root extracting.

They also solved some problems that were principally associated with multiple burner boilers. There were a lot more multiple burner boilers in the middle of the 20th century because they were either converted from firing coal or designed to be convertible to coal. Coal fired designs use a reasonably square furnace, not the long skinny ones we're used to on most boilers today. The shorter furnace required use of multiple burners.

Inferential metering is accomplished by considering the fuel delivery systems as an orifice with a pressure drop that can be measured and comparing that with the air side pressure drop. These systems were only applied to oil and gas fired boilers and they used the burner header pressure as a variable that equated to fuel flow. After all, the oil burner tip is an orifice or group of them and a gas ring or spud has orifices in it, and the pressure in the furnace (downstream of the orifice) was relatively close to zero so it is reasonable to treat the burner header pressure as a value of differential.

Some gas fired systems used gas at such low pressures it was essential to include a furnace pressure input to the measuring device so the changes in furnace pressure didn't upset the flow signal although they did experience some difficulty with pressure fluctuations (see draft control).

Modern instruments have erased the cost advantages of inferential metering systems so you will see fewer of them. When inferential metering is used today the differential is treated as a flow signal and the square root is extracted by the transmitter or controller to produce a linear flow signal. One of the more serious problems with inferential metering systems was their lack of linearity. The control response was normally tuned for the high end of the boiler operation and swings accepted at low loads.

In dealingx with those multiple burner boilers they had a distinct advantage, even over today's full metering systems. The fuel flow based on the burner header didn't account for the number of burners in service and the differential from windbox to furnace didn't account for the number of registers open. If a burner tripped the control backed down to restore the header pressure, effectively decreasing the flow so the air to fuel ratio at the other burners was restored. Later the operator could close the register and the air control would restore the

windbox to furnace differential to restore the air to fuel ratio again. The only problem came when someone put a burner in service and forgot to open the register.

FIRING RATE CONTROL— STEAM FLOW/AIR FLOW

Inferring fuel flow by pressure worked fairly well for oil and gas but it didn't help with coal firing. Steam flow/air flow systems were developed for coal firing and are basically inferential metering systems because the steam flow could be equated to fuel flow. If the boiler efficiency and steaming conditions were constant then a fixed relationship between steam flow and fuel flow would exist because the fuel would generate a proportional amount of steam. The systems eliminated the problems with, or impossibility of, measuring the coal flowing to the fire. Coal fired boilers larger than about 90,000 pph can justify the expense of metering the coal but smaller units still use steam flow/air flow control.

One problem with steam flow/air flow is the lag in response associated with load changes. If the plant master output increases there is a delay associated with the inertia of the boiler. It takes a little time for the higher coal flow rate to heat up the boiler a little more and increase steam flow rate. If the system was set up so air flow followed fuel flow the boiler would probably smoke on a load increase.

The systems normally use a parallel positioning control methodology where plant master changes produce a proportional change in fuel feed, primary air flow (on pulverized coal fired boilers) and combustion air flow and maintain the ratio of fuel and combustion air flow signals with the steam/air flow ratio on a slow reset. Some engineers refer to steam flow/air flow systems as parallel positioning with steam flow trim because the steam flow is used to trim the ratio between fuel and air.

It's the timing problem that dictates how tight air to fuel ratio can be maintained with a steam flow/air flow system. Gas and oil fired systems could actually run a little tighter than a system with air flow metering added because changes in fuel input produced a rapid change in steam rate. Pulverized coal fired boilers have a delay in load changes associated with changes in coal inventory in the pulverizer so they typically operate with excess air rates around 30%. Stoker fired boilers have a larger inventory change effect and have to operate closer to 50% excess air to eliminate fuel rich firing conditions (and smoking) during load changes.

Of course you don't run all boilers at that rate; the

wise operator will let one boiler take the load swing and set others (if they're needed) to fire at a constant load and much tighter excess air rates. The steam flow/air flow controls can then respond to variations in fuel quality to maintain the appropriate air to fuel ratio.

FIRING RATE CONTROL—FULL METERING

As the title indicates, full metering control systems measure the flow of fuel and air. Similar to labeling the steam flow/air flow metering systems some engineers will call the systems parallel positioning with flow tie-back. The advent of microprocessor based controls (which have drastically reduced the cost of control systems) and continued reductions in device costs allow for smaller and smaller boiler control systems of the full metering type. As of the writing of this book I recommended any oil or gas fired boiler that consistently operates at loads above 25,000 pounds of steam per hour (25 million Btuh) be equipped with full metering controls; they will return their cost in fuel savings in a matter of two or three years. Any step between a jackshaft system or parallel positioning and full metering (with the possible exception of adding oxygen trim which is covered later) is, in my judgment, a waste of money. The currernt gas bubble may not provide enough savings for boilers in the 20,000 to 25,000 range today.

The full metering system does use flow as feedback to the controls but I prefer to think that the controllers control the flow of the fuel and the flow of the air to produce a heat flow into the boiler that matches the load. The plant master signal which maintains a pressure at the common boiler header is proportional to the heat load. The boiler masters in a steam system pass the plant master signal plus or minus any bias at the boiler master to the firing rate controls. Hot water and fluid heating boilers each will have their own temperature control or, in large sizes, a load indication based on fluid flow and temperature differential to produce a boiler master signal for the firing rate controls.

The firing rate controls respond to the boiler master signal by changing their outputs until their respective fluid flow transmitters send back a signal that matches the boiler master. Modern full metering systems automatically include what we call cross limiting to prevent fuel rich firing conditions. There was a time when you added the term "cross-limiting" to your definition because it required additional control devices. Today cross limiting is simply a couple of extra instructions in the software.

The full metering system is shown in Figure 11-36 without the plant master controller. The lower of the master signal or air flow signal become the setpoint for the fuel flow controller. The symbol < in the diagram identifies a low signal selector, its output is the lower of the two inputs. This is part of the cross limiting because the fuel controller can't see a demand for fuel flow greater than the air flow signal. The fuel flow controller adjusts its output using PID algorithms until the fuel flow signal matches the lower of the air flow or master signal.

The air flow controller's setpoint is the higher of the master or fuel flow signal to provide the other part of cross limiting. The symbol > in the diagram identifies selection of the higher signal. The air flow controller will adjust its output until the air flow signal coming back to it is equal to the higher of fuel flow or master.

Just to make sure you understand what's happening, let's take a look at the system performance when a load change occurs. Say someone opened up a steam valve to a process in the facility so we have an increase in load. The plant master will detect a drop in pressure and change to increase its output. The increase in plant master output is passed through the boiler master to the firing rate controls. Since the air flow signal matched the previous boiler master signal it is lower than the master

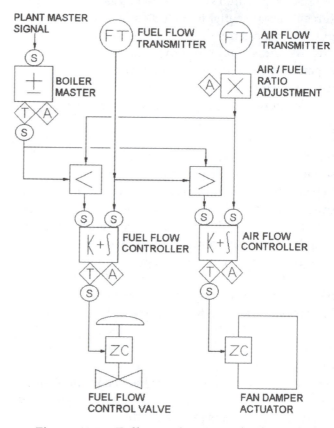

Figure 11-36. Full metering control schematic

so the fuel flow controller doesn't see any change in its remote setpoint.

The master signal is higher than the fuel flow signal so it passes through the high signal selector to become the remote setpoint of the air flow controller. The air flow controller then responds, changing its output to increase air flow. As the air flow increases the transmitted flow signal increases to raise the setpoint of the fuel controller. If the master signal stops changing the air flow signal will eventually come up to match the master and the fuel flow signal will follow it. Look at the diagram and see how the air will follow the fuel on a decrease in master signal. That's how the system with cross limiting works.

On a decrease in load the fuel flow goes down and the air flow controller follows the fuel flow signal (as its remote setpoint) down. Therefore, cross limiting prevents a fuel rich condition. Some engineers try to think of these as lead-lag systems because the air leads the fuel going up and lags it going down. They're incorrect because we've had lead-lag systems for years and it has nothing to do with fuel and air.

Since all the control signals have to match we have a problem when the air to fuel ratio has to change. Any change in the master signal between air and fuel controllers will upset the cross limiting. To resolve that problem we modify the air flow signal to indicate an air flow that is less or more than what it actually is. A typical method is to insert a ratio control between the air flow transmitter and the fuel and air controls with their signal selectors as shown in Figure 11-36.

When we had systems that used our ratio totalizer we used a special one with a threaded shaft through the pivot point extended to a knob on the panel. By turning the knob we changed the totalizer pivot position, sort of adjusting the gain, so the flow signal to the controllers was equal to the air flow transmitter times the totalizer gain.

Despite the fact that a full metering control eliminates many of the variables of pressure effects (people opening and closing windows and doors and other situations), there is one serious problem with full metering controls that you must be aware of. If the fuel flow signal is lost the controller will drive the fuel valve wide open almost instantly! If the air flow signal is lost the air controller will drive the damper wide open and can blow the fire out or produce a lot of unburned fuel by quenching the fire.

Either situation is hazardous but the loss of fuel flow signal is the most dangerous. Many system designers incorporate differential sensing devices that will shut down the boiler if the fuel and air flow signals don't match within limits; I don't favor shutting the boiler down. The choice we made was to compare the fuel flow signal with a prescribed minimum and drive the boiler to low fire if the signal was less than that value. It doesn't result in a boiler shutdown and gives an operator a chance to correct the situation or fire the boiler in hand rather than running around trying to get another boiler on line. The limit also prevents a shift above low fire in the event of loss of the control signal after start-up. We don't worry about loss of an air flow signal because we haven't had it happen... yet.

FIRING RATE CONTROL—DUAL FUEL FIRING

First let me explain that dual fuel firing means firing two fuels at the same time and under control. Boilers that can fire gas or oil are two fuel boilers, they can fire gas or they can fire oil but they can't fire both at the same time. Low fire changeover systems are discussed in the section on operating wisely and aren't dual fuel firing either.

To fire two fuels at once you have to have a full metering system. In addition you need a fuel flow summer that combines the two fuel flow signals so the total fuel flow is the feedback signal to the fuel controllers and to the high selector of the air flow controller. One of the two fuels has to be considered the primary fuel and the other fuel flow signal has to be adjusted with a gain so it produces an output that equates to the air flow demand of the primary fuel. Some engineers call the summer a Btu summer because it takes about the same amount of air to produce a Btu whether you're firing oil or gas. The rest of the controls don't know that they're looking at two fuels so they operate normally.

When dual fuel firing you're usually switching fuels. There are other operating conditions that favor dual fuel firing but the common one is switching fuels. A dual fuel firing system is the ultimate in control for a boiler and you should have it unless you only fire one fuel or almost never switch. I believe it's the best way to transfer fuels because you're always operating with an inert furnace environment. It's safer than stopping then restarting the boiler and a lot safer than the low fire changeover systems.

The standby fuel is brought on the burners at low fire then manually adjusted upward until the fuel flow controller output equals the manual output for the standby fuel; the controller will automatically reduce the firing rate of the leading fuel to compensate for the added standby fuel. When the two fuel flow signals are equal you switch the standby fuel controller to automatic and then switch the leading fuel to manual.

It doesn't require an instant transfer because the controller will simply adjust the two fuels in parallel. Control action will not be smooth with both fuels in auto because every change in output produces twice the change in flow compared to firing one fuel; don't leave both fuel controls in automatic unattended. It's possible to control two fuels in auto at once but, why? If you're dual fuel firing there are other reasons, not one of which involves auto operation of both fuels to maintain steam pressure. To complete the transfer you reduce the firing rate of the lead fuel manually until it is at low fire then shut it down.

There's nothing preventing you firing both fuels continuously as long as one is in manual control. It's convenient for burning down an oil tank while still firing natural gas or firing natural gas at the maximum rate allowed by your supplier.

I should have titled this section multi-fuel burning. I've put in a couple of projects where we burned three fuels simultaneously, gas, oil and a solid fuel. There are very few opportunities to do that so I stuck with the "dual fuel" label. Don't let that prevent you from considering firing more than two fuels on one burner; just keep in mind that only one can be operating on any one automatic control signal.

FIRING RATE CONTROL—CHOICE FUEL FIRING

It is possible to fire two fuels and have both on automatic control, just different automatic controls. Modern microprocessor based controls allow dynamic changes in controller gain so a fuel controller could fire oil and gas together. The question is, why would you want to do that? Choice fuel firing is the incorporation of additional controls to meet fuel supplier's criteria. You could have a system that calculates the rate gas should be fired based on the amount of gas you are allowed to burn and the number of hours left in the month. The gas glut created by fracking (a process of fracturing gas laden rock in the earth that has produced the gas glut) will eventually diminish and this type of firing may return.

It's easy with computerized control, you input the amount of gas you're allowed each month and the controls do the rest. There are several parameters that the control system needs to make a decision about the gas firing rate at any time including the need to burn a minimum amount of fuel oil. Some history of facility performance during that month can be used to predict situations when less gas could be burned and increase the current rate so the gas is consumed by the end of the month.

Choice fuel firing can also be associated with burning other combinations of fuels including a fuel produced by a process in your facility. Some of it may be sold to another facility. That case could involve burning the excess fuel the other facility does not use. When the other fuel is gas it is handled by pressure control, sensing the pressure of the generated gas and varying the flow of gas to your fired equipment to maintain a constant pressure at the output of the process. It could also include a variable setpoint for the gas pressure developed by the process that generates the gas. When the other facility uses more of the generated gas the pressure will attempt to drop so the controls reduce the amount fired in your equipment to maintain the pressure. When the other facility uses less generated gas the pressure will attempt to rise with the result of your burning more.

So, what happens when neither facility needs more of the generated gas? Your firing rate controls will first reduce the firing rate of the main fuel which is the only one normally controlled to maintain boiler output. When the main fuel firing rate is at minimum then the generated gas control is overridden to maintain the boiler output, reducing its use in your boiler. If there is still too much the generated gas pressure will increase and a relief valve will send it to a flare to simply burn it and waste the energy. You should be able to note this occurring (actually production should advise you that it's going to happen) and look for ways to use the generated gas in other equipment to offset purchased fuel cost by utilizing all the generated gas that's possible to use.

If the production generates a liquid fuel then deviations in supply of the generated fuel can be absorbed in storage tanks. Automatic control for such applications is unusual. Production should be able to tell you of any changes they make in production and you can determine an optimum firing rate for the generated fuel based on production data and the level in the storage tanks.

Burning of waste gases, oils, and solid fuels generated by the plant process operations can also require analysis of the fuel to determine its proper air to fuel ratio so it burns cleanly and efficiently. Liquid and solid fuels can be stored and tested with the ability to adjust air fuel ratio before firing it. Gas fuels, on the other hand, are not easily stored. I had questioned one customer, a refinery, with why they didn't burn gas that was flared in the boilers and was told the waste gas varied considerably in its hydrogen to carbon ratio so it wasn't possible. I encouraged them to use an online analyzer that could be used to automatically adjust the fuel air ratio for the gas being burned and incorporate that into an equivalent fuel value applied to the firing rate controls but I don't know if it was ever done. One tricky part of that application would be providing a means of accounting for the time delay between the analyzer's response to the waste gas values and the time it takes for the gas to get from the analyzer to the burner.

In earlier times we had a separate setpoint generator for the gas controller with a low signal selector that would reduce the gas firing rate when the fuel flow was at minimum. These systems are more dependent on the contracts with your fuel suppliers than any other parameter so, if you have one, realize that you're trying to use up fuel you've paid for without using any more—which normally costs an arm and a leg.

FIRING RATE CONTROL—OXYGEN TRIM

Oxygen trim controls actually measure and control excess air. The oxygen content of the flue gas is controlled but it's an indicator of excess air. An analyzer samples the flue gas in the furnace or at the outlet of the boiler to determine the amount of oxygen in the gas. The analyzer transmits a proportional signal to a controller which then changes the firing rate controls to alter the fuel to air ratio to maintain the oxygen content at a setpoint.

Almost any oxygen trim system you encounter will not have a simple oxygen setpoint because the amount of excess air required does vary with load. In most boilers the excess air can be held constant at loads over 50% of maximum but it has to increase almost exponentially as the load decreases (see the section on burners for reasons). The common approach is to generate an oxygen setpoint that, for all loads up to about 50% is a function of the boiler master signal or the steam flow signal.

I prefer steam flow because it produces higher oxygen requirements when increasing the firing rate of a cold boiler. It's when more excess air is needed to complete combustion because the furnace and refractory are not as hot so the flame temperatures are lower. The common approach is to use a function generator which allows the technician setting up the control to produce an output that bears no mathematical relationship to the input.

Data collected during firing tests on the boiler (to determine the necessary amount of excess air at each load) can be used to determine how to cut a cam in the function generator. Modern digital controllers have a similar application except you simply enter numbers instead of measuring a plastic or aluminum plate and cutting it to get the desired shape to produce the output. It's another blessing of microprocessor based controls that you can easily change them, you don't have to cut a new cam if you made a mistake at one point.

On jackshaft controlled boilers the trim is accomplished by adjusting the linkage connecting the fan damper to the shaft so changes in the relative position of damper and jackshaft alter the air to fuel ratio. The adjustment has to be made in a manner that maintains some relationship to firing rate because a change in damper position near maximum fire that would be considered minimal can be major change in air flow when the burner is at low fire. Once again microprocessor based controls serve to recognize those problems and correct for them but, if you have an older system, you should be aware that the same correction at high fire has to have a much smaller effect on air flow at low fire; the same rules for linearity exist.

Oxygen trim control of parallel positioning systems (including steam flow/air flow, inferential and full metering) should use a multiplier to change the relationship of fuel valve and fan damper position for oxygen trim control. That way any change in the two signals is proportional to load. Multipliers are not an easy device to make for pneumatic systems so many use a simple bias adjustment, adding to or subtracting from the signal to the damper positioner to trim the air to fuel ratio and maintain an oxygen setpoint. On inferential and full metering controls the air flow signal is modified by the oxygen trim so the output of the transmitter should be multiplied by the correcting output of the oxygen trim controller to change it proportionally over the load range.

These controls became acceptable during my later years in the business and represented another step forward in technology and reduced manufacturing costs. Originally only utility boilers could be equipped with oxygen trim control because the analyzers required almost constant maintenance and recalibration. Hot wire analyzers which combined a flue gas sample with some hydrogen and heated it until the hydrogen burned were the first analyzers to prove partially reliable and low enough in cost to be used in industrial plants.

The paramagnetic analyzer which used the difference in oxygen content of a gas to disturb a magnetic field then followed. Both required drawing a sample of the flue gas from the boiler or stack and conditioning it before analysis. They used water systems to cool the gas which always introduced a problem when there was any amount of oxygen in the water. The sampling systems had to operate at high velocity to reduce the time between analysis and a response to a change in burner operation so leaks in the sample piping was always a concern.

The advent of the zirconium oxide analyzer made oxygen trim possible on even small commercial boilers because the analyzer can be mounted in the boiler or stack to achieve fast sampling and analysis. There were a few made with sampling systems, some integral to the analyzer, and I installed a few before the "in-situ" analyzers came out.

The in-situ zirconium oxide analyzer doesn't measure the oxygen content of the flue gas. Before you start arguing with me you should read on because it really doesn't. The analyzer measures the difference between the oxygen in the flue gas and a reference gas. Often the reference gas is the air around the analyzer and, if the boiler casing, ductwork, or stack leaks that reference can vary in its oxygen content. Many units still use a compressed air source as a reference gas and that can be complicated by particles or droplets of oil in the compressed air.

To work the zirconium oxide cell (which is a ceramic substrate coated with the metal oxide) must be heated to a temperature around 1500°F. At that temperature any oil in the compressed air will burn and deplete some of the oxygen in the reference gas. If your analyzer(s) use compressed air I suggest you provide a separate compressor for them, one of the inexpensive oil-free compressors that only has to produce air at 10 psig or so. Besides, it's a real waste to dump air you compressed all the way up to 100 or 150 psig for use as a reference gas. You can size the little compressor to match your analyzer needs plus a little for calibration, get far less expensive air, and it's oil free.

If you fire oil regularly I suggest you incorporate a procedure to prevent damage to your analyzer while blowing tubes. Steam soot blowers add a considerable amount of moisture to the flue gas when they're operating. The problem with that is that steam has a much higher specific heat than air and the heater in a zirconium oxide analyzer has to really put out to push the gas temperature up to 1500°F. It's not the going up that's the problem, it's when the soot blower shuts off and all of a sudden that heat isn't needed; the analyzer overheats and parts burn out.

I solved a problem with repeated failures of an early model of zirconium oxide analyzer by inserting a soot blower header pressure switch in the heater power

circuit. The analyzer didn't work very well while we were blowing tubes and indicated low oxygen so the air flow went a bit high but the analyzer quit failing every month.

Regular failures of the analyzers and drifting of the calibration compelled me to provide an air fuel ratio adjustment independent of the oxygen trim control and really limit the trim control range so an analyzer failure didn't produce a hazardous situation or a lot of waste. Figure 11-37 shows a schematic of the air flow loop with this configuration. The summer is set to apply a gain of 0.1 to the input so the full range of output of the oxygen trim controller is reduced to a multiplier adjustment of ±5%. That not only limits the extent the trim controller can adjust excess air, it also uses the full range of the trim controller output. When the oxygen controller output is at 50% the multiplier for the fuel air ratio is 100%, basically one, so the air flow signal flows directly to the controls without modification.

That's where the output of the oxygen trim controller should be when everything is initially set up, right in the middle of the range so it can act to increase or decrease the excess air. It's something to look for after your controls are tuned. If the output is zero the technician didn't leave the control system any way to decrease the excess air. If it was set up in the summer a reduction will probably be necessary in the winter when the fan starts pumping colder air. On the other hand, if the output is well above 50% it limits the amount the system can increase the excess air. Lacking any reasonable explanation from the technician, the output of the controller

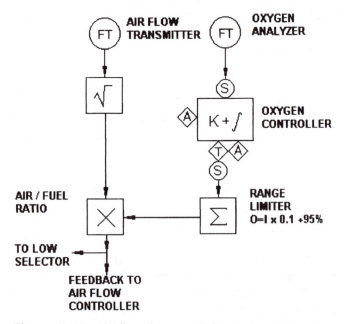

Figure 11-37. Air flow loop with limited oxygen trim

should be right at 50% at any firing rate immediately after it's set up.

I can recall many an operator that was confused because the trim controller didn't seem to be doing anything because only about 10% of its output range influenced the air to fuel ratio; talk about reset windup! Bias in the summer is set to 0.95 so the output is anything from 0.95 to 1.0 and the air flow transmitter is set for a range 16.87% higher than the actual differential (8.11% more flow) at full load at normal operating conditions so the output of the multiplier is 100% under those conditions.

Recall that the oxygen content had to increase as firing rate decreased because of loss of turbulence? An oxygen trim control has to provide for that and it does so by introducing a variable set-point that's fed to the oxygen controller. Two methods are used, one adjusts the oxygen set point as a function of firing rate (basically the boiler master signal) and the other adjusts the set point as a function of load using steam flow. I prefer the latter. A boiler needs more excess air during cold starts because the fire doesn't get the benefit of the heat reflected back from a hot burner throat and furnace refractory. Using steam flow to establish a variable oxygen set point produces a higher set point for the trim controls while the boiler is warming up. Once steam flow is established tighter control can be accommodated and the result is the same as using the master.

The variable set point is generated by a control device called a function generator. A function generator permits a control technician, or a savvy boiler operator, to produce preset outputs at multiple points of input. Some function generators provide for four points while others allow eleven. Its setup is simple with new microprocessor controls because you can set an output for each input by simply entering the input and output values into the controller. In my early days we had to do it with a cam, like on a valve positioner, to produce the desired output. I believe we did a better job back then, even though it took a lot of time and tooling to determine how then cut the cam. Why? Because the cam was smooth and a function generator that's set with specific points has sharp changes at the points that can introduce instability in the control.

In addition to the function generator the oxygen set point generator should include a bias adjustment or independent summer that allows the operator to increase the oxygen set-point to accommodate upsets or unusual fuel conditions and decrease it when the boiler is firing at a fixed rate (manually set) to optimize operation under that condition. The result is a set-point that follows the same slope for excess oxygen that's produced by the offset of the air flow zero.

With this method the oxygen trim controller is limited and it could easily wind up or down to the end of its output range if there was a considerable change in the fuel or some other factor. If the operator notes that condition the first action should be to check the fires to see if their appearance indicates a condition indicated by the analyzer output. If the fires appear normal then the analyzer should be checked for calibration. It's uncommon to need any more adjustment than that plus or minus 5%.

With our experience getting LNG (liquefied natural gas) added to our normal there was no guarantee that 5% would always be enough. Rather than allow the trim controls more latitude, I added a manual station so a boiler operator could put another signal on the summer. Giving that input a gain of 0.1 and changing the summer's bias from 95% to 90% allowed the trim control a ±5% and the boiler operator ±5%. If the fires indicate the analyzer is right the operator can adjust the manual air to fuel ratio adjustment (slowly) to restore the trim controller's output to 50%.

Note that I used a gain on those inputs so the controller or operator could adjust their respective outputs over the full control range, from zero to 100%. A large number of trim controls and similar control loops use limits that will not allow that. You might be able to adjust the output from 0 to 100% but the limits only allow it to work from 40% to 60%, the rest of the time nothing happens. Operators complained to me when they were exposed to limits on the output so I modified the control parameters so they got full range. Now they're happy. A control technician shouldn't set up a control system to impose limits that don't make sense.

A 2-1/2% swing in air to fuel ratio is not something that I would expect to see in the short term so regular checks and adjustment of the manual ratio adjustment (to restore the trim output to 50%) are not likely to happen. Any significant changes should be discussed with your control technician because it may indicate problems with the controls or (more likely) one of the flow transmitters.

A boiler with a high stack temperature (over 500°F) will benefit from oxygen trim control. Low pressure heating boilers and boilers with economizers or air heaters have to burn a fair amount of fuel to justify oxygen trim. Single burner boilers can be fired with stack oxygen content of 1/2 to 1% with a combination of full metering and oxygen trim but don't expect much in fuel savings from oxygen trim if you have low stack temperatures.

Also don't expect oxygen trim to be a cure all. There is a definite lag in time between the change of an air to fuel ratio on a burner and the appearance at the analyzer cell of a gas sample that is the result of that change. If the controls aren't aligned properly to maintain an air to fuel ratio with load changes don't expect the oxygen trim controls to correct that. If you see the trim controller output change with load that's what it's trying to do. You should also be aware that the oxygen measured at the analyzers didn't necessarily come through the burners; this is particularly true on induced draft and balanced draft boilers where the furnace and boiler passes are at a lower pressure than atmospheric and air can leak in at several points after the burners.

FIRING RATE CONTROL—CO TRIM

One solution to problems with oxygen analyzers sensing oxygen that didn't come through the burners is to control based on another parameter. Carbon monoxide, the result of incomplete combustion and a gas that is always present in some minute quantities in flue gas can be sensed and controlled. The original oxygen analyzer problems still hold for analyzing CO and there's the problem of its finite quantity.

We normally control at about 50 parts per million, that is 0.005%, a very little amount of gas in the whole and hard to measure accurately. Large utility boilers frequently use it to resolve the problems with casing and ductwork leakage. The operating modes are the same as for oxygen trim. I've never installed one but more modern analyzers could change all that.

DRAFT CONTROL

Many small boilers use natural draft and a natural means of draft control. The gas fired hot water heater in a house is one. Most use a draft hood, nothing more than an open box over the outlet of the boiler. Natural draft up the stack produces a difference in pressure between the bottom of the hood and the rest of the room so the space under the hood is negative with respect to the room. You know, and usually check to ensure, that there is a negative there by holding a match or lighter near the bottom edge of the hood. If there's a draft, air flowing from the room into the hood will pull the flame into the hood. I check draft on my wood stove before lighting it by holding a flame near the top of the charging door and open the door a little; the draft almost always pulls the flame down into the stove. If it doesn't, I use a newspaper fire with all the stove air inlets shut to heat up the stack before starting a wood fire.

The air that is pulled into the draft hood from the room goes up the stack. It cools the stack gases which lowers the natural draft until there isn't a significant difference between the pressure in the room and the pressure under the hood. Since the boiler outlet is under the hood the pressure at the outlet and the boiler inlet differs by the natural draft through the boiler. If the hood wasn't there the pressure at the boiler outlet could vary so much that it could blow out the fire, as on cold days, or be so high that you wouldn't get enough air for clean combustion. That draft hood stabilizes a fixed fire operation to ensure maintenance of the air to fuel ratio.

I still have friends asking me about problems with their gas fired appliances and discover that they went a little overboard with one of those insulating blankets that the home stores are pushing. They plugged up the opening to the hood!

Instead of a draft hood we'll occasionally use a barometric damper. That's a single bladed, usually round, pivoted just above its center, damper that separates the boiler room and the stack. These usually have stops on them so the damper will not swing out at the bottom into the boiler room. As the stack draft increases the difference in boiler room pressure and the stack base forces the damper open and cold air from the boiler room slips into the stack to cool the stack gases and reduce the draft. If the draft gets too low the damper closes down to restrict the flow of cold boiler room air into the stack. The stack temperature then increases to raise the draft. Those dampers usually have a weight mounted on a stud to adjust them, by screwing the weight in and out you change the pressure required to open the damper.

Barometric dampers or some other means of controlling draft is essential on systems with two or more boilers attached to the same stack. Draft is always a balance between the differential pressure produced by natural draft and the resistance to gas flowing up the stack. Double the flow of gas, by firing two boilers instead of one, and the resistance to flow up the stack will increase by a factor of four. It's obvious that the pressure at the base of the stack will differ considerably so the flow of combustion air and flue gases through the boilers will too. It's impossible to maintain air to fuel ratios in boilers with a common stack unless you have draft or metering controls.

Barometric dampers do a fair job but they also require a lot of additional air supplied to the boiler room. In the winter you have to heat that air or worry about freezing some pipes. If you can't get exhaust air from other sources and there's a lot needed to control that draft, other means of draft control, more expensive means, will lower the operating cost and pay for that expensive control.

In addition to accounting for a variation in flue gas flow, draft controls can maintain a parameter in the boiler, such as furnace pressure, a requirement for balanced draft boilers. Many operators believe that's the only place you can control the pressure with draft controls but nothing could be further from the truth. If you have two or more boilers capable of pressurized firing you can control the draft anywhere between the furnace and the outlet for individual boiler control.

If you're controlling the common draft (at all boiler outlets) it has to be controlled there or at a central point in the breaching where there's little difference in pressure as flows change. I don't recommend a common control because it can fail to prevent operation of all the boilers and it's very difficult to get the large damper in a common stack to handle all the turndown that's required of it.

Balanced draft boilers require a means of controlling the induced draft fan to keep a constant pressure in the furnace, something slightly less than atmospheric pressure. A typical control loop looks no different than any other control loop; a transmitter senses furnace pressure and sends a signal to a controller which alters its output to an actuator for a boiler outlet damper or a damper at the inlet or outlet of the induced draft fan. It can also vary the speed of the induced draft fan.

The control isn't that simple because there are a number of factors that influence it. First the pressure in the furnace of the boiler should only be slightly less than the pressure in the boiler room outside the furnace. That way any air that leaks into the furnace and boiler passes is kept at a minimum. That air is heated to stack temperature and thrown away just like excess air so it's a loss that should be minimized. The furnace pressure transmitter is really a differential pressure transmitter comparing furnace and boiler room pressure and it should have a maximum range of six inches water column and, preferably, have a range of one inch.

Don't do like one plant I checked where they mounted the transmitter in the control panel and sensed the pressure using draft gage piping. The control panel was in a conditioned room in another building. The boiler pressurized regularly, blowing smoke and soot out into the operating area. Of course it never blew any into the remote control room.

The differential that transmitter measures is so low it needs a large diaphragm to accurately measure it in the required range (less than two inches of water column). The larger diaphragm transmitter costs a lot more than the standard differential pressure transmitter (like about three to four times as much) so many a plant is fitted with one that saved the contractor a lot of money but doesn't work worth a darn. The desired operating point for a furnace pressure is 0.05 to 0.2 inches of water column below the boiler room pressure. Transmitters with a wide range, like 50 inches or so, become too unstable for good control. Also, at the low pressures we're dealing with for draft control any pressure fluctuations due to a noisy fire create a very noisy pressure signal and considerable filtering is necessary to get a steady output. Frequently the location of a furnace pressure sensing connection has to be moved because the selected spot just happens to be where heavy pressure waves from combustion noise strike it. Of course there's also the problem of incomplete combustion that can create a noisy signal.

Once any problems with the furnace pressure signal are resolved there's the problem of load changes. In the old days when controls were expensive we lived with that unless the boiler loads were constantly changing more than ten percent or so. When necessary we added cascade control where the output to the boiler outlet damper became the output of the air flow controller plus or minus the output of the furnace pressure controller. The summer which combined the air flow controller output and the furnace pressure controller output also needed a bias spring to subtract fifty percent so the furnace pressure controller output would end up at mid range just like two element and three element feedwater controls. Those draft control systems got tuned with changes in gain applied to the air flow controller input and the bias to satisfy control requirements which varied between boiler start-up and operating conditions. Modern microprocessor controls can use the stack temperature as an input to help compensate for the variation in conditions.

The use of balanced draft allows the boiler manufacturer to use open inspection doors and joints that aren't exactly gas tight in furnace construction. The result is there are plenty of places for atmospheric air to enter the furnace. I've visited many a plant where the furnace controller setpoint was at two tenths of an inch negative or more. The leakage at two tenths is three times as much as the leakage at five hundredths where the setpoint should be. Operating at five hundredths will allow an occasional puff of furnace gases into the boiler room, especially during start-up, but will provide far more efficient operation. With modern microprocessor based controls using the stack temperature could permit varying the setpoint to minus two tenths for start-up increasing to minus five hundredths for normal operation.

FEEDWATER PRESSURE CONTROLS

I decided to add this control consideration because it is unique to boiler plants. In many cases I consider it to be done improperly so I'll cover what's been done, why it was needed, and what you should consider for your plant.

Why even control the feedwater pressure? If you read the chapter on pumps you know the differential follows the pump curve and as long as the discharge pressure is less than the maximum pressure rating of the pump and piping there's no way the pressure can get too high. Some pumps do have a rather steep curve so we may choose to do something about the pressure getting too high but most of the time the problem is with the feedwater control valves.

A pneumatic or electro-hydraulic actuated feedwater control valve can be selected with an adequate diaphragm or enough hydraulic pressure to keep the valve closed under conditions of the maximum feed pump discharge pressure and no pressure in the boiler. The thermo-hydraulic and thermo-mechanical valves described earlier had limited power and in most cases couldn't operate with a pressure differential greater than thirty to fifty pounds per square inch.

Another reason for pressure control was to improve operating efficiency; turbine driven boiler feed pumps could be controlled using feedwater header pressure or the difference between feedwater and steam header pressures to throttle the steam to the turbine. It reduced steam flow through the turbine to save energy. Actually it saved by allowing operation of more auxiliary turbines to eliminate motor operating costs.

The normal practice for maintaining a constant feedwater header pressure, or a differential between feedwater and steam headers, consisted of installation of a control valve that dumped feedwater back into the deaerator or boiler feedwater tank. I'll admit that I designed and installed a lot of systems that did that before I became more energy conscious and started questioning why we did what we did.

Today I know that method maintains a header pressure or differential but it also wastes a lot of energy. I think the first time it was apparent to me was in an industrial plant where the operators had two feed pumps running (in case one failed) in the summertime when one was four times larger than the actual load. Maintaining the header pressure by recirculating the water ensures that the pump runs at full capacity (maximum horsepower) all the time.

That industrial plant was running one pump at 30 horsepower for nothing but the mental comfort of the operators, in case the other one failed. Total cost for that pump operation was 52.5 horsepower more than necessary, equal in 1997 energy costs to about $5,000 per year. That's money that never became a bonus for the operators. If you have one of those systems, don't abuse the power bill further by lowering the pressure setpoint one psi lower than it has to be.

If it's necessary to control feedwater header pressure with electric motor driven feed pumps try to get an evaluation of the application of one or more variable speed drives (one for each pump preferably) because they can be used to maintain pressure by slowing the pump down and saving on the horsepower. There's a practical limit to how slow they can go but most of the time they will provide all the pressure control that's necessary. As with other things technology improvements and manufacturing cost reductions has made such controls a wise investment.

All constantly operating boiler feed pumps have a potential problem with overheating, cavitation, and pump damage that can occur if all the feedwater control valves shut off. Temporary upsets in plant operations can result in high water levels in all the boilers so that happens. If the water doesn't flow through the pump then it just sits there and churns; all the energy put in by the motor is converted to heat that raises the temperature of the water. The water that left the deaerator was nearly at saturation conditions so the additional heat will most certainly result in steam generation, cavitation, and pump damage (see the discussion on pumps).

To prevent the damage from such an incident some feedwater circulation is provided. You could argue that the recirculation provided by pump discharge pressure control solves this problem, and it does, but at a very significant operating expense. The standard practice in my early days was to install a small recirculating line on each pump that returned enough water to the deaerator to prevent overheating of the water.

An orifice nipple is made for those recirculating lines and the recirculating lines were usually 3/4-inch pipe size so we could make the orifice out of a 1-inch diameter steel rod. One was installed on each pump to provide protection in the event we were so dumb as to start a pump with the discharge valve shut then forget to open it. Of course there were times when we forgot to open an isolating valve on that recirculating line too! Why, I don't know, because we should have left them open.

Another feature of those recirculating lines that I used in later designs was combining all the recirculating lines into one line returning to the deaerator with another orifice (sized for all pumps) so some of the recirculating water would flow backwards through the other, idle, pumps to keep them warm (the recirc line was connected before the discharge check valve). Knowing what I now know about the effect of piping stress on pumps (see piping flexibility) I think it also prevented damage to the pumps.

During work on a problem with some boiler feed pumps in 1999 I also discovered that higher pressure boilers require so much feed pump energy that the recirculating flow represented a significant amount of extra horsepower and, more importantly for that particular customer, a significant reduction in the amount of water that could be supplied to the boiler.

Plants operating at pressures of 250 psig and higher have had a solution for this problem for many years, it's a self contained check and recirculation valve which consists of a spring loaded check valve that checks the main fluid flow and an integral recirculating valve that opens as the main flow decreases. The problem with those valves is they are very expensive and most plant owners scream at the cost, they can cost almost as much as the pump! Regardless, they work and they pay for themselves in power savings.

There is another solution that I haven't had an opportunity to try yet and I hope to compare to those expensive recirculating valves. The cost of controls has dropped so much I believe you can justify installing two temperature sensors on the pump, one in the suction piping and one that senses the liquid near the pump discharge (without interfering with the flow patterns) to detect a rise in the fluid temperature in the pump. It would control a small pneumatic valve in the recirculating line. As long as the temperature differential is low enough there's no need to recirculate water. It eliminates the pressure loss attributable to the spring on the automatic recirculating valve check and, by using temperature differential, is oblivious to any changes in deaerator pressure that would change the temperature at which steam would start to form.

Another system I am trying is using the integrated controllers of automation systems to create a feedwater flow calculation based on the control signals to all the feedwater control valves in the plant and the number of

pumps in operation to determine when recirculation is necessary and open a solenoid valve in the recirculating line when necessary. I made it a point to have the solenoid valves supplied as normally open and the controls energize it to close. That way it will fail to recirculating to prevent damage to the pump. You should still use orifices but they can be a little larger.

Goofy Controls

The advent of microprocessor controls allows us to add more and more features to a control system. I like to think of them as ways to help the boiler operator. There are, however, some features added that make it more difficult for the operator.

I've encountered some fairly stupid concepts in recent years. Not because they were dangerous or wrong, they just made life difficult for the operator. One I ran into involved a boiler control system where the designer decided that any time any variable got out of range (more than 100% or lower than zero) all the controllers should switch to manual. With that logic, on a balanced draft boiler, every time the boiler was started and the purge commenced all the controls switched to manual because the boiler outlet dampers couldn't close down enough to prevent the furnace pressure dropping below the low end of the furnace pressure transmitter's range so everything switched to manual. The boiler hung up in manual until the operator switched everything back to automatic. Even though I explained what was happening and that there was no need to switch controls to manual simply because the measured variable was out of range the designer insisted that his system had to work that way. Hopefully it was finally resolved but I wouldn't be surprised to visit that plant and watch the operators switching all the controls back to automatic after every start-up.

I also recently had a situation where a technician insisted the drum level transmitter had to be set where mid range was the center of the drum. There are very few boilers out there which have a normal water level at the center of the drum, most are lower by two to four inches. I finally had to insist that the middle of the range was at the center of the gage glass, that it had been that way for ages, and he had better set it there if he wanted to get paid.

On a recent excursion to California to look at a system that had nothing to do with a boiler other than use steam I got frustrated with the programmable controller logic. It prevented starting a pump while the water level in a tank was high. It was high because the pump had been shut off! I had to drain the tank to get the pump started and if I drained it too far the system would shut down on low tank level... and lock out. I've recommended that the designer consider simply alarming some of the conditions and, provided the operator acknowledges the alarm, let the system start when there's no reason not to.

If you run into something that becomes a real pain don't hesitate to grab the engineer or technician and register a complaint. Of course if all you do is complain they may not do anything about it. If, on the other hand, you suggest an alternative approach and explain your reasons they may just go along with it.

INSTRUMENTATION

I have been in one or two boiler plants that honestly had no instrumentation. It was a violation of their State law but that's how they were. No pressure gauge, no thermometers, nothing! Their argument was that nobody ever looked at them anyway so why did they need them. One of those plants had a fuel bill equal to one third of the prior year when I finally got them to understand the need for instrumentation and how to use it.

On the other hand I've been in plants with all the requisite instrumentation and a log book where they recorded many readings and discovered they gave no thought to interpreting what they had. The instruments are there to provide the operator with information on the status of the plant and provide a history of the plant's performance. The wise operator knows how to use those instruments.

Instrumentation varies in sophistication and precision from an indicating light to a fully compensated fuel gas flow recorder. Some, like the indicating light, give an immediate perception of the status of the plant while others, like a flow totalizer reading, have to be subjected to study before the status is determined. One key to the use of instrumentation is—it isn't worth anything if it isn't recorded. Many of the reasons for recording data are explained in the section on boiler logs. The purpose of this section on instrumentation is to convey some points on interpreting readings and understanding the effect of other conditions on the instruments.

An indicating light provides you information on two states or conditions, on and off, right? Well that's a maybe; if the light is off it could be because the bulb is blown or the power is shut down to that piece of equipment. If the light is on, well you know there's voltage and current at the indicating light but that doesn't necessarily mean the status it's indicating exists. That's

very true for pumps, fans, etc., that are powered out of a motor control center with a common control power transformer or where there's a motor area disconnect.

It's possible for the motor starter to pull in, making a contact and energizing a motor running indicating light, and the motor to be sitting there powerless because the power circuit breaker or the disconnect at the motor is open. I watched an operator get very frustrated, throw tools and everything else one evening when he couldn't get a boiler to fire on oil. The light at his control panel said the oil pump was running; it wasn't. I do hope the obvious question came to your mind, what about the low oil pressure switch?

Meters and other electrical devices are directional. Recently some operators blew up a rather expensive set of electrical switchgear because the phases were not realigned after some maintenance. They thought "alternating current goes both ways so there isn't any direction" but it's important to remember that it's different for each phase and single phase power can come from a transformer on one of the other two phases so they don't parallel.

Pressure gauges show pressure but a steam pressure gauge that's mounted at the operating level and has connecting piping to a steam drum twenty feet or more above the gauge also has a standing leg of water on it. To properly indicate the pressure in the boiler the gauge has to be calibrated to read zero when it has that standing head of water.

Thermometers read the temperature at their bulb. That doesn't mean that the fluid is at the same temperature just a few inches away from the bulb. Use the steam tables in the appendix to find the temperature of the steam in your boiler and compare it to the stack temperature. I'm always amazed when someone tries to tell me the stack temperature on a boiler operating at 250 psig (406°F) is 350°F and there's no economizer or air heater. Either the temperature reading is wrong or there's a lot of tramp air leaking into that boiler.

Thermometers in the top of a pipeline can fail to indicate the temperature of the liquid flowing underneath the bulb. Similarly air in the top of piping or a vessel can insulate the thermometer from the heat of the liquid. Part of using instrumentation is realizing when a reading has to be wrong.

Steam flow recorders, unless compensated, are calibrated for a certain operating pressure. If the header pressure is higher or lower then the recorder then the readings are wrong. I've encountered many a plant that thought they had saved a lot by lowering the steam pressure, the recorders indicated they were making more steam per gallon of oil or hundred cubic feet of gas than they used to. They called me in to help them find out why they weren't saving any fuel because, for some strange reason, their steam consumption had increased. I hope you got it, their steam consumption hadn't increased, they just introduced a recorder error by lowering the pressure and had saved almost nothing.

If the steam pressure varies at the recorder (more than plus or minus two or three psi) and you want it to be accurate it needs to be compensated. Compensated recorders for steam use a steam pressure and/or temperature input that allows calculation of the density of the steam at the orifice for accurate measurement. Superheated steam flow recorders need both pressure and temperature inputs to determine the density of the steam, saturated steam only needs one of them.

Fuel gas flow recorders are subject to the same errors from pressure and temperature fluctuations as steam flow recorders. By maintaining the pressure constant there's usually little variation between actual and recorded flow so it's suitable to use a simple recorder. Normally fuel gas flow is recorded at each boiler because we have the flow instruments to provide a control signal for the firing rate controls and it doesn't cost much more to add the recorder. For purposes of control we can live with little errors in the gas flow recordings. Besides, we have a way of correcting them to the purchased values.

We do? Yes, you do. If you don't compare the readings of your fuel gas recorders with the gas company's meter you're missing a real bet. You'll also have some smart ass engineer like me come into the plant and demonstrate to your boss that you don't know what's going on. On the one hand you can catch problems with your metering. On the other, well, I could tell you about two jobs where customers were being billed for far more gas than they were actually using.

You should also track your inventory and manage it. When I was operating we burned heavy fuel oil. Since it had to be heated we always burned more oil than we had. Now that I have you confused I'll explain why. We knew how much oil we had by sounding the fuel oil tanks. The oil in the tanks was maintained at a temperature much lower than the temperature at which we burned the oil. The oil we burned was measured by a fluid meter after the oil was heated for firing. The oil expanded as it was heated so a gallon of fuel burned was always less mass than the gallon in the tank and less mass than the gallon of oil that was delivered.

You have to correct for temperature to keep a good accounting of your oil inventory. If you aren't watching your oil inventory then your employer has a good chance of being stung for a major cleanup cost. The oil in the tanks should match a calculation of what you had plus what was delivered less what you burned. If it's a little less or more you simply show an inventory adjustment, you must have burned it or not burned it depending on which way you're off. If the calculation says you should have a lot more than what's in the tanks you've got a leak or an oil thief. If it's a leak you have to call the local Coast Guard office and inform them, that's federal law.

Fuel oil or gas is measured by the supplier and the user has to pay for what they measured. The plant meter readings, values from your instrumentation, should be corrected to match the supplier's numbers so your data are considered accurate. Divide the combined fuel meter readings for all boilers by the fuel supplier's number to produce a correction factor then multiply that result by your meter readings at each boiler to get the actual fuel consumed in the each boiler. Keep track of the correction factor and ask yourself "why?" if it changes significantly. When firing oil you would use the oil drawn from a particular tank as the supplier's number since you normally verify each delivery with a sounding.

What's a sounding? It's a measurement of the depth of liquid in a tank. The term comes from taking an ullage reading. (Just like those darn engineers, use one confusing word to define another)… I'll clarify. When we measure the depth of heavy fuel oil in a tank we don't like to drop a tape all the way to the bottom then clean it off. We use a probe at the end of a tape that looks like a brass rod with an upside down cup on the end. When we lower that into the tank it makes a plop sound when it hits the surface of the oil. Using the tape measurement from the top of the pipe and subtracting the depth of the tank from the reading gives us the depth of the oil. Since the process involves making a sound (the probe going "plop" when it hits the oil we called it sounding the tank. The actual measurement is called an "ullage" when it's the distance down to the top of the oil.

The sounding of light oil storage tanks doesn't require wiping off a lot of black sticky oil so we usually take soundings where the probe is simply a pointed brass rod or wood stick that drops to the bottom of the tank. We read the level where the liquid coats the rod or stick and wipe the thin coat of oil off the rest of it. That stick you drop into the oil tank is an instrument too. The tip can be torn off (there's usually a brass button on the bottom) or, as in one case I encountered, someone can need a piece of wood about that size and cut a few inches off. Also, just like the meter readings, you can get strange results when the temperature of the oil in the tank and the oil delivered differ considerably. Sometimes it pays to take another reading on a tank a day later to ensure the change in volume is accounted for.

One of the most valuable and important instruments in any steam plant is the drum level gauge on the boiler. It's also one that can go wrong with disaster close on its heels. The most important thing I can say about that instrument is that if you don't trust its indication, shut the boiler down. Either leg can plug and present a false water level indication. Keep in mind that the only force that produces the level indication in that gauge glass is the level in the boiler and you're measuring something in inches of water column.

The steam side can be plugged to the point that only a small opening remains and the steam condenses in the glass faster than it can get through that small opening. The result is the level rises, compared to the level in the boiler, until the condensing matches the amount that can get through the opening. If there's nothing but a small opening in the water leg the level in the glass may rise to produce the additional pressure needed to force the condensate through the small opening. Any leak on the steam side of the glass has to be fed by steam flowing from the boiler. There is a pressure drop in the connection and piping associated with the friction of that steam flowing so the pressure in the glass is lower than it is in the drum and the result is a false high level indication.

Notice that all those potential problems produce a false high level. It can look pretty normal but be wrong. Only a liquid side leak in a gauge glass assembly will produce a false low level indication. I could tell you several stories about false drum level indications but all I really have to tell you is, if you don't think it looks right, it probably isn't and it's normally higher than what's really there!

A common instrument that doesn't get the attention it deserves is the draft gauge. Many plants today don't even have them. Typical vertical draft gauges provide an indication of the pressures in the air and flue gas flow streams of a boiler and are valuable for indicating soot formation and damage to baffles, seals, and dampers. If installed properly draft lines will not plug; the best connection for sensing draft with a draft gauge is shown in Figure 11-38.

You probably won't see many connections like it but it's the best way to do it. The large pipe is sloped where it penetrates the boiler wall so soot and dust that tries to accumulate in it can roll out. The cap at the end allows easy access to clean the boiler penetration when necessary. Every change in direction of the sensing pip-

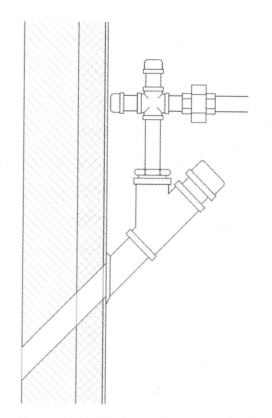

Figure 11-38. Draft sensing connection

ing is made with a cross closed with nipples and cap. Plugs in those crosses will be next to impossible to remove after a year or two. Note that I show pipe, usually no smaller than 3/4 inch. That's to allow a lot of room for dust to pile up before it fouls up the indication.

In addition to allowing for removal of the piping a union close to the sensing point is a great place to insert an orifice. You see, there's always problems with draft gauges because they're measuring such low pressures and the flame can make a lot of noise. In some cases you'll have to relocate a connection because it's looking right at the fire which can produce a very noisy pressure signal. When I say noise I mean the needle on the draft gage is just jumping up and down like crazy. Use a thin (1/16 inch) piece of copper with a small (1/8 inch) hole drilled off center in it and insert it in the union closest to the boiler. If the signal at the draft gage is still noisy take the piece to a vise and hammer around the hole to close it down some then try it again. If, on the other hand, the indication seems sluggish, you can ream the hole out some. It's always a good idea to hang a tag on the union with this orifice so it's the first thing somebody checks if the gage line acts like it's plugged.

Sensing lines for pressure gages can affect the quality of their reading and, in some cases, can produce some operating problems if not installed and maintained properly. First there is the matter of size of the sensing connection; none should be smaller than 1/2 inch NPS. I've broken a few 1/4 and 3/8 connections in my day and had to repair damage to a lot of them. A 1/2-inch schedule 80 pipe nipple and valve is strong enough for most people to stand on without damage; anything smaller is simply looking for trouble. I once spent twenty minutes with my finger pressed over a broken 1/4-inch nipple while someone else was machining a plug for it. On the other side was 300 psig heated Bunker C fuel oil at 220°F.

Sensing connections should be made at the side or top of process lines to limit any debris settling into the smaller line and blocking it. The connection should be isolated with a valve as close as reasonable; only provide enough room for a hand to get at the valve handle and make allowances for insulation. After the isolating valve you can install smaller piping or tubing from the connection to the gauge. If it gets broken you can quickly shut the valve.

I mentioned Bunker C, see the section on fuels, and the fact it was heated. If it isn't heated heavy fuel oil doesn't flow well and below a certain temperature it becomes quite solid. To prevent blockages in sensing lines for heavy fuel we don't put heavy fuel oil in them. There are two approaches to the problem of sensing pressure of heavy fuel oil and they are dependent on the fill liquid. You can use a light fuel oil, like Number 2, or a heavy mineral oil such as Nujol. One is lighter (floats on) the heavy fuel oil and one is heavier.

When using light oil the process connection and all pipe and tubing connected to the process line has to be flooded in such a manner that the light oil is trapped above the heavy oil. When using a heavy mineral oil the process connection should be on the side of the piping and turn down immediately into a separating chamber. Thereafter the sensing piping can be routed however you need it.

With both systems the separating oil must be injected into the sensing lines at regular intervals to refresh it because it will gradually mix with the heavy fuel oil. Since both burn it is best to inject the separating oil while the burner is in operation. Most heavy fuels are fired with steam atomizing and the atomizing steam differential control valves have a chamber filled with oil to sense the burner oil pressure; it's best to inject the separating oil at the valve chamber to flush the piping and tubing all the way to the process line; a valve for that purpose should be provided at the chamber or at the sensing line connection to the chamber. Pump it slowly so you don't blow the fire out.

A fuel oil sensing line can produce a hazardous condition. I encountered this one recently where the piping from the burner manifold to a pressure gauge in the control room was not properly vented. Since the line was full of air it compressed every time the burner operated allowing more than half the line to fill with fuel oil. When the burner shut down the air expanded forcing the contents of the sensing line into the furnace through the burner tip. In most instances the oil simply burns off but keep in mind that a tablespoon of fuel oil properly atomized and mixed with air to form an explosive mixture can blow a boiler casing off.

Always bleed the air out of piping when the accumulating effect of air is not desirable. Provide vent valves at the high points of the piping and keep a piece of the appropriate sized pipe bent with a 180° turn to insert in the outlet of those vent valves so you can cleanly and safely bleed the air and catch any liquid spill in a bucket.

On the other hand, some sensing lines and gauges are protected by air trapped in the sensing lines. The air can serve as a cushion to limit the impact of noise on the gauge. A gauge line for a heavy fuel gear pump can use the air to quiet the effect of the bump each time a gear squeezes out its oil. Centrifugal pumps can produce fluctuations in the line that are associated with the vanes passing the cutoff. Some acid and caustic processes provide for the air to separate a process fluid and a pressure gauge that would be destroyed by that fluid. When you have situations where it's desirable to have the gauge sensing piping full of air the sensing lines should be fitted with vent and drain valves to allow removal of any liquid that may absorb the air.

Note I didn't mention putting air in the sensing piping. Why not? If you do you could blow up your boiler or splash someone with a hazardous liquid. There's also the guy that filled his compressed air storage tank with lube oil.

Pressure and flow transmitters, hell—any transmitter, should be installed where it's convenient to get at for checking and calibration. I still don't understand why contractors insist on putting them ten feet above the floor, down in pits, or inside a maze of piping where you have to be a contortionist to get at them. I know why they do it, to avoid extra cost, because that's where the engineer showed it, or that's where the workman installing it could see the girls going in and out of the next building. I never allowed such inconsiderate locations when I was in charge of their going in because I had to operate with many such crappy installations.

I insist every transmitter has to be mounted at an

elevation four feet above a floor or platform and readily accessible to a person standing on that floor or platform. Sometimes it requires extra piping and installations where the operator may have to blow down the sensing lines a little more frequently. That's okay though, I don't mind doing something a little more frequently if I don't have to climb all over things to do it.

Pressure and differential flow transmitters require piping connecting them to the process line. Some of those lines require long runs of sensing lines and they should be installed in a manner that limits problems with the instruments. The most common problem I encounter (Figure 11-39) is a transmitter installed at the bottom of a sensing line. Any scale, rust, or sediment that comes drifting down the line ends up inside the transmitter.

Liquid pressure and differential pressure transmitters should be installed as shown in Figure 11-40 so the only thing that flows to the transmitter is liquid; the rust and sediment ends up in the drop leg where it can be

removed by blowing down the line through the drain valve. It's almost impossible for the dirt to get up into the transmitter (it will if the transmitter is vented too fast) and, despite some arguments to the contrary, steam will not get into the transmitter when a steam pressure sensing unit is blown down.

Where the transmitter is located and the fluid sensed has a lot to do with how a transmitter is piped. The diagram in Figure 11-41 is recommended for dirty liquid systems making it more difficult for solids and debris in the system getting into the transmitter.

The piping routed to the process sensing connection should always run vertically or at least slope up to the connection so any gas that may form in the sensing piping will naturally rise to the process connection and be replaced by liquid. A little air in a liquid sensing line for flow measurement will introduce a considerable error.

If the transmitter is sensing a non-condensing gas (just about anything but steam) the transmitter should be mounted above the process sensing connection and run in such a manner that anything condensing out of the gas will run back out of the sensing lines into the process line. When it's absolutely necessary to install

Figure 11-39. Improper transmitter installation

Figure 11-40. Proper transmitter installation

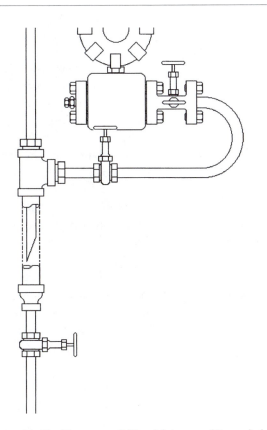

Figure 11-41. Steam and liquid transmitter piping

the transmitter below gas piping (especially for compressed air and, in some parts of the country, fuel gas) the arrangement shown for liquids should be used and a schedule prepared for regular draining of the dirt legs. Otherwise, install it above the line so everything can drain away.

Installation of oxygen analyzers and their sampling locations has varied with the type of instrument over the years. The in-situ analyzers eliminate problems with sampling lines but introduced other problems. The analyzer has to be installed where it senses a representative sample of the flue gas (that's engineer for taking a reading of what's really flowing). It also has to be where the wiring will not be overheated, and in a manner that ensures the reference gas isn't contaminated. See the discussion under oxygen trim.

Some in-situ analyzers have been installed at the furnace outlet which will work well on boilers with low heat release rates. If the temperature of the flue gas at the sampling point is above 1500°F then the gas will be too hot to control its temperature and the analyzer will produce erroneous signals. I recommend installation of the analyzer so the probe is centered in the upper third of the smallest gas passage (in cross section) at the boiler outlet. If the boiler is equipped with an economizer or air heater it should be installed before that equipment.

Thermometers and temperature transmitters are occasionally installed in such a manner that they're useless. The temperature sensing portion of the instrument must be in the process fluid where it's flowing. One measurement that is always a problem is boiler stack temperature. I've encountered situations where the stack thermometers had stems so short that they didn't penetrate the stack. On others the thermometer bulb was located in a zone where the flue gas was idle, a stagnant zone where the gas was much cooler than the flowing flue gas. Here's one spot where modern technology has created some problems because the two common measures used for temperature detection, RTDs and thermocouples, are point instruments, they only sense temperature at one point.

We used to have these wonderful capillary type temperature transmitter elements that allowed us to stretch the probe back and forth across the stack or boiler outlet several times in a pattern that insured we had an average reading of the gas temperature. The problem is they were filled with mercury. I wouldn't recommend an RTD for stack temperature service because they can't take high temperatures that can occasionally occur in a stack.

When doing it right I specify a multipoint thermocouple with an element that spans the stack and has several terminations in it along with several reference junctions outside the stack so it will provide an average reading. I prefer a single point bi-metal thermometer for the local instrument because the large dial makes it easy to read from floor or access platform level. I just make certain the stem is long enough so it will always be in the center of the gas flow. I have encountered stack thermometers with a bulb so short that the sensitive tip wasn't in the stack. A boiler stack should have 3" to 4" of insulation and a nipple and coupling extending through to mount the transmitter. A standard 2-1/2" bulb is too short.

With the possible exception of stack and air duct temperature measurements all thermometers and temperature transmitter elements should be installed in thermowells. That way, if you do question an instrument's accuracy you can remove it and have its calibration checked, or check it yourself if you have the right equipment.

Stacks and air ducts may simply contain air at ambient temperatures or be under negative pressure so there is no hazard associated with removal of the thermal element and a thermowell isn't necessary. Sometimes, however, the well is essential to support the thermal element. Thermowells tend to slow the response of the instrument to changes in temperature because they have to heat up before the thermal element so there's no reason to install them where they aren't necessary. Some process applications don't use thermowells to achieve faster response time. Many thermowells are filled with a grease or other compound to improve heat transfer between the well and the element.

I prefer temperature transmitters to recorders or controllers that are directly connected to the sensing element. Both RTDs and thermocouples require more expensive wiring than the typical twisted shielded pair required for a transmitter. Exposing that wiring to electromagnetic fields in the plant can also produce erroneous outputs.

By installing local transmitters you eliminate an inventory of special wire and a lot of running back and forth when trying to check the calibration of the instrument. A local reading of what the transmitter is sensing can be provided by adding a relatively inexpensive meter on a transmitter. The only caveat with local trans-

mitters is they are not designed to be mounted on hot ductwork and piping. Unless I'm certain the fluid in the piping will not be too hot and the transmitter will not be heated by another source I insist on mounting the temperature transmitter away from the probe on another support attached to the building structure.

That requires the temperature element be fitted with extension leads long enough to reach the transmitter. I have long specified three feet as a requirement for extension leads (except stack temperature elements where I double that) so there's enough lead to conveniently locate the transmitter at a platform or grade where it's readily accessible, four foot above just like for pressure and flow transmitters.

There are other stories in this book that address problems with instrumentation. These comments will, hopefully, give you the ability to know when the information you are looking at is flawed and what you might do about it.

Chapter 12

Why They Fail

When a boiler or related equipment fails it's usually due to a lack of attention. While modern control systems normally manage to ensure a failure in a safe manner, i.e. a shutdown, the news media frequently has headlines involving catastrophic failures. Some of those catastrophes involve human suffering and death. Although not at the frequency and numbers of a century ago it still generates grave concerns when an incident does occur.

WHY THEY FAIL

A Little Bit of History

The last year of the twentieth century was a disappointing one for those of us who believed we were making a difference in the industry. Despite maintaining an average of less than ten people killed by boiler accidents 1999 produced 21 deaths. Six died in what has been described as the most expensive single accident ever; by itself bearing losses in excess of one billion dollars. Despite the horror of September 11, 2001, (which wasn't an accident) a boiler explosion holds the record for the most deaths from a single accident.

It happened in 1865, at the end of the Civil War, shortly after Lee surrendered to Grant. Over 1900 union soldiers clambered aboard the riverboat Sultana heading north to Cincinnati. Shortly after leaving the dock the boilers exploded. Some died immediately, others suffered from burns and shrapnel wounds until succumbing weeks later. About 1800 people, including women and children, died in that accident. With little left of the ship the actual cause remains undetermined.

In the early 1900's thousands died each year from boiler accidents. That's why the ASME proceeded to produce the boiler construction codes at the beginning of the twentieth century. The dramatic improvements that reduced injuries up until the end of that century should continue but 1999 started a new trend.

Recent history is depicted by the charts in Figure 12-1 and Figure 12-2 which show the swing in primary cause from low water to operator error and poor maintenance plus an increase in incidents. You'll notice the data are very old. That's because the lawyers for the

National Board of Boiler Inspectors advised the board in 2003 to not publish that information because they might be subjected to a lawsuit regarding it. I consider it shameful because the public has no other source for information regarding boiler accidents. I can only hope that the trend to fewer accidents is continuing and that this book is contributing to that trend.

Boilers seldom wear out. The effects of wear that you associate with machinery and automobiles are nowhere near as significant with boilers. Most of the time the boiler just sits there. There is rubbing associated with movement as it heats up and cools down but, in a normal plant, it doesn't happen often enough to be important. Don't confuse the boiler with the burner. They really are separate items. I will admit that burners wear because there are so many moving parts associated with most of them.

I've worked on many a boiler that was over fifty years old and had no evidence that it was nearing the end of its life. I recently provided engineering assistance to rebuild three boilers that were thirty years old and will undoubtedly last another thirty. Boilers usually fail by incident and the most common incidents have to do with lack of, or improper, water treatment.

Water Treatment

Improper water treatment or the lack of it contributes to most of the failures that I have encountered. The boiler fails because scale builds up until some metal overheats, the metal fails to allow the steam and water to escape where the water then flashes into steam so quickly that it violently blows the boiler apart.

There's a whole chapter in this book on water treatment and opportunities for you to learn more at the treatment supplier's school or other sources. If a boiler operator is comfortable with his water treatment, the likelihood that the boiler will fail is very low.

LOW WATER

For years we could count on the reports of boiler failures to list low water as the primary reason the boiler failed. Even today, with special systems and all our knowledge, low water always stands out as a significant

RECENT HISTORY OF BOILER ACCIDENTS ACCORDING TO NATIONAL BOARD DATA					
POWER BOILERS	1996	1999	2000	2001	2002
SAFETY VALVE	1	1	1	4	8
LOW WATER	356	67	183	161	137
LIMIT CONTROLS	16	27	22	8	4
IMPROPER INSTALLATION	5	14	15	2	5
IMPROPER REPAIR	6	24	16	1	14
DESIGN OR FABRICATION	8	22	8	2	6
OPERATOR ERROR OR POOR MAINT.	125	140	193	82	90
BURNER FAILURE	40	27	10	29	16
HEATING BOILERS - STEAM					
SAFETY VALVE	5	2	14	2	2
LOW WATER	490	397	437	519	359
LIMIT CONTROLS	27	33	66	17	17
IMPROPER INSTALLATION	14	10	22	10	5
IMPROPER REPAIR	7	36	23	11	2
DESIGN OR FABRICATION	14	33	34	31	54
OPERATOR ERROR OR POOR MAINT.	125	258	412	406	262
BURNER FAILURE	59	20	19	29	16
HOT WATER HEATING BOILERS					
SAFETY VALVE	5	5	7	6	7
LOW WATER	112	221	258	195	96
LIMIT CONTROLS	24	68	69	19	23
IMPROPER INSTALLATION	15	31	68	13	11
IMPROPER REPAIR	3	87	28	10	2
DESIGN OR FABRICATION	20	67	40	30	60
OPERATOR ERROR OR POOR MAINT.	221	314	406	260	215
BURNER FAILURE	70	31	30	26	28

Figure 12-1. Chart of reasons for boiler failures in prior years

BOILER ACCIDENTS, INJURIES AND DEATHS ACCORDING TO NATIONAL BOARD					
	1996	1999	2000	2001	2002
SAFETY VALVE	11	8	22	12	17
LOW WATER	958	685	878	875	592
LIMIT CONTROLS	67	128	157	44	44
IMPROPER INSTALLATION	34	55	105	25	21
IMPROPER REPAIR	16	147	67	22	18
DESIGN OR FABRICATION	42	122	82	63	120
OPERATOR ERROR OR POOR MAINT.	471	712	1011	748	567
BURNER FAILURE	169	78	59	84	60
TOTAL INCIDENTS	1768	1935	2381	1873	1439
INJURIES	56	63	24	66	16
DEATHS	4	15	8	8	3

Figure 12-2. Chart of accidents, injuries and deaths, late 1990s to 2002

cause for boiler failures. Taking all the precautions and conducting the regular testing should prevent them but they continue to occur.

It doesn't matter if it's a hot water boiler or a steam boiler, it should have a low water cutoff; steam boilers should have two. In the last century the most consistent reason for a boiler failure, accounting for about one third of the incidents, was loss of water. You should check the cutoffs as often as possible and under different situations to be certain they are reliable. Low water cutoffs come in two basic forms, float and conductance. Float operated cutoffs, as their name implies, use a float to detect the water level and a lever connected to the float keeps the float in position and actuates the electrical contacts that open to stop burner operation.

Conductance cutoffs use probes, looking something like a spark plug, to detect water level by the difference in conductivity of water and steam or air. Low water cutoffs should be installed to prevent burner operation in the event the boiler water drops below a safe level where the heating surfaces are exposed to steam. Normally the lowest safe operating level in a boiler is the bottom of the gauge glass so the cutoff should prevent burner operation near it. Cutoffs are installed in two forms, external and internal. There are arguments for each installation and you should encounter some boilers with both.

The failures of boilers due to low water continues despite the provisions of extra low water cutoffs and regular testing of them. Perhaps one principle reason is the failure to test them regularly so a problem is detected before a failure occurs. Whatever else you choose to let go, never fail to test the low water cutoffs immediately after arriving on the job. They can fail because mud builds up in the piping connecting the cutoff to the boiler, or an accumulation of mud in the cutoff housing. The mud is dirt that enters with the makeup and accumulates in the boiler water. It's usually suspended in the boiler water by the rapid circulation but will settle out in the water column and cutoff piping and chambers because the water moves slowly in them.

Float operated low water cutoff failures include the normal problems of mud collecting in the piping between boiler and the float housing where the float chamber can't drain so the level is higher than that in the boiler, (This happens if either the water leg or the steam leg is plugged, the chamber fills with condensate and can't drain) mud filling the bellows and hardening to resist transmission of the float position, friction preventing operation of magnet actuated switches, also the stiffening with age of wiring connected to magnet actuated switches, fusing of contacts due to excessive electrical current, freezing of the switch actuating mechanism due to corrosion from boiler water leaks or leakage into the switch housing.

Probe types, using conductance, can fail because deposits coat the probe to simulate the presence of water. The opportunities for a low water cutoff to fail are so many that regular testing (to detect problems) is the most important thing you can do.

Remember that, despite the many schemes for testing the low water cutoff, the only sure proof that the low water cutoff works is gradually dropping water level with the burner operating until the cutoff shuts the burner down. Do it as often as possible while keeping a close eye on the water level. Other tests to check it, explained in the normal operation description, should be performed with the recommended frequency. Always watch the level until cutoff occurs because the odds are rather high that it will not work.

Since incorporating timing of low water cutoff testing into my burner management systems there have been no failures of the boilers with those systems. There were, however, three incidents of the testing revealing a problem with a low water cutoff!

THERMAL SHOCK

Of all the modes of boiler failure thermal shock seems to be the one that can happen at any time. I've seen boilers that didn't make it past their initial week's operation without failing as a result of thermal shock and boilers that failed after years of operation due to an incident of thermal shock. I also saw one that was replaced and repaired by the manufacturer under warranty three times before the manufacturer found an installation mistake that allowed them to refuse additional repairs.

It's important to understand exactly how thermal shock destroys a boiler because there are several situations that are called thermal shock that aren't consistent with the normal perception. Thermal shock can destroy a boiler in a single incident or it can take several shocks to produce evident damage. There is a specific combination that must exist for thermal shock damage. First the metal of the boiler (or refractory) must be exposed to a change in temperature that's enough to produce a range of stress in the material.

The best example of thermal shock is pouring water over ice cubes fresh out of the freezer. What happens to the ice cubes? They crack! Even if you use cold water stored in the refrigerator they crack. When you consider the fact that steel is only about 7% stronger than ice (ever try to chop a fishing hole with a plain piece of steel?) you can understand that thermal shock can destroy a boiler. The reason for the ice cracking can be explained by noting how the cracks form. When the water hits the ice there's a rapid transfer of heat from the water to the surface of the ice. Keep in mind that ice contracts as it is heated, and the operation is just the opposite for steel. The inside of the cube remains cold because the heat doesn't transfer through the ice as fast as the outside is warmed by the water flowing over it. Because it's warmed and tends to shrink the outer layer of the ice cube is placed in tension, as if something was trying to pull it apart. The result is it is pulled apart, cracks form and as the rest of the cube shrinks the crack continues.

The second important element of thermal shock is thickness of the material. Shaved ice doesn't crack when cold water is poured over it. When the metal is thin enough the difference in temperature across it is not adequate to produce enough stress to produce cracking. The thicker parts of a boiler, tube sheets, shells, and drums are more susceptible to thermal shock than the tubes.

The third element is frequency. One violent shock may not be good for a boiler but hundreds of little ones repeatedly occurring will eventually result in failure because tiny microfissures (very little cracks) that form in thinner metals or where the temperature differences are not dramatic will, if constantly bombarded with thermal shock conditions, eventually grow into large cracks that finally result in boiler failure.

Many people don't realize that thermal shock doesn't have to happen on the water side of a boiler. I normally differentiate it by calling it firing shock but it's really thermal shock. Any boiler that trips while running at high fire and immediately goes into a purge is subjected to thermal shock because the metal of the boiler's heating surface is immediately subjected to contact with cold purge air right after it was exposed to the hottest flue gas of normal operation. Add to that the trip occurring near the maximum operating temperature (and related pressure) and there's potential for failure.

I've never seen a significant indication in water tube boilers but that doesn't mean they can't experience it. The most common failure in this mode occurs with the ends of the fire tubes at the inlet of the second pass of a fire tube boiler. The reason they fail is because they're sticking out into the hot flue gas where their temperature is elevated by the high fire glue gases and then they suddenly encounter the cold purge air. That failure is usually one that results in gradual growth of microfissures in the ends of the tubes and will even happen in tubes that are welded to the tube sheet. The primary reason for this type of failure is improper adjustment of the firing rate controls such that the boiler cycles off while the modulating controls are still at high fire or just left high fire.

Hydronic heating systems can operate to produce significant thermal shock by returning water from idle sections of the system (where the water got very cold) to the hot boiler. A slug of cold water is directed against the boiler heating surfaces. In some cases this can be caused by automatic controls operation, especially day/night controls. Sources of the problem are usually close to the boiler because any slug of cold water in a remote system will be heated by the metal in the piping as it returns to the boiler. Service water heating with a hydronic boiler has a high potential for thermal shock if the heating water to the service water heater is cycled on and off. It's better to use a constant flow to the service water heat exchanger with other provisions to prevent overheating the service water. See the section on service water heating for more on thermal shock in that application.

CORROSION AND WEAR

Nothing lasts forever and that's very true for boilers. You will be hard pressed to find a boiler in operation that is more than fifty years old. I know where there are a few but they're few and far between and, when they've had good care and water treatment, it's predominantly because of corrosion and wear. I could argue that a boiler doesn't have any moving parts so it can't wear out. I just finished a project where we replaced all the tubes and casings in thirty year old boilers and I have every reason to believe that they'll last another thirty years but they're an exception because they're well cared for. The normal end of a well maintained boiler's life is almost always due to a decision to replace them, not wear.

There are areas in a boiler that can't be reached to monitor and prevent corrosion, sometimes they're due to installation and sometimes to manufacturing but they're there. In many cases, as in the project I just finished, the only way to address those spots is a major rebuild of the boiler to reach them and clean, protect, and recover them to extend the boiler's life. That's a sound decision in many cases but many boiler owners just won't do it.

I've seen failures due to rubbing in a boiler where each time it heats up and cools off metal to metal rubbing resulted in cutting through a tube. In one case I discovered three boilers were lost in a matter of three years due to lack of adequate combustion air openings, completely destroyed by alternating corrosion and reduction. Those, however, are the unusual cases. Most of the time the problems with wear are all at the burner.

When the control valve on a boiler has run from high fire to low fire six or seven times a day, 365 days a year for 10 years it's run over 21,000 cycles. Did you know that the ASME Code has factors for operating cycles with no additional allowance for boilers that are expected to cycle less than 7,000 times in their lifetime? Now wonder how many engineers allow for more than that many cycles. Exactly how long do you expect that system to run without failing? A major revamping of a boiler's burner and controls on a five year cycle should prevent failures due to wear but they never seem to happen.

OPERATOR ERROR AND POOR MAINTENANCE

Regrettably the National Board statistics, which are quoted here, don't provide enough breakdown to clearly indicate why trends exist or to detect reasons for trends. I've seen a considerable increase in the elimination of central plants with licensed boiler operators. Their replacement multiple low pressure heating plants are maintained by individuals without a license so the increasing contribution of operator error to boiler failures isn't really surprising. Until such time that the National Board chooses to differentiate between licensed individuals and the janitor there's no way for them to determine if that's the case. In my judgment it's the perception that licensed operators cost too much and actions taken to replace them that has resulted in increased losses and loss of life. When the person maintaining a boiler has all the training and skill of a janitor that was handed a broom and told where the boiler room is it's no wonder this facet of failures is showing an increase.

Is it that increase the operators' fault? Hell no! When I encounter problems that are attributable to operator error or poor maintenance I always find an attitude on the part of the plant management that promotes or enforces the improper action or lack of action. I've recommended training for upper management in many plants since the 1970's and have yet to do any. All that plant manager wants to hear from me is how screwed up the operators are and when I tell that manger that the problem originates at a higher level than the operators they go look for another consultant that will tell them what they want. I hope a lot of plant managers read this book but my experience indicates they won't.

Frequently it's not the operator that contributes to poor maintenance. The operator manages to keep the plant running by a growing mountain of temporary fixes that accumulate until nothing can keep the boiler running. The reason is management's attitude about maintenance. In some cases operators simply have to allow the boiler to fail or shut it down due to unsafe operating conditions. One of the advantages of a license is that license gives you the authority to do just that, shut it down and refuse to operate it. Of course there's a potential for being fired but you may get a supporting position from another source and after a hearing you will be reinstated. When you don't have the confidence to shut the boiler down you do have the option of reporting the condition to the State Chief Boiler Inspector who will send a deputy inspector to look at the boiler. If the problem is one that threatens failure the deputy will 'red tag' the boiler and instruct you to shut it down. There's absolutely no way you can be dismissed under those circumstances.

And, just because an insurance company inspector passed your boiler don't believe you have no recourse. I know of several instances where a State Deputy Inspector red tagged a boiler that was reported safe by an insurance company inspector. That's especially true if nobody sees the inspector but a new certificate to operate suddenly appears. There are situations where an insurance inspector has inspected the boiler while sitting in front of the television at his house several miles away. They aren't supposed to do it, but it's done.

The National Board's data doesn't break down maintenance problems either. The most likely is loss due to lack of proper water treatment but we simply don't know. I think that's highly probable because a large number of boilers are installed and operated with no consideration of water treatment beyond an initial charge of chemicals, especially hot water boilers.

If it isn't broke don't fix it! How often we've heard those words in one form or another. I'm always told that it hasn't broke yet so it must be okay. If there's no log, no record of maintenance, and no repair history I'm there because the plant is frequently shutting down for unknown reasons and fuel bills seem to be much higher. Just because it's working doesn't mean it's working right. People that use that excuse are costing their employer a lot of money and exposing themselves to increased risk of injury or death.

It's true that a licensed boiler operator could make a mistake with disastrous consequences, a license is no guarantee and neither is training. However, I've had many opportunities to observe individuals without a license and have no doubt that the lack of the discipline involved in training and preparing for the exam leaves lots of room for error. If you don't have a license that doesn't mean you're more likely to make a mistake because I'm reasonably confident that the operator that chooses to read this book is far less likely to do something that will result in an accident with loss of life or serious injury than one who believes it's a waste of time.

Part of the business of acquiring a license includes the development of respect for the profession and greater understanding of the responsibility so you should attempt to get a license even if you don't need to have one. It's more a matter of attitude than the actual license. When a state licensing program exists the wise operator seeks to obtain the license to support a professional perception of his role.

Attitude and perception seem to be the key to operator error. When a boiler is damaged, and I've investigated several cases of damage that never reached the status of a National Board investigation and report; any failure in operation is usually attributable to an attitude. The most disconcerting one is "the boss doesn't care so why should I?" Since I have the opportunity to get to know operators in several boiler plants I eventually learn a lot about their perception of their job and their attitude. It's the ones that seem to believe that they can get away with doing the minimum and the company should be happy that they even show up that eventually make the mistakes that result in damage. Usually that same attitude also protects them from exposure to the failure and eventual injury as well, an undeserved result. I know many operators who I'm certain will eventually do something, or not do something, that will result in failure and possible injury or death.

If you don't have some fear, fear that a boiler failure could occur if you did the wrong thing, then you are potentially one of those people that will make a mistake. You shouldn't be afraid of the plant but you do have to respect the potential for a boiler or furnace explosion and act accordingly. It's the people without fear, with an attitude that they're infallible, that take unnecessary risks with everything from shortening purge periods to skipping boiler water analysis which eventually result in a failure.

Over the years I've screwed up. In some cases it was a royal screw up. You'll never know how many of those operators described in this book were really the author. I give you all I can to prevent your making those mistakes and I hope you've learned something and even enjoyed that learning experience a little. I also hope you learned those priorities and acquired a respect for the equipment you're operating. God bless you all, the devil doesn't need any more help with his furnaces.

Appendix A

Properties of Water and Steam

Vacuum in. Hg	TEMP. °F	CU. FT. PER LB.		HEAT IN BTU PER POUND		
		LIQUID	STEAM	LIQUID	LATENT	STEAM
29.75	40	.01602	2423.7	8	1071	1079
29	79	.01608	652.3	47	1049	1096
25	134	.01626	143.2	102	1018	1119
20	161	.01640	74.8	129	1001	1131
15	179	.01650	51.1	147	991	1138
10	192	.01658	39.4	160	983	1143
5	203	.01666	31.8	171	976	1147
PRESS. PSIG	TEMP. °F	CU. FT. PER LB.		HEAT IN BTU PER POUND		
		LIQUID	STEAM	LIQUID	LATENT	STEAM
0	212	.01672	26.8	180	970	1150
2	218	.01675	24.1	186	966	1152
5	227	.01682	20.1	196	961	1156
8	234	.01688	17.9	202	955	1158
9	237	.01690	17.2	205	954	1159
10	240	.01692	16.6	207	953	1160
11	241	.01693	16.0	209	951	1161
12	243	.01695	15.4	212	950	1162
15	250	.01700	13.9	218	945	1163
15	250	.01700	13.9	218	945	1163
20	258	.01707	12.1	227	940	1167
25	266	.01715	10.6	235	934	1169
30	274	.01721	9.5	243	929	1172
45	281	.01727	8.59	250	924	1174
50	298	.01743	6.68	257	911	1179
60	307	.01753	5.89	277	904	1182
70	316	.01761	5.18	286	898	1184

PROPERTIES OF WATER AND STEAM (*Continued*)

PRESS. PSIG	TEMP. °F	CU. FT. PER LB.		HEAT IN BTU PER POUND		
		LIQUID	STEAM	LIQUID	LATENT	STEAM
80	324	.01770	4.65	294	892	1186
90	331	.01778	4.25	302	886	1188
100	338	.01785	3.90	309	881	1189
110	344	.01792	3.59	315	875	1191
120	350	.01800	3.34	322	871	1192
125	353	.01802	3.22	325	868	1193
150	366	.01819	2.74	339	857	1195
175	377	.01833	2.40	350	847	1197
200	388	.01847	2.13	362	837	1199
225	397	.01860	1.92	372	828	1200
250	406	.01873	1.74	382	820	1201
275	414	.01885	1.59	390	812	1202
300	422	.01897	1.47	399	804	1203
400	448	.01940	1.11	428	756	1204
600	489	.02020	0.73	474	728	1203
750	513	.02070	0.61	500	700	1200
900	534	.02130	0.49	529	665	1195
1200	574	.02233	0.39	587	624	1183

Notes:
- Data is for gage pressure at one standard atmosphere (14.696 psia)
- Entropy and internal energy shown on standard steam tables is not included for clarity.
- Only pressures commonly used are shown, use of tables from another source is recommended if precision is desired.

PROPERTIES OF SUPERHEATED STEAM

28" Hg	Temperature	150	200	250	350	600
(101)	Volume	2.01	2.05	2.08	2.10	2.27
	Heat	1128	1150	1173	1219	1336
26" Hg	Temperature	150	200	250	350	600
(126)	Volume	1.94	1.97	2.01	2.06	2.19
	Heat	1125	1150	1172	1227	1336
11" Hg	Temperature	250	300	350	450	700
(200)	Volume	1.80	1.84	1.87	1.92	1.99
	Heat	1169	1193	1216	1263	1335
0 psig	Temperature	250	300	350	450	700
(212)	Volume	1.78	1.81	1.84	1.90	2.02
	Heat	1169	1193	1216	1263	1383
5 psig	Temperature	250	300	350	450	700
(227)	Volume	1.75	1.78	1.82	1.87	1.98
	Heat	1167	1192	1134	1263	1383
10 psig	Temperature	300	350	400	500	750
(240)	Volume	1.75	1.78	1.81	1.87	1.98
	Heat	1186	1214	1238	1286	1407
15 psig	Temperature	300	350	400	500	750
(250)	Volume	1.73	1.76	1.79	1.85	1.96
	Heat	1209	1213	1238	1286	1400
60 psig	Temperature	350	400	450	550	800
(307)	Volume	1.67	1.70	1.73	1.78	1.89
	Heat	1207	1233	1268	1307	1430
120 psig	Temperature	400	450	500	600	850
(350)	Volume	1.61	1.65	1.67	1.72	1.83
	Heat	1222	1259	1276	1327	1453

PROPERTIES OF SUPERHEATED STEAM (*Continued*)

150 psig	Temperature	400	450	550	600	850
(366)	Volume	1.58	1.62	1.67	1.70	1.81
	Heat	1217	1245	1299	1324	1452
200 psig	Temperature	450	500	550	650	900
(388)	Volume	1.58	1.61	1.64	1.69	1.80
	Heat	1241	1267	1304	1346	1476
250 psig	Temperature	450	500	550	650	900
(406)	Volume	1.55	1.59	1.62	1.66	1.77
	Heat	1231	1262	1301	1344	1474
300 psig	Temperature	450	500	550	650	900
(422)	Volume	1.53	1.56	1.59	1.64	1.75
	Heat	1236	1256	1286	1340	1472
400 psig	Temperature	500	550	600	700	950
(448)	Volume	1.52	1.56	1.58	1.64	1.74
	Heat	1243	1271	1306	1362	1495
600 psig	Temperature	550	600	650	750	1000
(489)	Volume	1.49	1.53	1.56	1.61	1.71
	Heat	1254	1289	1320	1380	1516
750 psig	Temperature	550	600	650	750	1000
(513)	Volume	1.45	1.49	1.52	1.57	1.68
	Heat	1238	1274	1312	1370	1512
900 psig	Temperature	600	650	700	800	1050
(534)	Volume	1.47	1.49	1.53	1.58	1.68
	Heat	1258	1297	1331	1393	1536
1200 psig	Temperature	600	650	700	800	1050
(574)	Volume	1.40	1.44	1.48	1.54	1.64
	Heat	1220	1268	1310	1378	1537

Note: Value in parenthesis is temperature of steam at saturation for that pressure.

Appendix B

Water Pressure per Foot Head

HEAD (Feet)	Psi produced by water at				HEAD (Feet)	Psi produced by water at			
	60°F	140°F	212°F	240°F		60°F	140°F	212°F	240°F
1	0.433	0.456	0.415	0.410	36	15.59	15.35	14.95	14.78
2	0.866	0.853	0.831	0.821	38	16.45	16.20	15.78	15.60
3	1.299	1.279	1.246	1.231	40	17.32	17.05	16.61	16.42
4	1.732	1.705	1.661	1.642	45	19.48	19.18	18.69	18.47
5	2.165	2.132	2.077	2.052	50	21.65	21.32	20.77	20.52
6	2.598	2.558	2.492	2.463	55	23.81	23.45	22.84	22.57
7	3.031	2.984	2.907	2.873	60	25.98	25.58	24.92	24.63
8	3.464	3.410	3.323	3.283	65	28.14	27.71	27.00	26.68
9	3.987	3.837	3.738	3.694	70	30.31	29.84	29.07	28.73
10	4.329	4.263	4.153	4.104	75	32.47	31.97	31.15	30.78
11	4.762	4.689	4.569	4.515	80	34.64	34.10	33.23	32.83
12	5.195	5.116	4.984	4.925	85	36.80	36.24	35.30	34.89
13	5.628	5.542	5.399	5.336	90	38.97	38.37	37.38	36.94
14	6.061	5.968	5.815	5.746	95	41.13	40.50	39.46	38.99
15	6.494	6.395	6.230	6.156	100	43.29	42.63	41.53	41.04
16	6.927	6.821	6.645	6.567	110	47.62	46.89	45.69	45.15
17	7.360	7.247	7.061	6.977	120	51.95	51.16	49.84	49.25
18	7.793	7.673	7.476	7.388	130	56.28	55.42	53.99	53.36
19	8.226	8.100	7.891	7.798	140	60.61	59.68	58.15	57.46
20	8.659	8.526	8.307	8.209	150	64.94	63.95	62.30	61.56
22	9.525	9.379	9.137	9.029	200	86.59	85.26	83.07	82.09
24	10.391	10.231	9.968	9.850	250	108.24	106.58	103.83	102.61
26	11.257	11.084	10.799	10.671	300	129.88	127.89	124.60	123.13
28	12.122	11.936	11.629	11.492	350	151.13	149.21	145.37	143.65
30	12.988	12.789	12.460	12.313	400	173.18	170.52	166.14	164.17
32	13.854	13.642	13.291	13.134	450	194.83	191.84	186.90	184.69
34	14.720	14.494	14.121	13.955	500	216.47	213.15	207.67	205.21

Appendix C

Nominal Capacities of Pipe

Size	Sch.	Water gpm	Air - scfm @ 30 psig	Air - scfm @ 100 psig	Steam pph @ 12 psig	Steam pph @ 30 psig	Steam pph @150 psig	Steam pph @ 250 psig
3/4	S	6.70	0.15	0.38	35	45	116	198
	XS	5.43	0.09	0.23	28	36	94	112
1	S	12.70	0.50	1.28	66	89	233	730
	XS	10.57	0.32	0.81	55	74	194	445
1-1/4	S	26.00	1.98	5.06	138	199	523	3,188
	XS	22.30	1.35	3.44	118	171	449	2,111
1-1/2	S	39.50	4.28	10.9	210	309	813	2,921
	XS	34.28	3.00	7.67	182	268	706	2,536
2	S	75.00	14.9	38.1	410	627	1,650	4,815
	XS	65.99	10.8	27.7	361	552	1,452	4,237
2-1/2	S	120.0	36.3	92.7	660	1,033	2,430	6,870
	XS	106.2	26.7	68.3	584	914	2,151	6,082
3	S	210.0	69.3	177.1	1,160	1,880	4,210	10,608
	XS	187.6	61.9	158.2	1,036	1,679	3,760	9,478
	LW	416.3	125.2	320.0	2,518	4,197	8,814	19,168
4	S	396.8	119.3	305.0	2,400	4,000	8,400	18,268
	XS	358.3	107.8	275.4	2,168	3,614	7,588	16,498
5	S	623.6	187.6	479.3	4,250	7,390	15,000	28,708
	XS	567.1	170.6	435.9	3,863	6,718	13,636	26,107
	LW	937.1	281.9	720.3	7,284	12,633	26,224	43,141
6	S	900.5	270.8	692.2	7,000	12,140	25,200	41,457
	XS	812.5	244.4	624.5	6,322	10,964	22,758	37,405
	LW	1,641	493.5	1,261	15,048	26,570	52,614	75,541
8	S	1,559	469.0	1,199	14,300	25,250	50,000	71,787
	XS	1,423	428.1	1,094	13,070	23,079	45,700	65,526
	LW	2,602	782.7	2,000	27,527	49,654	95,285	119,798
10	S	2,458	739.3	1,889	26,000	46,900	90,000	113,153
	XS	2,327	700.0	1,789	23,660	42,680	81,901	107,137
	LW	3,674	1,105	2,824	41,684	77,115	161,526	169,124
12	S	3,525	1,060	2,710	40,000	74,000	155,000	162,291
	XS	3,380	1,017	2,598	38,338	70,925	148,559	155,599
	LW	4,461	1,342	3,429	50,608	93,625	196,106	205,400
14	S	4,298	1,293	3,304	48,751	90,189	188,910	197,863
	XS	4,137	1,244	3,180	46,929	86,818	181,848	190,466
	LW	5,881	1,769	4,521	66,714	123,420	258,515	270,767
16	S	5,693	1,712	4,376	64,579	119,471	250,243	262,103
	XS	5,508	1,657	4,234	62,479	115,586	242,106	253,580
	LW	7,497	2,255	5,763	85,041	157,325	329,533	345,150
18	S	7,284	2,191	5,599	82,628	152,862	320,185	335,359
	XS	7,075	2,128	5,438	80,251	148,464	310,971	325,709
	LW	9,308	2,800	7,155	105,589	195,340	409,159	
20	S	9,071	2,728	6,973	102,899	190,364	398,735	417,632

Size	Sch.	Water gpm	Air - scfm		Steam pph			
			@ 30 psig	@ 100 psig	@ 12 psig	@ 30 psig	@150 psig	@ 250 psig
	XS	8,837	2,658	6,793	100,244	185,451	388,445	406,854
	LW	11,316	3,404	8,698	128,359	237,465	497,393	
22	S	11,054	3,325	8,497	125,392	231,975	485,893	508,920
	XS	10,796	3,247	8,298	122,459	226,549	474,527	497,016
	LW	13,519	4,066	10,392	153,351	283,699	594,235	
24	S	13,233	3,980	10,172	150,106	277,695	581,659	609,225
26	S	15,607	4,694	11,997	177,041	327,526	686,034	718,546
28	S	18,178	5,468	13,973	206,198	381,466	799,016	836,883
30	S	20,944	6,300	16,099	237,576	439,516	920,607	964,237
32	S	23,906	7,191	18,376	271,176	501,675	1,050,806	
34	S	27,064	8,140	20,803	306,997	567,945	1,189,614	
36	S	30,418	9,149	23,381	345,040	638,324	1,337,029	
42	S	41,654	12,529	32,018	472,497	874,119	1,830,925	
48	S	54,653	16,439	42,010	619,947	1,146,902	2,402,295	

As stated in the title, these are nominal capacities. The pipe can always handle less than the indicated flow and will handle much more than the indicated flow with increasing pressure drop. These capacities are approximately what a piping designer would allow through the pipe.

Appendix D

Properties of Pipe

Schedule / Weight	Thickness Inches	I.D. Inches	Surface Inside sq. ft.	Cross-sectional		Weight of		
				Metal Area	Flow Area	Pipe pounds	Water pounds	Pipe & Water
0.405" O.D.				**1/8" NPS**		Outside surface area 0.106 sq. ft.		
40 / S	0.068	0.269	0.070	0.072	0.057	0.245	0.025	0.270
80 / XS	0.095	0.215	0.056	0.092	0.036	0.314	0.016	0.330
0.540" O.D.				**1/4" NPS**		Outside surface area 0.141 sq. ft.		
40 / S	0.088	0.364	0.095	0.125	0.104	0.425	0.045	0.470
80 / XS	0.126	0.302	0.079	0.157	0.072	0.535	0.031	0.566
0.675" O.D.				**3/8" NPS**		Outside surface area 0.177 sq. ft.		
40 / S	0.091	0.493	0.129	0.167	0.191	0.568	0.083	0.651
80 / XS	0.126	0.423	0.111	0.217	0.140	0.739	0.061	0.800
0.840" O.D.				**½" NPS**		Outside surface area 0.220 sq. ft.		
40 / S	0.109	0.622	0.163	0.250	0.304	0.851	0.132	0.983
80 / XS	0.147	0.546	0.143	0.320	0.234	1.088	0.101	1.189
160	0.187	0.466	0.122	0.384	0.171	1.304	0.074	1.378
XX	0.294	0.252	0.066	0.504	0.050	1.715	0.022	1.737
1.050" O.D.				**3/4" NPS**		Outside surface area 0.275 sq. ft.		
40 / S	0.113	0.824	0.216	0.333	0.533	1.131	0.231	1.362
80 / XS	0.154	0.742	0.194	0.434	0.432	1.474	0.187	1.661
160	0.218	0.614	0.161	0.570	0.296	1.937	0.128	2.065
1.315" O.D.				**1" NPS**		Outside surface area 0.344 sq. ft.		
40 / S	0.133	1.049	0.275	0.494	0.864	1.679	0.374	2.053
80 / XS	0.179	0.957	0.250	0.639	0.719	2.172	0.311	2.483
160	0.250	0.815	0.213	0.836	0.522	2.844	0.226	3.070
XX	0.358	0.599	0.157	1.076	0.282	3.659	0.122	3.781

PROPERTIES OF PIPE (*Continued*)

Schedule / Weight	Thickness Inches	I.D. Inches	Surface Inside sq. ft.	Cross-sectional Metal Area	Cross-sectional Flow Area	Weight of Pipe pounds	Weight of Water pounds	Weight of Pipe & Water
1.660" O.D.			**1 1/4" NPS**			Outside surface area 0.434 sq. ft.		
40 / S	0.140	1.380	0.361	0.668	1.496	2.273	0.648	2.921
80 / XS	0.191	1.278	0.334	0.881	1.283	2.997	0.555	3.552
160	0.250	1.160	0.304	1.107	1.057	3.765	0.458	4.223
XX	0.382	0.896	0.234	1.534	0.630	5.215	0.273	5.488
1.900" O.D.			**1 ½" NPS**			Outside surface area 0.497 sq. ft.		
40 / S	0.145	1.610	0.421	0.799	2.036	2.718	0.882	3.600
80 / XS	0.200	1.500	0.939	1.068	1.767	3.632	0.765	4.397
160	0.281	1.337	0.350	1.431	1.404	4.866	0.608	5.474
XX	0.400	1.100	0.288	1.885	0.950	6.409	0.411	6.820
2.375" O.D.			**2" NPS**			Outside surface area 0.622 sq. ft.		
40 / S	0.154	2.067	0.541	1.074	3.356	3.653	1.453	5.106
80 / XS	0.218	1.939	0.508	1.477	2.953	5.022	1.278	6.300
160	0.343	1.689	0.442	2.190	2.240	7.445	0.970	8.425
2.875" O.D.			**2 ½" NPS**			Outside surface area 0.753 sq. ft.		
10S	0.120	2.635	0.690	1.039	0.545	3.531	2.361	5.892
40 / S	0.203	2.469	0.646	1.704	4.790	5.794	2.073	7.867
80 / XS	0.276	2.323	0.608	2.254	4.240	7.662	1.835	9.497
160	0.375	2.125	0.556	2.945	3.550	10.01	1.540	11.550
XX	0.552	1.771	0.464	4.028	2.460	13.70	1.070	14.770
XX	0.436	1.503	0.393	2.656	1.774	9.030	0.768	9.798

PROPERTIES OF PIPE (*Continued*)

Schedule / Weight	Thickness Inches	I.D. Inches	Surface Inside sq. ft.	Cross-sectional Metal Area	Flow Area	Weight of Pipe pounds	Water pounds	Pipe & Water
3.500" O.D.			**3" NPS**			Outside surface area 0.916 sq. ft.		
10S	0.120	3.260	0.853	1.274	8.35	4.33	3.61	7.94
40 / S	0.216	3.068	0.803	2.228	7.39	7.58	3.20	10.78
80 / XS	0.300	2.900	0.759	3.016	6.60	10.25	2.86	13.11
160	0.437	2.626	0.687	4.205	5.42	14.30	2.35	16.65
XX	0.600	2.300	0.602	5.466	4.15	18.58	1.80	20.38
4.500" O.D.			**4" NPS**			Outside surface area 1.178 sq. ft.		
10S	0.120	4.260	1.115	1.651	14.25	5.61	6.17	11.78
LW	0.188	4.124	1.080	2.550	13.36	8.66	5.78	14.44
40 / S	0.237	4.026	1.054	3.170	12.73	10.79	5.51	16.30
80 / XS	0.337	3.826	1.002	4.410	11.50	14.99	4.98	19.97
120	0.437	3.626	0.949	5.580	10.33	18.96	4.47	23.43
160	0.531	3.438	0.900	6.620	9.28	22.51	4.02	26.53
XX	0.674	3.152	0.825	8.100	7.80	27.54	3.38	30.92
5.563" O.D.			**5" NPS**			Outside surface area 1.456 sq. ft.		
40 / S	0.258	5.047	1.321	4.30	20.01	14.62	8.66	23.28
6.625" O.D.			**6" NPS**			Outside surface area 1.734 sq. ft.		
5S	0.109	6.407	1.667	2.23	32.2	7.58	13.96	21.54
10S	0.134	6.357	1.664	2.73	31.7	9.29	13.74	23.03
LW	0.219	6.187	1.620	4.41	30.1	14.99	13.02	28.01
40 / S	0.280	6.065	1.588	5.58	28.9	18.98	12.51	31.49
80 / XS	0.432	5.761	1.508	8.40	26.1	28.58	11.29	39.87
120	0.562	5.501	1.440	10.7	23.8	36.40	10.29	46.69
160	0.718	5.189	1.358	13.32	21.1	45.30	9.16	54.46
XX	0.864	4.897	1.282	15.64	18.8	53.17	8.16	61.33
80 / XS	0.375	4.813	1.260	6.11	18.19	20.78	7.88	28.66

PROPERTIES OF PIPE (*Continued*)

Schedule / Weight	Thickness Inches	I.D. Inches	Surface Inside sq. ft.	Cross-sectional Metal Area	Cross-sectional Flow Area	Weight of Pipe pounds	Weight of Water pounds	Weight of Pipe & Water
8.625" O.D.				**8" NPS**		Outside surface area 2.258 sq. ft.		
5S	0.109	8.407	2.201	2.92	55.5	9.91	24.04	33.95
10S	0.148	8.329	2.180	3.94	54.5	13.4	23.59	36.99
LW	0.219	8.187	2.143	5.78	52.6	19.66	22.94	42.60
20	0.250	8.125	2.127	6.58	51.8	22.37	22.45	44.82
30	0.277	8.071	2.113	7.26	51.2	24.7	22.15	46.32
40 / S	0.322	7.981	2.089	8.40	50.0	28.56	21.68	50.24
60	0.406	7.813	2.045	10.48	47.9	35.6	20.8	56.4
80 / XS	0.500	7.625	1.996	12.76	45.7	43.4	19.8	63.2
120	0.718	7.189	1.882	17.84	40.6	60.6	17.6	78.2
XX	0.875	6.875	1.800	21.30	37.1	72.4	16.1	88.5
10.750" O.D.				**10" NPS**		Outside surface area 2.81 sq. ft.		
5S	0.134	10.482	2.74	4.47	86.3	15.2	37.4	52.6
10S	0.165	10.420	2.73	5.49	85.3	18.7	36.9	55.6
20	0.250	10.250	2.68	8.25	82.5	28.0	35.7	63.7
30	0.307	10.136	2.65	10.07	80.7	34.2	34.9	69.1
40 / S	0.365	10.020	2.62	11.91	78.9	40.5	34.1	74.6
60 / XS	0.500	9.750	2.55	16.10	74.7	54.7	32.3	87.0
80	0.593	9.564	2.50	18.92	71.8	64.3	31.1	95.4
160	1.125	8.500	2.23	34.02	56.7	115.7	24.6	140.3
12.750" O.D.				**12" NPS**		Outside surface area 3.34 sq. ft.		
5S	0.156	12.438	3.26	6.17	121.5	21.0	52.6	73.6
10S	0.180	12.390	3.24	7.11	120.6	24.2	52.2	76.4
20 / LW	0.250	12.250	3.21	9.82	117.9	33.4	51.0	84.4
40S / S	0.375	12.000	3.14	15.58	113.1	49.6	49.0	98.6
40	0.406	11.938	3.13	15.74	111.9	53.5	48.5	102.0
30	0.330	12.090	3.17	12.88	114.8	43.8	49.7	93.5

PROPERTIES OF PIPE (*Continued*)

Schedule / Weight	Thickness Inches	I.D. Inches	Surface Inside sq. ft.	Cross-sectional		Weight of		
				Metal Area	Flow Area	Pipe pounds	Water pounds	Pipe & Water
12" NPS (continued)								
80S/XS	0.500	11.750	3.08	19.24	108.4	65.4	47.0	112.4
80	0.687	11.376	2.98	26.04	101.6	88.5	44.0	132.5
160	1.312	10.126	2.65	47.14	80.5	160.3	34.9	195.2
14" O.D.				**14" NPS**		Outside surface area 3.67 sq. ft.		
5S	0.156	13.688	3.58	6.78	147.2	23.1	63.7	86.8
10S	0.188	13.624	3.57	8.16	145.8	27.7	63.1	90.8
10 / LW	0.250	13.500	3.53	10.80	143.1	36.7	62.0	98.7
20	0.312	13.375	3.50	13.44	140.5	45.7	60.8	106.5
30 / S	0.375	13.250	3.47	16.05	137.9	54.6	59.7	114.3
40	0.438	13.125	3.44	18.66	135.3	63.4	58.6	122.0
XS	0.500	13.000	3.40	21.21	132.7	72.1	57.5	129.6
80	0.750	12.500	3.27	31.22	122.7	106.1	53.1	159.2
120	1.093	11.814	3.09	44.32	109.6	150.7	47.5	198.2
160	1.406	11.188	2.93	55.63	98.3	189.1	42.6	231.7
16" O.D.				**16" NPS**		Outside surface area 4.19 sq. ft.		
5S	0.169	15.670	4.1	8.21	192.9	27.9	83.5	111.4
10S	0.188	15.624	4.09	9.34	191.7	31.8	83.0	114.8
10 / LW	0.250	15.500	4.06	12.37	188.7	42.1	81.7	123.8
20	0.312	15.375	4.02	15.4	185.7	52.4	80.4	132.8
30 / S	0.375	15.250	3.99	18.41	182.7	62.6	79.1	141.7
40 / XS	0.500	15.000	3.93	24.35	176.7	82.8	76.5	159.3
60	0.656	14.688	3.85	31.62	169.4	107.5	73.4	180.9
80	0.843	14.314	3.75	40.14	160.9	136.5	69.7	206.2
120	1.218	13.564	3.55	56.56	144.5	192.3	62.6	254.9

PROPERTIES OF PIPE (*Continued*)

Schedule / Weight	Thickness Inches	I.D. Inches	Surface Inside sq. ft.	Cross-sectional Metal Area	Flow Area	Weight of Pipe pounds	Water pounds	Pipe & Water
16" NPS (continued)								
160	1.593	12.814	3.35	72.1	129.0	245.1	55.8	300.9
18" O.D.				**18" NPS**		Outside surface area 4.71 sq. ft.		
5S	0.165	17.670	4.63	9.24	245.2	31.4	106.2	137.6
10S	0.188	17.624	4.61	10.52	243.9	35.8	105.6	141.4
20	0.312	17.375	4.55	17.36	237.1	59.0	102.7	161.7
S	0.250	17.250	4.52	20.76	233.7	70.6	101.2	171.8
30	0.438	17.124	4.48	24.17	230.3	82.2	99.7	181.9
XS	0.500	17.000	4.45	27.49	227.0	93.5	98.3	191.8
40	0.562	16.876	4.42	30.79	223.7	104.7	96.9	201.6
60	0.750	16.500	4.32	40.64	213.8	138.2	92.6	230.8
80	0.937	16.126	4.22	50.23	204.2	170.8	88.4	259.2
120	1.375	15.250	3.99	71.81	182.7	244.2	79.1	232.3
160	1.781	14.438	3.78	90.75	163.7	308.5	70.9	379.4
20" O.D.				**20" NPS**		Outside surface area 5.24 sq. ft.		
5S	0.188	19.624	5.14	11.7	302.5	39.8	131.0	170.8
10S	0.218	19.564	5.12	13.55	300.6	46.1	130.2	176.3
10 / LW	0.250	19.500	5.11	15.51	298.6	52.7	129.3	182.0
20 / S	0.375	19.250	5.04	23.12	291.0	78.6	126.0	204.6
30 / XS	0.500	19.000	4.97	30.6	283.5	104.1	122.8	226.9
40	0.593	18.814	4.93	36.2	278.0	122.9	120.4	243.3
60	0.812	18.376	4.81	48.9	265.2	166.4	114.8	281.2
80	1.031	17.938	4.70	61.4	252.7	208.9	109.4	318.3
120	1.500	17.00	4.45	87.2	227.0	296.4	98.3	394.7
160	1.968	16.064	4.21	111.5	202.7	379.1	87.8	466.9

PROPERTIES OF PIPE (*Continued*)

22" O.D.				**22" NPS**		Outside surface area 5.76 sq. ft.		
10 / LW	0.250	21.500	5.63	17.1	363	58.1	157.2	215.3
S	0.375	21.250	5.56	25.5	355	86.6	153.6	240.2
XS	0.500	21.000	5.50	33.8	346	114.8	150.0	264.8
24" O.D.				**24" NPS**		Outside surface area 6.28 sq. ft.		
5S	0.218	23.564	6.17	16.3	436	55.4	188.8	244.2
10 / LW	0.250	23.500	6.15	18.7	434	63.4	187.8	251.2
20 / S	0.375	23.250	6.09	27.8	425	94.6	183.8	278.4
XS	0.500	23.000	6.02	36.9	415	125.5	179.9	305.4
40	0.687	22.626	5.92	50.3	402	171.1	174.1	345.2
60	0.986	22.064	5.78	70.0	382	238.1	165.6	403.7
80	1.218	21.564	5.65	87.2	365	296.4	158.1	454.5
120	1.812	20.376	5.33	126.3	326	429.4	141.2	570.6
160	2.343	19.314	5.06	159.4	293	542	126.9	668.9
26" O.D.				**26" NPS**		Outside surface area 6.81 sq. ft.		
S	0.375	25.250	6.61	30.2	501	102.6	216.8	319.4
XS	0.500	25.000	6.54	40.1	491	136.2	212.5	348.7
28" O.D.				**28" NPS**		Outside surface area 7.33 sq. ft.		
S	0.375	27.250	7.13	32.5	583	110.7	252.5	363.2
XS	0.500	27.000	7.07	43.2	573	146.9	247.9	394.8
30" O.D.				**30" NPS**		Outside surface area 7.85 sq. ft.		
5S	0.250	29.500	7.72	23.4	683	79.4	296.0	375.4
10 / LW	0.312	29.376	7.69	29.1	678	98.9	293.5	392.4
S	0.375	29.250	7.66	34.9	672	118.7	291.0	409.7
20 / XS	0.500	29.000	7.59	46.3	661	157.6	286.0	443.6
30	0.625	28.750	7.53	57.7	649	196.1	281.1	477.2
32" O.D.				**32" NPS**		Outside surface area 8.38 sq. ft.		
S	0.375	31.250	8.18	37.3	767	126.7	332.1	458.8
XS	0.500	31.000	8.12	49.5	755	168.2	326.8	495.0

PROPERTIES OF PIPE (*Continued*)

Schedule / Weight	Thickness Inches	I.D. Inches	Surface Inside sq. ft.	Cross-sectional Metal Area	Flow Area	Weight of Pipe pounds	Water pounds	Pipe & Water
34" O.D.				**34" NPS**		Outside surface area 8.9 sq. ft.		
S	0.375	33.250	8.70	39.6	868	134.7	376.0	510.7
XS	0.500	33.000	8.64	52.6	855	178.9	370.3	549.2
36" O.D.				**36" NPS**		Outside surface area 9.42 sq. ft.		
S	0.375	35.250	9.23	42.0	976	142.7	422.6	565.3
XS	0.500	35.000	9.16	55.8	962	189.6	416.6	606.2
42" O.D.				**42" NPS**		Outside surface area 11.0 sq. ft.		
S	0.375	41.250	10.80	49.0	1336	166.7	578.7	745.4
XS	0.500	41.000	10.73	65.2	1320	221.6	571.7	793.3
48" O.D.				**48" NPS**		Outside surface area 12.56 sq. ft.		
S	0.375	47.250	12.37	56.1	1753	190.8	759.2	950.0
XS	0.500	47.000	12.30	74.6	1735	253.7	751.2	1004.9
LW	0.219	10.310	2.7	7.28	83.5	24.7	36.1	60.8

Notes applicable to properties of pipe:

- Metal area and flow area are in square inches.
- Surface area and weights of pipe and water are per foot of length
- There are other standard sizes of pipe based on schedule numbers that are not shown and other wall thicknesses. Check with your local pipe supplier for more information if it is needed.
- LW, S, XS, XX refer to Light Wall, Standard, Extra Strong, and Double Extra Strong commercial sizes.
- 5S and 10S sizes apply specifically to corrosion resistant materials.

Appendix E

Secondary Ratings

SECONDARY RATINGS OF JOINTS, FLANGES, VALVES, AND FITTINGS

SOLDER JOINTS - MAXIMUM WORKING PRESSURE				
Solder used in joints	Maximum working temperature	1/4" thru 1"	1-1/4" thru 2"	2-1/2" thru 4"
50-50 Tin - lead	100	200	175	150
	150	150	125	100
	200	100	90	75
	250	85	75	50
95 - 5 Tin - Antimony	100	500	400	300
	150	400	350	275
	200	300	250	200
	250	200	175	150

These ratings apply to any valve or fitting joined by soldering. They are also limited to water and other non-flammable liquids and gases. For steam, the pressure cannot exceed 15 psig saturated (maximum temperature of 250°F) These values limit all other ratings.

BRONZE VALVES AND FITTINGS						
Pressure Class & Conn	125	150		200	300	
	Screwed	Screwed	Flanged	Screwed	Screwed	Flanged
-20 to 150°F	200	300	225	400	1000	500
200°F	185	270	210	375	920	475
250°F	170	240	195	350	830	450
300°F	155	210	180	325	740	425
350°F	140	180	165	300	650	400
400°F				275	560	375
406°F	125	150	150			
450°F				250	480	350
500°F				225	390	325
550°F				200	300	300

Note: The 406°F value, saturation temperature for 250 psig steam, is normally the limit for all bronze valves according to the boiler construction code.

SECONDARY RATINGS OF JOINTS, FLANGES, VALVES, AND FITTINGS (*Continued*)

	125 LB CLASS			150 LB	250 LB CLASS		300 LB
Temp.	2" thru 12"	14" to 24"	30" to 36"	DUCTILE	2" thru 12"	14" thru 24"	DUCTILE
-20 to 100	200	150	150	250	500	300	500
150	200	150	150	250	500	300	500
200	190	135	115	235	460	280	480
225	180	130	85	230	440	270	470
250	175	125	65	225	415	260	460
275	170	120	50	220	395	250	450
300	165	110		215	375	240	440
325	155	105		212	355	230	430
350	150	100		208	335	220	420
375	145			204	315	210	410
400	140			200	290	200	400
425	130			190	270		390
450	125			180	250		380
500				170			360
600				140			320
650				125			300

Note: The values in the table are restricted to valves that are rated for steam at the class pressure. There are certain constructions that cannot use these secondary ratings.

SECONDARY RATINGS OF JOINTS, FLANGES, VALVES, AND FITTINGS (*Continued*)

CARBON STEEL VALVES AND FLANGES						
Temperature °F	PRESSURE CLASS					
	150	300	400	600	900	1500
-20 TO 100	275	720	960	1440	2160	3600
200	240	675	930	1400	2100	3500
300	210	655	910	1365	2050	3415
400	180	635	890	1330	2000	3330
500	150	600	835	1250	1875	3125
600	130	550	740	1110	1660	2770
650	120	535	690	1030	1550	2580
700	110	535	640	960	1440	2400
750	95	505	590	890	1330	2220
800	80	410	545	815	1225	2040
850	65	270	495	745	1115	1860
900	50	170	450	670	1010	1680
950	35	105	400	600	900	1500
1000	20	50	285	425	635	1065

Notes:
- The secondary ratings can vary by manufacturer, valve material, and valve construction. These are the lowest of secondary ratings listed by two manufacturers and can be considered reasonable for most valves. When operating a system with pressures close to these ratings the manufacturer's ratings for the valves and fittings you're using should be consulted.
- When piping is designed for operation at temperatures less than saturated steam temperature these tables should be consulted to ensure the hydrostatic test pressure does not exceed the rating of the valves and flanges.

Appendix F

Pressure Ratings for Various Pipe Materials

PIPE PRESSURE RATINGS							
SIZE NPS	PVC 75°F	CPVC 180°F	COPPER	BRASS	STEEL		
					STD	XS	XX
1/8	100	100		1,989	330	2,058	
1/4	100	100	1,530	1,984	1,178	3,115	
3/8	100	100	1,212	1,712	1,049	2,400	
1/2	100	100	954	1,625	1,390	2,626	8,485
3/4	100	100	900	1,355	1,200	2,248	6,903
1	100	100	691	1,188	1,344	2,288	6,575
1-1/4	70	100	560	1,079	1,164	1,980	5,460
1-1/2	60	100	524	961	1,080	1,844	4,969
2	50	100	460	793	954	1,658	4,318
2-1/2	50		425	784	1,221	1,895	4,731
3	50		409	751	1,091	1,722	4,202
4	40		380	663	960	1,539	3,659
5	35		365	531	868	1,412	3,291
6	35		367	444	809	1,401	3,215
8			391	425	740	1,269	2,445
10			391	409	691	1,010	
12			391	343	600	847	
14					545	769	
16					476	671	
18					423	595	
20					380	534	
24					316	444	
36					210	295	
48					157	220	

Notes:
- These are nominal pressure ratings for the common materials and standard wall thicknesses
- Steel pipe data includes a 1/16 inch corrosion allowance.

Appendix G

Square Root Flow Curve

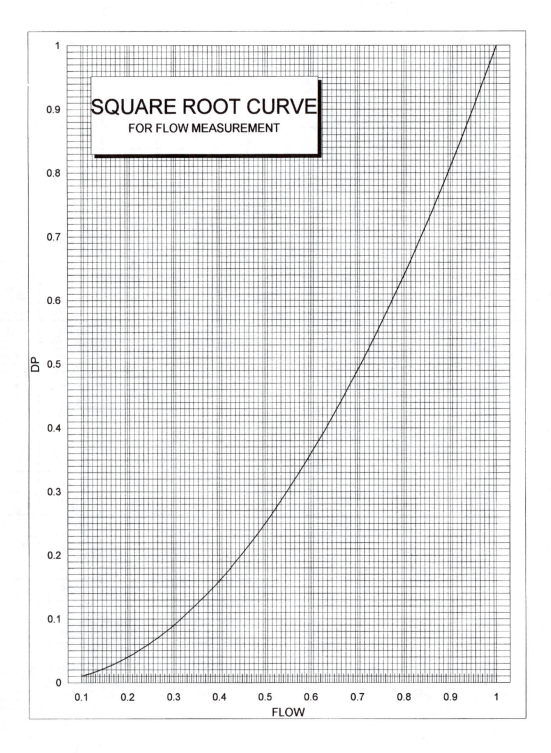

Appendix H
Square Root Graph Paper

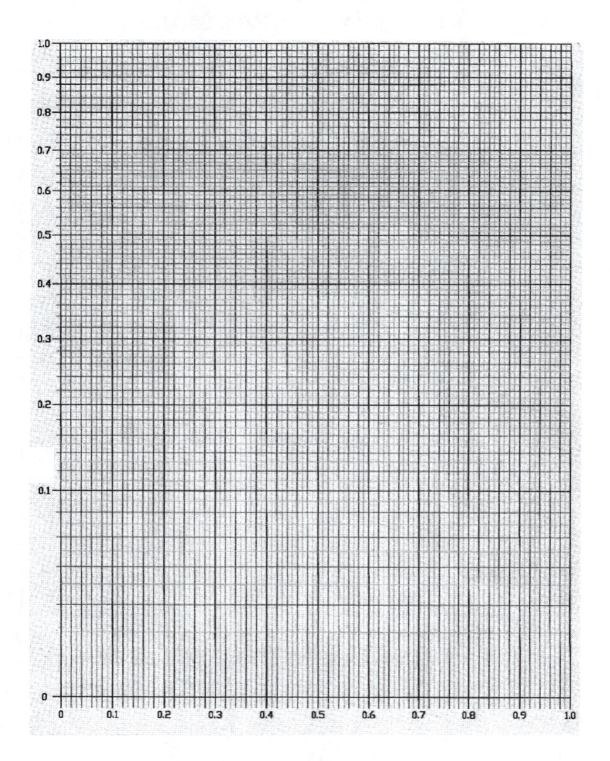

Appendix I

Viscosity Conversions

VISCOSITY CONVERSIONS					
SSU	SSF	Kinematic	SSU	SSF	Kinematic
40	6.4	0.042	1100	112	2.419
50	7.2	0.074	1110	113	2.441
60	8.2	0.103	1120	114	2.463
70	9.1	0.130	1130	115	2.485
80	10	0.156	1140	116	2.507
90	10.9	0.182	1150	117	2.529
100	11.9	0.207	1160	118	2.551
110	12.8	0.230	1170	119	2.573
120	13.7	0.253	1180	120	2.595
130	14.6	0.276	1190	121	2.617
140	15.6	0.298	1200	122	2.639
150	16.5	0.321	1300	132	2.859
160	17.5	0.344	1400	142	3.079
170	18.4	0.366	1500	152	3.299
180	19.4	0.389	1600	162	3.519
190	20.3	0.411	1700	172	3.739
200	21.3	0.433	1800	182	3.959
250	26.2	0.545	1900	192	4.179
300	31.2	0.656	2000	202	4.399
400	41.2	0.877	2100	212	4.619
500	51.2	1.097	2200	222	4.839
600	61.4	1.318	2300	232	5.059
700	71.5	1.538	2400	242	5.279
800	81.7	1.758	2500	252	5.499
900	91.8	1.979	2600	262	5.719
1000	102	2.199	2750	277	6.050

Notes: Conversion is for viscosity in Saybolt Universal Seconds at 100°F, Saybolt Seconds Furol at 122°F, and Kinematic viscosity in Centistokes.

FUEL FIRING TEMPERATURE CALCULATOR[12]

To determine correct burning temperature draw a diagonal line parallel to the one on the chart through the viscosity at temperature reported for the oil. Note the temperature for where that line intersects the line for the correct viscosity for firing

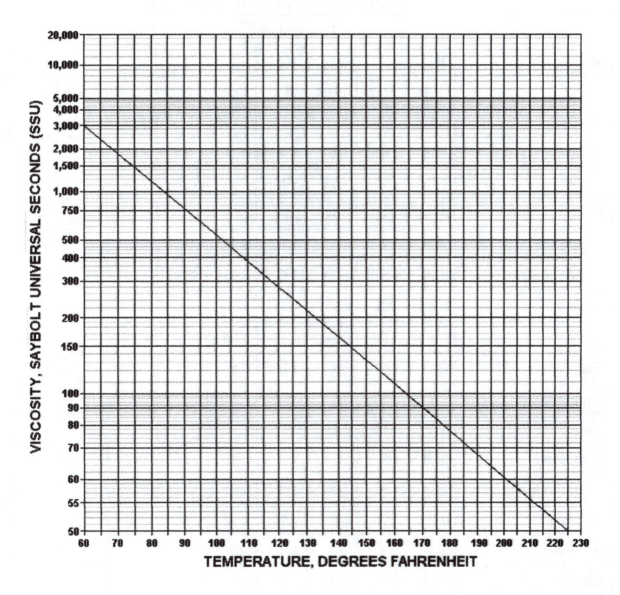

Appendix J

Thermal Expansion of Materials

EXPANSION OF MATERIALS, INCHES PER 100 FEET									
	Temperature range, from 70°F to:								
Material	200	300	400	500	600	700	800	900	1000
Air	22.08	39.06	56.04	73.02	90.00	107.0	124.0	140.9	157.9
Aluminum	2.00	3.66	5.39	7.17	9.03				
Austenitic stainless	1.46	2.61	3.80	5.01	6.24	7.50	8.80	10.12	11.48
Brass	1.52	2.76	4.05	5.40	6.80	8.26	9.78	11.35	12.98
Brick	0.78	1.44	2.14	2.87					
Bronze	1.56	2.79	4.05	5.33	6.64	7.95	9.30	10.68	12.05
Carbon steels	0.99	1.82	2.70	3.62	4.60	5.63	6.70	7.81	8.89
Concrete	1.26	2.31	3.43						
Chrome steels	0.94	1.71	2.50	3.35	4.24	5.14	6.10	7.07	8.06
Cast iron	0.90	1.64	2.42	3.24	4.11	5.03	5.98	6.97	8.02
Copper-nickel	1.33	2.40	3.52						
Glass	0.78	1.44	2.14	2.87					
Glass, pyrex	0.28	0.52	0.77	1.03	1.31	1.61	1.91		
High Chrome stainless	0.86	1.56	2.30	3.08	3.90	4.73	5.60	6.49	7.40
Titanium	0.52	0.96	1.42	1.90	2.42	2.95			
Water (liquid)	42.6	104	193	324	563	1557			
Wood	0.47	0.31	0.46	1.72					
Wrought iron	1.14	2.06	3.01	3.99	5.01	6.06	7.12	8.26	9.36

Values are approximate, for information only, and do not indicate the material is suitable for use at the temperatures indicated.

Appendix K

Value Conversions

To obtain	multiply	by	To obtain	multiply	by
atmospheres	ft. of water	0.0295	knots	miles per hour	0.8684
atmospheres	in. mercury	0.0334	liters	cubic feet	28.316
atmospheres	psi	0.0680	liters	gallons (US)	3.7853
barrels	gallons (US)	0.0238	horsepower	Btuh	0.00039
Btu	calories	252	horsepower	kW	1.341
Btu	hp-hr	2545	meters	feet	0.3048
Btu	kW-hr	3413	meters	inches	0.0254
Btu	watt-hr	3.413	meters	nautical miles	1852
Btuh	horsepower	2545	meters	miles	1609.34
Btuh	kW	3413	microns	inches	25.4
Btuh	refrigeration ton	12,000	miles	feet	5280
centimeter	inches	2.54	miles	meters	0.00062
cubic feet	gallons (US)	0.1337	miles	nautical miles	1.151
cubic meters	cubic feet	0.0283	miles, nautical	kilometers	0.54
feet of H_2O	atmospheres	33.899	miles, nautical	miles	0.8690
feet/minute	miles per hour	88	milliliters	microns	0.001
feet/second	gravity	32.174	mils	centimeters	393.7
foot-pounds	Btu	778	mils	inches	1000
foot-lbs/min.	horsepower	33,000	mils	microns	0.03937
gallons (US)	barrels	42	ounces	grains	0.00228
gallons (US)	cubic feet	7.4805	ounces	grams	0.03527
gallons (US)	Imperial gallons	1.201	ounces, liquid	gallons (US)	128
gallons (US)	Liters	0.2642	parts/million	grains/gallon	17.118
grains	grams	15.432	percent grade	ft. per 100 ft.	1.0
grains	ounces	437.5	pounds	grains	0.00014
grains	pounds	7000	pounds	grams	0.00220
grains/gallon	parts per million	0.0584	pounds	kilograms	2.2046
grams	grains	0.0648	pounds	long tons	2240
grams	ounces	28.35	pounds	metric tons	2204.6
Grams	pounds	453.59	pounds	short tons	2000
Inches	centimeters	0.3937	lbs.ice melt/hr.	refrigeration ton	83.711
inches	microns	0.00004	pounds/cu.ft.	grams/cu.cm.	62.428
in. mercury	feet of water	0.88265	pounds/cu.ft.	pounds/gallon	7.48
inches water	psi	27.673	psi	atmospheres	14.696
kilograms	pounds	0.45359	psi	feet of water	0.43352
kilometer	mile (US)	1.6093	psi	inches water	0.0361
km/hr	mph	1.6093	quarts	cubic feet	29.922
kW	Btu/minute	0.01758	quarts	liters	1.057
kW	horsepower	0.7457	Tons, metric	Tons, short	0.9072
kW-hour	Btu	0.00029			

Appendix L

Combustion and Efficiency Calculation Sheets

The combustion calculation sheet on the following page provides a means for comparing two fuels for their air to fuel ratio requirements and percent moisture in the flue gas. Those values are then used in the boiler efficiency calculation that follows.

The first requirement for an accurate analysis is an "ultimate analysis" of the fuel to provide the data to fill in the box on the top right of the combustion calculation sheet. If you're firing a gas fuel and receive a volumetric analysis these first two worksheets below can be used to produce an ultimate analysis.

Constituent	% Vol	Mol. Wt.	Density	#/C cu.ft.	% by Wt.
Methane (CH_4)		16.0400	0.0424		
Acetylene (C_2H_2)		26.0400	0.0697		
Ethylene (C_2H_4)		28.0500	0.0746		
Ethane (C_2H_6)		30.0700	0.0803		
Propylene (C_3H_6)		42.0800	0.1110		
Propane (C_3H_8)		44.0900	0.1196		
Butylene (C_4H_8)		56.1000	0.1480		
Butane (C_4H_{10})		58.1200	0.1582		
Pentene (C_5H_{10})		70.1300	0.1852		
Pentane (C_5H_{12})		72.1500	0.1904		
Benzene (C_6H_6)		78.1100	0.2060		
Hexane (C_6H_{14})		86.1700	0.2274		
Hydrogen (H_2)		2.0200	0.0053		
Ammonia (NH_3)		17.0300	0.0456		
Hydrogen sulfide (H_2S)		34.0800	0.0911		
Carbon Dioxide (CO_2)		44.0100	0.1170		
Carbon Monoxide (CO)		28.0100	0.0740		
Oxygen (O_2)		32.0000	0.0846		
Nitrogen (N_2)		28.0200	0.0744		
Moisture (H_2O)		18.0200	0.0476		
TOTALS	100.00%	Mixture total			100.00%

yours will not contain all of them. Simply skip lines that don't apply. Multiply those values by the values in the density column and enter the result in the #/C cu. ft. column. If your analysis includes gases labeled "iso-"

ignore that and simply combine the percentages. That result is pounds per hundred cubic feet (the large C represents 100). Total the results in that column to get a mixture total weight number. Divide each of the results

This first worksheet converts the gas constituents from volumetric to gravimetric portions. In other words, it changes it from percent by volume to percent by weight. Insert the percent by cubic foot values for each of the gases in the first open column. This tabulation is provided to accommodate a wide variety of gases and in the # / C cu. ft. column by the total and place the result in the % by weight column. Now that you know the percent by weight of each gas you can convert those values to pounds of carbon, hydrogen, etc. to develop the gravimetric analysis using the next worksheet.

Constituent	% by wt.	x =	C	x =	H_2	x =	O_2	x =	N_2	x =	S_2
Methane	%	0.7487		0.2513		0	0	0	0	0	0
Acetylene	%	0.9226		0.0074		0	0	0	0	0	0
Ethylene	%	0.8563		0.1437		0	0	0	0	0	0
Ethane	%	0.7989		0.2011		0	0	0	0	0	0
Propylene	%	0.8563		0.1437		0	0	0	0	0	0
Propane	%	0.8171		0.1829		0	0	0	0	0	0
Butylene	%	0.8563		0.1437		0	0	0	0	0	0
Butane	%	0.8266		0.1734		0	0	0	0	0	0
Pentene	%	0.8563		0.1437		0	0	0	0	0	0
Pentane	%	0.8323		0.1677		0	0	0	0	0	0
Benzene	%	0.9226		0.0774		0	0	0	0	0	0
Hexane	%	0.8362		0.1638		0	0	0	0	0	0
Hydrogen	%	0		1.0000		0	0	0	0	0	0
Ammonia	0	0		0.1776		0	0	0.8224		0	0
H_2S	%	0		0.0592		0	0	0	0	0.9408	
CO_2	%	0.2729		0		0.7271		0	0	0	0
CO	%	0.4288		0		0.5712			0	0	0
O_2	%	0		0		1.0000		0	0	0	0
N_2	%	0		0		0	0	1.0000		0	0
Totals	%										

Transfer the % by weight values from the first worksheet to the column in the second one. Multiply each value by the factors for carbon, hydrogen, oxygen, etc. in the succeeding columns then add them up. Add up all the values for each element to get the totals for the bottom of the worksheet and transfer those values to the combustion calculation sheet on the next page. Note that moisture in percent by weight is included in the first worksheet.

Always check your math by adding up the percentages. They will seldom total 100% precisely but should be very close to it.

The combustion calculation form has space for describing the fuel and indicating its source, and higher heating value. The predicted stack temperature and excess air percentage are used to determine the volume of the flue gas. Additional instructions on its use follow the combustion calculation form.

COMBUSTION CALCULATION

% by weight	
Carbon : _____	%
Hydrogen: _____	%
Oxygen: _____	%
Sulphur: _____	%
Nitrogen: _____	%
Moisture: _____	% [1]
Ash: _____	%
100.00 %	

FUEL: _____

Source: _____

HHV: _____

Predicted:

Stack temperature: _____ [2]

Excess air (percent): _____ %

Element	Weight	Oxygen Determination			Products	
	#/#	factor	weight	weight	factor	volume @70F
A	B	C	D = B x C	E = B + D	F	G = E x F
C		2.6640			8.7930	scf/# [4]
H_2		7.9360		[3]		
O_2		-1.0000				
S		0.9980			6.0410	scf/ #
N_2					13.8150	scf / #

Weight O_2 required = _____ #/#   Dry gas = _____ scf / # [5]

N_2 in air = 3.31 x O_2 = _____ #/#

Weight of air required = _____ #/# [6]   Theoretical CO_2 = [4/5]= _____ %

Theoretical air [6] x 13.33 = _____ scf / #

PREDICTED AIR/FUEL CALCULATION

Excess air   % x [6] = _____ [7] #/#

      + _____ [6] #/#

TOTAL AIR = _____ [8] #/#

COMBUSTION AIR = [8] x 13.33 = _____ scf / #

Volume excess air = [7]x 13.33 = _____ [9] scf / #

Volume dry product = [5] + [9] = _____ scf / # [10]

% CO2 (dry basis) =[4] x 100 / [10]= _____ %

Volume dry gas @ [2] = [10]*(460+[2])/530 = _____ [11] cu.ft./ #

Moisture in flue gas = [1] + [3] + H x [8] = _____ [12] #/#

M = Pounds of H_2O / pound of dry gas = [12]/([8] +1) = _____ #/#

Volume H_2O / volume dry gas = M/(M + 0.62133) = _____ [13] cu.ft./ cu. ft.

PREDICTED FLUE GAS VOLUME @ [2] = [11]/(1-[13]) = _____ [14] cu.ft./#

%O_2 (dry basis) = ([7]x12.094)/4.31x[10]) = _____ % [15]

%O_2 (wet basis) = [15] x [11]/[14] = _____ %

H = Moisture in Combustion air = _____ #/#

To convert percentages to pound per pound (#/#) numbers simply divide the percentage by 100. The values in column B should add up to 1 or less, less if there's water and/or ash in the fuel. All the calculations on this worksheet are based on one pound of fuel and all results are per pound of fuel.

Multiply the pounds of combustible in the fuel (column B) by the factor in column C to determine the pounds of oxygen required for each combustible and record it in column D. Add up all the results in column D to determine the pounds of oxygen required per pound of fuel, entering it after the "pounds of O_2 required." Calculate the amount of nitrogen in the air by multiplying that result by 3.31 and entering it in the space provided.

The weight of the products of combustion for each combustible is determined by adding the pounds per pound of fuel to the pounds of oxygen required and recording that result in column E. The pounds of nitrogen required in the air must be added to the weight of the nitrogen in the fuel to get the total pounds of nitrogen per pound of flue gas in column E. Combine the weight of water from the hydrogen in the fuel and it's oxygen with the moisture in the fuel [1] to get the total moisture from fuel [3].

Determine the volume of the dry gas by multiplying the weights in column E by the factors in column F and enter the result in column G. Note that we only calculate the volume of carbon dioxide, sulfur dioxide, and nitrogen because the oxygen from the theoretical air is consumed. The water volume isn't calculated because we're determining the volume of dry gas.

Combine the dry gas volumes to get the total theoretical volume of dry gas [5]. Add the weight of oxygen and weight of nitrogen from the air to get the theoretical weight of air required [6]. Divide the theoretical volume of carbon dioxide [4] by the theoretical volume of dry gas [5] to determine the maximum possible percentage of carbon dioxide in the flue gas. Multiplying the theoretical air weight by 13.33 produces the volume of combustion air to burn one pound of fuel in standard cubic feet.

The bottom box of the combustion calculation sheet is set up for determining actual firing conditions. Multiply the weight of air required [6] by the percent of excess air and divide by 100 to determine the weight of excess air [7]. Add [6] and [7] to get total air required for normal combustion [8]. Calculate the volume of excess air [9] by multiplying the excess air weight [7] by 13.33. Add the volume of excess air [9] and the theoretical product

[5] to get the volume of dry flue gas [10]. Perform the indicated calculation to determine what the percent of carbon dioxide should be at the normal firing condition.

The formula for calculating the actual volume of the dry flue gas is developed by adding 460 and the predicted (or actual) stack temperature, dividing that result by 530 then multiplying by the standard volume of dry product [10]. Add the result

To determine the total volume of flue gas we have to calculate the volume of the water. This sheet approximates it by using the formulas shown. Determine the pounds of water in the flue gas per pound of fuel by dividing the percent of water in the fuel [1] by 100, adding the water produced by the combustion of hydrogen in column E [3], and the moisture in the air which is equal to the total air [8] multiplied by the fraction of water that's in the air. You can obtain that information from a psychometric chart or use 0.009 which is a typical value for pounds of moisture per pound of dry air.

If you're using these calculations to get as precise a value as possible for a given operating condition you should make it a point to get the moisture in air value as precise as possible because that moisture can carry a lot of heat up the boiler stack. It can make a big difference in calculating the boiler efficiency for two different operating conditions like summer versus winter.

Add 0.62133 to the pounds of moisture then divide by the pounds of moisture to get the volumetric ratio of moisture [13]. Divide the actual volume of dry gas [11] by the percent of dry gas to wet gas (which is 100 minus the moisture ratio) to get the actual volume of wet flue gas [14].

Formulas for the percent oxygen give results on a dry basis [15] and a wet basis. Percent oxygen on a dry basis is what you would get using a fyr-rite or an orsat analyzer because the moisture is condensed from the flue gas to get the measurements. The oxygen content indicated by an in-situ analyzer, such as a zirconium oxide analyzer, measures the gases with the moisture as steam so it's included in the volume of the flue gases.

EFFICIENCY CALCULATION WORKSHEET

The last worksheet in this appendix uses the information developed in the earlier ones to predict, or calculate, the efficiency of a boiler burning the fuel having the ultimate analysis used for the combustion calculations.

If used for calculating an operating efficiency you have to use the stack temperature measured and adjust the excess air to match the actual operating condition. Some help in determining the excess air is obtained by using the graph in Appendix M.

Space is provided for the boiler name or number, the date, and the fuel to separately identify each worksheet. That's because you may be considering several fuels or have collected operating data on several boilers or you're comparing the boiler's performance to what is was at another time.

The excess air value (a) is the same as used in the combustion calculation sheet. You may have run the boiler at different values of excess air when collecting operating data so you can compare the difference in boiler efficiency. The air/fuel ratio (b) is the one calculated for the operating condition on the combustion calculation sheet [8].

Combustion air (c) and flue gas temperature (d) are recorded and can be adjusted for special applications. For example, you might want to compare the performance of the boiler to the performance of the boiler without its economizer. You could make one worksheet up for the boiler flue gas exit temperature and another with the economizer flue gas exit temperature to get that comparison of efficiencies.

For purposes of calculating efficiencies it's simply easier, and produces more meaningful numbers, if you calculate the results based on therms (100,000 Btu). To determine the amount of fuel required per therm (e) divide the higher heating value of the fuel into 100,000. The matching quantity of air (f) is determined by multiplying the fuel quantity (e) by the air/fuel ratio (b).

Moisture brought in with the combustion air (g) is determined by multiplying the ratio (H on the combustion

---

EFFICIENCY CALCULATIONS

Boiler: _____   Date: _____

Fuel: _____

Excess air:	_____ %	(a)
Air/fuel ratio:	_____ #/#(b)	
Combustion air temperature:	_____ °F (c)	
flue gas temperature:	_____ °F (d)	
Quantities per therm input:		
Fuel:	_____ #/Therm	(e)
Air:	_____ #/Therm	(f)
H_2O in air:	_____ #/Therm	(g)
Wet flue gas:	_____ #/Therm	(h)
H_2O fuel:	_____ #/Therm	(i)
H_2O in flue gas:	_____ #/Therm	(k)
Dry flue gas:	_____ #/Therm	(l)
Heat losses:		
Sensible heat:	_____ %	(m)
H_2O in flue gas:	_____ %	(n)
CO loss:	_____ %	(o)
Radiation:	_____ %	(p)
Total losses:	_____ %	(q)
Efficiency by difference:	%	

calculation sheet) by the quantity of air (f). Wet flue gas (h) is the sum of the fuel, air, and moisture in the air, (e) + (f) + (g). The water in the flue gas that is produced by burning the fuel (i) is determined by multiplying the fuel (e) by the calculated ratio [3] on the combustion calculation sheet.

The total moisture in the flue gas (k) is the sum of the moisture from the air (g) and the moisture from combustion (i). Subtract that from the wet flue gas (h) to get the quantity of dry flue gas (l).

Now you're ready to determine the boiler efficiency by the heat loss method. The loss in sensible heat, where you're just heating up the fuel, air, and water then throwing it away is called sensible heat loss and is calculated by multiplying the wet flue gas quantity (h) by the difference between the flue gas and combustion air temperatures (subtract (c) from (d)) and multiplying the result by the specific heat of the flue gas. The specific heat varies according to the ratio of carbon and hydrogen and you can get a more precise value from Figure 2-3 in PTC-4.1[13] but a value of 0.25 for gas or 0.245 for coal or oil is close enough for most calculations. The result of that calculation is a loss in Btu for the fuel burned so divide the result by 1,000 to get the loss in percent. (That's the same as dividing by a therm then multiplying by 100 to get percent)

The loss due to the moisture content of the flue gas is determined by multiplying the moisture in and from combustion of the fuel (i) by the difference between the enthalpy of the steam and the enthalpy of liquid water at room temperature. You can look up the value for steam at stack temperature and at one pound absolute pressure and subtract the value for water at the

combustion air temperature or simply use 1040 which is usually close enough. Divide that result by 1,000 to get percent.

You should not plan on having a carbon monoxide loss when predicting boiler efficiency and normally carbon monoxide (CO) is so small that its loss is insignificant but occasionally a problem occurs where there is significant loss so you want to determine it. Multiply the dry flue gas (l) by the CO measurement in ppm and divide by 230,200 to get the percent loss from incomplete combustion.

Radiation losses (p) are difficult to determine and frequently much of the heat lost to radiation is recovered in the combustion air making a true analysis even more difficult. You have the option of ignoring the radiation loss (that's what all those modern analyzers do) or using the boiler manufacturer's predicted radiation loss. If using the manufacturer's number it's important to consider that value is at full boiler load. When calculating efficiency at partial loads you should use a radiation loss equal to the manufacturer's prediction divided by the percent load on the boiler when the data was taken and multiply by 100 to get percent loss.

Add up all the losses to get the total losses (q) subtract that result from 100 to get the boiler efficiency in percent. This is a more precise determination than those made with charts and electronic analyzers because it considers the moisture in the fuel and air plus the moisture from combustion of hydrogen in the fuel for the fuel you burn. A few calculations with different fuel analysis will show that the moisture loss is a significant consideration.

Appendix M

Excess Air/O$_2$ Curve

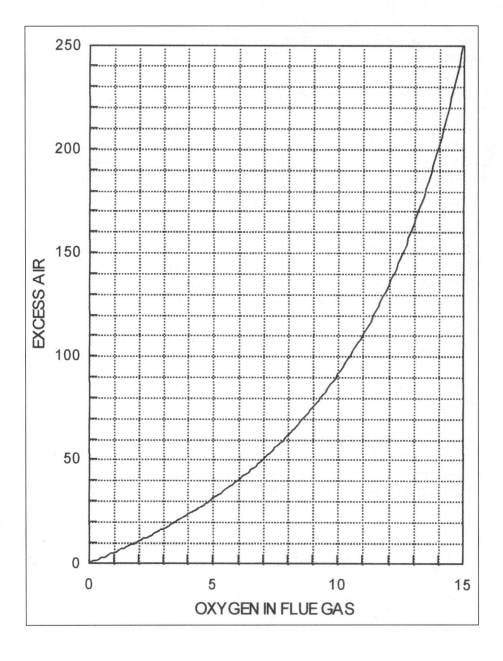

Appendix N

Properties of Dowtherm A

Vacuum Pressure	TEMP. °F	LB. PER CU. FT.		HEAT IN BTU PER POUND		
		LIQUID	VAPOR	LIQUID	LATENT	VAPOR
29.92 in. Hg	60	66.54	0.0000	2.4	174.4	176.8
29.92 in. Hg	80	65.82	0.0000	9.9	172.0	181.9
29.92 in. Hg	100	65.27	0.0000	17.6	169.6	187.2
29.92 in. Hg	120	64.72	0.0001	25.5	167.2	192.7
29.91 in. Hg	140	64.16	0.0002	33.5	164.9	198.4
29.90 in. Hg	160	63.60	0.0004	41.6	162.7	204.3
29.87 in. Hg	180	63.03	0.0007	49.9	160.4	210.3
29.83 in. Hg	200	62.46	0.0012	58.3	158.3	216.6
29.74 in. Hg	220	61.88	0.0021	66.9	156.2	223.1
29.60 in. Hg	240	61.30	0.0034	75.7	154.0	229.7
29.40 in. Hg	260	60.71	0.0055	84.5	152.0	236.5
29.09 in. Hg	280	60.11	0.0086	93.6	149.9	243.5
28.65 in. Hg	300	59.50	0.0129	102.7	147.9	250.6
27.97 in. Hg	320	58.89	0.0191	112.1	145.8	257.9
27.06 in. Hg	340	58.28	0.0274	121.5	143.8	265.3
25.80 in. Hg	360	57.65	0.0385	131.2	141.7	272.9
24.11 in. Hg	380	57.02	0.0532	140.9	139.8	280.7
21.87 in. Hg	400	56.37	0.0720	150.9	137.6	288.5
19.00 in. Hg	420	55.72	0.0959	160.9	135.6	296.5
15.31 in. Hg	440	55.06	0.1258	171.1	133.5	304.6
10.71 in. Hg	460	54.38	0.1626	181.5	131.3	312.8
5.01 in. Hg	480	53.70	0.2076	192.0	129.1	321.1
0 psig	494.8	53.18	0.2470	199.9	127.4	327.3
0.00 psig	494.8	53.18	0.2470	199.9	127.4	327.3
0.95 psig	500	53.00	0.2618	202.7	126.9	329.5
5.08 psig	520	52.29	0.3267	213.5	124.5	338.0
10.01 psig	540	51.56	0.4037	224.5	122.1	346.6
15.85 psig	560	50.82	0.4943	235.7	119.5	355.3
22.68 psig	580	50.06	0.6003	247.1	116.9	364.0
30.64 psig	600	49.29	0.7237	258.6	114.1	372.7
39.81 psig	620	48.49	0.8667	270.2	111.3	381.5
50.33 psig	640	47.67	1.032	282.0	108.3	390.4
62.30 psig	660	46.82	1.223	294.0	105.2	399.2
75.86 psig	680	45.94	1.442	306.1	102.0	408.1
91.10 psig	700	45.03	1.695	318.3	98.6	416.9
108.3 psig	720	44.08	1.988	330.7	95.0	425.8
127.4 psig	740	43.09	2.327	343.4	91.2	434.6
152.5 psig	760	42.04	2.723	356.2	87.1	443.3
198.6 psig	800	39.74	3.749	382.7	77.6	460.2

Reference state for heat of the fluid is zero at the freezing temperature of 53.6°F

Appendix O

Properties of Dowtherm J

Vacuum / Pressure	TEMP. °F	LB. PER CU. FT.		HEAT IN BTU PER POUND		
		LIQUID	VAPOR	LIQUID	LATENT	VAPOR
29.92" Hg	0	55.64	0.0000	-32.8	175.1	142.3
29.92" Hg	20	55.14	0.0000	-24.5	172.3	147.8
29.92" Hg	40	54.64	0.0001	-16.0	169.6	153.6
29.90" Hg	60	54.13	0.0003	-7.4	167.0	159.6
29.88" Hg	80	53.61	0.0006	1.3	164.5	165.8
29.81" Hg	100	53.09	0.0011	10.2	162.1	172.3
29.72" Hg	120	52.56	0.0021	19.3	159.7	179.0
29.55" Hg	140	52.02	0.0037	28.6	157.3	185.9
29.29" Hg	160	51.47	0.0063	38.0	155.0	193.0
28.86" Hg	180	50.92	0.0103	47.6	152.7	200.3
28.19" Hg	200	50.36	0.0161	57.4	150.4	207.8
27.21" Hg	220	49.79	0.0246	67.4	148.1	215.5
25.81" Hg	240	49.21	0.0364	77.6	145.7	223.3
23.83" Hg	260	48.62	0.0525	87.9	143.5	231.4
21.16" Hg	280	48.01	0.0739	98.5	141.1	239.6
16.48" Hg	300	47.40	0.1018	109.3	138.6	247.9
12.90" Hg	320	46.77	0.1375	120.3	136.2	256.4
6.91" Hg	340	46.13	0.1826	131.4	133.6	265.0
0 psig	358.4	45.53	0.2340	141.9	131.2	273.1
0.32 psig	360	45.48	0.2385	142.8	131.0	273.8
4.93 psig	380	44.80	0.3072	154.4	128.3	282.7
10.57 psig	400	44.11	0.3907	166.3	125.4	291.7
17.39 psig	420	43.40	0.4911	178.4	122.4	300.7
25.55 psig	440	42.67	0.6110	190.7	119.2	309.9
35.19 psig	460	41.91	0.7533	203.2	115.9	319.1
46.50 psig	480	41.12	0.9215	215.9	112.5	328.4
59.63 psig	500	40.30	1.1198	228.9	108.9	337.7
74.79 psig	520	39.44	1.3533	242.0	105.0	347.1
92.16 psig	540	38.53	1.6285	255.4	101.0	356.4
111.96 psig	560	37.58	1.9542	269.1	96.6	365.7
134.42 psig	580	36.56	2.3425	283.1	91.8	374.9
159.80 psig	600	35.46	2.8110	297.4	86.6	383.9

Appendix P

Chemical Tank Mixing Table

CHEMICAL TANK MIXING TABLE					
Inches tank Diameter	Gallons per inch of depth	Chemicals to be added for % solutions in pounds per inch			
		1%	2%	5%	10%
12	0.49	0.04	0.08	0.20	0.41
18	1.10	0.09	0.18	0.46	0.92
24	1.96	0.16	0.33	0.82	1.63
30	3.06	0.25	0.51	1.27	2.55
36	4.41	0.37	0.73	1.84	3.67
42	6.00	0.50	1.00	2.50	5.00
48	7.83	0.65	1.31	3.26	6.53
54	9.91	0.83	1.65	4.13	8.26
60	12.24	1.02	2.04	5.10	10.20
66	14.81	1.23	2.47	6.17	12.34
72	17.63	1.47	2.94	7.34	14.68
78	20.69	1.72	3.45	8.62	17.23
84	23.99	2.00	4.00	9.99	19.98
90	27.54	2.29	4.59	11.47	22.94
96	31.33	2.61	5.22	13.05	26.10
102	35.37	2.95	5.89	14.73	29.47
108	39.66	3.30	6.61	16.52	33.03
114	44.19	3.68	7.36	18.40	36.81
120	48.96	4.08	8.16	20.39	40.78
132	59.24	4.93	9.87	24.67	49.35
144	70.50	5.87	11.75	29.36	58.73
192	125.34	10.44	20.88	52.20	104.41
216	158.63	13.21	26.43	66.07	132.14
240	195.84	16.31	32.63	81.57	163.13

Use the data on this table to create your own table for your specific size of chemical tank and solution strength to be maintained. On a fresh piece of paper, write or print the range of levels of the tank from bottom to top then multiply the values of those levels by the pounds of chemical per inch numbers from the table above. You can list the levels and chemical requirements in multiple columns to produce a table like the one on the following page.

For this example the tank is 48 inches deep so we need forty-eight different level readings and the matching quantity of chemical to produce a 5% solution. After laying out the table so we have all 48 inches accounted for, we multiply each level by the 1.27 pounds per inch to determine the number of pounds that must be added to produce a 5% solution at each level.

Here's the table you made. Once you have the table made, it would pay to find some laminating plastic and cover it then mount it next to the tank. With this table you subtract the level in the tank (before you fill it with water) from the level after it's filled then find the pounds of chemical to add from the table.

Level	Chemical	Level	Chemical
1	1.27	25	31.75
2	2.54	26	33.02
3	3.81	27	34.29
4	5.08	28	35.56
5	6.35	29	36.83
6	7.62	30	38.1
7	8.89	31	39.37
8	10.16	32	40.64
9	11.43	33	41.91
10	12.7	34	43.18
11	13.97	35	44.45
12	15.24	36	45.72
13	16.51	37	46.99
14	17.78	38	48.26
15	19.05	39	49.53
16	20.32	40	50.8
17	21.59	41	52.07
18	22.86	42	53.34
19	24.13	43	54.61
20	25.4	44	55.88
21	26.67	45	57.15
22	27.94	46	58.42
23	29.21	47	59.69
24	30.48	48	60.96

You could also create a chart based on the level before you filled it when you fill the tank to a consistent level. If you always raise the level to the 48 inches, subtract the level values in the table from 48 and replace them with the result.

Appendix Q

Suggested Mnemonic Abbreviations

The use of mnemonic abbreviations to simplify communications and labeling of devices in a boiler plant is a common practice. This is a recommended list for identifying plant devices on logs, equipment lists, maintenance records, reports, etc. A plant can have many identical devices that are numbered sequentially (although earlier numbers may no longer exist) as indicated by the pound sign (#). Also, some devices are redundant (such as two safety shutoff valves in series) so the number can be followed by a letter, indicated by the asterisk (*). The two symbols (# and *) are shown only where the inclusion of a number/letter is common.

This list does contain duplicate abbreviations where it is necessary to determine which one is correct by how and where it is used.

3VBP	Three valve bypass
AACV	Atomizing air control valve
AAPS	Atomizing air pressure switch
ABC	Automatic blowdown control
ABCV	Automatic blowdown control valve
AFT	Air flow transmitter
AIS	Automatic Interruptible System
ALWCO	Auxiliary low water cutoff
ASBOV	Atomizing steam blow-out valve
ASCV	Atomizing steam control valve
ASPS	Atomizing steam pressure switch
ASSV	Atomizing steam shutoff valve
ASV	Anti-siphon valve
AT	Analysis transmitter
AW	Acid waste
BD	Blowdown (piping)
BDFT	Blowdown flash tank
BDHX	Blowdown heat exchanger
BDQT	Blowdown quench tank
BF	Boiler feedwater (piping)
BFP#	Boiler feed pump
BFV	Butterfly valve
BGV#*	Burner gas safety shutoff valve
BLR#	Boiler
BO	Blowoff (piping)
BOQT	Blowoff quench tank
BOS	Blowoff separator
BV	Ball valve
BVV#	Burner gas vent valve
CAFS	Combustion air flow switch
CF	Chemical feed
CHX	Condensing heat exchanger
CO	Carbon monoxide
CO_2	Carbon dioxide
COND	Condensate
CP	Circulating pump
CP#	Condensate polisher
CPMP	Condensate pump

CPMS	Circulating/condensate pump motor starter
CR	Control relay
CW	City water (piping)
DA	Deaerator
DEGAS	Degassifier
DBB	Double block and bleed (valve arrangement)
DI	Draft indicator
DI	Demineralized (water)
DLT	Drum level transmitter
DT	Draft transmitter
FC	Flow controller
FD	Forced draft
FDF	Forced draft fan
FFT	Feedwater flow transmitter
FIC	Flow indicating controller
FMS	Fan motor starter
FOR	Fuel oil return
FOP#	Fuel oil pump
FOS	Fuel oil supply/suction
FOT#	Fuel oil tank
FPSC	Frost proof sill cock
FR	Flame relay
FR	Flow recorder
FW	Boiler feedwater (piping)
FWCV	Feedwater control valve
FWHTR	Feedwater heater
FY	Flow totalizer
GCV	Gas flow control valve
GFT	Gas flow transmitter
GOS	Gas - off - oil selector (switch)
GPR	Gas pressure regulator
GT#	Gas turbine
GV	Gate valve
H_2	Hydrogen
HFPS	High furnace pressure switch
HGP	High gas pressure (limit switch)
HIGP	High ignitor gas pressure (limit switch)
HOT	High oil temperature (limit switch)

HPC..............High pressure condensate
HPS..............High pressure steam
HPS..............High pressure switch
HTS..............High temperature switch
ID.................Induced draft
IDF..............Induced draft fan
IGV#*..........Ignitor gas safety shutoff valve
IPS...............Intermediate pressure steam
IT.................Ignition timer
IT.................Ignition transformer (see IX)
IVV#............Ignitor gas vent valve
IX.................Ignition transformer
LAAD..........Low atomizing air differential pressure
                    (limit switch)
LAAP...........Low atomizing air pressure switch
LAF..............Low air flow (limit switch)
LASD...........Low atomizing steam differential pressure
                    (limit switch)
LASP   Low atomizing steam pressure (limit switch)
LC................Level controller
LDS..............Low draft switch
LG................Level glass
LGP..............Low gas pressure (limit switch)
LI.................Level indicator
LIC...............Level indicating controller
LIGP.............Low ignitor gas pressure (limit switch)
LOP..............Low oil pressure (limit switch)
LOT..............Low oil temperature (limit switch)
LPC..............Low pressure condensate
LPHTR.........Low pressure heater
LPS..............Low pressure switch
LPS..............Low pressure steam
LR................Level recorder
LS................Level switch
LSH..............Level switch, high level
LSL..............Level switch, low level
LT................Level transmitter
LTS..............Low temperature switch
LWCO.........Low water cutoff
LWFS...........Low water flow switch
LWL..............Low water level
MAFS..........Minimum air flow switch (limit switch)
MBDI...........Mixed bed demineralizer
MGPR..........Minimum gas pressure regulator
MGV#*.........Main gas safety shutoff valve
MOPR..........Minimum oil pressure regulator
MS...............Motor starter
MU...............Makeup water (piping)
MVV#..........Main gas vent valve

N_2..................Nitrogen
NG................Natural gas
NRV..............Non-return valve
O_2..................Oxygen
O_2T.............Oxygen transmitter
OCV..............Oil flow control valve
OF................Overflow
OFT..............Oil flow transmitter
OPMS...........Oil pump motor starter
OPR..............Oil pressure regulator
OV#*............Oil safety shutoff valve
PC................Pressure controller
PC................Pumped condensate
PI.................Pressure indicator
PIC...............Pressure indicating controller
PPT..............Post purge timer
PR................Pressure recorder
PR................Pressure regulator
PRV..............Pressure reducing valve (station)
PT................Purge timer
PT................Pressure transmitter
PV................Plug valve
RO................Reverse osmosis
ROW.............Reverse osmosis water (permeate)
ROV.............Recirculating oil valve
RV................Recirculating valve
RV................Relief valve
SAN.............Sanitary sewer
SOFT#.........Softener
SPT...............Steam pressure transmitter
STM.............Steam
STRNR.........Strainer
SV................Safety valve
SW...............Softened water
TC................Temperature controller
TE................Temperature element
TI.................Temperature indicator
TIC...............Temperature indicating controller
TR................Temperature recorder
TSTAT..........Thermostat
TT................Temperature transmitter
TV................Globe valve (throttling valve)
VC................Vent condenser
VTR..............Vent through roof
ZC................Position controller (valve positioner)

NOTE: A mnemonic is a device to help someone remember. The letters used in an alphabetic abbreviation help one remember the device that is referred to.

Appendix R

Specific Heats of Some Common Materials

It takes less heat to raise the temperature of most substances than it does to raise the temperature of water. To determine how much steam or hot water is needed to heat another substance, multiply the temperature rise of the substance by it's specific heat and the quantity in pounds. The result is the number of Btus needed. For heating products continuously use pounds per hour of the substance to get the result in Btuh.

SPECIFIC HEATS OF COMMON MATERIALS[14]

Acetic Acid	0.51		Hydrochloric acid	0.60
Acetone	0.54		Ice	0.47
Air	0.24		Iron oxide	0.17
Alcohol	0.58		Kerosene	0.50
Alumina (aluminum oxide)	0.18		Lead	0.03
Aluminum	0.23		Lead oxide	0.06
Asbestos	0.20		Limestone	0.22
Ashes	0.20		Magnesia	0.22
Bakelite	0.35		Marble	0.21
Basalt (Lava)	0.20		Nickel	0.07
Benzol	0.40		Nickel steel	0.10
Borax	0.23		Oil, fuel	0.50
Brass, Bronze	0.09		Oil, machine	0.40
Brick	0.22		Oil, olive	0.40
Carbon	0.20		Petroleum	0.50
Chalk	0.22		Rubber	0.37
Charcoal	0.20		Salt, rock	0.21
Chloroform	0.23		Sand	0.19
Cinders	0.18		Sandstone	0.22
Coal	0.30		Silica	0.19
Concrete	0.16		Silver	0.10
Copper	0.09		Soil	0.44
Copper oxide	0.11		Solders	0.04
Cork	0.48		Steel	0.11
Dolomite	0.22		Sulfur	0.18
Ether	0.54		Sulphuric acid	0.34
Ethylene glycol	0.60		Talc	0.21
Flint glass	0.12		Titanium	0.13
Gasoline	0.50		Toluene	0.40
Glass	0.20		Turpentine	0.42
Glycerin	0.58		Water, sea	0.94
Gold	0.03		Wax	0.69
Granite	0.20		Wood, oak	0.57
Graphite	0.20		Wood, pine	0.67
Gypsum	0.26		Zinc oxide	0.12

Appendix S

Design Temperatures and Degree Days

Design outdoor winter temperature and the number of degree days are provided below for a number of North American cities.[15] More precise values should be available for your plant from the local weather service.

Alabama
Anniston 5 2806
Birmingham 10 2611
Mobile 15 1566
Montgomery 10 2071

Alberta
Calgary .... -29 9520
Edmonton .... -33 10320
Lethbridge .... -32 8650
Medicine Hat .... -35 8650

Arizona
Flagstaff .... -10 7242
Phoenix 25 1441
Yuma 30 1036

Arkansas
Fort Smith 10 3226
Little Rock 5 3009

British Columbia
Prince George .... -32 9500
Prince Rupert 8 6910
Vancouver 11 5230
Victoria 15 5410

California
Eureka 30 4758
Fresno 25 2403
Los Angeles 35 1391
Sacramento 30 2680
San Diego 35 1596
San Francisco 35 3137
San Jose 25 2823

Colorado
Denver .... -10 5839
Grand Junction .... -15 5613
Pueblo .... -20 5558

Connecticut
Hartford 0 6113
New Haven 0 5880

Delaware
Wilmington 0

District of Columbia
Washington 0 4561

Florida
Apalachicola 25 1252
Jacksonville 25 1185
Key West 49 59
Miami 35 185
Pensacola 20 1281
Tampa 30 571
Tallahassee 25 1463

Georgia
Atlanta 10 2985
Augusta 10 2306
Macon 15 2338
Savannah 20 1635

Idaho
Boise .... -10 5678
Lewiston 5 5109
Pocatello .... -5 6741

Illinois
Cairo 0 3957
Chicago .... -10 6282
Peoria .... -10 6004
Springfield .... -10 5446

Indiana
Evansville 0 4410
Fort Wayne .... -10 6232
Indianapolis .... -10 5458

Iowa
Davenport .... -15 6252
Des Moines .... -15 6375
Dubuque .... -20 6820
Keokuk .... -10 5663
Sioux City .... -20 6905

Kansas
Concordia .... -10 5425
Dodge City .... -10 5069
Topeka .... -10 5075
Wichita .... -10 4664

Kentucky
Lexington 0 4792
Louisville 0 4417

Louisiana
New Orleans 20 1203
Shreveport 20 2132

Maine

Eastport	-10	8445
Presque Isle		9644
Portland	-5	7377

Manitoba

Brandon	-32	10930
Churchill	-42	16810
Winnipeg	-29	10630

Maryland

Baltimore	0	4487

Massachusetts

Boston	0	5936
Fitchburg	0	6743

Michigan

Alpena	-10	8278
Detroit	-10	6560
Escanoba	-15	8777
Grand Rapids	-10	6702
Lansing	-10	7149
Marquette	-10	8745
Sault St. Marie	-20	9307

Minnesota

Duluth	-25	9723
Minneapolis	-20	7966
Saint Paul	-20	7985

Mississippi

Meridian	10	2330
Vicksburg	10	2069

Missouri

Columbia	-10	5070
Kansas City	-10	4692
Saint Louis	0	4596
Saint Joseph	-10	5596
Springfield	-10	4569

Montana

Billings	-25	7213
Havre	-30	8416
Helena	-20	7930
Kalispell	-20	8032
Miles City	-35	7981
Missoula	-20	7604

Nebraska

Lincoln	-10	5980
North Platte	-20	6384
Omaha	-10	6095
Valetine	-25	7197

Nevada

Reno	-5	5621
Tonopah	5	5812
Winnemucca	-15	6357

New Brunswick

Fredericton	-6	8830
Moncton	-8	8700
Saint John	-3	8380

Newfoundland

Corner Brook	-1	9210
Gander	-3	9440
Goose Bay	-26	12140
Saint Johns	1	8780

New Hampshire

Concord	-15	7400

New Jersey

Atlantic City	5	5015
Newark	0	5500
Sandy Hook	0	5369
Trenton	0	5256

New Mexico

Albuquerque	0	4517
Roswell	-10	3578
Santa Fe	0	6123

New York

Albany	-10	6648
Binghamton	-10	6818
Buffalo	-5	6925
Canton	-25	8305
Ithaca	-15	6914
New York City	0	5280
Oswego	-10	7186
Rochester	-5	6772
Syracuse	-10	6899

North Carolina

Asheville	0	4236
Charlotte	10	3224
Greensboro	10	3849
Raleigh	10	3275
Wilmington	15	2420

North Dakota

Bismark	-30	8937
Devils Lake	-30	10104
Grand Forks	-25	9871
Williston	-35	9301

Northwest Territories

Aklavik	-46	17870
Fort Norman	-42	16020

Nova Scotia

Halifax	4	7570
Sydney	1	8220
Yarmouth	7	7520

Ohio

Cincinnati	0	4990
Cleveland	0	6144
Columbus	-10	5506

Dayton	0	5412
Sandusky	0	6095
Toledo	-10	6269

Oklahoma

Oklahoma City	0	3670

Ontario

Fort William	-24	10350
Hamilton	0	6890
Kapuskasing	-30	11790
Kingston	-11	7810
Kitchener	-3	7380
Ottawa	-15	8830
Toronto	0	7020

Oregon

Baker	-5	7197
Portland	10	4353

Pennsylvania

Erie	-5	6363
Harrisburg	0	5412
Philadelphia	0	4739
Pittsburgh	0	5430
Reading	0	5232
Scranton	-5	6218

Prince Edward Island

Charlottetown	-3	8380

Quebec

Arvida	-10	10440
Montreal	-9	8130
Quebec City	-12	9070
Sherbrooke	-12	8610

Rhode Island

Providence	0	5984

Saskatchewan

Prince Albert	-41	11430
Regina	-34	10770
Saskatoon	-37	10960
Swift Current	-33	9660

South Carolina

Charleston	15	1866
Columbia	10	2488
Greenville	10	3059

South Dakota

Huron	-20	7940
Rapid City	-20	7197

Tennessee

Chattanooga	10	3238
Knoxville	0	3658
Memphis	0	3090
Nashville	0	3613

Texas

Abilene	15	2573
Amarillo	-10	4196
Austin	20	1679
Brownsville	30	628
Corpus Christi	20	965
Dallas	0	2367
El Paso	10	2532
Fort Worth	10	2355
Galveston	20	1174
Houston	20	1315
Palestine	15	2068
Port Arthur	20	1352
San Antonio	20	1435

Utah

Modena	-15	6598
Salt Lake City	-10	5650

Vermont

Burlington	-10	8051

Virginia

Cape Henry	10	3538
Lynchburg	5	4068
Norfolk	15	3364
Richmond	15	3922
Roanoke	0	4075

Washington

North Head	20	5367
Seattle	15	4815
Spokane	-15	6138
Tacoma	15	5039
Tatoosh Island	15	5857
Walla Walla	-10	4910
Yakima	5	5585

West Virginia

Elkins	-10	5800
Parkersburg	-10	4928

Wisconsin

Green Bay	-20	7931
La Crosse	-25	7421
Madison	-15	7405
Milwaukee	-15	7079

Wyoming

Cheyenne	-15	7536
Lander	-18	8243
Sheridan	-30	7239

Yukon Territory

Dawson	-56	15040

Appendix T

Code Symbol Stamps

Your boiler or boilers will have one or more of the these ASME Code Symbol stamps applied to the construction.

The letter within the symbol identifies the product and quality of construction. These stamps can only be applied by manufacturers authorized by ASME to use their respective stamp. Under no circumstances should you remove, alter, or obliterate the symbol stamp and the lettering next to it (which is also required by the Code). The definition of the stamps and the general scope of the authorization, including those you will find on pressure vessels (not shown above) are as follows:

A - Assembly, to assemble boilers.
E - Electric boiler, to manufacture electric boilers.
H - Heating boiler, to manufacture heating boilers.
M - Miniature boiler, to manufacture miniature boilers.
PP - Power piping, to manufacture boiler external piping.

S- Steam boiler, to manufacture power boilers, high temperature hot water and organic fluid heating boilers.
U - Unfired pressure vessel, to manufacture pressure vessels.
UM - Miniature unfired pressure vessel, to manufacture small pressure vessels
UV - Safety valves, manufacture of safety valves for unfired pressure vessels
V - Safety valves, manufacture of safety valves for high pressure boilers

Note that the manufacturer's certificate will also define the locations where the manufacturing can be done, either in the shop named on the Certificate of Authorization, or (also) in the field.

BIBLIOGRAPHY

1. Klaus Scheiss, P.E., C.E.M., *Strategic Planning for Energy and the Environment* Vol 15, No. 2

2. National Board Bulletin/Summer 2003

3. WADITW = We Always Did It That Way, K.E. Heselton, *Strategic Planning for Energy and the Environment*, Vol. 17, No. 2 - 1997.

4. American Society of Heating and Air Conditioning Engineers "1999 HVAC Applications" Handbook, page 48. 10, Figure 11 "Residential Hourly Hot Water Use - 95% Confidence Level."

5. "Anatomy of a Catastrophic Boiler Accident" David G. Peterson, National Board Bulletin, Summer 1997

6. Combustion Engineering, Inc. Fuel Burning and Steam Generating Handbook, 1973.

7. *Power* magazine, January/February 2003

8. PG-27.2.2 of Section I of the ASME Boiler and Pressure Vessel Code, "Rules for Construction of Power Boilers."

9. *Thermodynamics*, Virgil Moring Faires, Fourth Edition - The Macmillan Company - New York.

10. *Steam, its Generation and Use* - The Babcock and Wilcox Company, New York, NY, at least 38 editions.

11. Edward J. Brown, *Heating Piping and Air Conditioning*, April 1960

12. Derived from the ASTM D341-43 chart labeled "Viscosity - temperature relationships for heavy fuel oils" of the American Society for Testing and Materials.

13. The ASME Power Test Code for "Steam Generating Units PTC-4. 1, American Society of Mechanical Engineers, New York, New York

14. *Standard Handbook for Mechanical Engineers*, seventh edition, Theodore Baumeister and Lionel S. Marks, Editors, McGraw Hill Book Company, New York, New York

15. "Carrier System Design Manual, Part 1, Load Estimating" Carrier Air Conditioning Company, Syracuse, New York, 1960 - seventh printing.

Index